International Conference on Energy Sciences (ICES 2016)

Journal of Physics: Conference Series Volume 877

Bandung, Indonesia
25 - 27 July 2016

Editors:

Abdul Waris
Byunga Shin
Masatoshi Kondo
Yose Fachmi Buys
Dwi Irwanto
Syeilendra Pramuditya

ISBN: 978-1-5108-4721-7
ISSN: 1742-6588

Printed from e-media with permission by:

Curran Associates, Inc.
57 Morehouse Lane
Red Hook, NY 12571

CURRAN ASSOCIATES INC.
proceedings
.com

Some format issues inherent in the e-media version may also appear in this print version.

Printed by Curran Associates, Inc. (2017)

For permission requests, please contact the Institute of Physics
at the address below.

Institute of Physics
Dirac House, Temple Back
Bristol BS1 6BE UK

Phone: 44 1 17 929 7481
Fax: 44 1 17 920 0979

techtracking@iop.org

Additional copies of this publication are available from:

Curran Associates, Inc.
57 Morehouse Lane
Red Hook, NY 12571 USA
Phone: 845-758-0400
Fax: 845-758-2633
Email: curran@proceedings.com
Web: www.proceedings.com

Table of contents

Volume 877

International Conference on Energy Sciences (ICES 2016)
25–27 July 2016, Bandung, Indonesia

Accepted papers received: 10 July 2017
Published online: 10 August 2017

Preface

011001
OPEN ACCESS
International Conference on Energy Sciences (ICES 2016)

011002
OPEN ACCESS
Peer review statement

Papers

012001
OPEN ACCESS
Experimental study on corrosion and precipitation in non-isothermal Pb-17Li system for development of liquid breeder blanket of fusion reactor

Masatoshi Kondo, Masaomi Ishii, Takayoshi Norimatsu and Takeo Muroga.....1

012002
OPEN ACCESS
From lab to full-scale ultrafiltration in microalgae harvesting

I G Wenten, S Steven, A Dwiputra, Khoiruddin and A N Hakim.....13

012003
OPEN ACCESS
Preliminary Study of Plutonium Utilization in AP1000 Reactor

Nailatussaadah, Puguh A. Prastyo, Abdul Waris, Rizal Kurniadi and Syeilendra Pramuditya.....19

012004
OPEN ACCESS
Effect of the Cerium Oxide (CeO_2) on the Structural and Electrochemical Properties of the $LaNi_5Ce$ Metal Hydride Anode

Ade Utami Hapsari, Anne Zulfia, Jarot Raharjo and Agustanhakri.....26

012005
OPEN ACCESS
Dye-sensitized solar cell based on TiO_2/MnO_2 composite film as working electrode

A Prasetio, A M Habieb, I Alkian, Z Arifin and H Widiyandari.....32

012006
OPEN ACCESS
The effect of temperature deposited on the performance of ZnO-CNT-graphite for supercapacitors

Alfin Darari, Istajib S. Hakim, Priyono, Agus Subagio, Pardoyo and Achmad Subhan.....39

012007
OPEN ACCESS
Conformation Analysis of T1 Lipase on Alcohols Solvent using Molecular Dynamics Simulation

A M Putri, T Sumaryada and S T Wahyudi.....45

012008
OPEN ACCESS
Effect of Acetylene Black Content in $Li_4Ti_5O_{12}$ Xerogel Solid-State Anode Materials on Half-Cell Li-ion Batteries Performance

N M Abdurrahman, B Priyono, A Z Syahrial and A Subhan.....53

012009
OPEN ACCESS
Effectiveness of Dry Cell Microscopic Simulation (DCMS) to Promote Conceptual Understanding about Battery

Firmanul Catur Wibowo, Andi Suhandi, Dadi Rusdiana, Achmad Samsudin, Dina Rahmi Darman, M. Noor Faizin, Wiyanto, Supriyatman, Anna Permanasari, Ida Kaniawati, Wawan Setiawan, Yudi Karyanto, Suharto Linuwih, Abdul Fatah, Bambang Subali, Aceng Hasani and Sholeh Hidayat.....62

012010
OPEN ACCESS
Superhydrophobic Membrane Contactor for Acid Gas Removal

Nurul Faiqotul Himma and I Gede Wenten.....72

012011
OPEN ACCESS
Modeling of various heat adapter plate 4 and 6 array for optimization of thermoelectric generator element using modified diffusion equation

Defrianto, W Tambunan and Lazuardi.....82

012012
OPEN ACCESS
Application of Implicit Finite Difference Method to Determine the 2D Patterns of Unsteady State Thermal Spreading of Geothermal Systems

Joko Sampurno, Apriansyah, Riza Adriat, Irfana Diah Faryuni and Wahyu Srigutomo.....89

012013
OPEN ACCESS
Analysis of Neutron Fission Reaction Rate in the Nuclear Fuel Cell Using Collision Probability Method with Non Flat Flux Approach

Mohammad Ali Shafii.....97

012014
OPEN ACCESS
Developing a Framework For The Assessment Of Pre-Service Physics Teachers' Energy Literacy

Muhamad Yusup, Agus Setiawan, Nuryani Y. Rustaman and Ida Kaniawati.....103

012015
OPEN ACCESS
Ni and Zn Substituted M-type Barium Hexaferrite Processed by Sol–Gel Auto Combustion Method

Widyastuti, Nia Sasria, A Marsha Alviani, M Dwi Febri R and P Vania Mitha.....111

012016
OPEN ACCESS
Modification of Blue LED using Organic-Inorganic Hybrid Polymer Doped with Nile Red for Artificial Lighting of Photosynthesis

Norman Syakir, Fahmi Syarifudin, Sahrul Hidayat and Fitrilawati.....119

012017
OPEN ACCESS
Synthesized $Li_4Ti_5O_{12}$ from Technical Grade Raw Material by Excess $LiOH.H_2O$ as Anode Lithium Ion Battery

S Priyono, R D Primasari, S A Saptari and B Prihandoko.....124

012018
OPEN ACCESS
Preparation of Heterogeneous CaO Catalysts for Biodiesel Production

W Widayat, T Darmawan, H Hadiyanto and R Ar Rosyid.....130

012019
OPEN ACCESS
Fuel Sustainability And Actinide Production Of Doping Minor Actinide In Water-Cooled Thorium Reactor

Sidik Permana.....137

012020
OPEN ACCESS
Synthesis of Carbon Nano Materials Originated from Waste Cooking Oil Using a Nebulized Spray Pyrolysis

A.A. Arie, L. Hadisaputra, R.F. Susanti, H. Devianto, M. Halim, R. Enggar and J.K. Lee.....143

012021
OPEN ACCESS
Study on Utilization of Super Grade Plutonium in Molten Salt Reactor FUJI-U3 using CITATION Code

Cici Wulandari, Abdul Waris, Syeilendra Pramuditya, Pramutadi AM Asril and Novitrian.....149

012022
OPEN ACCESS
Random Matrix Theory Approach to Indonesia Energy Portfolio Analysis

Alifian Mahardhika and Acep Purqon.....156

012023
OPEN ACCESS
Self-Assembly of ZnO-Nanorods and Its Performance in Quasi Solid Dye Sensitized Solar Cells

A Aprilia, A Erdienzy, A Bahtiar, L Safriani, N Syakir, Risdiana, T Saragi, S Hidayat, Fitrilawati, R Hidayat and R E Siregar.....162

012024
OPEN ACCESS
The effect of surface morphology in copper oxide nanostructure to photo detector characteristics

Angga Virdian, Christoforus Dimas Satrya, Eka Nurfani and Yudi Darma.....169

012025
OPEN ACCESS
Simulation of Heat Transfer in Husk Furnace with Cone Geometry Based on Conical Coordinate System

Iman Noor, Faozan Ahmad, Irzaman and Husin alatas.....175

012026
OPEN ACCESS
First Principles Study of Molybdenum Disulfide Electronic Structure

Imam Abdul Rahman and Acep Purqon.....182

012027
OPEN ACCESS
Electronic Properties of Rare-Earth Doped α-GaN

Muhammad Widianto and Acep Purqon.....188

012028
OPEN ACCESS
Study of Evaluation OSH Management System Policy Based On Safety Culture Dimensions in Construction Project

Yusuf Latief, Rossy Armyn Machfudiyanto, Rosmariani Arifuddin, R. Mahendra Fira Setiawan and Yoko Yogiswara.....194

012029
OPEN ACCESS
Scale Build-Up Prediction of FeS and $FeCO_3$ in Gas Production Pipes

R. K. Santoso, S. D. Rahmawati, A. Gadesa and D. Wahyuningrum.....202

012030
OPEN ACCESS
Modeling the Thermal Response of $Al_{0.3}Ga_{0.7}As$/GaAs/Ge Multijunction Solar Cells

T Sumaryada, N E Damayanti, S Rohaeni, H Syafutra, Irzaman, A Maddu, H Hardhienata and H Alatas.....212

012031
OPEN ACCESS
Comparison of Several Thermal Conductivity Constants for Thermal Hydraulic Calculation of Pebble Bed Reactor

Dwi Irwanto, Topan Setiadipura and Asril Pramutadi.....217

012032
OPEN ACCESS
Probabilistic Assessment of Severe Accident Consequence in West Bangka

Sunarko and Zaki Su'ud.....222

012033
OPEN ACCESS
Looking at groundwater research landscape of Jakarta Basin for better water management

Dasapta Erwin Irawan, Adhi Priyambodho, Cut Novianti Rachmi and Dimas Maulana Wibowo.....232

012034
OPEN ACCESS
The Effects of Hydration on Protein of Azurin using Coarse-Grained Method and The Free-Energy Analysis

Dian Fitrasari and Acep Purqon.....237

012035
OPEN ACCESS
Accuracy and Numerical Stabilty Analysis of Lattice Boltzmann Method with Multiple Relaxation Time for Incompressible Flows

Pradipto and Acep Purqon.....243

012036
OPEN ACCESS
Structural and optical properties analysis of MoS_2 nanoflakes on quartz substrate as prepared by mechanical exfoliation

Sandy Jaka Adilla, Eka Nurfani, Robi Kurniawan, Christoforus Dimas Satrya and Yudi Darma.....252

012037
OPEN ACCESS
Biodiesel Production by Using CaO Catalyst and Ultrasonic Assisted

W Widayat, T Darmawan, R Ar Rosyid and H Hadiyanto.....258

012038
OPEN ACCESS
Heat exchange studies on coconut oil cells as thermal energy storage for room thermal conditioning

I M Sutjahja, Widya A Putri, Z Fahmi, S Wonorahardjo and D Kurnia.....266

012039
OPEN ACCESS
Second Vertical Derivative Using 3-D Gravity Data for Fault Structure Interpretation

E J Wahyudi, Y Kynantoro and S Alawiyah.....272

012040
OPEN ACCESS
Dielectric-function analysis of metals for plasmonic-device application

Christoforus Dimas Satrya and Yudi Darma.....280

012041
OPEN ACCESS
Construction Performance Optimization toward Green Building Premium Cost Based on Greenship Rating Tools Assessment with Value Engineering Method

Yusuf Latief, Mohammed Ali Berawi, Van Basten, Riswanto and Rachmat Budiman.....290

012042
OPEN ACCESS
Modelling of Mechanical Coupling for Piezoelectric Energy Harvester Adapted to Low-Frequency Vibration

T Untoro, S Viridi, Suprijanto and E Ekawati.....301

012043
OPEN ACCESS
The Temperature Effect on the Working Characteristics of Solar Cells Based on Organometal Halide Perovskite Crystals

Tanti Dewinggih, Shobih, Lia Muliani, Herman and Rahmat Hidayat.....310

012044
OPEN ACCESS
Construction of Maximum Tortuosity of Single Fluid Path in Grid-based 2-d System (5×20) and 3-d system (5×20×3) for Certain Value of Porosity

S Viridi, FDE Latief and SN Khotimah.....315

012045
OPEN ACCESS
First Principle Calculation : Investigation on interaction of Pt/Graphene as Catalyst

Nuning Anugrah Putri Namari and Suprijadi.....324

012046
OPEN ACCESS
Influences of Precursor Solution Concentration and Temperature on $CH_3NH_3PbI_3$ Perovskite Layer Morphology and the Unconverted PbI_2 Proportion to their Perovskite Solar Cell Characteristics

Mhd. Ikhsan Alturisa, Joe Wira, Mardiyati, Herman and Rahmat Hidayat.....332

012047
OPEN ACCESS
Study of Natural Convection Passive Cooling System for Nuclear Reactors

Habibi Abdillah, Geby Saputra, Novitrian and Sidik Permana.....339

012048
OPEN ACCESS
Comparison of Pretreatment Methods on Vetiver Leaves for Efficient Processes of Simultaneous Saccharification and Fermentation by *Neurospora* sp.

E Restiawaty and A Dewi.....344

012049
OPEN ACCESS
Comparative Studies on UO_2 Fueled HTTR Several Nuclear Data Libraries

Anni N. Hidayati, Puguh A. Prastyo, Abdul Waris and Dwi Irwanto.....351

012050
OPEN ACCESS
Review of Membrane Oxygen Enrichment for Efficient Combustion

Danu Ariono and Anita Kusuma Wardani.....355

012051
OPEN ACCESS
The Calculation Study of Electronic Properties of Doped RE (Eu, Er and Tm)-GaN using Density Functional Theory

Aflah Zaharo and Acep Purqon.....362

012052
OPEN ACCESS
Effect of Acetylene Black Content to Half Cells Li-ion Battery Performance Based on $Li_4Ti_5O_{12}$ using Li_2CO_3 as Lithium Ion Source with Hydrothermal Mechanochemical Process

B Priyono, Faizah, A Z Syahrial and A Subhan.....372

012053
OPEN ACCESS
Understanding Kendal aquifer system: a baseline analysis for sustainable water management proposal

A Lukman, M D Aryanto, A Pramudito, A Andhika and D E Irawan.....379

012054
OPEN ACCESS
Ab-Initio Calculation of Electronic Structure of Lead Halide Perovskites with Formamidinium Cation as an Active Material for Perovskite Solar Cells

E D Indari, T D K Wungu and R Hidayat.....389

012055
OPEN ACCESS
Resistivtiy Structure in Tangkuban Parahu Area Drived from CSAMT Data

Enjang Jaenal Mustopa, Nurhasan, Wahyu Srigutomo and Doddy Sutarno.....395

012056
OPEN ACCESS
Simulation of Fluid Flow and Heat Transfer in Porous Medium Using Lattice Boltzmann Method

Imam Wijaya and Acep Purqon.....401

012057
OPEN ACCESS
Synthesis and Characterization of $BaFe_{12}O_{19}$/Poly(aniline, pyrrole, ethylene terephthalate) Composites Coatings as Radar Absorbing Material (RAM)

Nia Sasria, H Ardhyananta, R Fajarin and Widyastuti.....411

012058
OPEN ACCESS
Preliminary Study of 20 MWth Experiment Power Reactor based on Pebble Bed Reactor

Dwi Irwanto, Sidik Permana and Syeilendra Pramuditya.....421

012059
OPEN ACCESS
Thermoeconomical Productivity Analysis in Manufacturing Sector in Indonesia

Widya Liana Aji and Acep Purqon.....425

012060
OPEN ACCESS
Thorium Fuel Utilization Analysis on Small Long Life Reactor for Different Coolant Types

Sidik Permana.....433

012061
OPEN ACCESS
Heat and Mass Transfer Model in Freeze-Dried Medium

Sayahdin Alfat and Acep Purqon.....439

012062
OPEN ACCESS
Understanding Passive Layer Formation for Further Corrosion Management in Gas Production Pipes

R. K. Santoso, S. D. Rahmawati, A. Gadesa and D. Wahyuningrum.....446

012063
OPEN ACCESS
Carbon material@Chitosan composite as catalyst on the synthesis of FAME from used-cooking oil with electrocatalytic process

Rudy Syah Putra, Yudi Antono and Kharis Pratama.....456

012064
OPEN ACCESS
Comparative Study on Various Geometrical Core Design of 300 MWth Gas Cooled Fast Reactor with UN-PuN Fuel Longlife without Refuelling

Ratna Dewi Syarifah, Zaki Su'ud, Khairul Basar and Dwi Irwanto.....464

012065
OPEN ACCESS
Preliminary Studies of Thermoluminescence Dosimeter (TLD) $CaSO_4$:Dy Synthesis

N Nuraeni, F Iskandar, A Waris and F Haryanto.....469

012066
OPEN ACCESS
1-D DC Resistivity Inversion Using Singular Value Decomposition and Levenberg-Marquardt's Inversion Schemes

M Heriyanto and W Srigutomo.....474

012067
OPEN ACCESS
Neutronics Analysis of SMART Small Modular Reactor using SRAC 2006 Code

Rahmi N. Ramdhani, Puguh A Prastyo, Abdul Waris, Widayani and Rizal Kurniadi.....484

012068
OPEN ACCESS
Reducing The Light Reflected by Silicon Surface Using ZnO/TS Antireflection Coating

Andi Suhandi, Yuyu R. Tayubi, Firmanul C. Wibowo, Pepen Arifin and Supriyatman.....490

012069
OPEN ACCESS
Optimation Pressure Loss to Oil Production System with Electrical Submersible Pump (ESP) at the Well A SW Field Bojonegoro, East Java

A M Ilmah and L Hendrajaya.....496

012070
OPEN ACCESS
Preliminary Study of 2-D Time Domain Electromagnetic (TDEM) Modeling to Analyze Subsurface Resistivity Distribution and its Application to the Geothermal Systems

Cahyo Aji Hapsoro, Acep Purqon and Wahyu Srigutomo.....504

012071
OPEN ACCESS
Calculation Study of Electric Properties on Molybdenum Disulfide By Using Density Functional Theory

Diah Angraina Fitri and Acep Purqon.....514

012072
OPEN ACCESS
Ferrous Ion and Medium Composition Effects on Acidogenic Phase in Biobutanol Production from Molasses

E Restiawaty and D Grinanda.....519

012073
OPEN ACCESS
Preliminary Study on Synthesis of Composite rGO/Ni by Microwave Assisted Method

Hafizh A. Fakhri, Miftahul Husnah, Akfiny Hasdi Aimon and Iskandar Ferry.....527

012074
OPEN ACCESS
Recent Advances on Bioethanol Dehydration using Zeolite Membrane

I G B N Makertihartha, P T Dharmawijaya and I G Wenten.....532

012075
OPEN ACCESS
Preparation and characterization of polysulfone/PEG heterogeneous ion exchange membrane for reverse electrodialysis (RED)

D Ariono, Khoiruddin, D Prabandari, R Wulandari and I G Wenten.....539

012076
OPEN ACCESS
Advances of zeolite based membrane for hydrogen production via water gas shift reaction

I G B N Makertihartha, M Zunita, Z Rizki and P T Dharmawijaya.....545

012077
OPEN ACCESS
Synthesis and Characterization of La, Sc, Yb and Nd co-doped Gadolinium doped Cerium (GDC) Composite Electrolyte for IT-SOFC

Damisih, Jarot Raharjo, Masmui, Raffty Setya Aninda and Novita Ami Lestari.....554

Preface: International Conference on Energy Sciences 2016 (ICES 2016)

The 1st International Conference on Energy Sciences 2016 (ICES 2016), which is organized by the Physics Department, Faculty of Mathematics and Natural Sciences, Bandung Institute of Technology (ITB), and supported by Institute for Research and Community Services ITB is aimed at summarizing recent research activities relevant to the innovation in Energy Sciences and its applications and facilitates communication among relevant experts.

In this scientific event the, latest research results regarding the state-of-the-art development in the field have been presented. It is also has been designed to offer the opportunity of making direct contacts for the Indonesian scientists and students with well-known scientists abroad, and thereby fostering the existing research collaborations and extending international research networking for the future.

The range of topics covered by ICES-2016 include: Batteries, Conventional Energies, Renewable Energies, Nuclear Energies, Energy Conversion, Energy Storage, Energy Technologies, Policies & Management, Environmental Aspect of Energy Systems, International Cooperation and Innovation in Energy Systems, Energy Education, and Energy Conservation.

The program of ICES 2016 features 6 invited speeches and 125 contributed oral presentations, which come from 37 institutions, and 5 different countries, namely: Japan, South Korea, Malaysia, Philippines, and Indonesia.

To all participants, we hope that you have learn new subjects, make new contacts, and have fruitful discussions with others.

We wish to express our sincere appreciation to all of the presenters for their valuable contributions and also to the members of the program committee for their excellent works so that this proceeding become reality.

Finally we also would like to thank IoP management for their great efforts in publishing this proceeding.

July 2017
Chairman of ICES 2016,

Abdul Waris

List of Editors

Prof. Abdul Waris
Dept. of Physics,
Institut Teknologi Bandung, Indonesia

Prof.Byunga Shin
Department of Materials Science and Engineering (MSE)
Korea Advanced Institute of Science and Technology (KAIST), South Korean

Prof. Masatoshi Kondo
Laboratory for Advanced Nuclear Energy
Tokyo Institute of Technology, Japan

Dr. Yose Fachmi Buys
Department of Manufacturing and Materials Engineering, Faculty of Engineering,
International Islamic University Malaysia

Dr.Eng. Dwi Irwanto
Dept. of Physics,
Institut Teknologi Bandung, Indonesia

Syeilendra Pramuditya, PhD
Dept. of Physics,
Institut Teknologi Bandung, Indonesia

Organized by:

Dept. of Physics
Faculty Mathematics and Natural Sciences
Institut Teknologi Bandung

Supported by:

LPPM-ITB

IOP Conf. Series: Journal of Physics: Conf. Series **877** (2017) 011002 doi:10.1088/1742-6596/877/1/011002

Peer review statement

All papers published in this volume of *Journal of Physics: Conference Series* have been peer reviewed through processes administered by the proceedings Editors. Reviews were conducted by expert referees to the professional and scientific standards expected of a proceedings journal published by IOP Publishing.

IOP Conf. Series: Journal of Physics: Conf. Series **877** (2017) 012001 doi:10.1088/1742-6596/877/1/012001

Experimental study on corrosion and precipitation in non-isothermal Pb-17Li system for development of liquid breeder blanket of fusion reactor

Masatoshi Kondo[1], Masaomi Ishii[2], Takayoshi Norimatsu[3], Takeo Muroga[4]

[1] Institute of Innovative Research, Tokyo Institute of Technology, 2-12-1-N1-18 Ookayama, Meguro-ku, Tokyo 152-8550, Japan
[2] Department of Nuclear Engineering, Tokai university, 4-1-1 Kitakaname, Hiratsuka-shi, Kanagawa 259-1292, Japan
[3] Institute of Laser Engineering, Osaka university, 2-6 Yamadaoka, Suita, Osaka 565-0871, Japan
[4] Fusion System Research Division, National Institute for Fusion Science, 322-6 Oroshi-cho, Toki-city, Gifu 509-5292, Japan

E-mail: kondo.masatoshi@lane.iir.titech.ac.jp

Abstract. The corrosion characteristics of RAFM steel JLF-1 in a non-isothermal Pb-17Li flowing system were investigated by means of the corrosion test using a non-isothermal mixing pot. The corrosion test was performed at 739K with a temperature gradient of 14K for 500 hours. The corrosion tests at a static and a flowing conditions in an isothermal Pb-17Li system were also performed at the same temperature for the same duration with the non-isothermal test. Then, the effect of mass transfer both by the flow and the temperature gradient on the corrosion behaviors was featured by the comparison of these results. The corrosion was caused by the dissolution of Fe and Cr from the steel surface into the flowing Pb-17Li. The specimen surface revealed a fine granular microstructure after the corrosion tests. A large number of pebble-shaped protrusions were observed on the specimen surface. This microstructure was different from the original martensite microstructure of the steel, and might be formed by the influence of the reaction with Li component in the alloy. The formation of the granular microstructure was accelerated by the flow and the temperature gradient. Some pebble-shaped protrusions had gaps at their bases. The removal of these pebble-shaped granules by the flowing Pb-17Li might cause a small-scale corrosion-erosion. The results of metallurgical analysis indicated that a large-scale corrosion-erosion was also caused by their destruction of the corroded layer on the surface. The non-isothermal mixing pot equipped a cold trap by a metal mesh in the low temperature region. The metal elements of Fe and Cr were recovered as they precipitated on the surface of the metal mesh. It was found that a Fe-Cr binary intermetallic compound was formed in the precipitation procedure. The overall mass transfer coefficient for the dissolution type corrosion in the non-isothermal system was much bigger than that in the isothermal system. This model evaluation indicated that the temperature gradient accelerated the corrosion.

1. Introduction

A liquid lead lithium alloy (Pb-17Li) is one of the candidate tritium breeders of fusion reactors [1, 2] due to its excellent nuclear characteristics and thermodynamic stability. The experimental studies on the fabrication of high-purity alloy [3], the control of non-metal impurities in the liquid alloys [4] and the thermophysical properties at high-temperature [5, 6] have been performed for the development of the fusion blanket.

A reduced activation ferritic martensitic (RAFM) steel is one of the candidate structural materials for the liquid breeder blanket. The corrosion of the RAFM steel in a flowing Pb-17Li is one of the important issues. The corrosion is caused by the dissolution of the steel components from the steel surface into the liquid Pb-17Li [7, 8]. The mechanism of this dissolution type corrosion is based on the mass transfer, which is created by the flow and the concertation gradient in the liquid metal system. Then, the flow may accelerate the dissolution of the steel elements from the steel surface [9]. The heavy density fluid flow may also cause a corrosion-erosion [10], in which the corroded surface is destructed by the flow. However, the information of the corrosion-erosion behavior is still limited.

The blanket system is a non-isothermal system, which consists of a high-temperature region and a low-temperature region. The corrosion is caused in the high-temperature region, and the dissolved elements transfer to the low temperature region on a flowing Pb-17Li. The solubility of the metal elements in the liquid alloy depends on the temperature, and it is larger at higher temperature. Therefore, the elements are precipitated in the low temperature region, when the concentration of the elements in the flowing Pb-17Li is higher than the solubility of the low temperature region. Then, the dissolution type corrosion is continuously caused in the high-temperature region without the saturation of the metal elements in the liquid metals. The previous articles reported about the precipitation of the metal elements in a flowing Pb-17Li [11] and a flowing Pb-Bi [12]. The information of the corrosion and the precipitation in these non-isothermal systems is still limited. The metal impurities dissolved in the flowing Pb-17Li should be recovered during the operation of the blanket system to suppress the loop plugging accident by the large precipitation in the low temperature region. The investigation on the effective way for the removal of the precipitates is also important issue.

The purpose of the present study is to make clear the corrosion characteristics of the RAFM steel and the precipitation behaviours of the metal elements in the non-isothermal system of the liquid Pb-17Li flow. The corrosion characteristics of RAFM steel JLF-1 in the non-isothermal system were investigated by means of the corrosion test using a non-isothermal mixing pot. The results were compared with those obtained from the isothermal corrosion tests to feature the effect of the temperature gradient on the corrosion characteristics.

2. Experimental apparatus and conditions

2.1 Static pot apparatus

Figure 1 (a) shows the schematic drawing of the isothermal static pot. The rectangular type specimen of JLF-1 steel steel [13] was installed in the crucible, which was also made of JLF-1 steel. The chemical compositions of JLF-1 (JOYO-HEAT) are summarized in Table 1. The crucible was filled with a liquid Pb-17Li (3cc). Then, the specimen was fixed to the crucible using Mo wire to suppress the floating of the specimen in a liquid Pb-17Li. The crucible was installed in the vessel made of SUS316 austenitic steel (Fe-18Cr-12Ni), which was filled with the cover gas of a high-purity Ar (99.999%). The vessel was heated by the heater installed on the outside of the vessel.

2.2 Isothermal mixing pot

Figure 1 (b) shows the schematic drawing of the isothermal mixing pot. The inventory of the liquid Pb-17Li in a crucible of this pot is 75cc. The crucible was made of Fe-12Cr steel. The cover gas was a high purity Ar (99.999%). Four cylindrical specimens were installed on the same circumferential position whose center was the shaft of the rotating impeller. The specimens were fixed to the specimen holder made of Fe-12Cr steel. The specimens were immersed to a rotational flow made by the co-rotating with the rotating impeller. The flow velocity around the specimens was roughly expressed by:

$$v=\pi dn \quad (1),$$

where *d* is the diameter [m] of the circumferential layout of the specimens and *n* is the rotating speed [s^{-1}] of the impeller. Then, the Re number for mixing is expressed by:

$$\mathrm{Re}_{\mathrm{mix}}=\frac{\rho n d^2}{\mu} \quad (2),$$

where ρ and μ are density [kg/m^3] and viscosity [Pa·s] of liquid Pb-17Li, respectively. In the current work, *d, n, υ* and $\mathrm{Re}_{\mathrm{mix}}$ are 0.033 m, 50 rpm, 8.7×10^{-2} m/s and 8564, respectively.

2.3 Non-isothermal mixing pot

Figure 1 (c) shows the schematic drawing of the non-isothermal mixing pot. The inventory of the liquid Pb-17Li is 200 cm^3. The sizes of the specimen and the specimen holder are the same with those of the isothermal mixing pot. Four cylindrical specimens of JLF-1 steel were installed in the high-temperature region as the same way with the isothermal mixing pot. As discussed in the previous study [7, 9], the size and the layout of the specimens in the test apparatus must influence on the test results. Therefore, the non-isothermal mixing pot was designed as it had a similar structure and a geometrical parameter with the isothermal pot apparatus. The geometrical parameter is the ration of the liquid volume (V, [m^3]) to the total surface area (A, [m_2]) for corrosion (V/A, [m]). The values of this geometrical parameter for the apparatuses used in the current work were presented in Table 2.

The apparatus consists of the high-temperature region and the low temperature region. The high-temperature region was heated by some heaters and the low temperature region was cooled by a flowing air. The vessel was made of SS316 steel (Fe-18Cr-12Ni). The cylindrical partition, which was made of Fe-12Cr steel, was installed in the vessel. A rotational flow of the liquid Pb-17Li was made by the rotation of an impeller inside the partition. In the same time, a descending flow passed the outside of the partition. Then, one part of the descending flow passed the metal mesh, which was installed as a cold trap in the low temperature region. The metal mesh is made of Fe-12Cr steel. The diameter of the wire of the metal mesh and the mesh opening are 0.22mm and 0.4mm, respectively. The cover gas was a high purity Ar (99.999%).

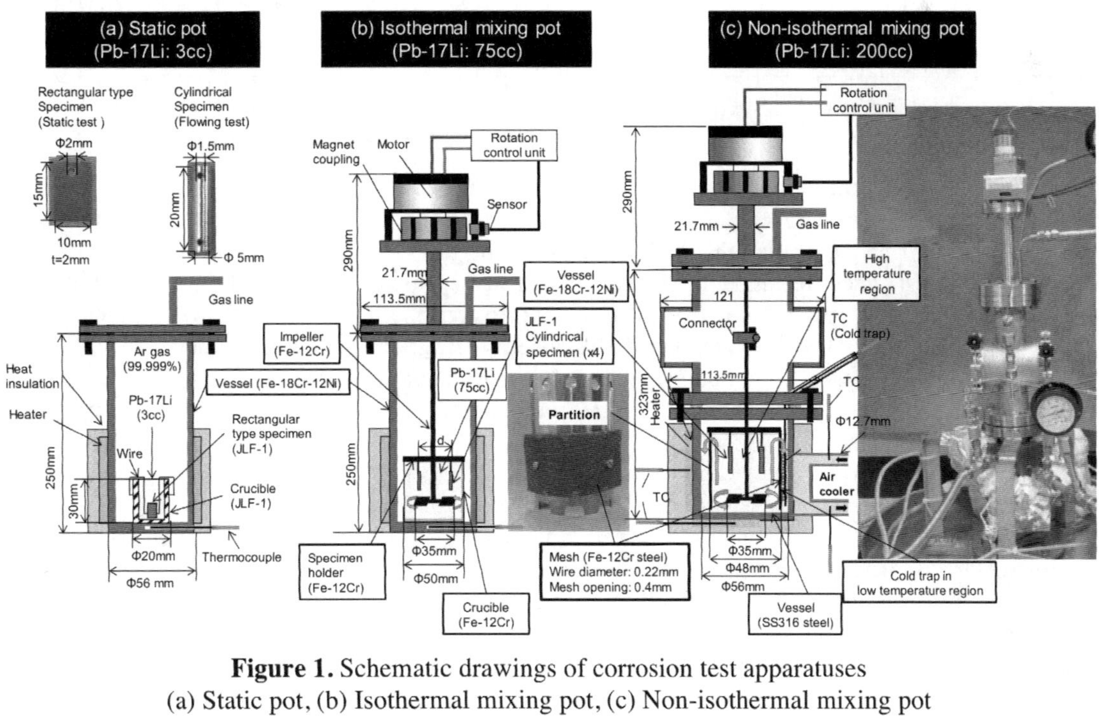

Figure 1. Schematic drawings of corrosion test apparatuses
(a) Static pot, (b) Isothermal mixing pot, (c) Non-isothermal mixing pot

3. Experimental conditions

The conditions of the corrosion tests are summarized in Table 2. The test duration and the temperature for the tests at a static and an isothermal conditions were 500hours and 773K, respectively. This temperature corresponds to the outlet temperature of a liquid Pb-17Li in fusion blanket systems. In the non-isothermal flowing test, the temperature of the high-temperature region was 739K. This temperature was slightly lower than the other tests, because the space for the heater installation was limited in the high-temperature region, and then the heater power was smaller than the other apparatuses. The temperature gradient in the system was approximately 14K. Re number for mixing in the corrosion test section was 8567. The flow velocity in the cold trap was estimated approximately 1mm/s from the heat balance of the system.

After the corrosion tests, the specimens were taken out from the liquid Pb-17Li. Some specimens were cleaned in a liquid Li at 350°C to remove an adhered Pb-17Li from the specimen surfaces. After the immersion to the liquid Li, an adhered Li on the specimens was removed by the immersion to ethanol. Then, the weight changes of the specimens by the corrosion were measured, and the surface microstructure was observed by field emission type scanning electron microscope/energy dispersive X-ray spectrometry (FE-SEN/EDX). The surface cross sections of the specimens without the removal of adhered Pb-Li were metallurgically analyzed by FE-SEM/EDX and electron prove micro analyzer (EPMA).

Table 1. Chemical compositions of RAFM steel JLF-1 (JOYO-HEAT) (wt%) [13]

	Cr	W	C	Mn	Other	Fe
JLF-1	9	1.94	0.09	0.49	V:0.2	Balance

Table 2. Experimental conditions

	Temperature [K] T_{high}	T_{Low}	ΔT [K]	Time [hour]	υ [m/s]	Re_{MIX}	V/A [m]
Static test	773	-	-	500	-	-	$2.1x10^{-3}$
Isothermal flowing test	773	-	-	500	$8.7x10^{-2}$	8567	$8.1x10^{-3}$
Non-isothermal flowing test	739	725	14	500	$8.7x10^{-2}$	8567	$4.6x10^{-3}$

4. Results and Discussions

4.1 Corrosion behavior of JLF-1 steel in static Pb-17Li at 773K

Figure 2 (a-1) shows the SEM image of the specimen surface after the test. The result of the EDX analysis for the specimen surface is presented in Table 3. The concentrations of Cr and W on the corroded surface was larger than that before the corrosion test. The Fe concentration was lower than that before the test. These results indicated that the corrosion was mainly caused by the dissolution of Fe. Though Cr and W were also dissolved by the corrosion, the dissolution of these elements were smaller than that of Fe. The specimen surface revealed small granular microstructure. The size of the granules is approximately 0.2μm. Some large-size granules having a diameter of approximately 0.5μm were observed on the surface. These large-size granules might be formed as the small granules are combined and coarsened.

Figure 2 (a-2) shows the tilt view of the corroded surface by FE-SEM. Small pebble-shaped protrusions were observed on the surface. The size of the protrusions was approximately 0.2μm, and almost the same with that of granules observed in the plan-view image by SEM (Figure 2 (a-2). This microstructure is different from the original martensite microstructure of the steel. However, this microstructure is similar to that observed after the immersion to a liquid Li [14, 15]. The pebble-shaped protrusions were formed in the liquid Li by the progress of the selective corrosion along the boundary of grains and sub-grains. Therefore, this microstructure might be formed by the influence of the reaction between the Li component in the alloy and the steel surface. Some gaps are observed at the base of the protrusions.

Figure 3 (a-1) shows the result of SEM observation and EDX analysis for the cross section of specimen surface. The solid arrow in Figure 3 (a-1) indicated the protrusions observed on the corroded surface in Figure 2 (a-2). From the results of EDX analysis shown in Figure 3 (a-1), the surface layer can be classified into four layers marked as (I) – (IV) in the Figure. The layer (I) is located in the Pb-Li alloy adhered on the specimen surface. The thickness of this layer is approximately 3μm. Fe and Cr are enriched in this layer. This enrichment was due to dissolution of Fe and Cr from the steel surface and their diffusion in a static Pb-17Li. The layer (II) is the Pb diffusion layer in the steel matrix. The thickness of the layer (II) is approximately 1.5μm. The layer (III) is the Fe depleted layer due to the dissolution of Fe into a static Pb-17Li, and its depth is approximately 3.5μm. The Pb concentration in the layer (IV) was higher than the other part of the JLF-1 matrix. This high concentration of Pb was possibly due to the Pb diffusion at the dent part of the surface as illustrated in Fig. 3 (a-2). This result indicated that the gaps, which were observed at the bases of the protrusions (Fig. 2(a-2)), might be formed by the removal of Pb by the cleaning procedure of the specimen.

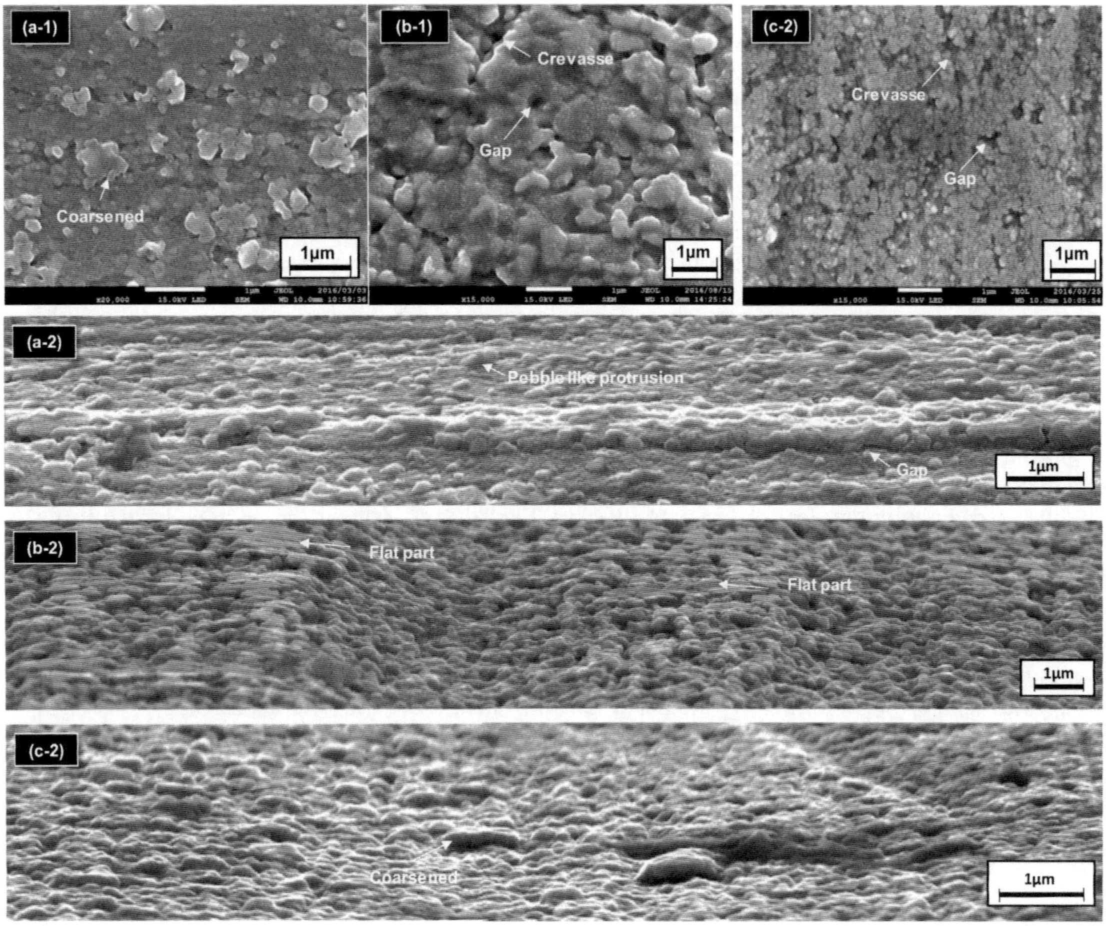

Figure 2. Plan-view and tilt-view SEM images of specimen surfaces, (a-1) & (a-2) surface after static test, (b-1) & (b-2) surface after isothermal flowing test, (c-1) & (c-2) surface after non-isothermal flowing test.

Table 3. Composition ratio of Fe, Cr and W of specimen surface after corrosion test by EDX (wt %)

	Fe	Cr	W	Total
Static test	88.33	9.52	2.15	99.97
Isothermal flowing test	88.74	7.82	3.43	100
Non-isothermal flowing test	87.24	10.40	2.36	100

4.2 Acceleration of corrosion by flow in isothermal system

Figure 2 (b-1) shows the SEM image of the specimen surface after the test. The surface revealed the granular microstructure having some crevasses and gaps. The crevasses and the gaps were formed along the boundaries of grains and sub-grains of the martensitic microstructure. Figure 2 (b-2) shows the tilt-view for the corroded surface by FE-SEM. The specimen surface was fully covered by the pebble-shaped protrusions. The number of the protrusions on the specimen surface was larger than that observed after the static test. The corroded surface had a flattened shape in several places, as the corroded surface was made flat by the shaving and/or the removal of the granules in the flowing Pb-17Li. The diameter of the protrusions is approximately 0.4μm. This size is twice larger than that observed after the static test. These results indicated that the protrusions were combined and coarsened by the corrosion under the flowing conditions. The results of EDX analysis for the corroded surface indicated that the depletion of Cr from the steel surface into Pb-Li alloy.

Figure 3 (b-1) shows the results of FE-SEM/EDX analysis for the cross section of the specimen surface. The surface had a multiple layer, which was formed by the corrosion. The layer is classified into three layers (I), (II) and (III). The layer (I) is the adhered Pb-17Li. The layer (II) is formed by the alloying of Pb, Fe and Cr. In the layer (III), the diffusion of Pb into the steel matrix and the depletion of Fe and Cr from the steel surface were observed. The eroded trace was observed, and its width and depth are approximately 50μm and 15μm, respectively. In the eroded part, the layers (II and III) were not observed. Therefore, this large-scale corrosion-erosion was possibly caused by the destruction of corroded surface as illustrated in Fig. 3 (b-2).

4.3 Corrosion and precipitation in non-isothermal Pb-17Li system

The corroded surface revealed a fine granular surface (Figure 2 (c-1)). The diameter of granules is approximately 0.1μm. Some crevasses and gaps are observed on the surface. The size of the gaps was larger than the diameter of the granules. Therefore, these gaps were possibly made by the removal of the granules in flowing Pb-17Li. The results of the EDX analysis on the corroded surface indicated the dissolution of Fe from the steel surface into Pb-17Li. Figure 2 (c-2) shows the tilt view for the surface. The surface was fully covered by the pebble-shaped protrusions. Some large size protrusions, which had a diameter of 0.5-1μm, were dotted on the surface.

Figure 3 (b-3) shows the result of FE-SEM/EDX analysis on the cross sections of the large-size protrusions. The size of the protrusions are approximately 2.6μm. The concentrations of Fe and Cr in the protrusions are lower than the steel matrix (L-1 and L-3 in Fig.3 (b-3)). This result indicated that the protrusions might be formed by the corrosion in flowing Pb-17Li. The depletion of Fe was also detected on the surface of steel matrix (L-1, L-2 and L-3 in Fig.3 (b-3)). There was the gap between the protrusions and the steel matrix, and Pb was detected in the gap. These results indicated that the protrusions were not firmly fixed on the steel surface, and these could be removed from the surface by flowing Pb-17Li. Then, the removal of the pebble-shaped protrusions may cause a small-scale corrosion-erosion.

Figure 4 shows the results of EPMA analysis on the cross sections of the metal mesh installed as the cold trap in the low temperature region. The black portions, which are dotted in Pb-17Li (Figure 4 (a-1)), are abrasive powders used for the polishing procedure of the cross section. The surface of the wire was fully covered by the metal elements of Fe and Cr as shown in Figures 4 (a-2) and (a-3). These metal elements were dissolved from the surface of the specimen and the wall of the vessel in the high-temperature region of the pot. These elements were transferred to the low temperature region by flowing Pb-17Li, and then they precipitated on the wire surface since their solubility in the low temperature region was lower than the concentration in flowing Pb-17Li.

The precipitates of Fe and Cr were observed as they formed a double layer on the wire surface. In the layer (I), Fe and Cr were precipitated in a solidified Pb-Li. The thickness of the layer (I) was approximately 10μm. The positions of the Cr precipitates overlapped with those of Fe as shown in Figures 4 (b-2) and (b-3). These trends indicated that Fe-Cr alloy was formed in the precipitation procedure. In the layer (II), the particulate precipitates of Fe and Cr were observed. However, some Fe precipitates in the layer (II) did not contained the element of Cr. The solubility of Fe in Pb-17Li and its temperature dependence were larger than those of Cr [16-18]. Therefore, the precipitation of Fe was larger than that of Cr. The metal element of Ni was not detected in the precipitates. The solubility of Ni in Pb-17Li is much larger than those of Fe and Cr [18]. Then, Ni was not precipitated since the Ni concentration in Pb-17Li did not reach the solubility at the temperature of the cold trap during the test duration.

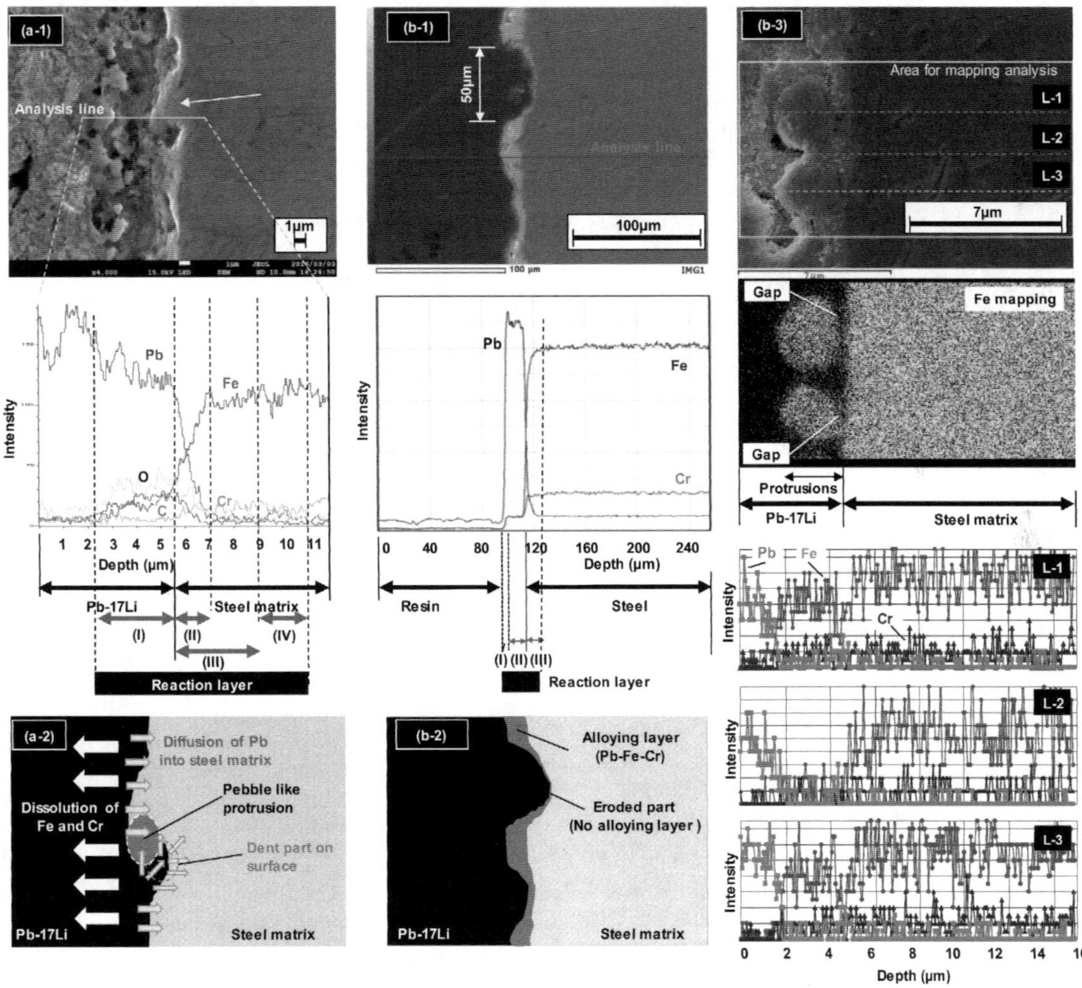

Figure 3. Results of FE-SEM/EDX analysis, (a-1) surface cross section of specimen exposed to static condition, (a-2) schematic drawing of corrosion and Pb diffusion around pebble-shaped protrusion, (b-1) surface exposed to non-isothermal system (low magnification), (b-2) schematic drawing of large scall corrosion-erosion, (b-3) surface cross section of specimen exposed to non-isothermal system (high magnification).

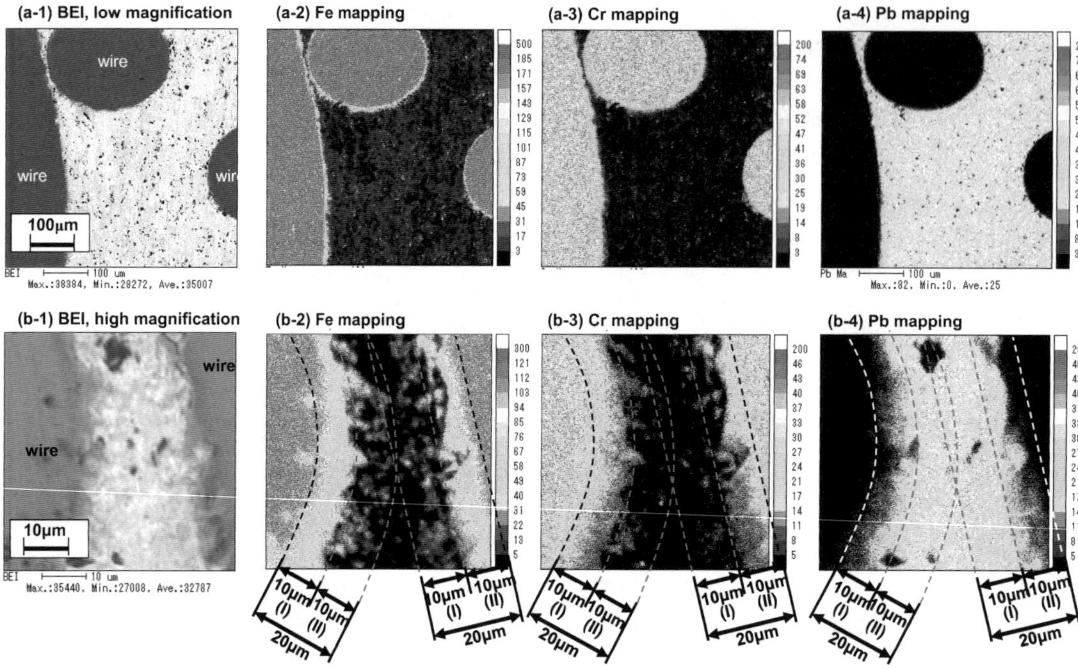

Figure 4 Results of EPMA analysis for cross section of metal mesh installed in cold trap, (a-1), (a-2), (a-3) and (a-4): BEI image and mapping analysis for Fe, Cr and Pb with low magnification, (b-1), (b-2), (b-3) and (b-4): BEI image and mapping analysis for Fe, Cr and Pb with high magnification.

4.4 Corrosion evaluation by mass transfer model in non-isothermal Pb-17Li system

The mass flux for the dissolution type corrosion on the specimen surface ($J_{corrosion}$, [g/m^2s]) is expressed by:

$$J_{corrosion}=h\rho(C_s-C) \quad (3),$$

where h, C and C_s are the overall mass transfer coefficient [m/s], the metal concentration and the metal solubility in flowing Pb-17Li, respectively. The solubility is the function of the temperature.

The mass flux for the metal precipitation in the low temperature region of the non-isothermal mixing pot ($J_{precipitation}$, [g/m^2s]) is given by:

$$J_{precipitation}=k\rho(C-C_s{}^*) \quad (4),$$

where k is the effective rate constant [m/s] for the precipitation including the trap of the precipitation on the surface of the cold trap. $C_s{}^*$ is the solubility of metal elements at the temperature of the cold trap. At the balance point [11] in the non-isothermal system, the mass fluxes for the corrosion and the precipitation are equal expressed as:

$$J_{corrosion}=J_{precipitation}=0 \quad (5).$$

The overall mass transfer coefficient for the dissolution type corrosion is derived from the Sherwood number, which is the non-dimensional number for mass transfer in fluids [9]. The coefficient increases with the Reynolds number. On the contrary, the effective rate constant for the precipitation decreases with the flow rate.

The weight losses per unit area of the specimens obtained by the corrosion experiments are summarized in Table 4. The total weight loss of the specimens and the crucible by the dissolution type corrosion in the isothermal system ($\Delta m_{\mathrm{all(isothermal)}}$, [g]) is derived as:

$$\Delta m_{\mathrm{all(isothermal)}} = \rho V C_s \left(1 - \exp\left(-\frac{h(A_{\mathrm{crucible}}+A_{\mathrm{specimens}})t}{V}\right)\right) \quad (6),$$

where V and t are the inventory of liquid Pb-17Li [m^3] and the exposure time [s], respectively. A_{crucible} and $A_{\mathrm{specimens}}$ are the fluid contact area of the crucible and the specimens, respectively. The material of the crucible had similar chemical compositions with the specimen. The flowing condition near the vessel wall was also the similar to that on around the specimen. Here, the corrosion on the wall of the crucible is assumed as the same with the specimens in the current work. Then, $\Delta m_{\mathrm{all(isothermal)}}$ is also given by the total weight loss of the specimens exposed to the liquid Pb-17Li ($\Delta m_{\mathrm{specimens(isothermal)}}$, [g]) in the corrosion test. This relation is expressed as:

$$\Delta m_{\mathrm{all(isothermal)}} = \frac{A_{\mathrm{crucible}}+A_{\mathrm{specimens}}}{A_{\mathrm{specimens}}} \Delta m_{\mathrm{specimens\ (isothermal)}} \quad (7),$$

The solubilities of Fe and Cr in liquid Pb-17Li at the temperature of 773K are assumed as 47wppm and 10wppm, respectively [17]. Then, the maximum weight loss of the specimens from these solubilities in the isothermal system is given by:

$$\Delta m_{\mathrm{specimens(isothermal),max}} = \rho V C_s \frac{A_{\mathrm{specimens}}}{A_{\mathrm{crucible}}+A_{\mathrm{specimens}}} \quad (8).$$

The overall mass transfer coefficients in the isothermal system are obtained from the weight loss of the specimens by eqs. (6) and (7). The results of the model evaluation are summarized in Table 4. The weight losses of the specimens obtained by the corrosion tests were smaller than the values given by eq. (8). It was indicated that liquid Pb-17Li was not saturated with the metal elements during the corrosion tests. The overall mass transfer coefficient at the flowing conditions are approximately 10 times larger than that at the static condition. These results indicated that the mass transfer on the corrosion was accelerated by the flow. However, it must be carefully considered that the occurrence of the corrosion-erosion is not separately evaluated in the current model evaluation.

In the non-isothermal Pb-17Li system, the concentration of the metal impurities in the fluid at the inlet of the high-temperature region can be the same with the solubility in the low temperature region due to the cold trap after the operation of a certain period. Then, the concentration of metal impurities at the inlet is constant. The mass flux for the corrosion in the high-temperature region of the non-isothermal system under the steady state condition is expressed by:

$$J^*_{\mathrm{corrosion}} = h\rho(C_s - C_s^*) \quad (9).$$

The weight gain in the cold trap by the precipitations of the dissolved metal elements in total is expressed by:

$$\Delta m_{\mathrm{precipitation}} = J_{\mathrm{precipitation}} A_{\mathrm{precipitation}} t \quad (10),$$

where $A_{\mathrm{precipitation}}$ is the effective surface area for the precipitation in the cold trap [m^2]. Based on the mass balance between the corrosion in the high-temperature region and the precipitation in the low temperature region, the total weight loss of the specimens and the vessel in the non-isothermal system is expressed:

$$\Delta m_{\text{all(non-isothermal)}}=\frac{A_{\text{vessel}}+A_{\text{specimens}}}{A_{\text{specimens}}}\Delta m_{\text{specimens (non-isothermal)}} \quad (11).$$

The weight loss of the specimens after the exposure for a certain period is expressed as:

$$\Delta m_{\text{specimens (non-isothermal)}}= J^*_{\text{corrosion}}A_{specimens}t \quad (12).$$

The overall mass transfer coefficient for the corrosion in non-isothermal system under the steady state condition is derived from eqs. (9) and (12) as:

$$h=\frac{\Delta m_{\text{specimens (non-isothermal)}}}{\rho A_{specimens}t(C_s-C_s^*)} \quad (13).$$

The overall mass transfer coefficient was obtained as 3.8×10^{-9} m/s by eqs. (9) and (12), when C_s* was assumed as a half of C_s due to the small temperature difference in the current system. The coefficient was obtained as 8.3×10^{-11} m/s by eqs. (6) and (7) under the assumption as the steady state condition was not achieved during the current corrosion test. This value is also approximately 8 times larger than that in the isothermal flowing condition. These results indicated that the temperature gradient might strongly accelerate the corrosion in liquid metal flowing system.

Table 4 Weight losses of specimens and mass transfer coefficient

	Weight loss of specimens per unit area by experiment [g/m²] $\left(\frac{\Delta m_{\text{specimens(isothermal or non-isothremal)}}}{A_{\text{specimens}}}\right)$	Maximum weight loss of specimens per unit area by eq. (8) [g/m²] $\left(\frac{\Delta m_{\text{specimens(isothermal),max}}}{A_{\text{specimens}}}\right)$	Mass transfer coefficient [m/s]
Static test	7.62×10^{-1}	1.24	1.0×10^{-12} by eq. (6)
Isothermal flowing test	1.23	4.81	1.1×10^{-11} by eq. (6)
Non-isothermal flowing test	2.03	(2.77)	3.8×10^{-9} by eq. (13) (8.3×10^{-11} by eq. (6))

5. Conclusion

The corrosion characteristics of RAFM steel JLF-1 in the non-isothermal Pb-17Li system were investigated by means of the corrosion test using the non-isothermal mixing pot. The corrosion test was performed at 739K with the temperature gradient of 14K for 500 hours. The corrosion tests in the isothermal Pb-17Li system were performed at the same temperature for the same duration with the non-isothermal test to feature the effect of mass transfer in the non-isothermal system on the corrosion behaviors. Major conclusions are follows;

(1) The corrosion was mainly caused by the dissolution of Fe and Cr from the steel surface into the Pb-17Li regardless of the flowing condition and the temperature gradient. The corroded surface of the specimens revealed the slight enrichment of Cr and W, and this behavior indicated that the dissolution ratios of Cr and W were smaller than that of Fe in Pb-17Li.

(2) The specimen surface revealed the granular microstructure after the corrosion test. The pebble-shaped protrusions were observed on the corroded surface. The size of the protrusions was 0.2μm - 1μm. The size of the protrusions, which were formed in the flowing Pb-17Li with the temperature gradient, was larger than that formed in the isothermal conditions. This microstructure was not based on the original martensite microstructure of the steel, and might be formed by the influence of the reaction with Li component in the alloy. The formation was accelerated by the flow and the temperature gradient. Some pebble-shaped protrusions had gaps

around their bases. This unstable structure indicated the protrusions could be removed by the flowing Pb-17Li.

(3) The trace of the large-scale corrosion-erosion was observed on the corroded surface after the exposure to the liquid Pb-17Li flow at the isothermal condition. The width and the depth of the trace was approximately 50μm and 15μm, respectively. The Fe-Cr-Pb layer was formed on the surface. The thickness of the layer on the eroded part was much thinner than that in the other part. These results indicated that the large-scale corrosion-erosion was caused on the surface by the partial destruction of the corroded layer.

(4) The metal elements of Fe and Cr were dissolved by the corrosion in the non-isothermal Pb-17Li system. These elements were precipitated on the surface of the metal mesh of the cold trap installed in the low temperature region of the pot. These elements precipitated as they formed Fe-Cr alloy on the surface of the wire.

(5) The overall mass transfer coefficient for the dissolution type corrosion in non-isothermal Pb-17Li system was larger than that in the isothermal system. This model evaluation indicated that the temperature gradient strongly accelerated the corrosion.

Acknowledgments
Authors acknowledge Mr. Ohe T, Mr. Nakano J and Mr. Awaji T in Marumoto Struers K. K for their technical advices on the cross sectional polishing procedure of the JLF-1 specimen and the solidified Pb-17Li. Authors also acknowledge Mr. Y. Tsuruoka in Tokyo Institute of Technology for his technical assistant for the corrosion experiments. This work was partially supported by NIFS budget code NIFS14KOBF028, NIFS13KUGK076 and Grant-in-Aid for Scientific Research (A) 19206100 (2007–2009), Young scientists (B) 25820437 (2013–2016) and Scientific Research (B) 16H04617 (2016).

References
[1] Malang S, Raffray A R and Morley N B 2009 *Fusion Eng. Des.* **84** pp 2145–57
[2] Kozaki Y, Norimatsu T, Furukawa Y, Hayashi T, Souman Y, Nishikawa M and Tomabechi K 2007 *J. Plasma Fusion Res.* **82** No.12 pp 19-27
[3] Nakajima Y, Kondo M and Nozawa T 2015 *Fusion Eng. Des.* **98–99** pp 2009–14
[4] Kondo M, Nakajima Y, Tanaka T, Nozawa T and Yokomine T 2016 *J. Plasma Fusion Res.* **11** 2405076
[5] Kondo M and Nakajima Y 2013 *Fusion Eng. Des.* **88** pp 2556-59
[6] Kondo M, Nakajima Y, Tsuji M and Nozawa T 2016 *Fusion Eng. Des.* **109-111** pp 1345-50
[7] Kondo M, Takahashi M, Teruya T and Tsisar V and Muroga T 2012 *Fusion Eng. Des.* **87** pp 1777-1787
[8] Kondo M and Muroga T 2013 *Plasma Fusion Res.* **8** 3405053
[9] Kondo M, Muroga T, Sagara A, Valentyn T, Suzuki A, Terai T, Takahashi M, Fujii N, Yokoyama Y, Miyamoto H and Nakamura E 2011 *Fusion Eng. Des.* **86** pp 2500–03
[10] Kondo M, Takahashi M, Suzuki T, Ishikawa K, Hata K, Qiu S and Sekimoto H 2005 *J. Nucl. Mater.* **343** pp 349–59
[11] Tortorelli P F 1991 *Fusion Eng. Des.* **14** pp 335-45
[12] Kondo M and Takahashi M, 2005 *Prog. Nucl. Energ.* **47**, No. 1-4 pp 639-47
[13] Ono H, Kasada R and Kimura A 2004 *J. Nucl. Mater.* **329–333** pp 1117–21
[14] Kondo M, Tsisar V, Muroga T, Nagasaka T and Yeliseyeva O 2010 *J. Plasma Fusion Res.* **9** pp 294-99
[15] Tsisar V, Kondo M, Xu Q, Muroga T, Nagasaka T and Yeliseyeva O 2011 *J. Nucl. Mater.* **417** pp 1205-09

[16] Balbaud-Celerier F and Barbier F 2001 *J. Nucl. Mater.* **289** pp 227-42
[17] Simon N, Terlain A and Flament T 1998 *J. Nucl. Mater.* **254** pp 185-90
[18] Baker M G and Sample T 1991 *Fusion Eng. Des.* **14** pp 219-26

From lab to full-scale ultrafiltration in microalgae harvesting

I G Wenten[1,*], S Steven[1], A Dwiputra[1], Khoiruddin[1] and A N Hakim[1]

[1] Department of Chemical Engineering, Institut Teknologi Bandung, Jl. Ganesha 10, Bandung 40132, Indonesia

*Email: igw@che.itb.ac.id

Abstract. Ponding system is generally used for microalgae cultivation. However, selection of appropriate technology for the harvesting process is challenging due to the low cell density of cultivated microalgae from the ponding system and the large volume of water to be handled. One of the promising technologies for microalgae harvesting is ultrafiltration (UF). In this study, the performance of UF during harvesting of microalgae in a lab- and a full-scale test is investigated. The performances of both scales are compared and analyzed to provide an understanding of several aspects which affect the yield produced from lab and actual conditions. Furthermore, a unique self-standing non-modular UF is introduced in the full-scale test. The non-modular UF exhibits several advantages, such as simple piping and connection, single pump for filtration and backwashing, and smaller footprint. With those advantages, the non-modular UF could be a promising technology for microalgae harvesting in industrial-scale.

1. Introduction

Ponding system is still generally used for microalgae cultivation [1]. *Nannochloropsis sp.* is one of the microalgae which has high economic value and mostly cultivated due to its nutritional content and potentiality in a wide range of usage. Additionally, *Nannochloropsis sp.* is easily cultured, does not cause toxicity or damage to the ecosystem in the tub larval rearing, growing relatively fast, and contain antibiotic substances. However, low cell density resulted from the ponding system is still the main challenge. This low cell density is attributed to a low light intensity and a low CO2 concentration in the water [2, 3]. The low cell density biomass and large volumes of water will become a burden on the harvesting process which is the largest cost in the production of biomass [4].

Basically, there are three main steps in the production of the biomass from microalgae which are cultivation, harvesting, and post-harvesting processes. Post-harvesting process is generally conducted after microalgae are successfully harvested and concentrated up to 10^9 cells/mL. Microalgae are usually harvested using the conventional process such as chemical flocculation and centrifugation. However, these conventional processes possess several disadvantages such as high chemical consumption, time-consuming, and expensive [5].

Recently, membrane technology is widely applied in almost every industrial sector due to advantages of this technology [6]. Energy efficiency, high separation capacity, selective separation, reduction in a number of unit process, and low investment cost are several advantages of membrane technology which made this technology is competitive compared to other technologies in various functions [7-18]. Ultrafiltration (UF) is a pressure-driven membrane process which is generally used in water treatment, wastewater treatment, biotechnology, food processing, etc. In microalgae harvesting, UF is the most promising techniques for concentrating microalgae [19-26]. The advantages of UF in microalgae harvesting include less damage to cell integrity due to cross flow filtration mode, low or no chemical consumption, easy to scale up, flexible operation, and low energy consumption. However, fouling is still the main problems which result in flux decline overtime. Many strategies have been proposed to control the fouling phenomenon such as backwashing, chemically enhanced backwash, chemical cleaning, selection of appropriate operation, pretreatment, and membrane modification [13, 27-33]. By using the appropriate strategies, membrane technology could be operated effectively.

In this study, the performance of UF during harvesting of microalgae in a lab- and a full-scale test is investigated. The performances of both scales are compared and analyzed to provide an understanding of several aspects which affect the yield produced from lab and actual conditions. Furthermore, a unique self-standing non-modular UF is introduced in the full-scale test. The performance and advantages of this technique will be presented.

2. Materials and method

2.1. Lab-scale experiment

Microalgae were grown in the medium called "Walne medium". It was made with the following composition: 100 grams of NH_4NO_3; 20 grams NaH_2PO_4; 33.6 grams of H_3BO_3; 0.36 grams $MnCl_2$; 1.3 grams $FeCl_3$; and 45 grams of EDTA were dissolved in 1 liter of demineralized water. This solution is called pro-analyze. Then, 1 ml of trace elements is added into 1 liter of pro-analyze solution. The trace elements used were 2.1 grams of $ZnCl_2$; 2 g $CoCl_2$; 2 grams of $CuSO_4$; and 0.9 grams of $(NH_4)_6MoO_{25}$ which were dissolved in 100 mL of water. The medium is then sterilized if necessary. One milliliter of Walne medium was used for 1 liter of microalgae's culture solution.

This study begins with acclimatization isolates of *Nannochloropsis sp.* in the laboratory by aerating the microalgae isolates for 2-3 days. Microalgae then cultivated in the Walne medium for about 2 weeks. Microalgae were aerated with 3 L/min compressed air and under artificial light of 2000 lux intensity. The light intensity is measured by lux meter.

Microalgae's growth was assayed by taking a sample and then viewed under a light microscope to identify the cells. *Nannochloropsis sp.* that already reached the stationary phase will be counted the total cells using a counting chamber.

For filtration experiment, hollow fiber polypropylene ultrafiltration membrane (effective area: 0.001 m^2) supplied by GDP Filter Indonesia was used. The UF was operated at 2 bar of trans-membrane pressure. UF process was held for about 90 minutes. Chemically enhanced backwash (CEB) was also conducted at 15 minutes duration with the interval of 30 and 45 minutes.

Table 1. Properties of full-scale UF membrane.

Properties	Value
Membrane material	Polypropylene
Hollow fiber OD (μm)	400
Membrane area (m^2)	120
Module diameter (inch)	12
Module length (m)	1,5
Feed pump type	Centrifugal
Head and impeller material	SS 316
Pump flowrate (m^3/h)	20
Pump head (m)	40
Motor power (kW)	4,5

2.2. *Full-scale UF plant*

The full-scale UF plant uses a unique self-standing non-modular UF membrane which is fabricated by GDP Filter Indonesia. The properties of the membrane are listed in table 1. Schematic diagram of full-scale UF plant for microalgae harvesting and the photograph of the plant are shown in figure 1 and 2, respectively. The plant consists of concentration tank (product tank) with capacity of 2.5 m^3 wherein cultivated microalgae from the pond will be concentrated, feed pump for pumping microalgae to UF module and backwashing of UF module when single batch of concentration cycle is completed, UF module, and permeate tank with capacity of 1 m3 for storing permeate water which is used for backwash. In the operation, cultivated microalgae from the pond are pumped to product tank then it is fed to UF membrane using UF circulating pump where permeate is stored in permeate tank while concentrate is recirculated until desired cell concentration is obtained. This system utilizes a single pump for circulation and backwash. Backwash is conducted at the end of concentration cycle using a suction action of circulation pump which draws pure water from permeate tank through UF module. Air scouring is also employed during backwash.

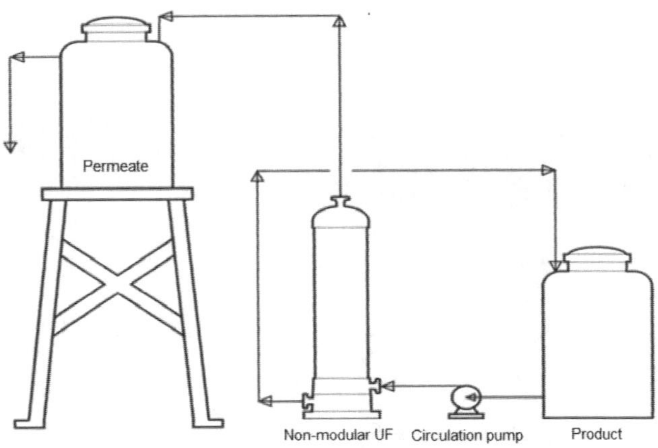

Figure 1. Schematic of the full-scale UF plant

(a) (b)

Figure 2. Full-scale UF: (a) Self-standing non-modular UF (12 inch in diameter) and (b) UF plant

3. Results and discussion

The performance of lab-scale UF in microalgae harvesting process is illustrated in figure 3 and 4. Figure 3 shows that UF is able to concentrate microalgae cells up to 1×10^9 cells/ml from the cultivated microalgae containing 4.6×10^7 cells/ml. Almost 98% of the microalgae cells are successfully rejected by UF membrane. However, rapid flux declining was observed during the filtration. Chemically enhanced backwash was then conducted to recover the flux. Even though the backwash successfully recovered the flux, this rapid decline would not applicable for the full-scale since it would require an intensive chemically backwash. Another alternative way for maintaining the flux stability is by operating the membrane under a relatively low flux. Under this condition, a drag force for solute deposition on membrane surface is decreased thus the fouling formation could be reduced [34, 35]. This strategy would be preferred for a large scale test.

For energy production purpose, open ponding system is the most suitable system for large-scale microalgae cultivation. Supported by tropical weather, it is considered a simple and low cost since uses direct sunlight intensity for growth. In this study, full-scale microalgae cultivation pond consists of 4 square pond with dimension of 200 x 200 m^2 each. The mature pond will have 1.4 x 10^7 cells/ml in average that is 3 times lower than lab cultivated culture (figure 3). Contamination from the atmosphere and lower light intensity especially at rainy season are identified as the root cause of the problem. Growth competition between microalgae and zooplankton is also noticed which negatively affects microalgae population.

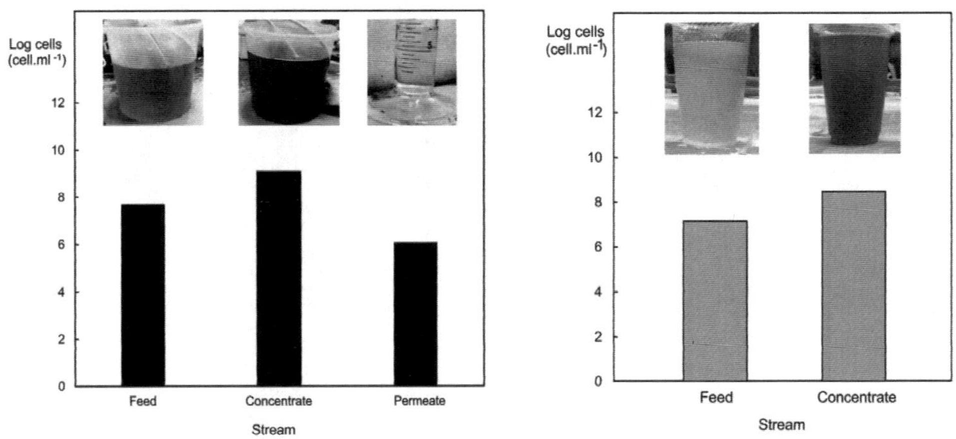

(a) (b)

Figure 3. Microalgae cells in (a) lab and (b) full-scale test.

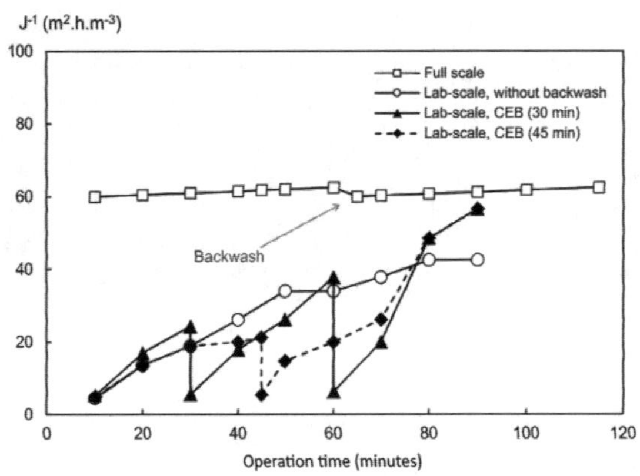

Figure 4. UF performance during filtration of microalgae in lab-scale experiment

Apart from its disadvantages, open ponding system is still attractive from economic and operation point of view while finding best harvesting system is the main challenge. Full-scale ultrafiltration plant for microalgae harvesting process has been successfully operated to concentrate cultivated microalgae from 1.4×10^7 cell/ml to 3×10^8 cell/ml (figure 3). According to experience in lab scale test, low flux (16,67 $l.m^{-2}.h^{-1}$) and high turbulence (Re = 16,121) was used in the full-scale operation to maintain performance stability while outside-in filtration mode was used to prevent fiber clogging. High degree of turbulence which is applied in this process prevent the formation of the film layer on membrane surface resulting stable permeation flux with insignificant decrease. Low flux and high turbulence operation on full-scale plant exhibit lower filtration resistance due to cake build up indicated by the lower slope of $1/J$ vs t as shown in figure 4.

Beside superiority in performance stability, the full-scale plant suffers disadvantage attributed to high circulation rate operation. At high circulation rate, heat is continuously generated which mainly due to intensive friction between the fluid and pump head (especially when using a centrifugal pump) and accumulated to the circulation fluid thus significantly increase it is temperature. A significant increase in temperature may have a negative effect on cell survival during storage. For better performance, positive displacement pump e.g. rotary screw pump is recommended since the friction is lower even at high circulation rate. According to our experience in full-scale operation, we found that there was an entrapped volume of fluid with high cell concentration inside ultrafiltration module as a dead volume at the end of concentration cycle which is proportional to module dimension. This fluid should be drained from module before backwashing UF module. To enhance draining process, pressurized air from is used to force the dead fluid out of UF module by flowing from lumen side to shell side (backwash flow).

4. Conclusion

In this study, the performance of UF during harvesting of microalgae in a lab- and a full-scale test is investigated. The performances of both scales are compared and analyzed to provide an understanding of several aspects which affect the yield produced from lab and actual conditions. Furthermore, a unique self-standing non-modular UF is introduced in the full-scale test. The non-modular UF exhibits several advantages, such as simple piping and connection, single pump for filtration and backwashing, and smaller footprint. With those advantages, the non-modular UF could be a promising technology for microalgae harvesting in industrial-scale.

Acknowledgments

The authors would like to express their gratitude to GDP Filter, Indonesia, for supporting the data and pictures of full-scale UF plant.

References

[1] Milano J, Ong H C, Masjuki H, Chong W, Lam M K, Loh P K and Vellayan V 2016 *Renewable and Sustainable Energy Reviews* **58** 180
[2] Jorquera O, Kiperstok A, Sales E A, Embiruçu M and Ghirardi M L 2010 *Bioresource technology* **101** 1406
[3] Makareviciene V, Skorupskaite V and Andruleviciute V 2013 *Reviews in Environmental Science and Bio/Technology* **12** 119
[4] Molina Grima E, Belarbi E H, Acién Fernández F G, Robles Medina A and Chisti Y 2003 *Biotechnology Advances* **20** 491
[5] Brennan L and Owende P 2010 *Renewable and sustainable energy reviews* **14** 557
[6] Wenten I G and Khoiruddin 2016 *Desalination* **391** 112
[7] Drioli E, Stankiewicz A I and Macedonio F 2011 *J. Membr. Sci.* **380** 1
[8] Drioli E and Curcio E 2007 *J. Chem. Technol. Biotechnol.* **82** 223
[9] Wenten I G 2002 *J Sci Technol Membrane Sci Technol* **24** 1010

[10] Drioli E, Curcio E and Di Profio G 2005 *Chem. Eng. Res. Des.* **83** 223
[11] Purwasasmita M, Nabu E B P, Khoiruddin and Wenten I G 2015 *Journal of Engineering and Technological Sciences* **47** 426
[12] Wenten I G, Julian H and Panjaitan N T 2012 *Desalination* **306** 29
[13] Wenten I G and Widiasa I N 2002 *Desalination* **149** 279
[14] Poulsen B R and Iversen J J L 1998 *Biotechnol. Bioeng.* **58** 633
[15] Rios G, Belleville M, Paolucci D and Sanchez J 2004 *J. Membr. Sci.* **242** 189
[16] Shu J, Grandjean B, Neste A v and Kaliaguine S 1991 *The Canadian Journal of Chemical Engineering* **69** 1036
[17] Judd S 2008 *Trends Biotechnol.* **26** 109
[18] Joscelyne S M and Trägårdh G 2000 *J. Membr. Sci.* **169** 107
[19] Discart V, Bilad M R, Vandamme D, Foubert I, Muylaert K and Vankelecom I 2013 *Bioresour. Technol.* **129** 18
[20] Merle T, Dramas L, Gutierrez L, Garcia-Molina V and Croué J-P 2016 *Water Res.* **93** 10
[21] Nguyen T, Fan L, Roddick F and Harris J 2009 *Desalination* **236** 208
[22] Ruiz-Martinez A, Garcia N M, Romero I, Seco A and Ferrer J 2012 *Bioresour. Technol.* **126** 247
[23] Zhang S and Santschi P H 2009 *Limnology and Oceanography: Methods* **7** 419
[24] Zhang X, Hu Q, Sommerfeld M, Puruhito E and Chen Y 2010 *Bioresour. Technol.* **101** 5297
[25] Zhang Y, Tang C Y and Li G 2012 *Water Res.* **46** 4783
[26] Zhang Y, Tian J, Liang H, Nan J, Chen Z and Li G 2011 *J. Environ. Sci.* **23** 529
[27] Yang X, Wang R, Fane A G, Tang C Y and Wenten I G 2013 *Desalin. Water Treat.* **51** 3604
[28] Aryanti P T P, Yustiana R, Purnama R E D and Wenten I G 2015 *Membr. Water Treat.* **6** 379
[29] Himma N F, Anisah S, Prasetya N and Wenten I G 2016 *J. Polym. Eng.* **36** 329
[30] Kwon B, Park N and Cho J 2005 *Desalination* **179** 203
[31] Liang H, Gong W, Chen J and Li G 2008 *Desalination* **220** 267
[32] Kanchanatip E, Su B-R, Tulaphol S, Den W, Grisdanurak N and Kuo C-C 2016 *Bioresour. Technol.* **209** 23
[33] Zhao F, Chu H, Tan X, Yang L, Su Y, Zhou X, Zhao J and Zhang Y 2016 *J. Membr. Sci.* **517** 30
[34] Wenten I and Widiasa I 2002 *Desalination* **149** 279
[35] Chen V, Fane A, Madaeni S and Wenten I 1997 *J. Membr. Sci.* **125** 109

Preliminary Study of Plutonium Utilization in AP1000 Reactor

Nailatussaadah[1], Puguh A. Prastyo[1], Abdul Waris[2,*], Rizal Kurniadi[2] and Syeilendra Pramuditya[2]

[1]Department of Physics, Faculty of Mathematics and Natural Sciences,
Institut Teknologi Bandung,
Jl. Ganesa 10 Bandung 40132, INDONESIA
[2]Nuclear Physics & Biophysics Research Division, Department of Physics,
Faculty of Mathematics and Natural Sciences,
Institut Teknologi Bandung,
Jl. Ganesa 10 Bandung 40132, INDONESIA
*E-mail: awaris@fi.itb.ac.id

Abstract. Preliminary study of plutonium utilization in AP1000 reactor has been conducted. This study evaluated the standard of Westinghouse AP1000 reactor and ZrB_2 as *Integral Fuel Burnable Absorber* (IFBA). Different fuel compositions of assembly were analyzed by using SRAC 2006 code system with JENDL 4.0 nuclear data library. This study aiming to compare the neutronics characteristics of UO_2 and $(U, Pu)O_2$ fuel assembly designs. Some results show that criticality can be accomplished by using 5% enrichment of U-235 for UO_2 fuel and 8.75% plutonium fraction for $(U,Pu)O_2$ fuel assembly.

1. Introduction

In the present and future the need of energy will increase along with the increasing of human population, the advanced technology and economic. This advances should be supported by an adequate of energy supply. However the avaibility of primary energy sources today, which is fossil fuels, become less and less and also unrenewable. Besides, the effect of fossil fuels on the environmental become an important issue due to its *green house effect* or CO_2 emission.

Other than that, learning from Chernobyl, Three Mile Island and Fukushima Daiichi accidents, the reactor should be designed with passive safety system. Passive safety system is a safety feature of nuclear reactor that does not require operator actions or electronic feedback in order to shutdown safely in the event of a particular type of emergency [1]. Also the development of nuclear technology requires some criteria such as the increasing of safety, economical aspects, less fuel waste and also non proliferation factor. The type of reactors with those requirements is from Generation IV reactors. But the Generation IV designs are still on the drawing board and will not be operational before 2020 at the earliest [2].

However the use of nuclear reactor also give us some new issue such as the accumulation of plutonium stockpile. This enforced the nuclear scientists and engineers to find another way to stabilize it. One of the enable way is to reprocessing and recycling it in the form of mixed oxide (MOX) fuel. We know that it is an established industry in several countries , like Japan, UK and France [3].

2. AP1000 reactor

AP1000 reactor is one of Generation III+ reactor designs from Westinghouse company. Its designs meet applicable safety requirements and goals defined for advanced light water pressurized water reactors with passive safety features. Different from Generation III PWR reactor, AP1000 designed by its simplicity so we can get cheaper cost, especially in its construction. Those reasons somehow make AP1000 one of the best candidates for nuclear power plant for nowadays.

This reactor has 3400 MW thermal power and 1117 MW electrical power outputs. The standard fuel is enriched UO_2 type with light water as moderator and coolant. The reactor core contains a matrix of fuel rods assembled into 157 identical fuel assemblies along with control and structural elements. The fuel assemblies are arranged in an approached circular cylinder. There are three radial regions in the core with different enrichments to establish a favorable power distribution. The enrichment used in this core is 2.35%, 3.34% and 4.45%. The temperature coefficient of reactivity of the core is highly negative. The core is designed for 18 months of fuel cycle [4].

Each of the fuel assemby consists of 264 fuel rods distributed in a square 17x17 array, 24 are guide tubes, one in the center is instrumentation tube. There are several kinds of burnable absorbers. One of them in the form of IFBA (*Integral Fuel Burnable Absorber*) [5].

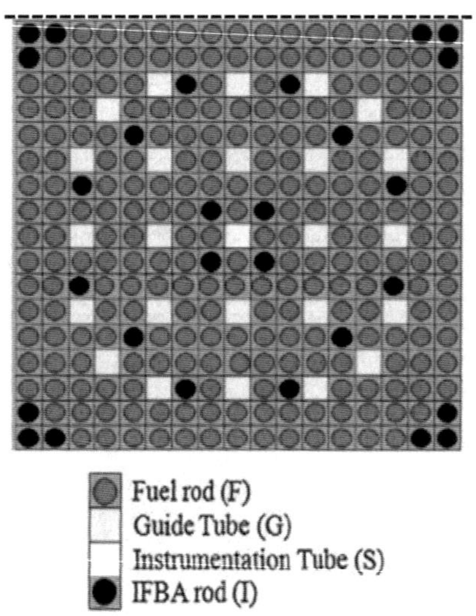

Figure 1. One of AP1000 fuel assembly configurations [4].

3. Methodology

3.1 Fuel assembly designs

There are two fuel assembly designs in this study. They are UO_2 and MOX (U, Pu)O_2 fuel assembly designs. The enrichment of U-235 in the fuel rod of UO_2 fuel range 4% - 7.5% while in the MOX one we used only natural uranium and plutonium fraction range 6.5% - 9.75%. This two kinds of fuel assembly has no difference configuration with the original one which showed in Figure 1.

In this study we have employed the reactor grade plutonium only. The composition of reactor grade plutonium is given in Table 1.

IOP Conf. Series: Journal of Physics: Conf. Series **877** (2017) 012003 doi:10.1088/1742-6596/877/1/012003

Table 1. Composition of Reactor Grade Plutonium.

Pu isotope	Percentage (%)
Pu-238	1.81
Pu-239	59.14
Pu-240	22.96
Pu-241	12.13
Pu-242	3.96

This data taken from the spent fuel composition of the 3 GWth PWR with 33 tons of annual loaded UO_2 fuel, 33 GWd/t burnup, and 10 years cooling [6,7].

3.2 Computational procedure

The burnup calculations in this study have been conducted by using SRAC 2006 code system [8] and JENDL 4.0 as nuclear data library [9]. SRAC (*Standard Reactor Analysis Code system*) was developed by JAERI (*Japan Atomic Energy Research Institute*). This code system was operated under UBUNTU system.

The fuel assembly burnup was calculated by using octant symmetric square with square array of pin rods. Then these fuel assemblies are arranged in reactor core for the core burnup calculation by using CITATION module of SRAC 2006 code. Flowchart of the calculation processes is shown in Figure 2.

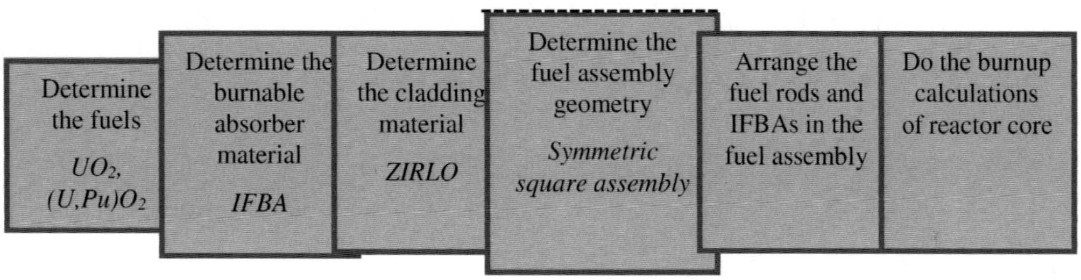

Figure 2. Flowchart of calculation processes

4. Results and Discussion

4.1. Effective multiplication factor

Figure 3 presents the neutron effective multiplication factor (*k-eff*) for UO_2 fuel type of AP1000 with several UO_2.enrichments. AP1000 can achieve its criticallity condition when U-235 enrichment in the fuel is 5.00% where *k-eff* is 1.0818 in the beginning and 1.0051 in the end of operation period. The nuclear reactor criticality condition means that the reactor can maintain the chain nuclear fission reactions undergoes continuously, which indicated by k-eff $\geq$ 1.0 during the whole operation time.

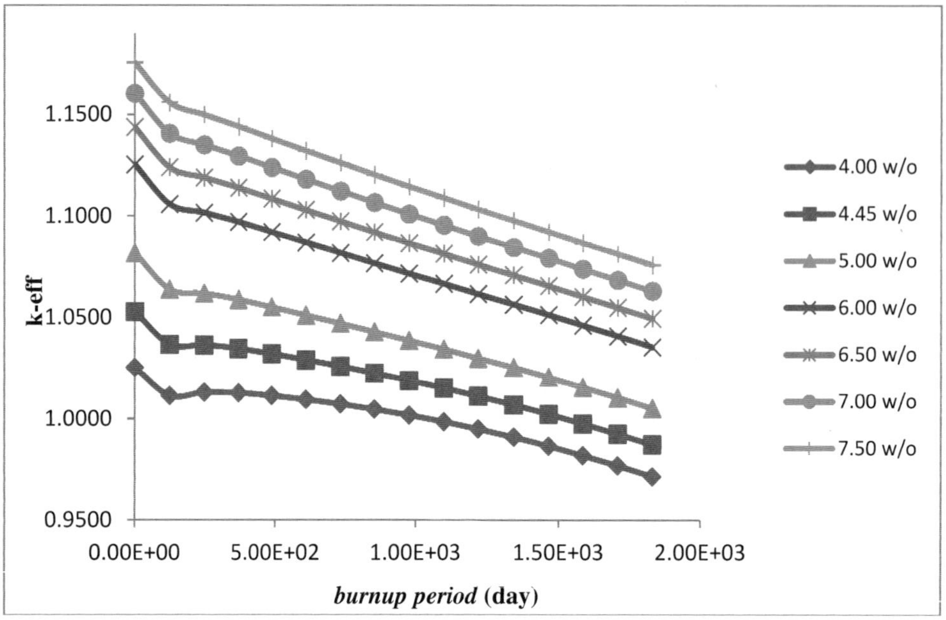

Figure 3. K-eff vs burnup period forUO$_2$ fuel with different U-235 enrichments.

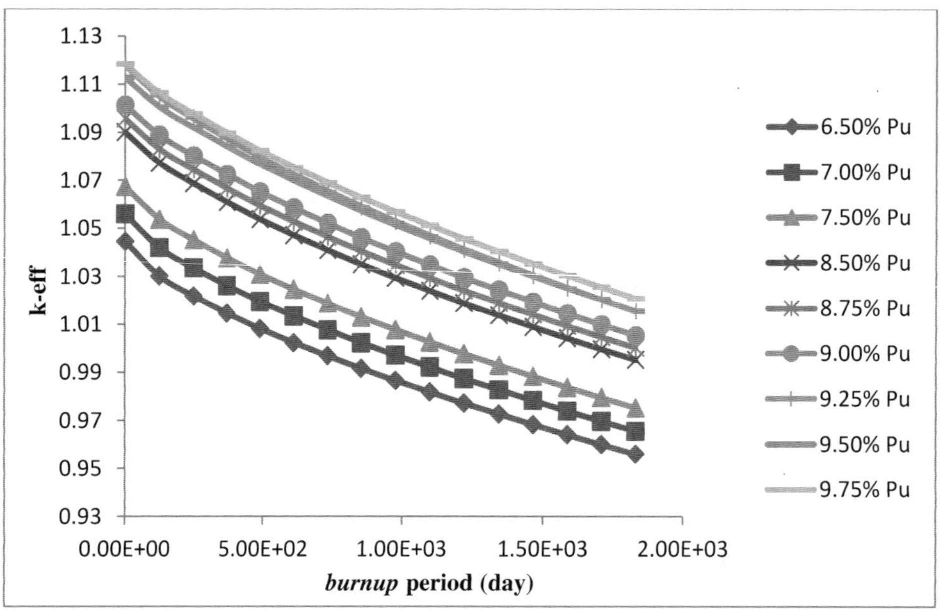

Figure 4. K-eff vs burnup period for (U,Pu)O$_2$ fuel with different plutonium fractions.

Figure 4 shows the effective multiplication factor for (U,Pu)O$_2$ fuel. In this case, AP1000 can obtain its criticality with 8.75% of plutonium fraction or more. The effective multiplication factor is 1.0958 in the beginning and 1.0003 in the end of operation period. As we know that plutonium composition dominated by Pu-239 which is a kind of fissile material. The increasing of plutonium fraction which means the increasing of the number of fissile materials will also increase the number of fission reaction in the reactor. So that the multiplication factor will become higher.

4.2. Conversion ratio

The conversion ratio in the fuel assembly for UO_2 fuel and $(U,Pu)O_2$ fuel are given in Figures 5 and 6, correspondingly.

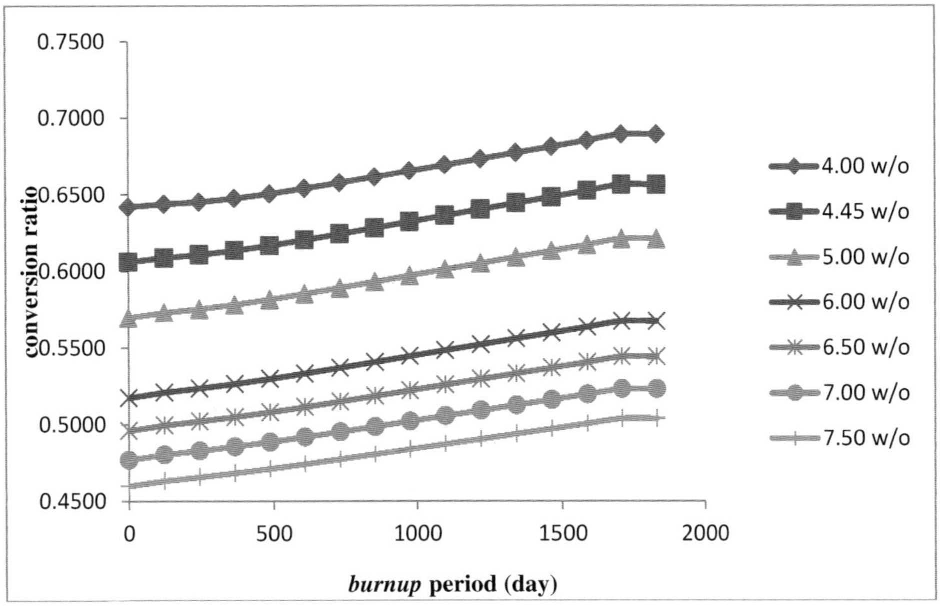

Figure 5. Conversion ratio vs burnup period for UO_2 fuel with different U-235 enrichments.

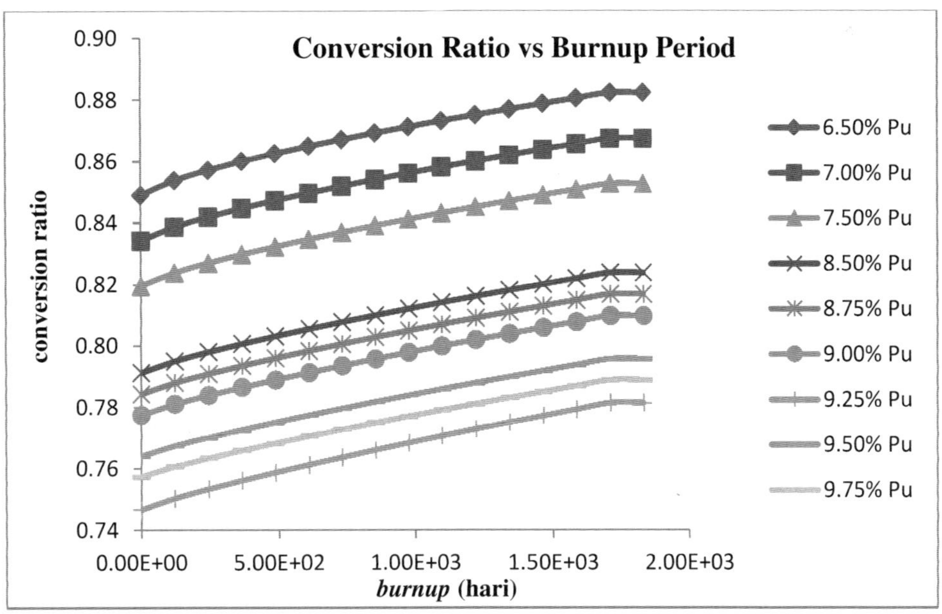

Figure 6. Conversion ratio vs burnup period for $(U,Pu)O_2$ fuel with different plutonium fractions.

The conversion ratio of the two kinds of fuels increase slowly from the beginning until the end of operation period. But the average value of the conversion ratios are below 1. It means that the number of fissile materials have been produced was lower than the fissile materials have been consumed. This also means that AP1000 reactor is a converter only not a breeder andit is a kind of thermal neutron reactor.

4.3. Neutron spectrum

The neutron spectra of the UO_2 and $(U,Pu)O_2$ fuels are given in Figures 7 and 8, respectively.

.

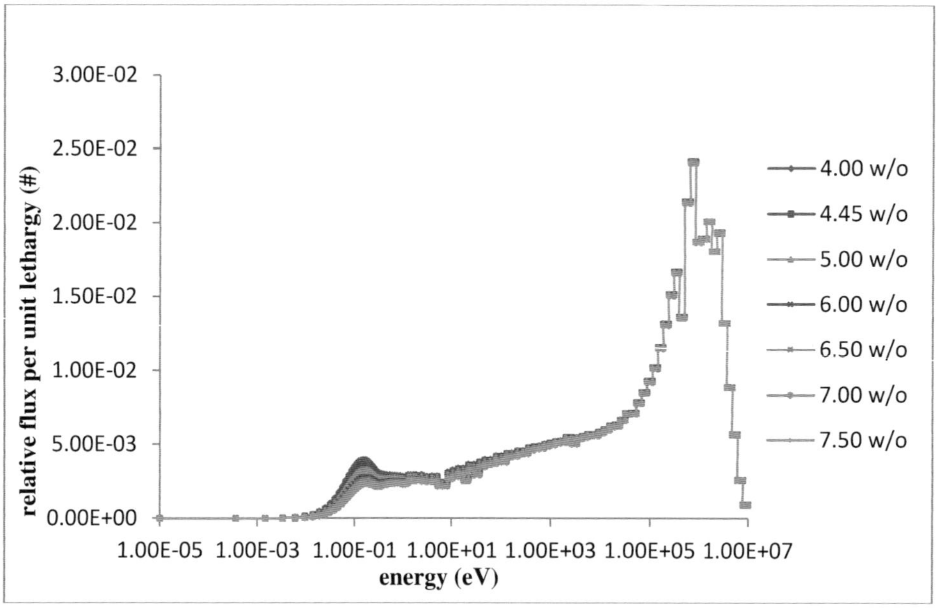

Figure 7. Neutron spectra for UO_2 fuel.

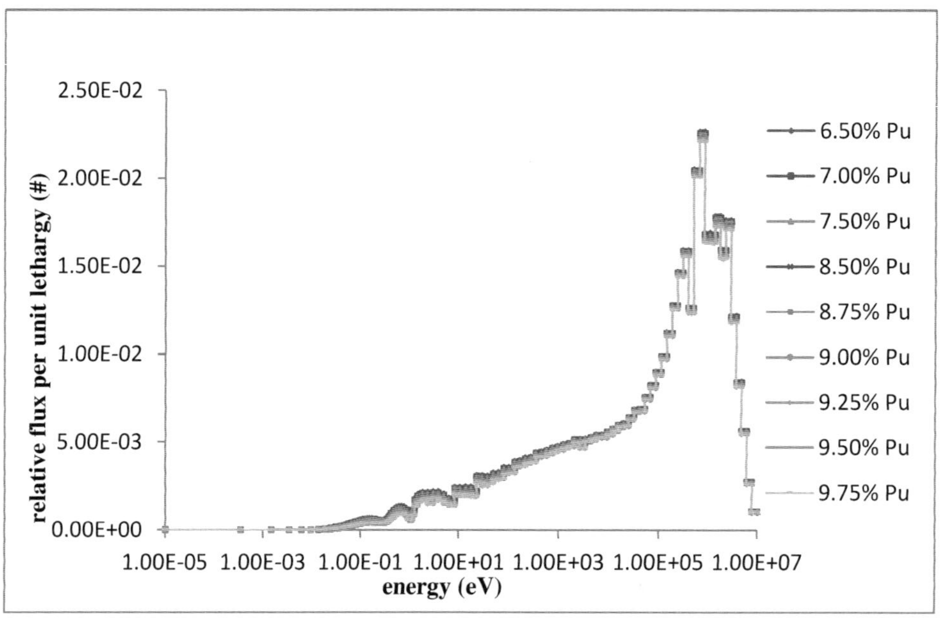

Figure 8 Neutron spectra for $(U,Pu)O_2$ fuel.

The neutron spectra for UO_2 fuel become slightly harder (shift to high energy region or peaks in thermal energy region (1e-3 eV to 1e+1 eV) become lower) with the increasing of U-235 enrichment. This fact due to more thermal neutron absorbed with the enlarging of U-235 amount in fuel. In case of the $(U,Pu)O_2$ fuel, the neutron spectrum has similar trend with the UO_2 one with the augmenting of plutonium fraction. Moreover, the neutron spectra become much harder for the $(U,Pu)O_2$ fuel since fissile isotopes of plutonium Pu-239 and Pu-241 have larger fission cross-section compared to that of U-235 in thermal energy region [7,10].

5. Conclusions

Preliminary study of plutonium utilization in Westinghouse AP1000 at fuel assembly level by using SRAC 2006 code with JENDL 4.0 nuclear data library has been conducted. The criticallity can be achieved by using $\geq 5\%$ enrichment of U-235 for the UO_2 fuel. For the $(U,Pu)O_2$ fuel, AP1000 can gain its criticality for at least 8.75% plutonium fraction. The neutron spectra become harder with the boosting of U-235 enrichment as well as plutonium fraction in loaded fuel.

References

[1] IAEA 1991 *IAEA-TECDOC-626* Safety Related Terms for Advanced Nuclear Plants ISSN 1011 4289 Vienna pp 1–20
[2] http://www.world-nuclear.org/information-library/nuclear-fuel-cycle/nuclear-power-reactors/advanced-nuclear-power-reactors.aspx
[3] Waris A, et.al. 2008 *Progress in Nuclear Energy. Influence of void fraction change on plutonium and minor actinides recycling in BWR with equilibrium burnup* 50 pp 295-298
[4] Elsawi M A and Hraiz A S B 2015 *Nuclear Engineering and Design. Benchmarking of the WIMS9/PARCS/TRACE Code System for Neutronic Calculations of The Westinghouse AP1000 Reactor* **293** pp 249-257
[5] Sanders C E and Wagners J C 2002 *Study of The Effect of Integral Fuel Burnable Absorber for PWR Burnup Credit* Tennessee Oak Ridge National Laboratory
[6] Broeders C H M et al 2002 *Nuclear Engineering and Design. Burning Transuranium Isotopes in Thermal and Fast Reactor* **202** pp 157-172
[7] Waris A, et.al. 2013 *Energy Conversion and Management. Preliminary study on plutonium and minor actinides utilization in thorims-nes minifuji reactor* 72 pp 27-32
[8] Okumura K, Kugo T, Kaneko K and Tsuchihashi K 2007 *SRAC2006: a Comprehensive Neutronics Calculation Code System,* JAEA
[9] Shibata K, et. al. 2011 *Journal of Nuclear Science and Technology. JENDL-4.0: A New Library for Nuclear Science and Engineering* **48(1)** pp 1-30
[10] Waris A, et.al. 2012 *Energy Conversion and Management, Effect of void-fraction on characteristics of several thorium fuel cycles in BWR* 63 pp 11-16

Effect of the Cerium Oxide (CeO_2) on the Structural and Electrochemical Properties of the $LaNi_5Ce$ Metal Hydride Anode

Ade Utami Hapsari[1], Anne Zulfia[2], Jarot Raharjo[1], Agustanhakri[1]

[1] Agency for the Assesment and Application of Technology, Center for Materials Technology, Bldg 224, PUSPIPTEK, Serpong – South Tangerang 15314, Indonesia
[2] Department of Metallurgy and Material Engineering, Faculty of Engineering, University of Indonesia, Kampus UI, Depok 16424, Indonesia

*ade.utami@bppt.go.id ; tam.hapsa@gmail.com

Abstract. One of negative electrode, AB_5-type alloy electrodes, have been extensively studied and applied in rechargeable Ni-MH batteries due to their excellent electrochemical characteristics. Some researchers have found that addition of rare earth oxides (La, Ce, Pr, Er, Tm, Yb) to AB_5-type alloy (MH) electrode improves battery performance significantly. Cerium Oxide (CeO_2) is a light rare earth oxide is widely obtained from the processing of tailings in mining activities. During this time, there is still little data for research applications of cerium oxide for electrode materials. In this paper, the effects of adding CeO_2 on the performance metal hydride electrode were investigated. In order to study the effects of CeO_2 on the performance of anode material, 1%, 2%, and 3% of weight ratio CeO_2 was mixed to $LaNi_5$ as an negative electrode. The powder mixtures were mechanically milled at a speed of rpm 240 for 2 hours using ball mill. The powder mixtures were characterized by X-Ray Diffraction (XRD) and Scanning Electron Microscope (SEM). Electrochemical characteristics were measured using electrochemical impedance spectroscopy (EIS). The powder mixing showed the presence of Ce atom substitution into $LaNi_5$ structures that affect the electrochemical properties of the material. The addition of cerium oxide at $LaNi_5$ increase of the value of impedance. However, the addition of the value of impedance at 1% CeO_2 is not significant when compared with the addition of 2% and 3% CeO_2 that actually make the electrochemical properties of $LaNi_5$ worst. Although the addition of 1% CeO_2 also slightly increases the impedance value of $LaNi_5$, but the addition of 1% CeO_2 showed increase the corrosion resistance than without the addition of CeO_2 and the addition of 2% and 3% CeO_2.

1. Introduction

Nickel-metal hydride (Ni-MH) batteries have been widely used for both consumer and industrial applications due to their advantages such as flexible design, excellent power, long cycle life and environmental friendliness. Although Ni-MH batteries are commercially available, their performance still needs to be improved by further research. It is well known that the capability of negative electrode is one of the key issues for Ni-MH batteries, depending on active materials and the preparation method of negative electrode. One of negative electrode, AB_5-type alloy electrodes, have been extensively studied and applied in rechargeable Ni-MH batteries due to their excellent electrochemical characteristics. Some researchers have found that addition of rare earth oxides (La, Ce, Pr, Er, Tm, Yb) to AB_5-type alloy (MH) electrode improves battery performance significantly [1-3].

Cerium Oxide (CeO_2) is a light rare earth oxide is widely obtained from the processing of tailings in mining activities. In Indonesia, mineral containing rare earth metal elements (La, Ce, Nd, Y) are as associated minerals in mineral gold, alluvial tin, bauxite, and copper. Despite the potential of rare earth metals in Indonesia is quite high, but during this time, there is still little data for research applications of cerium oxide for electrode materials. In this paper, the effects of adding CeO_2 on the performance metal hydride electrode were investigated. In order to study the effects of CeO_2 on the performance of anode material, 1%, 2%, and 3% of weight ratio CeO_2 was mixed to $LaNi_5$ as an negative electrode.

2. Experiment Detail

Hydrogen storage grade $LaNi_5$ samples from Aldrich were used for the study. The powder $LaNi_5$ then mixtures with 1%, 2%, and 3% of weight ratio powder CeO_2 and alcohol to complete the mixture. The powder mixture then was mechanically milled using RETSCH ball mill at a speed of rpm 240 with a 5 steel ball. Samples were prepared with milling time 2 hours and 30 min pause-time to avoid the heating of the sample during the milling process. The powder mixtures were drying with oven drying at 80°C for 17 hours.

The powder mixtures were characterized by X-Ray Diffraction (XRD) and Scanning Electron Microscope (SEM). Electrochemical characteristics were measured using electrochemical impedance spectroscopy (EIS) in 6 M KOH aqueous solution using a three-electrode cell. The working electrode was composed of 2 g composite powder and 0.1 ml PEG, and then was cold pressed into a pellet of 15 mm in diameter under 10 Metric Ton. Carbon was the counter electrode, and Hg/HgCl was the reference electrode. Electrochemical impedance spectroscopy (EIS) studies of the electrodes were performed in a frequency range between 5 mHz – 100 kHz. Tafel polarization curves were measured at a scanning rate of 1 mV/s from -1.5 to -0.5 V.

3. Result and Discussion

From the results of phase identification by XRD Figure 1, the addition of CeO_2 concentration in the powder mixture $LaNi_5$ would changes in the peak of the phase. The dominant phase before and after the addition of the oxide to $LaNi_5$ look slightly formed a new peak. The addition of CeO_2 concentration showed peaks that began appearing on the addition of 1%, 2%, or 3%. Although the peak of CeO_2 which appears not very significant between each concentration is added, but the peak of CeO_2 that appears with the addition of 1%, 2%, and 3% showed the presence of a new phase.

The formation of a new phase identified by XRD analysis, reinforced by the analysis using method GSAS. Table 1 shows the results of the analysis phase identified and size of cell volume with variations in concentration of 1%, 2%, and 3% CeO_2 using methods GSAS. With GSAS method is known that there are atom substitution La and Ce. The existence of substitution atoms La and Ce increases with increasing concentration of CeO_2 added. La and Ce atom substitution also affect the cell volume. Cell volume increased with the addition of oxide concentration of 1% and the cell volume decreases with the addition of oxide concentration of 2% and 3%. Contraction of volume cell indicates the size of the interstitial space for the hydrogen atoms to the alloy. This will cause the process of diffusion of hydrogen atoms on the surface of the alloy becomes more difficult [4].

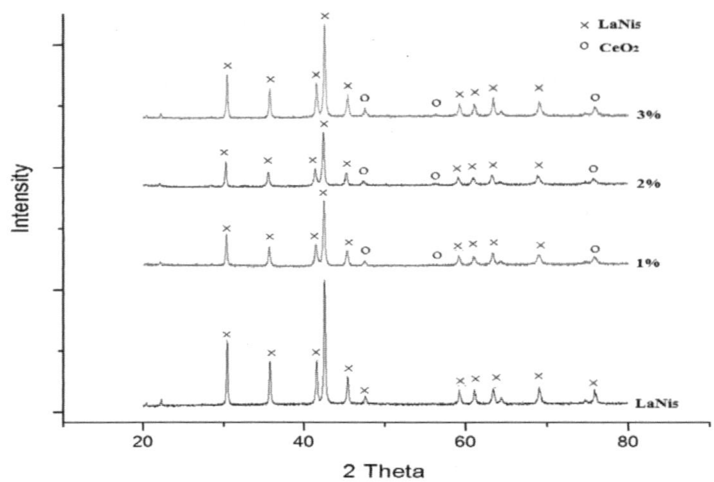

Figure 1. The results of XRD analysis of the powder mixture $LaNi_5$ and CeO_2 with various concentration of 1%, 2%, and 3% CeO_2.

Table 1. Results of the analysis phase identified and size of cell volume with variations in the concentration of 1%, 2%, and 3% CeO_2 using GSAS methods.

Sample	a	b	c	Volume Cell	Fasa Identifikasi
LaNi5	5.014567	5.014567	3.98548	86.792	$LaNi_5$
+ 1%CeO_2	5.021173	5.021173	3.98548	87.157	$La_{0.99}Ce_{0.01}Ni_{4.922}$
+ 2%CeO_2	5.020298	5.020298	3.990819	87.107	$La_{0.98}Ce_{0.02}Ni_{5.202}$
+ 3%CeO_2	5.013347	5.013347	3.986693	86.776	$La_{0.97}Ce_{0.03}Ni_{5.562}$

Analysis of the morphology of the SEM is intended to look at the effect of the milling process of the morphology of the particles of the powder. $LaNi_5$ particle morphology before and after the addition of cerium oxide concentration of 1%, 2%, and 3% are shown in Figure 2. Observations made on 500x magnification. In general, from the observation with SEM, reflecting the particle size reduction of about 50 μm to 5-10 μm from before the addition and after the addition of cerium oxide concentration in all variations. Particle size reduction occurs because of the process of mixing with cerium oxide (CeO2) using the method of milling for 2 hours at 240 rpm SEM results showed that the particles spread evenly on all additional concentration. This shows that the process of mixing with milling method successfully leveled the cerium oxide particles in $LaNi_5$.

Figure 2. The morphology of the powder $LaNi_5$ and mixture $LaNi_5$ (a) with variation CeO_2 concentration (b) 1%, (c) 2% and (d) 3%

Profile EIS shown from measurements $LaNi_5$ material with varying concentrations of cerium oxide is shown in Figure 3. It looks a diffusion process in areas with high frequency which is then followed by the formation of an arc at a frequency being in every variation of oxide concentration. The addition of oxide concentration in $LaNi_5$, will increases large arc pattern. The profile area that forms the greater arc at a frequency of charge-transfer medium indicates that the greater the resistance. The formation of small semicircle at low frequency area indicates the resistance area between the current collector material / pellet alloy powder [5]. In the measurement results, profile EIS of $LaNi_5$ material with the addition of 1% CeO_2 which showed the formation of small semicircle at low frequency area. This shows that the material $LaNi_5$ with the addition of 1% CeO_2 can be used as an anode material.

As is known from previous studies, the addition of rare earth metal oxides into metal hydride electrode will improve battery performance which is to reduce the rate of corrosion and increases the longevity of battery life [1-3]. To determine the corrosion rate of a material can be calculated by using the polarization curves. Linear polarization curves of LaNi5 with CeO_2 concentration variation shown in Figure 4. The polarization curves show a pattern of increasing current density (Io) when given in addition to the concentration of cerium oxide material $LaNi_5$.

Calculation shows the highest E_{corr} was -0.6432 V which when added 1% cerium oxide. And the lowest E_{corr} -1.0411 V when added 3% cerium oxide. The greater value of E_{corr} indicates better corrosion resistance on the electrode and can improve the cycle stability of the electrode [6].

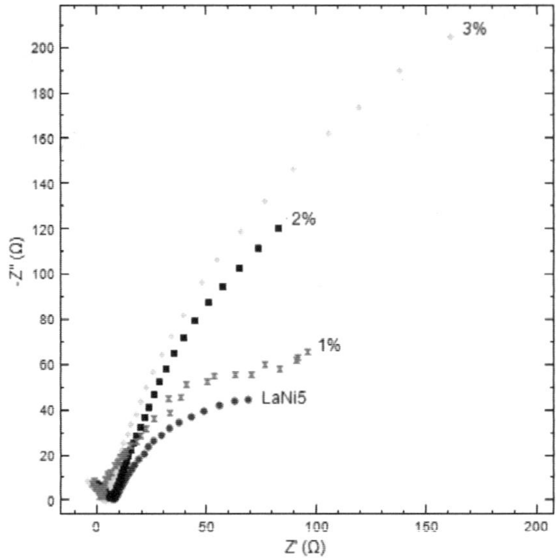

Figure 3. AC impedance spectra at room temperature of $LaNi_5$ with varying concentrations of cerium oxide.

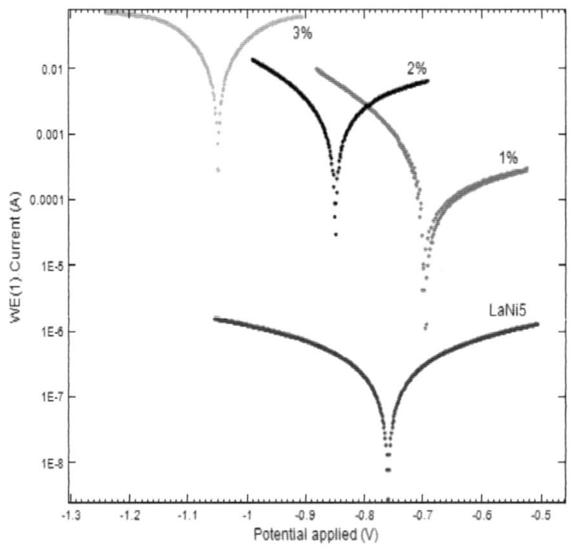

Figure 4. Linear Polarization curves of $LaNi_5$ with variations CeO_2 concentration.

4. Conclusion

The powder mixing showed the presence of Ce atom substitution into $LaNi_5$ structures that affect the amount of cell volume. The amount of cell volume affects the electrochemical properties of the material. In general, the new phased are formed with addition of CeO_2 concentration were $La_{1-x}Ce_x Ni_5$ (x =0.01 – 0.03). The addition of cerium oxide at $LaNi_5$ shows the increase of the value of impedance. However, the addition of the value of impedance at 1% CeO_2 is not significant when compared with the addition of 2% and 3% CeO_2 that actually make the electrochemical properties of $LaNi_5$ worst. It is showed by impedance value that is higher on the addition of 2% and 3% CeO_2. Although the addition of 1% CeO_2 also slightly increases the impedance value of $LaNi_5$, but from the polarization curve, the addition of 1% CeO_2 showed increase the corrosion resistance than without the addition of CeO_2 and the addition of 2% and 3% CeO_2

Acknowledgments
The authors would like to thank the Centre for Materials Technology for the financial support and Department of Metallurgy and Material Engineering University of Indonesia for laboratories facilities

References

[1] Maurela, F, P.Leblanc, B. Knosp, M.Backhaus-Ricaoult 2000 *Journal of Alloys and Compounds* **309** 88–94
[2] H. Kaiya, and T. Ookawa 1995 *Journal of Alloys and Compounds* **231** 598-603
[3] Tanaka, Toshiki, Minoru Kuzuhara, Masaharu Watada, Masahiko Oshitani 2006 *Journal of Alloy and Compounds* **408-412** 323 -326
[4] J.Chen, S.X.Dou, H.K.Liu 1996 *Journal of Power Sources* **63** 267 – 270
[5] N.Kuriyama, T.Sakai,H. Miyamura, I.Uehara, H. Ishikawa, and T.Iwasaki 1993 *Journal of Alloys and Compounds* **202** 183-197
[6] Pu, Zenggan, Yungfeng Zhu, Jinyu Zhu, Jianguang Yuan, Jiguang Zhang, Wei Chen, Jiaojiao Fang, Liquan Li 2014 *International Journal of Hydrogen Energy* **39** 3887-3894
[7] Su, Geng, He, Yue-hui, Liu, Kai-yu 2012 *International journal of hydrogen energy* **37** 11994 - 12002
[8] Joseph, B, B.Schiavo, G.D. Ali Staiti, B.R.Sekhar 2011 *International Journal of Hydrogen Energy* **36** 7914 – 7919

Dye-sensitized solar cell based on TiO_2/MnO_2 composite film as working electrode

A Prasetio[1], A M Habieb[1], I Alkian[1], Z Arifin[2] and H Widiyandari[1]

[1] Department of Physics, Diponegoro University, Jl. Prof. Soedharto SH, Semarang, Indonesia.
[2] Department of Mechanical Engineering, Sebelas Maret University, Jl. Ir. Sutami 36A, Surakarta, Indonesia.

E-mail: h.widiyandari@undip.ac.id

Abstract. The purpose of this study is to develop DSSC based on TiO_2/MnO_2 composite film as working electrode. The working electrodes were prepared by the doctor blade method and their characteristics were investigated by using scanning electron microscopy (SEM), X-ray diffraction (XRD) and UV-Vis spectroscopy. Moreover, the performance of DSSCs was performed under simulated AM 1.5G solar illumination using 100 mW/cm^2. The indirect band gap of TiO_2 pure, TiO_2/MnO_2 and TiO_2/MnO_2 6% films are 3.29, 2.98 and 2.81 eV, respectively. The highest power conversion was 0.018 % which corresponds to TiO_2/MnO_2 6% composite film. The results suggest that the MnO_2 added to the semiconductor TiO_2 can improve the efficiency of the DSSCs.

1. Introduction

Energy is the biggest problem will be faced by human due to global energy consumption was growth. The sun is an obvious the largest of clean and sustainable energy source. Photovoltaic (PV) technology, the direct conversion of sunlight to electricity, was classified into three generations of solar cell. The first generation is the silicon wafer based solar cell which categorized into mono-crystalline and poly-crystalline solar cell [1]. The second generation cells are most of thin film solar cells and amorphous (a-Si) solar cell. The third generation cells are the new promising technologies investigated in detail [2]. Dye-sensitized solar cell (DSSC) is one of the third generation PV developed by O'Regan and Grätzel in 1991. The DSSC has attracted attention due to its simple fabrication process, low fabrication cost and promising efficiency in converting energy from solar energy to electricity [3]. Commonly, the DSSC consists of a platinum as counter electrode, an electrolyte containing iodide/ triiodide (I^-/I_3^-), a sensitizer adsorbed on semiconductor surface, and conducting glass usually FTO or ITO coated with nanocrystalline titanium dioxide (TiO_2) semiconductor as working electrode [4]. The working electrode is an important component of the DSSC which its function in injecting the light generated electrons into conduction band of semiconductor and transferring them to the conducting glass [5]. TiO_2 semiconductor has a key role as working electrode due to its high transparency for visible light, chemically stable, non-toxic, and controllable electric properties. However, TiO_2 has a relatively high energy gap (Eg=3.2 eV for anatase) and works at a small fraction of the solar spectrums (ultra-violet region) [6-8]. The further development in terms of both mechanisms DSSC fabrication and material synthesis are needed to improve the power efficiency. One promising approach is based on adding other semiconductor materials with lower energy gap into TiO_2 semiconductor. Pure manganese oxide (MnO_2) semiconductor has direct energy gap of about 2.5 eV which can decrease the energy gap of TiO_2 and increase the power efficiency of DSSC [9-11]. Thus, the main purpose of this research is to improve performance of the DSSC through band gap engineering on working electrode by adding MnO_2 to TiO_2.

In this paper, TiO_2/MnO_2 composite films were synthesized and deposited onto FTO glass substrates using the doctor blade method. The characteristics of TiO_2/MnO_2 composite films were characterized by using scanning electron microscopy (SEM), energy dispersive microanalysis (EDS), X-ray diffraction (XRD) and ultraviolet-visible (UV-Vis) spectroscopy. The performance of the DSSCs was measured under AM 1.5G solar illumination at 100 mW/cm^2 light intensity.

2. Method

2.1. Materials

TiO_2 P25 degusa powder (Sigma Aldrich, USA) was used as working electrode layer. $KMnO_4$ powder (Merck, Germany) was used as precursor of MnO_2 powder. Platinum paste (Dyesol, Australia) was used as counter electrode layer. Fluorine-doped tin oxide (FTO) glass substrate (15Ω/sq, 2.3 mm thickness, Dyesol, Australia) was used as conducting glass. Electrolyte HSE containing iodide and triiodide (Dyesol, Australia) solution was used as the transport for redox mediator between two electrodes. Ruthenium dye N719 (Dyesol, Australia) was used as sensitizer and sunlight harvester. Sealant (Dyesol, Australia) was used for sealing spacer DSSC device for longer term application.

2.2. MnO_2 powder synthesis

To synthesis MnO_2 powder, 0.75 gr $KMnO_4$ was solved in 20 ml aqua DM and 10 ml ethanol (Merck, Germany) and stirred for 15 min at room temperature. The $KMnO_4$ solution was filtered by using filter paper for 24 h. The solution precipitate into blackish MnO_2 sediment as a result of chemical reaction. The MnO_2 sediment was dried at 90°C for 5 h and milled for 15 min.

2.3. Preparation of the working electrodes

The FTO glass substrate with area 2.5 x 2.5 cm^2 was washed with ethanol and aquades several times. The MnO_2 powder was added into TiO_2 powder and milled for 15 min in high energy milling 3D (Nanotech Indonesia, Indonesia). This process was enough to provide homogeneous powder. The TiO_2 P25 powder with various concentrations of 0, 2 and 6% of the MnO_2 powder were named as TiO_2 pure, TiO_2/MnO_2 2% and TiO_2/MnO_2 6%, respectively.

To prepare TiO_2/MnO_2 paste, the mixture of TiO_2/MnO_2 powder (MnO_2 content: 0, 2 and 6%) was dispersed into the binder which prepared from our previous research [12]. TiO_2/MnO_2 paste was deposited onto the FTO glass substrate by using the doctor blade method with thickness of about 5 μm and active area 1 x 1 cm^2. After that, the working electrode was sintered at 500°C for 45 min then immersed into a 0.3 mM ruthenium dye N719 in a mixture of tert-butanol and acetonitrile solution for 24 h at room temperature.

2.4. Preparation of the Pt-counter electrode

To prepare the counter electrode, two holes (1 mm diameter on active surface) were drilled through for each FTO glass substrate by using diamond bor. The FTO glass substrate was washed with ethanol and aquades. Platinum paste was deposited onto the FTO glass substrate by using the doctor blade method with thickness is equal to the thickness of the scotch tape with active area 1 x 1 cm^2. After drying at 80°C, the Pt-counter electrode was sintered at 450°C for 60 min.

2.5. Assembly of the DSSC

The working electrode was assembled into a sandwich like structure with the Pt-counter electrode (Figure 1). The sealant was put on as sealing spacer between two electrodes, the inner dimensions 1.5 mm wider than TiO_2/MnO_2 layer. The working electrode and the Pt-counter electrode assembled were secured using a binder clip on each side. The electrolyte solution was injected through one of the holes. Finally, the holes were sealed with silicon glue to prevent evaporation.

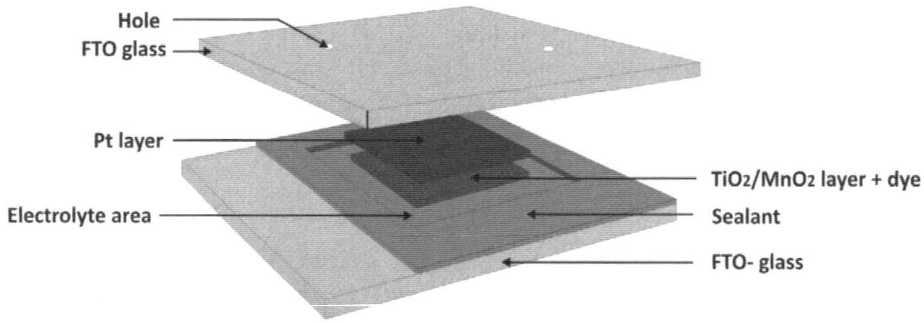

Figure 1. A sandwich like structure for DSSC.

2.6. Chacterization and measurement

The structural analysis of the TiO_2/MnO_2 films was characterized using XRD (Rigaku-Denki Corp, Japan) with a CuKα source (λ= 0.154 nm). The surface morphology and elemental composition of the TiO_2/MnO_2 films were observed using SEM-EDS (JEOL JSM-6360 LA, Japan). The optical properties of the TiO_2/MnO_2 films were obtained using Shimadzu 1240 SA UV-Vis spectroscopy (Shimadzhu, Japan). The photovoltaic performance of the DSSCs was measured using a Keithley 2602A sourcemeter (Keithley Instrument, USA) under AM 1.5G solar illumination at 100 mW/cm^2 light intensity.

3. Results and discussion

High-magnification of SEM images of the TiO_2/MnO_2 samples are shown in Figure 2. By comparing particles with the scale bar, we determined that the average diameter of the TiO_2 particles was approximately 200 nm. Large particle sizes could be attributed to a strong light scattering and less dye adsorption leading to lower efficiency of the DSSCs. These are should be caused by small surface area of the samples. In other hand, both MnO_2 and TiO_2 particles have a similar shape as shown in Figure (b) and (c).

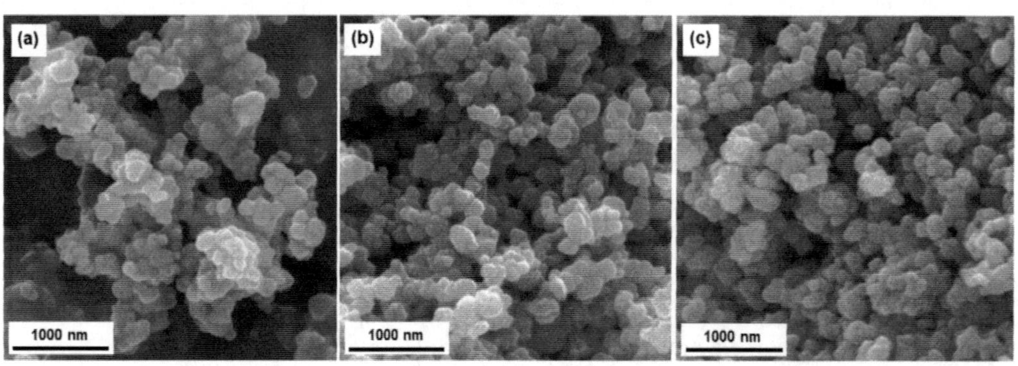

Figure 2. High-magnification SEM images of the(a) TiO_2 pure; (b) TiO_2/MnO_2 2% and (c) TiO_2/MnO_2 6% samples.

Figure 3 shows chemical composition of the samples which obtained using EDS. Peaks at 0.525, 4.508 and 5.894 keV agree with O, Ti and Mn, respectively. It proves that the samples consisted of only three elements.

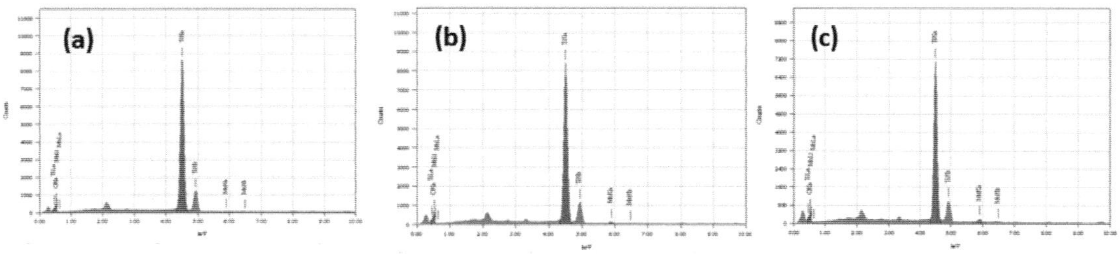

Figure 3. EDS observation of the (a) TiO_2 pure; (b) TiO_2/MnO_2 2% and (c) TiO_2/MnO_2 6% samples.

The XRD patterns for TiO_2, TiO_2/MnO_2 2% and TiO_2/MnO_2 6% samples are shown in Figure 4(a). Similar patterns were obtained from the samples. The XRD patterns agree with the Joint Committee on Powder Diffraction Standards (JCPDS) database no. 21-1272. The strongest peaks at 25.4, 37.8 and 48.0° correspond respectively to the (101), (004) and (200) planes of anatase phase. Figure 4(b) shows that d-spacing of (101) lattice plane of TiO_2 decreases should be caused by increasing of MnO_2 content. However, MnO_2 patterns are not shown clearly in the XRD patterns.

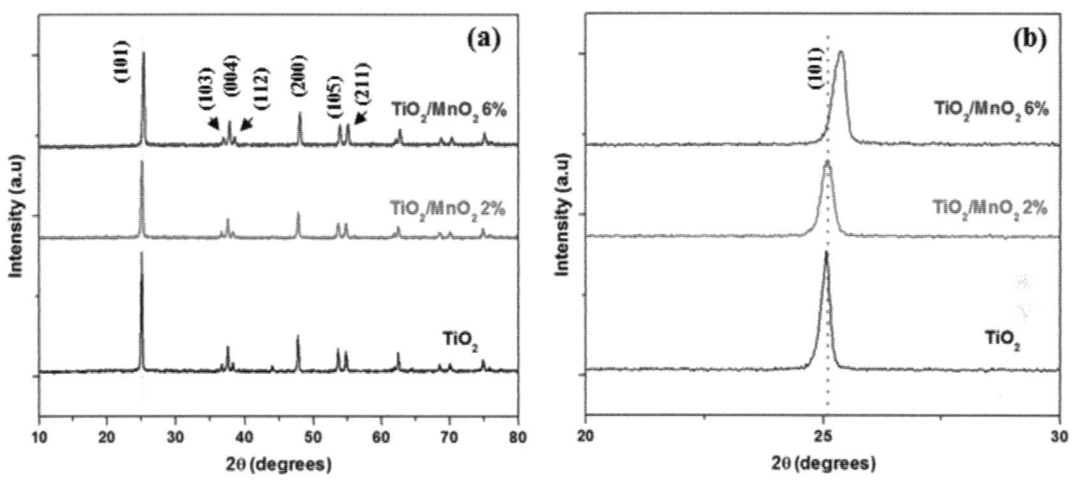

Figure 4. (a) The XRD patterns of the TiO_2, TiO_2/MnO_2 2% and TiO_2/MnO_2 6% and (b) details of the XRD patterns around from 25° to 30° 2θ values.

The optical energy gap of each sample can be calculated and estimated by using Tauc's plot and equation (1):

$$(\alpha h\nu) = A(h\nu - E_g)^n \quad (1)$$

where α is the optical absorption coeffcient, $h\nu$ is photon energy, A is the constant, E_g is optical energy gap and the exponent n is depending on the nature of the transition (n = 2 corresponds to indirect energy gap of anatase TiO_2) [13]. Therefore, the plot of $(\alpha h\nu)^{1/2}$ vs $h\nu$ represent E_g, as shown in Figure 5.

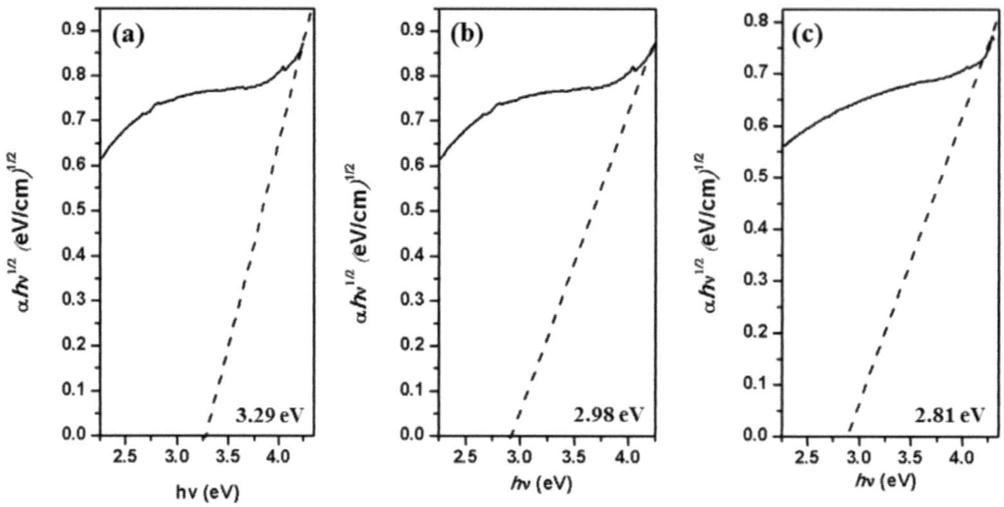

Figure 5. *Plot* $(\alpha h\nu)^{1/2}$ vs $h\nu$ of (a) TiO_2, (b) TiO_2/MnO_2 2% and (c) TiO_2/MnO_2 6% samples.

The energy gap of TiO_2, TiO_2/MnO_2 2% and TiO_2/MnO_2 6% films of approximately 3.29, 2.98 and 2.81 eV, respectively. The energy gap of various films is found decrease along with increases of MnO_2 concentration. Lower energy gap to be responsible to enhancement of electron injection and transport [14].

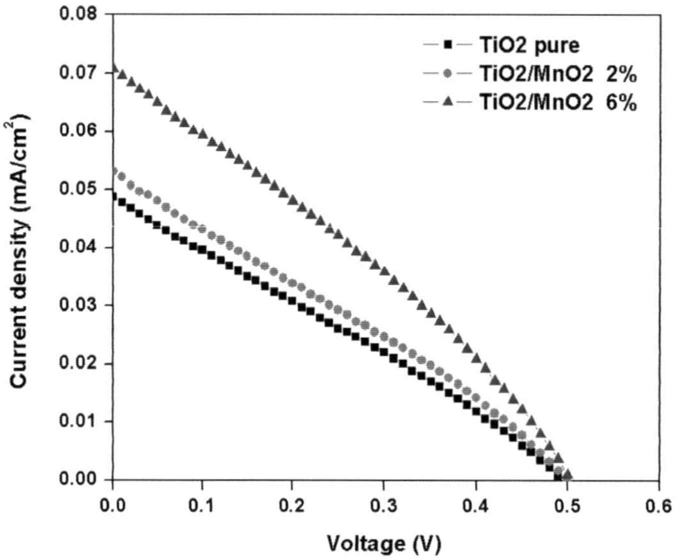

Figure 6. J-V curves of DSSCs for TiO_2, TiO_2/MnO_2 2% and TiO_2/MnO_2 6%.

Figure 6 shows the current density-voltage (J-V) curve of the DSSCs with various concentrations of the MnO_2 which were performed under simulated AM 1.5G solar illumination using 100 mW/cm^2. The conversion efficiency (η) of the DSSCs is determined by short-circuit current density (J_{sc}), open-circuit potential (V_{oc}) and the intensity of the incident light per unit area (P_{in}). The overall conversion efficiency is shown in equation (2).

$$\eta = \frac{J_{sc} \cdot V_{oc} \cdot FF}{P_{in}} \times 100\% \tag{2}$$

The fill factor (FF) can be calculated according to the equation (3) below:

$$FF = \frac{J_{max} \cdot V_{max}}{J_{sc} \cdot V_{oc}} \tag{3}$$

where J_{max} and V_{max} are the current density and potential at the point of the maximum power respectively [15]. The photovoltaic performance of DSSCs based on TiO_2/MnO_2 electrodes with different MnO_2 contents are listed in Table 1.

Table 1. Photovoltaic properties of DSSCs with various working electrodes

Photoanode	J_{sc} (mA.cm^{-2})	V_{oc} (V)	Fill Factor	Efficiency (%)
TiO_2	0.0487	0.4948	0.2773	0.0067
TiO_2/MnO_2 2%	0.0531	0.4997	0.2811	0.0075
TiO_2/MnO_2 6%	0.0707	0.5049	0.3008	0.0108

Both the curve and the table show that the DSSC with TiO_2/MnO_2 6% photoanode has the highest value of J_{sc} as well as efficiency. Meanwhile, the DSSC with TiO_2 pure photoanode has the lowest value of J_{sc} as well as efficiency. The results show that the MnO_2 added to the TiO_2 semiconductor can improve both the efficiency and J_{sc} of the DSSCs.

4. Conclusion

The DSSCs using TiO_2/MnO_2 films as working electrode have been successfully fabricated by using the doctor blade method. By adding of MnO_2 to TiO_2 semiconductor can decrease energy gap of the TiO_2, which means the DSSCs can work under visible light and harvest energy optimally. Additionally, lower energy gap responsibles to enhance injection as well as transport of electron. The enhancement of electron injection and transport leads to the improvement of J_{sc}. Consequently, FF and efficiency of the DSSC increases significant enough.

Acknowledgement

Financial support from Ministry of Research, Technology and Higher Education of Indonesia is gratefully acknowledged. In addition, we would thank to La Ode Yusni Malik for valuable discussion.

References

[1] Barham K W J, Mazzer M and Clive B 2006 *Nat. Mater.* **5** 161
[2] Sharma S, Jain K K and Sharma A 2015 *Mater. Sci. Appl.* **6** 1145
[3] O'Regan B and Grätzel M 1991 *Nature* **353** 737
[4] Nazeeruddin Md K, Baranoff E and Grätzel M 2011 *Solar Energy* **85** 1172
[5] Grätzel M 2003 *J. Photochem. Photobio. C: Photochem. Rev.* **4** 145
[6] Sedghi A and Miankushki H N 2012 *Int. J. Elecrochem. Sci.* **7** 12078
[7] Qin W, Lu S, Wu X and Wang S 2013 *Int. J. Elecrochem. Sci.* **8** 7984
[8] Lin Y, Jiang Z, Zhu C, Hu X, Zhang X, Zhu H, Fan J and Lin S H 2013 *J. Mater. Chem.* A **1** 4516
[9] Li T, Wu J, Xiao X, Zhang B, Hu Z, Zhou J, Yang P, Chen X, Wang B and Huang L 2016 *RSC Adv.* **6** 13914
[10] Toufiq A M, Wang F, Javed Q, Li Q and Li Y 2014 *Appl. Phys.* A **116** 1127

[11] Brus V V, Pidkamin L J, Abashin S L, Kovalyuk Z D, Marynchuk P D and Chugai O M 2012 *Optic. Mater.* **34** 1940

[12] Prasetio A, Purwanto A, Subagio A and Widiyandari H 2016 *6th AIP Conf. Proc. Nanosci. Nanotech. Symp. 2015(Surakarta)* vol 1710 (College Park: Maryland/AIP Publishing) 030054

[13] Buchholz D B, Liu J, Marks T J, Zhang M and Chang R P H 2009 *ACS Appl. Mater. Interfaces* **1** 2147

[14] Sua H, Huanga Y T, Changa Y H, Zhaia P, Haua N Y, Cheunga P C H, Yehc W T, Weid T C and Fenga S P 2015 *Electrochimica Acta* **182** 230

[15] Nazeeruddin M K, Kay A, Rodicio I, Humphry-Baker R, Muller E, Liska P, Vlachopoulos N and Gratzel M 1993 *J. Am. Chem. Soc.***115** 6382

The effect of temperature deposited on the performance of ZnO-CNT-graphite for supercapacitors

Alfin Darari[1], Istajib S. Hakim[1], Priyono[1], Agus Subagio[1], Pardoyo[2] and Achmad Subhan[3]

[1]Department of Physics, Faculty of Science and Mathematics, Diponegoro University, Indonesia
[2]Department of Chemistry, Faculty of Science and Mathematics, Diponegoro University, Indonesia
[3]Center of Physics Research, LIPI, Indonesia

E-mail : alfindarari@st.fisika.undip.ac.id and agus_fadhil@yahoo.com

Abstract. Carbon nanotubes (CNTs), graphite are now widely studied as the electrodes of supercapacitor, owing to their high conductivity, large surface area, chemical stability, etc. A lot of research has been focused on Carbon/metal oxide nanocomposite electrode for Electrode supercapacitor because it will increase the total capacitance. In this research, ZnO nanoparticles were deposited onto substrate CNT:Graphite in different temperatures such as 300°, 350°, and 400°C. The characterization of the crystal size using *X-Ray Diffraction (XRD)* patterns showed ZnO material peak was detected a ZnO crystallite. The size of ZnO crystallite in 300°, 350°, and 400°C consecutively is 101.1; 103.4; and 116.7 nm. The test results are *Electrochemical impedance spectrometry* (EIS) high electrical conductivity values obtained on the composition of ZnO-CNT-graphite with a temperature of 350°C 4.6 (S/m); and (2) the highest value of capacitance in 300°C is 1.23 F /g.

1. Introduction

The energy crisis towards the end of the 20th century cannot offset the high demand. Since electric energy becomes one of the best alternatives for future energy supply, research on renewable energy devices continues to grow. However, the constraints now are related to the energy storage. Supercapacitors are considered as an ideal energy storage device due to their long cycle-life, high power and energy density characteristics [1-2]. Currently, supercapacitors are developed for various functions such as military technology, aerospace, and electric cars [3]. There are two types of supercapacitors based charger principle namely electrochemical double layer supercapacitors (EDCL) and pseudocapacitors. Carbon nanotubes (CNTs) are now widely studied, the most notably for EDCL. Kim et al. [4] report supercapacitors based CNTs and graphite have electron storage principle which can increase conductivity, because it has large surface area. Metal oxides, such as RuO_2 , TiO_2 and MnO_2 have a redox reaction principle which were studied for supercapacitor, namely pseudocapacitor. It can increase a capacitance, but it has a weakness. It takes time for chargering compared to EDCLs type [5]. In recent years, there is a method that combines both these supercapacitors which can improve the performance of the charge transfer. In addition, the ability of material that is filled over and over again becomes more stable. Moreover, it has a high capacity and the characteristics of fast charger [6].

Zinc oxide (ZnO) is a semiconductor material that has a high power density, which is 650 Ah/g for supercapasitor. It happens because of the significant advantages of low cost, abundant availability and eco-friendly. Recently, ZnO has been used as supercapacitor electrodes with carbon nanocomposite material types [7]. In this research, the focus is on the exploration of supercapacitor electrode-based ZnO-CNT-graphite nanocomposites. The CNT-graphite were screen-printed on the stainless steel foil (SS foil) substrate and ZnO source precursors (zinc acetate) were deposited on CNT films by spraying pyrolysis of different temperatures. The purpose of this research is to look for good temperature deposited and to inverstigate the relation between ZnO crystal size with capacitance and conductivity supercapacitors.

2. Experimental

The materials used in this study is foil substrates (MTI, Malaysia). They were only 2 cm^2 used in this research. Zinc Acetat Tetrahydrate (Merck, Australia) was used as electrodes material. Electrolyte which consist of KCl (Merck, Germany) was used as the transport mechanism for the redox mediator of the electrode. The CNT powder which prepared using *spray pyrolysis method* as previous reported by Subagio, et al., 2009 [8-9] was used as electrode material. Ethanol (Merck, Germany), terpineol (Sigma Aldrich, USA), ethylcellulose (Sigma Aldrich, USA), monoethanolamine (Merck, Germany), and ISO-Propanol (Merck, Germany) were used without purification.

0.2 g of ethylcellulose was dissolved in 4 mL ethanol which is stirred for 10 minutes, followed by mixing 1.6 g terpineol into the binder. The CNT-graphite (composition of mass 1:1) was dispersed in the binder with stirrer for 10 minutes. Then, CNT paste was deposited onto the SS Foil substrate by using *doctor blade method* and drying in the temperature of 150 °C for 2 hours. Then, ZnO nanoparticles was synthesized by sol-gel method 0.5 M using $Zn(CH_3COO)_2.2H_2O$ as sources of Zn. In addition, isopropanol (IPA: (CH_3) $_2CHOH$), monoethanolamine (MEA: $HOCH_2CH_2NH_2$) were stirred for 30 minutes. ZnO liquid was sprayed onto CNT-graphite substrate in different temperatures, such as in 300, 350, 400 and, 450 °C. The surface structure of electrodes were characterized by X-ray diffraction spectroscopy (XRD, Philip Analytical X-Ray B.V) with Cu Kα radiation (λ=1.5418 Å) at 40 kV and 20 mA of Electrochemical impedance spectrometry (EIS, Hioki 3522-50) was performed to study the electrodes by using a standard double cell containing 1 M KCl aqueous solution.

3. Result and discussion

Supercapacitor electrode of the samples summarized in Table 1 shows the name of samples in this study.

Table 1. Samples of supercapacitors electrode

Sampel	Composition
S1	CNT-graphite
S2	CNT-graphite ZnO 300 °C
S3	CNT-graphite ZnO 350 °C
S4	CNT-graphite ZnO 400 °C
S5	CNT-graphite ZnO 450 °C

There were 4 samples (S1, S2, S3 and S4) can be tested, while sample S5 cannot be tested because the S5 sample was broken. This is due to high temperatures in coating so that the coating of CNT-graphite-ZnO cannot be included. The 4 samples test by XRD of ZnO layer coated on the CNT-graphite nanocomposite is shown in Figure 1.

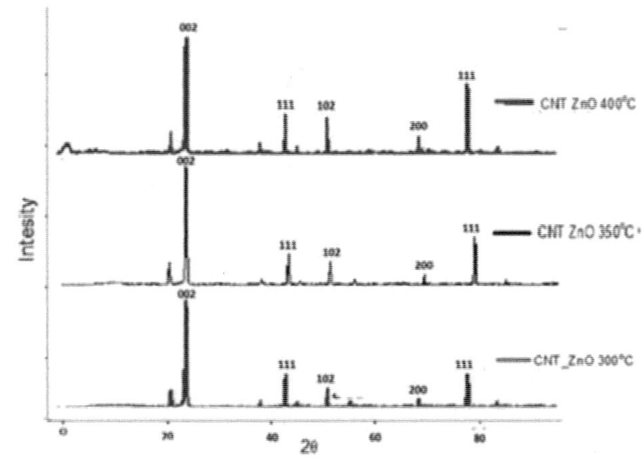

Figure 1. The XRD results of ZnO-CNT-graphite electrode at various temperatures of 300, 350 and 400 °C

Figure 1 shows that a diffraction pattern is dominated by diffraction peak of carbon. Those are adjusted with reference to the program *Powder Diffraction Standarts*. The carbon type detected in reference (JCPDS number 020 456) looks to dominate the summit between 26.29 - 26.39°. It shows the type of carbon with (002) orientation of the hexagonal structure of graphite [10]. Diamond structure was dominated by the peak between 43.44 - 43.91° with reference of JCPDS number 060 675 to the orientation (1 1 1). ZnO material and the peak were detected at peak 66.20 – 66.77° in JCPDS number 011 136 with the (200) orientation. Peak areas of the crystal orientation is used in the calculation to find crystallite size because the dominant peak is sharp and strong with a high intensity. Crystallite size was calculated using the scherer equation shown in figure 2.

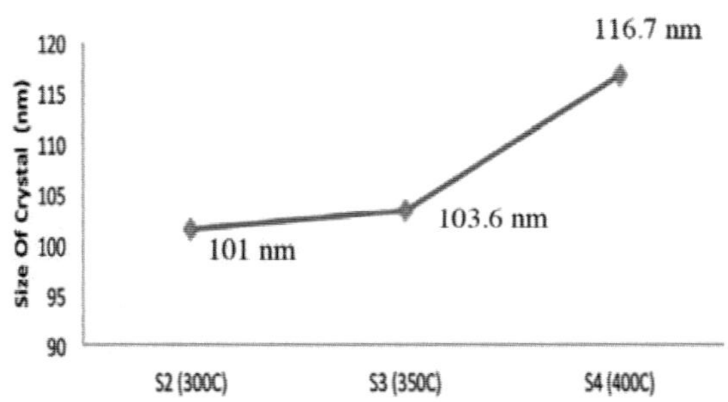

Figure 2. Crystalite size relationship chart with temperature variation of ZnO coating

ZnO resulted from the source of Zn acetate solution carried spray with high temperature will vaporize CH3COO and ZnO oxidizes with the air binds atoms into ZnO [11].

$$Zn(CH_3COO)_2.2H_2O_{(solid\ near\ substrate)} \xrightarrow{heat} Zn(CH_3COO)_2.2H_2O_{(gass)} \rightarrow ZnO + CH_3COO_{(gass)} \quad (1)$$

The size of ZnO crystallite in 300, 350 and 400 °C consecutively is 101.1; 103.4 and 116.7 nm. This indicates that the increasing of temperature coatings can increase the particle size of ZnO material. Crystallite size becomes larger when temperature increases during the ZnO deposition. The size of the crystal lattice parameter values increases due to the temperature increases. It is to produce a modified configuration for new crystal grain boundaries, so it looks as if the grain size becomes larger [12].

The purpose of testing Electrochemical Impedance Spectroscopy (EIS) is to find the value of the conductivity and capacitance of CNT-graphite-ZnO electrodes mode at frequency range that is given between 0.1 KHz to 100 KHz. It aims to further evaluate the electrochemical behavior of the fabricated electrodes. The S1, S2, S3, and S4 Nyquist curves exhibited a semicircular arc in the high-frequency region and a straight line in the low frequency region. Electrical conductivity value can be calculated using the equation (2). The results of the electrical conductivity of each sample sheet are shown in Figure 3 :

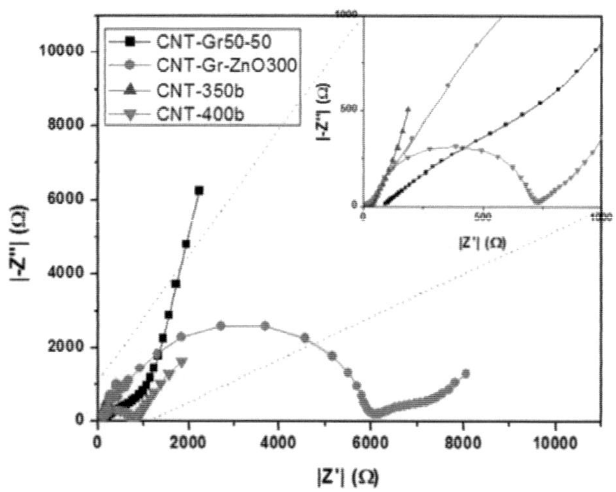

Figure 3. Nyquist impedance plots for electrode of different deposition temperature.

Figure 3 shows a graph of the nyquist impedance for samples S1, S2, S3 and S4. The graph explains the value of the charge transfer resistance (R_{ct}). The 45° slope portion in the low-frequency regions demonstrated the Warburg impedance of the electrolyte into the interior of the electrode surface and ion diffusion/transport into the electrode surface [13]. When ZnO material is add, the visible arc size enlarges. This can be saw in the sample S2, S3, and S4. Magnification corner semicircle described in the graph shows the capacitance properties of materials [14]. The height of conductivity is show in Table 2. When there is no addition of ZnO, it will increase conductivity.

Table 2. Conductivity (σ)and interfacial electron transfer resistance (R_{ct}) of the electrodes

	S1	S2	S3	S4
$R_{ct}(\Omega)$	974.04	6113.34	34.32	751.74
σ(S/m)	0.16	0.22	4.6	1.8

Results consecutive R_{ct} value sample respectively is 751.74; 34.32; 6113.33 and 974.04 Ω. The conductivity can be calculated from:

$$\sigma = \frac{t}{R.A} \quad (2)$$

t, R and A mean the thickness, resistance and surface area of electrode. For the cell with CNT-graphite (S1), the value of conductivity (Table 2) is 0.16 S/m. The increasing of conductivity with addition ZnO happened for the samples of S2, S3 and S4. However, there is a decrease in the conductivity of sample S4 to be 1.8 S/m, but it is still higher without the addition of ZnO. The addition of ZnO has a significant influence on the electrical conductivity. The smallest crystal sizes lead to facilitate the mobility of electrons. ZnO Crystals will produce a set of items denser. Moreover, it facilitates the electron transfer distance so that the conductivity of the electricity generated will be greater [15]. However, the deposition in temperature of 400 °C degrades the conductivity to be 1.8 S/m. This is because a large amount of ZnO covering the carbon destroys the surface structure of carbon morphology. The overall addition of ZnO increases a significant multiple value of conductivity without addition ZnO which is higher than multiple value addition MnO_2/CNT [6, 16] and TiO_2/CNT [17].

To investigate a specific capacitance calculation, it can be calculated with the equation (3). Specific capacitance can be defined as the ability of a capacitor to store charge at a certain frequency per unit

From the data processing results, EIS is formulated as follows:

$$C = \frac{1}{2\pi f Z m} \quad (3)$$

f, Z and m are the frequency, impedance and mass electrode. The result of calculation represented in diagram (Figure 4). Specific capacitance (Cs) on several frequencies obtained on the sample S2 has the largest capacitance is 1.23 F/g at a frequency of 0.1 Hz. Then sample S1 has a capacitance of 1.7 x 10^{-2} F/g at a frequency 0.5 Hz. Sample S3 has a frequency of 4.09 x 10^{-1} F/g at a frequency of 0.2 Hz. Then, sample S4 has a capacitance of 1.95 x 10^{-1} F/g at a frequency of 0.5 Hz.

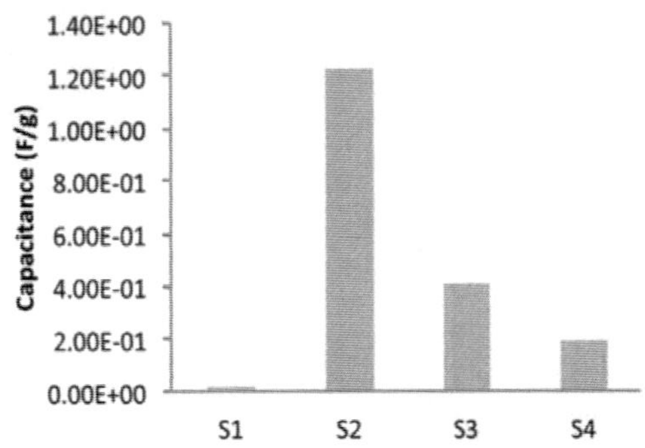

Figure 4. Diagram capacitance value analysis results on testing Electrochemical Impedance Spectroscopy

In addition, ZnO coating in the manufacture of ZnO-CNT-graphite nanocomposite electrode can improve conductivity and capacitance values. The high capacitance in S2 sample was caused by effect of the small crystallite size of ZnO. This effect will increase redox reaction and support electrons into electrolyte in the ZnO-CNT-graphite electrode and connect the contact between electrodes and electrolyte [14]. They can improve the utilization ratio of the electrode active material in storing an electric charge [18]. The increasing of the ZnO crystallite size will cover structure of carbon material, so it will intrude carbon to storage of the redox reaction from Zn. Value of conductivity increases until deposition temperature of 350 °C, the same result has been obtained by Aravinda [6] and Zhang [7]. The high deposition temperature will increases a crystallite size of ZnO and carbon materials, however above a temperature of 350 °C, ethylcellulose as a material binder of carbon is damage. It results carbon material inovolved broken and affect decrease of conductivity values.

4. Conclusion

Based on the results of the study, it can be concluded that manufacturing of ZnO-CNT-graphite nanocomposite with a different temperature of ZnO deposition has been successfully. The addition of ZnO with given temperature can affect the value of conductivity and capacitance at the electrodes. The studies reveal that the supercapacitors have good capacitance and conductivity. Hence, values obtained the best results with the highest capacitance in temperature 300°C and high conductivity in 350°C deposition.

Acknowledgments

We would like to thank Department of Physics, Faculty of Science and Mathematics Diponegoro University and Center of Physics Research Indonesia (LIPI) for funding our research. We are also indebted to Sulistiyono Susilo for his valuable editing assistance.

References

[1] L Qi, Han K, M R Gadinski, G Zhang, and Q. Wang, 2014 *Advanced Materials* Vol **26**(36), pp.6244-6249.

[2] Boyea J.M., Camacho, R.E., Turano, S.P. and Ready, W.J, 2007 Nanotech. L. & Bus Vol **4**(1), pp.585–593.

[3] Du, Chungseng., and Pan Ning, 2006 *Journal of Power Source* Vol **4**(1), pp.1487-1494.

[4] Kim, H.I., Kim, H.J., Morita, and Park, S.G., 2010 *Bull. Korean Chem.* Soc, **31**(2), pp.423-428.

[5] Wang, J.G., Yang, Y. and Kang, F., 2012 *Electrochimica Acta* **75**, pp.213–219.

[6] Zhang, Y., Sun, X. and Pan, L., 2009 *Solid State Ionics*, **180**(32), pp.1525–1528.

[7] Aravinda, L.S, Nagaraja, K.K., Nagaraja, H.S. and Udaya, B., 2013 *Electrochimica Acta* Vol **95**, pp.119–124.

[8] A. Subagio, Priyono, Pardoyo, Aswardi, R Yudiantu, 2009 *Materias Science Forum* Vol **827,** pp.113 - 118

[9] A. Subagio, Pardoyo, K. NgurahAyu ,V. Gunawan, Sony, and Rowi J, 2009 Nanosains Nanotek. Vol **2**(1), pp. 13 – 15.

[10] Rohman, M. and Subagio, Agus., 2013 *Youngster Physics Journal,* Vol **1**(3), pp.35-40.

[11] Karakaya, S. and Ozbas,, 2013 *Journal of Natural and Applied Science* Vol **17**(3), pp.9-51.

[12] Djelloul, A. and Khedidja B., 2008 Turk J Phys Vol **32**, pp.49-58.

[13] Frackowiak, E., Abbas, Q. and Béguin, F., 2013 *Journal of Energy Chemistry* Vol **22**(2), pp.226–240.

[14] R. Kotz, M. Carlen, 2000 Electrochimica Acta Vol 45, pp 2483 - 2498.

[15] Esmaili, S., Illadis, A.A., Samataki, A. and Konofaos, M., 2010 *Solid-State Electronics* Vol **54**, pp.1150–1154.

[16] Ramadoss, A. and Jae, S., 2014 *International Journal of Hydrogen Energy*, Vol **39**(23), pp.12201–12212.

[17] J.B. Chen, C.-W. Wang, J.Wang, Y. Li, R.-S. Guo, B.-H.Ma, F. Zhou, and W.-M. Liu, 2009 Appl. Surf. Sci Vol **256**(1), pp.39–42.

[18] R.B. Rakhi, W. Chen, D. Cha, H.N. Alshareef, 2012 Advance Energy Materials Vol **2** (3), pp. 381-389.

IOP Conf. Series: Journal of Physics: Conf. Series **877** (2017) 012007 doi:10.1088/1742-6596/877/1/012007

Conformation Analysis of T1 Lipase on Alcohols Solvent using Molecular Dynamics Simulation

A M Putri[1], **T Sumaryada**[2,3], **S T Wahyudi**[1]*

[1]Biophysics Division, Department of Physics, Bogor Agricultural University, Jalan Meranti Kampus IPB Dramaga Bogor 16680, Indonesia

[2] Computational Biophysics and Molecular Modeling Research Group (CBMoRG), Department of Physics, Bogor Agricultural University, Jalan Meranti Kampus IPB Dramaga Bogor 16680, Indonesia

[3]Tropical Biopharmaca Research Center Jalan Taman Kencana No. 3 Bogor 16128, Indonesia

E-mail: stwahyudi@apps.ipb.ac.id

Abstract. Biodiesel usually is produced commercially via a transesterification reaction of vegetable oil with alcohol and alkali catalyst. The alkali catalyst has some drawbacks, such as the soap formation during the reaction. T1 Lipase enzyme had been known as a thermostable biocatalyst which is able to produce biodiesel through a cleaner process. In this paper the performance of T1 lipase enzyme as catalyst for transesterification reaction in pure ethanol, methanol, and water solvents were studied using a Molecular Dynamics (MD) Simulation at temperature of 300 K for 10 nanoseconds. The results have shown that in general the conformation of T1 lipase enzyme in methanol is more dynamics as shown by the value of root mean square deviation (RMSD), root mean squared fluctuation (RMSF), and radius of gyration. The highest solvent accessible surface area (SASA) total was also found in methanol due to the contribution of non-polar amino acid in the interior of the protein. Analysis of MD simulation has also revealed that the enzyme structure tend to be more rigid in ethanol environment. The analysis of electrostatic interactions have shown that Glu359-Arg270 salt-bridge pair might hold the key of thermostability of T1 lipase enzyme as shown by its strong and stable binding in all three solvents.

Keywords: T1 Lipase enzyme, molecular dynamics simulation, alcohol solvents, biodiesel

1. Introduction

One type of renewable energy sources is biodiesel. Biodiesel is basically a diesel engine fuel made from natural products (vegetable oil) and consists of alkyl esters of fatty acids. Production of biodiesel can be obtained through many routes of reaction, for example by using alkali (alkali metal) catalyst. Biodiesel that produced via transesterification of vegetable oils with methanol and alkali catalyst has some drawbacks, such as saponification which contaminates the end products [1]. One alternative to avoid this problem is by using other type of catalyst i.e lipase enzyme, which can be used as a biocatalyst to accelerate the transesterification reactions through a clean and thermostable reaction.

A good biocatalyst has to maintain its structure and should not dissolve in a solvent while directing a specific reaction to produce the desired product. Lipase enzymes become popular nowadays due to its ability to maintain its structure, even at relatively high (industrial) temperature, so the separation process between the biodiesel and side products become easier [2,3]. However, the use of lipase as biocatalyst leaves one problem, alcohol such as methanol might destabilize lipase enzyme quickly and halt the biodiesel production process. Therefore, the study of lipase and alcohol solvent interaction is very important to optimize and speed up the biodiesel production process.

One of the thermostable lipases that already have a 3D structure is T1 lipase with PDB code 2dsn.PDB. T1 lipase is a thermostable enzyme and was isolated from Palm Oil Mill Effluent (POME) in Malaysia [4,5]. Besides resistance to high-temperature, T1 lipase also has resistance to alkalies solvent. This is due to its environment contains many alkalies. T1 lipase comprises of two independently chains, A and B. Each of them consists of 388 amino acid with missing the first residue. T1 lipase belonging of lipase that has two lids that cover active site [4]. According to Rahman *et al* [6], the lids are alpha-helix 6 and 7, but according to Wang *et al* [7], the lids are alpha-helix 6 and 9. The activity of lipase depends on the flexibility of lids covering the active site.

Researchers had previously conducted a study to see the opening mechanism of lid that covering active site [7] and examined the T1 lipase flexibility in the variation of temperature [8]. This paper is aimed to analyze the stability of the T1 lipase enzyme [5] in some alcohol solvents (ethanol and methanol) and water by observing the change in their geometrical and energetical parameters by using a Molecular dynamics simulations of T1 lipase enzyme in various solvents at temperature of 300 K and 10 ns simulation.

2. Material and Methods

2.1. Material

The materials used in this study is the crystallography data from *Geobacilus Żalihae* lipase (PDB code 2DSN) known as T1 lipase [4], which can be downloaded from the Protein Data Bank website (http://www.rscb.org/pdb). The equipment used in this study consist of hardware with the specification of a computer with 8 GB RAM, Quad Core Processor (Intel Corei7), NVIDIA Card GTX-760, Intel Corei7 and LINUX operating system Ubuntu version 14.04. The software used for the simulation process is AMBER 12 [9] and VMD (Visual Molecular Dynamics Program) version 1.9.2 [10] for visualization and analysis of the simulation result. The moving average technique for data smoothing were done using VBA program. For plotting the results, MS Excel 2010 is used to create the graphs.

2.2. Methods

T1 lipase was then solvated in water, ethanol, and methanol box using Packmol program were used for solvent preparation. 14658, 4542, and 6514 particle of water, ethanol and methanol respectively was used to build 10 Å boxes. Amber force field ff12SB was used parameterize amino acid. TIP3P was used to modelling the water [11] and the parameter for alcohol solvent, ethanol and methanol were obtained from R.E.D Server. The periodic boundary condition (PBC) was also used to eliminate the effect of surface tension and to achieve a condition with uniform density and pressure [12]. The SHAKE algorithm with a tolerance of 10^{-5} is applied to the system to curb the entire bond containing hydrogen atoms [13]. The electrostatic energy systems were calculated using the particle mesh Ewald (PME) [14] whereas the van der Waals interaction was calculated using the Lennard-Jones potential with a cutoff distance of 12 Å [15]. To control the stages of minimization, heating and the production run the NVT ensemble (constant number, volume and temperature) was used to the desired temperature where the system is coupled to the thermostat velocity rescaling the coupling constants of 0.1 [16]. To control the equilibration stage maintained at NPT Ensemble (constant number, pressure and temperature) and using langevin dynamics for temperature regulation [17].

Molecules that have solvated then minimized to avoid the van der Waals contacts that do not match (bad contact) and to minimize the steric effects of high energy. Each simulation starts with minimization process which is aimed to put the protein in its lowest energy (native conformation). In the heating process, protein was heated gradually until it reach the intended simulation temperature, 300K. To secure the stable conformation of protein, the equilibration processes were done using Langevin protocol. The final process is the production run at 300 K for 10 ns in three different solvents environment, where protein free to move without any constraint. The molecular dynamics results then analyzed and discussed.

3. Results and Discussions

3.1. Global Descriptor Analysis

The stability of a protein can be analyzed by a number of parameters such as, root mean square deviation (RMSD), root mean square fluctuation (RMSF), radius of gyration (Rg), solvent accessible surface area (SASA), and Secondary Structure. Figure 1 shows the fluctuations of RMSD, and RMSF value at 300 K and for 10 ns simulation in water, ethanol, and methanol environment.

RMSD is the average deviation between the protein conformation at a particular time and the reference structure. A seen in figure 1(a), at 5 ns first movement of RMSD in the water is relatively less than in Ethanol and Methanol. But after that, the movement of RMSD molecule in ethanol is relatively as the same as RMSD in the water. The RMSD molecule in methanol up to 10 ns which keep moving is higher than in ethanol and water as shown in figure 1(a). Dynamics of RMSD in methanol solvent which is higher than in the water and ethanol, allegedly is because of relative methanol polarity (0.762) is closer to water polarity (1) than ethanol polarity (0.654). The relative methanol polarity allegedly has hydrophilic and hydrophobic residues to keep it moving. The high movement in ethanol solvent at 5 ns allegedly is the movement of polar residues on the suppressed surface encountering non-polar solvent.

High peaks of amino acid residues show on RMSF graph showed the residue flexibility. The amino acid residue in water and methanol showed higher flexibility than in ethanol (Fig. 1b). Residue of Thr74, Thr93, Gly104, Ser201, Arg230, and Ser333 is polar residue, which located on the surface and has high flexibility when it was in the water than in methanol and ethanol. However, there were two hydrophobic residues Pro217 and Val294 which have higher flexibility in water than in methanol and ethanol. Although Pros217 is a hydrophobic residue but the position is squeezed by surrounding 3 polar residues, so the RMSF value became higher in water than in methanol and ethanol. The hydrophobic residue Val294 is on helix 9th which adjacent to active center lid and the position Val294 is shoved by Arg179 residue which polar residue. The RMSF result also showed that the fluctuation of amino acid atomic in ethanol solvent is generally more rigid than in water and methanol (figure 1b).

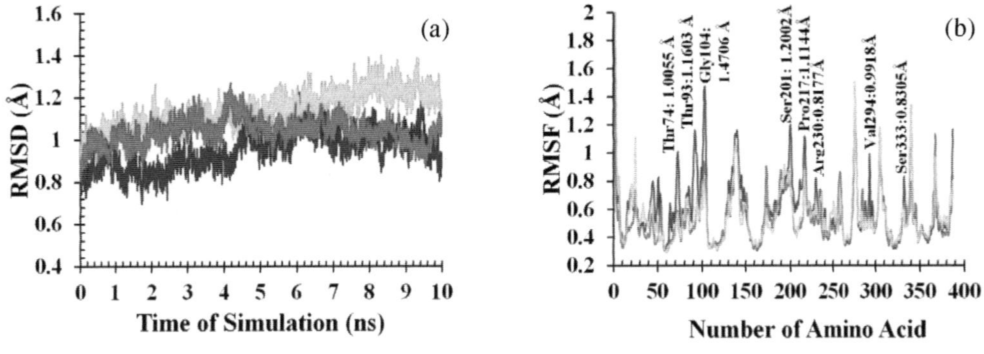

Figure 1. The pattern of RMSD (a) and RMSF (b) at 300K of T1 lipase molecule in water (blue line), methanol (green line) and ethanol solvents (red line).

Initially at simulation of 4 ns, SASA total molecule T1 lipase in methanol is higher than in water and ethanol. After 4 ns until end of simulation, SASA total in methanol and water is relatively balanced showed in figure 2(a). Different result of SASA total showed in figure 2(a) when the molecule was in ethanol solvents. The profile tends to decline and less than SASA total in water and methanol as seen in figure 2(a). The highest average value of SASA total is when the molecule in methanol, 16600 Å^2, followed by SASA Total in water, 16343 Å^2, and the last one in ethanol 16185 Å^2. The high value of SASA total when solvated in methanol is contributed by SASA non-polar which also the highest of the three solvents as seen in figure 2(b). The average value of SASA non-polar is respectively 5264, 5112 and 4729 Å^2 in methanol, ethanol and water. Calculation of SASA polar for the T1 lipase when in methanol obtained middle value among other solvents that is equal to 11336 Å^2, while SASA polar in water is 11614 Å^2 and in ethanol is 11074 A^2, showed in figure 2(c). The result of SASA analysis agree with the RMSD data as shown in figure 1(a), which the higher dynamic of T1 lipase in methanol is due to the non-polar residue movement and the quiet enough of polar residue movement. The decreasing of RMSD T1 lipase when in ethanol, showed in figure 1(a), allegedly is that of the SASA polar get shrink and the SASA non polar didn't changed much after 4 ns.

The radius of gyration of protein which shows the compactness of the protein structure did not show any significant difference in the three solvents. T1 lipase just a little bit experience shrinkage when in ethanol than in water or methanol. This shrinkage show that this structure more rigid compared to structure in water or ethanol. The Radius of gyration not so different when immersed in methanol and water as seen in figure 2(d).

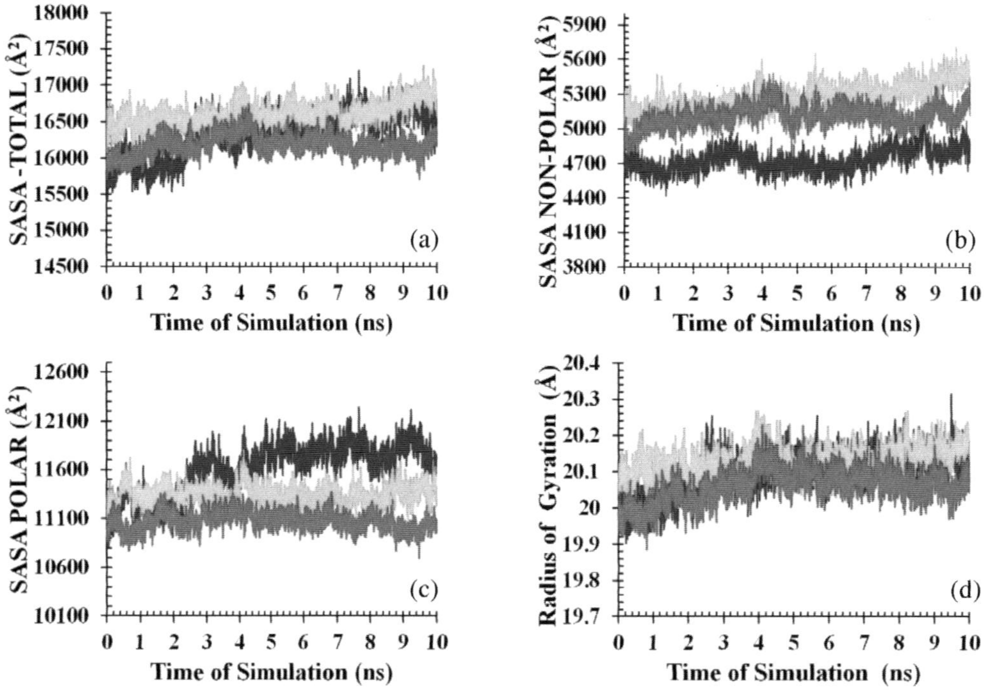

Figure 2. The fluctuation of Solvent Accessible Surface Area (SASA) during the simulation at 300 K for 10 ns (a) SASA total, (b) SASA non-polar, (c) SASA polar, & (d) radius of gyration. Blue, red and green line indicated the molecule immersed in water, ethanol and methanol respectively.

3.2. Salt bridge Analysis

Electrostatic interactions are often referred to as ion-pair interactions or a salt-bridge is an interaction between charged amino acid residues. The importance of electrostatic interactions in protein structure were first reported by [18] which stated that this interaction contributes significantly in stabilizing the protein.

The results of our simulation have indicated the occurrence of 14 salt-bridge pairs that appeared in all solvents. There are also 3 salt-bridge that appeared only in water, 2 salt-bridge that appeared in ethanol and 2 salt-bridge that appeared in methanol solvents. Of 14 salt-bridge that appeared in all solvents, four salt-bridge that might play an important role in the rigid conformation of T1 lipase molecule when solvated in ethanol and methanol are Glu132-Lys84, Glu149-Arg134, Asp205-Arg92 and Glu250-Arg330.

The Glu132-Lys84 salt-bridge pair as seen in figure 3(a) appeared to be stable in ethanol and methanol environment. This salt-bridge pair slightly away at 2 to 4 ns in ethanol solvents but showed come up again until the end of simulation. This is indicates that this salt bridge is strong and responsible to the stability of T1 lipase enzyme in ethanol and methanol. Glu149-Arg134 salt bridge pair have intermittent relation since the beginning of simulation until the end of simulation in water environment, showed in figure 3(b). This salt bridge also have a bond distance is far in ethanol but very close enough after 6 ns, as shown in figure 3(b). When this molecule immersed in methanol, this salt bridge pair has a moderate energy around -40 kJ/mol since the beginning of simulation until 8ns. After that this salt bridge has similar profile with salt bridge pair when immersed in ethanol, as shown in figure 3(b). This is indicates that this salt bridge strong enough to make the molecule structure stable in methanol and even more rigid in ethanol.

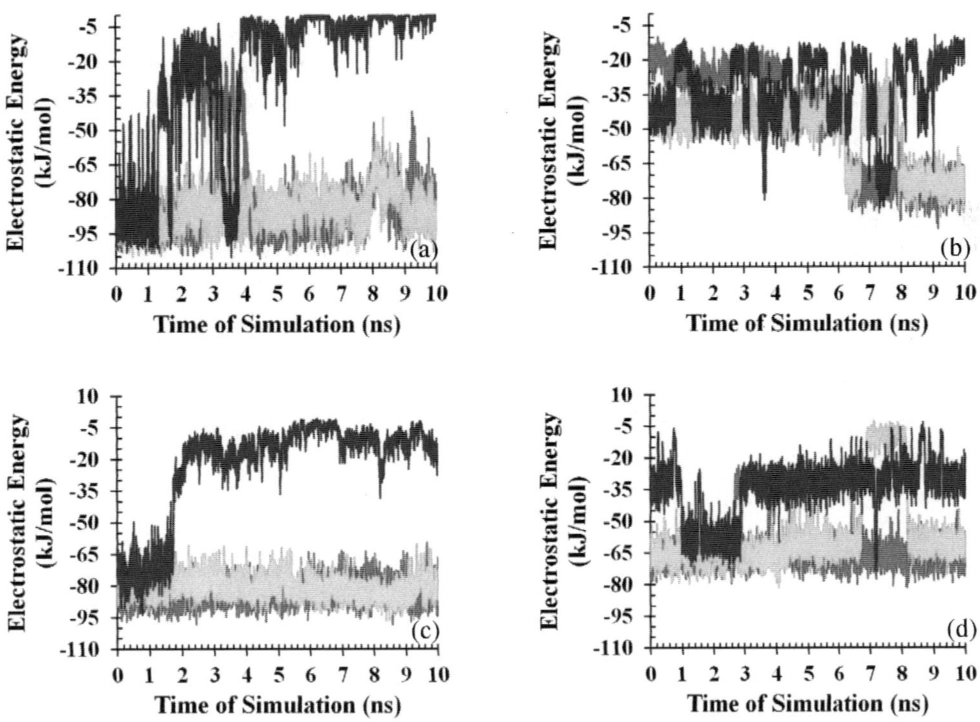

Figure 3. Electrostatic interaction of Glu132-Lys84 salt bridge pair (a), Glu149-Arg134 salt bridge pair (b), Asp205-Arg92 salt bridge pair (c) and Glu250-Arg330 salt bridge pair (d) Blue, red and green line indicated the molecule solvated in water, ethanol and methanol respectively.

The Asp205-Arg92 salt bridge pair have a strong energy binding in ethanol and methanol solvents and also show weak energy binding after breaking at 1.6 ns in water solvent (Figure 3c). This is indicates that this salt bridge is very strong to make the structure more rigid in ethanol and methanol. The Glu250-Arg330 salt bridge pair also showed intermittent distance in water environment (Figure 3d). From 0-1 ns the pair was away then closed for next 2 ns and then far again until the end of simulation. This salt bridge has electrostatic energy around -30 kJ/mol as shown in figure 3(d). Different situation was happened when this salt bridge appear in ethanol and methanol. In this situation, this salt bridge has energy around -70 kJ/mol. In ethanol, this Glu250-Arg330 pair was experienced a bit far distance at 3-4 ns, also in methanol this salt bridge has a bit far distance at 6.8 to 8 ns as seen in figure 3(d). This salt-bridge is weak in the water, but quite strong in ethanol and methanol environment.

3.3. Secondary Analysis

(a) (b) (c)

Figure 4. Secondary structure evolution during the simulation at 300 K for 10 ns in (a) water, (b) ethanol (c) methanol; α-helix (Purple), β-sheet (Yellow), Turn (Blue) and Coil (Black); α-helix (Purple), β-sheet (Yellow) 3_{10}-helix (Blue), Turn (Green) and Coil (White)

Analysis of secondary structures shows that during the simulation, α-helix in ethanol environment only suffered a relatively very small damage as compared to other solvents as seen in figure 4(b). The β-sheet structure relatively stable and constant in all three solvents, showed in Figure 4. At this point, it is too early to state the type of unfolding mechanism in this simulation since we need a longer simulation time (probably in the range of millisecond), but initial indication tend to support the zipper model which start with α-helix destruction [19].

The structure of lipase T1 after 10 ns simulation solvated in ethanol and methanol show a bit conformation change compared to structure in water. The changes occur on polar residue in surface area as expected before, as seen in figure 5.

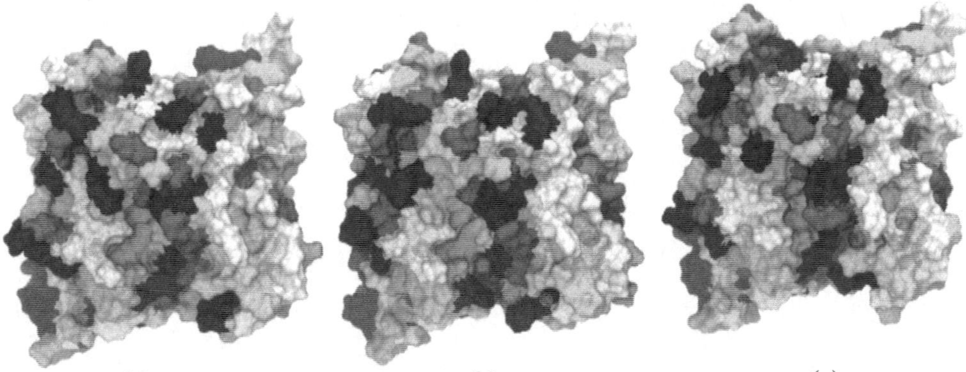

(a) (b) (c)

Figure 5. Conformation of lipase T1 enzyme at 300K after 10 ns solvated in ethanol (a), in water (b), and in methanol (c). Red and Blue indicated negative and positive charge residues respectively, green indicated polar neutral residues and grey indicated as hydrophobic residues.

4. Conclusion

The molecular dynamic simulations have indicated that T1 lipase enzyme is able to maintain its structure integrity in various solvents i.e 100% pure ethanol, methanol, and water. This result is important since the previous assumption stated that T1 lipase enzyme would be dissolved easily in pure alcohols environment. In alcohol environment, our results indicates that the enzyme is more dynamics in methanol, whereas more rigid in ethanol environment. When the conformation is too rigid, the functionality of the enzyme will be reduced, so it is important to let the T1 lipase enzyme structure to have dynamics. This result also suggest that the usage of methanol solvent in biodiesel reaction will be expected well than ethanol solvent. Varying the composition of mixed water and alcohol will be our next tasks.

Acknowledgment

The authors express their gratitude for the funding from The Ministry of Research Technology and Higher Education of The Republic of Indonesia through the scheme of Penelitian Unggulan Perguruan Tinggi, Penelitian Unggulan Divisi of contract No.079/SP2H/LT/DRPM/I/2016. The authors also thank to Dr. Husin Alatas for the access to High Performance Computing Facility (HPC) in The Theoretical and Computational Physics Laboratory at the Department of Physics and to Dr. Wisnu Ananta Kusuma for the access to HPC facility in The Department of Computer Science Bogor Agricultural University, Indonesia.

References

[1] Knothe, G 2005 *Fuel processing Technology* volume **86**, Issue 10, 1059-1070
[2] Kumar A, Dhar K, Kanwar S S and Arora P K 2016 *Biological Procedures Online* **18** 2
[3] Fan X, Niehus X and Sandoval G 2012 *Methods in Molecular Biology* **861** 471
[4] Matsumura H, Yamamoto T, Leow T C, Mori T, Salleh A B, Basri M, Inoue T, Kai Y and Rahman R N Z R A 2007 *Proteins* **70** 592
[5] Rahman R, Leow T C, Salleh A B and Basri M 2007 *BMC Microbiology* **7** 77
[6] Rahman R, Salleh A B, Rahman R N Z R A, Rahman M A B, and Leow T C 2012 *Protein Science* **21** 1210 - 21
[7] Wang Y, Wei D Q and Wang J F 2010 *J. Chem. Inf. Model* **50** 875–78
[8] Abdul Rahman M B, Karjiban R A, Salleh A B, Jacobs D, Basri M, Thean Chor A L, Abdul Wahab H and Rahman, R N 2009 *Protein and Peptide Letters* **16** 1360-70

[9] Case D.A., T.A. Darden, T.E. Cheatham, III, C.L. Simmerling, J. Wang, R.E. Duke, R. Luo, R.C. Walker, W. Zhang, K.M. Merz, B. Roberts, S. Hayik, A. Roitberg, G. Seabra, J. Swails, A.W. Götz, I. Kolossváry, K.F. Wong, F. Paesani, J. Vanicek, R.M. Wolf, J. Liu, X. Wu, S.R. Brozell, T. Steinbrecher, H. Gohlke, Q. Cai, X. Ye, J. Wang, M.-J. Hsieh, G. Cui, D.R. Roe, D.H. Mathews, M.G. Seetin, R. Salomon-Ferrer, C. Sagui, V. Babin, T. Luchko, S. Gusarov, A. Kovalenko and P.A. Kollman (2012), AMBER 12, University of California, San Francisco

[10] Humprey W, Dalke A and Schulten K 1996 *J.Molec.Graphics* **14** 33

[11] Jorgensen W L, Chandrasekhar J, Madura J D, Impey R W and Klein M L 1983 J. Chem.Phys.**79** 926

[12] Allen M P and Tildesley D J 1989. *Computer Simulation of Liquids*, Oxford University Press,USA.

[13] Ryckaert J, Ciccotti G and Berendsen H. 1977 *J Comput. Phys*.**23**: 327

[14] Essmann U, Perera L, Berkowitz M L, Darden T, Lee H, and Pedersen, L G. 1995 *J. Chem.Phys*., **103**(19) 8577

[15] Bui J M, Gsponer J, Vendruscolo M and Dobson C M. 2009 *Biophyics Journal*. **97**: 2513

[16] Bussi G, Donadio D and Parrinello M. 2007 *J. Chem. Phys*. **126** 014101

[17] Izaguirre J A, Catarello D P, Wozniak J M, and Skeel, R D 2001 *J. Chem. Phys* **114** 2090 - 98

[18] Perutz M. 1978 *Science*. **201**(4362) 1187

[19] Muñoz V, Thompson P A, Hofrichter J and Eaton W A 1997 *Nature* **390** 196

Effect of Acetylene Black Content in $Li_4Ti_5O_{12}$ Xerogel Solid-State Anode Materials on Half-Cell Li-ion Batteries Performance

N M Abdurrahman[1], B Priyono[1], A Z Syahrial[1], A Subhan[2]

[1]Departement of Metallurgical and Materials Engineering, Universitas Indonesia, Depok, 16424, Indonesia
[2]Research Center for Physics – LIPI, PUSPIPTEK, Tangerang, Banten, 15310, Indonesia

E-mail: bambang.priyono@ui.ac.id

Abstract. The effect of Acetylene Black (AB) additive contents in lithium titanate/$Li_4Ti_5O_{12}$ (LTO) anode on Li-ion Batteries performance is studied in this work. The LTO active material for Li-ion batteries anode was successfully synthesized using sol-gel method to form TiO_2 xerogel continued by mixing process with LiOH in ball-mill and then sintered to obtain spinel LTO. The LTO powder is characterized by X-Ray Diffraction (XRD), scanning electron microscopy-Energy Dispersive Spectroscopy (SEM-EDS), and Brunauer–Emmett–Teller (BET) . The spinel LTO and TiO_2 rutile were detected by XRD diffractogram. The LTO powder is in the form of agglomerates structure. This powder then was mixed with PVDF binder (10%wt) and AB additives with various amount from 10%wt (LTO2 Ac-1), 12%wt (LTO2 Ac-2), and 15%wt (LTO2 Ac-3) of total weight solid content to form electrode sheet. Half-cell coin battery was made with lithium metal foil as a counter electrode. Cyclic voltammetry (CV), Electrochemical-impedance spectroscopy (EIS), and charge discharge (CD) test used to examine the battery performance. The highest resistance value is obtained in LTO2 Ac-3 sample with 15%wt of AB. It might be caused by the formation of side reaction product on electrode surface at initial cycle due to high reactivity of LTO2 Ac-3 electrode. The highest initial capacity at CV test and CD test was obtained in LTO2 Ac-1 (10%wt AB) sample, due to the best proportion of active material content in the compound. While, in the charge-discharge test at high current rate, the best sample rate-capability performance belongs to LTO2 Ac-3 sample (15%wt AB), which still have 24.12 mAh/g of discharge capacity at 10 C with 71.34% capacity loss.

1. Introduction

Battery is one of the energy storage medium that have a good capacity and energy densities performance. It should have a good rechargeable performance and more safe to use as well. However, Li-ion Batteries (LiB) to be used for electric vehicle (EV) needs to have high power density as one of the basic properties [1].

On their development, graphite is selected as anode materials in LIB. However the poor cyclic life performance of the graphite, makes the LIB with graphite based anode is less durable. It can be caused by unavoidable solid electrolyte interface (SEI) film formation on electrode surface and thickening on the graphite anode, lithium plating while charging at low temperature, and other side reaction [2].

Recently, $Li_4Ti_5O_{12}$ (LTO) based anode materials is considered to substitute the graphite based anode, due to high safety, and excellent cyclic performance, compared to graphite based anode materials [3]. LTO based anode has a zero strain properties, no SEI film formation, fast charging ability, and good thermal stability, therefore makes LTO based anode has a long cycle life, high safety, and can be implemented for EV LIB anode materials [3]. However, LTO has a low intrinsic conductivity (10^{-9} $S.cm^{-1}$) [4].Thus, a conductive agent additives must be added into LTO mixing slurry to improve the conductivity of anode materials. Acetylene Black (AB) is often used for conductive agent additives to improve the conductivity of electrode.

In this research, the LTO compound prepared by sol gel process to produced TiO_2 anatase and then by a solid-state process to form $Li_4Ti_5O_{12}$ mixed with AB with various contents, then it is assembled into a half cell LiB to observed the influence of Acetylene Black (AB) content to the performance of the battery cell.

2. Experimental

2.1 $Li_4Ti_5O_{12}$ powder preparation and characterization

LTO powder is synthesized through a sol-gel process, calcination, ball-milling, and sintering process. Sol gel process is started by making primary solution that contain titanium tetra-n-butoxide (Kanto chemical, Co. Inc.) and ethanol (Merck) and mixed that with secondary solution that contains ethanol and distillated water. Mixing process is carried out in magnetic stirring for about 2 h until gel formed. The gel is then allowed to dry in room temperature for about 12 h to form xerogel. Calcination is carried out at 300 °C for 2 h to make anatase TiO_2. The high energy ball milling (HEBM) is used to mix anatase TiO_2 with LiOH for 1 h in an intermittent way. Sintering process is then carried out at 750 °C for 3 h to form spinel $Li_4Ti_5O_{12}$ (LTO) that ready to be used as an active electrode material.

The morphology of sample was observed by scanning electron microscope (SEM). The particle size of sample was measured with Image J software. Phase identification was examined by X-ray diffraction analysis (XRD) using Cu Kα radiation in an angular range 2θ from 10° to 100°.

2.2 Cell assembly and testing

The half-cell coins battery were assembled and tested to observe the electrochemical properties and performance of LTO compound, in which LTO was prepared as cathode active material, lithium foil as an anode, and $LiPF_6$ 1M as an electrolyte. While, DMAC, Acetylene Black (AB) and PVDF were used as solvent, conductive additive and binder, respectively. The compositions were prepared with fixed binder content of 10%wt, while the LTO and AB contents were 10%wt for LTO2 Ac-1, 12%wt for LTO2 Ac-2, and 15%wt for LTO2 Ac-3. All of these material then mixed by magnetic stirrer at 300 rpm at 70°C for 2 h to obtain slurry.

The slurry was coated at copper foil current collector by electrode coater (doctor blade) with 200 μm of wet thickness to form 100 μm of dry thickness. Coated Cu foil was then dried at doctor blade at 80 °C and cut to form coin shape. The assembly process was carried out in a glove box filled with high purity argon gas.

Cyclic voltammetry (CV), Electro-impedance spectroscopy (EIS), and Charge-discharge test were performed to evaluate the cell electrochemical performance.

3. Results and Discussions

3.1 Phase analysis of the LTO

X-ray Diffraction analysis was performed to analyze phases in LTO powder after sintering process. An expected phase in XRD pattern was spinel $Li_4Ti_5O_{12}$, with smoothing by X-pert High Score Plus software corresponds to JCPDS no 26-1198 as shown in Figure 1. From the result, a spinel LTO phase was successfully identified in six angle positions of 2Θ.

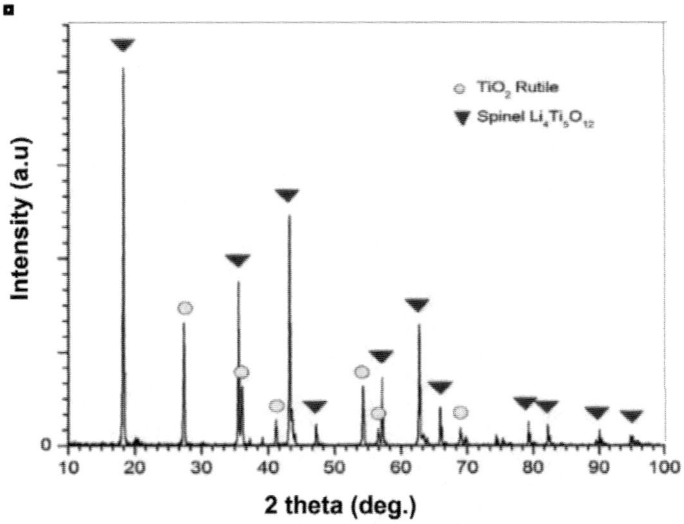

Figure 1. The XRD result pattern of the $Li_4Ti_5O_{12}$ powder by solid state synthesis.

The XRD result was not only identified a LTO pattern, but also TiO_2 rutile was presence in LTO powder. TiO_2 rutile identified with JCPDS no 01-076-0318. The presence of TiO_2 rutile in XRD pattern was caused by the unreacted TiO_2 anatase due to the loss of the Li^+ ion source during sintering process, and transform to TiO_2 rutile in high temperature of sintering process [5].

3.2 Structure and morphology of LTO
Structure and morphology analysis of LTO powder were conducted using scanning electron microscopy. The magnifications used in this work were 5000X, 10000X and 20000X to see the particle size and the presence of agglomerates.

Morphology of this LTO powder can be seen in Figure 2, with 5000X magnification, an agglomerate structure was seen. With the increasing size and quantity of agglomerates, it could lead to decreasing in the reactivity and capacity of sample as a consequence of decreasing the sample's surface area. The presence of more agglomerates is caused by a quite high sintering temperature (750°C). Sintering process at high temperature could induce diffusion phenomenon and promote the LTO particle clustering generated agglomerates formation [6].

LTO particle size measurement was conducted on the SEM micrograph of microstructure by Image-J software. An average size of LTO agglomerates particle is 4.22 μm as obtained from Figure 2a. While for the average of all LTO particles shown in the micrograph is 0.73 μm as obtained from Figure 2b and Figure 2c.

The material contain in EDS test was Titanium, Chlor, Carbon, Oxygen with Titanium compared Oxygen contain 21,9% : 78,1%.

Figure 2. Microstructure of LTO by SEM test with magnification of a) 5000X, b) 10000X, and c) 20000X.

3.3 EIS characterization of LTO electrode

Conductivity is one of the most important properties of active materials for use in the battery electrodes. Effect of AB on the conductivity is measured by EIS to obatin the impedance (resistivity) level of the electrode. The EIS results could be seen in Table 1 and Figure 3. The results show that LTO2 Ac-3 (15%wt) has the highest resistivity value, the Nyquist semicircle of LTO2 Ac-3 has the longest diameter corresponds the highest value of R_{ct}. Meanwhile, the LTO2 Ac-2 has the lowest value of R_{ct} that would results in highest level of conductivity.

Table 1. Impedance result of the samples.

Sample	R_e (Ω)	R_{ct}(Ω)
LTO2-Ac1	4,440	51,660
LTO2-Ac2	12	41
LTO2-Ac3	10,100	62,500

The high value level of R_{ct} in LTO2 Ac-3 sample might be caused by side reaction product formed in the early cycle of LTO2 Ac-3 sample electrode surface. The presence of side reaction product in electrode surface would block lithium ion to reach electrode surface that makes resistivity value increased.

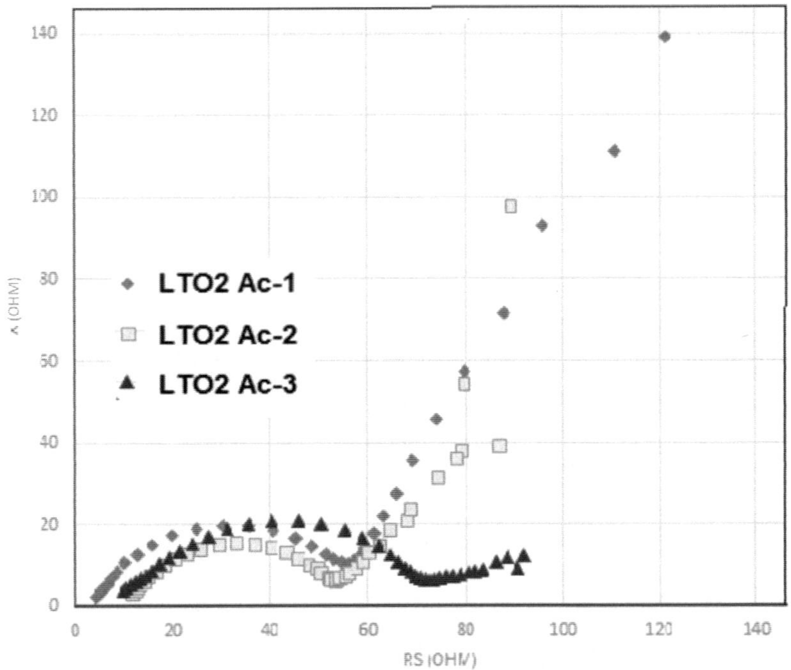

Figure 3. Nyquist semicircle-curve of the samples.

Side reaction products that formed at low cycle in LTO Ac-3 sample electrode surface might be caused by improving in electrode's reactivity due to highest content of AB. As the results, increasing reactivity would lead to a side reaction product formed even at low cycle and low rate. That phenomenon was also observed by Yan-Bing He, et al.[7]. He observed that resistivity value indicated more surface area and reactivity of electrode, the side reaction product became easier to form in early cycle and low current rate even at voltage value above 1V. This would be the reason to increased R_{ct} value at EIS test for LTO2 Ac-3 sample. This result is supported by the charge-discharge testing results.

3.4 Chemical performance by cyclic voltammetry

The cyclic voltammetry test is conducted to examine the electrochemical performance of the sample, such as: capacity, working voltage, and other chemical performance properties of electrochemical cell. Theoretically, LTO has 1.55V of cathodic voltage peak value vs. Li with 175 mAh/g of capacity [8]. Cathodic peak relates to the intercalation process of lithium to $Li_4Ti_5O_{12}$ to form $Li_7Ti_5O_{12}$. Meanwhile, the anodic peak relates to the de-intercalation process of lithium ion from $Li_7Ti_5O_{12}$ to form $Li_4Ti_5O_{12}$ ($Li_4Ti_5O_{12} + 3Li^+ \leftrightarrow Li_7Ti_5O_{12}$) [9]. The results of CV test as shown in Figure 4 were obtained with 100 µV/s scan rate and presented in Table 2 for more detailed data.

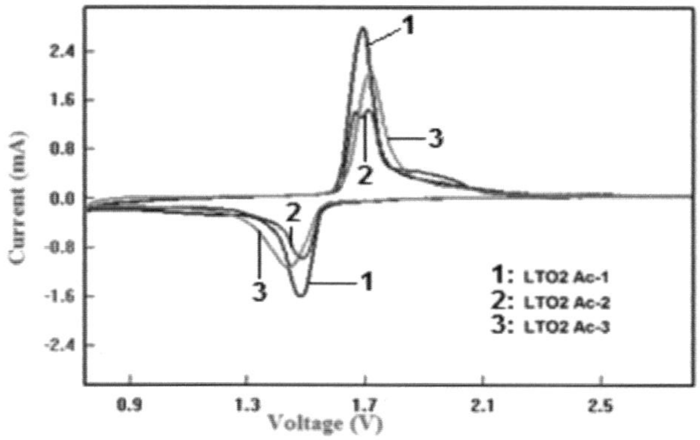

Figure 4. Cyclic voltammetry curve from sample.

Table 2. Capacity and working voltage of the samples.

	Capacity (mAh/g)	**E^{O} (V)**
LTO2 Ac-1	129,710	1,585
LTO2 Ac-2	94,750	1,575
LTO2 Ac-3	83,350	1,573

For working potential, it was obtained by calculation using Eq.1 as follows [10]:

$$E^0 = \frac{E_{p,c} + E_{p,a}}{2} \quad (1)$$

The result in CV test shows that all of the sample has a similar working potential that similar to a LTO theoretical working potential, i.e: 1.55V for lithiation process and 1.56V for de-lithiation [11]. Thus, AB content has no effect improved the working potential of the LTO electrodes.

The graph in Figure 4 also informs a reversibility of the reactions of this LTO. The results show that all of the samples have a pair of peaks, reduction and oxidation peak of a reversible reaction[12].

Specific capacity data is also provided by CV test. As shown in Table 2, LTO2 Ac-1 sample who has smallest AB content, has the highest value in the specific capacity, i.e: 129.71 mAh/g. Besides, the LTO2 Ac-3 which contains highest AB content has the lowest value of the specific capacity, i.e: 83.35 mAh/g. The decrease of active material content (LTO) as increasing of AB content and formation of side reaction product is predicted to be the reasons of this phenomenon that consume a part of capacity in more reactive electrode [7]. This data was supported by the EIS results.

3.5 Charge-discharge test results

Charge-discharge test was performed to analyze battery performance at different given current rate. This test began from the current rate of C/3 until 10C. The results of the influence of AB content in CD test results in LTO's electrode are presented in Figure 5 and Figure 6.

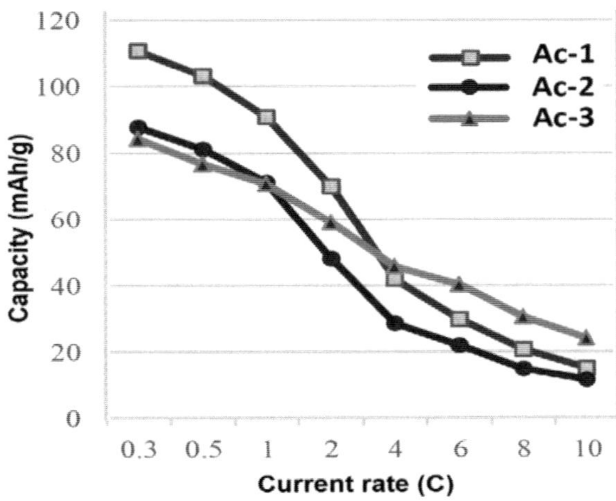

Figure 5. Discharge capacity vs. current rate.

Increased current rate would decrease the instant capacity of the battery. As shown in Figure 5, there is a relation between AB content to rate-capability in battery. The increased of AB content would increase the rate capability of battery sample. LTO2 Ac-3 sample, with highest AB content, has the best rate-capability. Although the sample has worst initial capacity, at 4C current rate, it will surpass all of capacity of other sample and less steep slope curve than the other samples. In 10C, LTO2 Ac-3 sample reaches 24 mAh/g, while, LTO2 Ac-1 and LTO2 Ac-2 only have the capacity of 15.14 mAh/g and 11.57 mAh/g, respectively. It shown that the optimum AB content is depend at the current rate that is applied to the electrodes. When 0.3C until 4C current rate applied in the electrodes, the optimum composition was LTO2 Ac-1. After 4C until 10C, the optimum composition was LTO2 Ac-3 who highest percentage value of AB. This could indicate that AB acts to increase stability, conductivity, and also electrode's reactivity so that sample has better rate-capability or high power density. It is also confirmed that LTO2 Ac-1 sample with smallest AB content has a steepest slope in discharge capacity vs. current rate curve indicating that this sample has the lowest rate-capability. Thus, LTO2 Ac-1 composition is for highest energy density. More detailed data was provided in Figure 6, Figure 7, Figure 8 and Table 3.

Table 3. Capacity percentage at actual discharge rate.

Current Rate	LTO2 Ac-1	LTO2 Ac-2	LTO2 Ac-3
4C	37.76%	32.47%	54.22%
10C	13.65%	13.17%	28.66%

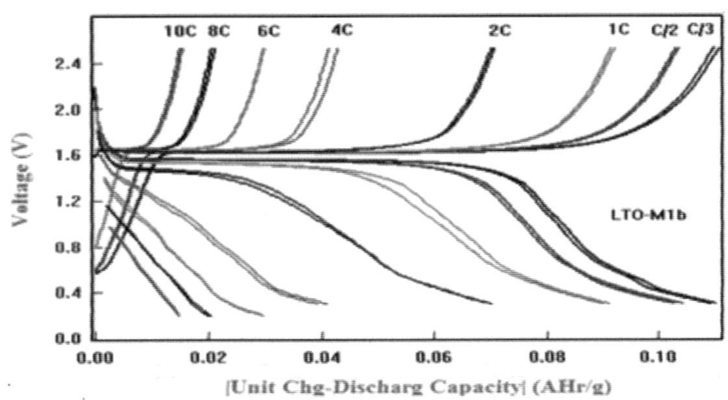

Figure 6. Charge-discharge curve of sample LTO2 Ac-1 (10wt% AB).

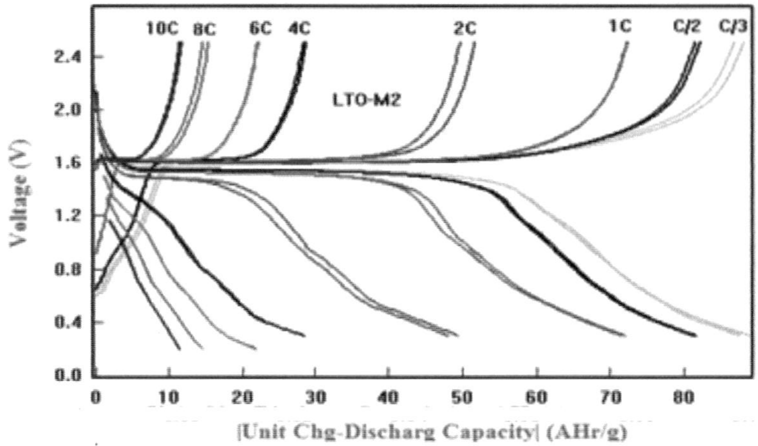

Figure 7. Charge-discharge curve of sample LTO2 Ac-2 (12wt% AB).

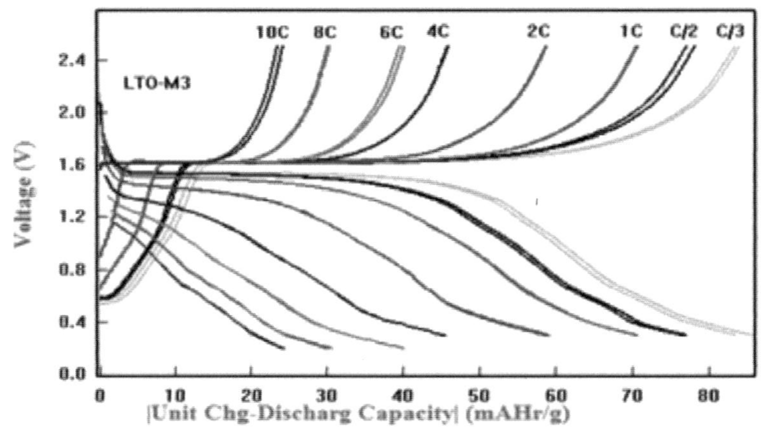

Figure 8. Charge-discharge curve of sample LTO2 Ac-3 (15wt% AB).

Low initial capacity in LTO2 Ac-3 sample might be caused by side reaction product at electrode's surface present at low cycle and low rate that consumes the initial capacity of the LTO2 Ac-3 sample. This phenomenon occurred because of AB content increased, which would increase the reactivity of the electrodes. With high reactivity, side reaction product would form easier at low rate and low cycle condition. Because of side reaction product formation has been ended at low cycle and low rate. They will not form at further rate and cycle anymore, and hinder the rate capability of the sample. Meanwhile, the less reactive electrode's is stable at initial rate and cycle, but less stable at further rate and cycle because of the formation of side reaction product at further cycle and rate that would decrease the capacity of samples in CD test [7].

4. Conclusion

Active material in the form of spinel $Li_4Ti_5O_{12}$ structures was successfully processed by sol-gel continued with solid-state methods, and possessed agglomerates micro structure. The phase obtained has spinel LTO pattern and with little amount of TiO_2 rutile analysed by XRD. Electrochemical performance obtained from EIS test show that LTO2 Ac-3 has the highest R_{ct} value, it is predicted that the side reaction product was formed at electrode's surface. In the CV test, it is found that LTO2 Ac-1 sample has the highest specific capacity value. Charge-Discharge test, show that LTO2 Ac-3 sample has the best rate-capability and supported by CV and EIS test result. Therefore, it concluded that the effect of acetylene black content on LiB battery's electrode composition is not significantly improved the conductivity of anode materials as the reactivity of the electrode is proportional to content of AB, and with the proper AB amount, it will obtained the best performance of LiB electrode composition. In this research, it is concluded that the optimum percentage of AB was depends on by current applied to the electrodes. For application at higher current rate (4C and above) or high power density, LTO Ac-3 has best performance. But for high density application in lower current rate (below 4C), LTO Ac-1 has best performance. Overall, considering the high initial specific capacity, LTO2 Ac-1 composition would be regarded as the optimum composition in this research. Finally, the power density or high-rate capability of this LTO compound is considered as well above the anode requirement of Li-ion batteries for electric vehicle.

Acknowledgements
The authors would like to thank the Direktorat Riset dan Pengabdian Masyarakat Universitas Indonesia (DRPM-UI)for the financial support of this research under the grant of Hibah PITTA with contract number: 2106 /UN2.R12/ HKP.05.00 /2016.

References
[1] X. Han, M. Ouyang, L. Lu, and J. Li, "A comparative study of commercial lithium ion battery cycle life in electric vehicle : Capacity loss estimation," *Journal of Power Sources*, vol. 268, pp. 658–669, 2014.
[2] X. Han, M. Ouyang, L. Lu, and J. Li, "Cycle life of commercial lithium-ion batteries with lithium titanium oxide anodes in electric vehicles," *Energies*, vol. 7, no. 8, pp. 4895–4909, 2014.
[3] I. Belharouak, G. M. Koenig, and K. Amine, "Electrochemistry and safety of $Li_4Ti_5O_{12}$ and graphite anodes paired with $LiMn_2O_4$ for hybrid electric vehicle Li-ion battery applications," *Journal of Power Sources*, vol. 196, no. 23, pp. 10344–10350, 2011.
[4] Y. Li, Z. Wang, D. Zhao, and L. Zhang, "Gd doped single-crystalline $Li_4Ti_5O_{12}/TiO_2$ nanosheets composites as superior anode material in lithium ion batteries," *Electrochimica Acta*, vol. 182, pp. 368–375, 2015.
[5] B. Li, F. Ning, Y. He, H. Du, Q. Yang, F. Kang, and C. Hsu, "Synthesis and Characterization of Long Life $Li_4Ti_5O_{12}$/C Composite Using Amorphous TiO_2 Nanoparticles," *International Journal of Electrochemical Science*, vol. 6, pp. 3210 – 3223, 2011.
[6] B. Priyono, A. Z. Syahrial, A. H. Yuwono, E. Kartini, M. Marfelly, and W. M. F. Rahmatulloh, "Synthesis of Lithium Titanate ($Li_4Ti_5O_{12}$) through Hydrothermal Process by using Lithium Hydroxide (LiOH) and Titanium Dioxide (TiO_2) Xerogel," *International Journal of Technology*, vol. 6, no. 4, pp. 555–564, Oct. 2015.
[7] Y. B. He, M. Liu, Z. D. Huang, B. Zhang, Y. Yu, B. Li, F. Kang, and J. K. Kim, "Effect of solid electrolyte interface (SEI) film on cyclic performance of $Li_4Ti_5O_{12}$ anodes for Li ion batteries," *Journal of Power Sources*, vol. 239, pp. 269–276, 2013.
[8] X. Sun, M. Hegde, Y. Zhang, M. He, L. Gu, Y. Wang, and J. Shu, "Structure and Electrochemical Properties of Spinel $Li_4Ti_5O_{12}$ Nanocomposites as Anode for Lithium-Ion Battery," vol. 9, pp. 1583–1596, 2014.
[9] Q. Wang, M. Lu, J. Miao, and Y. Hao, "Synthesis and electrochemical performance of nanodiamond supported $Li_4Ti_5O_{12}$-Fe_2O_3 composites as anode materials for lithium ion batteries," *Materials Letters*, vol. 164, pp. 1–4, 2016.
[10] D. K. Gosser, *Cyclic Voltammetry Simulation and Analysis of Reaction Mechanisms*, 1st ed., no. 9. New York: Wiley-VCH, 1993.
[11] N. Nitta, F. Wu, J. T. Lee, and G. Yushin, "Li-ion battery materials: Present and future," *Materials Today*, vol. 18, no. 5, pp. 252–264, 2015.
[12] D. Linden and T. B. Reddy, *HANDBOOK OF BATTERIES*, 3rd ed. McGraw-Hill, 2002.

Effectiveness of Dry Cell Microscopic Simulation (DCMS) to Promote Conceptual Understanding about Battery

Firmanul Catur Wibowo[1], Andi Suhandi[2], Dadi Rusdiana[3], Achmad Samsudin[4], Dina Rahmi Darman[5], M. Noor Faizin[6], Wiyanto[7], Supriyatman[8], Anna Permanasari[9], Ida Kaniawati[10], Wawan Setiawan[11], Yudi Karyanto[12], Suharto Linuwih[13], Abdul Fatah[14], Bambang Subali[15], Aceng Hasani[16], Sholeh Hidayat[17]

[1,5,14,15,16,17] Department of Physics Education, Universitas Sultan Ageng Tirtayasa, Indonesia
[2,3,4,10] Department of Physics Education, Universitas Pendidikan Indonesian, Indonesia
[1,12] Department of Automotive Safety Engineering, Politeknik Keselamatan Transportasi Jalan, Indonesia
[6] Junior High School, Sekolah Menengah Pertama Negeri 2 Kudus City, Indonesia
[7,13,15] Department of Physics, Universitas Negeri Semarang, Indonesia
[8] Department of Physics Education, Universitas Tadulako, Indonesia
[9] Department of Chemistry Education, Universitas Pendidikan Indonesian, Indonesia
[11] Department of Computer Science Education, Universitas Pendidikan Indonesian, Indonesia

E-mail: firmanulcaturwibowo@untirta.ac.id

Abstract. Electricity is a concept that is abstract and difficult to see by eye directly, one example electric shock, but cannot see the movement of electric current so that students have difficulty by students. A computer simulation designed to improve the understanding of the concept of the workings of the dry cell (battery). This study was conducted to 82 students (aged 18-20 years) in the experimental group by learning to use the Dry Cell Microscopic Simulation (DCMS). The result shows the improving of students' conceptual understanding scores from post test were statistically significantly of the workings of batteries. The implication using computer simulations designed to overcome the difficulties of conceptual understanding, can effectively help students in facilitating conceptual change.

1. Introduction

The battery is a device that can store chemical energy and make electricity when needed. The battery has been widely known in its use as an energy source electronic items such as toys, flashlights, and others. Excellence battery as a source of electrical energy is the ease to carry. Electricity generated by a battery arising from potential difference electrical energy both pieces of the electrodes. The potential difference is known as the cell potential or electromotive force. To complete the reaction within the battery needed a charge transfer media and an external circuit as the electric flow path.

Natural phenomena are formed when there is an interaction between matter (objects) with energy. So really it was studying the physics of matter, energy, and the interaction between them. For example in everyday live are observable phenomenon that is increasing the temperature of a substance when heated. This event is actually an interaction between substances with the effect of heat energy on the rising temperature of the substance. To be able to answer the questions above is necessary to study more fundamental to the microscopic level of the object. "Macroscopic system to answer and understand how important of the questions we ask about macroscopic systems" [16]. The simulation may be used as an alternative instructional tool, in order to help students to confront and constructed their basic of scientific conception and developed their understanding [14]. If the student has a certain conception and that conception is different from the scientific conception, then said to the student experience misconceptions [20]. Especially for materials which are microscopic physics. For example, a battery can work if the interactions between its constituent molecules.

The low level of understanding of the concept which achieved the detection of students and students who have misconceptions (misconceptions) allegedly closely associated with learning physics implemented is still conventional. Based on observations in the classroom and learning plan documents study, data showed that the learning process is carried still using the traditional pattern, which centered learning lecturer lecturing dominated by the verbal information so that students tend to be a listener.

To visualize the phenomenon of microscopic physics simulation or visualization media needed to model the phenomenon of microscopic invisible becomes as if viewable. For example, the temperature of a substance that microscopically illustrated by the level of the velocity of constituent particles. Simulation of the battery as well as that of the circuit being powered Allows determination of high performance and minimum energy operational modes [20]. The faster the movement of electrons moving from the positive pole to pole negative then the current generated is also getting stronger. Because of the size of the constituent particles of substances is very small and cannot be observed senses of sight, the movement also cannot be observed much less understood. If the movement of the particles of matter can be visualized it will be easy to understand. [2] states that "An instructional simulation Often involves concrete and abstract representations of objects". Learning physics by using virtual simulations can visualize abstract concepts become real.

Media simulated been learning physics concepts microscopic phenomena is needed both for learning the physics-oriented conception of construction and reconstruction of learning-oriented conception. Characteristics of learning physics oriented conception construction is learning physics in which there are the build process from the initial conception of physics do not understand the concept of being understood concept. Therefore, it takes a real step in the learning process to construct knowledge and correcting (reconstruct) the concept if not in accordance with the scientific context [21]. Evidence has been going on in a person's conception of construction can be seen from the level of understanding (level of understanding). Construction conception performed on the learning physics simulation using virtual calls for construction-related conception comprehension level (level of understanding) students and conception in reducing misconceptions [27]. Stages of construction of learning begins with the conception of preliminary activities, which consists of core activities (Prediction, Experience and Reflection) and cover. In the learning stages of construction conception of lecturers serving phenomena or events, which aired in the form of microscopic phenomena. Microscopic phenomena using the virtual simulation carried out in several stages of construction conception of learning. While the material studied and served in the learning-oriented conception construction on the concept of the battery.

However, look at the battery that requires media to be able to see the movement of electrical current needed a computer simulation designed to enhance understanding of the concept of working from dry cells (batteries). Because the particle size that size microscopic, small chance conceptual changes can be made by direct experience. So that the required simulation as a tool to visualize the concept microscopic, through virtual simulation. Virtual multimedia development stage that is done in this study included the analysis phase, the planning or design, production, and evaluation [15].

One software application that can be used to create a virtual simulation is Macromedia Flash 8.0. Some of the advantages possessed by this simulation can combine text, animation, sound and color.

Based on the description that has been stated above, then the formulation of the problem in general is: "What Dry Cell Microscopic Simulation (DCMS) to Promote Conceptual Understanding about Battery?". Research problems above can be translated into research questions as follows: (1) How DCMS characteristics developed for learning the physics-oriented conception construction ?. (2) Is the use of learning physics DCMS has developed highly effective in constructing scientific conception in the minds of students? "

2. Microscopic Virtual Simulations in Physics Learning

Modern service systems are complex, multifactor systems of a distributed nature. For teaching and studying how interactions, decisions, and collaboration of different parties are managed when designing and deploying complex systems, and simulations for learning approaches are gaining popularity [31], [28].

Some virtual media related teaching materials have been developed partly physics at the University of Collorado PhET Simulation in the form that can be accessed and downloaded for free by the public. The use of virtual simulation of microscopic phenomena in physics learning, some researchers have published the results of its investigations both learning physics oriented conception construction or reconstruction of conception. [3], [4] the use of virtual media can change the conception of students who have misconceptions through in-depth restructuring concept. Besides virtual media can also give way scientific reasoning in studying the concept of thermal expansion and heat transfer. research results [17] and [9] the use of virtual media can assist students in building-related conception velocity and acceleration in the motion of projectiles. [1] the use of computer simulations in learning Newton's laws of motion can help remediate misconceptions that occur on college campuses. [5] that the use of simulation media on material climate change could facilitate construction of the transformation in the conception in the minds of students. Results of research [11] learning with media simulation can reduce the quantity of student misconceptions in the optical material. [34] the use of media simulation of light and color dispersion is effective in helping students to construct its conception so that the ability to understand can be improved; [21] that the use of virtual simulation media on the concept of light waves to enhance understanding of the concept of wave and help the process of changing the conception (Conceptual Change) students to the scientific conception. Simulation battery performance of a digital circuit [19].

3. Understanding of the Concepts

In recent decades, research have publicized that students move toward to science lessons by instructional conceptions and ideas concerning the phenomena and concepts to be academic that are not in harmony amid knowledge views. Besides, conceptions and ideas are firmly held and are resistant to change [12], [13]. The first was [10] dictum that the most important single factor influencing learning is what the learner already knows and hence to teach the learner accordingly. The second theoretical perspective was Piaget's idea of the interplay of assimilation and accommodation.

His clinical interview method deeply influenced research on investigating students' conceptions [33].

By the end of the 1970s and the beginning of the 1980s preliminary conceptual change ideas addressing students' conceptions were revealed in the various studies that developed. [32] identify four possible strategies within this theoretical framework:

1. Integrating students' existing conceptions with scientific conceptions,
2. Differentiating existing conceptions into separate but more clearly defined conceptions to show the students that their alternative ideas are not plausible in related situations (not general).
3. Exchanging the entire alternative conception for the scientific conception after demonstrating the explanatory and predictive power of the scientific conception, and
4. Conceptual bridging (described as bridging analogies above).

Most conceptual change strategies were designed based on students' alternative conceptions [7]. Conceptual change texts that can be used to overcome some common misconceptions regarding "sound intensity" and "how fast sound travels [22]. It is suggested that combining different conceptual change methods such conceptual change text/refutation text, argumentation with the intervention may be more effective in reducing student alternative conceptions [6]. Development of conceptual change in the teaching of aspects important the nature of science [25].

However, teaching with analogies is an effective teaching method for higher learning achievement and in preventing misconceptions [26]. Conceptual change strategy was designed based on students' alternative conceptions. Consistency conception of the students in this study includes the pattern of student answers using the same concept models in answering a series of questions asking the same concept. This data was obtained through the conception consistency test instruments tested one by asking questions with the same concept more than once.

In this research, Conceptual change based on students' alternative conceptions reduce with DCMS. Because DCMS of battery is the ability to make normally unobservable occurrences plainly visible for the student. These may be processes that:

1. are microscopic such as particle,
2. have many complex sub-processes such as moved electron
3. computer simulations can be credible representations of reality, and
4. remediation produced significant conceptual change in those students holding alternative conceptions.

4. Method

4.1. Research Design

The purpose of the present study is to investigate the effects of Dry Cell Microscopic Simulation (DCMS) to Promote Conceptual Understanding about Battery. Quantitative study design used is one-group pretest posttest design with a scheme according to [8]. Tests of understanding the concept of the experimental class applied using One group posttest only design as shown in Figure 1.

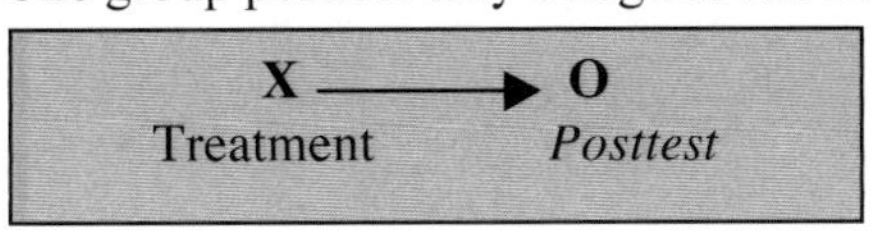

Figure 1. Design One Group Posttest Only

Information:
O: The test conceptual understanding of physics of microscopic phenomena
X: Treatment in the form of Interactive Learning Lecture Demonstration using DCMS experimental group.

4.2. Study Participants

A total of 82 student-respondents in their eleventh grade, age ranges from 18 to 21 years. The sample was randomly selected from second year student teachers undertaking an introductory physics course. A few parts of these students have already studied electricity and magnetism in physics lessons in high school; but the other students took this subject for the first time in their physics courses at university physics education, Universitas Sultan Ageng Tirtayasa, Serang City, Banten, Indonesia.

4.3. Instrument and Data Collection Process

An achievement test, developed by the researchers, about conceptual understanding about battery composed of 3 open-ended questions or items, was used as a data-gathering instrument. In the first part of the test, an everyday example or sample case was presented to students.

Table 1. Learning Models and Instruments

Learning	Variables
Construction of conception using the model Interactive Lecture Demonstration (ILD)	1. Level of understanding 2. Model of understanding

They were asked questions about the "sample case" that probed their ability to explain the sample case (Item A), determine the physics law explaining the sample case (Item B) and provide an additional sample case about related law (Item C). Overall, three questions were asked about conceptual understanding about battery, totaling nine questions in all. In the second part of the test, participants were asked three questions about defining conceptual understanding about battery (Item D). We introduce the test items below:

Problem 1

A. In everyday life often find dry battery, the flashlight, a toy children and remote. So we often define as an element batteries (cells) source of direct current which work by converting chemical energy into electrical energy. The dry battery components as shown in the figure.

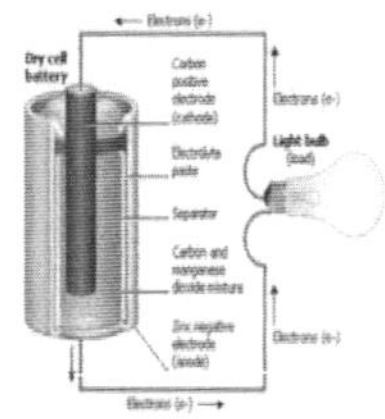

Try to explain why up to a microscopic level when a convert chemical energy into electrical energy?

B. The concept of physics what can be used to explain the relationship between chemical energy can be transformed into electrical energy is?

C. Try to put forward other examples in everyday life related to the physics concepts?

D. Based on the answers to the items of the previous question please you define the concept of physics?

Problem 2

A. When we use the dry battery mounted on a flashlight, when we are to the ON switch, the light goes on. The process of the lights on a chemical oxidation reaction occurs in the battery.

$$Zn + 2MnO_2 + 2NH_4Cl \longrightarrow ZnCl_2 + Mn_2O_3 + 2NH_3 + H_2O + \text{electrical energy}$$

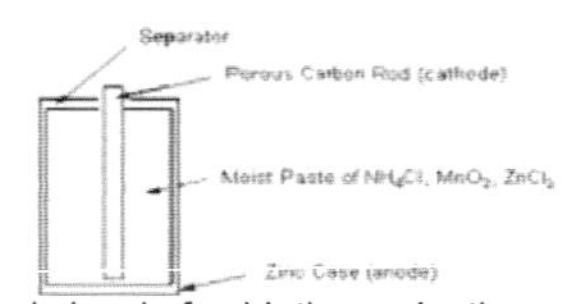

Try to explain to the microscopic level of oxidation-reduction reaction process on a dry battery?

B. The concept of physics what can be used for explain the events mentioned above?

C. Try to put forward other examples in everyday life related to the physics concepts?

D. Based on the answers to the items of the previous question please you define the physics concept?

Problem 3

A. When a dry battery has a battery inside the electric energy then there is the electron charge will always move.

B. The concept of what physics can be used to explain the above phenomenon?

C. Try to put forward other examples in everyday life related to the physics concepts?

D. Based on the answers to the items of the previous question please you define the concept of physics?

One researcher completed the data collection during introductory physics courses. Students were given 40 minute to complete the test and they were encouraged to freely express their thoughts. They were assured that it was not an examination and their answers were not to be used in order to evaluate their academic levels.

4.4. Data Analysis

Student responses were analyzed in two ways. The first identified the student level of understanding and the second identified the student model of understanding. The analysis of understanding level is a suitable analysis method, which gives a general view of the academic level of participants about each question on an achievement test. On the other hand, the student model of understanding helps to analyze a student's answers to all questions about a subject, in other words, helps to analyze students answers individually. In that way, one can find a link between the students knowledge about a subject. In other words, one can reach some clues about students mental model. Because mental models are used to construct new knowledge, it is important to reach some understanding about them. As there are mental models composed by the students behind the meaningful learning [12]. Data obtained from the instrument was analyzed to understand levels of understanding as suggested by [7]. Model of understanding were adapted from the Student's Model of Understanding in Typology of Perceived Knowledge as suggested by [27].

5. Analysis

5.1. Development Dry Cell Microscopic Simulation (DCMS)

Virtual multimedia development stage that is done in this study included the analysis phase, the planning or design, production, and evaluation [15]. Analysis of the curriculum successfully selected thermal expansion of solids material with consideration of the material contains some abstract concept, so that very precise when made interactive multimedia that can simulate events that are difficult to observe by students and can simulate the dry cell (battery). Result it showed that the use of virtual media can change the conception of learners who have conceptual understanding through restructuring concept in depth [4].

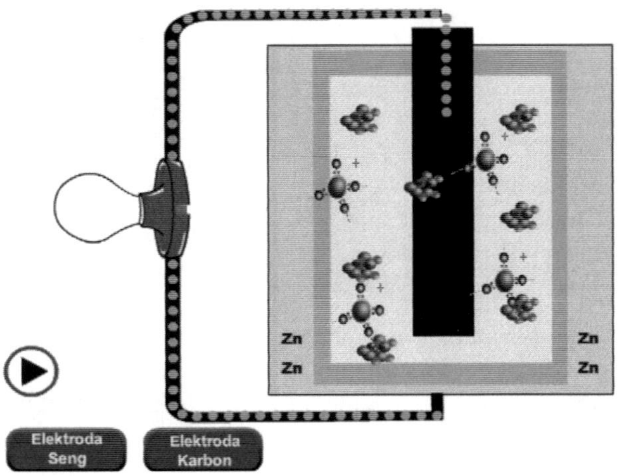

Figure 2. Result Development of Dry Cell Microscopic Simulation of understanding Concept of battery

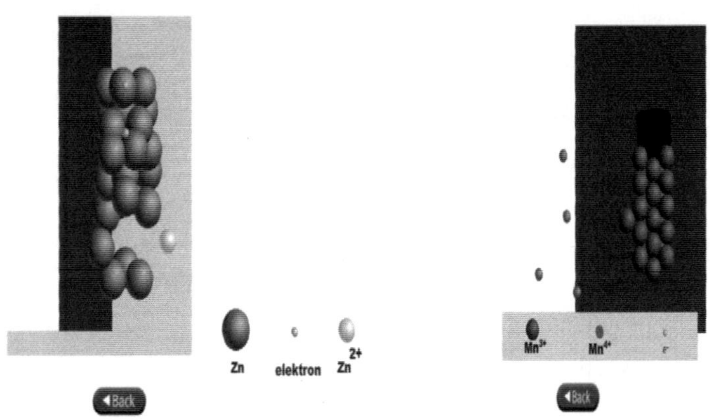

Figure 3. Reaction of zing electrode and carbon electrode for Dry Cell Microscopic Simulation

5.2. Students' level of understanding

The findings from conceptual understanding test are presented using the headings students' level of understanding and students' model of understanding. The findings from the achievement test are presented below after analyzing each item about battery.

Table 2. Students' level of understanding

Level Understanding	Indicator	Concept Battery						Average
		1		2		3		
		Total	%	Total	%	Total	%	
[4] Understanding the Whole (UW)	Correctly answered all the questions in a matter of	40	49	38	46	36	44	46
[3] Understanding the Majority (UM)	Answering the questions correctly in part on a matter	24	29	25	30	22	27	29
[2] In Erroneous Understanding (MSK)	With all the questions answered on a question but the answer is not clear, or the answer is wrong	10	12	11	13	13	16	14
[1] Not Understand (UN)	Answered all the questions on a problem but the answer given is not appropriate (irrelevant) with any questions.	6	7	6	7	8	10	8
[0] No Answer (NA)	Does not provide answers to all the questions on an items	2	2	2	2	3	4	3

Based on Table 2 obtained information that in consecutive average level of understanding of each concept battery from high to low after the application of learning physics conception construction by using DCMS. The average percentage of Understanding the Whole (UW) of level medium.

5.3. Students' Model of understanding

The findings from the achievement test are presented below after analyzing each item about battery.

Table 3. Students' Model of Understanding

Model Understanding	Characteristics Answer Students	Concept Battery		
		1	2	3
		Σ & %	Σ & %	Σ & %
Optimal Model (OM)	Correctly answered the questions A, B, C, D on a matter	30 (37)	28 (23)	26 (32)
Model Not Creative (MNC)	Correctly answered only the questions C course at a problem	19 (23)	21 (26)	23 (28)
Theoretical models (TM)	Correctly answered only the questions A, B, D only on a matter.	13 (16)	9 (11)	13 (16)

Model Understanding	**Characteristics Answer Students**	**Concept Battery**		
		1	**2**	**3**
		Σ & %	**Σ & %**	**Σ & %**
Practical Model (PM)	Correctly answered only the questions A and C only in a matter of a question on an items	10 (12)	13 (16)	8 (10)
Memorizing Model (MM)	Correctly answered only the questions D alone on a matter	8 (10)	9 (11)	9 (11)
Not Have Model (NHM)	Not answered correctly all the questions A, B, C, D on a matter	2 (2)	2 (2)	3 (4)

Σ: Total of students who experienced levels of understanding (the total number of students)
%: The percentage of students who experienced the level of understanding
1: Battery Energy Concept
2: The concept of oxidation reduction reaction in the battery
3: The concept of battery function

5.4. Effectiveness of Using DCMS in Learning Electric Concept

Effectiveness of the media virtual simulation developed in learning physics student oriented conception construction on the concept of battery, the quantity is determined by the percentage of students Understanding the Whole (UW) concept battery. Meanwhile, the effectiveness of the use of DCMS on construction of learning physics oriented conception in helping students construct understanding on the concept of battery.

Table 4. Percentage of quantity Understanding Whole (UW) concept of battery

Label Conception	**Quantity Understanding Whole (UW) %**	**Category**
KI	49	Medium
KII	46	Medium
KIII	44	Medium
Average	46	Medium

Information:
KI: The concept of Battery Energy
KII: Concepts of oxidation reduction reaction in the battery
KIII: Concept of battery function

Based on Table 4. obtained information that the average percentage of the quantity of understanding as a whole (MSU) of electrical material after use DCMS in physics oriented learning concept of battery conception construction is 46 %. This percentage when with DCMS effectiveness criteria according to research [21] obtained the criterion of being. It can be concluded that the progression DCMS developed in the battery material "more effective" in the construction of student understanding.

Based on the data, obtained information that that students construct sensible and coherent understandings of phenomena and concepts as seen through Reviews their own eyes that do not match the views that are Universally accepted by the scientific community [14]. Media simulation is effective in helping students to construct its conception so that the ability to understand can be improved [34]. Use of virtual simulation media on the concept improve understanding of the concept of process of changing the conception (Conceptual Change) students towards scientific conception. Learning physics simulation that uses electric magnetic media can change the conception of a student who is not scientific to the scientific conception.

IOP Conf. Series: Journal of Physics: Conf. Series **877** (2017) 012009 doi:10.1088/1742-6596/877/1/012009

6. Conclusion

This study demonstrates Effectiveness of Dry Cell Microscopic Simulation (DCMS) to Promote Conceptual Understanding about Battery helped students successfully construct scientific understanding of concept battery. The results suggested that incorporation of learning by DCMS has the potential for the development of students' conceptual understanding in science through the mechanical Process of conceptual change. Based on the results of research and discussion of the conclusions of this research are the results presented here show that students working with virtual simulation exhibited significantly higher scores. Our findings strongly support that DCMS can be used as an alternative instructional tool, in order to help students develop an understanding of physics concepts.

References

[1] Atasoy and A. R. Akdeniz. Developing and applying a thesis related to appearing misconceptions about Newtonian laws of motion. *Journal of Turkish Science Education*, 4 (1), pp. 45-50, 2007.

[2] C. Buckley. Interactive Multimedia and Model-based Learning in Biology. *International Journal of Science Education*, 22 (9), pp. 895-935, 2000.

[3] C. She. DSLM Instructional Approach to Conceptual Change Involving Thermal Expansion. *Research in Science and Technological Education*, 21 (l), pp. 43-54, 2003.

[4] C. She. Fostering Radical Conceptual Change through Dual-Situated Learning Model. *Journal of Research in Science Teaching*, 41 (2), pp. 142-164, 2004b.

[5] C. Trundle, and R. L. Bell. The Use of a Computer Simulation to Promote Conceptual Change: A Quasi-Experimental Study. *Journal Computers & Education*, 4, pp. 1078-1088, 2010.

[6] Çalik, A. Kolomuç, and Z. Karagölge. The Effect of Conceptual Change Pedagogy on Students' Conceptions of Rate of Reaction. *Journal of Science Education and Technology*, 19 (5), pp. 422-433, 2010.

[7] Coştu, A. Ayas, M. Niaz, S. Unal, and M. Calik. M. Facilitating Conceptual Change in Students' Understanding of Boiling Concept. *Journal of Science Education and Technology*. 16, pp. 524–536, 2007.

[8] Cresswell. *Research Design: Qualitative, Quantitative and Mixed methods Approach*. London: Sage publication, lnc, 2014.

[9] D. Finkelstein, W. K. Adams, C. J., Keller, P. B., Kohl, K. K., Perkins, N. S. Podolefsky, S., Reid., and R. LeMaster. When Learning about the Real World is better done Virtually: A study of Substituting Computer Simulations for Laboratory Equipment. *Physical Review Special Topics - Physics Education Research*. 1 (1), pp. 1-8, 2005.

[10] D. P. Ausubel. *Educational psychology: A cognitive view*. New York: Holt, Rinehart and Winston. 1968.

[11] Djanett, C., Fouad, and K. Djamel. What Thinks' the University's Students about Propagarion of Light in the Vacuum?. *European Scientific Journal*, 9 (24), pp. 197-213, 2013.

[12] Duit, and D. Treagust. Conceptual Change: A Powerfull Framework for Improving Science Teaching and Learning. *International Journal of Science Education*. 25, pp. 671-688, 2003.

[13] Duit. STCSE – Bibliography: Students' and teachers' conceptions and science education. Kiel, Germany: IPN – Leibniz Institute for Science Education (http://www.ipn.uni-kiel.de/aktuell/stcse/stcse.html), 2006.

[14] F C Wibowo, A Suhandi, D Rusdiana, D R Darman, Y Ruhiat, Y R Denny, Suherman, A Fatah. Microscopic Virtual Media (MVM) in Physics Learning: Case Study on Students Understanding of Heat Transfer. *J. Phys.: Conf. Ser.* 739 012044, 2016.

[15] F. C. Wibowo, A. Suhandi, D. Rusdiana, Y. Ruhiat, and D. R. Darman. Microscopic Virtual Media (MVM) in Physics Learning to Build a Scientific Conception and Reduce Misconceptions: A Case Study on Students' Understanding of the Thermal Expansion of Solids. International Conference on Innovation in Engineering and Vocational Education (ICIEVE 2015). Atlantis Press, pp. 239-244, 2016.

[16] H. Gould and J. Tobochnik. *Thermal And Statistical Physics*. Princeton University Press, 2006.
[17] Jimoyiannis, and Komis. Computer Simulations In Physics Teaching And Learning: A Case Study On Student's Understanding Of Trajectory Motion. *Journal Computers and Education* 36, pp. 183-204, 2001.
[18] Kaewkhong, A. Mazzolini, N. Narumon Emarat and C. Arayathanitkul. Thai High-School Students' Misconceptions about and Models of Light Refraction through a Planar Surface. *Physics Education Journal,* 45 (l), pp. 91-101, 2010.
[19] Kulkarni dan V. D. Agrawal. A Tutorial on Battery Simulation-Matching Power Source to Electronic System " in Proc. 14th IEEE VLSI Design and Test Symp. 2010.
[20] N. Kartal, Ozturk, and H.G. Yalvac. *Misconceptions of Sciences Teacher Candidats about Heat and Temperature*. Science Direct. Procedia Social and Behavioral Sciences, 2011.
[21] N. Srisawasdi, and S. Kroothkeaw. Supporting Students Conceptual Development of Light Refraction by Simulation-Based Open Inquiry with Dual-Situated Learning Model. *Journal Computer & Education,* 1 (1), pp. 49-79, 2014.
[22] Özkan, and G. Sezgin Selçuk. The use of conceptual change texts as class material in the teaching of "sound" in physics. In Asia-Pacific Forum on Science Learning and Teaching, 14 (1), Article 11, 2013.
[23] Q. B. Macabebe, I. B Culaba, and J. T. Maquiling. Pre-Conceptions of Newton's Laws of Motion of Students in Introductory Physics. *AIP Conference Proceedings*, 1263 (1), pp. 106-109, 2010.
[24] Ramadesigan, Paul, Northrop, and Sumitava. (2012). Modeling and Simulation of Lithium-Ion Batteries from a Systems Engineering Perspective. *Journal of The Electrochemical Society*. 159 (3) 31-45.
[25] S. Cepni, and E. Cil. The Effectiveness of the Conceptual Change Approach, Explicit Reflective Approach, and Course Book by the Ministry of Education on the Views of the Nature of Science and Conceptual Change in Light Unit. *Educational Sciences: Theory and Practice*, 12 (2), pp. 1107-1113, 2012.
[26] Ş. Pekmez.Using analogies to prevent misconceptions about chemical equilibrium. *Asia-Pacific Forum on Science Learning and Teaching*, 12 (2), 2010.
[27] Saglam-Arslan, and Y. Uevecioglu. Student Teachers' Levels of Understanding and Model of Understanding about Newton's Laws of Motion. *Asia-pacific Forum on science learning & Teaching,* 11 (1), pp. 1-20, 2010.
[28] Saunders, A. F. Rutkowski, M. van Genuchten, D. Vogel, and J. Molina Orrega. Virtual space and place: Theory and test. MIS Quarterly, 3 (4), pp. 1079-1098, 2011.
[29] Savinainen, and P. Scott. Using the Force Concept Inventory to Monitor Student Learning and to Plan Teaching. *Physics Education Journal*, 37, pp. 53-58, 2002.
[30] Spyrtou, E. Hatzikraniotis, and P. Kariotoglou. Educational software for improving learning aspects of Newton's Third Law for student teachers. *Educational and Information Technologies*, 14 (2), pp. 163-187, 2009.
[31] T. M. Corsi, S. Boyson, A. Verbraeck, S. P. A. Van Houten, C. Han, and j. R. MacDonald. The real-time global supply chain game: New educational tool for developing supply chain management professionals. *Transportation Journal*, 45 (3), pp. 61-73, 2006.
[32] W. Hewson. A case study of conceptual change in special relativity: The influence of prior knowledge in learning. *European Journal of Science Education*, 4, pp. 61-78, 1982.
[33] White and R. Gunstone. *Probing understanding*. London, UK: The Falmer Press, 1992.
[34] Z. Zacharias, G. Olympiou, and T. de Jong. Making the Invisible Visible: Enhancing Students' Conceptual Understanding by Introducing Representations of Abstract Objects in a Simulation. *Instruction Science,* 41, pp. 575-596, 2013.

Superhydrophobic Membrane Contactor for Acid Gas Removal

Nurul Faiqotul Himma[1] and I Gede Wenten[2]*

[1] Department of Chemical Engineering, Universitas Brawijaya, Jl. Mayjen Haryono 167, Malang 65145, Indonesia
[2] Department of Chemical Engineering, Institut Teknologi Bandung, Jl. Ganesha 10, Bandung 40132, Indonesia

*Corresponding author (Email: igw@che.itb.ac.id)

Abstract. Gas–liquid membrane contactor has gained a great attention as an alternative to conventional absorption columns in acid gas removal from natural gas or post-combustion. The membrane contactor offers high mass transfer area and excellent operational flexibility. However, hydrophobic microporous membranes commonly used are still susceptible to wetting by liquid absorbents, leading to the deterioration of absorption performance in long-term operation. Therefore, many studies were recently directed to improve the membrane wetting resistant by endowing superhydrophobicity. This article then presents a review on superhydrophobic membrane development and its application for acid gas removal using membrane contactor. An overview of gas–liquid membrane contactor is firstly presented, followed by the preparation of superhydrophobic membranes. The performances of superhydrophobic membranes in acid gas absorption are then discussed, and the recommendation for future research is finally outlined. This review may provide an insight into the further development of superhydrophobic membrane contactor.

1. Introduction

Acid gases such as carbon dioxide (CO_2), hydrogen sulfide (H_2S), and other sulphuric components are frequently contained in raw natural gases or resulted from combustion of fossil fuels. The acid gases can lead to corrosion and plugging problems. Besides, the global emissions of CO_2 are believed as the main factor causing climate change and global warming, which become a global concern [1]. Whereas sulfur dioxide (SO_2) that can be resulted from H_2S oxidation is one of the major source causing acid rain. Therefore, the acid gases are necessary to be removed from the gas stream either for the operational reason or to meet the environmental emission.

Technologies for acid gases removal include absorption using solvent or solid sorbents, pressure- and temperature-swing adsorption using various solid sorbents, cryogenic distillation, and membranes. Among these methods, absorption using solvent is the most promising current method due to the high selectivity and capacity, and easy regeneration of solvent [2, 3]. Although absorption is well-established process that traditionally uses packed columns, this process has some disadvantages such as difficulty in obtaining an accurate estimate of the gas–liquid mass transfer area and a limited range of gas and liquid flow rates due to operational problems including flooding, loading, channeling, and foaming [4, 5].

Gas–liquid membrane contactor can be an alternative technology to overcome the drawbacks. This is a hybrid process combining conventional gas absorption in liquid solvent and additional membrane as contacting device between gas and liquid phases, offering some advantages such as high and constant specific interfacial area, independent gas/liquid control with absence of entrainment flooding, downflow flooding or weeping, modularity, and low cost [6, 7]. However, the risk of membrane wettability is a major drawback for its long-term operation. Therefore, wetting phenomena and prevention in gas–liquid membrane contactor is widely investigated recently [8-12]. Different methods have been used to improve the membrane wetting resistant, including optimization of operating condition [13, 14], solvent and membrane selection [14-17], and membrane development by control the membrane pore structure [18, 19] or the membrane surface properties [20-23]. Among these methods, development of membrane with superior hydrophobicity is an attractive way with the expectation of high wetting resistant without sacrifice the high permeability and selectivity with the maintained large pore size and the selected absorption liquid.

Gas–liquid membrane contactor for acid gases removal have been reviewed in some published papers [5, 24, 25]. However, the use of superhydrophobic membrane in such application has not been reviewed in elsewhere. This paper aims to provide a summary and direction of superhydrophobic membrane contactor development for acid gas removal. A brief overview of gas–liquid membrane contactor is firstly given, followed by superhydrophobic membrane preparation method. The performances of superhydrophobic membranes in acid gas absorption are then discussed, and the future outlook is finally outlined. This review may provide an insight into the further development of superhydrophobic membrane contactor.

2. Gas–liquid membrane contactor for acid gas removal

2.1. Process Principle

Gas absorption using gas–liquid membrane contactor is an integrated technology of membrane separation and conventional absorption. The membrane used acts as a support to keep in contact gas and liquid phases and to promote the mass transfer between them [6]. A membrane with hollow fiber configuration is preferentially used [5, 24, 26-28]. The hollow fiber modules give high specific surface area, i.e. in the range of 1500–3000 m^2/m^3, whereas in conventional contactor it is in the range of 100–800 m^2/m^3 [5]. The process of gas–liquid membrane contactor is depicted in figure 1a. Gas flows on one side and the absorbent liquid flows on the other side of the membrane. Mass transfer occurs when the solute gas diffuses through the membrane and absorbed into the liquid solvent. This is nonwetted mode. Besides in nonwetted mode (gas-filled pores), the operation mode can be in wetted mode (absorbent liquid-filled pores) [29]. The former one uses hydrophobic membrane, whereas the latter one uses hydrophilic membrane. However, the nonwetted mode is preferentially used due to the lower resistance to mass transfer [29, 30].

The membrane used is a non-selective barrier, but the selectivity is provided by the absorbent liquid. Thus, the selection of solvent is same with the conventional absorption in term of solvent capability in absorbing the acid gases and easiness of regeneration. The solvents can be chemical solvent, physical solvent, or mixed-solvent [31]. Monoethanolamine (MEA) is a chemical solvent that is commonly used for CO_2 absorption due to its absorption capability in low partial pressure of CO_2 and low cost. Investigation on various absorbents showed that MEA is superior solvent in term of CO_2 removal efficiency, compared with other absorbents such as methyldiethanolamine (MDEA), 2-amino-2-methyl-1-propanol (AMP), and distilled water [16, 32]. However, AMP can be a good choice considering the energy requirement for regeneration [32]. Moreover, MEA can be degraded in the presence of SO_2 and O_2 [3]. Ammonia has been used by Resnik et al. [33] as the solvent to capture CO_2, SO_2, and NO_x simultaneously from flue gas. Besides the benefit of a single process, the ammonia can give CO_2 loading capacity three times higher than MEA. Compared with CO_2, investigation of removal of other acid gases such as H_2S and SO_2 is rarer. It was reported that high SO_2 removal efficiency above 80% can be achieved using NaOH and Na_2SO_3 aqueous solutions [34]. Regarding the selection of solvent, additional criteria emerged in membrane contactor are surface tension and chemical compatibility with the membrane material [24, 25]. A solvent with lower surface tension gives a higher tendency to membrane wetting [17, 35]. Meanwhile, a solvent with low compatibility may deteriorate the membrane structure and increase the membrane wetting [10].

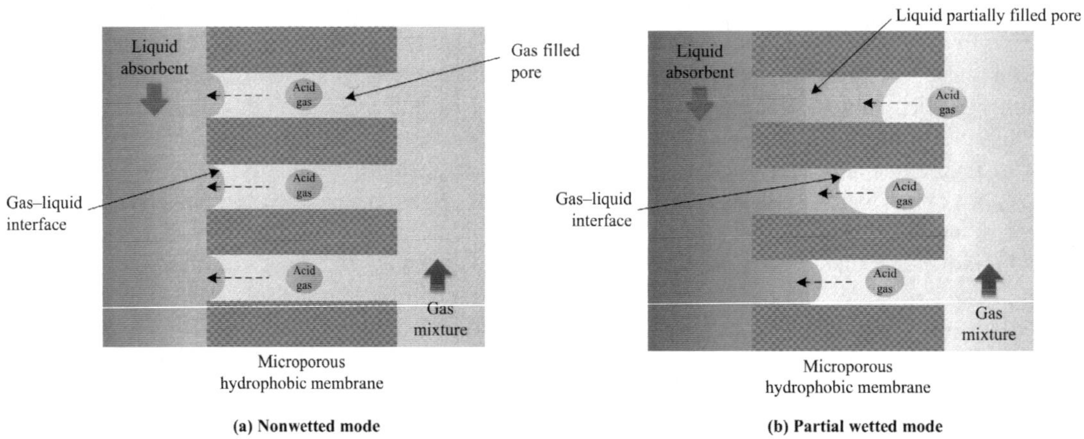

Figure 1. Scheme of hydrophobic membrane-based gas–liquid contactor. Adapted from [29, 30].

2.2. Membrane wetting

The properties of membranes used highly determine the membrane contactor performance. In the case of nonwetted mode, hydrophobic membranes from polypropylene (PP), polytetrafluoroethylene (PTFE), and polyvinylidenefluoride (PVDF) are generally used [6, 16, 34]. These membranes are commercially available. Among these membranes, PTFE membrane is more resistant to wetting [16, 36, 37]. The water contact angles of PP, PVDF, and PTFE membranes are about of 100–127°, 92–130°, and 113–139°, respectively [4, 37, 38]. Even though the membrane used is intensive hydrophobicity, such as PP, PTFE, and PVDF, the absorbents, especially aqueous solution of organic compounds such as alkanolamines, can penetrate into partial pores of the hydrophobic membrane, that is partial wetting mode (figure 1b) [11, 30, 39]. In this case, the gas–liquid interface is pushed to inside the membrane phase. The partial wetting may change the membrane morphology, decrease the membrane hydrophobicity, and in turn deteriorate the membrane performance.

Some studies have reported the morphological changes due to the intrusion of absorbent into membrane pores. Barbe et al. [40] showed that after exposed to water for 72 h, the microporous PP membranes have overall increases in porosity, pore length, and pore equivalent diameter. The increase of pore size becomes more significant with the immersion using alkanolamine solution such as MEA and MDEA [10]. Besides, the interaction between the membrane and absorbent also reduces the surface hydrophobicity which indicated by a reduction of contact angles. The PP hollow fiber membranes have a significant reduction of contact angles, i.e. from 121.6° to 90.8°, when immersed in 30% MDEA for 60 days. The increase of membrane pore size and the decrease of membrane hydrophobicity lead to wetting in a greater extent.

The membrane wetting even the partial wetting can increase the membrane resistance significantly due to the liquid phase occupying the membrane pores [11, 41]. Rangwala [41] found that even below 2% liquid-filled pores, the membrane resistance could be as high as 60% of the total mass transfer resistance. This would result in significant decrease of acid gas transfer rate. It was reported that partial wetting of PP membrane pores by MEA absorbent caused a significant decrease in mass transfer rate of CO_2 (i.e. decreased to 59% of initial value after 14 days operation time) [42].

The membrane pores may be wetted by absorption liquid when the transmembrane pressure (pressure difference between the liquid and the gas) exceeds the critical transmembrane pressure or breakthrough pressure. Following Laplace–Young equation [43], the breakthrough pressure depends on membrane properties (i.e. pore size), absorption liquid properties (i.e. surface tension), and interaction between the membrane material and absorption liquid (i.e. contact angle). Using membrane with smaller pore size may reduce the membrane wetting. However, it may also result in decreased permeability. Meanwhile, increase the absorption liquid surface tension by reducing the solution concentration may reduce the selectivity. Increasing the contact angle by using superhydrophobic membranes is considered to be an effective way to prevent the penetration of absorption liquid.

3. Superhydrophobic membrane preparation and characterization

The term of a superhydrophobic surface is recently emerged although the investigation on water repellent or highly hydrophobic surfaces has been conducted since a long time ago. In a common agreement, a superhydrophobic surface is characterized by water contact angle (WCA) above 150° and sliding angle (SA) and hysteresis contact angle (HCA) below 10° [44, 45]. To obtain the high WCA, using low surface energy material only without concerning the surface morphology is insufficient, but it is necessary to be combined with the appropriate rough surface. Using this concept, the superhydrophobic membrane preparation is widely developed. In this paper, a review is more focused onto superhydrophobic membrane for membrane contactor that the membrane is usually microporous.

3.1. Inorganic membranes

Among the inorganic materials, ceramics from metal oxides such as alumina, zirconia, and silica are the most used materials for membrane preparation. These materials are hydrophilic in nature due to the presence of hydroxyl (–OH) groups on the surfaces. On the membrane surface, the liquid water can pass through at a fast rate [46]. Therefore, the ceramic membranes need to be surface modified to improve the hydrophobicity, making them are suitable for membrane contactor. For this purpose, surface modification by direct grafting using low surface energy materials such as fluoroalkyl silane (FAS) is commonly used [46-49]. Usually, the ceramic membrane is immersed in FAS solution for a certain time to allowing the grafting process. During the grafting, the OH groups on the ceramic membrane surface react with Si–O–alkyl groups of the silane [50]. Modified ceramic membranes having WCA higher than 150° with enhanced liquid entry pressure of water (LEPw) have been successfully obtained [47, 49]. Studies of the superhydrophobic ceramic membrane are summarized in table 1.

Table 1. Superhydrophobic ceramic membranes prepared by FAS grafting method for membrane contactor

Ceramic Membrane	Grafting parameter	WCA (°)	LEP_w (bar)	Process	Ref.
alumina tube with ZrO_2 layer	FAS: 2 wt.% grafting time: 24 h	153	6.5	CO_2 absorption using MEA	[47]
kaolin-alumina hollow fibre	FAS: 0.01 mol/L grafting time: 4 h	142	2.5	CO_2 absorption using water	[48]
γ-alumina membranes on α-alumina support	FAS: 0.1 mol/L eroding time*: 5 min grafting time: 24 h multiciplity: 4 times	164.5	-	-	[49]

*The membrane was eroded in NaOH

IOP Conf. Series: Journal of Physics: Conf. Series **877** (2017) 012010 doi:10.1088/1742-6596/877/1/012010

3.2. Polymeric membranes

Differ from inorganic membranes that the superhydrophobization is achieved by surface modification, the polymeric membranes can be superhydrophobized either by surface modification or by direct processing [51, 52]. In direct processing, the superhydrophobic modification can be achieved during the preparation process. Meanwhile, the surface modification can be performed through one-step or more, depending on chosen material and method.

3.2.1. Improvement of membrane preparation process. In this direct processing method, hydrophobicity can be enhanced by improvement of phase separation process, blending method, or electrospinning. In the former way, the enhancement of hydrophobicity is attributed to the surface morphology which is controlled by solution and process parameters. It has been reported that highly hydrophobic membrane with porous structure can be obtained by using coagulant which led to delay demixing phase separation [53], by using long vapor exposure time facilitating crystallization process [54], and by using high air temperature and low PVDF content [55]. Blending method is a simple way to enhance the hydrophobicity of membrane by blending low surface energy material such as surface modifying macromolecule (SMM) [20, 21, 56-60] or by blending nanoparticle [61] into casting solution. However, there isn't a published work reporting superhydrophobic membrane prepared using the methods. Electrospinning method has been widely investigated in superhydrophobic membrane preparation, either from melt or solution of polymer, with or no addition of nanoparticles. This method has been used for superhydrophobic membrane preparation from the solution of polystyrene (PS) [62, 63], PVDF [64], and PP [65]. This method has also been used for superhydrophobic inorganic-polymer composite membrane preparation from PVD/SiO_2 [66-68], polyurethane (PU)/terminal fluorinated polyurethane (FPU)/carbon nanotubes (CNTs) [69], and polyvinylidene fluoride-co-hexafluoropropylene (PcH)/CNTs [70].

3.2.2. Surface modification of existing membrane. Some methods of surface modification have been proposed to obtain a superhydrophobic membrane. Plasma treatment, coating with roughened hydrophobic polymer film, and chemical vapor deposition (CVD) are methods that can render superhydrophobicity on membrane surface through simultaneous surface roughening and hydrophobization using low surface energy materials. These methods are also called one-step surface modification. The rough surface of plasma-treated membrane is resulted from an etching process during the plasma treatment, while the low surface energy is due to fluorine deposition. Plasma treatment using CF_4 [71, 72], PTFE [73, 74], and benzene [75] have been successfully used to prepare superhydrophobic polymer membranes. Besides choosing of plasma reagent, other crucial parameters in this method are plasma treatment time and working power. Excessive treatment could result in no enhancement or even some deterioration in superhydrophobicity [71, 73, 76, 77], thus the optimization of both process parameters is important. Coating with roughened hydrophobic polymer film is also an attractive method for superhydrophobic polymer membrane preparation, which the surface roughening process can be accomplished by using non-solvent [38, 78-81] or nanoparticles [82, 83]. A few works reported superhydrophobic polymer membrane preparation by CVD method [84]. Besides the one-step surface modification, the superhydrophobic membrane can also be produced by two-step surface modification, which the sequence can be surface roughening followed by hydrophobization using low-surface-energy material or vice versa. In this regard, surface coating with nanoparticles followed by coating with fluorosilane compounds is often used. Superhydrophobic membrane preparation with surface roughening by using TiO_2 [85], silver [86], and silica [87] nanoparticles have been reported. Studies on preparation of superhydrophobic membrane from polymeric materials for membrane contactor are summarized in table 2.

Table 2. Superhydrophobic polymeric membranes for membrane contactor

Membrane Base Material	Preparation/Modification Method	WCA (°)	LEP_w (bar)	Process	Ref.
PVDF	SMM blending method	92	3	CO_2 absorption using water	[21]
PVDF	SMM blending method	96.2	8	CO_2 absorption using water	[60]
PDMS/PVDF	Electrospinning and FAS modification	150	-	CO_2 absorption using AMP/PZ	[88]
PP	CF_4 plasma treatment	143	-	CO_2 absorption using PZ, MDEA, and AMP	[77]
PP	Plasma treatment using PTFE	151	0.7*	CO_2 absorption using MEA	[73]
PVDF	Coating with rough LDPE layer	152	-	CO_2 absorption using MEA	[80]
PP	Coating with rough PP layer	158	-	CO_2 absorption using MEA	[78]

*The liquid entry pressure of 2-propanol

4. Performances of superhydrophobic membrane contactor in acid gas removal

4.1. Performances

It has been proved that superhydrophobic modification increases the WCA and LEP, thus it is expected that the superhydrophobic membrane has a higher resistant to wetting than a hydrophobic one. As the superhydrophobic membrane isn't wetted, the gas–liquid interface should be maintained at the pore mouth, which means that there is no additional membrane resistance imposed by the stagnant liquid phase. With this condition, a high mass transfer rate of the acid gas can be achieved.

Investigations of membrane contactor in acid gas removal showed that either superhydrophobic polymeric membranes or superhydrophobic ceramic membranes exhibited a higher and more stable flux than commercial hydrophobic membrane [47, 74, 77, 78, 80]. It was reported by Lin et al. [77] that the CO_2 absorption flux of the superhydrophobic PP membrane contactor only dropped from 7.4 x 10^{-4} mol/(m^2 s) to 7.1 x 10^{-4} mol/(m^2.s), even after 30 days. Whereas hydrophobic PP membrane and hydrophobic PVDF membrane exhibited a significant decrease of CO_2 absorption flux from 6.84 x 10^{-4} mol/(m^2 s) to 7.4 x 10^{-4} mol/(m^2 s) and from 7.05 x 10^{-4} mol/(m^2 s) to 5.52 x 10^{-4} mol/(m^2 s), respectively. The high and stable absorption flux could be attributed to the decrease in wetting degree with the increase of membrane hydrophobicity [73, 77]. An estimation of wetting degree showed that superhydrophobization of PP membrane could decrease the pore wetting degree more than 40% with the increase of membrane mass-transfer coefficient about 207% [73].

However, one matter should be noted is that the benefit of superhydrophobicity rendered in the membrane isn't offset by the negative effect of the superhydrophobic modification to the membrane characteristics such as pore size, porosity, and membrane thickness. As studied by Lv et al. [78], superhydrophobic PP membrane prepared by roughened polymer film coating using solvent/non-solvent method exhibited a slight decrease in CO_2 flux within the first 7 days, approximately 86% of the initial value. Moreover, the CO_2 flux of modified membrane was lower than that of unmodified membrane in the first 6 days, as shown in figure 2. This is due to an increase in the membrane thickness and a decrease in the surface porosity after the superhydrophobic modification. Thus, the superhydrophobic modification that maintains the membrane thickness and porosity is highly desired.

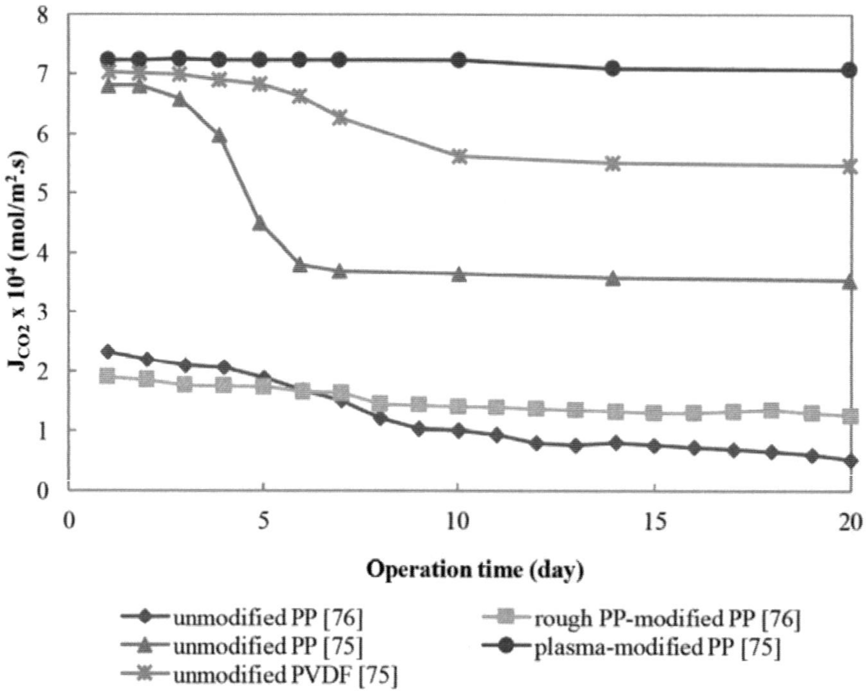

Figure 2. Long-term performances of superhydrophobic membranes compared with hydrophobic membranes [77, 78].

4.2. Self-cleaning function

Fouling is one of the major problems in porous membrane processes such as microfiltration and ultrafiltration [89-92]. In membrane contactor, the fouling isn't significant as much as in membrane filtration since there is no convective flow through the membrane pores. However, in industrial applications such as coal-fired power plant, gas and liquid streams with suspended particles can cause plugging due to the small membrane contactor diameter. Superhydrophobic surface has been acknowledged for its self-cleaning ability [93, 94]. Therefore, the use of superhydrophobic membrane is intended to mitigate the membrane fouling in membrane contactor. Regarding this case, only a few reports that can be found. Yu et al. [47] showed that ceramic membrane with superhydrophobicity exhibited a better anti-fouling ability than PP membrane without superhydrophobicity. The superhydrophobic ceramic membrane could be easily cleaned easily carbon powder, whereas for the hydrophilic ceramic membrane and hydrophobic membrane, the carbon powder could not be effectively removed. Moreover, one-month operation using flue gas from thermal power plant showed that on the superhydrophobic ceramic membrane, there was no obvious dust. Meanwhile, the PP membranes were fouled by dust in the flue gas.

5. Conclusion and recommendation

Membrane contactor offers some advantages over conventional technologies which makes it gained a great attention for acid gas removal. However, membrane wetting by liquid absorbent becomes a drawback which deteriorates the membrane performance. To address it, superhydrophobic membrane contactor was then developed, either from polymer or inorganic materials. Superhydrophobic polymer membrane can be prepared by either direct processing or surface modification. However, surface modification of hydrophobic polymer membrane by plasma treatment or solvent/non-solvent coating method is more likely to be used to prepare superhydrophobic membrane contactor for acid gas removal. For inorganic membrane, surface modification is obviously required to render superhydrophobicity due to the hydrophilic nature. It has been reported that the increase of hydrophobicity led to a higher and more stable flux compared to the unmodified membrane. However, additional of membrane thickness and pore blocking due to the surface modification need to be concerned because these offset the benefit of superhydrophobicity. Besides offering wetting resistant, superhydrophobic modification also gives self-cleaning ability. It was newly investigated in membrane contactor which is important for industrial flue gas streams. For future improvement of membrane contactor, development of superhydrophobic microporous membrane with main concern on the stability of superhydrophobicity, thickness addition, loss in pore size and porosity is very necessary. As the flue gas from the combustion process is in high temperature, and various absorbents are being developed, the thermal stability and chemical stability of the superhydrophobic membrane also need to be investigated.

References

[1] Huisingh D, Zhang Z, Moore J C, Qiao Q and Li Q 2015 *J. Cleaner Prod.* **103** 1-12
[2] Aaron D and Tsouris C 2005 *Sep. Sci. Technol.* **40** 321-48
[3] Wang M, Lawal A, Stephenson P, Sidders J and Ramshaw C 2011 *Chem. Eng. Res. Des.* **89** 1609-24
[4] Ahmad A L, Sunarti A R, Lee K T and Fernando W J N 2010 *Int. J. Greenhouse Gas Control* **4** 495-8
[5] Mansourizadeh A and Ismail A F 2009 *J. Hazard. Mater.* **171** 38-53
[6] Drioli E, Curcio E and di Profio G 2005 *Chem. Eng. Res. Des.* **83** 223-33
[7] Klaassen R, Feron P H M and Jansen A E 2005 *Chem. Eng. Res. Des.* **83** 234-46
[8] Malek A, Li K and Teo W K 1997 *Ind. Eng. Chem. Res.* **36** 784-93
[9] Goyal N, Suman S and Gupta S K 2015 *J. Membr. Sci.* **474** 64-82
[10] Lv Y, Yu X, Tu S-T, Yan J and Dahlquist E 2010 *J. Membr. Sci.* **362** 444-52
[11] Zhang H-Y, Wang R, Liang D T and Tay J H 2008 *J. Membr. Sci.* **308** 162-70
[12] Mosadegh-Sedghi S, Rodrigue D, Brisson J and Iliuta M C 2014 *J. Membr. Sci.* **452** 332-53
[13] Simons K, Nijmeijer K and Wessling M 2009 *J. Membr. Sci.* **340** 214-20
[14] Yan S-p, Fang M-X, Zhang W-F, Wang S-Y, Xu Z-K, Luo Z-Y and Cen K-F 2007 *Fuel Process. Technol.* **88** 501-11
[15] Albrecht W, Hilke R, Kneifel K, Weigel T and Peinemann K V 2005 *J. Membr. Sci.* **263** 66-76
[16] deMontigny D, Tontiwachwuthikul P and Chakma A 2006 *J. Membr. Sci.* **277** 99-107
[17] Dindore V Y, Brilman D W F, Geuzebroek F H and Versteeg G F 2004 *Sep. Purif. Technol.* **40** 133-45
[18] Rajabzadeh S, Yoshimoto S, Teramoto M, Al-Marzouqi M, Ohmukai Y, Maruyama T and Matsuyama H 2013 *Sep. Purif. Technol.* **108** 65-73
[19] Nguyen P T, Lasseuguette E, Medina-Gonzalez Y, Remigy J C, Roizard D and Favre E 2011 *J. Membr. Sci.* **377** 261-72
[20] Bakeri G, Ismail A F, Rana D and Matsuura T 2012 *Chem. Eng. J.* **198–199** 327-37
[21] Rahbari-Sisakht M, Ismail A F, Rana D and Matsuura T 2012 *J. Membr. Sci.* **415–416** 221-8
[22] Zhang Y, Wang R, Zhang L and Fane A G 2012 *Sep. Purif. Technol.* **101** 76-84
[23] Wongchitphimon S, Wang R and Jiraratananon R 2011 *J. Membr. Sci.* **381** 183-91
[24] Li J-L and Chen B-H 2005 *Sep. Purif. Technol.* **41** 109-22
[25] Zhang Y and Wang R 2013 *Curr. Opin. Chem. Eng.* **2** 255-62
[26] Gabelman A and Hwang S-T 1999 *J. Membr. Sci.* **159** 61-106
[27] Pabby A K and Sastre A M 2013 *J. Membr. Sci.* **430** 263-303
[28] Wenten I G and Widiasa I N 2004 *Indonesia Patent* **046.1440 A**
[29] Karoor S and Sirkar K K 1993 *Ind. Eng. Chem. Res.* **32** 674-84
[30] Lu J-G, Zheng Y-F and Cheng M-D 2008 *J. Membr. Sci.* **308** 180-90

[31] Rivas O R and Prausnitz J M 1979 *AlChE J.* **25** 975-84
[32] Kim Y-S and Yang S-M 2000 *Sep. Purif. Technol.* **21** 101-9
[33] Resnik K P, Yeh J T and Pennline H W 2004 *Int. J. Environ. Technol. Manage.* **4** 89-104
[34] Jeon H, Ahn H, Inho S, Jeong H-K, Yongtaek L and Hyung-Keun L 2008 *Desalination* **234** 252-60
[35] Kumar P S, Hogendoorn J A, Feron P H M and Versteeg G F 2002 *Chem. Eng. Sci.* **57** 1639-51
[36] Nishikawa N, Ishibashi M, Ohta H, Akutsu N, Matsumoto H, Kamata T and Kitamura H 1995 *Energy Convers. Manage.* **36** 415-8
[37] Khaisri S, deMontigny D, Tontiwachwuthikul P and Jiraratananon R 2009 *Sep. Purif. Technol.* **65** 290-7
[38] Franco J A, Kentish S E, Perera J M and Stevens G W 2008 *J. Membr. Sci.* **318** 107-13
[39] Chabanon E, Roizard D and Favre E 2011 *Ind. Eng. Chem. Res.* **50** 8237-44
[40] Barbe A M, Hogan P A and Johnson R A 2000 *J. Membr. Sci.* **172** 149-56
[41] Rangwala H A 1996 *J. Membr. Sci.* **112** 229-40
[42] Lv Y, Yu X, Tu S-T, Yan J and Dahlquist E 2012 *Appl. Energy* **97** 283-8
[43] Kim B-S and Harriott P 1987 *J. Colloid Interface Sci.* **115** 1-8
[44] Roach P, Shirtcliffe N J and Newton M I 2008 *Soft Matter* **4** 224-40
[45] Celia E, Darmanin T, Taffin de Givenchy E, Amigoni S and Guittard F 2013 *J. Colloid Interface Sci.* **402** 1-18
[46] An L, Yu X, Yang J, Tu S-T and Yan J 2015 *Energy Procedia* **75** 2287-92
[47] Yu X, An L, Yang J, Tu S-T and Yan J 2015 *J. Membr. Sci.* **496** 1-12
[48] Abdulhameed M A, Othman M H D, Ismail A F, Matsuura T, Harun Z, Rahman M A, Puteh M H, Jaafar J, Rezaei M and Hubadillah S K *J. Cleaner Prod.*
[49] Lu J, Yu Y, Zhou J, Song L, Hu X and Larbot A 2009 *Appl. Surf. Sci.* **255** 9092-9
[50] Picard C, Larbot A, Guida-Pietrasanta F, Boutevin B and Ratsimihety A 2001 *Sep. Purif. Technol.* **25** 65-9
[51] Himma N F, Anisah S, Prasetya N and Wenten I G 2016 *J. Polym. Eng.* **36** 329-62
[52] Crick C R and Parkin I P 2010 *Chem. - Eur. J.* **16** 3568-88
[53] Kuo C-Y, Lin H-N, Tsai H-A, Wang D-M and Lai J-Y 2008 *Desalination* **233** 40-7
[54] Peng Y, Fan H, Dong Y, Song Y and Han H 2012 *Appl. Surf. Sci.* **258** 7872-81
[55] Peng Y, Fan H, Ge J, Wang S, Chen P and Jiang Q 2012 *Appl. Surf. Sci.* **263** 737-44
[56] Essalhi M and Khayet M 2012 *J. Membr. Sci.* **417–418** 163-73
[57] Suk D E, Matsuura T, Park H B and Lee Y M 2010 *Desalination* **261** 300-12
[58] Bakeri G, Matsuura T, Ismail A F and Rana D 2012 *Sep. Purif. Technol.* **89** 160-70
[59] Rahbari-Sisakht M, Ismail A F, Rana D and Matsuura T 2012 *Sep. Purif. Technol.* **99** 61-8
[60] Mansourizadeh A, Aslmahdavi Z, Ismail A F and Matsuura T 2014 *Int. J. Greenhouse Gas Control* **26** 83-92
[61] Efome J E, Baghbanzadeh M, Rana D, Matsuura T and Lan C Q 2015 *Desalination* **373** 47-57
[62] Kang M, Jung R, Kim H-S and Jin H-J 2008 *Colloids Surf. Physicochem. Eng. Aspects* **313–314** 411-4
[63] Lee M W, An S, Latthe S S, Lee C, Hong S and Yoon S S 2013 *ACS Appl. Mater. Interfaces* **5** 10597-604
[64] Liao Y, Wang R, Tian M, Qiu C and Fane A G 2013 *J. Membr. Sci.* **425–426** 30-9
[65] Patel S U and Chase G G 2014 *Sep. Purif. Technol.* **126** 62-8
[66] Wang S, Li Y, Fei X, Sun M, Zhang C, Li Y, Yang Q and Hong X 2011 *J. Colloid Interface Sci.* **359** 380-8
[67] Liao Y, Wang R and Fane A G 2014 *Environ. Sci. Technol.* **48** 6335-41
[68] Liao Y, Loh C-H, Wang R and Fane A G 2014 *ACS Appl. Mater. Interfaces* **6** 16035-48
[69] Li Y, Zhu Z, Yu J and Ding B 2015 *ACS Appl. Mater. Interfaces* **7** 13538-46

[70] Tijing L D, Woo Y C, Shim W-G, He T, Choi J-S, Kim S-H and Shon H K 2016 *J. Membr. Sci.* **502** 158-70
[71] Yang C, Li X-M, Gilron J, Kong D-f, Yin Y, Oren Y, Linder C and He T 2014 *J. Membr. Sci.* **456** 155-61
[72] Yang C, Tian M, Xie Y, Li X-M, Zhao B, He T and Liu J 2015 *J. Membr. Sci.* **482** 25-32
[73] Franco J A, Kentish S E, Perera J M and Stevens G W 2011 *Ind. Eng. Chem. Res.* **50** 4011-20
[74] Franco J A, deMontigny D D, Kentish S E, Perera J M and Stevens G W 2011 *Ind. Eng. Chem. Res.* **51** 1376-82
[75] Lee S H, Dilworth Z R, Hsiao E, Barnette A L, Marino M, Kim J H, Kang J-G, Jung T-H and Kim S H 2011 *ACS Appl. Mater. Interfaces* **3** 476-81
[76] Bae B, Chun B H and Kim D 2001 *Polymer* **42** 7879-85
[77] Lin S-H, Tung K-L, Chen W-J and Chang H-W 2009 *J. Membr. Sci.* **333** 30-7
[78] Lv Y, Yu X, Jia J, Tu S-T, Yan J and Dahlquist E 2012 *Appl. Energy* **90** 167-74
[79] Ahmad A L, Mohammed H N, Ooi B S and Leo C P 2013 *Caspian Journal of Applied Sciences Research* **2** 52-7
[80] Ahmad A L, Mohammed H N, Ooi B S and Leo C P 2013 *Pol. J. Chem. Technol.* **15** 1-6
[81] Himma N F, Wardani A K and Wenten I G 2017 *Polymer-Plastics Technology and Engineering* **56** 184-94
[82] Zhang J, Song Z, Li B, Wang Q and Wang S 2013 *Desalination* **324** 1-9
[83] Zhou H, Shi R and Jin W 2014 *Sep. Purif. Technol.* **127** 61-9
[84] Zheng Z, Gu Z, Huo R and Ye Y 2009 *Appl. Surf. Sci.* **255** 7263-7
[85] Razmjou A, Arifin E, Dong G, Mansouri J and Chen V 2012 *J. Membr. Sci.* **415–416** 850-63
[86] Liao Y, Wang R and Fane A G 2013 *J. Membr. Sci.* **440** 77-87
[87] Fang J, Wang H, Wang X and Lin T 2011 *The Journal of The Textile Institute* **103** 937-44
[88] Lin Y-F, Ye Q, Hsu S-H and Chung T-W 2016 *Chem. Eng. J.* **284** 888-95
[89] Howe K J and Clark M M 2002 *Environ. Sci. Technol.* **36** 3571-6
[90] Aryanti P T P, Subagjo S, Ariono D and Wenten I G 2015 *Journal of Membrane Science and Research* **1** 41-5
[91] Wenten I, Koenhen D, Roesink H, Rasmussen A and Jonsson G 1994 *Proc. of Engineering of Membrane Processes, II Environmental Applications* (Ciocco, Italy)
[92] Wenten I G 1995 *Filtr. Sep.* **32** 252-3
[93] Xue C-H, Jia S-T, Zhang J and Ma J-Z 2010 *Sci. Technol. Adv. Mater.* **11** 033002
[94] Bhushan B, Jung Y C and Koch K 2009 *Langmuir* **25** 3240-8

Modeling of various heat adapter plate 4 and 6 array for optimization of thermoelectric generator element using modified diffusion equation

Defrianto[1,a], W Tambunan[1] and Lazuardi[1]

[1]Department of Physics, FMIPA University of Riau
Kampus Bina Widya Simpang Baru, Pekanbaru 28293, Indonesia
[a]Corresponding author: defriaanto@gmail.com

Abstract. The use of waste heat from exhaust gas and converting it to electricity is now an alternative to harvest a cheap and clean energy. Thermoelectric generator (TEG) has the ability to directly recover such waste heat and generate electricity. The aim of this study is to simulate the heat transfer on the aluminum adapter plate for homogeneity temperature distribution coupled with hot side of TEG type 40-40-10/100 from Firma Eureka and adjust their high temperatures to the TEG operating temperature to avoid the element damage. Modelling was carried out using MATLAB modified diffusion equation with Dirichlet boundary conditions at defined temperature which has been set at the ends of the heat source at 463K and 373K ± 10% on the hot side of the TEG element. The use of nylon insulated material is modeled after Neumann boundary condition in which the temperature gradient is $\partial T/\partial n = 0$ out of boundary. Realization of the modelling is done by designing a heat conductive plate using software ACAD 2015 and converted into a binary file format of Mathlab to form a finite element mesh with geometry variations of solid model. The solid cubic model of aluminum adapter plate has a dimension of 40mm length, 40mm width and also 20mm, 30mm and 40mm thickness arranged in two arrays of 2x2 and 2x3 of TEG elements. Results showed a temperature decrease about 40.95% and 50.02% respectively from the initial source and appropriate with TEG temperature tolerance.

1. Introduction

Micro scale energy sources that produce an order of milliwatt electrical power from solar, vibration, thermal and biological resources has been a part of research on the current electronics design. A new device in micro energy generation systems nowadays is a thermoelectric generator element (TEG), known as Peltier element which in principle generates electric power based on the Seebeck effect utilizing the temperature difference. The power generated is relatively small and proportional to the magnitude of the temperature difference between the two sides of Peltier elements [1].

Some applications use the waste heat from an industrial process and automotive, like waste heat from a car muffler or the temperature difference between the two parts. In automotive vehicles, there is a large power dissipation resulted from fuel combustion in automobile exhaust chamber and as an energy source is still relatively rarely used [2]. A single auto muffler can produce high temperature from 600K up to 900K then it release to the free air. However, the waste heat of car muffler is not in accordance with the characteristics of a commercially available thermoelectric generators, so there needs to be an adjustment between a heat source in the exhaust plate with the surface temperature of a TEG element, so that an optimum power can be generated without damage to the device [3].

Heat transfer in the elements can be expressed mathematically in a diffusion equation. Transient heat conduction (or diffusion) equations with nonlinear source terms have been used as model equations for heat transfer problems in many different areas of mathematical physics, applied science, and engineering. In numerical methods, the boundary element method has become increasingly attractive to scientists and engineers since it has certain advantages over other methods such finite difference method (FDM). One of the advantages is its ability in reducing the dimensionality of a problem by one [4-5].

This study simulated the heat distribution in various form of the adapter plates using a modified diffusion equation. The TEG element used has rectangular shape and was purchased from Fa. Eureca Germany with technical data as given in Table 1.

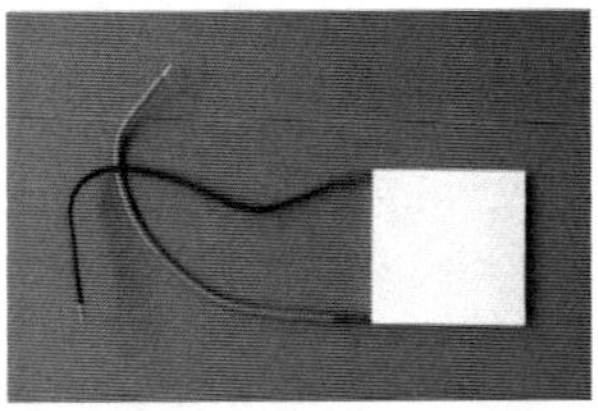

Figure 1. The thermoelectric generator element of Fa. Eureka

Table 1. Thermal, electrical and mechanical properties of TEG type 1-40-40-10/100 [6]

No	Parameters	Symbols	
1	thermal force	α	0.0823 V/K
2	resistance	ρ	1.67Ω
3	thermal conductivity	γ	1.56 W/K
4	maximum operating temperatures	Tmax	150°C
5	size of cold/ hot side	L x B x H	40.0x 40.0 x3.20 mm
6	weight	m	24g

Table 1 shows that the element TEG has maximum operating temperature around 150°C (under temperature of muffler). However, based on previous experiments, devices which operate near the operating temperature can damage TEG terminal served as a current wire. It is due to melting of soldering tin and inhomogeneity of heat distribution on the surface of the TEG resulted instability of the generated electrically output [7]. It is necessary to simulate the optimum homogeneous temperature on the surface of TEG and to be within the tolerance limits and to avoid permanently damage to the TEG elements. The arrangement of 2x2 array (4 elements) and 2x3 (6 elements) provided in this study aims that TEG element can be arranged in series and parallel to provide a large output results.

2. Temperature distribution model

2.1 Model overview

The waste heat generated from auto muffler can be utilized as a heat source for the thermoelectric generator and be connected using a conductive plate as shown in Figure 2. For this purpose, a modified diffusion equation is used to model the heat distribution in steady state.

The heat from auto muffler at 190°C was feds through aluminum base and adapter cubic plate with different thickness of 20mm, 30mm and 40mm, so that the gradient temperature only in the direction of the z axis (heat deep penetration). Heat leakage between adapter plates is terminated using isolated materials such as pertinaks filled in gaps.

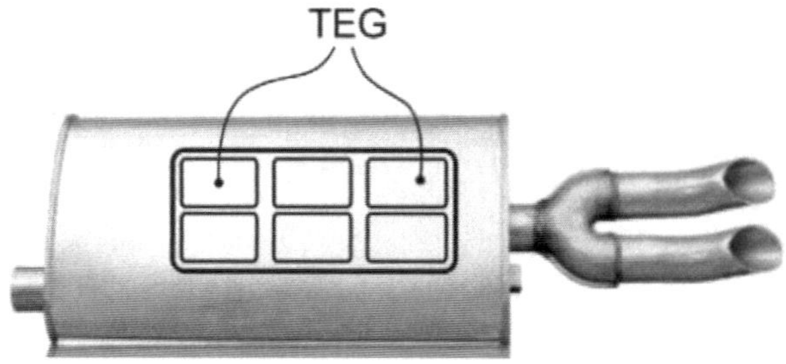

Figure 2. Array arrangement of TEG elements for simulating the muffler heat distribution

2.2. Modified diffusion equation

In this study, a thermoelectric generator consists of a base plate connected to the waste heat source from an auto muffler is modeled to simplify the simulation which is used to decrease the source temperature to 100 C°± 10% below the maximum operating temperature of TEG elements. This goes to the hot side of TEG surfaces. On the cold side, the heat will be removed in the form of convection and radiation, thus the temperature difference between both sides of the TEG (ΔT) becomes larger and the efficiency of TEG converted heat into electricity is higher. Figure 3 illustrates the schematic of the model and flow of heat distribution in three ways.

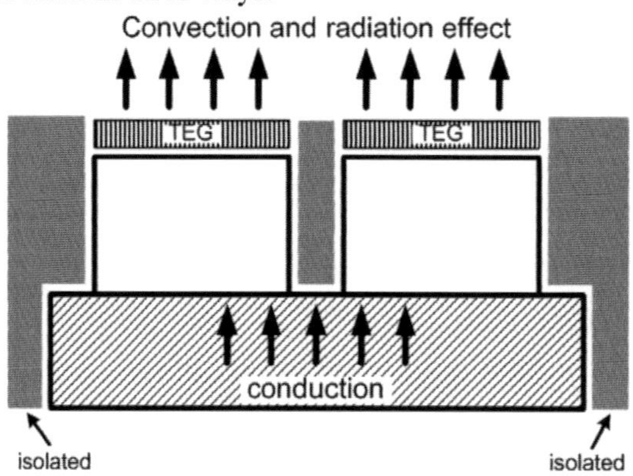

Figure 3. Components of heat transfer element TEG, conduction, convection and radiation

The TEG element as thermoelectric generator works using two different medium temperatures on its surfaces, as heat source and heat dissipation. Heat transfer is the exchange of thermal energy between physical systems in form of conduction, convection and radiation. This can be modeled using a heat conduction equation and in general is expressed as diffusion equation [8]:

$$k\nabla^2 T + q = \rho c \frac{\partial T}{\partial t} \tag{1}$$

Equation (1) is then modified by inserting a heat source q as wasted heat due to convection and radiation proces:

$$k\nabla^2 T - h(T - T_a) - \in \sigma(T^4 - T_a^4) = \rho c \frac{\partial T}{\partial t} \tag{2}$$

In steady state, the diffuse equation become Poisson equation by eliminating the time component:

$$k\nabla^2 T - h(T - T_a) - \in \sigma(T^4 - T_a^4) = 0 \tag{3}$$

Equation (3) is used to the calculate model at steady state condition, thus the function of time can be ignored. Using simulation, the heat energy reached to the hot side of the TEG device adjusted about 373K ± 10%, which is remaining under maximum operating temperature of the element, and can be tolerated, this does not cause permanent damage to the TEG devices.

There are three boundary conditions for the model in Figure 3. First condition is the temperature maintained at T = Ts for the base plate coupled to the heat source, which is stated by the Dirichlet boundary conditions. The temperature at Dirichlet boundary condition is applied choose to any heat source. Second condition is for insulated walls which is reject the heat propagation in the x-axis and y-axis and apply $\partial T/\partial n = 0$ the Neumann boundary condition, and last condition is on the cold side of TEG element. The heat is released in the appearance of convection and radiation using equation $-k\frac{\partial T}{\partial n} = h(T - T_a) + \in\sigma(T^4 - T_a^4)$ which is Neumann boundary conditions and Ts = 463K from sources temperature.

The previous TEG model discussed in Figure 3 is computationally solved using Finite Difference Method (FDM). To determine the temperature distribution of T (x, y, z) in a computational domain using the FDM, the Poisson equation (3) must be discrete towards the x, y and z plane and can be written in equation (4), (5) and (6).

$$\frac{\partial^2 T}{\partial x^2} = \frac{T(x+\Delta x, y, z) - 2T(x, y, z) + T(x-\Delta x, y, z)}{\Delta x^2} \tag{4}$$

$$\frac{\partial^2 T}{\partial y^2} = \frac{T(x, y+\Delta y, z) - 2T(x, y, z) + T(x, y-\Delta y, z)}{\Delta y^2} \tag{5}$$

$$\frac{\partial^2 T}{\partial z^2} = \frac{T(x, y, z+\Delta z) - 2T(x, y, z) + T(x, y, z-\Delta z)}{\Delta z^2} \tag{6}$$

Taking the value of Δx = Δy = Δz = h, then the steady state temperature distribution become discrete near the points of (i, j, k) as written in equation (7).

$$T_{i+1,j,k} + T_{i-1,j,k} + T_{i,j+1,k} + T_{i,j-1,k} + T_{i,j,k+1} + T_{i,j,k-1} - 6T_{i,j,k} = \left(\frac{h}{k}\right) q_{i,j,k} \tag{7}$$

By inserting the all values of i, j and k, the matrix equation A.T = Q will be obtained.

After modeling heat distribution connected to hot side surface of Peltier element which is simulated using a diffusion equation, the distribution of color patterns will be resulted which describes the rate of heat transfer from the source and depict the penetration of heat transfer from source through an aluminum heat plate adapter with specific geometries. Optimization is obtained by changing the parameters of the equation (7) which are the boundary conditions and the geometries of heat source.

3. Results and Disscusion

The termoelectric plate model was designed using Autocad software which consisted of two parts, namely a base plate and an adapter plate [9]. The base plate has dimensions of 110mm length, 110mm width for 4 element array of TEG and 160mm width for 6 elements array, each having a thickness of 20mm. The adapter plate laid on the base plate formed a solid rectangular cube and connected to the hot side of the TEG Eureca with dimension of 40mm length and 40mm width. Waste heat from the auto muffler at 463K will spread to the base and the adapter plate and reached a final temperature around 373K ± 10%, lower than the maximum operating temperature of TEG. The slit between the adapter plates was isolated to cancel the heat transfer between them, so that final temperature could be reduced by adjustingthe thickness of the adapter plate.

A computer aided design program, ACAD was used to create the geometric design plate in the format of stereo lithography and then imported into MATLAB program for meshing process using a triangular pattern connected elements on the plate [10]. Figure 4a presented the meshing adapter plate for 4 arrays and 6 arrays and the results of the modeling. Figure 4b showed the heat distribution on the hot side of the TEG elements originating from the auto muffler waste heat. Each of the 4 and 6 elements illustrate a homogeneous blue color on the top of adapter cubic plate at the same temperature around 373K or 100°C.

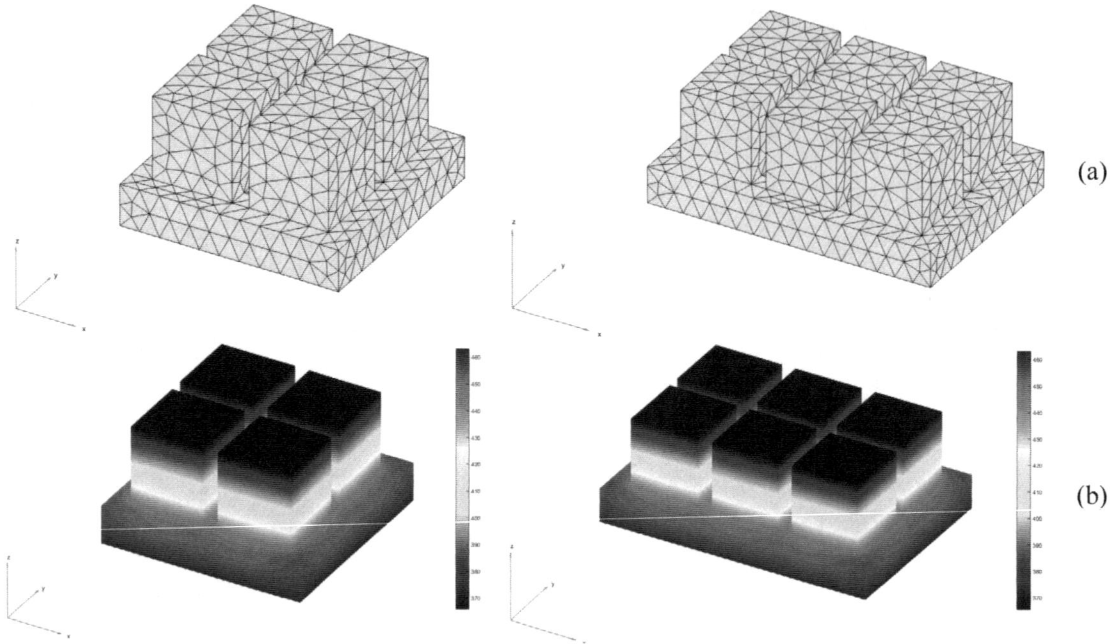

Figure 4. (a) Design and meshing adapter plate and (b) heat distribution simulation using MATLAB

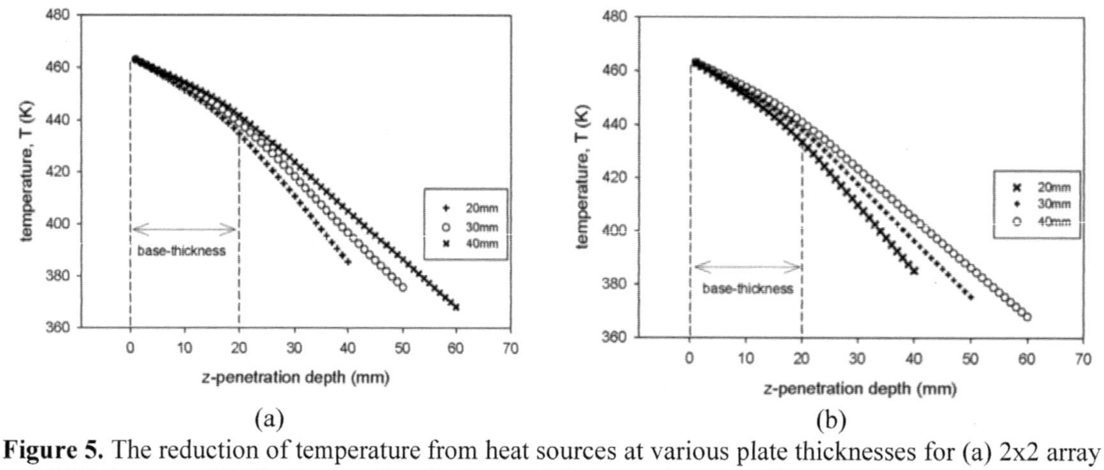

Figure 5. The reduction of temperature from heat sources at various plate thicknesses for (a) 2x2 array and (b) 2x3 array of TEG element. The thickness of the base plate was 20mm each.

As can seen from Figure 5 and Figure 6, the initial temperature for 2x2 arrays of TEG element was 463K and decreased about $\Delta T = 77{,}81$K which is proportional to 40.95% heat losses intended for a plate thickness of 20mm and 50.02% heat losses for 40mm. For 2x3 arrays, a minimum temperature was achieved for 40mm thickness as well as 49.98% heat losses. Temperature was decreasing gradually toward the z-axis, while on the direction of the x-axis and y-axis, the temperature was remains constant. Based on the computer simulation, the received heat temperature at the hot side of the TEG elements could be adjusted within operating range by means of plate thickness.

The cold side of TEG element would undergo two heat transfer process; convection and radiation, while the surface temperature kept constant at T_{cold} = 300K. From the described results, the temperature difference between the hot side and cold side, can be calculated and resulted $T_{hot} - T_{cold} = 68$K. The greater temperature difference arised between the two sides of the elements; more electrical energy will be produced by the TEG element.

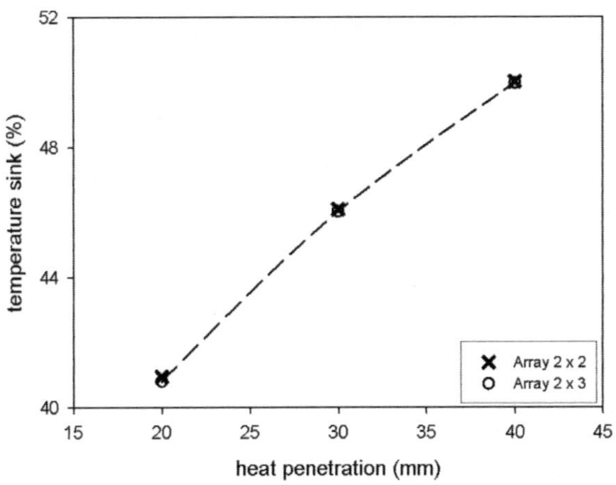

Figure 6. Temperature decreased in the z-axis of the initial temperature

Theoretically, there is no heat transfer between plate adapters since it has been isolated and the heat will only flow through the z-axis. Due to the thermal conductivity γ factor of the element, a small quantity of heat from hot side as amount of 1.56W goes through cold side, thus decreasing the temperature difference. Based on these results, the simulated temperature is complied with the TEG temperature by adjusting the adapter thickness. This can be used to achieve the efficiency and optimization of the electrical energy generated without deleterious effects on the device.

4. Conclusion

Heat distribution on an adapter plate for TEG element has been successfully simulated. Based on the results, the geometry and shape of adapter plate affects the temperature decrease in at the hot side of TEG element. Plate made of conductive material such as aluminum with high conductivity and high emissivity is able to reduce the muffler heat temperature with difference of 77,81K for 2x2 array and 94,97K for 2x3 array configuration. The temperature difference between two surfaces TEG, hot side and cold side, can be calculated and result as $T_{hot} - T_{cold} = 68K$. Greater the temperature difference arises between the two sides of the elements; more electrical energy will be produced by the TEG element. The simulated temperature is complied with the TEG temperature for optimum output and avoids damage from overheating.

Acknowledgement

This work was supported by Project of Hibah Bersaing 2016, DRPM Indonesia Ministry of Research, Technology and Higher Education.

References

[1] Date A, Date A, Dixon C and Akbarzadeh A 2014 *Renewable and Sustainable Energy Reviews* **33** 371
[2] Zhang X, Chau K T and Chan C C 2008 *Journal of Asian Electric Vehicle*, **6**(2) 1119
[3] Salerno D 2010 *Journal of Analog Innovation*, **20**(3)
[4] Chang C Y and Ma C C 2001 *International Journal of Heat and Mass Transfer* **44**(13) 2423
[5] Hu K Q and Chen Z T 2013 *International Journal of Heat and Mass Transfer* **62** 445
[6] Eureka 2010 EURECA Messtechnik GmbH, http://www.eureca.de
[7] Edwards M and Brinkfeldt K 2013 *Proc. 14th International Conference on Thermal, Mechanical and Multi-Physics Simulation and Experiments in Microelectronics and Microsystems, EuroSimE 2013*
[8] Kontogeorgis G M and Folas G K 2010 *Thermodynamic Models for Industrial Applications: From Classical and Advanced Mixing Rules to Association Theories* (Wiltshire: Wiley) p710

[9] Autocad version 2015 http://www.autodesk.com
[10] Matlab version R2015a https://www.mathworks.com

Application of Implicit Finite Difference Method to Determine the 2D Patterns of Unsteady State Thermal Spreading of Geothermal Systems

Joko Sampurno[1, *)], Apriansyah[2], Riza Adriat[1], Irfana Diah Faryuni[1], Wahyu Srigutomo[3]

[1]Department of Physics, Faculty of Mathematics and Natural Sciences, Tanjungpura University

[2]Department of Marine Sciences, Faculty of Mathematics and Natural Sciences, Tanjungpura University

[3]Department of Physics, Faculty of Mathematics and Natural Sciences, Institut Teknologi Bandung, Jalan Ganesa 10, Bandung 40132, Indonesia

[*)]Corresponding author: jokosampurno@physics.untan.ac.id

Abstract. In this study, we built a thermal numerical modelling scheme using finite difference method based on the implicit Crank-Nicholson algorithm. The aim of the study is to describe the process of heat transfer, which is calculated using a thermal diffusion equation (2D vertical) at the unsteady-state conditions, in the geothermal area. The physical parameter used as the input is the thermal diffusivity of the rocks. This parameter was calculated from the other parameters (thermal conductivity, heat capacity, and density). The model was applied to the Wapsalit geothermal area whose geological structure has been known. The model successfully describes the heat flow that spread from the bottom layer to the surface by adjusting to the conditions of the rock.

1. Introduction

In a geothermal system, temperature distribution of the subsurface is formed by conduction process, convection process, or both [1]. The pattern of the temperature distribution is directly controlled by the variation of rock thermal conductivity in the area. The composition of the rock thermal conductivity will affect the pattern of heat distribution from the reservoir layer to the surface of the earth.

Geothermal system generally consists of four main parts: heat source, reservoir rock (permeable rock), cap rock and fluid circulation [2]. The most important part of this system is the reservoir. In this part processes of heat transfer and mass transfer take place. The processes that occur in this reservoir will affect the manifestations that appear at the surface.

Within the reservoir, the energy is stored at certain geological conditions. This energy is obtained from the system that has unique characteristics. The main characteristic of the system is the pattern of heat distribution. Therefore, the pattern of the heat distribution becomes a very important to be investigated.

In this study, we built a numerical modeling scheme using finite difference method based on the implicit Crank-Nicholson algorithm to describe the conduction process of heat transfer at the unsteady-state condition in the geothermal subsurface area. This modeling scheme was applied to some synthetic geological models (for testing the sensitivity of the models) and the subsurface structure of the Wapsalit geothermal area. The modeling result is used to interpret the subsurface thermal structure of the study area.

2. Method

2.1. The Equation Generator and Implicit Finite Difference Approach

Convection and conduction are the heat transfer processes that occur in the earth. In a geothermal system, the convection process occurs in the recharge area and in the permeable reservoir layer. On the other hand, conduction process occurs when heat migrates from the heat source to the surface through an impermeable rock layer. The 2D heat conduction equation in unsteady state condition is given by [3]:

$$\frac{\partial T}{\partial t} = \kappa\left(\frac{\partial^2 T}{\partial x^2} + \frac{\partial^2 T}{\partial z^2}\right) \tag{1}$$

With

$$\kappa = k/\rho c_p$$

Where:

κ : Coefficient of thermal diffusivity (m^2/s)
k : Thermal conductivity of the rock (*W/mK*)
ρ : Density of the fluid and the rock (*kg/m*^{3})
c_p : Specific heat capacity (*kJ/kg*).

To solve the equation, we applied a numerical approach using the finite-difference method based on the implicit Crank-Nicholson (CN) algorithm. The implicit Crank-Nicholson algorithm produces two sets of tridiagonal simultaneous equations which have to be solved in sequence. The first step is solving the equation on the *x*-axis, while the second step is solving the equation on the *y*-axis [4].

For *x*-axis, the numerical solution of the Equation (1) is:

$$\frac{\kappa_{i,j}\Delta t}{4\Delta x^2}T_{i-1,j}^{n+\frac{1}{2}} + \left(1 + 2\frac{\kappa_{i,j}\Delta t}{4\Delta x^2}\right)T_{i,j}^{n+\frac{1}{2}} + \frac{\kappa_{i,j}\Delta t}{4\Delta x^2}T_{i+1,j}^{n+\frac{1}{2}} = T_{i,j}^{n} + \frac{\kappa_{i,j}\Delta t}{4\Delta x^2}\left(T_{i+1,j}^{n} - 2T_{i,j}^{n} + T_{i-1,j}^{n}\right) \tag{2}$$

While the solution for y-axis is:

$$\frac{\kappa_{i,j}\Delta t}{4\Delta y^2}T_{i,j-1}^{n+1} + \left(1 + 2\frac{\kappa_{i,j}\Delta t}{4\Delta y^2}\right)T_{i,j}^{n+1} + \frac{\kappa_{i,j}\Delta t}{4\Delta y^2}T_{i,j+1}^{n+1} = T_{i,j}^{n+\frac{1}{2}} + \frac{\kappa_{i,j}\Delta t}{4\Delta y^2}\left(T_{i,j+1}^{n+\frac{1}{2}} - 2T_{i,j}^{n+\frac{1}{2}} + T_{i,j-1}^{n+\frac{1}{2}}\right) \tag{3}$$

2.2. Stability analysis of Crank Nicholson's Method

The derivative approximation of Equation (1) using Crank-Nicholson's method is:

$$\frac{T_{i,j}^{n+1} - T_{i,j}^{n}}{\Delta t} = \frac{1}{2}\kappa\left(\frac{T_{i+1,j}^{n+1} - 2T_{i,j}^{n+1} + T_{i-1,j}^{n+1}}{\Delta x^2} + \frac{T_{i+1,j}^{n} - 2T_{i,j}^{n} + T_{i-1,j}^{n}}{\Delta x^2} + \frac{T_{i,j\ 1}^{n+1} - 2T_{+,j}^{n+1} + T_{i,j-1}^{n+1}}{\Delta z^2} + \frac{T_{i,j+1}^{n} - 2T_{i,j}^{n}\quad T_{i,j\ \vdash}^{n}}{\Delta z^2}\right) \tag{4}$$

By defining that $d_1 = \frac{1}{2}\frac{\kappa\Delta t}{\Delta x^2}$ and $d_2 = \frac{1}{2}\frac{\kappa\Delta t}{\Delta z^2}$ we obtain:

$$T_{i,j}^{n+1} = T_{i,j}^{n} + d_1\left(T_{i+1,j}^{n+1} - 2T_{i,j}^{n+1} + T_{i-1,j}^{n+1} + T_{i+1,j}^{n} - 2T_{i,j}^{n} + T_{i-1,j}^{n}\right) + d_2\left(T_{i,j+1}^{n+1} - 2T_{i,j}^{n+1} + T_{i,j-1}^{n+1} + T_{i,j+1}^{n} - 2T_{i,j}^{n} + T_{i,j-1}^{n}\right) \quad (5)$$

The Fourier Component was defined as:

$$T_{i,j}^{n} = T^{n} e^{I(\theta i + \phi j)};$$
$$T_{i,j}^{n+1} = T^{n+1} e^{I(\theta i + \phi j)};$$
$$T_{i\pm1,j}^{n} = T^{n} e^{I(\theta(i\pm1) + \phi j)};$$
$$T_{i\pm1,j}^{n+1} = T^{n+1} e^{I(\theta(i\pm1) + \phi j)};$$
$$T_{i,j\pm1}^{n+1} = T^{n+1} e^{I(\theta i + \phi(j\pm1))} \quad (6)$$

So, by substituting the Equations (6) into Equation (5), we obtain:

$$T^{n+1} e^{I(\theta i+\phi j)} = T^{n} e^{I(\theta i+\phi j)} + d_1\begin{pmatrix} T^{n+1} e^{I(\theta(i+1)+\phi j)} - 2T^{n+1} e^{I(\theta i+\phi j)} + T^{n+1} e^{I(\theta(i-1)+\phi j)} + \\ T^{n} e^{I(\theta(i+1)+\phi j)} - 2T^{n} e^{I(\theta i+\phi j)} T_{i,j}^{n} + T^{n} e^{I(\theta(i-1)+\phi j)} \end{pmatrix}$$
$$+ d_2\begin{pmatrix} T^{n+1} e^{I(\theta i+\phi(j+1))} - 2T^{n+1} e^{I(\theta i+\phi j)} + T^{n+1} e^{I(\theta i+\phi(j-1))} + \\ T^{n} e^{I(\theta i+\phi(j+1))} - 2T^{n} e^{I(\theta i+\phi j)} + T^{n} e^{I(\theta i+\phi(j-1))} \end{pmatrix} \quad (7)$$

After doing some eliminations, we obtain:

$$T^{n+1}\left(1 + d_1 + d_2 - d_1\left(\frac{e^{I\theta} + e^{-I\theta}}{2}\right) - d_2\left(\frac{e^{I\phi} + e^{-I\phi}}{2}\right)\right) = T^{n}\left(1 - d_1 - d_2 + d_1\left(\frac{e^{I\theta} + e^{-I\theta}}{2}\right) + d_2\left(\frac{e^{I\phi} + e^{-I\phi}}{2}\right)\right) \quad (8)$$

$$T^{n+1} = T^{n}\frac{\left(1 - d_1 - d_2 + d_1\left(\frac{e^{I\theta} + e^{-I\theta}}{2}\right) + d_2\left(\frac{e^{I\phi} + e^{-I\phi}}{2}\right)\right)}{\left(1 + d_1 + d_2 - d_1\left(\frac{e^{I\theta} + e^{-I\theta}}{2}\right) - d_2\left(\frac{e^{I\phi} + e^{-I\phi}}{2}\right)\right)} \quad (9)$$

Next, using the identities $\cos\theta = \left(\frac{e^{I\theta} + e^{-I\theta}}{2}\right)$ and $\sin\theta = \left(\frac{e^{I\theta} - e^{-I\theta}}{2I}\right)$, we get:

$$T^{n+1} = T^{n}\left[\frac{(1 - d_1 - d_2 + d_1\cos\theta + d_2\cos\phi)}{(1 + d_1 + d_2 - d_1\cos\theta - d_2\cos\phi)}\right] \text{ Or}$$

$$G = \left[\frac{(1 - d_1 - d_2 + d_1\cos\theta + d_2\cos\phi)}{(1 + d_1 + d_2 - d_1\cos\theta - d_2\cos\phi)}\right] \quad (10)$$

A stable solution is fulfilled when $|G| \le 1$ or:

$$\left|\frac{(1 - d_1 - d_2 + d_1\cos\theta + d_2\cos\phi)}{(1 + d_1 + d_2 - d_1\cos\theta - d_2\cos\phi)}\right| \le 1 \quad (11)$$

Since $-2 \le \cos\theta - 1 \le 0$ and $-2 \le \cos\phi - 1 \le 0$, then Equation (11) will be fulfilled for every $d_1 \in R$ and $d_2 \in R$. As the conclusion, we can define that stability of the 2D heat conduction equation in unsteady state condition using Crank Nicholson's scheme is unconditionally stable.

2.3. Wapsalit Geological Models

Figure 1 is the geological model of Wapsalit geothermal area taken from Idral [5]. The Wapsalit geothermal area is located in the district of Apo Wae, Buru, Maluku. This area is one of the prospective zones of the geothermal energy in Indonesia.

The subsurface structure shown in Figure 1 is represented by the rock resistivity. Based on the rock resistivity, it can be known that the Wapsalit geothermal area is composed of several layers. The top layer consists of sandy soil that stretches from the southwest to the northeast. Marble (fresh metamorphic rock) can be found at some depth. Slate (weathered metamorphic rock) is found under the marble layer. Granite (as the altered rock) is found at the top of cap rock layers. The cap rock layers are composed of shale. Reservoir layer is composed of limestone. The heat source is located under the reservoir layer.

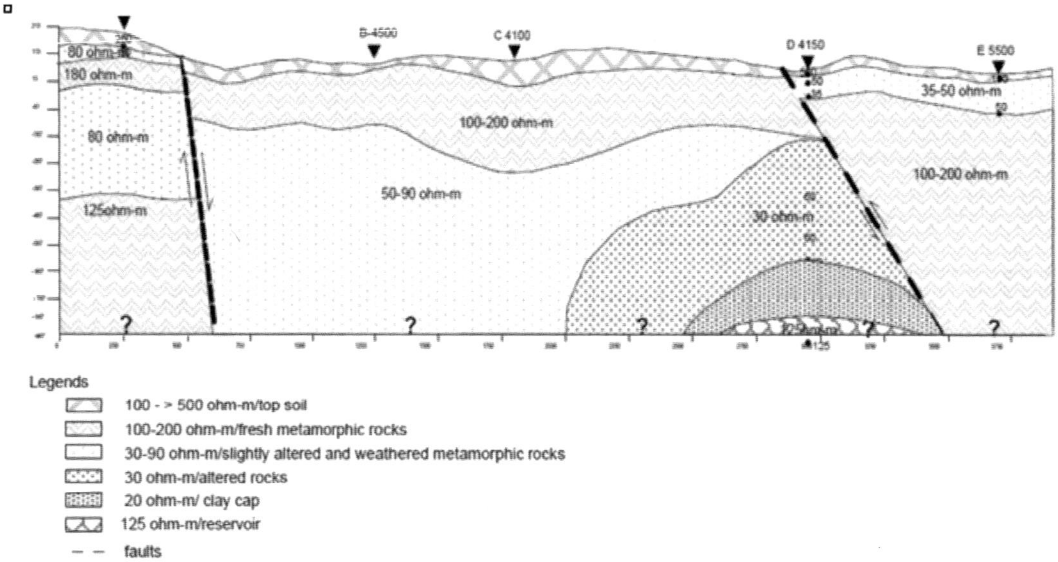

Figure 1. Resistivity cross-section of Wapsalit geothermal area [5]

In this study, we solved the thermal distributions for the Wapsalit geological models. The geological model was interpreted to obtain the diffusivity coefficient of rocks. The diffusivity coefficient was calculated based on the thermal conductivity, specific heat capacity, and density of rocks (see Table 1). The geological model was used as the input for the modeling of the thermal distributions.

Figure 2 depicts the geometric domain of Wapsalit geological models. The geometry consists of 12 rows × 40 columns grid points. The grid spacing of the model is 100 meters, so the total dimension is 1200 × 4000 meters. The initial value for this model is zero ($T(x, y) = 0$) at all points. The boundary condition at the left side, right side, and the top of the model is Neumann boundary condition. The heat source is located under the reservoir rock. The temperature of the heat source is assumed about 600^{0}C to 900^{0}C.

In this modelling, the iteration process will be stopped if the surface temperature of the model has reached about 60.7^{0}C. This criterion was chosen based on the real temperature at the study site [6].

IOP Conf. Series: Journal of Physics: Conf. Series **877** (2017) 012012 doi:10.1088/1742-6596/877/1/012012

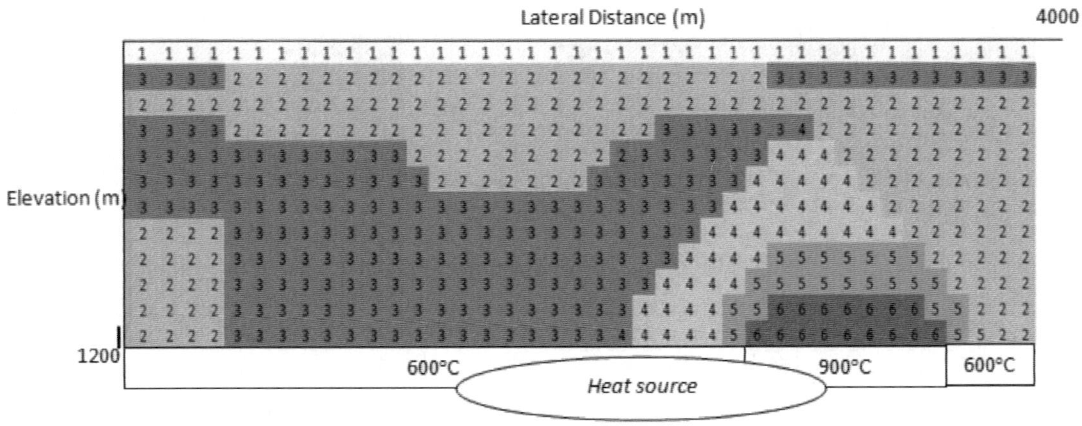

Figure 2. Geometry of the modelling domain of Wapsalit geological model (numbers and colours within the cells correspond to the types of rocks shown in Table 1)

Table 1. Thermal diffusivity of the rock types used in Wapsalit geological model [7]

No	Color Index	Parameter	
		Rock types	Thermal diffusivity (10^{-3} cm^2/s)
1		Sandy Soil/top soil	3
2		Marble/fresh metamorphic rock	10
3		Slate/weathered metamorphic rock	11
4		Granite/altered rock	16
5		Shale/cap rock	8
6		Limestone/ reservoir rock	11

3. Results and Discuss

3.1. *Sensitivity Study*

Before applying the model to the Wapsalit geothermal area, firstly we conducted a sensitivity test of the model to see the effect of the geometry, the coefficient of thermal diffusivity parameters (presented in Table 2) and the boundary condition using some simple synthetic models. The simple models for sensitivity study consist of a homogeneous earth model (shown in Figure 3a), an anomalous body embedded in a homogeneous earth model (shown in Figure 3b) and an anomalous body trapped within a layered earth model (shown in Figure 3c). Colors depicted within the cells correspond to the type of rocks listed in Table 2.

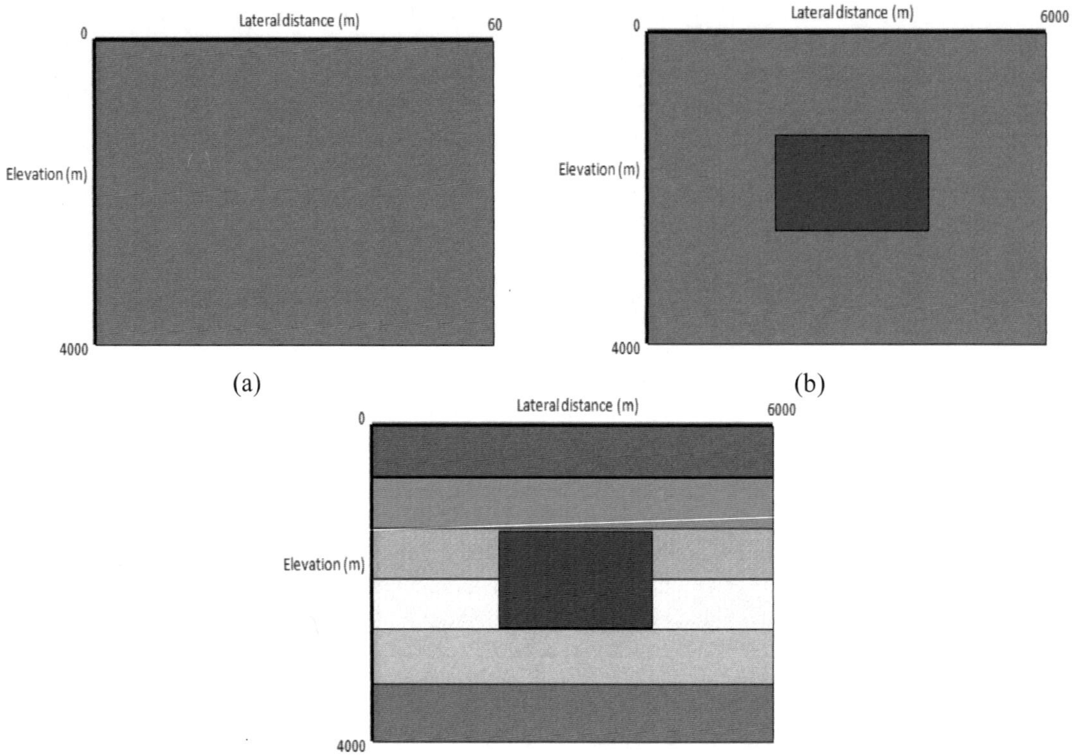

(c)

Figure 3. Simple Models for sensitivity study consist of homogeneous earth model (a); anomalous body embedded in homogeneous earth model (b); anomalous body within layered earth model (c). Colours within the cells correspond to the types of rocks listed in Table 2 [8].

Table 2. Thermal diffusivity of rocks used in the sensitivity study [7]

No	Color Index	Parameter	
		Rock types	Thermal diffusivity (10^{-3} cm^2/s)
1		Quartzite	26
2		Peridotite	17
3		Serpentine	13
4		Slate	11
5		Basalt	9
6		Sandy Soil	3
7		Steel	173

The geometry models shown in Figure 3 have a size of 4000 × 6000 meters, divided into 20 rows × 30 columns. The initial temperature is zero ($T(x, y) = 0$) for all points in the grid. The boundary condition at the left, the right, and the top of the model are Neumann boundary condition. The temperature of the heat source is 500^0C and it is located at the bottom of the model.

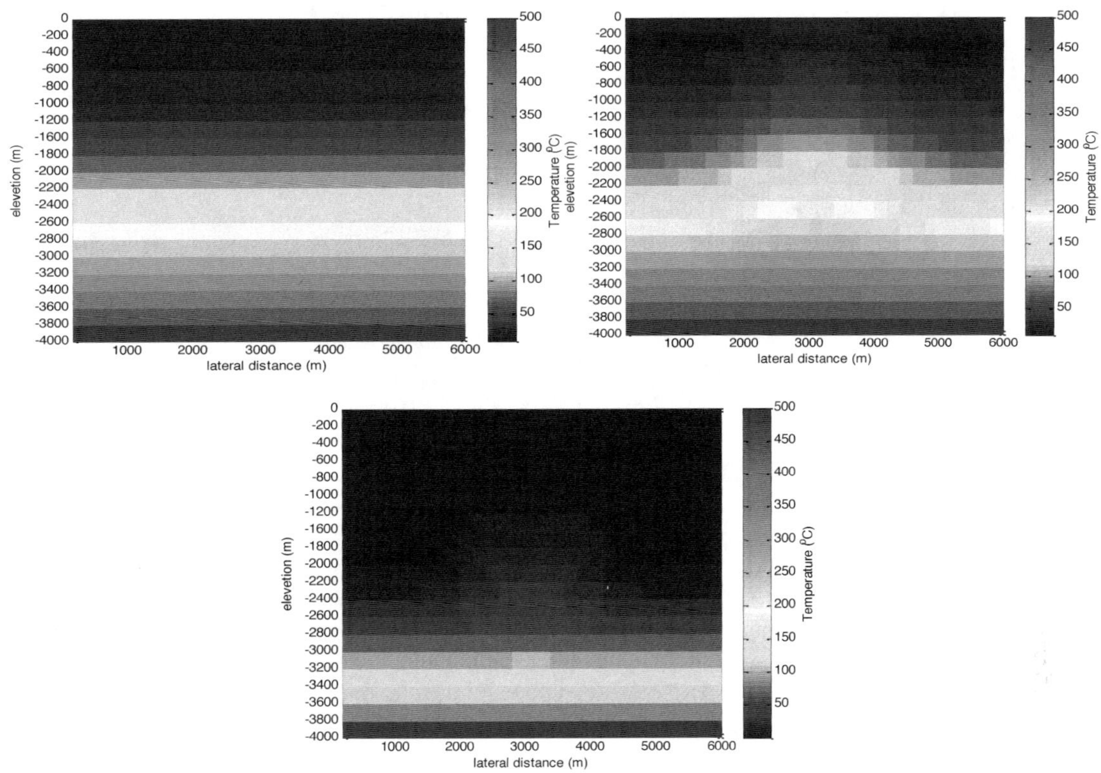

Figure 4. Thermal distribution of a homogeneous earth model (a); an anomalous body embedded in a homogeneous earth model (b); an anomalous body embedded within a layered earth model (c).

Modeling result for the first model (Figure 4a) shows that the temperature is decreased (evenly) when it gets closer to the surface. This shows that there is no difference in conductivity at each depth (homogeneous layers). From the second model (shown in Figure 4b), we can see a change in temperature in the anomalous rock. The rock causes the changes of the temperature distribution to be faster. This is because the thermal diffusivity of the anomalous rock has greater value than the surrounding rocks.

From the third model (shown in Figure 4c) we can see that the distribution of temperature flow is disturbed by rock anomalies. The model shows that the temperature flows from the bottom layer to the surface slowly. This is because each layer in this model has a different thermal diffusivity. However, because there is an anomalous rock in the middle of the model, which has a high thermal diffusivity, then there is a heat flow disorder in that layer. Yet, the contours of the thermal distribution models still show a good distribution pattern.

The results of sensitivity studies of three models of the earth in Figure 4 show that the pattern of heat distribution corresponds to the synthetic geological models. The sensitivity study of the model also shows a good performance. The variation of the model structure can be successfully represented by the heat distribution. Within this perspective, the Crank-Nicholson algorithm can be used to determine the distribution of heat in the Wapsalit geothermal area.

3.2. *Thermal Numerical Modelling using Wapsalit Geological Model*

Figure 5a shows the results of numerical modeling of 2D unsteady state thermal diffusivity at the Wapsalit geothermal area. The heat distribution pattern of the model shows that heat flow spreads by adjusting the conditions of the rock layers. The temperature of the reservoir layer is higher than surrounding due to direct contact with the heat source (probably derived from magma intrusion).

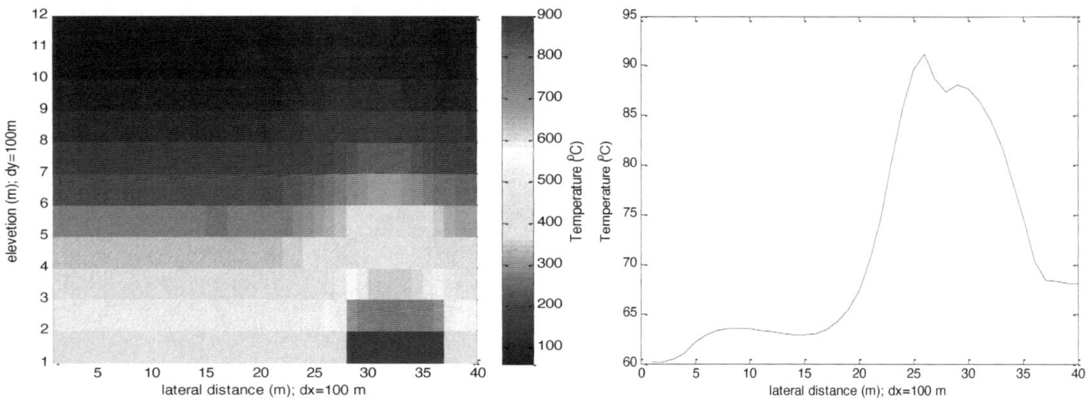

Figure 5. (a) Thermal distribution model of Wapsalit geothermal area; (b) Profile of temperature distribution at the surface

Figure 5a shows the result of the thermal spreading model in the Wapsalit geothermal area. The heat distribution pattern shows that the heat flow adjusts to the rock layers in its path. On the other hand, Figure 5b depicts the profile of surface temperature in the Wapsalit geothermal area. It can be seen that for this geothermal area the temperature tends to be higher in zone that contain a reservoir in the subsurface. Based on this pattern, we have an opportunity to build an inversion method to investigate the subsurface in the future.

4. Conclusions

The finite difference method based on the Implicit Crank-Nicholson algorithm has successfully applied to describe the distribution of 2D heat diffusion in the unsteady state condition of geothermal systems. The sensitivity study shows that the modeling scheme exhibits acceptable performance. The model of the Wapsalit geothermal area successfully exhibits the pattern of heat diffusion, which is in accordance with the conditions of rock layers.

References

[1] Putra SDH, Fajar SJ and Srigutomo W 2014 *Proc. Int. Conf. on Physics 2014 (ICP-14)* (Amsterdam: Atlantis Press) pp 51-61

[2] Abebe, T 2000 *Proc. World Geothermal Congress* (Kyushu-Tohoku, Japan: International Geothermal Association) pp 2025-2030

[3] Pletcher RH 2000 *Handbook of Numerical Heat Transfer, Second Edition,* eds W J Minkowycz, E M Sparrow and J Y Murthy (Hoboken, USA: John Wiley & Sons, Inc.)

[4] Hoffmann KA and Chiang ST 1989 *Computational Fluid Dynamics for Engineers. 4th ed.* (Austin, TX: Engineering Education System) pp 165-172

[5] Idral A 2010 *Proc. World Geothermal Congress* (Bali-Indonesia: International Geothermal Association)

[6] Zarkasyi A, Hermawan D and Kasbani *Proc. Pemaparan Hasil Kegiatan Lapangan dan Non Lapangan Tahun 2007 Pusat Sumber Daya Geologi* (http://psdg.bgl.esdm.go.id/kolokium%202007/PANASBUMI/Penyelidikan%20Terpadu%20daerah%20panas%20bumi%20Wapsalit%20Maluku.pdf)

[7] Clauser C and Huenges E 1995 *"Thermal conductivity of rocks and minerals." Rock physics & phase relations: A handbook of physical constants* (USA: American Geophysical Union) pp 105-126.

[8] Hardiansah CP, Dharmawan IA, Susanto K and Hilman I 2012 *Lokakarya Komputasi dalam Sains dan Teknologi Nuklir* (Jakarta-Indonesia: BATAN) pp 123-141

IOP Conf. Series: Journal of Physics: Conf. Series **877** (2017) 012013 doi:10.1088/1742-6596/877/1/012013

Analysis of Neutron Fission Reaction Rate in the Nuclear Fuel Cell Using Collision Probability Method with Non Flat Flux Approach

Mohammad Ali Shafii

Department of Physics, Andalas University
Kampus Limau Manis Padang West Sumatera Indonesia

Email: mashafii@fmipa.unand.ac.id

Abstract. Neutron fission reaction rate in the nuclear reactor depends on macroscopic cross section and neutron flux distribution. The macroscopic cross section depends on the type of nuclide, the type of reaction, and the group energy of the neutrons relative to the nuclides. Flux distribution is very important in a nuclear reactor, because it is closely related to power distribution. In general, the integral neutron transport equation is solved using a collision probability (CP) method with a flat flux (FF) approach. Consequently, the CP matrix is also assumed constantly, therefore, the distribution of the neutron flux throughout the cell becomes flat. In the non-flat flux (NFF) approach, the neutron flux is modellled by linear interpolation as a function of mesh in the cylindrical nuclear fuel cell of a fast reactor type. This study uses the CP method with a NFF approach and it is applied to analyze the neutron fission reaction rate of a cylindrical nuclear fuel cell of a fast reactor type. Nuclear data library that is used in this study is JFS-3-J33 which belongs to the SLAROM computer code. Calculation results of the fission reaction rate shows that it is decrease in the high energy region due to the events of elastic collision that caused the neutron easier to lose of energy. The same fission reaction rate pattern occurs in the FF and NFF approaches.

1. Introduction

The most difficult part of the nuclear reactor analysis is to solve the neutron distribution system as it describes the integral transport equation with variable of energy, space and time [1]. In general, the integral transport is solved using collision probability (CP) method with flat flux (FF) approach [2]. CP method has advantage for simple and symmetrical geometry such as cylindrical cell. Flat flux means that the neutron flux is every points in the region of nuclear fuel cell is considered to be constant and the neutron cross section in all region of the cell is homogeneous. Consequently, the CP matrix is also assumed constantly, therefore, the distribution of the neutron flux throughout the cell becomes flat. Transport of neutron in this study covers two perspectives. First, calculation of homogenization of the nuclear fuel cell is performed using CP method with the FF approach [3]. Second, the points of mesh in cell region is assumed not flat, rather they follow neutron interpolation model based on non-flat flux (NFF) approach [4]. The selected flux interpolation is linear, quadratic or spline cubic fuction and it is applied to analyze flux distribution of a cylindrical nuclear fuel cell of fast reactor type. The method itself requires fine spatial subdivision of the geometrical domain into so called flat-flux zones where the material properties are assumed to be constant [5].

Neutron interaction with the nuclear fuel cell is presented through the CP process from one nucleus to another until it is absorbed or escapes the boundary of a system. One of the important and characteristic features of neutron interactions with matter that proceed through a compound nucleus formation is that cross sections exhibit maximum values at certain incident neutron energies [5].

Fission process represents a type of nuclear interactions that produces neutrons and energy and as such is a basic principle of nuclear power generation. In a nuclear reactor, neutrons not only have mono energetic, it covers a wide range of energy, and this is referred to as multi group energy. The reaction rates as well as neutron flux and cross sections are energy dependent. At a given neutron energy, the neutron flux at a given time varies with the spatial position in the reactor core. The spatial distribution of fissile material is not entirely uniform initially and is not uniform after the reactor has been operating for a certain time. In order to determine a fission rate at a given time, the neutron transport equation must be integrated over all neutron energies and spatial positions in the reactor. In thermal reactors, the majority of fissions occur in the thermal energy region where flux and macroscopic cross sections are both very large. The fission rate can be roughly estimated assuming the average values of space and energy for flux and cross section. A representative geometry of a complex reactor assembly is selected to show the distribution of neutron flux and reaction rates as a function of neutron energy group and spatial coordinates. Neutron flux at a certain point in the reactor core will depend on the distribution of nuclear properties such as cross section and the position where the flux is calculated. The reaction rate for various types of interactions is found from the appropriate cross section type: fission, absorbtion and total. In order to get a good fission reaction rate, cross section of fission must be dominant.

This study uses the CP method with a NFF approach and it is applied to analyze the neutron fission reaction rate of a cylindrical nuclear fuel cell of a fast reactor type for 70 energy group. Library data that is used in the research is the JFS-3-J33 JAEA (Japan Atomic Energy Agency) of group constants for the SLAROM code.

2. Theoretical Background

Probability of neutron that uniformly spreading out and isotropic in the *region i* and suffer first collision in *region j* of cylindrical nuclear fuel cell is follow [6]

$$P_{ij} = \frac{2}{\Sigma_i V_i} \int_0^{r_i} d\rho \{ K_{i3}(\lambda^1_{ij}) - K_{i3}(\lambda^1_{ij} + \lambda_i) - K_{i3}(\lambda^1_{ij} + \lambda_j) + K_{i3}(\lambda^1_{ij} + \lambda_i + \lambda_j) + K_{i3}(\lambda^2_{ij}) \\ - K_{i3}(\lambda^2_{ij} + \lambda_i) \quad K_{i3}(\lambda^2_{ij} + \lambda_j) + K_{i3}(\lambda^2_{ij} + \lambda_i + \lambda_j) \} \tag{1}$$

where $\lambda_k = \Sigma_k (x_k - x_{k-1}), \lambda^2_{ij} = \sum_{k=1}^{i-1} \lambda_k + \sum_{k=1}^{j-1} \lambda_k$ and $\lambda^1_{ij} = \sum_{k=i-1}^{j-1} \lambda_k$. $K_{ij}(\lambda)$ is Bickley-Nayler function of third order. Therefore, if energy range of neutron is divided into multi group energy, then the average flux in the energy interval ΔE_g is described by φ_{ig}, so the neutron transport equation is performed as follow [7]

$$\Delta E_g \Sigma_{jg} V_{jg} \varphi_{jg} = \sum_i P_{ijg} V_{ig} \left[\sum_{g'} \Delta E_{g'} \Sigma_{sig' \to g} \varphi_{ig'} + \Delta E_{g'} S_{ig} \right] \tag{2}$$

where ΔE_g and $\Delta E_{g'}$ are energy width in the group g and g', $\Sigma_{sig' \to g}$ is scattering cross section in the region i from the group g' to g. For the computational process, equation (2) can be rewritten as multi group energy [2]

$$\Sigma_{jg} V_j \varphi_{jg} = \frac{1}{k_{eff}} \sum_i V_i P_{ijg} S_{ig} \tag{3}$$

Neutron reaction rate proportional to neutron flux and target area. The reaction rate for various types of interactions is found from the appropriate cross section type: Fission, absorption and total. In order to get a good fission rate, cross section fission must be the dominant in the interaction. Reaction rate of neutron fission of through shielding material of fuel in the region i with energy group g is defined by

$$R_{fig} = \varphi_{ig} \Sigma_{fig} \tag{4}$$

where φ_{ig} is neutron flux and Σ_{fig} is macroscopic cross section. In the case of NFF, neutron flux in the equation (4) is taken from [3]

$$\varphi_{ig} = \frac{1}{\alpha_{i,g}}\left(Q_{ig} - \beta_{i,g}\varphi_{i-1g}\right) \tag{5}$$

where $$\alpha_{ig} = \left\{\Sigma_{jg}\left(\frac{1}{2}x_j^2 x_i - \frac{1}{2}x_{j-1}^2 x_i - \frac{1}{3}x_j^3 + \frac{1}{3}x_{j-1}^3\right) + \sum_{i,g} P_{ijg}\Sigma_{sig}\left(\frac{1}{6}x_i^3 - \frac{1}{2}x_{i-1}^2 x_i \quad \frac{1}{3}x_{i-1}^3\right)\right\}$$

$$\beta_{ig} = \left\{\Sigma_{jg}\left(\frac{1}{2}x_{j-1}^2 x_i - \frac{1}{2}x_j^2 x_{i-1} - \frac{1}{3}x_{j-1}^3 + \frac{1}{3}x_j^3\right) + \sum_{i,g} P_{ijg}\Sigma_{sig}\left(\frac{1}{6}x_{i-1}^3 - \frac{1}{2}x_i^2 x_{i-1} \quad \frac{1}{3}x_i^3\right)\right\}$$

$$Q_{ig} = \frac{1}{2}\sum_{i,g} P_{ijg} S_{ig}\left(x_i^3 - x_i^2 x_{i-1} - x_{i-1}^2 x_i \quad x_{i-1}^3\right).$$

3. Design and Computational Model

The geometry of cylindrical nuclear fuel cell is divided into three regions: region 1 is fuel, region 2 is cladding and region 3 is coolant [2]. The composition of mesh follows [3]. Specification design of nuclear fuel cell is shown in Table 1. Library data that is used in the research is the JFS-3-J33 of group constants for the SLAROM code in 70 energy group.

Table 1. Specification design of nuclear fuel cell.

Parameters	Specification
Type of nuclear fuel cell	Uranium-Plutonium Nitride
Geometry of cell	Cylindrical 1D
Material Structure	Stainless steel
Coolant	Lead-Bismuth (Pb-Bi)
Diameter of pin cell	1.134 cm
Cladding thickness	0.11 cm
Temperature Average	1183 K
Volume fraction: fuel	61.73%
structure	19.40%
coolant	18.87%

The computational program to calculate fission reaction rate in the nuclear fuel cell using CP method with FF and NFF approach has been written in Delphi. The procedure to get fission reaction rate are follow the mechanism, after reading input data, microscopic cross section is calculated for each nuclide by considering temperature and homogeneous background cross section. Macroscopic cross section is calculated using corrected background cross section for width of cell. The collision probability matrix p_{ij} is calculated for all regions that divided into several mesh uses equation (1). For case NFF approach, equation (5) is inserted to equation (3) to get the *k*-eff and then neutron flux in the equation (5) is inserted to equation (4) to get the fission reaction rate. The iteration is executed for 6 meshes and 70 group of energy. The detail of mesh composition, the radius of each region cell and number of nuclide in each region are shown in [3].

4. Results and Discussion

Figure 1 shows that the fission reaction rate has high value in the region of high energy and then decreased in the intermediate energy region and increase in the low energy region for FF and NFF aproach. Fast reactor is working at high energy, between 0.1 eV to 10 MeV (energy group 1 to 30). Fission reaction rate has decreased in the region of high energy due to elastic collision events, therefore neutrons easier to lose energy so the fission reaction rate also decreased. Energy group increase to right, but the energy decreases. The number 1 to 19 are fast neutron energy groups covers the cross section data for neutron energy range from 0.11109 to 10 MeV, while the 20 to 70 are thermal neutron energy group cover the neutron energy range from 0.32242 eV to 86.517 keV [8]. This is in accordance with the fact that the formation of neutron flux spectrum only occur at a high energies, because the fast region occurred in a high energy, while the thermal region occurred at low energy. Since the fast reactor is used, then the influence of thermal region can be neglected. It is noticed that the behavior of thermal region, in order to calculate constant group for fast reactor, occasionally avoids the differentiation.

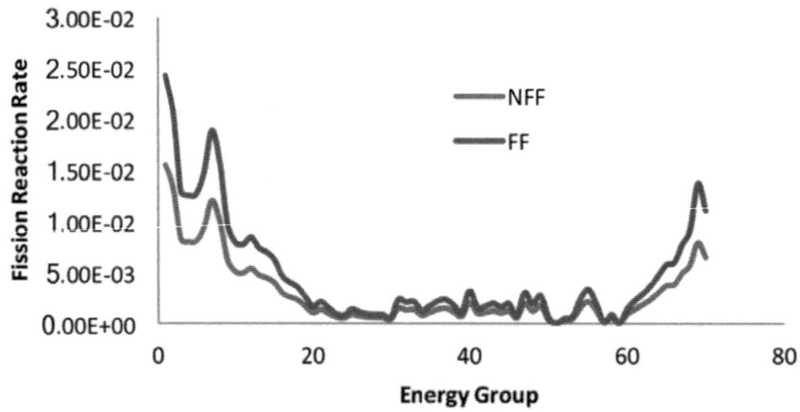

Figure 1. Neutron fission reaction rate in each energy group for FF and NFF aproach.

In the intermediate energy range or in the resonant energy region, for all reactors, neutrons fission is absorbed in the resonant energy region. Resonant energy region occurs in the energy group 30 to group 65, or are in the energy range of 1 eV to 1 keV. Fission resonant region is where the energy characteristics of fission nuclides easily excited that indicated the presence of the cross. As may be seen, the NFF approach result is in good agreement with the FF approach, both have same pattern. In the region of low energy or often referred to as the resolved resonance region will depend on the nature of the radionuclide used. For fertile nuclides resolved resonance region until a few keV energy, while for the fissile nuclides in the region of about 50 eV. In this incident is demonstrated how the scattering dominates in the MeV range for most of the fission neutrons have an energy that is just above 1 MeV. For commercial fast reactors, most of the neutrons will have a population in the region around 100 keV. For heavy nuclides, energy is lost in the event of elastic collisions is much smaller than in the inelastic collision. Neutron in the event of elastic collisions with the heavy nuclei, energy is almost equal to the initial energy. In the process of inelastic collision, neutron energy mostly used for excitation.

Figure 2 shows that in the fast energy region, the fission reactions rate occurs only in the fuel region, whereas the cladding and the coolant region does not occur because the both region does not occur fission. For thermal energy region, same pattern of fission reactions rate are shown in the Figure 3. This pattern follows the neutron flux distribution in each mesh using NFF and FF approach. Neutron flux spectrum behavior in the fuel cell region for 6 meshes distribute neutron flux in three regions, i.e. fuel, cladding and coolant. In the NFF and FF approach, neutron flux distribution tends to form linearly in the fuel region, then slowdown in the cladding and in the coolant. Neutron flux distribution tends to decrease in the cladding and coolant for fast energy region, in accordance situation come from the thermal energy region, although in the thermal energy region, flux distribution tends to increase in the cladding and coolant region [3].

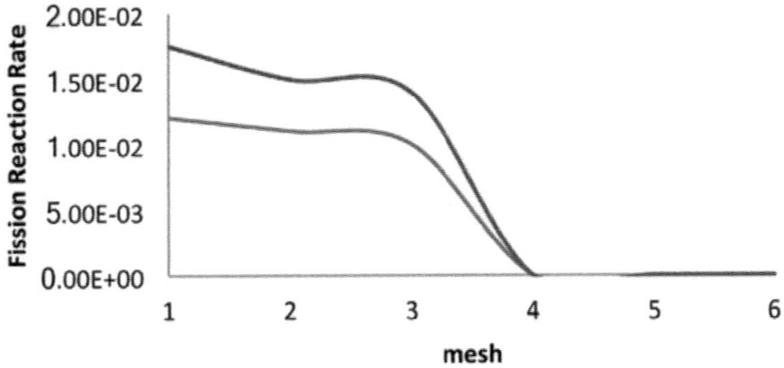

Figure 2. Fission reactions rate of the nuclear fuel cell for 6 mesh in the fast region for FF and NFF approach.

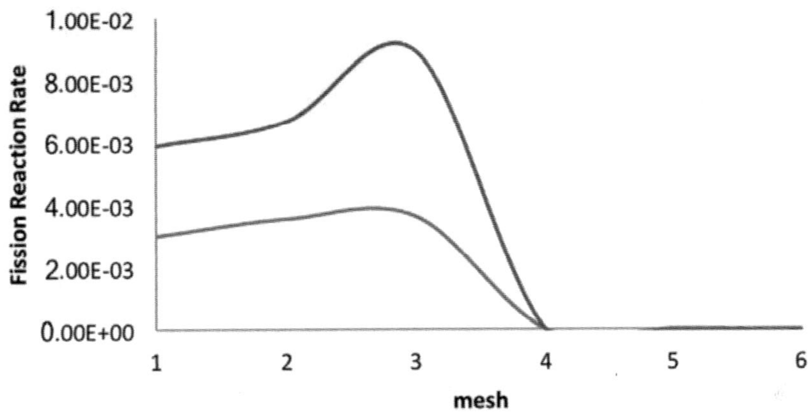

Figure 3. Fission reactions rate of the nuclear fuel cell for 6 mesh in the thermal region for FF and NFF approach.

5. Conclusion

Solution of neutron transport equation using the collision probabilities method with NFF approach that implemented in cylindrical cell calculation yields fairly good results comparing with FF approach. The neutron fission reaction rate has high value in the region of high energy and then decreased in the intermediate energy region and increase again in the low energy region for FF and NFF aproach. The decreasing of neutron fission reaction rate in the high energy region due to the events of elastic collision that caused the neutron easier to lose of energy. The neutron fission reactions rate in the fast energy region, occurs only in the fuel region, whereas the cladding and the coolant region does not occur because the both region does not occur fission. Same situation also occurs for thermal energy region using FF and NFF approaches, because the pattern follows the neutron flux distribution in each mesh of region.

Acknowledgments
The author wishing to thanks for financial support from the Directorate General of Strengthening of Research and Development, Ministry of Research, Technology and Higher Education RI with Contract No : 020/SP2H/LT/DRPM/II/2016 for *Stranas* Research.

References

[1] Duderstadt, J.J., and Hamilton,L.J. 1976, *Nuclear Reactor Analysis*, John Wiley and Sons, NY.
[2] Shafii M.A., Su'ud, Z. Waris, A., Kurniasih, N., Ariani, M. dan Yulianti, Y. 2010, Neutron Flux Interpolation with Finite Element Method in the Nuclear Fuel Cell Calculation using Collision Probability Method, *AIP Conf. Proc.* Vol. **1325**, p. 253.
[3] Shafii, M.A., Su'ud, Z. Waris, A., dan Kurniasih, N. 2012, Nuclear Fuel Cell Calculation using Collision Probability Method with Linear Non Flat Flux Approach, *World Journal of Nuclear Science and Technology*, Vol. **2** p. 49.
[4] Shafii, M.A., Meidianti, R, Wildian, Fitriyani, D, Tongkukut, SHJ and Arkundato, A, 2014, Theoretical analysis of integral neutron transport equation using collision probability method with quadratic flux approach, *AIP Conf. Proc.* Vol. **1615**, p. 91.
[5] Jevremovic, T, 2005, *Nuclear Principle in Engineering*, Springer, NY.
[6] Nakagawa, M and Tsuchihashi, K, 1984, SLAROM: *A Code for Cell Calculation of Fast Reactor,* JAERI, Japan.
[7] Okumura, K, Kugo, T, Kaneko, K. and Tsuchihashi, K, 2007, SRAC 2006: *A Comprehensive Neutronics Calculation Code System,* JAEA, Japan.
[8] Hazama, T., Chiba, G., dan Sugino, K. 2006, Development of a Fine and Ultra-Fine Group Cell Calculation Code SLAROM-UF for Fast Reactor Analyses, *J. Nucl. Sci. Technol.*, Vol. **43**, No. 8, p. 908–918.

Developing a Framework For The Assessment Of Pre-Service Physics Teachers' Energy Literacy

Muhamad Yusup[1,3], Agus Setiawan[2], Nuryani Y. Rustaman[2], Ida Kaniawati[2]

[1] Doctoral student of Postgraduate School of Universitas Pendidikan Indonesia
[2] Postgraduate School of Universitas Pendidikan Indonesia
[3] Physics Education Department of Universitas Sriwijaya

Email: yusufunsri@yahoo.com.

Abstract. Energy concept is one key component in physics education. Pre-service physics teachers are expected have energy literacy as their learning experience and for preparing their professional teaching. Measuring their energy literacy is therefore very important and its results are useful not only for faculty but also for national curriculum developers and policy makers. Yet there is no assessment instrument to measure how literate they are. In this paper, we develop an assessment framework to measure pre-service physics teachers' energy literacy.

1. Introduction

Energy literacy is a topic in science education research that has been risen over the last few decades. Most of the research have focused either on determining attitudes toward energy conservation and education or on trying to effect a change in attitude toward energy conservation and education through courses and workshops [1]. Reseachers then broadened the domain of energy literacy. DeWaters, Powers, and Graham [2] developed energy literacy scale for a written survey, named Energy Literacy Qusetionnaire, The scale measures energy literacy in three domain: cognitive, affective, and behavioral outcomes. This scale used by many reseachers in different country [3–7].

By adopting and extending the framework from DeWaters and Powers [3], other researchers [8] developed framework for energy literacy in four domain: energy concepts, reasoning of energy issues, low carbon lifesyle, and civic responsibility for sustainable society. Similar to the former, this frameworks were intended use adequately for the assessment of energy literacy among middle and high school students. So far, we have not found a assessment framework to assess energy literacy of physics pre-service teacher.

Pre-service teachers is a key intervention into the future generation of citizens facing sustainability challenges [9]. Therefore, addressing energy literacy as part of preparation of future teachers is important. Teacher preparation education can transform human behavior towards the rational use of energy and increase energy literacy [10]. Pre-service teacher education is a promising means to achieve this transformation [9].

The National Science Teachers' Association outlined standards for Science Teacher Preparation [11]. One of the standards stated that teachers of science should recognize that informed citizens must be prepared to make decisions and take action on contemporary science-and-technology-related issues of interest to the general society. For this reason, we viewed that assessment framework to measure physics pre-service teacher energy literacy is needed. The existing frameworks are not adequate to be used fot this purpose because of their intended purpose for middle and high school students.

2. Method

We developed the framework assessment in the following steps. Firstly, we started with determining definition of energy literacy and individual characteristics who has it. In developing a working definition of energy literacy that can be used as the basis for designing an assessment, we used definitions of literacy, several national frameworks for energy education, and pertinent research. Secondly, we developed a model of framework by matching the criteria with a model of taxonomy of educational objectives we selected. In selecting the taxonomy of educational objectives we applied the following procedures. (1) We considered cognitive processes and knowledge domains that former researchers [8], [12], [13] identified as relevant for energy literacy. (2) We searched among taxonomies of educational objectives [14–20] and chose one that appropriate to be applied for energy literacy assessment.

3. Energy Education and Energy Literacy

Energy education receives much attention for its importance in instilling knowledge, making connections with environment and society, developing responsibility, and shaping behavior regarding energy issues [5]. There is no agreement among researchers or curriculum developers about energy education definition. One definition of energy education stated as a vehicle to help students respond to present and future energy-related concerns, which include the political, social, economics, and environmental dimension [1]. The broad objectives of energy-education for students are as follows [21-22].

a) Becoming acquainted with various forms of energy and their interconversion.
b) Learning about the role of energy in their daily lives.
c) Becoming aware that energy is not infinitely available—through this the methods of conserving energy, augmenting it.
d) Developing an awareness about the nature, cause of energy crises and methods of overcoming it.
e) Making aware of various types of nonrenewable and renewable sources of energy, their resource potential, existing technologies to harness them, economics and energetics of these technologies, and their socio-cultural and environmental aspects.
f) Making the students appreciate the consequences of various energy-related policy measures.
g) Making the students appreciative of the energy-environment nexus and enable them to evolve holistic solutions to ensure sustainability

The goal of energy education is to develop energy-literate citizens. Energy literacy is an understanding of the nature and role of energy in the universe and in our lives. Energy literacy is also the ability to apply this understanding to answer questions and solve problem [23]. By reviewing literature [12], [23], we defined an energy-literate person as one who:

a) can trace energy flows and think in terms of energy systems;
b) knows how energy is used in everyday life;
c) can assess the credibility of information about energy;
d) can communicate about energy and energy use in meaningful ways;
e) is able to make informed energy and energy use decisions and take action based on an understanding of impacts and consequences;
f) understands the impacts that energy production and consumption have on all spheres of environment and society;
g) is aware of the need for energy conservation and the need to develop renewable energy resources; and
h) continues to learn about energy throughout his or her life.

4. Choosing an appropriate taxonomy for assessment framework

We did not use those aforementioned energy literacy-related assessment framework [8], [12], because of three reason. Firstly, they did not represent a model or a theory of human thought as opposed to a taxonomy [24]. Secondly, they did not give clear information about in what level of thinking the item of assessment would be addressed. Thirdly, they did not clear about how knowledge type and context are integrated in assesment item.

We examined among the taxonomies [14–20] that could be categorized as framework, because of their broad in use in the world. The criteria we used to select the taxonomy that appropriate for our purpose were:

(1) addresses cognitive, as well as affective, behavioral, and knowledge domain in a one integrated model;
(2) makes a clear distinction between the thinking processes and the knowledge; and
(3) be able to predict phenomena of energy behavior.

Employing the above criteria, we found that only Marzano's taxonomy [20], named The New Taxonomy that fulfilled both criteria (1) and (2). The New Taxonomy also gave us a model of behavior that satisfied our criteria (3) and was in line with our definition of energy literacy.

Briefly, The New Taxonomy is a two-dimensional framework having three systems of thinking as one dimension and three types of knowledge as the other dimension. The three systems of thinking are ordered in the following six levels:

Level 6: Self-system
Level 5: Metacognitive system
Level 4: Knowledge utilization (cognitive system)
Level 3: Analysis (cognitive system)
Level 2: Comprehension (cognitive system)
Level 1: Retrieval (cognitive system)

For the purpose of our work, we did not include all three types of knowledge in The New Taxonomy, that are information, mental procedures, and psychomotor procedures. We substituted them with system knowledge, action-related knowledge, effectiveness knowledge [25], and pedagogical content knowledge (PCK) [26]. The first three are forms of declarative knowledge, that is akin to information knowledge in The New Taxonomy. We included PCK in this framework because PCK is a type of knowledge that is unique to teachers, as this framework is intended. The elaboration of these systems of thinking and types of knowledge are presented in the following section.

5. Organizing the domains

The way the domain of energy literacy is organized determines the assessment design, including the test items. As we presented in the previous section, we decided to use The New Taxonomy as a basis of the framework we will develop. The framework comprises five interrelated components: self-system, metacognitive system, cognitive system, knowledge, and contexts. Figure 1 presents these components.

5.1. Knowledge domain

Knowledge is needed to solve certain task. A review of the knowledge domain from existing frameworks for environmental literacy provided the basis for this framework. We adopted the forms of environmental knowledge proposed by Frick et al [25]. The first form of knowledge is system knowledge. System knowledge is defined as "knowing what". This knowledge usually relates to the question of how energy system operate or knowledge about energy issues. A typical example is knowledge of the relationship between carbon dioxide (CO_2) and global climate change.

The second form of knowledge is action-related knowledge, defined as ''knowing how'' or knowledge of behavioral options and possible courses of action. Unlike factual knowledge, action-related knowledge is more likely to affect behavior. For example, if people know that CO_2 contributes to global warming, they may still not know what actions they can take to reduce their CO_2 emissions. The third form of knowledge is effectiveness knowledge, addresses the relative gain or benefit (i.e. the relative conservational effectiveness) that is associated with a partiçular behavior. With this form of knowledge, the focus in action-related knowledge has obviously been extended from a mere knowing how to conserve to knowing how to get the greatest environmental benefit. For example, buying an energy-efficient light bulb is a better way to reduce energy consumption than an incandescent light bulb.

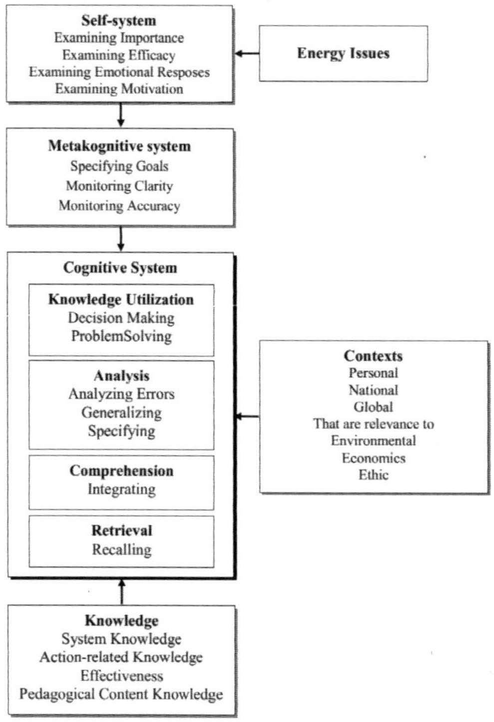

Figure 1. A framework for assessing energy literacy of pre-service physics teachers (adapted from [20]). Some components from the original framework are exclude due to the purpose of our framework.

5.2. *Contexts for the assessment item*

Individuals confront their everyday life situations in which they must use knowledge. Context refers to these situation in which knowledge about energy issues must be applied, ranging from personal to global, that are combined with environmental, economics, and ethics aspects. The combinations are shown in the Table 1. Energy literacy assessment does not assess contexts, rather, it assesses competencies and knowledge in particular context.

Table 1. Contexts for the assessment of energy literacy of preservice physics teachers.

	Personal	Local/National	Global
Environment	Environmental impact of energy use	Renewable energy technologies	Climate change
Economics	Energy efficiency	Energy policy	Energy resources and global development
Ethics	Green lifestyle, energy conservation	Energy exploration and production	Global consumption of energy

5.3. *Thinking system and competencies*

Competencies are defined as clusters of skills and abilities that may be called upon and expressed in real-world and assessment settings for a specific purpose [27]. Tabel 2 describes the competencies of energy literate person relate to the six levels of thinking system in The New Taxonomy.

Table 2. The six levels of thinking and competencies relate to each level.

System of thinking	Competencies
Level 1: Retrieval	Recognizing nonrenewable and renewable energy resources.
Level 2: Comprehension	Explaining that energy dissipation occur in every energy transfer. Counting energy consumption of electrical equipment.
Level 3: Analysis	Identifying logic error of an information provided about energy. Analyzing the environmental impact of fossil fuel usage.
Level 4: Knowledge utilization	Using information to make a decision about energy use and purchase. Using information to solve problem about energy. Proposing personal action to conserve energy.
Level 5. Metacognition	Specifying goals of conserving energy. Specifying learning objective of energy concept.
Level 6. Self-system	Examining importance of energy conservation. Identifying beliefs about one ability to conserve energy. Identifying own emotional response related energy use. Identifying overall level of motivation to take action in energy conservation.

6. Sample items

In this section, examples of item of energy literacy assessment for pre-service physics teachers is presented.

Question 1 : Air Conditioner (AC)

I feel annoyed to find AC in the classroom is on whereas no people there.

A. Always B. Often C. Sometimes C. Never

The categorization for sample Question 1 above is presented in Table 3.

Table 3. Framework categorization for sample Question 1.

Categories	Framework
Knowledge type	Action-related knowledge.
Competency	Identifying own emotional response related energy use.
Context	Personal, economics.

Question 2 : Air Conditioner (AC)

Human body uses energy at the rate of approximately 100 W when at rest. This energy, from chemical energy in our body's stores, is ultimately converted entirely to thermal energy, which is then transferred as heat to the environment. Estimate what BTU/hour of AC you need for a lecture room if its capacity about a hundred people in order to to take account of transferred thermal energy.

The categorization for sample Question 2 is presented in Table 4.

Table 4. Framework categorization for sample Question 2.

Categories	Framework
Knowledge type	System knowledge.
Competency	Using information to make a decision about energy use and purchase.
Context	Personal, economics.

Question 3 : Air Conditioner (AC)
One day, you are asked by your father, who is with no science background, to accompanying him to buy an AC that will be used in the small guest room of your home. He tell you that he want to buy it cash and he has IDR 3.000.000,- for its budget. At electronic store, you find there are many kind of AC with similar specification but differ in power consumption and price, as shown in the table below.

Brand	Power (watts)	Price (IDR)
A	795	2.900.000
B	840	2.700.000
C	900	2.600.000
D	925	2.500.000

Based on table above, and only consider on both its power consumption and price, which one brand of AC you will recommend to your father to buy? Give your consideration.

The categorization for sample Question 3 above is presented in Table 5.

Table 5. Framework categorization for sample Question 3

Categories	Framework
Knowledge type	Effectiveness knowledge
Competency	Using information to make a decision about energy use and purchase.
Context	Personal, Economics

7. Conclusion

The purpose of this paper is to develop an assessment framework to measure pre-service physics teachers' energy literacy. Due to an energy literate person is who not only has energy knowledge but also is be able to use his/her knowledge, we chose The New Taxonomy of Educational Objectives [20], which satisfied our criteria, as a basis for the framework. The framework comprises knowledge domain, context, and three systems of thinking. For knowledge domain, we substituted the original in The New Taxonomy with system knowledge, action-related knowledge, and effectiveness knowledge, that are akin to declarative knowledge. Assessment items are constructed in the personal, local/national, and global context related to environment, economics, and ethics aspect. Competencies are assessed with relation to the six levels of thinking system (i.e., self-system thinking, metacognition, knowledge utilization, analysis, comprehension, and retreval). Sample items show that the framework developed applicable for assessment item to measure energy literacy of pre-service physics teachers.

References

[1] J. T. Morrisey and L. Barrow, "A Review of Energy Education : 1975 to NEED 1981," vol. 68, no. 4, pp. 365–379, 1984.
[2] J. E. DeWaters, S. E. Powers, and M. Graham, "Developing an Energy Literacy Scale," in *Proceedings of the 114th Annual ASEE Conference & Exposition (Honolulu HI, June, 2007, session AC 2007-1069, on CD)*, 2007.
[3] J. E. Dewaters and S. E. Powers, "Energy literacy of secondary students in New York State (USA): A measure of knowledge , affect , and behavior," *Energy Policy*, vol. 39, no. 3, pp. 1699–1710, 2011.
[4] Y. L. C. Khoo and D. F. Treagust, "Assessing secondary school students ' u nderstanding of the relevance of energy in their daily lives," *Int. J. Sci. Educ.*, vol. 8, no. 1, pp. 199–215, 2013.
[5] S. Chen, Y. Chou, and H. Yen, "Investigating and structural modeling energy literacy of high school students in Taiwan," *Energy Effic.*, 2015.
[6] L. S. Lee, Y. F. Lee, J. W. Altschuld, and Y. J. Pan, "Energy literacy: Evaluating knowledge, affect, and behavior of students in Taiwan," *Energy Policy*, vol. 76, pp. 98–106, 2015.
[7] A. M. Bodzin, Q. Fu, T. E. Peffer, and V. Kulo, "Developing Energy Literacy in US Middle-Level Students Using the Geospatial Curriculum Approach," *Int. J. Sci. Educ.*, vol. 35, no. 9, pp. 1561–1589, 2013.
[8] K. Chen, S. Huang, and S. Liu, "Devising a framework for energy education in Taiwan using the analytic hierarchy process," *Energy Policy*, vol. 55, pp. 396–403, 2013.
[9] A. E. Foley, Rider W., Archambault, Leanna M., Warren, "Building sustainability literacy among preservice teachers: An initial evaluation of a sustainability courses designed for K-8 educator.," in *Educating science teachers for sustainability*, S. K. Stratton, R. Hagevik, A. Feldman, and M. Bloom, Eds. New York: Springer, 2015, pp. 49–67.
[10] N. Zografakis, A. N. Menegaki, and K. P. Tsagarakis, "Effective education for energy efficiency," *Energy Policy*, vol. 36, pp. 3226–3232, 2008.
[11] National Science Teachers Association, "Standards for Science Teacher Preparation." 2003.
[12] J. Dewaters and S. Powers, "Establishing Measurement Criteria for an Energy Literacy Questionnaire," *J. Environ. Educ.*, vol. 44, no. 1, pp. 38–55, 2013.
[13] E. O. James, M. Robinson, and R. R. Powell, "Beyond STS : An Energy Education Curriculum Context for the 21st Century," *J. Sci. Teacher Educ.*, vol. 5, no. 1, pp. 6–14, 1994.
[14] B. S. Bloom, *Taxonomy of Educational Objectives, Handbook I: The Cognitive Domain*. New York: David McKay, 1956.
[15] B. B. Krathwohl,D. R., Bloom, B. S., Masia, *Taxonomy of Educational Objectives Book 2 Affective Domain*. New York: Longman, 1964.
[16] T. M. Haladyna, *Writing Test Items to Evaluate Higher Order Thinking*. Boston, MA: Allyn & Bacon, 1997.
[17] L. S. Hannah and J. U. Michaelis, *A Comprehensive Framework for Instructional Objectives: A Guide to Systematic Planning and Evaluation*. Reading, MA.: Addison Wesley, 1977.
[18] J. Biggs and K. Collis, *Evaluating the Quality of Learning: The SOLO taxonomy*. New York: Academic Press, 1982.
[19] L. W. Anderson, D. R. Krathwohl, P. W. Airasian, K. A. Cruikshank, R. E. Mayer, P. R. Pintrich, J. Raths, and M. C. Wittrock, *A Taxonomy for Learning, Teaching and Assessing. A Revision of Bloom's Taxonomy of Educational Objectives*. New York: Addison Wesley, 2001.
[20] R. J. Marzano and J. S. Kendall, *The New Taxonomy of Educational Objectives*, 2nd ed. Thousand Oaks, California: Corwin Press, 2007.
[21] D. R. Baluragi, "Teaching Methodology in Science Education," in *Energy Resources in Science Education*, D. F. Kirwan, Ed. Oxford, England: Pergamon Press Ltd, 1987, pp. 21–22.
[22] T. C. Kandpal and H. P. Garg, "Energy education," vol. 64, pp. 71–78, 1999.
[23] U.S. Department of Energy, "Energy literacy: Essential principles and fundamental concepts for energy education." The US Department of Energy., Washington, DC, 2012.

[24] J. R. Anderson, *The adaptive character of thought*. Hillsdale, New Jersey: Lawrence Erlbaum, 1990.
[25] J. Frick, F. G. Kaiser, and M. Wilson, "Environmental knowledge and conservation behavior : exploring prevalence and structure in a representative sample," *Pers. Individ. Differ.*, vol. 37, pp. 1597–1613, 2004.
[26] L. S. Shulman, "Those who understand: Knowledge growth in teaching," *Educ. Res.*, vol. 15, no. 2, pp. 4–14, 1986.
[27] K. S. Hollweg, J. R. Taylor, R. W. Bybee, T. J. Marcinkowski, W. C. McBeth, and P. Zoido, "Developing a Framework for Assessing Environmental Literacy." North American Association for Environmental Education, Washington, DC, 2011.

Ni and Zn Substituted M-type Barium Hexaferrite Processed by Sol–Gel Auto Combustion Method

Widyastuti[1,2], Nia Sasria[1], A Marsha Alviani[1], M Dwi Febri R[1] and Vania Mitha P[1]

[1] Department of Materials and Metallurgical Engineering, Faculty of Industrial Technology, Sepuluh Nopember Institute of Technology (ITS), Surabaya, Indonesia

Email wiwid@mat-eng.its.ac.id

Abstract. Barium hexaferrite ($BaFe_{12}O_{19}$) or BaM was synthesized by sol-gel auto combustion method with the new dopan elements, Ni and Zn, with mole fraction (0.2, 0.4, 0.6), pH (7, 9, 11) and sintering temperature (850, 950, 1050 °C) for 3 hours. Ni - Zn ions were substituted to Fe^{3+} ion of BaM in order to modify the magnetic properties such as magnetic saturation (Ms) and coercivity (Hc). Magnetic properties and microstructure were characterized by various instruments; XRD, SEM, FTIR, and VSM. It was obtained that the largest crystal size of BaM (588.049 nm) was found at X=0.6, pH 11 and sintering temperature of 850 °C. It was found that from the VSM graphs that magnetic saturation (Ms) has reached a maximum of 95 emu/g at X=0.6 pH 11, sintering temperature at 1050 °C and coercivity (Hc) at 0.13 T. Considering these properties, BaM was qualified as a candidate for stealth technology materials.

1. Introduction

Stealth technology has been widely used in many national defense systems. The aim of this technology is to minimize the response of the electromagnetic wave fired by radar so the object undetectable. The ability to absorb these waves is influenced by a material called Radar Absorbent Material (RAM). This material is expected to have ability to absorb high energy waves and sensitive to the coming waves. The ability to absorb the wave is called magnetic saturation (Ms) and the sensitivity of the material when receive a wave is called magnetic coercivity (Hc). So that the material has to have a high value of magnetic saturation and low magnetic coercivity. Barium Hexaferrite (BaM) has been known as one of the material that is used as RAM. Zainuri reported that, BaM had a value of Hc = 6700 Oe while Ms = 72 emu / g and a curie temperature of 450 °C [1]. These properties indicated the high value of Ms and Hc. Producing BaM materials with high Ms and low Hc value, the Fe ions inside BaM have to be substituted with another ferromagnetic metal cations which have a similar atomic radius and electron configuration to Fe such as Ti, Zn, Mn, Ni, Co, etc [2]. M Jazireh Pour has dopped BaM with Mg-Ti within various dopan amount 0,1,2,3,4,5 using sol-gel method and it resulted the unsufficient magnetic saturation at 30 emu/gr [3]. The similar research using La-Ni as doppan was carried out by Talwinder and it was obtained the better magnetic saturation at 60.73 emu/gr and coercivity at 2,617 T [4].

In this research, new combination of dopant elements, Ni and Zn were added into BaM by sol - gel auto combustion method. This method had several advantages such as less processing time and easier to obtain the reagent. The variations that were used in this research were the Ni-Zn mole fraction, pH value and sintering temperature.

2. Experimental analysis

2.1. Materials

All chemicals used to prepare the Barium hexaferrite ($BaFe_{12}O_{19}$) were reagent grade and used without further purifications. Barium nitrate ($Ba(NO_3)_2$), ferric nitrate ($Fe(NO_3)_3.9H_2O$), zinc nitrate ($Zn(NO_3)_2.4H_2O$), nickel nitrate ($Ni(NO_3)_2.6H_2O$), citric acid ($C_6H_8O_7.H_2O$) and ammonium hydroxide (NH_4OH) were supplied by Merck. Demineralized water was used during all synthesis and treatment process.

2.2. Preparation of Ni-Zn Dopped Barium Hexaferrite (BaM)

Barium Hexaferrite (BaM) was produced by mixing $Ba(NO_3)_2$ and ($Fe(NO_3)_3.9H_2O$ solutions using a magnetic stirrer for 15 minutes. Ni $(NO_3)_2.6H_2O$ and Zn $(NO_3)_2.4H_2O$ (Ni-Zn X 0.2, 0.4, 0.6) were added then as doping materials. Citric acid was added as a reducing agent and then stirred for 15 minutes. After that, ammonia was added to obtain pH at 7, 9, and 11 and then stirred for 1 hour at room temperature. The next process was stirring at 80 °C for 3 hours to produce wet gels of BaM. Then the samples were heated in furnace at 250 °C for 3 hours to produce a dry gel of BaM (Auto combustion).Then, dry gel were crushed into a fine powder and sintered at temperature of 850, 950, 1050 °C for 3 hours.

2.3. Characterization

Phase characteristic, composition, and crystal structure were determined using X-Ray Diffraction (XRD) method. Scanning Electron Microscope (SEM) with maximum magnification up to 5000 times was carried out to analyze the morphological structure. Chemical bond of the BaM was examined using Fourier Transform Infrared (FTIR) method, while magnetic properties (magnetic saturation, coercivity and remanence) of BaM were analyzed using Vibrating Sample Magnetometer (VSM).

3. Result and discussion

3.1. XRD (X-Ray Diffraction) analysis

Figure 1a and Table 1 show that each sample has a crystal size range from 70.29 nm to 588.049 nm. The largest crystal size of BaM (588.049) nm is found at pH 11, X=0.6 and sintering temperature of 850 °C. The crystal size has increased significantly compared with the initial crystal size of BaM at 33.58 nm [5]. The temperature increase reduces the crystal size and the increase in pH affects $BaFe_{12}O_{19}$ crystal with Ni-Zn doping fraction of 0.6 to become much larger as showed in Table 1.

From figure 1b, it can be seen that the pH increases the intensity of peak and the formation of

Figure 1b showed that the enhancement in pH increased the intensity of the peak or increased $BaFe_{12}O_{19}$ single phase. In Figure 1c, based on PDF card 01-075-9113, at a temperature of 850 °C Barium Monoferrite ($BaFe_2O_4$) phase is formed and at 950 and 1050 °C, BaM ($BaFe_{12}O_{19}$) single phase is formed. However, the impurities, Fe_2O_3 phase, was found when heating temperature reached at 800°C and 850 °C. This condition was similar with the result of research conducted by Sozeri in the BaM material[6]. At sintering temperature of 950 and 1050 °C, the single phase of BaM was more formed and the the fewer impurity phases found.

3.2. SEM - EDX (Scanning Electron Microscopy) analysis

Figure 2 illustrates that all specimens have pores on the surface and the number increase along with the Ni-Zn mole fraction. With the addition of Ni-Zn mole fraction, the particles become more irregular and more pores on the surface. This particle characteristics was also obtained in BaM investigation carried out by L. Junliang which produced the poligonal shape [6]. The pores are formed due to the elongated crystal structure that formed needle-shaped crystals or tube. The irregular particle shape is caused by the stress in Barium Hexaferrite crystal structure due to the difference in atomic radius between the doping elements Ni-Zn and Fe [7].

The influence of pH to particle sice can be seen in figure 2a. From this figure, the increase in pH affects the particle size to become larger. Figure 2d-2f shows that with sintering temperature of 850 °C, $BaFe_{10,8}Ni_{0,6}Zn_{0,6}O_{19}$ has more pores compared with the sintering temperature at 1050 °C. Therefore, the increase in sintering temperature stimulates the crystal size growth of $BaFe_{10,8}Ni_{0,6}Zn_{0,6}O_{19}$. From the EDX result in Table 2, it can be observed that the substitution of Ni-Zn elements has been successfully carried out and the increment of wt% of Ni and Zn due to the increase of mole fraction and sintering temperature.

3.3. FTIR (Fourier Transform Infrared) analysis

Figure 4 and Table 3 show that all the samples have M – O (metal-oxygen) bond and the C = O bond. This proves that there is a bond between the Fe, Ba, Ni, and Zn metals with oxygen in the specimen. C = O bond shows the chemical bond between oxygen for combustion with carbon during the formation of Barium Hexaferrite phase. C – H bond was also found in Figure 4 indicates the presence of citric acid reaction. The wavenumber range 1080-1300 cm^{-1} shows the C – O bond and this is formed when the combustion reaction in BaM happened. The rest of product and water are shown by H-O bond at wavenumber of 2161.45 and 2360.68.

3.4. VSM (Vibrating Sample Magnetometer) analysis

From figure 5, it can be known that the enhancement of the mole fraction decreases the Ms, Hc, and Mr value . Regarding the doping mechanism, Ni has a tendency to subtitute the Fe^{3+} ions at $4f_2$ position (spin down) if the mole fraction is at a value of 0.1. When the mole fraction was larger than 0.1, it will replace the Fe^{3+} ions in the 12k position (spin up) which will reduce the value of Ms and vice versa. Thus, the value of Ms has declined. [8]. The reduction of coercivity happened because the magnetic anisotropic value become lower. This properties significantly affect the magnetic performance and direction of magnetic domain [9].

Figure 5 also shows that the value of Hc decreases when pH increased, but at X=0.6 and pH 9 the value of Hc was higher than at pH 11. Although, the value of Ms does not indicate the similar trend with pH enhancement. The enhancement of pH value will increase with the value of Ms [4]. This is because the formation rate of single phase $BaFe_{12}O_{19}$ becomes higher. It also affects the crystal size and magnetic domain to be much bigger and it leads to a higher amount of spins with uniform direction that increases the capacity of the magnetization [10]. This condition is obtained when the addition of dopping element has reached X=0.6. However, the impurities in $BaFe_2O_4$ are still found at X=0.2 and X=0.4 and these will reduce the magnetic characteristics.

The ffect of sintering temperature on the magnetic properties is generally related with the effect of sintering temperature as reported in research by Sozeri [6]. The difference is at the value of coercivity that obtained by Sozeri [6] increased up to 1100 °C and then slightly declined at temperature of 1200 °C. These properties value increase is supported by the investigation in BaM materials conducted by Mosleh that used sintering temperature from 750 to 950 °C [11]. In this research, it has obtained Ms and Hc optimum at 95 emu/g and 0,13 Tesla, respectively, at X=0.6, pH 11 and sintering temperature of 1050 °C. Based on high Ms and low Hc value of BaM, this material can be used in stealth technology.

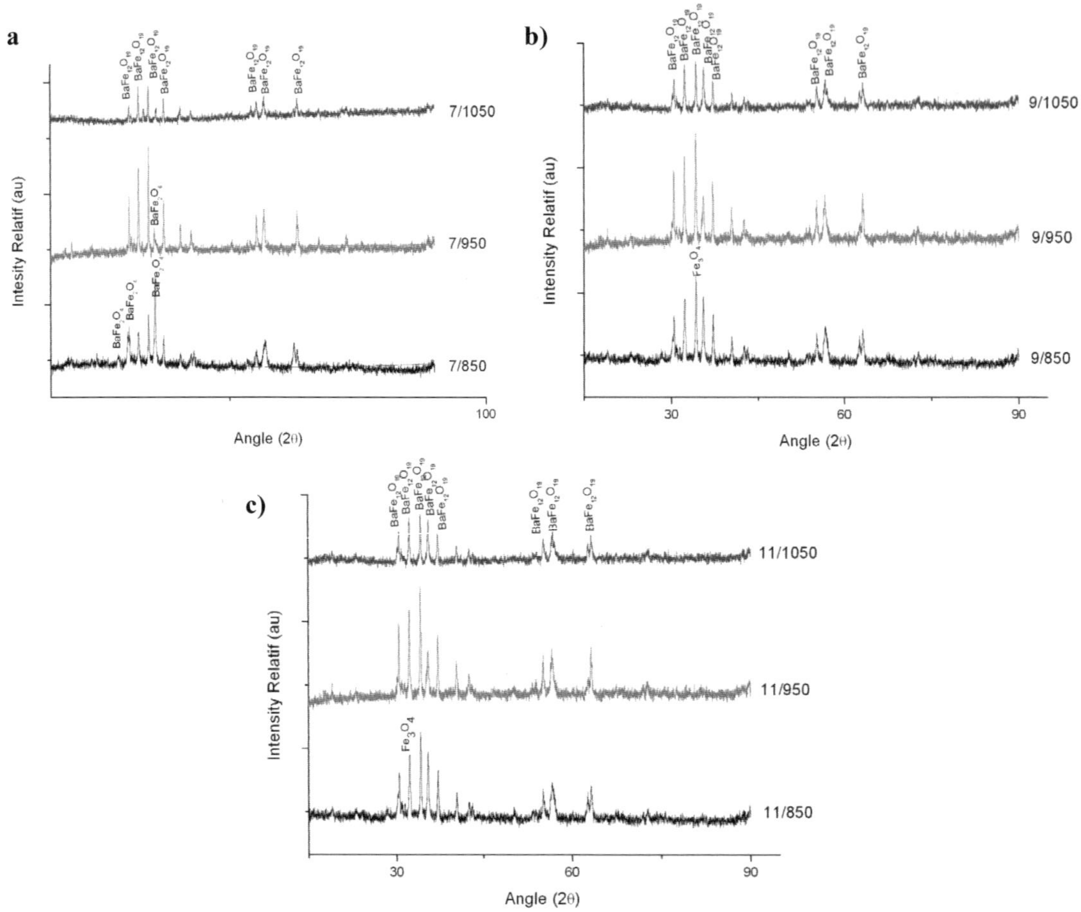

Figure 1. XRD at X=0.6 T=850, 950, 1050 °C (a) pH 7, (b) pH 9, (c) pH 11

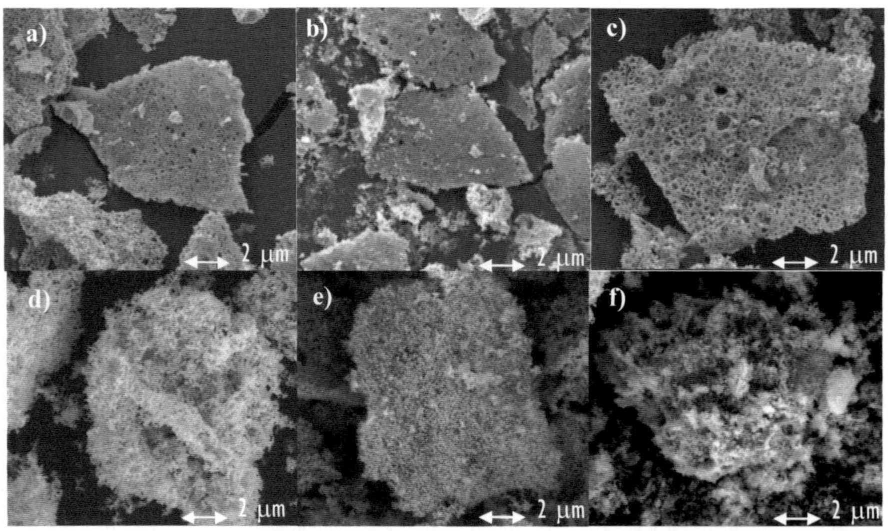

Figure 2. SEM at 5000x Magnification; T=950 °C pH 7 (a) X=0.2, (b) X=0.4, (c) X=0.6 and X=0.6 pH 11 (d) T=850 °C, (e) T=950 °C, (f) T=1050 °C

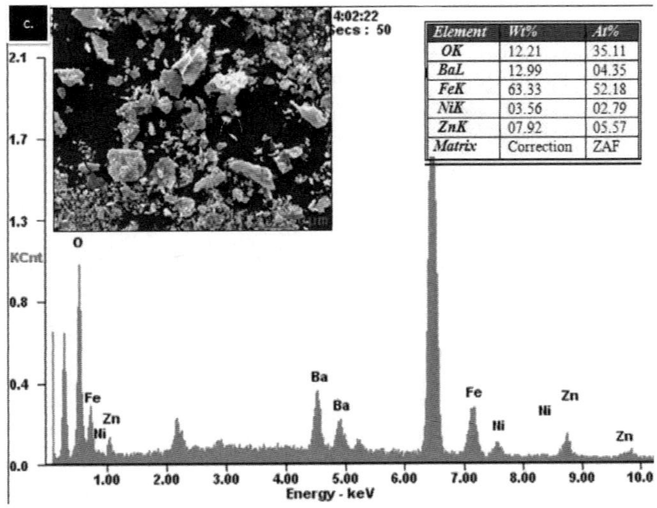

Figure 3. EDX at Specimen pH 7 X=0.6

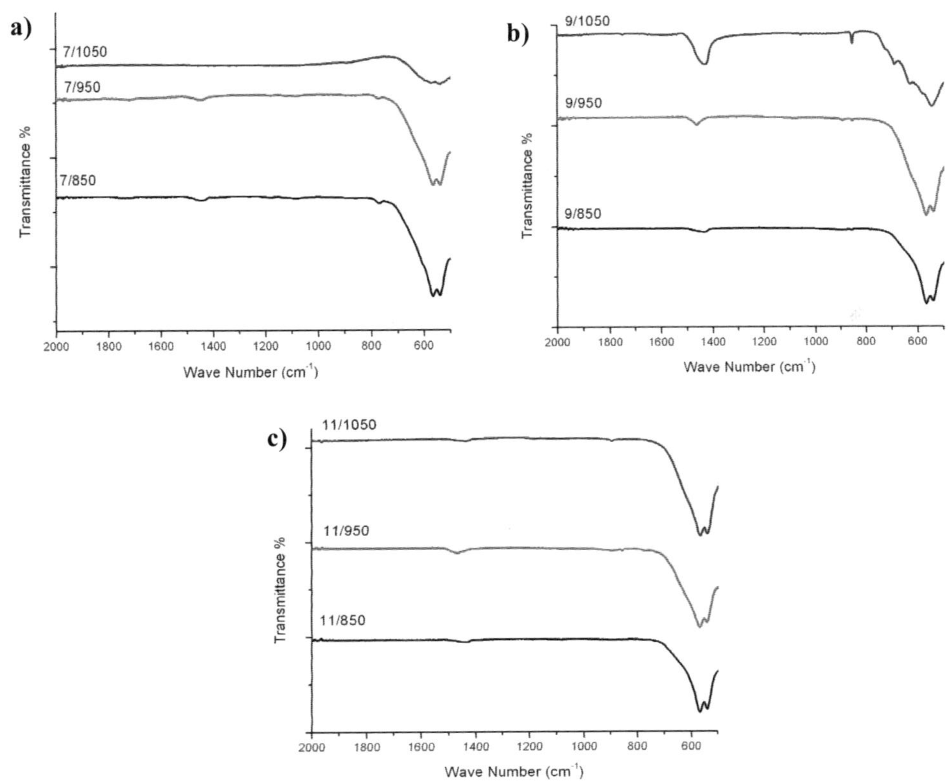

Figure 4. FTIR at X=0.6 T=850, 950, 1050 °C (a) pH 7, (b) pH 9, (c) pH 11

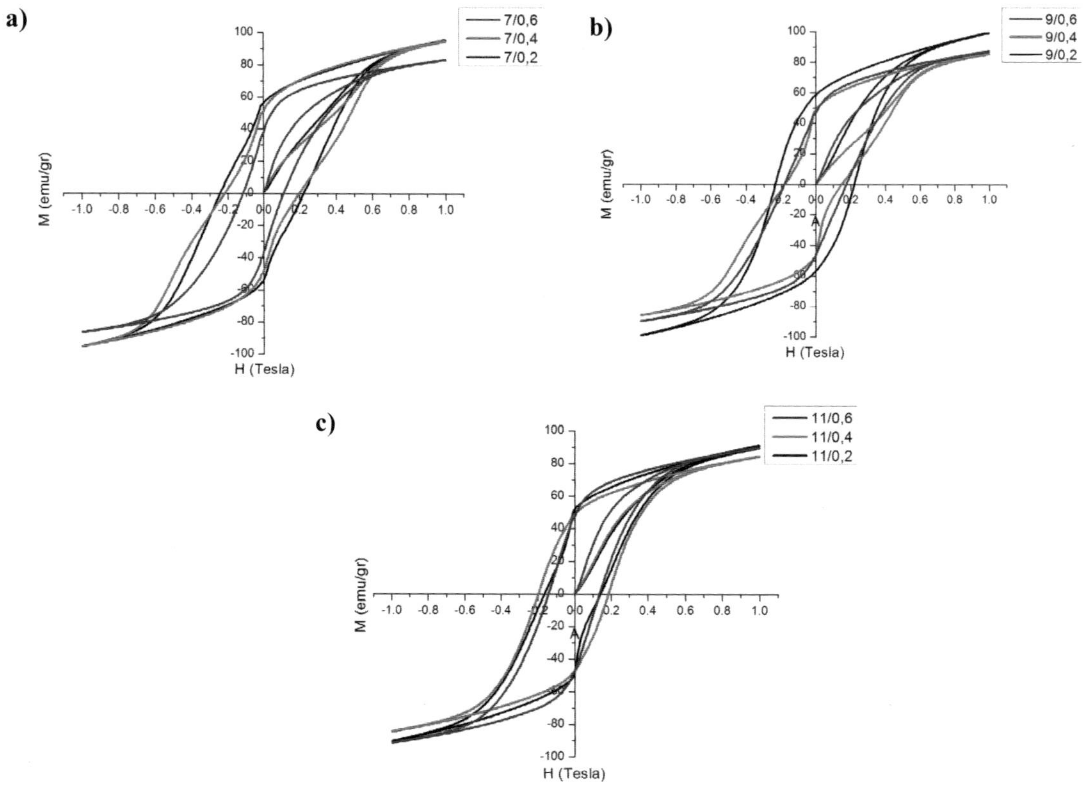

Figure 5. VSM at T=950 °C X=0.2, 0.4, 0.6 (a) pH 7, (b) pH 9, (c) pH 11

Table 1. Crystal Size of Barium Hexaferrite

Mol Fraction	pH	Sintering Temperature (°C)	Crystal Size (nm)
0.6	7	850	199,65
		950	102,05
		1050	189,19
	9	850	199,65
		950	199,65
		1050	70,29
	11	850	588,05
		950	199,65
		1050	83,13

Table 2. Wt% of Ni and Zn Elements

pH	Mol Fraction	Ni (Wt%)	Zn (Wt%)	pH	Sintering Temperature (°C)	Ni (Wt%)	Zn (Wt%)
	0,2	01,01	01,99		850	2,01	4,08
7	0,4	02,49	04,90	7	950	3,90	6,93
	0,6	03,56	07,92		1050	1,80	2,17
	0,2	02,08	01,82		850	2,08	4,41
9	0,4	06,28	05,65	9	950	2,66	2,31
	0,6	02,70	02,35		1050	2,66	2,31
	0,2	02,77	03,84		850	1,11	2,00
11	0,4	03,32	05,94	11	950	6,00	7,38
	0,6	06,00	07,38		1050	6,11	5,43

Table 3. Peak of Barium Hexaferrite

pH	Mol Fraction	Peak List			pH	Sintering Temperature (°C)	Peak List		
		M – O Stretching	M – O	C = O Stretching			C – H Stretching	M – O	C = O Stretching
	0.2	566.98	538.01	1457.05		850	773.61	566.04	1438.84
7	0.4	565.81	538.95	1457.95	7	950	767.90	539.04	1445.70
	0.6	567.07	538.60	1448.02		1050	700.05	570.19	1098.88
	0.2	568.86	540.23	1467.02		850	859.10	567.61	1438.52
9	0.4	569.01	538.44	1406.86	9	950	852.77	568.13	1457.52
	0.6	568.13	540.03	1457.52		1050	857.04	545.50	1423.42
	0.2	567.08	538.60	1448.02		850	865.44	567.61	1435.36
11	0.4	567.53	539.95	1463.85	11	950	853.96	569.81	1469.59
	0.6	569.81	541.71	1469.59		1050	906.60	567.32	1096.57

Table 4. Hc, Ms, and Mr of Barium Hexaferrite

pH	Mol Fraction	Magnetic Properties			pH	Sintering Temperature (°C)	Magnetic Properties		
		Hc (Tesla)	Ms (emu/gr)	Mr (emu/gr)			Hc (Tesla)	Ms (emu/gr)	Mr (emu/gr)
	0,2	0.24	95.48	55.90		850	0.15	94	51
7	0,4	0.22	95.45	51.29	7	950	0.12	85	40
	0,6	0.12	84.89	39.62		1050	0.11	74.5	38
	0,2	0.23	99,27	58.22		850	0.12	80.7	39.7
9	0,4	0.18	86.42	49.25	9	950	0.19	88.4	48.2
	0,6	0,19	88.45	48.20		1050	0.03	90.9	35.6
	0,2	0.16	91.18	52.19		850	0.18	80.3	43.4
11	0,4	0.20	84.54	48.84	11	950	0.15	90.7	48.6
	0,6	0.15	90.69	48.59		1050	0.13	95	49

Conclusions

1. The addition of doping ions Ni - Zn leads to decrease in the value of coercivity (Hc) and magnetic saturation (Ms). The lowest coercivity (0.12 T) is found at pH 7 and X=0.6.
2. The enhancement in pH increases the magnetic saturation (Ms) and the crystal size of BaM. Otherwise it leads to an impairment of the coercivity (Hc). The largest crystal size (588.049 nm) is found at X=0.6, pH 11 and sintering temperature at 850 °C.
3. The addition of sintering temperature produces the impairment in the crystal size of $BaFe_{12}O_{19}$, otherwise it increases the coercivity (Hc) and magnetic saturation (Ms) of Barium Hexaferrite that has been doped by Ni and Zn. The smallest crystal size is 70.29 nm observed at X=0.6, pH 9 and sintering temperature of 1050 °C. The highest value of magnetic saturation (Ms) is 95 emu/g with coercivity (Hc) of 0.13 T observed at X=0.6, pH 11 and sintering temperature of 1050 °C.

Acknowledgments

The author thank to ITS (Sepuluh Nopember Institute of Technology) and PSTBM (Pusat Sains dan Teknologi Bahan Maju) – BATAN Tangerang for support of this research.

References

[1] Zainuri M 2010 *Ristek* (Surabaya: Physics ITS)
[2] Mali. A. Ataie 2005 *Scripte Materials* **53** 1065 – 1070
[3] Jazirehpour, M.H Shams and O. Khani 2012 *Journal of Alloys and Compound*.**545** 32-40
[4] Talwinder 2013 *International Journal of Research in Mechanical Engineering and Technology* **3** 171 – 173
[5] Chauhan P 2010 *Dissertation* (Patiala: Materials and Metallurgical Engineering Thapar University) pp 1-72
[6] Sozeri H, Durmus Z, Baykal A and Uysal E 2012 *Mat. Sci. and Eng. B*. **177** 949-55
[7] Junliang Liu, et.all 2010 *European Ceramics Society*. **30** 993 – 997
[8] Kanagesan S, Hashim M, Jesurani S, Kalaivani T and Ismail I 2013 *J. Supercond. Nov. Magn.*
[9] Nobuyoshi Koga 2007 *Journal of Magnetism and Magnetic Materials* **323** 36-41
[10] Kaur T and Srivastava AK 2013 *Int. J. of Res. in Mech. Eng & Technol*. **3** 171-3
[11] Mosleh Z, Kameli P, Ranjbar M and Salamati H 2014 *Ceram. Int*. **40** 7279-84

Modification of Blue LED using Organic-Inorganic Hybrid Polymer Doped with Nile Red for Artificial Lighting of Photosynthesis

Norman Syakir[1], Fahmi Syarifudin[1], Sahrul Hidayat[1], Fitrilawati[1]*

[1]Department of Physics, Faculty of Mathematics and Natural Sciences, Universitas Padjadjaran, Jalan Raya Bandung-Sumedang KM 21. Jatinangor, Sumedang 45363, West Java, INDONESIA

*fitrilawati@phys.unpad.ac.id

Abstract. The photosynthesis process of chlorophyll absorbs only the light with wavelength in the blue and red ranges. The absorption peak of the chlorophyll-A is at 428 nm and 660 nm, while absorption peak of chlorophyll-B is at 453 nm and 643 nm. We report the modification of blue LED using hybrid polymer doped with Nile Red. In order to match the total absorption spectra of chlorophyll-A and chlorophyll-B, the emission spectrum of the modified blue LED was taken out by using the wavelength conversion material. We modified the blue LED by covering the blue LED of 450 nm as excitation source with precursor of red wavelength conversion material. The red wavelength conversion material was prepared by doped precursor of TMSPMA hybrid polymer with organic phosphor of Nile Red. The precursor of hybrid polymer was synthesized using sol-gel process and then it was doped with 0.1% Nile Red. In order to freeze the precursor of these conversion material, we employed UV photo-polymerization process. The modified blue LED has two emission peaks, which are at 448 nm (blue emission) and at 651 nm (red emission). The optimum spectrum profile of the modified blue LED has similar range as the total absorption spectra of chlorophyll-A and chlorophyll-B that obtain using Nile Red with the mass of 2.9 μg and the driven current of 60 mA. This result has a potential application for the artificial lighting in the photosynthesis process of horticultures at indoor plantation.

1. Introduction.

Currently it has been developed a horticultures arrangement in the room (indoor) that using modern horticultures technology. For this purpose, an artificial lighting was needed to fulfill requirement of photosynthesis process. The sufficient light is desperately needed to synthesize water and carbon dioxide into glucose. In the photosynthesis process, chlorophyll of the plants is functioned as a light absorber. The chlorophyll does not absorb the whole range of the sun light spectrum, but only absorb a particular wavelength range, that covers of red and blue wavelengths [1]. Based on chlorophyll absorption, it has been developed lights that can be used for plant growth. Many owned manufacturers have lighting products for plant growth from incandescent lamps, fluorescent lamp and a CFL, SSL lights up. Especially for SSL most products still use a combination of red LEDs and blue LEDs to meet the conditions of light emission of red and blue that will be absorbed by the plants, but the efficiency of the red LED is very small compared to the efficiency of blue LEDs [2].

Another way to get a combination of desired light emission color can be achieved by the color conversion using a wavelength conversion material. Besides being able to overcome the problem of efficiency red LEDs, the color conversion method also can streamline the distribution of color more evenly and especially more economical because it only uses a single excitation. Conversion materials most widely used in LED phosphor is an inorganic or organic. Phosphor organic hybrid polymer is integrated with one of the alternative material for converting the wavelength of the LEDs.
To meet the emission color red and blue light is required, has been created as an alternative hybrid LED manufacturing LEDs for plant growth. The hybrid LED using a combination of a blue LED with a peak emission of 450 nm and an organic chromophore Nile Red (9- (diethylamino) benzo [a] phenoxazin-5 (5H) -one) having absorption at 552 nm and emission at 636 nm **[3]**. Selection of the chromophore Nile Red LED and aims to produce a hybrid that has a pattern of LED emission curve approaching the curve pattern of absorption of chlorophyll-A and chlorophyll-B. In this paper, we reported the results of the fabrication and characterization of LED hybrid using single conversion approach for artificial lighting during photosynthesis process.

2. Experimental Methods

In this experiments, we used monomer TMSPMA (*3-(Trimethoxysilyl) propyl methacrylate*), organic chromophores NR (Nile Red) as photoluminescence material, Irgacure-819 as initiator on UV photopolymerisation process the network of organic link in hybrid polymer matrix, and a blue LED 450 nm as the excitation and source light. Other chemicals used are ethanol, DI water, hydrochloric acid and chloroform. The first step of experiment is to synthesize hybrid polymer precursors of monomer TMSPMA using sol-gel techniques. Sol-gel process is a part of inorganic phase polymerization of monomer, which is the main chain bonding (-Si-O-Si) formation. Precursors are made using a volume ratio of monomer: ethanol: DI water as 1 : 2 : 4. This ratio is based on the optimization of previous studies **[4]**. Sol-gel process results in the form of colored translucent and transparent gel. NR chromophore doping process on the hybrid polymer precursors begins by dissolving the precursor and the chromophore in chloroform. Furthermore, the chromophore solution (concentration of 0.1% by weight) were mixed with the precursor solution and stirred to form a homogenous solution. Before deposited, hybrid polymer precursor solution containing chromophore mixed with photoinitiator Irgacure-819 at a concentration of 0.1% by weight of the mass of the precursor. The second step is to make sample of modified blue LED as Hybrid LED, which is made by coating the blue LED with the mixture using a solution coating techniques. Finally, Hybrid LED pre-baked in an oven 50°C for 30 minutes to remove residual solvent following by UV photopolymerisation for 15 minutes and finishing with post-baked in an oven 50°C for 18 hours to complete the whole polymerisation process. The emission spectra of this Hybrid LED were characterized for several variations electric current drive from 10 mA to 80 mA and variation of the mass of the chromophore Nile Red performance films ranged 1.4 μg to 2.9 μg. Emission spectrum measurements has been done using Fluorescence USB4000FL Spectrometer and Spectra suite software from Ocean Optics.

3. Results and Discussions

The resulting Hybrid LED has a film on it's surface containing Nile Red in organic groups of hybrid polymer TMSPMA that have been polymerized. Thin film hybrid polymers containing Nile Red looks red as shown in **Figure 1.(a).** There emission spectrum of the blue LED and Nile Red (included its absorption spectrum) shown in **Figure 1.(b).**
In principle, this Hybrid LED is combining blue light emission of the blue LED as the excitation source then most of emission blue light is converted into red light by the chromophore Nile Red, which is indicated in **Figure 2.** Mass of thin films in Hybrid LED for each sample LED hybrid consist of 0.1 % Nile Red. Since we know the value of the thin film mass, Nile Red mass chromophore contained in each sample will be known, as shown in **Table 1**.

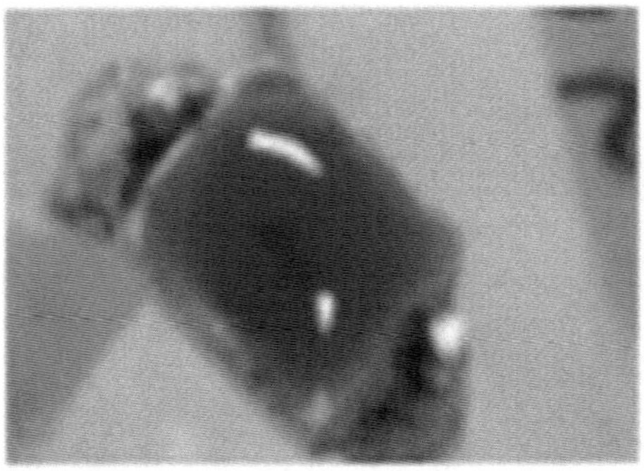

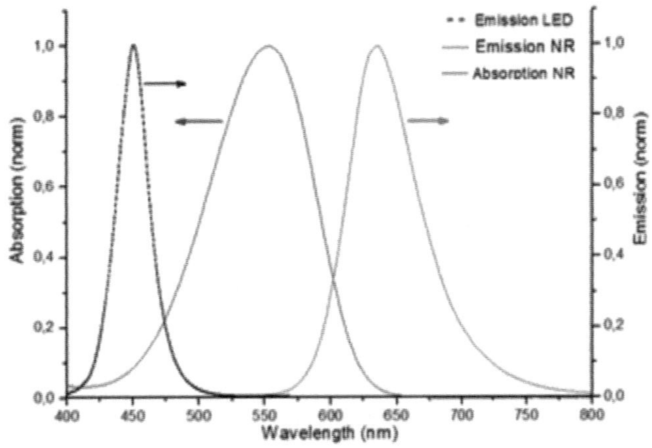

(a) (b)

Figure 1. (a) Photograph of the modified blue LED as Hybrid LED, (b) the emission spectrum of blue LED and chromophores Nile Red (blue line indicate the absorption spectrum of its Nile Red).

Table 1. The mass of the coated wavelength conversion material on the surface of Hybrid LED and Nile Red mass that present in each sample.

No.	Sample Code of Hybrid LED	Coated mass in mg	Nile Red mass in μg
1	H-LED_1	1.4	1.4
2	H-LED_2	1.9	1.9
3	H-LED_3	2.0	2.0
4	H-LED_4	2.4	2.4
5	H-LED_5	2.9	2.9

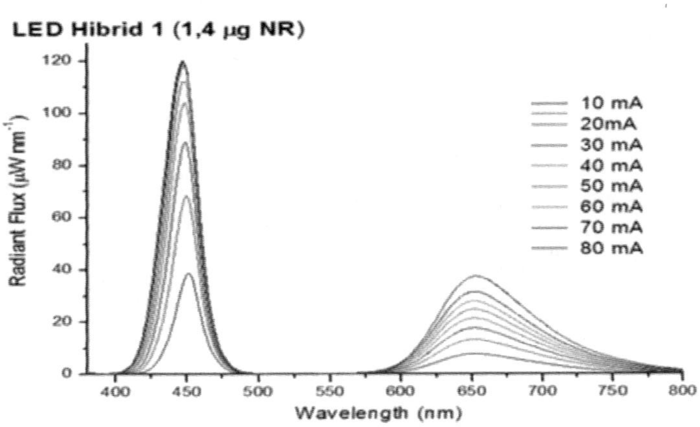

Figure 2. The emission spectra of H-LED_1 at various driven electric current from 10 mA to 80 mA.

The Hybrid LED emission curve profile has match the total profile of absorption curve of chlorophyll-A and chlorophyll-B. The incorporation of Hybrid LED emission curve normalized with the normalization absorption curve of chlorophyll-A and chlorophyll-B. Its normalization absorption curve spectra was using an absorption curve of chlorophyll-A and B in the solvent diethyl ether **[5,6]**.

All the five samples Hybrid LED has yield emission spectra curve mostly close to the total absorption curve of chlorophyll-A and B. The optimum emission spectra depend on NR mass content and driven current to the LED. Sample H-LED_3, H-LED_4, and H-LED_5 have optimum emission spectra with condition as shown in **Table 2**.

Table 2. Optimum parameter for Hybrid LED

No.	Sample Code of Hybrid LED	Nile Red mass in µg	Driven current in mA
1	H-LED_3	2.0	80
2	H-LED_4	2.4	60
3	H-LED_5	2.9	60

The effectiveness of the emission curve Hybrid LED with optimum conditions as shown above has meet the absorption curve of chlorophyll in the blue and red wavelength region. This comparation results shown in **Figure 3**. The efficiency of Hybrid LED that meet the chlorophyll absorption curves in the blue and red wavelength region can be seen in **Table 3**.

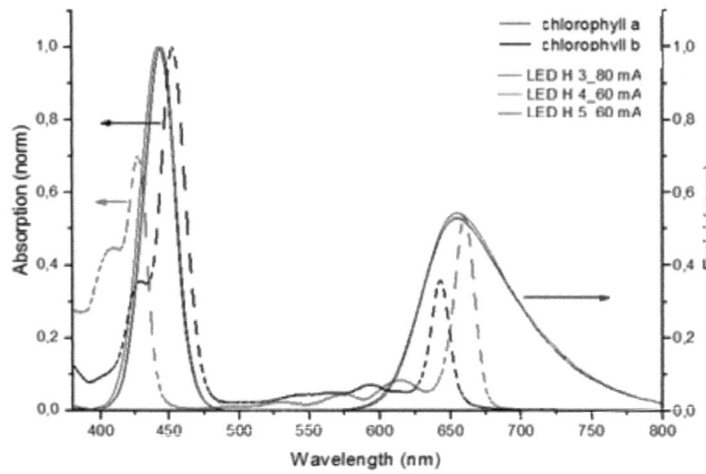

Figure 3. The Optimum Hybrid LED emission curve that able to covering superposition absorption curve of chlorophyll-A and B.

Table 3. Efficiency of Hybrid LED

No.	Sample Code of Hybrid LED	Driven current in mA	Efficiency in %
1	H-LED_3	80	49.1
2	H-LED_4	60	44.3
3	H-LED_5	60	47.0

The performance of Hybrid LEDs were observed over eight hours at optimum condition of operation. The optimum operation condition, namely for samples H-LED_3 and H-LED_5 were driven with electric current of 80 mA and 60 mA, respectively. The measurement results are shown in **Figure 4.** From this figure, it can be observed that when Hybrid LED lighting continuously for eight hours, the total emission intensity of Hybrid LED were almost constant, it means Hybrid LED has good stability. Except in the first half hours after its turn on, this is due to the burnout process of the wavelength conversion material itself.

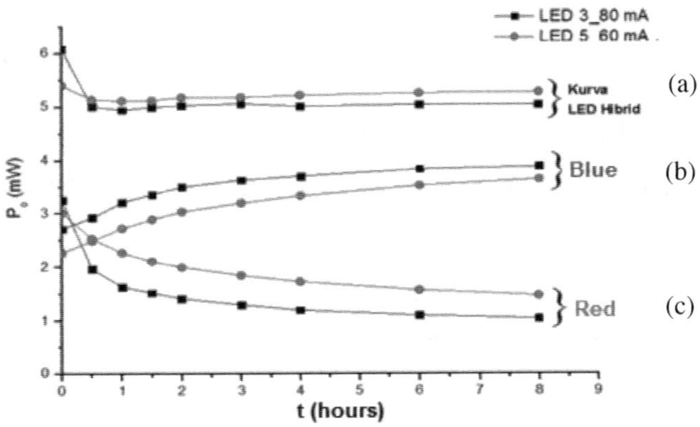

Figure 4. Hybrid LED performance, (a) total emission intensity of Hybrid LED, (b) emission intensity of blue light component, and (c) emission intensity of red light component.

The emission stability of blue and red light components, above two hours after Hybrid LED was turned on, tend to stable up to eight hours. Indicate the good stability on blue and red light emission of Hybrid LED. At the beginning the intensity changing on the blue region were slightly increases, however, on the red region seen the emission intensity slightly decreases when constantly lit. But after two hours is turned on, the emission intensity of blue and red LED hybrid tends to be stable. Also can be observed that when the supplied current to Hybrid LED has a great value, it made red light component curve was more lower as shown in H-LED_3 with driven current of 80 mA. Compare to H-LED_3, H-LED_5 with driven current of 60 mA has the red light component more higher and more stable.

4. Conclusion

Hybrid LEDs have been succesfully made as an artificial lighting source that meet the purpose for the photosynthesis process. The emission spectrum of Hybrid LED has a peak at 450 nm in the blue region and a peak at 651 nm is the red region. Characteristics of the resulting hybrid LED has the best results for the mass of Nile Red 2.9 micrograms with the driven current of 60 milliAmperes

Acknowledgement
This work was partly supported by Hibah Riset Unggulan Strategis Nasional DIKTI tahun 2012.

References

[1] Campbell, Neil. Reece, Jane. 2009. *Biology 7th Ed.* San Francisco: Pearson Benjamin Cummings.
[2] Fitrilawati, Norman Syakir, Linda Pebriani, Yolla Sukma Handayani, Sahrul Hidayat, Togar Saragi, Rustam Siregar, Pina Pitriana, Rahmat Hidayat. *Pengembangan phosphor organik sebagai bahan konversi warna emisi untuk aplikasi lampu SSL.* Jurnal Material dan Energi Indonesia Vol. 01, No. 03 (2011) 180 – 195.
[3] Sackett, D. L. and J. Wolff. 1987. *Nile red as a polarity-sensitive fluorescent probe of hydrophobic protein surfaces.* Anal. Biochem. 167, 228-234.
[4] Sinuhaji, Monica P. 2012. *Karakteristik Poli(3-(trimethoxysilyl) prophyl metahcrylate) dengan doping DCM (4-(dicyanomethylene)-2-methyl-6-(p-dimethylaminostyryl)-4H-pyran) sebagai bahan pengkonversi sepktrum LED biru.* Skripsi, Universitas Padjadjaran.
[5] Strain, H. H., M. R. Thomas and J. J. Katz. 1963. *Spectral absorption properties of ordinary and fully deuteriated chlorophylls a and b.* Biochim. Biophys. Acta 75, 306-311.
[6] Vernon, L. P. and G. R. Seely. 1966. *The chlorophylls.* Academic Press, New York.

IOP Conf. Series: Journal of Physics: Conf. Series **877** (2017) 012017 doi:10.1088/1742-6596/877/1/012017

Synthesized $Li_4Ti_5O_{12}$ from Technical Grade Raw Material by Excess $LiOH.H_2O$ as Anode Lithium Ion Battery

S Priyono[1], R D Primasari[2], S A Saptari[2], and B Prihandoko[1]
[1]Research Center for Physics – LIPI
Kawasan PUSPIPTEK Gedung 442 Serpong, Tangerang Selatan 15314, Banten, Indonesia
[2]Faculty of Science and Technology, State Islamic University
Jl. Ir. H. Juanda No.95 Ciputat 15412 Jakarta, Indonesia

Corresponding author: slam013@lipi.go.id

Abstract. $Li_4Ti_5O_{12}$ powder as anode lithium ion battery was synthesized via solid state reaction with excess $LiOH.H_2O$. Technical grades raw materials like $LiOH.H_2O$ and TiO_2 were used as starting materials. $LiOH.H_2O$ excess was varied from 0; 2.5; 5 and 7.5% to get higher optimum phases and capacity of $Li_4Ti_5O_{12}$. All raw materials were mixed stoichiometry then followed by calcination and sintering process to get final products. The obtained products were characterized by XRD, SEM, and PSA to get properties of active materials and the electrochemical properties were done by cyclic voltametry and charge-discharge test. The XRD test showed that 5% excess have highest $Li_4Ti_5O_{12}$ phases. All samples have same in morphology, agglomerate and same in particle size distribution. Sample with 5% excess showed good reversible process and charge-discharge test showed that increasing $Li_4Ti_5O_{12}$ phase can improve specific capacity.

1. Introduction

Lithium-ion batteries have been widely used as the major power sources for portable electronic devices are regarded as a promising candidate of power sources for electric vehicles and large scale electricity storage stations [1].Lithium ion battery is also one important device to store energy, it converts electrical energy to chemical energy when charge and vice versa when it uses. The lithium ion battery research has been done until now which focusing to increase capacity and capability. One application of lithium battery is for electric vehicle, electric energy can reduce CO_2 emission [2-3]. Lithium battery more confortable used in electric vehicle than others electrochemical devices. Today, there are two lithium battery technology i.e primary battery (not-rechargeable) and secondary battery (rechargeable). Battery that used for electric vehicle as energy storage system belong to secondary battery.

Battery consists of four components like cathode, anode, electrolyte and separator. Each component was made by different materials. Cathode materials that common used are $LiFePO_4$ and $LiMnPO_4$[4] while Anode material is graphite[5]. $LiPF_6$ [6] or LIBOB [7] are used as electrolyte as ion transfer media from cathode to anode and vice versa. Separator used to separate cathode and anode was made from polyethylene.

In this paper, we focus the research on anode materials. The conventional anode for lithium ion battery is graphite and others carbon based materials. But all these materials weren't suitable for electric vehicle battery because of safety reason. Graphite and carbon based materials have very low working potential, it cause electrolyte decomposition when charge-discharge process especially in high rate.

Electrolyte will decompose become sharp dendrite lithium on anode surface and will wreck the separator [8][9].

Recently, Lithium titanate had chemical formula $Li_4Ti_5O_{12}$ or sometime called LTO becomes popular and more suitable candidate to replace carbon as anode material for lithium battery because it has some advantages. LTO has high insertion potential at around 1.55 V (vs. Li/Li+), which can avoid the dendrite lithium growth on the anode surface at high charging rate. It can be almost fully charged within 10 min, exceptional cycle, 'zero-train' effect and the flat Li insertion voltage [10]. Besides, LTO has excellent Li-ion insertion and extraction reversibility and structure stability with almost negligible volume change during charge-discharge process [11]. So, LTO is indeed a potential candidate material for high-rate LIBs.

On previous work, it had synthesized $Li_4Ti_5O_{12}$ with raw materials such as $LiOH.H_2O$ or Li_2CO_3 and TiO_2 via solid state method. The XRD results show that $Li_4Ti_5O_{12}$ phase only about 85-96% and TiO_2 phase about 4-15%. These result aren't optimum. Herein, in this work $Li_4Ti_5O_{12}$ will be synthesized by vary addition $LiOH.H_2O$ to get optimum phases [12][13].

2. Experiment

The $Li_4Ti_5O_{12}$ powders were synthesized by solid state reaction. $LiOH.H_2O$ and TiO_2 were as Li and Ti sources. The experiment was done by varying $LiOH.H_2O$ added 0; 2.5; 5; and 7.5 % weight. Stoichiometric raw materials were mixed by ball milling for 15 h. then all samples were calcined at 750°C for 2 h in an air atmosphere to remove organic compound. Finally, the products were sintered at 850°C for 4 h to obtain $Li_4Ti_5O_{12}$ powders. The crystal structure of the prepared samples was identified by XRD using Cu-K-alpha radiation ($10^o \leq 2\theta \leq 80^o$). SEM and PSA were used to investigate the morphologies and particle size of the as-prepared powders, respectively.

The electrochemical characteristics were evaluated by half-cell coin cell. The sample slurry was prepared by mixing active materials powders with conductive carbon (acetylene black) and binder (PVDF) at a weight ratio of 85:5:10 in DMAC solution. Subsequently, the slurry was coated on a copper foil using the doctor blade technique and dried at 80°C for 1 h to evaporate the DMAC solvent. The electrode foil was punched to 12 mm diameter disc, which were used to assemble the coin cell in an Ar glove box. Li metal was used as the counter and reference electrode in the cell. Polyethyelene was the separator and 1M LiPF6 in EC:DEC solution with composition of 1:1 v/v was used as electrolyte. PE/PP-based film Celgard 2000 was used as a separator between the electrodes.

The charge-discharge experiments were conducted by using battery testing instrument (Wonatech WBCS3000, Automatic Battery Cycler). The cut-off voltage was set to be 1.0 V and 2.6 V for discharge and charge, respectively.

3. Result and Discussion

Fig 1 showed that Each XRD pattern consists of two dominant phases like $Li_4Ti_5O_{12}$ and rutile TiO_2 are respectively marked triangles and circles, respectively. Fig 1 also showed the formation of the phase $Li_4Ti_5O_{12}$ that the addition of $LiOH.H_2O$ of 0%, 2.5%, 5% and 7.5%. Another phase formed is an impurity in the form of rutile TiO_2. Analysis by using highscore software obtained that composition Lithium Titanate and rutile TiO_2 as shown in Table 1. Results showed that the data processing phase $Li_4Ti_5O_{12}$ $LiOH.H_2O$ formed on the addition of 0% is 94.50%, in increments of 2.5% was 94.70%, in increments of 5% is 95.90% and the addition of 7,5% is 95.50%. Diffraction peaks (2θ) $Li_4Ti_5O_{12}$ occur at 18°, 37°, 43°, 47°, 58°, 63°, 66°, 75°, 76° and 79°. Although the main phase have been obtained, but the rutile TiO_2 peaks at diffraction angles 27° caused final powder is not a single phase but it is a mixture $Li_4Ti_5O_{12}$ and rutile TiO_2.

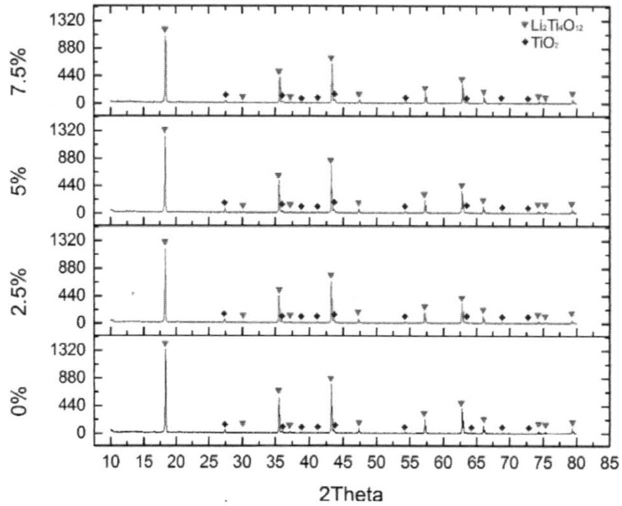

Figure 1. XRD pattern of $Li_4Ti_5O_{12}$ synthesized by varying $LiOH.H_2O$ excess.

Table 1. Phases composition of $Li_4Ti_5O_{12}$ with $LiOH.H_2O$ Excess from HIGHSCORE Software analysis.

Sampel	$Li_4Ti_5O_{12}$ (%)	TiO_2 (%)
0%	94,50	5,50
2,5%	94,70	5,30
5%	95,90	4,10
7,5%	95,50	4,50

$LiOH.H_2O$ additions in the manufacture of active ingredients $Li_4Ti_5O_{12}$ managed to cause a phase rutile TiO_2 formed tend to decline until the addition of $LiOH.H_2O$ as much as 5% and increased again in addition $LiOH.H_2O$ as much as 7.5%. This shows that the addition of $LiOH.H_2O$ in the synthesis cause initial mixture more evenly and more homogenous. Each item starting material of TiO_2 mixed with $LiOH.H_2O$ so $Li_4Ti_5O_{12}$ phase is expected to increase.

Table. 2. The particle size of the powder $Li_4Ti_5O_{12}$ synthesized with $LiOH.H_2O$ excess

$LiOH.H_2O$ excess	**Particle Size (μm)**
0 %	26,46
2,5 %	32,93
5,0 %	25,29
7,5 %	33,13

Figure 2 showed that SEM analysis results also showed a smooth texture on each sample. $Li_4Ti_5O_{12}$ particles shaped like small pebbles (small particles) that are agglomerate with size uniformly distributed (homogeneous). Whereas small particles have a variety of shapes such as round, oval, square and irregular. Addition of $LiOH.H_2O$ does not affect the shape, texture, and particle size of $Li_4Ti_5O_{12}$.

IOP Conf. Series: Journal of Physics: Conf. Series **877** (2017) 012017 doi:10.1088/1742-6596/877/1/012017

Figure 2. SEM image of $Li_4Ti_5O_{12}$ powder with a magnification of 10k SE with $LiOH.H_2O$ excess (a) 0% (b) 2.5% (c) 5% (d) 7.5%.

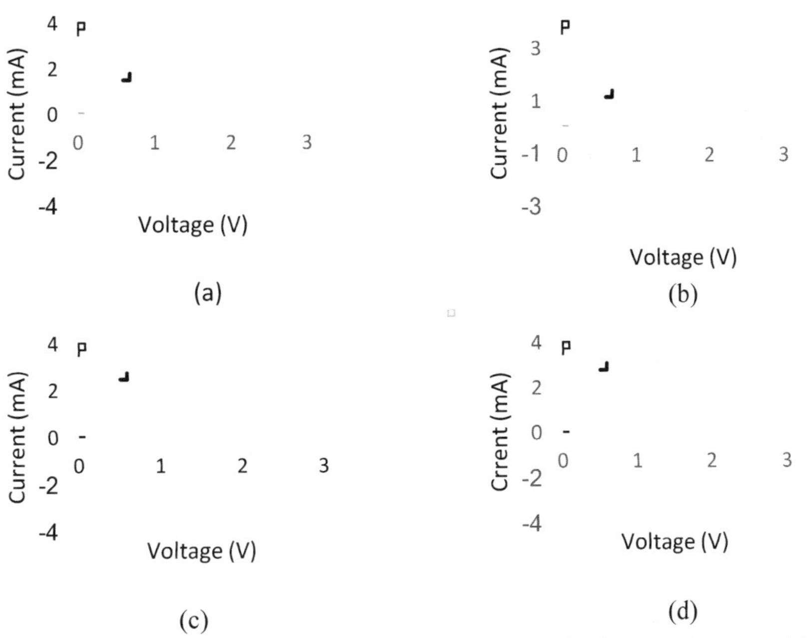

Figure 3. Cyclic Voltametry of $Li_4Ti_5O_{12}$ powder with $LiOH.H_2O$ excess (a) 0% (b) 2.5% (c) 5% and (d) 7.5%.

Table 2 showed larger particle size as measured by CILAS Particle Size Analyzer Brand (1190). The more the addition $LiOH.H_2O$ tend to increase the particle size, despite the increase in the size of the particle is not too significant. Particle size tends to be greater due to the total mass of the material during the milling process become more numerous.

Cyclic voltamograms of all sample obtained at a scan rate of 0.1 mV/s between 1 V and 2.5 V are shown in Figure 3. There are a pair reversible redox peaks for each sample, indicating all sample are rechargeable. Sample a and d have wide redox peak indicating bad electrode kinetic. While, sample b and c show sharp redok peak indicating rapid electrochemical process. It can be seen in Figure 3 that the CV of all sample are similar, suggesting that LiOH.H2O excess in $Li_4Ti_5O_{12}$ syintesize does not change the electrochemical reaction process of $Li_4Ti_5O_{12}$.

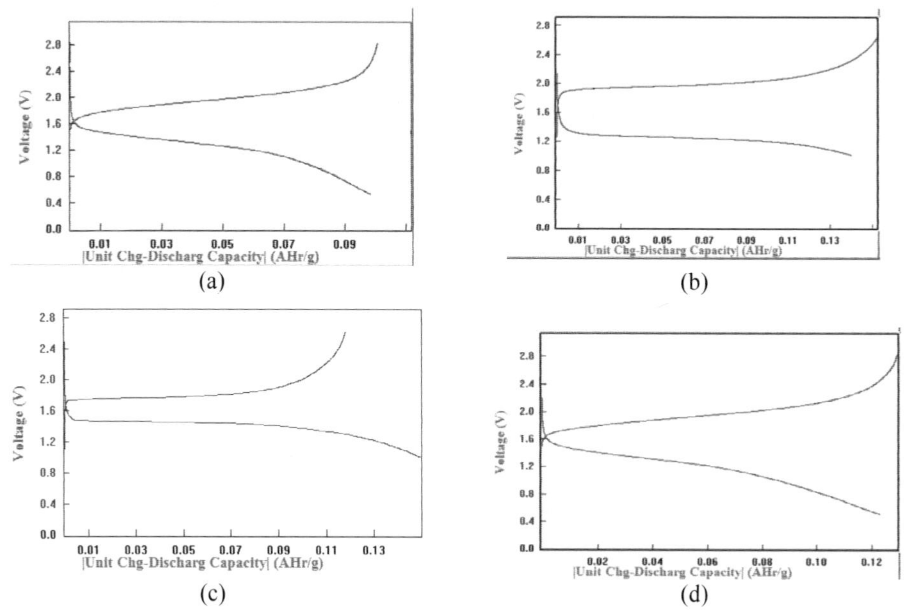

Figure.4. Charge-Discharge Curve of $Li_4Ti_5O_{12}$ powder with LiOH.H_2O excess (a) 0% (b) 2.5% (c) 5% and (d) 7.5%.

Table 3. Half-cell battery capacity of $Li_4Ti_5O_{12}$ were synthesized by LiOH.H_2O excess

Sample	Specific capacity (mAHr/g)
0%	85
2,5%	140
5%	150
7,5%	110

Diagrams of voltage Vs capacity from charge-discharge test results are shown in Figure 4. All samples showed good charge-discharge curves. Samples 5% LiOH.H_2O excess showed best curve with flat charge-discharge plateau. Difference between charge-discharge plateau showed polarization. It can be seen that the charge-discharging test results indicate the presence of only one phase with a curve that is not marked with staircase. This is because the amount of TiO_2 rutile phase at each sample is less than 10%. The curve will staircase (mark the 2 phases) if the amount of rutile TiO_2 phase of more than 30%.

Results of analysis of the capacity of the battery cell with charge-discharging test are shown in Table 3. Along with the raw materials LiOH.H_2O capacity battery cells from each sample tended to increase with increasing $Li_4Ti_5O_{12}$ phase is formed. This is caused by increasing of $Li_4Ti_5O_{12}$ phase and an impurity phase TiO_2 decreased.

4. Conclusion

Synthesis $Li_4Ti_5O_{12}$ with $LiOH.H_2O$ exccess has been successfully carried out by using solid state reaction. Crystal structure analysis by XRD test equipment showed the lowest rutile TiO_2 phase composition is shown by samples with the addition of $LiOH.H_2O$ as much as 5%. Addition $LiOH.H_2O$ in 4 variants of the sample does not alter the crystal structure on the results of XRD analysis but only affects the amount of phase composition and does not affect morphology and particle size. Morphology of each samples were agglomerate and form small particles with diverse forms. $LiOH.H_2O$ additions increase the capacity of the battery cell because $Li_4Ti_5O_{12}$ phase increase and phase impurities (rutile TiO_2) decrease.

References

[1]. Yurui G, Zhaoxiang W, Liquan C **2014** Stability of spinel $Li_4Ti_5O_{12}$ in air *J Power Sources* **245** 684-690.

[2]. Robert F N **2000** Power Requirements for Batteries in Hybrid Electric Vehicles *J Power Sources* **91** 2-26.

[3]. Armand M, Jean M T **2008** Building Better Batteries Nat **451** 652-657.

[4]. Zagrib K, Trudeau M, Guerfi A, Trottier J, Mauger A, Veillette R, Julien C M **2012** New advanced cathode material: $LiMnPO_4$ encapsulated with $LiFePO_4$ *J Power Sources* **204** 177-181

[5]. Aojun B, Li W, Jiaoyang L, Xiangming H, Jixian W, Jianlong W **2015** Composite of graphite/phosphorus as anode for lithium-ion batteries *J Power Sources* **289** 100-104

[6]. Andreas H, Thomas H **2015** Novel electrolyte mixtures based on dimethyl sulfone, ethylene carbonate and $LiPF_6$ for lithium-ion batteries *J Power Sources* **298** 322-330

[7] Noor I S, Majid S R, Arof A K **2013** Poly(vinyl alcohol)-LIBOB complexes for lithium-air cells *Electrochim Acta* **102** 149-160

[8]. Feng N, Yan B H, Baohua L, Hongda D, Dengyun Z, Feiyu K **2012** Effects of TiO_2 crystal structure on the performance of $Li_4Ti_5O_{12}$ anode material *J. Alloys Compd* **513** 524-529

[9]. Yan B H, Ming L, Zhen D H, Biao Z, Yang Y, Baohua L, Feiyu K, Jang K K **2013** Effect of solid electrolyte interface (SEI) film on cyclic performance of $Li_4Ti_5O_{12}$ anodes for Li ion batteries *J Power Sources* **239** 269-276

[10]. Euh D J, Hyun J H, Ok S J, Myoung G H, Chil H D, Min J H, Ho S Y, Hong K S **2012** Characterizations and electrochemical performance of pure and metal-doped $Li_4Ti_5O_{12}$ for anode materials of lithium-ion batteries *Mater Res Bull* **47** 2847-2850

[11]. Qianyu Z, Chengli Z, Bo L, Shifei K, Xi L, Yangang W **2013** Preparation and electrochemical properties of Ca-doped $Li_4Ti_5O_{12}$ as anode materials in lithium-ion battery *Electrochim Acta* **98** 146-152

[12]. Guo Q Z, Wenjuan L, Hongwei Y, Yahui W, Sowjanya B R, Yanli C, Chunbao Z, Keqiang D, Zhanhu G **2013** Influence of Preparation Conditions on the Properties of Lithium Titanate Fabricated by a Solid-state Method *J. New Mat. Electrochem Systems* **16** 025-032

[13]. Qian Y, Hailei Z, Jie W, Jing W, Chunmei W, Xinmei H **2014** Synthesis parameter dependence of the electrochemical performance of solvothermally synthesized $Li_4Ti_5O_{12}$ *Mater Renew Sustain Energy* 3:24

Preparation of Heterogeneous CaO Catalysts for Biodiesel Production

W Widayat[1,2,*], T Darmawan[2], H Hadiyanto[2] and R Ar Rosyid[2]

[1]Laboratory of Advanced Material Center of Research and Service Diponegoro University
[2]Department of Chemical Engineering Diponegoro University
Jl. Prof. Soedarto SH, Tembalang, Semarang, 50275

[*]Email. widayat@live.undip.ac.id

Abstract. The objective of this research was to develop heterogeneous catalysts from three CaO sources for biodiesel synthesis. The CaO catalyst were prepared from limestone, calcium hydroxide and calciun carbonate with thermal processing in a muffle furnace at 900°C.. The results showed that CaO catalyst from limestone has better characteristic than catalyst from Calcium Hydroxide and Calcium Carbonate. From morphology testing, the CaO catalyst derived from limestone formed a crystal, while The X-ray difraction analysis showed that the amount of CaO contained in limestone was the highest among the others. The yield of biodiesel obtained from the experiment was 89.98% for the catalyst from limestone; 85.15% for the catalyst Ca $(OH)_2$; and 78.71% for $CaCO_3$ catalyst.

1. Introduction

Catalyst is a substance that enables a chemical reaction to proceed at a usually faster rate. Commonly, catalyst was developed in homogenous or heterogenous phases. In biodiesel production, the homogenous catalyst generally use acid and base catalysts such as sulphuric acid, sodium hydroxide (NaOH) and potasium hydroxide (KOH). The sulphuric acid is used in esterification reaction in which free fatty acid (FFA) is reacted to methanol. The sodium hydroxide and potasium hydroxide are mostly used in transesterification reaction where triglyceride is reacted with methanol[1,2]. The disadvantages of utilization of homogeneous catalysts in biodiesel production are saponification reaction [1] and regeneration issue after the transesterification process is completed as well as producing toxic wastewater[3] and corosive in reactor[4]. Hence, the heterogeneous base catalysts was introduced to overcome these problems[5].

The heterogeneous catalysts which have been widely used include KI/KIO_3/H-zeolite[6], calcium hydroxide (CaO)[7-12], zirconia[13-14] and zinc oxide[15]. To saving cost of biodiesel production, the catalyst must be economically feasible. One option is by developing catalyst from natural resourcessuch as bone, waste and natural rock like limestone. Limestone can be found as a raw material for cement production. Limestone contains a calcium carbonate and calcium hydroxide / $Ca(OH)_2$[16]. If the calcium carbonate ($CaCO_3$) and $Ca(OH)_2$ were calcined in high temperature, then they will decompose to CaO which is the main source of heterogeneous catalyst. CaO is a strong basic oxide which has high catalytic activity [12]. The objective of this research was to prepare CaO catalyst from multi resources of CaO, like limestone, $Ca(OH)_2$ and $CaCO_3$..

2. Methods

2.1. *Material*
Limestone was obtained from local material store. Methanol has industrial specification and sulphuric acid, calcium hydroxide and calcium carbonate were obtained from Merck Ltd. Palm oil was used as feedstock in performance testing and it was purchased from local market. Palm oil must has FFA content under 0.5%. Pretreatment for reducing of FFA was done by using esterification reaction. The esterification reaction was carried out in a three-neck bottle reactor equipped with a condenser and a thermometer. A 800 ml cooking oil was mixed with 10 ml sulphuric acid and methanol by mole ratio to FFA 40:1, which all of the material had already been heated to 60°C. The reactor was placed in a water bath and heated on a hotplate. The agitating speed was kept at 600 rpm to ensure the efficient mixing. The reaction kept going for 2 hours. FFA contents was analyzed by using acidi-alkalimetry.

2.2. *Catalyst preparation*
The limestone were washed by deionized water to remove dust and impurities, and then dried overnight in an oven at 105°C. The acquired limestone as well as commercial $CaCO_3$ were crushed and sieved, and then calcined at 900°C for 1 ½ h. The same process was carried out for calcium carbonate and calcium hydroxide.

2.3. Catalyst characteristic
The catalyst product was analyzed of characterization consist of morphology and crystallite. The analysis crystallography was used x-ray diffractometer, XRD-7000S model Shimadzu brand with X-ray tube target Cu, voltage 30 kV, current 30 mA, and Kα radiation. XRD data was analyzed with PCXRD software. Morphology of catalyst was analyzed with JEOL PC Scanning Electron Microscope (PCSEM) model JSM-6510LA with magnification x5000. The analysis process was conducted in Center of Research and Service Diponegoro University (CORES DU).

2.4 *Catalytic testing for biodiesel production*
The transesterification reaction was carried out in a batch reactor equipped with a propeller. The reactions were performed using Branson (USA) ultrasonic processor (42 kHz) with a full power of 200 W. 100 ml of cooking oil filled into reactor and require in ultrasonic reactor. Methanol was added with mole ratio methanol to oil 9:1 and catalyst loading 2.6 wt%. The transesterification reaction has been done for 40 minute and CaO catalyst was separated by filtration. The filtrate was deposited until the glycerol (lower layer) and the methyl ester (upper layer) separate. FAME product was analyzed with Gas chromatography (GCMS) for composition of methyl ester[17].

3. Result and discussions

3.1 Characterization of CaO catalysts
The results SEM analysis was presented in Figure 1. Figure 1a is a result of SEM of limestone before calcination. The surface is still amorphous because there are many impurities contents. Limestone contains $Ca(OH)_2$ and $CaCO_3$ whilelimestone looks fragile. After calcination process the limestone convert to CaO and carbon dioxide and water gases. Figure 1 b) is a CaO catalyst after calcination of limestone so called as CaO catalyst 01. This catalyst has formed crystal on it's surface and hence will act as an active site and speeding up the reaction time. Figure 1c) is the catalyst of Ca $(OH)_2$ after calcination so called as CaO catalyst 02. The surface of the particles is similar to Figure 1 b) which has formed crystal on its surface. While Figure 1 d) is the catalyst of the raw material $Ca(CO)_3$ so called as CaO catalyst 03 The surface of CaO catalys 03 has particles that looks different compared to the other catalyst. It could be concluded that no crystal was formed on its surface. This could be caused because the calcination process is not perfectly occured. Thus, the formation of crystals of CaO

catalyst of limestone and $Ca(OH)_2$ is better than $Ca(CO)_3$ because morphologically it has already formed crystals on its surface.

Calcination temperature of 900°C is chosen because at these temperatures, will be produced an optimized CaO and it will be indicated by how much H_2O and CO_2 evaporates, the surface area formed, and the catalyst activity[11].

a) Limestone before calcination
b) Limestone after calcination
c) $Ca(OH)_2$ after calcination
d) $Ca(CO)_3$ after calcination

Figure 1. The result of SEM Analysis with x10.000 magnifications.

The graphics shown in Figure 2 are based on the XRD analysis. CaO catalyst 01 has higher intensity than the other subtances. XRD pattern of limestone was diferrent with CaO catalysts. Limestone was contained of CaCO3 and $Ca(OH)_2$ as indentified in 2 theta 29; 34, 48 and 52. CaO catalyst 02 was also identified at peaks of 2 theta 29, 34, 48 and 52. CaO catalyst 02 was CaO catalyst that prepared from $Ca(OH)_2$. It shows that calcium hydroxide can not be decomposed to calcium oxide and water vapor at 900°C during 1,5 hours. CaO catalyst 03 has the same peaks ash CaO catalyst 01, but intensity CaO catalyst 03 is more less than intensity CaO catalyst 01.

The results of XRD analysis was presented in Table 1. Table 1 shows that limestone was contained $Ca(OH)_2$, $CaCO_3$, and $Cr_{0,5}Li_2O_3Sb_{0,5}$. CaO catalyst 01 was contained CaO and small amount of CaD_2O_2. CaO catalyst 02 was contained CaO and few $Ca(OH)_2$ remaining. CaO catalyst 03 was contained mostly $CaCO_3$, and a few of $Ca(OH)_2$ and CaO. This shows that calcination processing of limestone was bettter than calcination of $Ca(OH)_2$ and $CaCO_3$. $Ca(OH)_2$ substance contains of hydrogen bond so the energy for decomposition of $Ca(OH)_2$ is more than energy for decomposition $CaCO_3$. Therefore $Ca(OH)_2$ is still detected in XRD analysis.

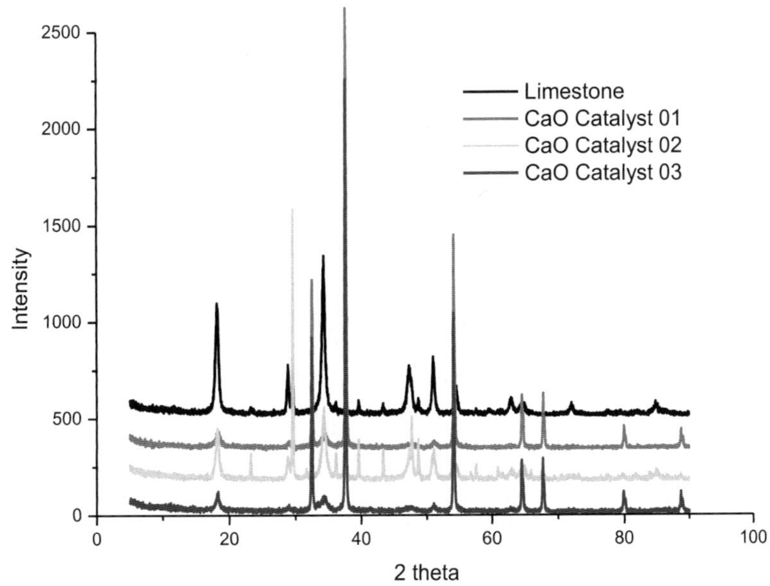

Figure 2. XRD pattern of limestone and CaO catalyst

Table 1. composition of the CaO catalyst and limestone.

Material	Compound	Formula	Composition (%)
Limestone	Portlandite	$Ca(OH)_2$	78,7
	Calcium Carbonate	$CaCO_3$	13,2
		$Cr_{0,5}Li_2O_3Sb_{0,5}$	8
CaO catalyst 01	Calcium oxide	CaO	88,1
		CaD_2O_2	11,9
CaO catalyst 02	Portlandite	$Ca(OH)_2$	19,4
	Calcium Oxide	CaO	80,6
CaO catalyst 03	Calcum carbonate	$CaCO_3$	37,7
	Calcium Oxide	CaO	18
	Portlandite	$Ca(OH)_2$	44,2

3.2 Catalytic testing for biodiesel production

Biodiesel from catalytic testing was analyzed by using Gas Chromatography Mass Spectrophotometry (GCMS) (Figure 3) Figure 3 shows typical methyl ester in a biodiesel product. Dodecanoic acid methyl ester was detected in a retention time (RT) of 26.62 minute, tetradecanoic acid methyl ester was detected in a RT of 32.43 minute, pentadenoic acid methyl ester was detected in a RT of 35.06 minute. All atty acid methyl ester (FAME) compounds found in the Chromatography is listed in Table 2 The largest compound found in FAME is 9-Octadecenoic acid methyl ester followed by hexadecanoic acid methyl ester and 9,12-Octadecadienoic acid methyl ester.

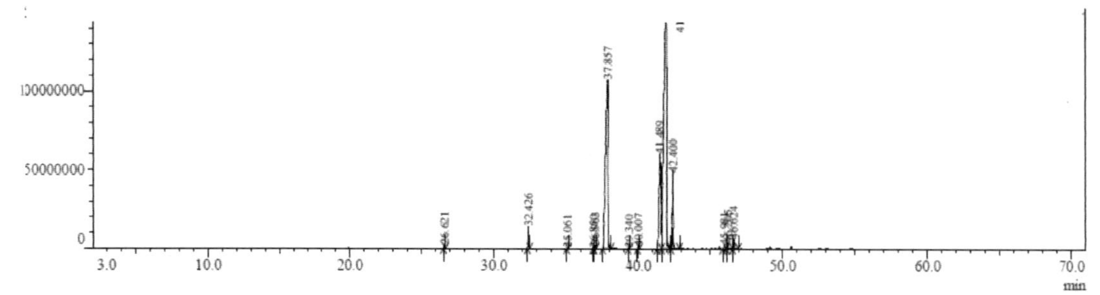

Figure 3. Chromatograph of biodiesel product

Table 2. Types of FAME in biodiesel product

Retention time (minute)	Compound	Composition (%)
26.62	Dodevanoic acid methyl ester	0.17
32.43	Tetradecanoic acid methyl ester	1.02
35.06	Pentadecanoic acid methyl ester	0.03
36.85	9-Hexadecenoic acid methyl ester	0.21
37.86	Hexadecanoic acid methyl ester	29.66
39.34	Cis-10 Heptadecenoic acid methyl ester	0.03
40.01	Heptadecanoic acid methyl ester	0.09
41.49	9-12 Octadecadienoic acid methyl ester	14.43
41.92	9-Octadecenoic acid methyl ester	47.79
42.40	Octadecanoic acid methyl ester	5.50
45.98	Cis 10-heptadecenoic acid methyl ester	0.22
46.21	Octadecanoic acid 9,10 dihydroxy methyl ester	0.35
46.62	Ecosanoic acid methyl ester	0.50

The yield of biodiesel was calculated by using Eq 1

$$Yield\ FAME = \frac{Mass\ of\ FAME}{Mass\ of\ Tryglyceride}\ x\ 100 \quad (1)$$

The calculated yield was shown in Figure 4. The best catalyst performance was shown at experiment with CaO catalyst 01 produces a yield of 89.98%. Yield of FAME by using CaO catalyst 02 was 78.71% and CaO catalyst 03 yield of FAME 85.15%.. In this study, the compound serves as the catalyst is a compound CaO based on following reaction:

$$Ca(CO)_3 \rightarrow CaO_{(s)} + CO_{2(g)}$$
(900°C, 90 minutes)

$$Ca(OH)_2 \rightarrow CaO_{(s)} + H_2O_{(g)}$$
(900°C, 90 minutes)

The three raw materials after calcination contains different amount of CaO. Based on XRD analysis and Table 1, CaO content of each raw material is 88.1% for limestone, 80.6% for $Ca(OH)_2$, 18% for $CaCO_3CaO$. The higher CaO contents give higher yield due to more catalytic activity in the [12]. $Ca(OH)_2$ shall be used in liquid phase so that it can act as catalyst in transesterification reaction.

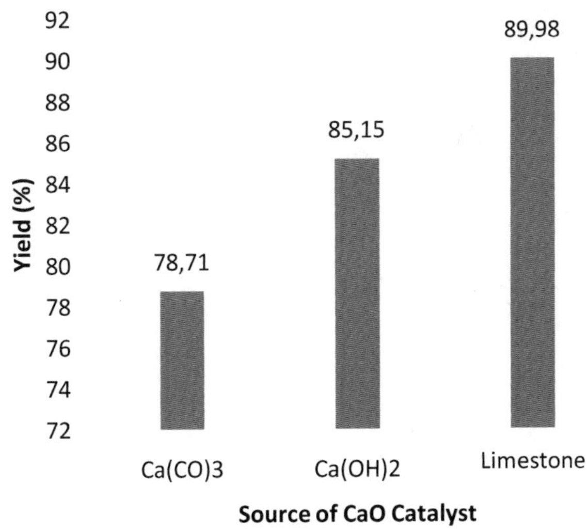

Figure 4. Yield amount of Methyl Ester produced by each different catalyst.

4. Conclusion

In summary, CaO catalyst has been prepared from limestone, calcium carbonate and calcium hydroxide by using calicination process. CaO catalyst prepared from limestone, give higher performance than other CaO catalysts from calcium hydroxide and calcium carbonate.

References

[1] Widayat W, Satriadi H, Nafiega FN, Alimin AJ and Ali MFM 2015 *AIP Conference Proceedings*, **1699, 030020**
[2] Widayat W, Satriadi H, Yuariski O and Murwono D. 2013 *Research Journal of Applied Sciences, Engineering and Technology*, **6 (10),** 1732-1739
[3] Endalew AK, Kiros Y, and Zanzi R 2011 *Biomass Bioenergy* **35** 3787 -809
[4] Vicente G, Martı´ nez M and Aracil J 2004 *Bioresour. Technol.* **92,** 297–305
[5] Talebian-Kiakalaieh A, Amin N A S and Mazaheri H 2013 *Appl. Energy* **104** 683–710.
[6] Widayat W, Wicaksono AR, Firdaus LH and Okvitarini N 2016 *IOP Conference Series: Materials Science and Engineering* **107** 012044
[7] Hadiyanto H, Lestari SP and Widayat W 2016 *Bulletin of Chemical Reaction Engineering and Catalysis* **11(1)** pp. 21-26
[8] Hadiyanto H, Lestari SP, Abdullah A, Widayat W and Sutanto H 2016 *International Journal of Energy and Environmental Engineering*
[9] Obadiah A, Swaroopa G A, Kumar S V and Jeganathan K R 2011 *Bioresource Technology* **116** pp. 512-516
[10] Granadosa M L, Alonsoa D M, Sádabaa I, Mariscala R and Ocón P 2009 *Applied Catalysis B: Environmental* **89** pp. 265-272.
[11] Kouzua M, Kasuno T, Tajika M, Sugimoto Y, Yamanaka S and Hidaka J 2008 *Fuel* **87,** pp. 2798-2806.
[12] Tanga Y, Menga M, Zhanga J and Lub Y 2011 *Applied energy* **88** pp. 2735-2739.
[13] Patel A, Brahmkhatri V and Singh N 2013 *Renewable Energy* **51** pp. 227-233
[14] Wan Omar WNN and Amin NAS (2011) *Fuel Processing Technology* **92** pp. 2397-2405
[15] Gurunathan B and Ravi A 2011 *Bioresources Technology* **190** pp. 424-428
[16] Stanmore BR and Gilot P 2005 *Fuel Processing Technology* **86** pp. 1707-1743

[17] Widayat W, Aulia H N, Hadiyanto and Sasongko S B 2015 *Journal of Engineering and Technological Sciences* **47** pp. 374-388.

Fuel Sustainability And Actinide Production Of Doping Minor Actinide In Water-Cooled Thorium Reactor

Sidik Permana [1]
[1]Nuclear Physics and Biophysics Research Division, Institut Teknologi Bandung
Jl. Ganesha 10, Bandung 40132, Gedung Fisika FMIPA ITB Indonesia

E-mail: psidik@fi.itb.ac.id

Abstract. Fuel sustainability of nuclear energy is coming from an optimum fuel utilization of the reactor and fuel breeding program. Fuel cycle option becomes more important for fuel cycle utilization as well as fuel sustainability capability of the reactor. One of the important issues for recycle fuel option is nuclear proliferation resistance issue due to production plutonium. To reduce the proliferation resistance level, some barriers were used such as matrial barrier of nuclear fuel based on isotopic composition of even mass number of plutonium isotope. Analysis on nuclear fuel sustainability and actinide production composition based on water-cooled thorium reactor system has been done and all actinide composition are recycled into the reactor as a basic fuel cycle scheme. Some important parameters are evaluated such as doping composition of minor actinide (MA) and volume ratio of moderator to fuel (MFR). Some feasible parameters of breeding gains have been obtained by additional MA doping and some less moderation to fuel ratios (MFR). The system shows that plutonium and MA are obtained low compositions and it obtains some higher productions of even mass plutonium, which is mainly Pu-238 composition, as a control material to protect plutonium to be used as explosive devices.

1. Introduction

Energy supply which can sustain its supply to secure world energy demand in order to maintain their prosperities and daily life activities are subtaintial to be pursued. In term of nuclear energy, its sustainability of nuclear energy comes from optimum fuel utilizations based on exixting fuel resources, fuel management of the reactor including fuel cycle program as well as fuel breeding program. Fuel cycle option becomes more important for fuel cycle utilization as well as fuel sustainability capability of the reactor. Those fuel optimization programs can be improved which has a similar trend with the renewable energies has been shown as a sustainable energy source [1]. In addition, a similar way, is to "re-use" spent fuel such as some depleted uranium, plutonium and minor actinides as a new fuel to be used for conventional or advance nuclear reactors. In term of recycle nuclear fuel program, one of the important issues is in conjunction with the nuclear proliferation resistance issues. The issue comes from the production of good quality of weaponable grade of spent nuclear fuel such as plutonium. To reduce the proliferation resistance level, some barriers were used such as matrial barrier of nuclear fuel based on isotopic composition. Some material barrier of isotopic composition can be estimated from critical mass, heat-generation rate, spontaneous neutron generation and radiation. Isotopic barrier of plutonium comes from the even mass number of plutonium isotope such as ^{238}Pu, ^{240}Pu and ^{242}Pu and in case of thorium cycle will depend on ^{232}U. For decades some thorium fuel technologies have being developed, for both conventional and advanced nuclear reactor as well as some thorium breeder reactor programs which was also developed for breeding program in uranium fuel technology. Thorium fuel cycle have some advantages including better fuel breeding

Published under licence by IOP Publishing Ltd

capability in thermal and epi-thermal neutron energy region, higher fuel stability and higher level proliferation resistance [2-6] in comparison with others fuel cycles.

Some analyses on nuclear fuel sustainability and actinide production composition have been done based on heaevy water-cooled thorium reactor system and will be shown in the present study. Fuel cycle flow scheme will be based on the fuel recycled composition which all actinide compositions except fission products are recycled into the reactor as a basic fuel cycle scheme. Two key important parameters are evaluated such as doping composition of minor actinide (MA) and volume ratio of moderator to fuel (MFR) which have some significant impact to the reactor performances as well as fuel composition behaviors.

2. Parameter and Method

Some analyses on nuclear fuel sustainability and actinide production composition have been done based on heaevy In the present study, well established reactor technology of water-cooled reactor based technology has been adopted as a basis for evaluation which is based on current LWR plant technology as general design parameter and using heavy water as coolant and moderator instead of light water. **Table 1** shows a basic design parameters which consist of some key parameters such as moderator to fuel ratio (MFR), burnup and doping rate of minor actinide (MA) based on large 3000 MWt power output and heavy water as coolant and moderator. Employed MFR of 1 and 1.5 are used to evaluate a moderator effect to the reactor performance in comparing to standard MFR of PWR (MFR =2) which show the neutron spectra are relatively harder than standard PWR. Minor actinide (MA) effecta are used based on no MA doping and with 10 % MA doping for fix discharged fuel burnup of 36 GWd/t. A 1.452 cm was used as fuel pin diameter based on the regular blanket fuel diameter of the Shippingport [2,6-7]. Fuel cycle scheme of closed cycle fuel that all actinides are returned to the reactor_and only FPs are sent to final disposal stream, are used for this analysis based on adopted thorium fuel based with additional MA are used for doping material. To evaluate fuel sustainability based on fuel conversion or fuel breeding capability and actinide production of the reactors, an equilibrium burn-up calculation code (Equilibrium Cell Iterative Calculation System: ECICS) was used which is coupled with SRAC code and JENDL3.2 nuclear data library [2,7-18]. Fuel sustainability based on fuel breeding and fuel conversion ratio ares based on the reaction rate of fissile and fertile nuclides and some small contributions of intermediate nuclides. When the conversion ratio is higher than unity than the definition becomes fuel breeding ratio and some surplus fuel from fuel breeding production is defined as fuel breeding gain.

Table 1. Reactor parameters

Parameters	Values/Type
Thermal Power Output [MWt]	3000
Coolant	D_2O
Fuel Cycle Option	All Actinides closed
Supply Fuel Composition	Th + MA
Cladding	Zircaloy-4
Fuel pellet diameter [cm]	1.31
Fuel pin outer diameter [cm]	1.45
Moderator to Fuel Ratio (MFR) [-]	1 and 1.5
Burn-up [GWd/t]	36
MA Content [%]	0 - 10

3. Results and Discussions

3.1. Fuel sustainability of breeding capability

Obtained results based on fuel conversion capability to evaluate the effect of doping MA in the reactor are shown in **figure 1.** Obtained fuel conversion capabilities are higher than unity, which means the results have fuel-breeding capability. As shown in Figure 1, doping material of MA is effective to increase the fuel breeding level because of some additional fissile materials are produced from converted MA material during reactor operation. To estimate some surplus from fuel breeding condition, a fuel breeding gain is adopted which is shown in **figure 2** for fuel sustainability from the effect of MA doping. Initial condition for thorium reactor without additional MA material as doping obtained higher than unity, which means fuel-breeding gain can be achieved. Additional fissile material based on breeding gain is estimated of about 1,2% additional fissile materials were produced during reactor operation. In addition, when some MA material are loaded into the reactor, the fuel breeding gain is increasing significantly from 1.2% up to about 2,2% for MA doping of 10%. Those materials of MA are estimated to produce more fissile material from fuel conversion process of MA into some plutonium fissile and fertile materials through neutron capture during reactor operation.

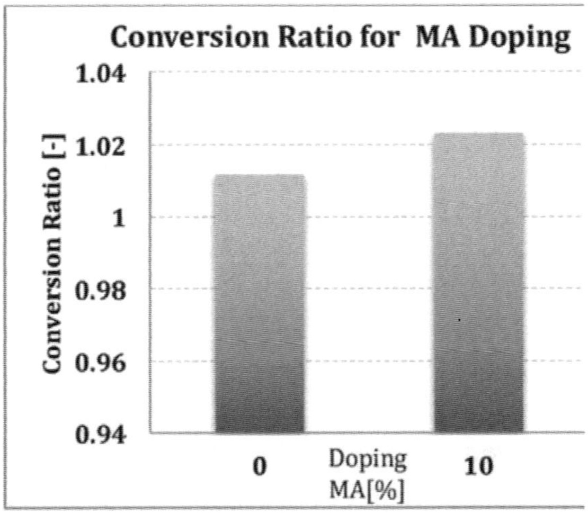

Figure 1 Fuel conversion rasio for MA content of 0 and 10 % as doping

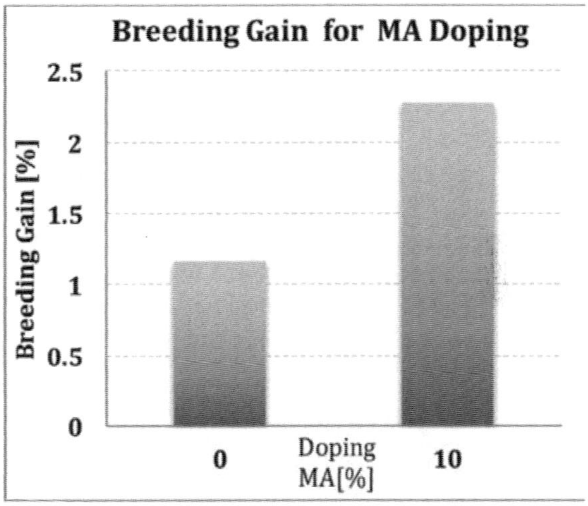

Figure 2 Fuel breeding gain for MA content of 0 and 10 % as doping

As mentioned in the previous section that less moderation ratio to fuel or MFR is expected to make the reactor has a harder neutron spectrum, which is estimated to have more fuel conversion ratio or fuel-breeding ratio based on higher production of fissile material, which is converted from fertile material by absorption neutrons. The effect of less moderator to the reactor performance capability in relation to fuel sustainability are shown in **figure 3** which shows MFR of 1 and 1.5 as representative of different MFR for less moderation ratios. MFR of 1.5 is difficult to have fuel conversion higher than unity in comparing to smaller MFR value of 1, which obtains fuel conversion ratio of higher than unity. MFR of 1 is good enough to produce more fissile materials and it obtained more than supply fuel. Meanwhile MFR of 1.5 is not sufficient to achieve more fissile material, which is higher than unity. It is estimated that MFR of 1 has good enough neutron spectrum to make the reactor having some fuel-breeding ratio because of harder neutron spectrum, which is used to convert some fertile material of thorium as well as additional MA as doping to fissile material. Neutron spectrum of MFR of 1.5 is no sufficient to convert more breeding fuel through neutron capture during reactor operation. As expected that, fuel-breeding gain only can be obtained by MFR of 1 while MFR of 1.5 is difficult to have fuel-breeding gain as mentioned in **figure 4**. Figure shows that two MFR values give a positive and a negative values for fuel breeding gain which shows a positive value of almost 2 % of additional fuel breeding is achieved by MFR of 1. A negative fuel breeding gain is obtained by MFR of 1.5 which means the fuel breeding gain of the reactor is not sufficient to achieved higher than unity, it request more than 5% additional breeding gain. Even though fuel conversion ratio of 0.95 is higher enough to be called as higher converter reactor. Based on the MFR effect to achieve fuel breeding gain, is estimated that between MFR 1 and MFR 1.5 has an optimum value to produce a breeding gain condition and its breeding gain is expected become higher for less MFR values. The results show that some feasible parameters can be used for obtaining breeding gain and it can be fulfilled by some additional MA doping as well as arranging some less moderation to fuel ratios (MFR), although higher MFR is expected to become a high conversion reactor.

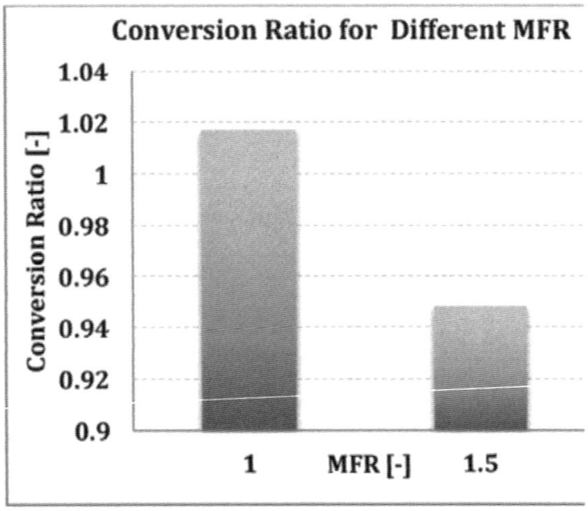

Figure 3 Fuel conversion rasio for different less MFR values.

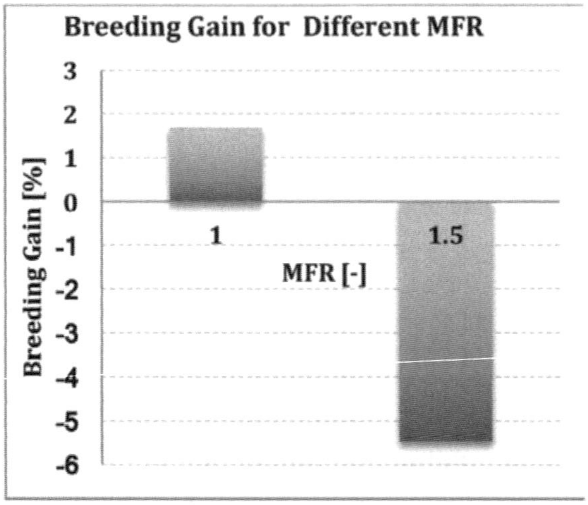

Figure 4 Fuel breeding gain for different less MFR values

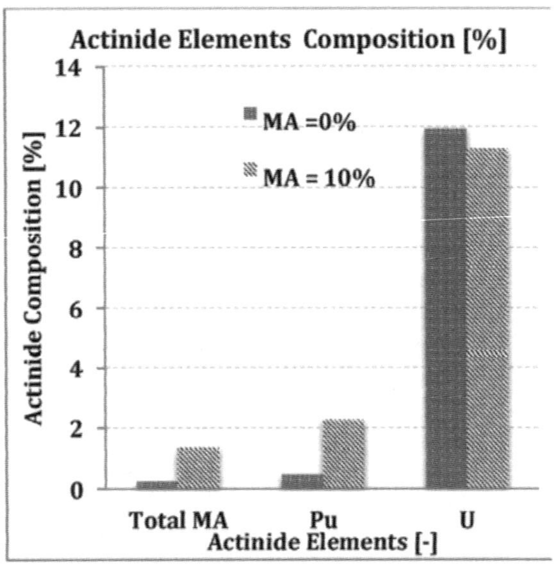

Figure 5 Actinide element composition for different MA doping rates

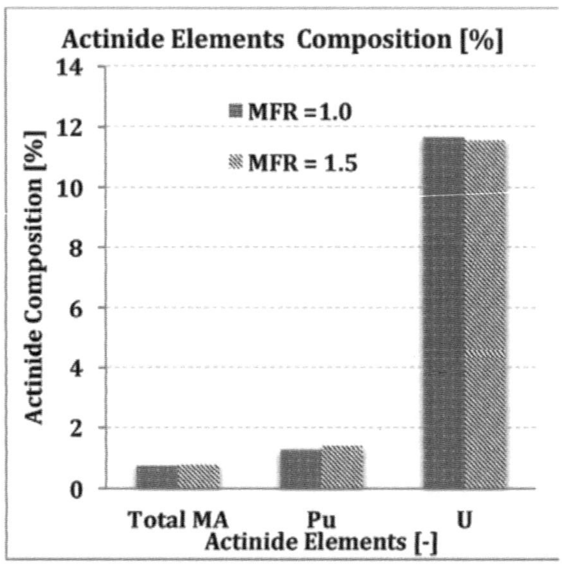

Figure 6 Actinide element composition for different less MFR values

3.2. Actinide element compositions

Some obtained results of actinide compositions are shown in **figures 5** through **8** for different MA doping and MFR values, which is based on equilibrium nuclide density composition. Some actinide composition based on actinide element of MA, plutonium and uranium are shown in figures 5 and 6 for different effects of MA doping and less MFR values. Doping MA is effective to increase plutonium production as well as MA, while uranium element is decreasing when MA is loaded. Plutonium element composition is produced from converted MA as well as some MA is still exist as or production from produced MA. Uranium production is coming mainly from thorium cycle, which is mainly from U-233. This decreasing uranium is estimated from less fraction of thorium as fresh fuel because some portions are replaced by MA material as doping materials at fresh fuel. The effect of MFR to produce actinide element is shown in Figure 6 form MFR 1 and 1.5. It shows that higher MFR (MFR of 1.5) produces slightly more MA and plutonium productions, while uranium is slightly decreasing. Changing the neutron spectra from both employed MFR are not so significant to the change in actinide element compositions. Both data show that plutonium compositions are obtained 2% or less from total actinides, while uranium compositions are obtained more than 10%.

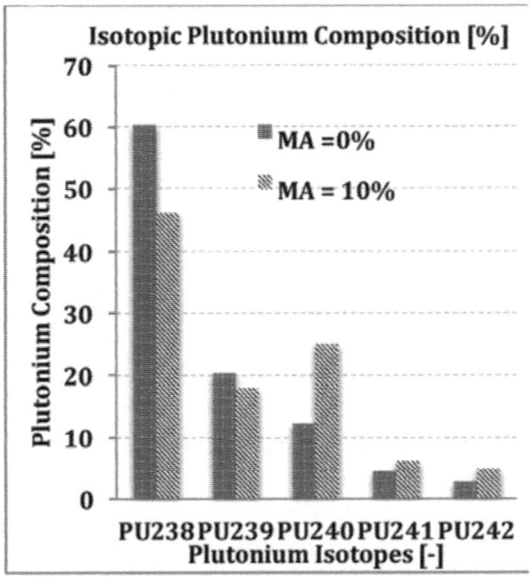

Figure 7 Isotopic Plutonium composition for different MA doping rate

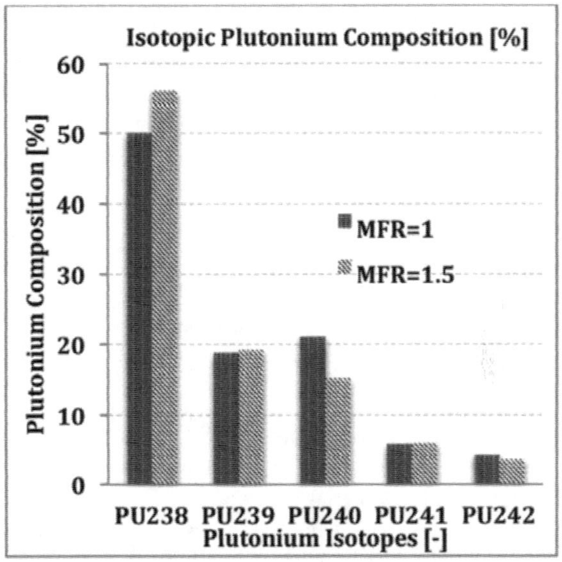

Figure 8 Isotopic Plutonium composition for different MFR Values

3.3. Plutonium isotopes compositions

Obtained produced plutonium composition based on isotopic plutonium compositions are shown in figure 7 and 8 for different MA doping rates and different MFR values. Data shows that both effect of MA doping and MFR values have a significant change in plutonium isotopes composition. The main contributions of total plutonium composition are coming from Pu-238, followed by Pu-239 or Pu-240 and Pu-241, Pu-242. MA doping is effective to increase the composition Pu-240 up to Pu-242 while it affects to decrease compositions of Pu-238 and Pu-239. While the opposite trend is shown by the effect of higher MFR that shows an increasing trend are obtained by Pu-238 and Pu-239 for higher MFR, while less composition for Pu-240 up to Pu-242. MA doping affects to reduce Pu-238 from 60% composition without doping up to about less than 50% when 10% doping is employed. While its Pu-238 composition for different MFR values are shown more than 50%. These Pu-238 values are higher enough based on protected plutonium composition in term of proliferation resistance point of view issue. The composition of Pu-238 is suitable to protect the plutonium to be used for explosive devices based on some even number mass of plutonium isotope which can be used to some criteria of plutonium characterization such as IAEA, Pellaud and Kessler criteria (IAEA, 1972; Pellaud, 2002; and Kessler, 2004) [19-21]. Trans uranium composition including plutonium and MA is expected to be less composition for thorium based reactor which is shown that very low MA compositions are obtained, while plutonium are achieved less than 2% in total composition for MA doping cases. In term of isotopic plutonium composition, it shows higher production of even mass plutonium mainly

Pu-238 which can be used as a control material to protect plutonium to be used as explosive devices which is higher enough based on Kessler criterion.

4. Conclusions

Nuclear fuel sustainability and actinide production composition analyses have been done based on heaevy water-cooled thorium reactor system which adopted all actinide compositions except fission products are recycled into the reactor as a basic fuel cycle scheme. Some important parameters are evaluated such as doping composition of minor actinide (MA) and volume ratio of moderator to fuel (MFR) which have some significant impact to the reactor performances as well as fuel composition behaviors. Some feasible parameters have been obtained to obtain some breeding gains such as some additional MA doping as well as arranging some less moderation to fuel ratios (MFR). Trans uranium composition including plutonium and MA for thorium based reactor obtained very low compositions although for MA doping cases. The system obtains some higher productions of even mass plutonium, which is mainly Pu-238 composition which can be used as a control material to protect plutonium to be used as explosive devices based on Kessler criterion.

Acknowledgments

We would like acknowledge and extend our gratitude to desentralisasi research program of ministry of research, technology and higher education and research innovation program of ITB for the grant and international conference publication

References

[1] President's meeting of G8 countries, Brazil, China, India and South Africa, 2006 Moscow April 19-20, Contribution of Coal and Nuclear to Sustainable Energy Supply: Perspectives and Problems, Paper presented at *the Russian Academy of Sciences, Academies of Sciences*
[2] P. Sidik, N. Takaki, H. Sekimoto 2007 "Feasible region of design parameters for water cooled thorium breeder reactor," *J. Nucl. Sci. Technol*, **44**, 7, 946-957
[3] T. K. Kim and T. J. Downar 2002 *Nuclear Technology,* **138**,[1] 17-29
[4] L. Michael, G. Otto 1998 *Nucl. Eng. Des.* **180,** 2, 133
[5] U. Turan 2000 *Prog. Nucl. Energy*, **37**, 137
[6] L. B. Freeman et al., 1989 *Nucl. Sci. Eng.*, **102**, 341
[7] P. Sidik, N. Takaki, H. Sekimoto, 2008 *J. Nucl. Sci. Technol.*,. **45,** 7, 589-600
[8] P. Sidik, N. Takaki, H. Sekimoto, 2006 *Ann. Nucl. Energy*, **33**, 561
[9] H. Sekimoto, N. Takaki, 1991 *J. Nucl. Sci. Technol.*, **28**, 941
[10] A. Waris, H. Sekimoto 2001 *J. Nucl. Sci. Technol.*, **38**, 7, 517
[11] A. Waris, H. Sekimoto 2001 *Ann. Nucl. Energy*, **28**, 153
[12] A. Mizutani, H. Sekimoto 1998 *Ann. Nucl. Energy*, **25**, 9, 623
[13] A. Mizutani, H. Sekimoto 1998 *Ann. Nucl. Energy*, **25**, 13, 1011
[14] A. Mizutani, H. Sekimoto 1998 *Prog. Nucl. Energy*, **32**, 713
[15] A. Mizutani, H. Sekimoto 1997 *J. Nucl. Sci. Technol.*, **34**, 6, 596 (1997).
[16] T. Seino, H. Sekimoto 1998 *Ann. Nucl. Energy*, **25**, 4–5, 223
[17] K. Okumura et al., 1996 *Japan Atomic Energy Research Institute (JAERI)*
[18] T. Nakagawa et al., 1995 *J. Nucl. Sci. Technol.*, **32**, 1259
[19] International Atomic Energy Agency, I972, Information Circular, INFCIRC/153
[20] B. Pellaud, 2002 *J. Nucl. Mater. Managements*, XXXI, No. 1
[21] G. Kessler, 2004 The First Int. Sci. Technol. Forum on Protected Plutonium Utilization for Peace and Sustainable Prosperity, March 1–3, Tokyo, Japan

Synthesis of Carbon Nano Materials Originated from Waste Cooking Oil Using a Nebulized Spray Pyrolysis

A.A. Arie [1*], L. Hadisaputra [1], R.F. Susanti [1], H. Devianto[2], M. Halim[3], R. Enggar[3] and J.K. Lee[3]

[1]Department of Chemical Engineering, Parahyangan Catholic University, Ciumbuleuit 94, Bandung, Indonesia

[2]Department of Chemical Engineering, Bandung Institute of Technology, Ganesha 10, Bandung, Indonesia

[3]Center for Energy Convergence Research, Korea Institute of Science and Technology, 5, Hwarang-ro 14-gil, Seongbuk-gu, Seoul, Republic of Korea

*arenst@unpar.ac.id

Abstract. Synthesis of nanocarbon on snake fruit-peel's activated carbon from waste cooking oil palm was conducted by a nebulized spray pyrolysis process (NSP) by varying the processing temperature from 650 to 750 ºC. Ferrocene was used as a catalyst with constant concentration of 0.015 g/ml of carbon source. The structure of nanocarbon was studied by using scanning electron microscope (SEM),x-ray diffraction (XRD), surface area analyzer and Raman spectroscopy. SEM results showed that the structures of carbon products was in the the form of carbon nanopsheres (CNS). XRD and Raman analysis confirmed the CNS structure. The carbon producs were then tested as electrode's materials for lithium ion capacitors (LIC) by cyclic voltammetry (CV) instruments. From the CV results the specific capacitance was estimated as 79.57 F / g at a scan rate of 0.1 mV / s and voltage range from 2.5 - 4 V. This study shows that the nano carbons synthesized from the waste cooking oil can be used as prospective electrode materials for LIC.

1. Introduction

Carbon is one of the most abundant elements in the world and it has many forms of allotropes. One type of carbon allotropes is the nano-carbon family. Nano-carbon is a carbon based material possessing a specified structure with nanometer-scaled particle size [1]. The application of nano carbons is very broad in various fields, such as construction, mechanical, electrical, renewable resources, etc. Because of the structure of nanocarbon is strong, high temperature resistance and have a good electrical properties, it has a good potential to applied in the electrochemical field, especially at energy storage such as supercapacitors [3-5]. Supercapacitor or also called ultracapacitor emerged after the innovation in enlarging the surface area and minimize the distance between the plates so the power and the capacity becomes much greater than conventional capacitors. Supercapacitor is expected to replace the battery as a media of electrical energy storage. This is due to the advantages of supercapacitors that power densities is much greater, the charging time is very short and longer life cycles [6].

Nanocarbon can be produced through various methods, such as chemical vapor deposition, laser ablation and many others. Simple method which can be used to synthesize nano-carbons from liquid or gas precursors is the nebulized spray pyrolysis (NSP) method. This method has several advantages compared with other methods, one of which is that it can produce large scale nanocarbon with good homogeneity carbon nanotubes (CNT) quality [7]. NSP technique requires a carbon source which can come from the organic solvent or various kinds of oil. The renewable resorces that can be used for the production of nano carbons is the palm oil. The use of palm oil in Indonesia is very high, up to 5 L/month from every household, producing a waste cooking oil. Hence, it is required to use the abundant waste cooking as raw materials for functional producs such as nano carbons.

In this work, nano-carbons were synthesized by a nebulized spray pyrolysis using waste cooking oil as carbon precursors, salacca peel based activated carbons as substrate and ferrocene as metal catalyst. The specific objectives of this works were to investigate the effects of pyrolysis temperatures on the structural characteristics of produced nano carbons. The utilization of cooking oil as carbon precursors for the production of nano carbons has been explored by some researchers [8-9]. Furthermore, the nano-carbons were then tested as cathode materials in lithium ion capacitors.

2. Experimental

The nebulised spray pyrolysis equipment is shown in Fig 1. One gram of salacca peel based activated carbon, which was used as substrate, was filled in the small quartz tube. Ferrocene with a concentration of 0.015 g/ml was dissolved in the waste cooking oil and then placed in the nebulisation container. The nebulisation container was then loaded in the nebulisation unit (D).The small tube was then loaded in the middle of the outside quartz tube (D). The outside quartz tube was then loaded in the furnace (C) and was connected with hose to water bubbler and with the container (C). The container (D) was also connected with flowmeter (F) which has a function to read the flowrate of nitrogen as carrier gas in the flow of 1.21 sscm. The furnace was then switched on and heated to various temperature (650, 700 and 750 ^{o}C After the furnace reached the operation temperature, the nebulisation unit was turned on in the frequency of 1.7 MHz for 20 minutes. After the reaction, furnace and nebulisation unit were allowed to cool down to room temperature

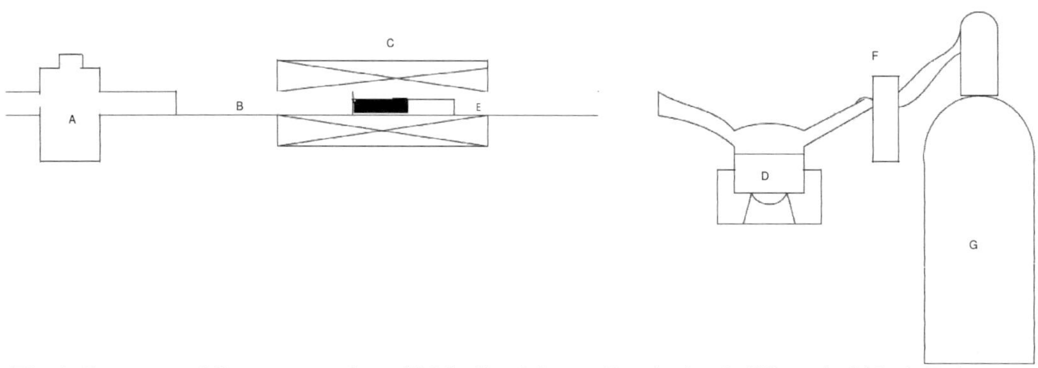

Fig 1. Process and Instrumentation of Nebulized Spray Pyrolysis, A (Water bubbler), B (quartz cylinder), C (Horizontal Furnace), D (Nebulizer), E (quartz reaction cylinder), F (Flowmeter), G (Nitrogen Gas)

The structural characteristics of carbon products were investiagted by scanning electron microscope (FEICompany, Nova 200), Raman spectroscopy (Renishaw, in Via Raman Microscope), BET analyzer and x-ray diffraction (Rigaku, Ultima IV). The surface texture and porosity of the carbon products were examined by N_2 adsorption at 77 K . The specific surface area was calculated from the N_2 adsorption isotherm using the BET equation.

The electrochemical performance of the nano-carbons was evaluated in a two-electrode cell Electrodes were prepared by mixing 90 wt% active material, 5 wt% carbon black (Mitsubishi), and 5 wt% polyvinylidene-fluoride in N-methyl pyrrolidone to form slurry. The slurry was painted in a 1 cm^2 area on Cu foil, with typically 2 mg carbon applied to each electrode. A sandwich type cell was constructed from two electrodes, with similar weights, facing each other and separated by glassy fiber paper. The 1 M $LiPF_6$ electrolyte was added to the cell under vacuum to reduce air contamination and improve wettability of the electrodes. Cyclic voltammetry (CV) was conducted using Maccor automated battery tester (series 4000). CV test was done at scan rate 0.1, 0.5 and 1 mV s^{-1}, with voltage window 2.5 - 4 V. Cell capacitance from CV test [F g^{-1}] was calculated by Eq. 1:

$$C_{CV} = \frac{(\sum|I|.\Delta t)}{2.m.\Delta V} \quad (1)$$

where $(\sum|I|.\Delta t)$ is the area of the current [A] against time [s] curve, *m* the mass of active material in the electrode [g], and ΔV the voltage window [V].

3. Results and Discussion

Fig 2 shows the SEM images of nano carbons produced at different reaction temperatures. It can be seen that the agglomeration of carbon nanospheres (CNS) occur at each of the temperature variations. The agglomers with the size of ~200 nm were observed. It seems that the amount of agglomerated CNS increses as the temperature raises from 550 to 750 ºC due to the high activity of ferrocene catalyst at higher temperature [10]

Fig 2. SEM images of nano carbons produced by a nebulized spray pyrolysis at reaction temperature of (a) 650 , (b) 700 and (c) 750 ºC

Table 1. Textural Properties of Carbon Products

Sample	Surface Area (m2/g)	Average Pores Size (Å)
Salacca Peel Based Activated Carbon (Substrate)	2500	22.65
Nano Carbons at 650 °C	983.35	21.46
Nano Carbons at 700 °C	1053.385	21.67
Nano Carbons at 750 °C	957.88	19.91

Table 1 shows the textural properties of carbon products synthesized at different reaction temperatures compared to the activated carbon substrate. It can be observed that the surface area was reduced after the growth of nano carbons on the surface of activated carbon substrate. However the average pore size remains in the range of 19 to 22 **Å**.

The XRD profiles of CNSs synthesized with different reaction temperatures are presented in Fig 2. Two distinct peaks normally observed in CNSs, those are (002) peak at ~23° and (100) peak at ~44°, can be clearly seen in all samples. It can be seen that those profiles are very different with the activated carbon supports. Peak C(002) indicates the graphitic structure formation in the sample. This also shown that the cristallinity of the material from NSP process is much higher than that of the activated carbon substrates..

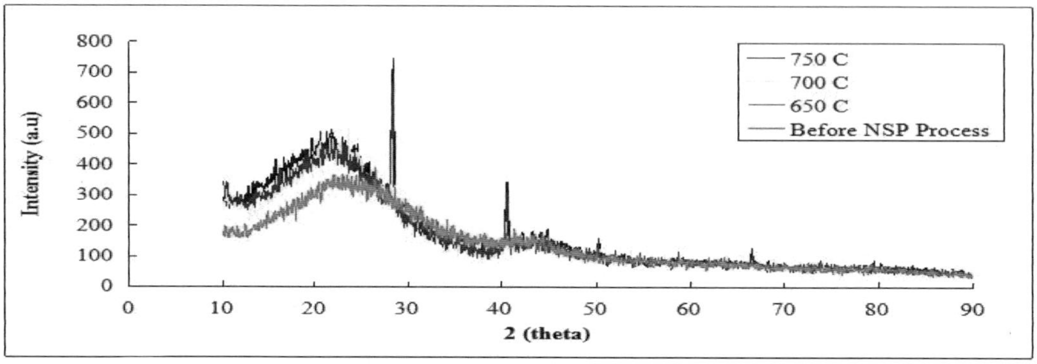

Fig 3. XRD profiles of nano carbons produced by a nebulized spray pyrolysis at reaction temperature

Table 2. I_D/I_G Value of Carbon Products

Reaction Temperature	Wave Length	ID	Wave Length	IG	ID/IG
650 °C	1333.7	5533.01	1593.73	6154.51	0.89901
700 °C	1341.63	9482.18	1587.19	10689.7	0.88703
750 °C	1342.87	4238.16	1592.74	4857.9	0.87242
Salacca peel Activated Carbon (substrate)	1341.63	2238.79	1586.1	2369.24	0.94494

The results of the analysis of the carbon structures using raman spectroscopy are shown in Table 2. It can be seen that the value of I_D/I_G ratio is reduced after the deposition of nano carbons on the surface of activated carbon substrates. It indicated that the NSP can increase the amount of graphitic carbons. The higher process temperatures will reduce the value of I_D/I_G ratio. At higher temperatures, the amount of carbon that binds with a grapitic process will increase along with the growth rate of nanocarbon.

Fig 4 shows the cyclic voltammetry profiles of nano carbons synthesized at temperature of 750 °C, measured at various scan rate. Voltammogram profiles shows a shape like a rectangle with a kind of horn/peak formed on cut-voltage above 3.5 V. This phenomenon appears on every scan rate. This indicates that the charge process will slower at the higher voltage value.

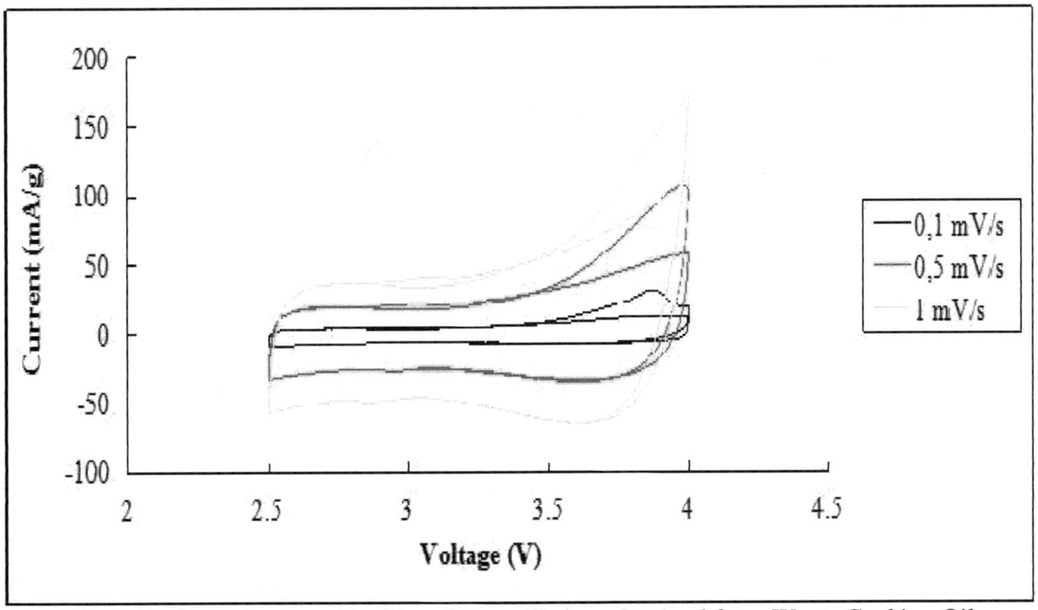

Fig 4. Cyclic Voltametry Profiles of Nano Carbons Derived from Waste Cooking Oil

Table 3 Specific capacitance of nano carbon samples

Cycle / Scan Rate	**0.1 mV/s**	**0.5 mV/s**	**1 mV/s**
1st Cycle	79.57 F/g	65.10 F/g	56.65 F/g
2nd Cycle	63.06 F/g	55.13 F/g	50.72 F/g

Table 3 present capacitancy data on every scan rate. It can be seen that the capacitance decreases with the increase in scan rate. This phenomenon occur because the decreasing of the rate of ion diffusion in pores at higher scan rate. Capacitance would be decreased in the second cycle. The capacitance reduction value will be smaller on the greater scan rate. This indicates that the stability of the electrode will be better on the greater scan rate. The capacitance value reached 79.57 F/g in the first cycle on scan rate 0.1 mV/s.

4. Conclusion

In summary, the nano carbon electrodes for lithium ion capacitors were prepared by a nebulized spray pyrolysis using waste cooking oil as carbon precursors by varying reaction temperatures from 650 to 750 °C. The structure of nano carbons were confirmed as carbon nano spheres (CNS) from the SEM observations. Raman and XRD analysis determined that the formation of graphitic carbon structures in the nano carbon samples. The specific capacitance of carbon products was obtained by cyclic voltammetry analysis conducted at various scan rate. The specific capacitance of carbon products was estimated as 79.75 at the first cycle at scan rate of 0.1 mV/s. This study shows the prospect of waste cooking oil based carbon products as electrode materials for electrochemical applications.

Acknowlagements

This study was supported by the program of International Research Collaboration and Scientific Publication provided by the High Education Council (DIKTI).

References

[1] Vivekchand RC, Cele LM, Deepak, FL, Raju AR, Govindaraj A 2004 Carbon Physics Letter 386 313-318
[2] Suriani AB 2010 Journal of the Ceramic Society of Japan 118 (10) 963
[3] Obreja VV 2008.Journal Physica E 40 2596-2605
[4] Xiaodong B 2005 Journal Tsinghua Science and Technology 10 (6) 729

[6] Marque AN, Romero R, Romero A, Valverde JL 2011 J. Mater. Chem., **21**, 1664-1672
[7] Ndwandwe S, Tshibangu P, Dikio ED 2011 Int. J. Electrochem. Sci., **6**, 749-760
[8] Suriani AB, Norhafizah J, Mohamed A, Mamat MH, Malek MF, Ahmad MK 2016 Journal of Material Science 27(11) 11599-11605.
[9] Suraini, AB, Dalila AR, Mohamed A, Mamat MH, Salina M, Rosmi MS, Rosly J, Nor RM, Rusop, M 2013 Materials Letters 101 61-64.
[10] Roberts AD, Li X, Zhang H 2014 Chem. Soc Rev., 43, 4341

Study on Utilization of Super Grade Plutonium in Molten Salt Reactor FUJI-U3 using CITATION Code

Cici Wulandari[1], Abdul Waris[2,*], Syeilendra Pramuditya[2], Asril Pramutadi AM[2], and Novitrian[2]

[1]Department of Physics, Faculty of Mathematics and Natural Sciences, Institut Teknologi Bandung, Jl. Ganesa 10 Bandung 40132, INDONESIA
[2]Nuclear Physics & Biophysics Research Division, Department of Physics, Faculty of Mathematics and Natural Sciences, Institut Teknologi Bandung, Jl. Ganesa 10 Bandung 40132, INDONESIA

*E-mail: awaris@fi.itb.ac.id

Abstract. FUJI-U3 type of Molten Salt Reactor (MSR) has a unique design since it consists of three core regions in order to avoid the replacement of graphite as moderator. MSR uses floride as a nuclear fuel salt with the most popular chemical composition is LiF-BeF_2-ThF_4-$^{233}UF_4$. ThF_4 and $^{233}UF_4$ are the fertile and fissile materials, respectively. On the other hand, LiF and BeF_2 working as both fuel and heat transfer medium. In this study, the super grade plutonium will be utilized as substitution of ^{233}U since plutonium is easier to be obtained compared to ^{233}U as main fuel. Neutronics calculation was performed by using PIJ and CITATION modules of SRAC 2002 code with JENDL 3.2 as nuclear data library.

1. Introduction

The energy consumption growth in the twentieth century demonstrated a twenty-fold increase in the use of fossil fuels, such as oil and coal, and it is followed by hydro power and renewable energy. In nuclear energy, as of 1 July 2016, there are 444 operable grid electric nuclear power reactors and 62 others under construction [1]. The majority of reactor in operation around the world are considered Generation II nuclear reactor system, and Generation I nuclear reactor system were retired and there is only a dozen of Generation III reactors in operation. Currently, Generation IV nuclear reactor system are a set of nuclear reactor designs and it is being researched for commercial applications with depend on the particular designs. The characteristic of Generation IV nuclear power system are highly economical, enhanced safety, minimal waste and proliferation resistant.

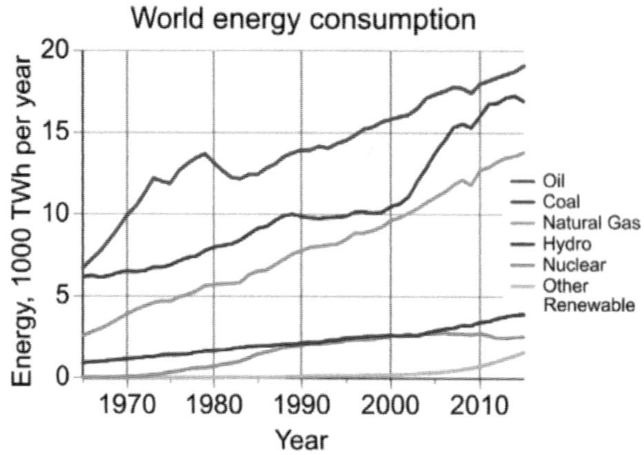

Figure 1. Graph of world energy consumption [2]

One of six of the Generation IV nuclear power systems is Molten Salt Reactor that utilize thorium as main fuel. MSR has many advantages such as safety improvement, proliferation resistance, resource sustainability, waste burning and ability for hydrogen production [3]. MSR was first studied in Oak Ridge National Laboratory in aircraft design [4]. In 1954, Aircraft Reactor Experiment (ARE) built and operated successfully with power output is 2500 kWt and temperature is 1150 K. ARE used NaF-ZrF_4-UF_4 and beryllium oxide as main fuel and moderator, respectively. After this project, ORNL proposed the secondary experimental of molten salt reactor, Molten Salt Reactor Experiment [4] from 1965-1969. This reactor used single core and graphite as moderator. MSRE was operated with power output 8 MWT in 1965 and this reactor was using thermal neutron spectrum for 4.5 years and shutdown in December 1969. The fuel salt for MSRE was LiF-BeF_2-ZrF_4-UF_4 and pyrolytic graphite as moderator.

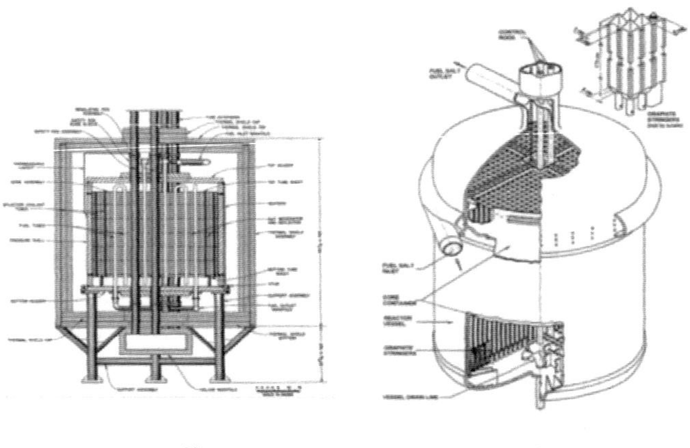

Figure 2. (a) Aircraft Reactor Experiment (ARE) diagram, (b) Molten Salt Reactor Experiment [4]

Based on the concept of MSRE, ORNL proposed Molten Salt Breeder Reactor with power output 1000 MWe, the fuel salt composed by LiF-BeF_2-ThF_4- UF_4 and NaF-$NaBF_4$ as the secondary coolant. In 1970, MSR project was closed from ORNL program. Currently, many conceptual design of Molten Salt Reactor was developed by several countries such as Russia (Molten Salt Actinide Recycler and Transmuter/ MOSART [5][6]), France (Molten Salt Fast Reactor/MSFR [6]), Japan (Fuel Irradiation for JNS and PSI/FUJI [7]) and Czech Republic (Spent Hot Fuel Incinerator by Neutron Flux/SPHINX [6]).

In this study, we focus on FUJI-U3 type of MSR that developed in Japan. In the previous study, this reactor was used Th/^{233}U as main fuel [8][9], but in this paper, the super grade plutonium will be utilized as substitution of ^{233}U because plutonium is easier to be obtained than ^{233}U.

2. Methodology

FUJI-U3 type of MSR has a unique design, it has three regions core and its reflector is made of boron carbide. Diameter and height of active core are 4.72 m and 4.66 m, respectively. Active core is consists of LiF, BeF_2, ThF_4, PuF_4 as fuel salts and graphite as moderator. The hexagonal graphite is modelled as a cylindrical element with radius 0.20 m. The lifetime of reactor is 30 years, it is depend on lifetime of graphite moderator. But in this paper, we calculate for 2000 Effective Full Power Days or 7.5 years for one loading fuel salt. The Neutronics calculation was performed by using PIJ and CITATION modules of SRAC 2002 code with JENDL 3.2 as nuclear data library. In PIJ modules, nuclear cross section of 107 groups are compressed into 30 groups, 24 fast neutron and 6 thermal neutron groups. In the diffusion calculation, CITATION, the core is divided into 65 radial and 32 axial zones. The following table is data of design parameters of FUJI-U3:

Table 1. Design parameters of FUJI-U3 [8]

Parameters	Specification
Electric Output	200 Mwe
Thermal Output	450MW(th)
Thermal Efficiency	44.40%
Reactor Vessel	
- Diameter / Height (inner)	5.40 m/5.34 m
- Thickness	0.05 m
Core	
- Diameter / Height	4.72 m/4.66 m
- Fuel volume fraction(av.)	36%
Fuel path / Duct	
- Width	0.04 m
- Fuel volume fraction	90vol%
Reflector	
- Thickness	0.30 m
- Fuel volume fraction	0.5 vol%
Power density	5.5 MW/m^3
Multiplication factor	ca. 1.01
conversion ratio (av.)	1.01
Temperature coefficient (av.)	- 2.7 x 10^{-3} 1/K
Fuel salt	
- composition	LiF-BeF_2-ThF_4-PuF_4
- mol %	71.76 - 16.00 -12.0 - 0.24*

The configuration of active core FUJI-U3 is shown in Figure 3, there are three regions core, among others core 1, core 2 and core 3. The fuel volume fraction in core 1, core 2, and core 3 are 0.39, 0.27 and 0.45, respectively. Fuel path (width:0.04 m) is region in the midst of reflector and core 3. The reflector in this reactor is used both for neutron absorber and reactor shielding. Data of radius and height of each region core is shown in Table 2.

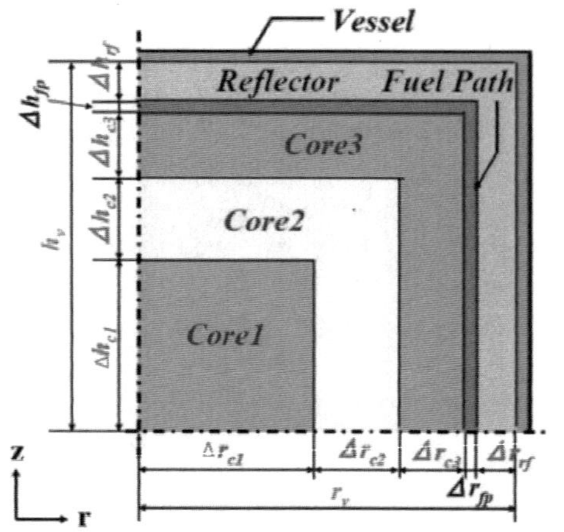

Figure 3. Configuration of active core FUJI-U3 [8]

Table 2. Data of configuration in active core of FUJI-U3 [8]

	core 1	core 2	core 3
Δr (m)	1.16	0.80	0.40
Δh (m)	1.23	0.70	0.40
Fuel volume fraction	0.39	0.27	0.45

The composition of fuel salt is varied to obtained the criticality of reactor. The percentage of LiF and BeF_2 in fuel salt is fixed is 87.76%, while the percentage of ThF_4 and PuF_4 is 12.24%. Super grade Plutonium is composed by 98% of ^{239}Pu as fissile nuclide and 2% of ^{240}Pu as fertile nuclide [10]. The composition of fuel salt is tabulated in Table 3.

Table 3. Composition of fuel salt

LiF	BeF_2	ThF_4	*PuF_4
71.76%	16%	11.64% -12%	0.24% - 0.60%

3. Results and Discussion

Figure 4(a) shows the effective multiplication factors as a function of burnup in each concentrations of super grade plutonium. The concentration of super grade plutonium is varied from 0.24% to 0.60%. Reactor can achieved their criticality with concentration of super grade plutonium of 0.60% in fuel salt with burnup 7.00 GWday/ton for 2000 EFPD. As shown in these figure, the effective multiplication factor is increased if the concentration of super grade plutonium in fuel salt is increased too. This is caused by the total of fissile material (^{239}Pu) in super grade plutonium is increased, so the fission in active core of FUJI-U3 is increased.

The conversion ratio of FUJI-U3 with thorium-super grade plutonium as fuel salt is shown in Figure 4(b). From these figure, if the concentration of super grade plutonium of 0.24% in fuel salt, the conversion ratio in the starting point of 0.79 and this value is fallen to 0.65 if the concentration of super grade plutonium of 0.60%. When the concentration of super grade plutonium is increased, the total of thorium as fertile material in fuel salt will be decreased and fissile materials produced will be reduced, so the conversion ratio of this reactor will be decreased.

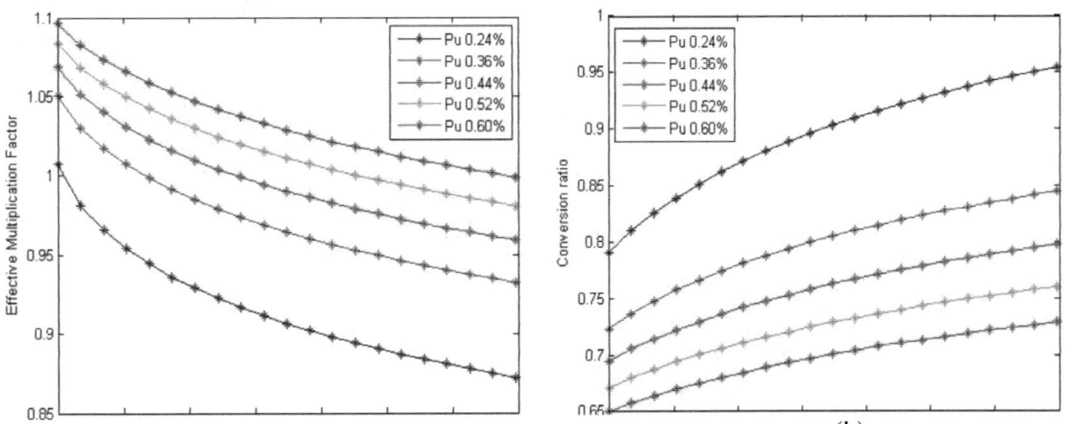

(a) (b)

Figure 4. (a) Effective multiplication factor vs burnup and (b) conversion ratio vs burnup for MSR type FUJI-U3 with Thorium-super grade plutonium as main fuel

The neutron spectra of each regions core are presented in figure 5. The trends of neutron spectra as a function of log energy are similar for each regions core. In the thermal energy range (10^{-5} eV- 10^{4} eV), neutron spectra become harder with the increasing of super grade plutonium in the fuel salt. In the previous study about the plutonium utilization in thermal reactor [11] [12], there is hardening of the neutron spectra like this case. The peak of neutron spectra in thermal energy range in figure 5(b) is higher than 5(a) and 5(c), because the fraction of moderator in core 2 is higher than others, so the thermal neutron in this region is more than other regions.

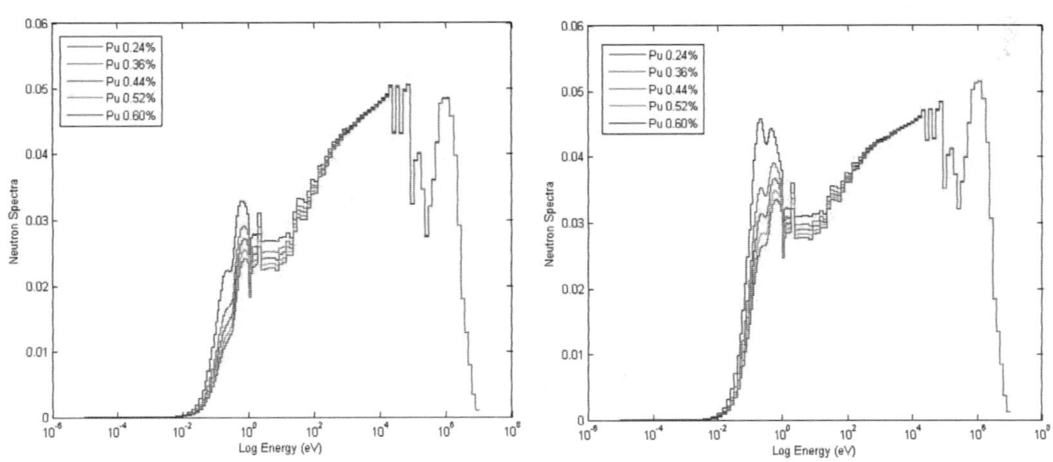

(a) (b)

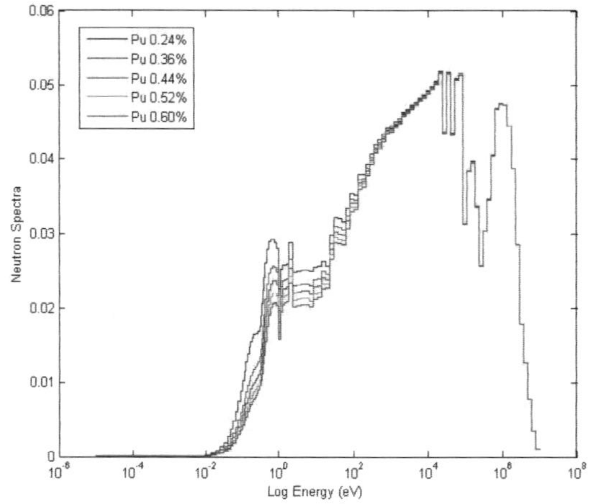

(c)

Figure 5. Neutron spectra vs log energy for FUJI-U3 using Thorium-super grade plutonium as main fuel, (a) for core 1, (b) core 2 and (c) core 3.

4. Conclusion

The study on super grade plutonium in molten salt reactor FUJI-U3 using CITATION code has been carried out. The criticality of this reactor can be achieved by loading 0.6% of super grade plutonium in fuel salt. When the concentration of super grade plutonium is improved, effective multiplication factor is increased, while conversion ratio is decreased as a function of burnup. In the thermal energy range, the neutron spectra become harder with the raising of super grade plutonium and the decreasing of moderator in active core.

References

[1] *World Nuclear Power Reactors & Uranium Requirements* (http://www.world-nuclear.org/information-library/facts-and-figures/world-nuclear-power-reactors-and-uranium-requireme.aspx, first visited August, 2016)

[2] *World Energy consumption* (https://en.wikipedia.org/wiki/World_energy_consumption-#cite_note-BP-Review-2013-1, first visited August,2016)

[3] Moir R.W, Dolan T. J, Sean M. M, Williams D. F, Greenspan E 2002 *Deep-burn molten-salt reactors* p 1

[4] Jerome S, *et al* 2014 The molten salt reactor (MSR) in generation IV: overview and perspectives *Prog. In Nuclear Energy* **77** 308-319

[5] Ignatiev V, Afonichkin V, Feynberg O, Merzlyakov A, Surenkov A, Subbotin V, 2012 Molten salt reactor: new possibilities, problems and solutions *At. Energy* **112** 157

[6] Endicott N 2013 *Thorium-fuelled molten salt reactors* (London: Weinberg Foundation) p 14-16

[7] Furukawa K, Erbay L.B, Aykol A 2011 *A study on a symbiotic thorium breeding fuel-cycle: Thorim-nes through fuji* (Istanbul: SET2011)

[8] Mitachi K, Yamamoto T, Yoshioka R 2007 *Self-sustaining core design for 200 MWe molten-salt reactor with thorium-uranium fuel: FUJI-U3-(0)* (Beijing: TU2007)

[9] Nagy K, Kloosterman J. L, Lathouwers D, Van der Hagen T. H. J. J 2012 The effects of core zoning on the graphite lifespan and breeding gain of a moderated molten salt reactor *Annals of Nulear Energy* **43** p 19-25

[10] Mark J. C, Hippel F.V, Lyman E 2009 Exploxive properties of reactor-grade plutonium *Science and Global Security* **17** p 170-185

[11] Waris A, Richardina V, Aji I. K, Su'ud Z 2013 Preliminary study on plutonium and minor actinides utilization in thorims-nes minifuji reactor *Energy Conversion and Management* **72** p 27-32
[12] Waris A, Aji I. K, Pramuditya S, Novitrian, Permana S, Su'ud Z 2015 Comparative studies on plutonium and minor actinides utilization in small molten salt reactors with various powers and core sizes *Energy Procedia* **71** p 62-68

Random Matrix Theory Approach to Indonesia Energy Portfolio Analysis

Alifian Mahardhika, Acep Purqon

Department of Physics, Faculty of Mathematics and Natural Sciences, Institut Teknologi Bandung

fiansmamda@gmail.com

Abstract. In a few years, Indonesia experienced difficulties in maintaining energy security, the problem is the decline in oil production from 1.6 million barrels per day to 861 thousand barrels per day in 2012. However, there is a difference condition in 2015 until the third week in 2016, world oil prices actually fell at the lowest price level since last 12 years. The decline in oil prices due to oversupply of oil by oil-producing countries of the world due to the instability of the world economy. Wave of layoffs in Indonesia is a response to the decline in oil prices, this led to the energy and mines portfolios Indonesia feared would not be more advantageous than the portfolio in other countries. In this research, portfolio analysis will be done on energy and mining in Indonesia by using stock price data of energy and mines in the period 26 November 2010 until April 1, 2016. It was found that the results have a wide effect of the market potential is high in the determination of the return on the portfolio energy and mines. Later, it was found that there are eight of the thirty stocks in the energy and mining portfolio of Indonesia which have a high probability of return relative to the average return of stocks in a portfolio of energy and mines.

1. Introduction

1.1. Background

According on local news, a lot of layoff happened on mining and energy employees both national and multinational companies. Domestically, PT Chevron Pacific Indonesia has sent a layoff letter of 1,200 employees plan to the Special Unit of Upstream Oil and Gas (SKK Migas). In the coal mining sector, more than 125 companies in East Kalimantan is not operating and causing thousands of people affected by layoffs. A wave of layoffs has potential to continue to happen given the price of energy commodities is still deteriorating. One of the companies is INPEX Indonesia, which decided to downsize personnel to 40% of the total personnel in Indonesia, downsizing personnel triggered mining companies and global oil and gas in the country to consider moving the portfolio from Indonesia. In other countries, the wave of layoffs hit the global enterprise. Heillong Longmay Mining, Schlumberger, Halliburton and Chevron each had to lay off 100,000 people, 34,000 people, 20,000 people, and 1,500 people.

Executive Director of the Indonesian Coal Mining Association (ICMA) stated, mining, oil and gas sector companies in global scale, began reviewing its business portfolio in Indonesia. If possible, they will focus on the mining sites in other countries which more profitable. Portfolio review has done considering the condition of the mining, oil and gas industry are worrying. The company's portfolio review also performed to ensure the company remains efficient, under employment and declining investment confidence. What can be done is to improve efficiency, control the cost of production, and optimization.

Imelda Adhisaputra (Director of Corporate Affairs BHP Billiton) said that the global mining industry is in poor condition, especially the coal mine. Therefore, it is only natural that as an international business entity, BHP Billiton is considering reviewing and reviewing back the economic value of their projects worldwide.

From the point of view mentioned before, we know that Indonesia energy portfolio need to be reviewed, then every mining, oil, and gas company portfolio can be optimized. Through this paper, we will try to analyze Indonesia energy portfolio using random matrix theory and then calculate the mean and variance of the portfolio to create Efficient Frontier graph then we can analyze the optimum return value of the portfolio.

1.2. Random Matrix Theory

Random matrix theory (RMT) was developed by Wigner, Dyson, Mehta, and others in order to explain the statistics of energy levels of complex quantum systems. RMT methods to analyze the properties of Correlation matrices (C), showed that 98% of the eigenvalues of C agree with RMT predictions, suggesting a considerable degree of randomness in the measured cross correlations. It is also found that there are deviations from RMT predictions for 2% of the largest eigenvalues. We find that the largest eigenvalue of C represents the influence of the entire market that is common to all stocks. Our analysis of the contents of the remaining eigenvalues that deviate from RMT shows the existence of cross correlations between stocks of the same type of industry, stocks having large market capitalization, and stocks of firms having business in certain geographical areas [1-3]. As stated above, our aim is to extract information about cross correlations from C. So, we compare the properties of C with those of a random cross-correlation matrix [4]. In matrix notation, the correlation matrix can be expressed as

$$C = \frac{1}{L} GG^T \tag{1}$$

where G is an $N \times L$ matrix with elements $\{g_{im} \equiv g_i(m\Delta t); i = 1,\cdots,N; m = 0,\cdots,L-1\}$ and G^T denotes the transpose of G. Therefore, we consider a random correlation matrix

$$R = \frac{1}{L} AA^T \tag{2}$$

where A is an $N \times L$ matrix containing N time series of L random elements $a_{im}$with zero mean and unit variance, that are mutually uncorrelated. Statistical properties of random matrices such as R are known [5,6]. Particularly, in the limit $N \to \infty, L \to \infty$ such that $Q = \frac{L}{N}$ (> 1) is fixed, it was shown analytically [6] that the probability density function of eigenvalues of the random correlation matrix R is given by

$$P_{rm}(\lambda) = \frac{Q}{2\pi} \frac{\sqrt{(\lambda_+ - \lambda)(\lambda - \lambda_-)}}{\lambda} \tag{3}$$

for λ within the bounds $\lambda_- < \lambda_i < \lambda_+$, where λ_- and λ_+ are the minimum and maximum eigenvalues of R, respectively given by

$$\lambda_{\pm} = 1 + \frac{1}{Q} \pm 2\sqrt{\frac{1}{Q}} \tag{4}$$

2. Methods

We constructed an Indonesia energy portfolio consists of 32 companies in energy, mining, oil, and gas sectors. We analyzed in the period of 11-26-2010 until 01-04-2016 (approximately 1378 rows of prices data from each companies). Then analyzed the probability density function (PDF) of the eigenvalues. From the result of analyzing the PDF, then we used the filtered eigenvalues to compare correlation before and after the RMT processed. After we've done the analysis of the correlation, we analyzed the eigenvector which represent each companies expected return, then we analyzed the efficient frontier of our energy portfolio. From the graphic of efficient frontier, we'll know the optimum expected return and risk value of our energy portfolio.

3. Results

As we can see at figure 1 that the contents of our C matrix mostly agree with the PDF from the random matrix and that only a few (largest) eigenvalues that deviate from the random PDF, these eigenvalues will be used to generate eigenvectors. From PDF analysis we've got $\lambda_+ = 2$ and $\lambda_- = 0.7224$.

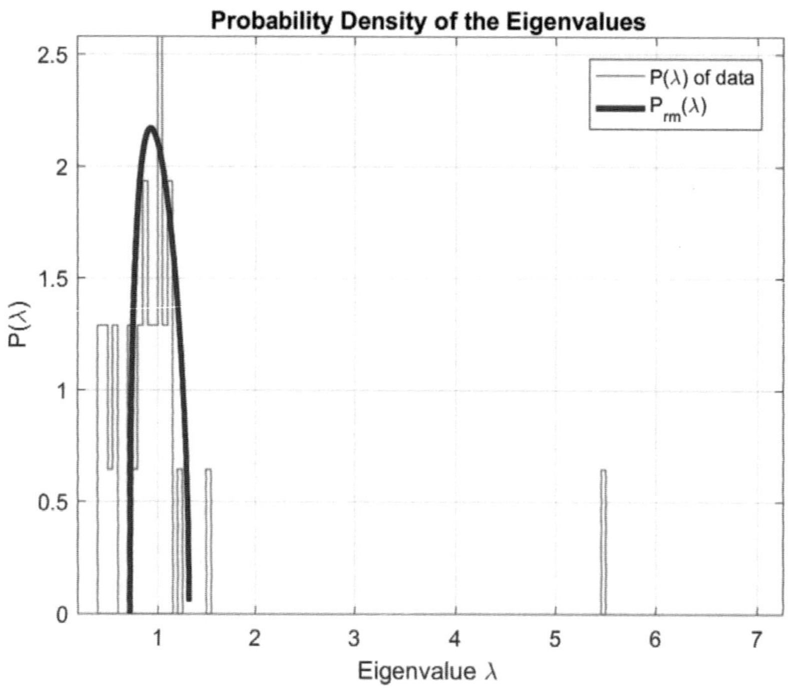

Figure 1. Probability Density Function (PDF)

Figure 2 and figure 3 shows the correlation between stocks at our energy portfolio before and after the RMT filtering. We've got that after the RMT filtering, the correlation between stocks are weaker than before we did the RMT filtering.

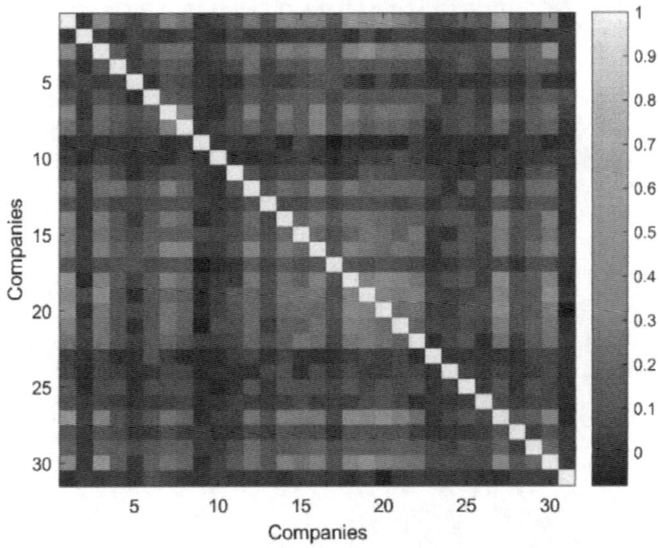

Figure 2. Correlation before RMT processed

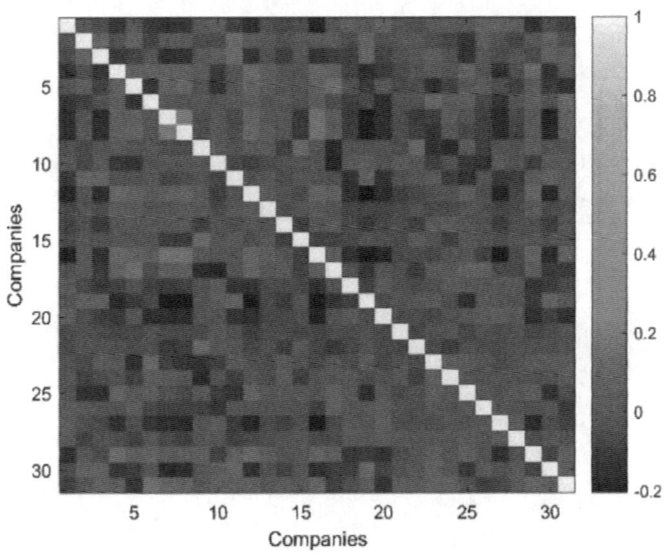

Figure 3. Correlation after RMT processed

After the RMT processed, we did a decomposition on our filtered C matrix, from the decomposition, we know that the market wide effect had a big enough potential to affect the return value of our portfolio (see figure 4).

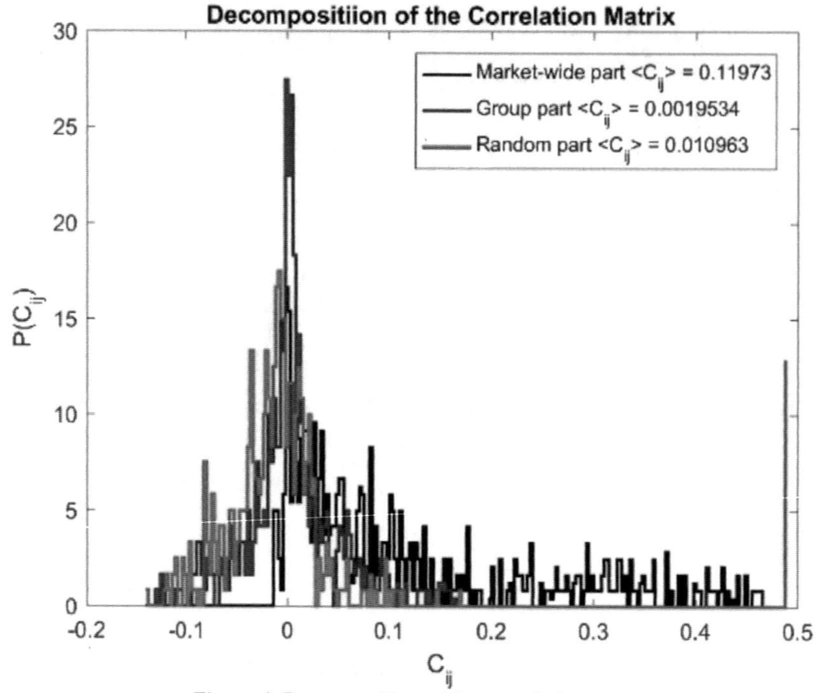

Figure 4. Decomposition of the correlation matrix

Below (figure 5) is the efficient frontier graphic, which show us about the return and risk value of our portfolio, this graph could help us to decide the risk we would take if we build the portfolio.

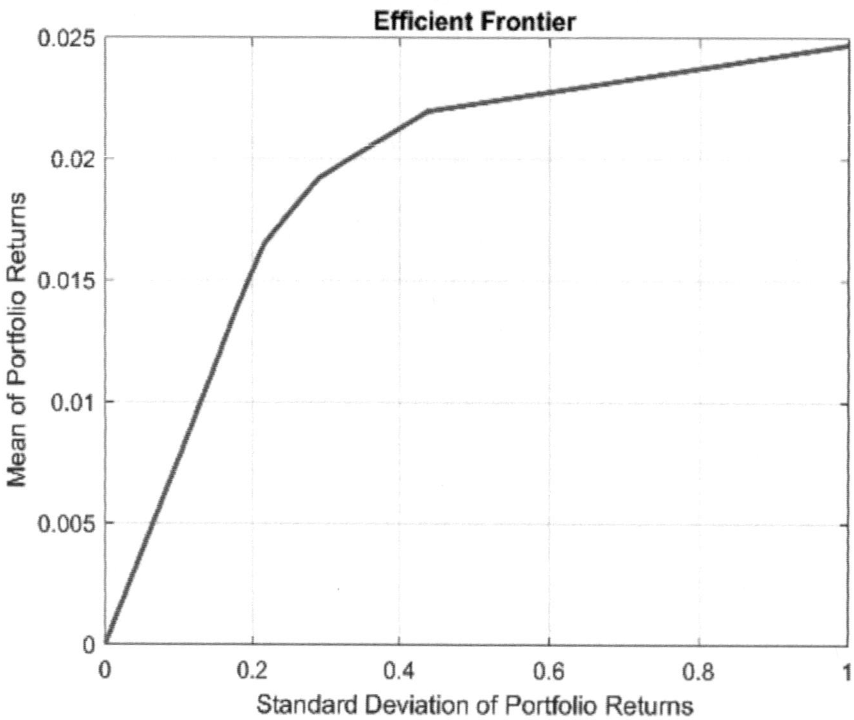

Figure 5 Efficient frontier graphic

4. Conclusions

According to our results, we can conclude that:

1. There are 8 out of 31 oil and mine stocks company that have higher return relative to the average of another stocks on the portfolio energy of Indonesia,
2. Portfolio energy of Indonesia have an optimum return approximately at 2.1 percent,
3. Portfolio energy of Indonesia will be more stable if there are just a few of another country intervention on Indonesia energy portfolio cause the 8 companies that have higher return value, the major owner of them are Indonesia.

References

[1] R.N. Mantegna, Eur. Phys. J. B 11, 193 (1999); L. Kullmann, J. Kertesz, and R.N. Mantegna, e-print cond-mat/0002238; an interesting analysis of cross correlation between stock market indices can be found in G. Bonanno, N. Vandewalle, and R.N.Mantegna, Phys. Rev. E 62, 7615 (2000).
[2] P. Gopikrishnan, B. Rosenow, V. Plerou, and H.E. Stanley, Phys. Rev. E 64, 035106(R)(2001).
[3] P. Ormerod and C. Mounfield, Physica A 280, 497 (2000). Use RMT methods to analyze cross correlations in the gross domestic product (GDP) of countries.
[4] L. Laloux, P. Cizeau, J.-P. Bouchaud, and M. Potters, Phys.Rev. Lett. 83, 1467 (1999).
[5] F.J. Dyson, Rev. Mex. Fis. 20, 231 (1971).
[6] A.M. Sengupta and P.P. Mitra, Phys. Rev. E 60, 3389 (1999).
[7] L. Laloux et al., Int. J. Theor. Appl. Finance 3, 391 (2000).
[8] F. Black and R. Litterman, Financial Analysts J. 28, (1992).
[9] J.D. Noh, Phys. Rev. E 61, 5981 (2000).
[10] M. Marsili, e-print cond-mat/0003241.

IOP Conf. Series: Journal of Physics: Conf. Series **877** (2017) 012023 doi:10.1088/1742-6596/877/1/012023

Self-Assembly of ZnO-Nanorods and Its Performance in Quasi Solid Dye Sensitized Solar Cells

A Aprilia[1*], A Erdienzy[1], A Bahtiar[1], L Safriani[1], N Syakir[1], Risdiana[1], T Saragi[1], S Hidayat[1], Fitrilawati[1], R Hidayat[2], R E Siregar[1]

[1] Department of Physics, Faculty of Mathematics and Natural Sciences, Universitas Padjadjaran, JL. Raya Bandung-Sumedang KM 21. Jatinangor, 45363, West Java, INDONESIA

[2] Department of Physics, Faculty of Mathematics and Natural Sciences, Institut Teknologi Bandung, Jl. Ganesha 10, Bandung 40132, West Java, INDONESIA

*a.aprilia@phys.unpad.ac.id

Abstract. Zinc oxide (ZnO) nanorods (NRs) were successfully prepared by self-assembly methods using zinc nitrate hexahydrate and hexamethylenetetramine as raw materials. ZnO-NRs were grown on FTO/ZnO seed layer and to enhance dye adsorption it was continued by deposition of titania (TiO_2) paste by screen printing method. Deposition time of ZnO-NRs were varied, for 120, 150 and 180 minutes and subsequently stacked with one layer of TiO_2 mesoporous. The resulting heterojunction layers of FTO/ZnO-Nrs/TiO_2 was then applied as a photoanode in quasi-solid dye sensitized solar cell (QS- DSSC) with polymer gel electrolyte (PGE) as a hole conductor. UV-Vis spectrometer was used to investigate the changes of dye adsorption in photoanode with/without inserting titania mesoporous. Characterizations of scanning electron microscopy (SEM) and X-ray diffraction was carried out and the results shows that increasing the deposition time produces a smaller average grain size, diameter and denser layer of ZnO-nanorods. From current-voltage measurement, higher efficiency (η = 2.53%) was obtained for 120 min ZnO nanorods with short circuit current density (J_{sc}) of 2.84 mA/cm^2 and open circuit voltage (V_{oc}) of 0.7 V. The combination of TiO_2 and ZnO-NRs shows a better performance in solar cells characteristics due to increases of dye adsorption on photoanode and high photogenerated electron transport rate. This work emphasizes an optimum condition of ZnO-NRs in combination with TiO_2 mesoporous as an alternative photoanode in QS- DSSC.

1. Introduction

Many researchers have been investigated the use of metal oxides such as zinc oxide and titanium dioxide as working electrode in many types emerging photovoltaic devices [1]. As semiconducting material with high energy exciton binding (~ 60 meV), direct wide band gap (~ 3.34 eV), and high electron mobility (200 – 300 cm^2 V/s), ZnO is attractively to use as an active electrode in hybrid/organic solar cells [2], meanwhile titanium dioxide (TiO_2) commonly use in dye sensitized solar cell (DSSC) systems [3]. A. K. Chandiran et.al. have been investigated the differences between ZnO and TiO_2 as working electrode in DSSC and their self-advantages. TiO_2 based photoanode generally introduce a high external quantum efficiency (EQE) due to high energy internal surface area, making a large contact with dye particles, while ZnO photoanode with high electron mobility will also produce a high recombination process on interface between ZnO and dye particle [4]. In order to utilizing the unique properties of ZnO in dye sensitized solar cells, many researchers have tried to modify the micro/nanostructures ZnO surface become nanocone, nanowire, nanorod or even nanoforest to produce a large surface area (in consequences with a better contact and high loading of dye particles) [5,6].

In dye sensitized solar cell (DSSC) systems, photoanode plays an important role in order to collect the regenerated electrons from photoexcitation of dye molecules and transfer it to external circuit through the transparent electrode, meanwhile the positive charges would be collected as cations in electrolyte side, contributed in redox cycle with incoming electrons from the external circuit thus electrical works were performed [7]. Besides that, there were several physical properties requirement as a photoanode layers in DSSC -systems such as highly transparent at visible area (400-750 nm), large surface area leading to high loading of dye particles, has a level conduction band lower than LUMO (lowest unoccupied molecular orbital) of sensitizer (dye) in order to compensate the generated charge injection at opposite side, high charge carrier mobility or even produces an efficient charge transport, and also has a good stability on redox electrolyte processes [8].

Almost physical properties of semiconductor metal oxides as photoanode correlated with its nano-crystalline morphology contribute to the efficient operation of the device. On a flat surface, only a few amount of light can absorb and used in photoexcitation in active layer (dye molecule) because of smaller cross section/interfacing area between dye molecule and photoanode. To enhance the incoming light absorbtion in active layer, one of methods is by surface micro/nanostructures engineering of photoanodes. Fine structures photoanode could be enhance the interfacing surface between metal oxide and dye molecules. According to this, we try to combine ZnO nanorod with titania mesoporous as working electrode in dye sensitized solar cell using polymer gel electrolyte based on polyethylene glycol and ionic liquid (mosalyte). It was already reported, the uses of this kind gel electrolyte in DSSC system shows a relatively stable photovoltaic performance [9, 10]. Figure.1 shows the energy band diagrams and schematic structure of the device during this research. Combination between zinc oxide nanorods and titania mesoporous were expected to enhance device performance due to the advantages properties of each metal oxides.

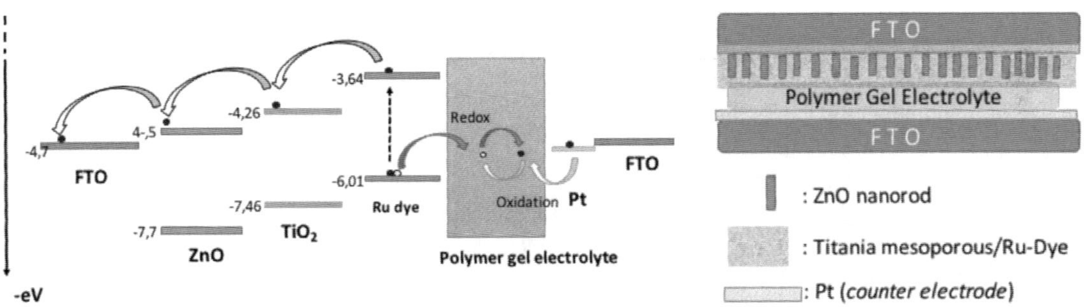

(a) (b)

Figure 1. (a) Energy band diagrams and (b) Schematic representation of FTO/ZnO-Nrs/TiO_2/Ru-dye/Pt/FTO device structure.

2. Experimental Methods

All chemicals used in this study were analytical grade without further purifications. Zinc acetate dehydrate ($Zn(CH_3COO)_2.2H_2O$) and zinc nitrate hexahydrate($Zn(NO_3)_2.6H_2O$) were purchased from Merck and used as raw materials for ZnO seed layer preparation and ZnO nanorods growth solution respectively. Titania mesoporous (Ti-Nanooxide T/SP) and dye molecules (Ruthenizer 535-bis TBA) purchased from Solaronix S.A. (Switzerland). Low iron 2.2 mm thick glass substrate, with a 7 ohm/square (FTO) fluorine doped tin oxide coating one side with and without platinum layer used as a chatode and anode (Solaronix S.A).

2.1. Photoanode preparation
Two layers of ZnO seed layer was deposited on FTO substrate using spin coating technique at 1500 rpm. Before used, FTO substrate were cleaned in ethanol and acetone by ultrasonic cleaner equipment. ZnO seed layer precursor made from 0.5mM zinc acetate dehydrate in 2-methoxyethanol anhydrous (Sigma-Aldrich) and diethanolamine (Sigma Aldrich) as a solvent and stabilizer respectively. A transparent precursor solution was achieved and continued with deposits it onto cleaned FTO substrate. Two step annealing treatment at 250 °C and 500 °C was followed to form the ZnO particle as seed layer. These all methods following the procedure from our previous research [11].

ZnO nanorods (ZnO-Nrs) were grown by self-assembly methods in isolated weighting bottle, with FTO/ZnO seed layer side was facing down at 45 °. The solution for growing ZnO nanorods was prepared by dissolving equimolar zinc nitrate hexahydrate and hexametylenetetramine (HMTA) in deionized water. Temperature of growth was fixed at 100 °C for 120 min, 150 min and 180 min inside the regularly laboratory oven. After that, ZnO nanorods was washed in ethanol and deionized water several times and then annealed at 500 °C for 30 min on isolated hotplate. For combination with titania mesoporous, one layer of Ti-Nanooxide T/SP was deposited by screen printing methods and gradually annealed until 500 °C for 30 minutes on hot plate.

2.2. DSSC assembly
The FTO with ZnO-Nrs and titania mesoporous on top is ready to use as working electrode (photoanode) of gel state (quasi solid) DSSC with structure FTO/ZnO-Nrs/TiO_2/Ru-dye/Pt/FTO. The photoanodes was immersed on dye solutions (Ruthenium) containing of 535-bisTBA (Solaronix SA) and chenodeoxolydacid (1:10) overnight. Before use, the FTO/ZnO-Nrs/TiO_2/Ru-dye was washed by acetonitrile and ethanol several times to remove the residues of dye molecules. Drilled platinum coated FTO substrate (Pt/FTO) was used as counter electrode and it was sandwiched with working electrode (FTO/ZnO-Nrs/TiO_2/Ru-dye) separated by hot melt surylin film (thickness 25 μm). Polymer gel electrolyte (PGE) was prepared by dissolves a polyethylene glycol (0.25 M) in polycarbonate and mixed with 0.2 ml of mosalyte. After that, PGE was injected to the hole at Pt/FTO side, and sealed with transparent tape to avoid a leaking.

2.3 Characterizations
The scanning electron microscopy (JEOL JSM-6510A) was used to identify the surface morphology of different time of growth ZnO-Nanorods. In order to study an addition of titania mesoporous on photoanode system, ultra-violet visible spectroscopy (T70+ PG Instrument) was carried out for each layers includes Ru-dye loaded on photoanode surfaces. To confirm the nanorod structure on each samples, X-ray diffraction measurement was used (X-pert Pro, PW3050/60 x-ray diffractometer) and continued with data analysis to determine crystals lattice of ZnO-Nrs by Debye-Scherrer method. Current voltage (J-V) characteristics was investigated using halogen lamp (power input 36.5 mW/cm^2) integrated with Yokogawa GS 200 DC voltage-current source and Yokogawa digital multimeter 7555.

3. Results and Discussions
The SEM images of ZnO-Nrs on FTO substrate was observed as shown in figure 2. The images show that a pore size (space between each ZnO-Nrs) was decreased as time of growth increased. A sample with 180 min time of growth denser than the others, with pore size approximately 50-150 nm. The 120 min pore size was bigger than other samples (200-500 nm). This results imply, when the deposition time increase, the space between nanorod would be reduced. From SEM images it is ensured that TiO_2 with particle size about 15-20 nm could be diffused easily and fill the ZnO-Nrs pores.

Figure 3 shows x-ray diffraction pattern of ZnO-Nrs on FTO substrate. All peaks of ZnO-Nrs were observed as hexagonal wurtzite zinc oxide structure consistent with the standard card Joint Committee on Powder Diffraction Standard (JCPDS) 36-1451. In order to investigate the influences of growing process, Debye –Scherrer formula was employed to determine the crystal lattice of each sample [12]. The result showed that the ZnO-Nrs samples prepared through self-assembly methods have strong peak located (2θ) between 34.1° and 34.5° corresponds with (002) plane and possessed pure hexagonal wurtzite structure with high *c*-axis orientation. The sharp peaks on XRD pattern indicate that the ZnO-Nrs possessed an excellent crystal quality. The average grain size of the ZnO-Nrs was estimated using Scherrer's formula that already reported elsewhere [11,12], and the sample with 180 min time of growth has a smallest average grain size than the others.

(a) (b) (c)

Figure 2. SEM images for ZnO Nanorods (a) 120 min (b) 150 min (c) 180 min

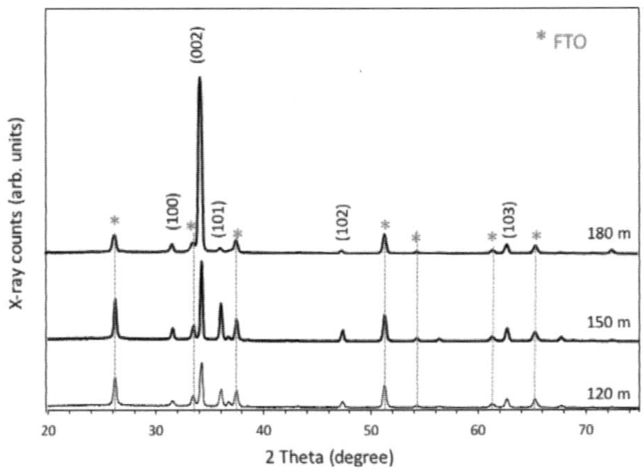

Figure 3. XRD patterns of ZnO-Nrs by variations of deposition time. Peaks marked by red star attributed to fluorine tin oxide film on glass substrate. In fact, there are several ZnO peaks that agreed with JCPDS data for hexagonal wurtzite zinc oxide structure. A dominant (002) preferred orientations was observed as time of growth increase, particularly for samples with 180 min time of growth. Is indicates that increasing a deposition time the nanorods with smaller structure were formed.

Table 1. X-ray diffraction parameters of ZnO-Nrs. *a* and *c* were crystal lattice in wurtzite (hcp) structure, and *D* is grain size of ZnO-Nrs

ZnO-Nrs time of growth (minutes)	2θ (100)	(002)	*a* (Å)	*c* (Å)	*D* (nm)
120	31.48	34.15	3.28	5.25	53.9
150	31.54	34.22	3.27	5.24	68.9
180	31.87	34.44	3.24	5.20	44.8
ZnO (JCPDS)	31.76	34.42	3.25	5.20	

Table 1. shows x-ray diffraction parameters of ZnO-Nrs and this results were supported by SEM images for 180 min time of growth that shows a smallest average diameter of ZnO-Nrs. These results

indicate that increases time of growth would reduce the grain size, diameters and producing a denser layer of ZnO-Nrs structures. Actually, this phenomenon come to an understanding with the size of nanorods strongly depend on [Zn^{2+}] and [OH^-] concentrations during the growth process. Thus a long time of growing process facilitates the supersaturation of ZnO leading to produce a smaller structure [13]. Lin et.al have been clarified that increasing temperature and time deposition on ZnO-nanorods growth by hydrothermal method, the aspect ratio (L/r) would be increased while the concentration of precursor keeps constants [14].

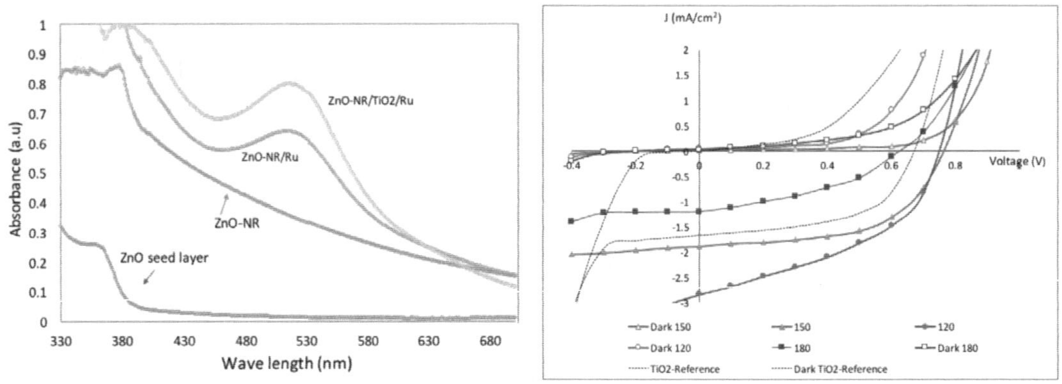

(a) (b)

Figure 4. (a) Ultra-violet visible spectra of each layers that composed of ZnO-Nrs/TiO_2 photoanodes for 120 min time of growth nanorods structure (b) Current-voltage characteristics curve of gel state dye sensitized solar cell (quasi solid DSSC) measured under irradiation of 36.5 mW/cm^2 halogen lamp.

TiO_2 material is already known to has a large surface area; thus a dye particle efficiently adsorb prominent higher dye loading compared to ZnO [4]. Therefore, in this work we combine the titania mesoporous and ZnO-Nrs as photoanode of the devices. Figure 4 (a) shows UV-Vis spectra for all samples, it reveals that ZnO-Nrs/TiO_2/Ru-dye heterojunction layer has a higher absorbance at visible region (correlated with Ru-dye absorbance area) compared with ZnO-Nr/Ru-dye. This result implies that an addition of titania mesoporous is effective to enhance dye loading in photoanode.

Table 2. Device parameters determined from current-voltage characteristics.

ZnO-Nrs Time of growth (minutes)	J_{sc} (mA/cm^2)	V_{oc} (volt)	FF (%)	Efficiency (%)
120	2.84	0.74	44	2.53
150	1.88	0.75	60	2.31
180	1.3	0.62	40	0.88
TiO_2 only	1.67	0.64	57	1.67

Current-voltage characteristics of all devices was carried out under irradiated by halogen lamp with 36.5 mW/cm^2 power input. The J-V (photocurret density – voltage) curves are shown in figure. 4 (b). Short circuit current density (J_{sc}), open circuit voltage (V_{oc}), fill factor (FF) and efficiency were observed and calculated as presented in table 2. Energy conversion efficiency obtained for quasi solid DSSC with 120 min, 150 min, 180 min time deposition of ZnO-Nrs are 2.53%, 2.31%, 0.88%

respectively. A lower efficiency for 180 min was induced by the low of short circuit current and the FF correlated with the ZnO nanostructures. ZnO-Nrs with 180 min time depositions has a smaller pores size thus the titania particles could not fill the pore efficiently and makes a self-layer on top of ZnO-Nrs. This situation produces an empty space between them (ZnO-Nrs/TiO_2) leading to a bad contact of photoanode systems consequently generating small current. By contrast, however the sample with 120 min time deposition of ZnO-Nrs has a large short circuit current density (J_{sc} = 2.84 mA/cm^2) but shows a lower fill factor (FF) compared with 150 min. It was attributed by higher photogenerated electron transport rate and better interfacing between ZnO-Nrs and titania mesoporous, but it could be produce an unbalance of charge carrier mobility between cathode and anode leading to small FF. ZnO material has a fast electron transfer compared to TiO_2 but also has a high probability for recombination process, therefore the relative small FF was obtained. Interestingly in this case, the 150 min ZnO-Nrs shows a large FF compared to other samples even for TiO_2 only as photoanode, indicates a better junction inside DSSC system, predominantly between photoanode and electrolyte gel. These results assume that ZnO-Nrs/TiO_2 heterojunction; with an optimum condition of controllable ZnO nanostructures, can be used as an alternative photoanode in QS-DSSC.

4. Conclusion

ZnO nanorods with fine structure was successfully prepared using self-assembly method, with time of growth was varied by 120 min, 150 min and 180 min, at 100 °C. From SEM and XRD measurements, it concluded that increasing time of growth in deposition process produce a smaller average grain size, diameter and denser layer of ZnO-nanorods. From current-voltage measurement, higher efficiency was obtained using 120 min ZnO nanorods with short circuit current density (J_{sc}) of 2.84 mA/cm^2 and open circuit voltage (V_{oc}) of 0.7 volt. Furthermore, the combination of TiO_2 and ZnO-NRs shows a better performance in solar cells characteristics due to increases of dye adsorption in photoanode leading to high photogenerated electron transport rate. A better fill factor was achieved for device with 150 min ZnO-Nrs indicates an optimum condition of nanostructures and a balance of charge collections at each opposite electrodes.

Acknowledgement

This work was financially supported by Universitas Padjadjaran under Academic Leadership Grant (ALG 1-1-6) year of 2016.

References

[1] Meidan Y, Xiaoru W, Mengye W, James I, Nan Z, Changjian L and Zhiqun L 2015 *Materials Today*, Volume **18**, Number **3**

[2] Mohammad V, Ahmad U and Yoong-Bong H 2010, Metal Oxide Nanostructures and Their Applications *ZnO Nanoparticles: Growth, Properties, and Applications* (vol 5) ed A Umar and Y-B Hahn (American Scientific Publisher) chapter 4 pp 1-36

[3] Yichuan L, Jason K C, Yi Y, Gongming W, Linda M, Hanyu W, Jin Z Z and Yat Li 2013 *Nano Energy* **2** 1373-1382

[4] Aravind Kumar C, Mojtaba A-J, Mohammad K.N. and Michael G 2014 *ACSNano* vol. **8** No.**3** 2261-2268

[5] Seung H K, Daeho L, Hyun W K, Koo H N, Joon Y Y, Suk J H, Costas P G, and Hyung Jin S 2011 Nano Lett. **11** 666–671

[6] Lori E G, Matt L, Dawud H T, Max M, Josh G, Gabor S and Peidong Y 2005 *Nano Letters* Vol.5 No. 7 1231-1236

[7] Michael Grätzel 2006 Thin film solar cells; Fabrication, characterization and applications *Nanocrystalline Injection Solar Cells* ed Jef P and Vladimir A (Jhon Wiley & Sons Publisher) chapter 9 pp 363-385

[8] Hari M U, S Senthilarasu, Min-Hung H, and D Kishore K 2013 *Solar Energy Materials & Solar*

Cells **119** 291-295

[9] G Anantharaj, J Joseph, M Selvaraj and D Jeyakumar 2015 *Electrochimica Acta* **176** 1403-1409

[10] Waode S A, Herman, Fitrilawati and Rahmat H 2015 *Advanced Material Research* vol. **1112** pp. 256-261

[11] Annisa A, Priastuti W, Veinardi S, Herman, Rahmat H, Akihiko F, and Masanori O 2013 *Solar Energy Materials & Solar Cells* **111** 181-188

[12] Kai long F, Uda H, Kashif M and Chun Hong V 2014, *Nanoscale Research Letters* **9**:429

[13] D Polsongkram, P Chamninok, S Purkird, L Chow, O Lupan, G Chai, H Khallaf, S Park and A Schulte 2008 *Physica B* **403** 3713-3717

[14] Suanzhi L, Hailong H, Weifeng Z, Yan Q and Fachun L 2013 *Nanosclae Research Letters 8:158*

IOP Conf. Series: Journal of Physics: Conf. Series **877** (2017) 012024 doi:10.1088/1742-6596/877/1/012024

The effect of surface morphology in copper oxide nanostructure to photo detector characteristics

Angga Virdian[1], Christoforus Dimas Satrya[1], Eka Nurfani[1], and Yudi Darma[1,a]

[1]Quantum Semiconductor and Device Laboratory, Department of Physics, Institut Teknologi Bandung, Bandung 40132, Indonesia.

Email: [a]yudi@fi.itb.ac.id

Abstract. We study the effect of structural properties of copper oxide nanostructure and its sensitivity for photo detector application. Copper oxide (CuO) nanostructures were prepared by combining the thermal evaporation with dry oxidation. We select CuO system at different thickness (around 90, 170, and 270 nm) and X-ray diffraction spectra confirms their polycrystalline structure. Scanning electron microscope images show the distinct CuO morphology surface for each sample on which the grain shape is changed and the size is grown with the increasing of the CuO thickness. Fourier-transform infra-red spectra reveals the high penetration depth of oxygen at ~90 nm thick sample indicated by large Si-O-Si bond intensity. Photo current characterization for each sample is carried out to compare its light sensing properties. We found that the thickness of CuO nanostructure system is related to the sensitivity of device during the light exposure in which thinner samples have better performance. Furthermore different band gap for each CuO sample is predicted from its sensitivity at continuous light exposure. We suggest that the grain boundary and the distinct morphology might promote a unique confined nanostructure effect, resulting in band gap widening. These studies demonstrate a new approach for tunable photonic device efficiency that could be beneficial in reducing energy loss during energy conversion.

1. Introduction

Computer has modernized every single aspect of people life, ranging from factory control to sophisticated scientific machinery. It is based on semiconductor material such as Silicon and Germanium which employ electron movement to carry and process information. Transistor which facilitates such control is characterized by its thickness; overly thick of the chip would not perform well while too thin chip might experience quantum tunneling that causes current leakage. The alternative solution is to combine with photonic technology that uses light as information carrier [1]. Such advancement can be achieved with development of Photo detector that works at low energy with minimum loss and detects wide range of photon wavelength. Alteration of certain aspect in material can be done to manage the requirement, and in this case is nanostructure [2]. Engineered diminutive structure of material can substantially change the electronic structure of material with an effect called Quantum confinement. With tunable electronic energy level, we can produce photo detecting device to suit our needs to detecting light property.

In this paper, we investigate the properties of copper oxide (CuO) light sensor with distinctive surface morphology. CuO nanostructures are fabricated by simple method of metallization and oxidation, and variations of surface are obtained through thickness difference. White light from LED that contains different wavelength will be used as light source. X-Ray Diffraction (XRD) and Fourier Transform Infrared (FTIR) spectra will be used to determine the differences in crystal orientation and atomic bonding at each nanostructure. Scanning Electron Microscope (SEM) is used to verify the nanostructure visually. Finally, photo current characterization of nanostructure was performed to analyze its sensing characteristic, and then a model derived from Brus equation is used to explain this phenomena [3].

2. Experimental Method

Device has been developed by using simple method of thermal annealing, metallization and dry oxidation. In the beginning stage, a thin layer of copper is deposited on RCA pre-cleaned silicon substrate. We use 99.999% pure copper wire to grow such layer and vary its length into 2 mm, 3 mm, and 4 mm in order to fabricate the samples with 3 different thicknesses. Then, these samples were oxidized in rich oxygen environment at atmospheric ambience (600^0C for 4 hours). Finally, metal contacts were prepared at the sides of each sample. Each sample was characterized with X-Ray diffraction spectra, Fourier transform infra-red spectra, Scanning electron microscope and photo current characterization.

3. Result and Discussion

In order to find the thickness of our structure, we used Spectroscopic Ellipsometry (SE) equipment in the energy ranges of 1.2 to 6.5 eV. For Cu films, Lorentz model was used in the fitting process ranging in the full energy spectrum of SE data and we found that the thickness of Cu films are 50 mm, 59 mm and 62 nm from 2 mm, 3 mm and 4 mm of length of copper wire respectively. In the fitting and analysis of CuO structure from SE data, Cauchy model that describes transparent material was used in energy ranges below energy bandgap (transparent area) for obtaining the thickness of CuO that are 90, 170 and 270 nm respectively.

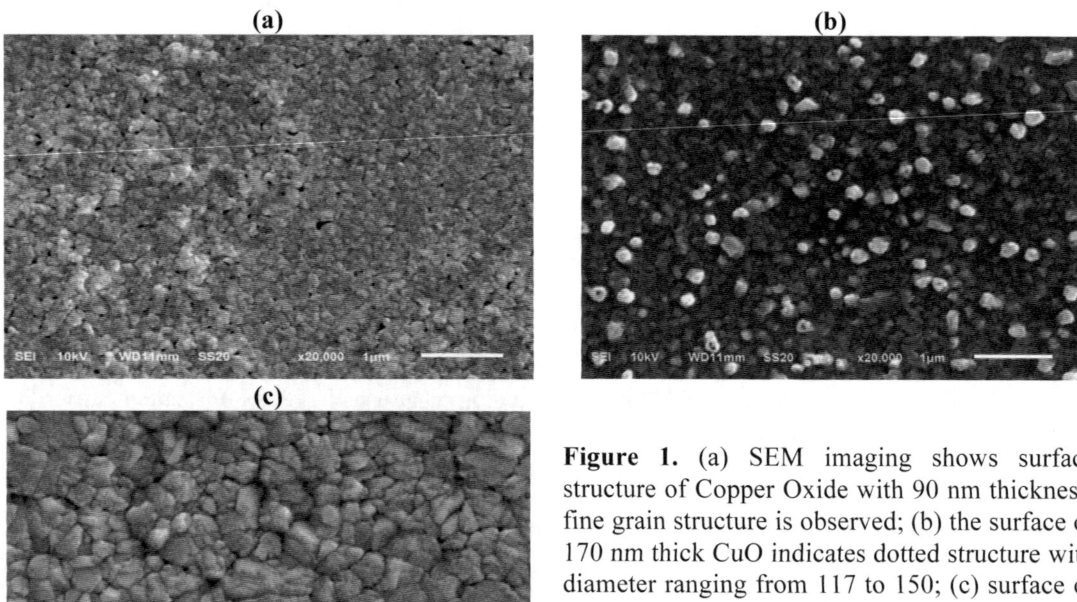

Figure 1. (a) SEM imaging shows surface structure of Copper Oxide with 90 nm thickness, fine grain structure is observed; (b) the surface of 170 nm thick CuO indicates dotted structure with diameter ranging from 117 to 150; (c) surface of the 270 nm thick CuO is flat with boundary of the grain even smaller, ranging from 59 to 100 nm.

Figure 1 shows visual morphology of CuO nanostructure with different thickness obtained by SEM imaging. Figure 1(a) depicts the surface of 90 nm thick sample which shows fine grain of CuO cluster that spread almost evenly. This image shows that there is no gap between each cluster and suggests that every single cluster is connected each other to form a CuO layer. Figure 1(b) shows the surface morphology of the sample with the thickness of 170 nm. Dark background and some white dots indicated a tiny cluster of CuO with diameter of approximately 117 to 150 nm is sitting on the top of CuO layer, and these tiny dots are small enough to quantum confinement effect to take place. Figure 1(c) is the surface of 270 nm CuO sample. Noting the relatively same coloration of image and small crack with distinct black color that separates the layer into cluster, there is a strong indication that small flat fraction of CuO layer has been formed with diameter of 58 nm to 100 nm. As these fractions are even smaller than the dots before, we can also expect that the confinement effect to be occurred.

Figure 2 shows XRD spectra of all CuO nanostructures. In general, all of the CuO samples have a good crystalline structure which is indicated with some sharp peaks at common points that correspond with crystal orientation of (111), (-111), and (-113) respectively [4]. However, some of the structure developed a unique orientation correlated to the thickness. At CuO sample with the thickness of 170, the new peaks emerge at points correlated with (110) orientation, meanwhile in the thicker sample of 270 nm, an additional new peaks indicated crystal with (-311) direction is found [4,5]. Comparing the intensity of spectra at these points in each sample, we can conclude that as the thickness increase, CuO structure forms a better crystal and promotes more diverse crystal orientation.

Figure 3 shows FTIR spectra of the CuO nanostructure systems, confirming several important bonding such as Cu-O, Si-Si, and Si-O-Si [6,7]. The copper and oxygen bonding are observed in all samples with distinction in absorbance intensity which depends on the thickness of each sample. At the thinner CuO sample, the FTIR spectra reveals that Si-O-Si bond was formed and shows the highest absorbance intensity compared to the thicker layer, which it indicates the decreasing quantity of this bond that is directly related with increasing copper oxide layer. These results confirm the oxygen penetration through the copper layer that easily bonds with Silicon substrate at relatively thin layer.

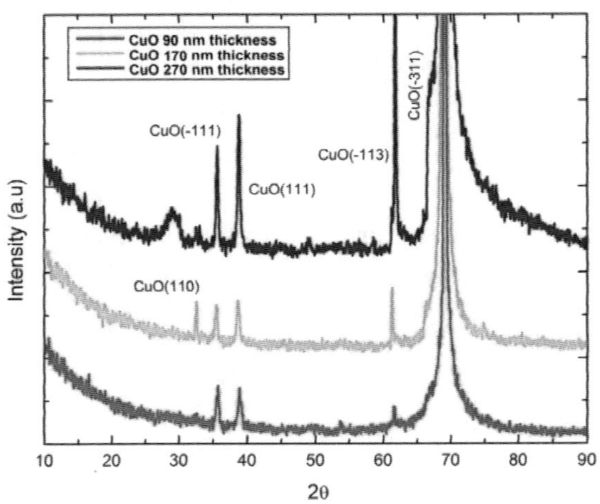

Figure 2. XRD spectra of all samples indicate all crystal orientation formed and crystalline of each.

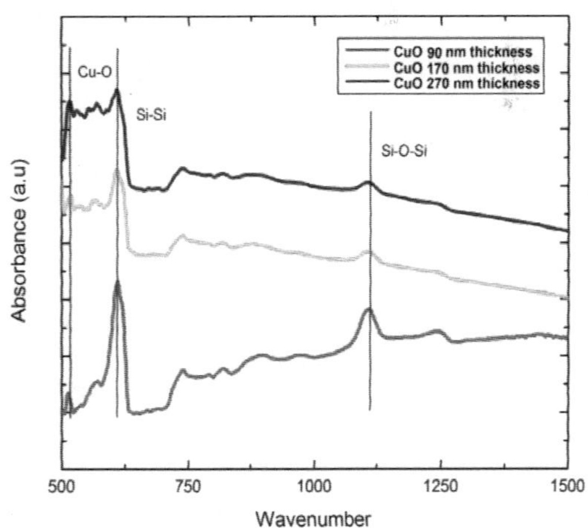

Figure 3. FTIR spectra show important atomic bonding between Oxygen, Copper, and Silicon.

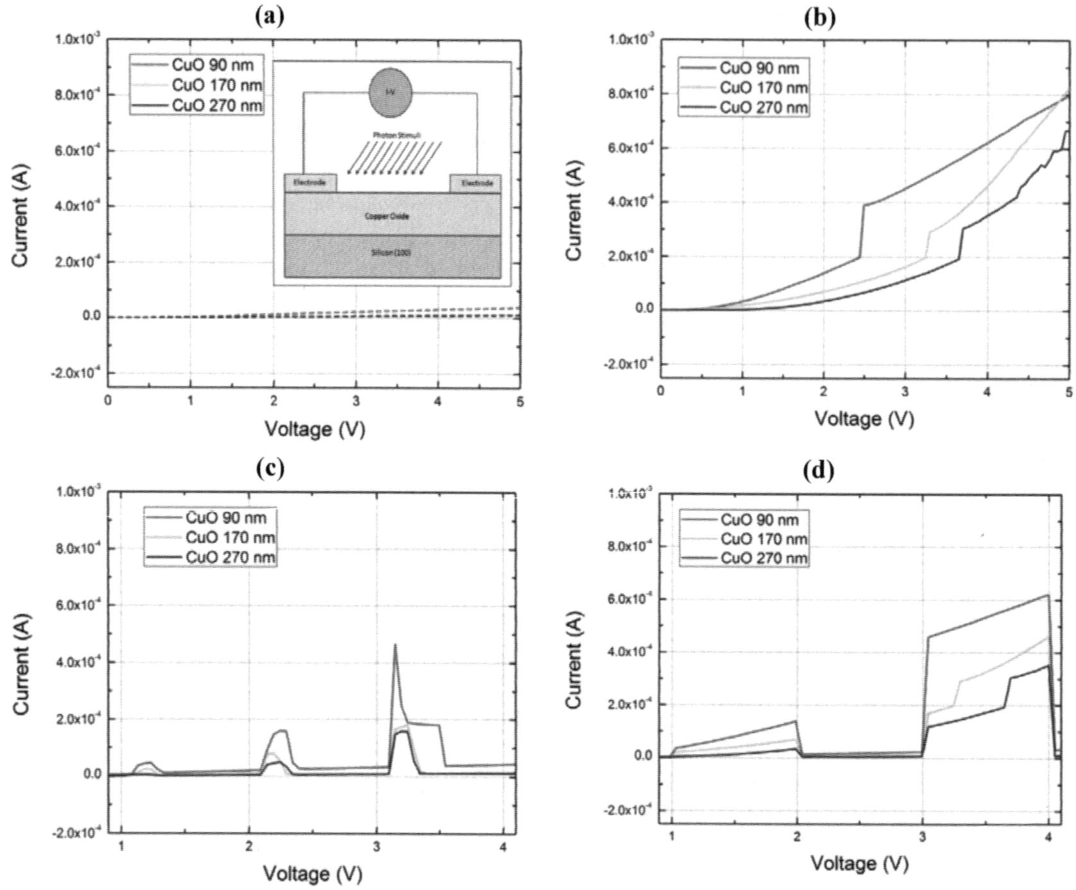

Figure 4. Figure (a) shows experimental setup for Photo current characterization and the current respond when no photon introduced in the system, notable feature is not observed, which is contrary with figure (b) that shows noticeable increment in current characteristics; figure (c) shows photo current character when photon only given when voltage have integer value; figure (d) shows IV characteristics when light stimulates sensor at certain interval in respect with voltage, in this case only at 1 to 2 volt and 3 to 4 volt the light switched on.

Using the experimental setup as shown above, the photo current relation is investigated. We measure the photo current characteristics at the maximum exposure of halogen light source that used measurement configuration. Figure 4(a) and (b) show the Photo current performance of the device. At the absence of light stimuli, electron current in sensors do not increase much even when the applied voltage is increased, in contrast, noticeable increment is observed when exposure is given, clearly photon energy given has interacted each unique nanostructure and yield different electron current properties. Since the voltage and light source is identical, there is a strong indication that these results are triggered by the structure of each nanostructure.

These experiments reveal a unique pattern which extrapolates from all of sensors, the electron current seems increasing in respect with voltage and increases rapidly after exceeding the current boundary value of 2.0×10^{-4} Ampere. The applied voltage where this phenomenon occurs also vary in increasing fashion as the sample getting thicker, and it means the energy used by electron to cross band gap of device to achieve the current boundary value is increasing. It also means that nanostructure developed alters the electronic structure therefore resulting in I-V shifting. SEM image analysis confirms structural differences between samples. As the confinement radius of structures is smaller, the band gap structure of material is expected to change as governed in Brus equation (equation 1).

$$\Delta E = E_{\text{qd}} - E_{\text{bulk}} = \frac{\pi^2 \hbar^2}{2R^2}\left(\frac{1}{\text{m}_\text{e}} + \frac{1}{\text{m}_\text{h}}\right) - \frac{1{,}8\text{e}^2}{4\pi\varepsilon_0\varepsilon_\text{r}R} \quad (1)$$

Where E_{qd} is band gap energy for confined material, E_{bulk} is band gap energy for bulk material (in CuO 1.6 eV [8]), m_e and m_h are respectively electron effective mass and hole with value of 0.4 m_o [9] and 9.7 m_o [10] (m_o is electron rest mass), ε_r is dielectric constant (10^4 for CuO [8]) and R is nanodot radius. Using this and radius information, calculated band structure widening is shown at table 1. The results conclusively explain the band gap alteration due to different confinement size which leads to phenomenon above. Besides the confinement effect, crystal orientation taking major role explaining the band gap difference, as mentioned before as Copper Oxide thicker, it is more diversified and will lead to a lot of lattice mismatch that interfaces different crystal orientation [11]. Space between Copper and Oxygen atoms would slightly move and change band structure entirely ranging from band gap widening to redirect valence state into another crystal point which means extra energy needed to excite electron [12,13].

Table 1. Calculation result with Brus equation.

Nanostructure Diameter	Band Gap Widening	Wavelength equivalent
58 nm	0.001 eV	824.6 nm
117 nm	0.00026 eV	825 nm

In the next part, we evaluate the response and sensitivity of Copper Oxide sensor by exposing the system with light stimuli at the intermittent mode. Figure 4(c) depicts the I-V properties that indicate current response to exposure. Copper Oxide sample with 90 nm thickness shows the high sensitivity, while the sensitivity decrease when CuO thickness increases. So in this study, CuO sensor with 270 nm thick Copper Oxide layer has the least sensitivity. In addition, figure 4(d) shows I-V response when light is applied at certain voltage interval. It shows the rapid changes and is interpreted as the effect of the light switching.

4. Conclusion

Nanostructured Copper Oxide photo detector has been developed successfully by varying the CuO layer thickness. Distinct surface structure has been confirmed by SEM image and leads to the structure that could introduce the confinement effect. From the structural point of view, our results show that thicker CuO layer promotes the good crystallinity with more crystal orientation. It is important to conclude that photo current characterization shows the different response as the effect of carrier confinement on the surface and also due to the distinct crystallinity. This research demonstrated the controlling of photo detector characteristics by surface morphology modification.

Acknowledgement
This work is partly supported by Desentralisasi DIKTI 2016 program and RUPT 2016 from Indonesian government. We also thank to SSLS-NUS Singapore for SE measurement facility.

IOP Conf. Series: Journal of Physics: Conf. Series **877** (2017) 012024 doi:10.1088/1742-6596/877/1/012024

References

[1] Dai D, Bauter J and Bower J E 2012 *Light: Science and Application* **1** e1.
[2] Furthmüller J *et al* 2005 *Physical Review B* **72** 20.
[3] Brus L 1986 *The Journal of Physical Chemistry* **90** 2555-2560.
[4] Etefagh R, Azhir E and Shahtahmasebi N 2013 *Scientica Iranica* **20** 1055-1058.
[5] Wang W, Zhan Y and Wang G 2001 *Chemical Communications* **8** 727-728.
[6] Zhao B *et al* 2013 *Journal of Material Chemistry A* **1** 367-373.
[7] Mansour N *et al* 2012 *Optical Material Express* **2** 740-748.
[8] Ching W Y, Xu Y N and Wong K W 1989 *Physical Review B* **40** 11.
[9] Brus L 1984 *The journal of Chemical Physics* **80** 4403-4409.
[10] Koffyberg F P and Benko F A *Journal of Applied Physics* **53** 1173-1177.
[11] Zhang P *et al* 2001 *Nature* **409** 69-71.
[12] Xiuying G *et al* 1992 *Crystal Research and Technology* **27** 1087-1096.
[13] Humayun M A *et al* 2012 *Journal of Russian Laser Research* **33** 387-394.

International Conference on Energy Sciences (ICES 2016) IOP Publishing
IOP Conf. Series: Journal of Physics: Conf. Series **877** (2017) 012025 doi:10.1088/1742-6596/877/1/012025

Simulation of Heat Transfer in Husk Furnace with Cone Geometry Based on Conical Coordinate System

Iman Noor[1a], Faozan Ahmad[2a], Irzaman[2b] and Husin alatas[2c]

[1] Department of Physics, Bogor Agricultural University, Jalan Meranti Kampus Dramaga IPB, 16880, INDONESIA
[2] Theoretical Physics Division, Department of Physics, Bogor Agricultural University, Jalan Meranti Kampus Dramaga IPB, 16880, INDONESIA

[1a] Email: iman_noor09@apps.ipb.ac.id
[2a] Email: faozan@ipb.ac.id
[2b] Email: irzaman@ipb.ac.id
[2c] Email: h.alatas@gmail.com

Abstract. Simulation of Heat Transfer in Husk Furnace with Cone Geometry Based on Conical Coordinates has been performed.This simulation aimed to study the heat distribution of temperature based on conduction and convection mechanism on conical coordinate system. Fluid dynamics inside the cone of husk furnace was obtained by solving the Navier - Stokes equations with laminar flow approach. The initial temperature in all parts of the cone is room temperature, except at the bottom of the cone is 700 °C. Through numerical calculation of heat conduction and convection equation by FDM method, we got that the velocity of fluid flow at the center cone is 13.69 m/s for 45 s, 11.90 m/s for 60 s, and 7.25 m/s for 120 s, with unfixed temperature condition in the cone.

1. Introduction

Husk furnace is a conical stove which uses fuel from rice husk. There are two component geometry in husk furnace that are cone part and cylinder part [1] (see Fig 1). Cylinder part function's used as heat source of a stove. While the cone function is the part of furnace that will ensure the heat is delivered well [2]. Heat distribution of temperature in the cone husk furnace can be determined theoretically by studying heat transfer based on conduction and convection process. Free convection flow in heat transfer can be solved by laminar model. The Graetz problem that we knew as the heat transfer problem of laminar fluid flow in ducts [3]. The Graetz problem could be studied with axial diffusion in circular tube, using a semi- infinite domain formulation with a specified inlet condition [4]. While the effects of axial diffusion in a infinite domain formed by an insulated preparation region followed by an isothermal wall was analyzed [5]. Noor [1] studied the heat transfer husk furnace in cylinder coordinate with laminar approach. Srinivasa [6] studied unsteady free convection flow and heat transfer from an isothermal truncated cone. Cone had been made by cylinder coordinate [7]. Except that to describe a cone can be modeled by conical coordinate system [8]. Fluid dynamics inside the cone husk furnace can be obtained by solving the Navier-Stokes equations to determines how convection works [9]. The pressure gradient of the fluid inside the cone can be derived from temperature gradient or fluid density gradient [1]. The aim of this research is to study the heat distribution of cone's temperature husk furnace based on conduction and convection mechanism in conical coordinate system with laminar fluid flow approach.

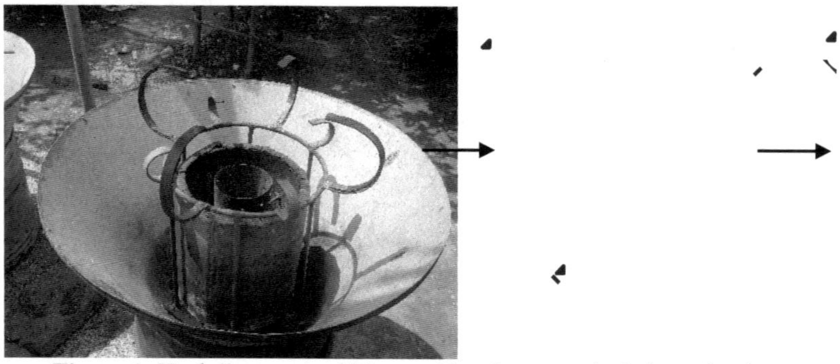

Figure 1. Husk Furnace and it's considered geometrical shape for the correspondingly numerical modeling

2. Result and Discussion

2.1 Mathematical Modelling of Heat Transfer in Conical Coordinate

Heat transfer on the cone husk furnace is associated with the phenomenon of conduction and convection [10]. The base governing equation for the heat transfer can be expressed as

$$\rho C_p \frac{\partial T}{\partial t} + \vec{U}\nabla T = k\nabla^2 T \tag{1}$$

With $\vec{U}$ is fluid flow velocity, ρ is air density, Cp is constant of material specific heat, k is constant of conductivity thermal matter, α is constant of diffusivity thermal matter, and T is temperature on conical [8].

The term on the right side of Equation (1) corresponds to the heat conduction. The Laplacian on the equation in conical coordinate is written as :

$$\begin{aligned}\nabla^2 T = {} & \frac{\partial^2 T}{\partial r^2} + \frac{2}{r}\frac{\partial T}{\partial r} \\ & + \frac{1}{r^2(\theta^2 - \lambda^2)}\left\{(\theta^2 - b^2)(c^2 - \theta^2)\frac{\partial^2 T}{\partial \theta^2} - \theta[2\theta^2 - (b^2 + c^2)]\frac{\partial T}{\partial \theta}\right. \\ & + (b^2 - \lambda^2)(c^2 - \lambda^2)\frac{\partial^2 T}{\partial \lambda^2} \\ & \left. + \lambda[2\lambda^2 - (b^2 + c^2)]\frac{\partial T}{\partial \lambda}\right\}\end{aligned} \tag{2}$$

While, the second term on left side of the equation corresponds to the heat convection and in conical coordinate the term can be written as:

$$\begin{aligned}\nabla T = {} & \overrightarrow{\boldsymbol{U}_r}\frac{\partial T}{\partial r} + \frac{1}{r(\theta^2 - \lambda^2)^{\frac{1}{2}}} X \left\{\overrightarrow{\boldsymbol{U}_\theta}(\theta^2 - b^2)^{\frac{1}{2}}(c^2 - \theta^2)^{\frac{1}{2}}\frac{\partial T}{\partial \theta}\right. \\ & \left. + \overrightarrow{\boldsymbol{U}_\lambda}(b^2 - \lambda^2)^{\frac{1}{2}}(c^2 - \lambda^2)^{\frac{1}{2}}\frac{\partial T}{\partial \lambda}\right\}\end{aligned} \tag{3}$$

With $\overrightarrow{U_r}$ is fluid flow velocity for r radius direction and $\overrightarrow{U_\theta}$ is fluid flow velocity for θ direction. $\overrightarrow{U_\lambda}$ is fluid flow velocity for λ degrees direction. Cause of fluid flow velocity that used is r radius direction than the others direction assumed is 0, so from equation (3) can simplified be :

$$\nabla T = \overrightarrow{\boldsymbol{U}_r}\frac{\partial T}{\partial r} \tag{4}$$

From equation (2) and (4), we combine it to equation (1), so the equation be :

$$\frac{dT}{dt} = \alpha\left(\frac{\partial^2 T}{\partial r^2} + \frac{2}{r}\frac{\partial T}{\partial r} + \frac{1}{r^2(\theta^2 - \lambda^2)}\left\{(\theta^2 - b^2)(c^2 - \theta^2)\frac{\partial^2 T}{\partial \theta^2} - \theta[2\theta^2 - (b^2 + c^2)]\frac{\partial T}{\partial \theta} + (b^2 - \lambda^2)(c^2 - \lambda^2)\frac{\partial^2 T}{\partial \lambda^2} + \lambda[2\lambda^2 - (b^2 + c^2)]\frac{\partial T}{\partial \lambda}\right\}\right) - \frac{1}{\rho Cp}\left(\overrightarrow{\boldsymbol{U}_r}\frac{\partial T}{\partial r}\right) \tag{5}$$

2.2 Fluid dynamics

Fluid dynamics inside the cone husk furnace was obtained by solving the Navier-Stokes equations with laminar flow approach. Analytical solution of the equation for laminar flow case is given by [10]:

$$U_r(r) = \frac{\beta}{4\mu}(a^2 - r^2) \tag{6}$$

Where boundary condition at the wall is :

$U_r(0) = finite\ ; \boldsymbol{r} = \boldsymbol{0}$

$U_r(a) = 0\ ; \boldsymbol{r} = \boldsymbol{a}$

With **a** is constant which have same value with **r**.

2.3 Numerical Calculation of Heat Conduction and Convection equation in Conical

Numerical calculation of heat conduction and convection equation in conical coordinates can be solved by Finite Difference Method (FDM) based on forward and center formulas [11]. Finite difference method is one of numerical technique to solved differential equation problem [1]. The initial temperature in all parts of the cone is at room temperature, except the bottom of the cone is at 700 °C. There are two times simulation that used is 45 and 60 second with unfixed cone blancket temperature case. For unfixed temperature assumed that the cone walls in no contact with low heat conductivity medium.

IOP Conf. Series: Journal of Physics: Conf. Series **877** (2017) 012025 doi:10.1088/1742-6596/877/1/012025

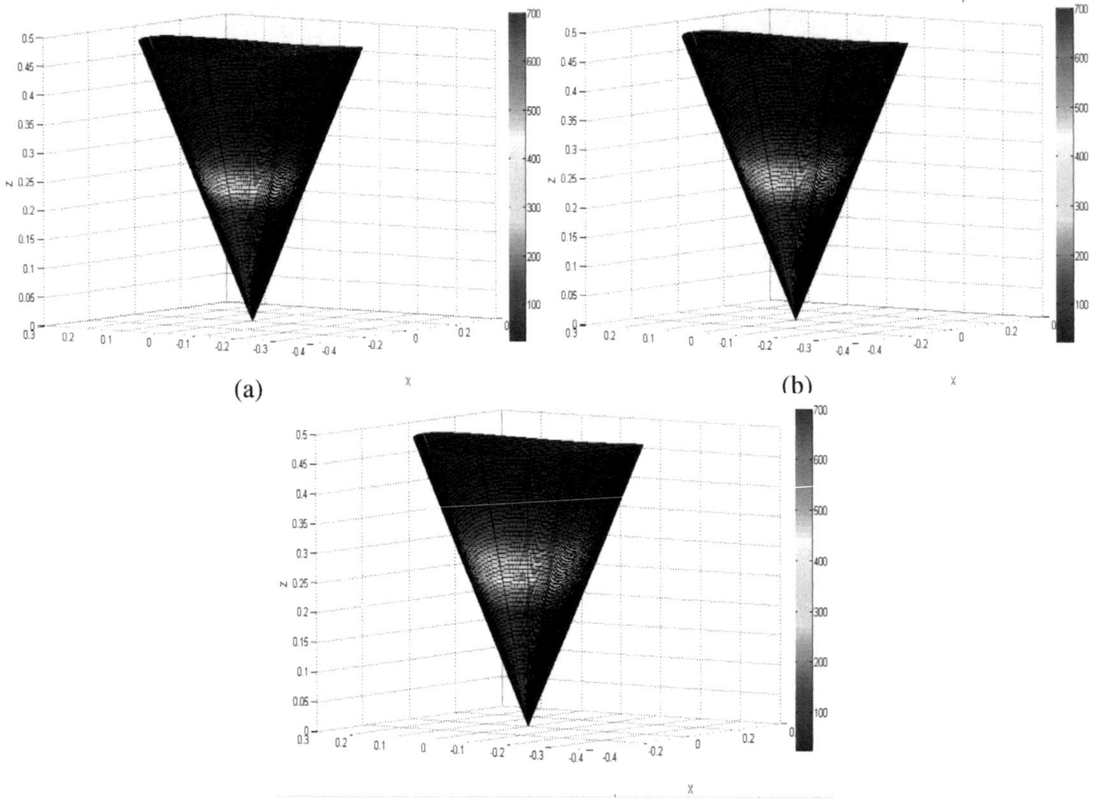

(a) (b) (c)

Figure 2. Heat distribution in the cone by conduction for unfixed blacket temperature at time 45 second (a) 60 second (b) and 120 second (c)

Figure 2 show the simulation of heat distribution of cone furnace caused by conduction process for 45 second and 60 second. It is appears that the rate of heat transfer in the (a) is lower than (b) unfixed temperature case. In this case, heat transfer in the cone is unsteady state condition. The temperature within the system does vary with time. While in unfixed temperature case, there is a contribution from the wall heat conduction with higher conductivity than fluid, so the temperature of cone edge is higher than its center.

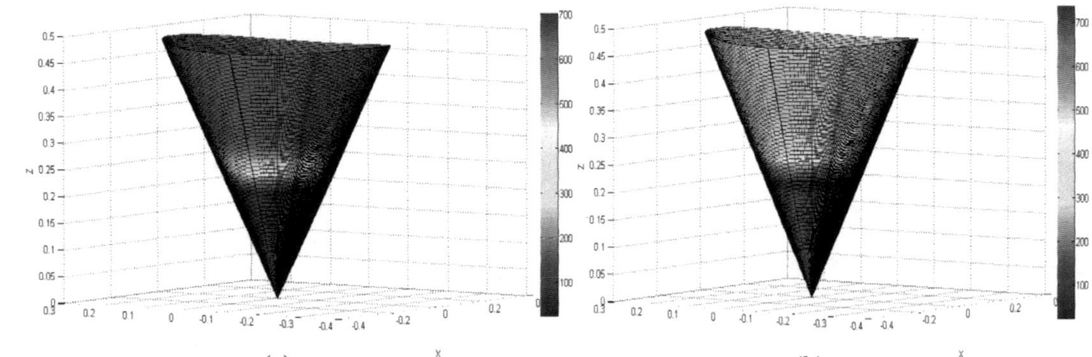

(a) (b)

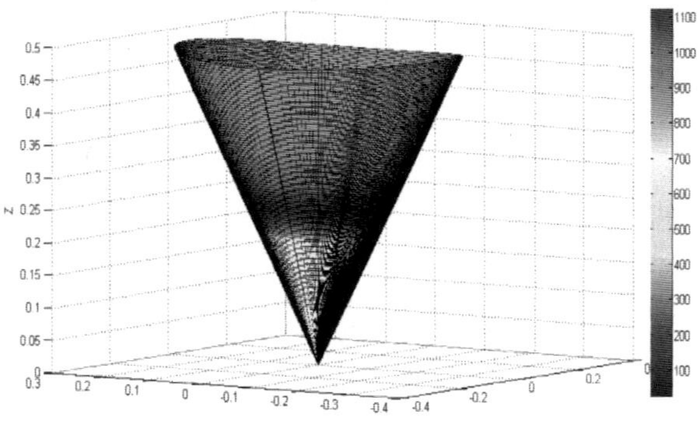

(c)

Figure 3. Heat distribution in the cone by convection for unfixed blacket temperature at time 45 second (a) 60 second (b) and 120 second (c)

Figure 3 show the simulation of heat distribution of cone furnace caused by conduction and convection process simultaneously for 45 second, 60 second, and 120 second. By comparing Fig 2 and Fig 3, it can be concluded that the heat transfer in the cone husk furnace is dominated by convection mechanism. On the mechanism of convection, the heat transfers through the movement of air particles due to the pressure gradient caused by the temperature gradient. While in conduction, there is only vibration between atoms of air due to a temperature gradient.

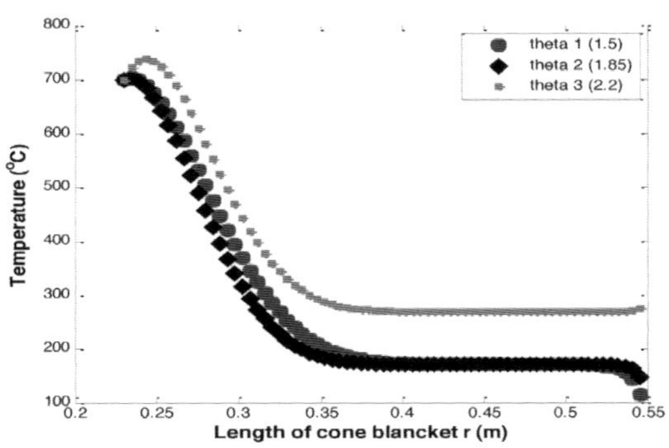

(a)

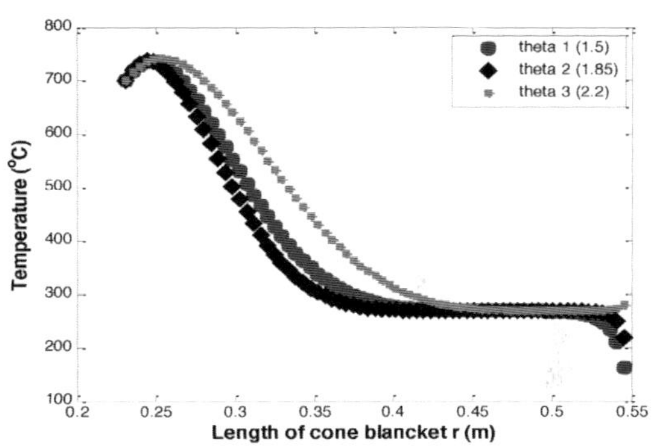

(b)

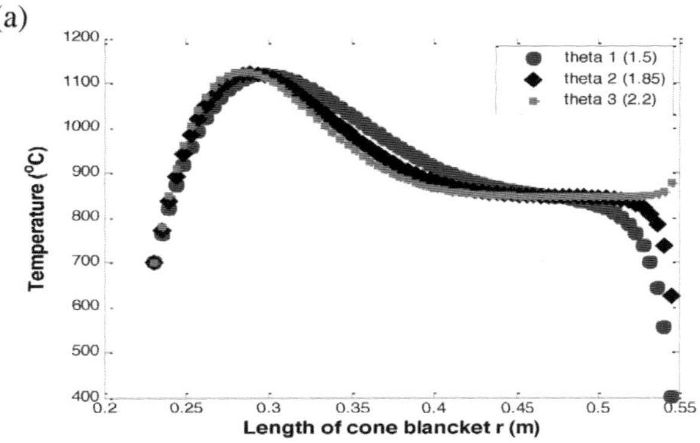

(c)

Figure 4. Graph of temperature vs length's cone blacket at time 45 second (a) 60 second (b) and 120 second (c)

Figure 4 show the graph temperature distribution by conduction and convection process simultaneously in the cone. There are three theta used, that is theta 1 (cone center), theta 2 (between center and cone edge) and theta 3 (cone edge) for 45 second, 60 second, and 120 second, respectively. For 45 second, temperature at the r=0.35 m in theta 1 is 201.94°C and 287.66 °C in theta 3. For 60 second, temperature at the r=0.35 m in theta 1 is 337.70 °C and 430.60 °C in theta 3. there is a contribution from the wall heat conduction with higher conductivity than fluid, so the temperature of cone edge (theta 3) is higher than its center (theta 1).

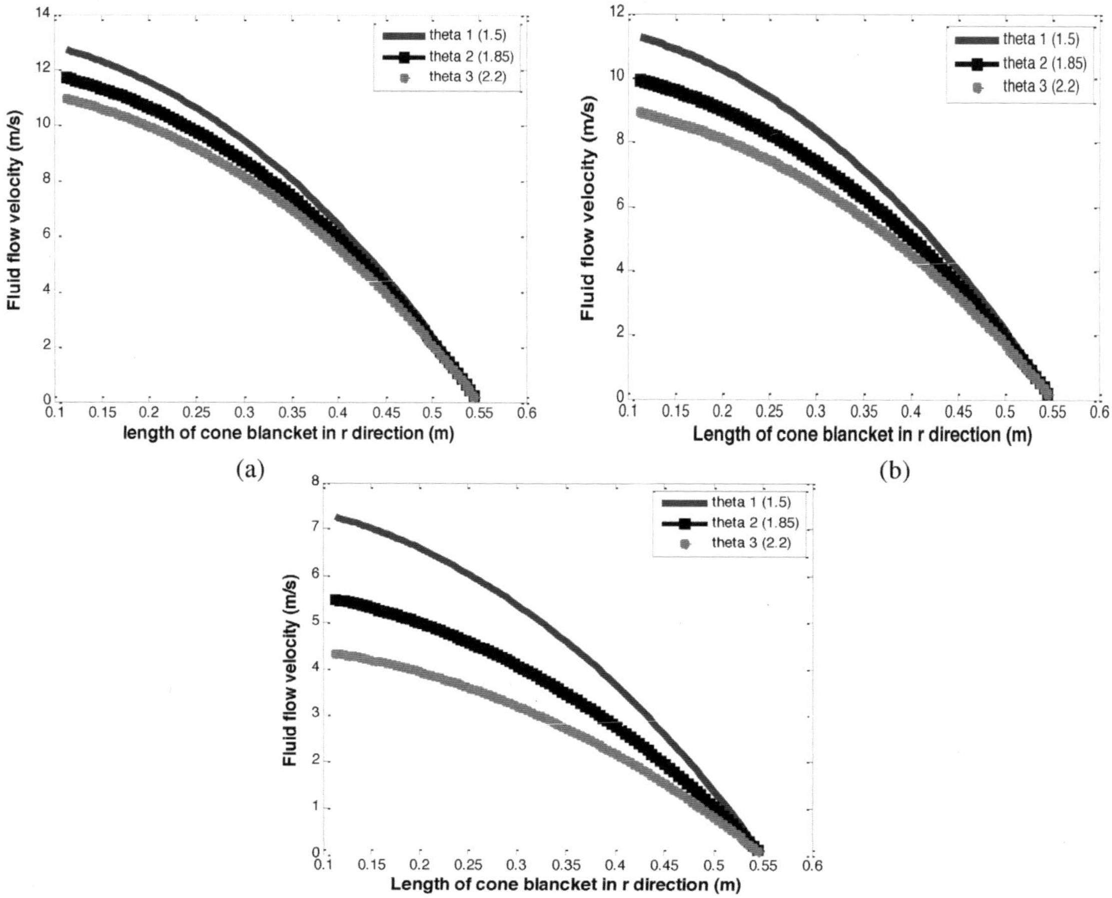

Figure 5. Graph of fluid flow velocity in **r** direction vs length's cone blacket at time 45 second (a) 60 second (b) and 120 second

Figure 5 above show the graph of fluid flow velocity in r direction at the time 45 and 60 second with unfixed temperature case. The fluid flow velocity at the center of the cone is bigger than fluid flow velocity at cone edge. The highest fluid flow velocity at r=0.11m is 13.69 m/s, 11.90 m/s, and 7.25 m/s for 45, 60, and 120 second, respectively. The different result is due to the pressure gradient for 45 second is the biggest than 60 second and 120 for unfixed temperature case.

3. Conclusion

We have performed simulation of heat transfer on cone husk furnace with FDM (Finite Difference Method). Based on numerical calculation with FDM for heat conduction and convection, temperature gradient convection is bigger than conduction. On convection, the heat transfer through the movement of air particles due to the pressure gradient caused by the temperature gradient. While in conduction, there is only vibration between atoms of air due to temperature gradient. The highest fluid flow velocity at r=0.11m is 13.69 m/s and 11.90 m/s for 45 and 60 second, respectively. The different result is due to the pressure gradient for 45 second is bigger than 60 second unfixed temperature case. The velocity with maximum speed is at the center of the cone.

References

[1] I. Noor, Irzaman, H. Syafutra, F. Ahmad. Simulation of heat transfer in cylinder husks furnace with finite difference method. IOP Conf. 31(2016) 012013

[2] Irzaman, H. Darmasetiawan, H. Alatas, M.N Indro, H. Hardhienata, K. Abdullah, T. Mandang, S. Tojo, Optimization of thermal efficiency of cooking stove with rice - husk fuel in supporting the proliferation of alternative energy in indonesia, Symposium Advanced Technological Development of biomass Utilization in Southeast Asia, Tokyo, Tokyo University of Agriculture and Technology (2009) 32 - 35.

[3] M. Ebadian, H. Zhang, An exact solution of extended Graetz problem with axial heat conduction, Int. J. Heat Mass Transfer 32 (9) (1989) 1709–1717.

[4] C.J. Hsu, Exact solution to entry-region laminar heat transfer with axial conduction and the boundary condition of the third kind, Chem. Eng. Sci. 23 (5) (1968) 457–468.

[5] M.L. Michelsen, J. Villadsen, The Graetz problem with axial heat conduction, Int. J. Heat Mass Transfer 17 (11) (1974) 1391–1402.

[6] A.H. Srinivasa, A.T. Eswara. Unsteady free convection flow and heat transfer from an isothermal truncated cone with variable viscosity, Int. J. Heat Mass Transfer 57 (2013) 411–420

[7] S. Ray, A. Loukou, D. Trimis. Evaluation of heat conduction through truncated conical shells, Int. J. Thermal Sciences 57 (2012) 183–191

[8] Moon, P. and Spencer, D. E. 1988. Field Theory Handbook, Including Coordinate Systems, Differential Equations, and Their Solutions, 2nd ed. "Conical Coordinates (r, θ, λ)." Table 1.09 in New York: Springer-Verlag, pp. 37-40.

[9] Lienhard, John H. 2005. A Heat Transfer Textbook .Third edition. Phlogiston Pressridge, Massachusetts, U.S.A.

[10] Williams, J. 2000. Engineering Heat Transfer. Second edition. CRC Press, New York, U.S.A

[11] Rectenwald, Gerald W. 2011. Finite-Difference Approximations to The Heat Equation. Mechanical Engineering Department Portland State University, Portland, Oregon.

First Principles Study of Molybdenum Disulfide Electronic Structure

Imam Abdul Rahman[1], **Acep Purqon**[2]
Departement of Physics, Institut Teknologi Bandung
E-mail: [1]imamar@students.itb.ac.id, [2]acep@fi.itb.ac.id

Abstract. The study of two dimensional material has gain interest due to unique properties which are different from the bulk precursors. Mono- and few-layered of Transition metal dichalcogenides (TMDCs) has band gap properties between 1-2 eV that suitable of FET devices or any optoelectronic devices. Among TMDCs, Molybdenum Disulfide (MoS2) has gain interest due to its promising band gap-tuning and transition between direct to indirect band gap properties depends on its thickness. First principles calculation by Density Functional Theory has been performed to study the characteristic of MoS2 electronic structure. Indirect band gap of MoS2 lies between point Γ to Γ-K in first brillouine zone, while the direct bandgap lies in point-K. The indirect band gap became larger while the number of layer decreased due to quantum confinement effect in c axis direction. In monolayer MoS2, the indirect band gap become larger than direct one, band gap properties transitioned from indirect to direct. The unique bandgap properties of MoS2 can lead into better application in energy devices such as solar cell [11], FET [10], and photoluminescence device.

1. Introduction

The study of two dimensional material has gain interest due to unique properties which are different from the bulk precursors either in its electrical properties also in its optoelectronical properties.This kind of thin material also allowed to make very thin devices since its already two dimensional.

Most viral two dimensional material is graphene that has outstanding electrical properties [9]. Graphene is semi-metal material that has zero band gap. It good electrical properties allow graphene be a suitable material for any conductor application. But since graphene has zero bandgap, it can not be applied as semiconductor devices. Whereas a lot of technology developed based on semiconductor properties of material included energy devices such as photoluminescence device, Solar Cell, etc.

TMDSc (Transition Metal Dichalcogenide) group of material can be suitable candidate as two dimensional material that has band gap, since its band gap in range 1 to 2 eV [12]. Among TMDCs, MoS_2 (Molybdenum Disulfide) is the most interesting for its fine electrical and optoelectronic properties in two dimensional state. In bulk, MoS_2 known has 1.29 eV [2] band gap with indirect one yet shifted into 1.8-1.9 eV [7] with direct band gap in two dimensional single layer. This value of band gap allow MoS_2 to be applied in semiconductor based devices, including any energy devices. Several study show MoS_2 application in solar cell [11], FET [10], and photoluminescence devices.

Since its wide use of MoS_2, in this work will be focused in basic properties of MoS_2. This work will show how electronic structure of MoS_2 shifted by its structure and number of layer. This preliminary study using Density Functional Theory (DFT) for identifying theoretical electronic structure shift along it structure change.

2. Method

Structure electronic of material defined by interaction of its all of electron and nuclei. Density Functional Theory calculate these interaction with simplified quantum calculation. In DFT, Kohn Sham Equation used instead of classic Schrdinger equation [3]. K-S Equation shown,

$$\left(-\frac{1}{2}\nabla + \int \frac{n(r')}{|r-r'|}\, d^3r' + \frac{\partial}{\partial n(r)} E_{xc} + v_{ext}\right)\phi_i = \lambda_i \phi_i \tag{1}$$

Since input of K-S Equation is electron density ($n(r)$) yet $n(r)$ it self is a square of solution wave function ϕ_i, this equation can be solved using Self Consistent Field algorithm. In this

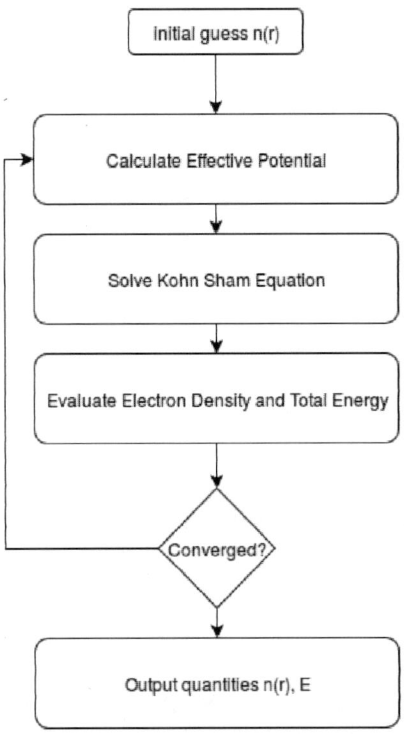

Figure 1. Self Consistent Field Flow Chart. First make initial guest in n(r), then calculate effective potential and solve Kohn Sham Equation. Final step in iteration is to evaluate total energy which are converged or not, if not, make output of electron density as new n(r) and do iteration again.

work PHASE0 package software used to calculate structure electronic of MoS_2. Structural optimization calculated with c/a optimization first then volume optimization. 10^{-9} Hartree used as converged energy criteria in SCF. K-poin sampling used is 8x8x8 for bulk and 8x8x1 for two dimensional, with 50 Rydberg Wave Function cutoff.

MoS_2 bulk model used as in figure 2. 2H structure is known most stable structure for MoS_2, since 3R structure can be reformed into 2H trough heating [14]. For two dimensional model, single layer and double layer calculated with 15 $\mathring{A}$ vacuum gap to make sure no interaction between layer. Model shown in figure 3.

3. Result and Discussion

Lattice constant of MoS_2 calculated shown in table 1. It shown also total energy per MoS_2 in 2H structure has lower than in 3R. This show that 2H has more stable structure rather

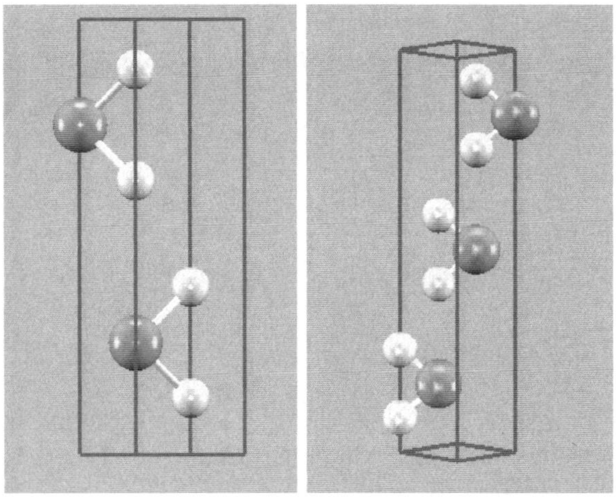

Figure 2. Bulk structure model of MoS_2 with 2H (left) and 3R (right). 2H has 2 molecule of MoS_2 in each cell while 3R has 3 molecule of MoS_2 in each cell.

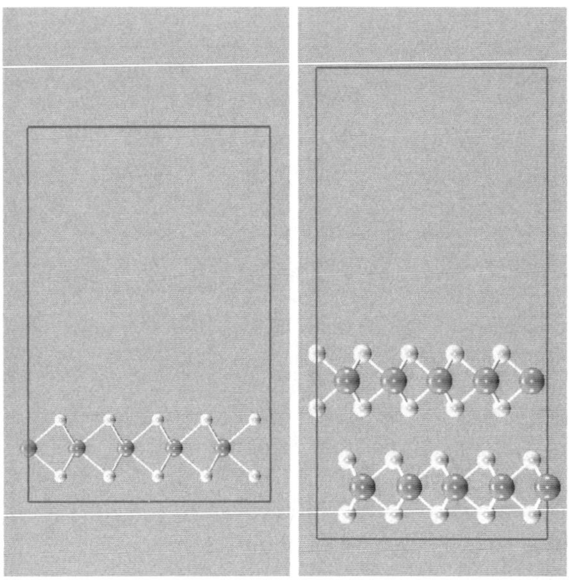

Figure 3. Model for monolayer (left) and dual layer (right) of MoS_2. 15 $\dot{A}$ used as vacuum layer for each model.

than 3R. Also provided the comparison table between this work and some reference in table 2. Optimization of lattice constant show very close result for each reference and this work. This show optimization work well enough and can be used to calculate electronic structure of each model.

Table 1. Calculated lattice constant and total energy

	2H	**3R**
a ($\dot{A}$)	3.22	3.24
c ($\dot{A}$)	13.26	18.92
E (Hartree)	68.021	68.017

Calculated electronic structure represented in band structure and density of states for each structure shown in figure 4, 5, 6, and 7. Comparison of band gap calculation result with some reference shown in table 3. Some of theoretical result lies very close with result of this work,

IOP Conf. Series: Journal of Physics: Conf. Series **877** (2017) 012026 doi:10.1088/1742-6596/877/1/012026

Table 2. Lattice constant on 2H-Bulk and MoS_2 single layer comparison with some reference

	MoS_2 Bulk		MoS_2 *Monolayer*	
	Result ($\dot{A}$)	Reference ($\dot{A}$)	Result ($\dot{A}$)	Referensi ($\dot{A}$)
Lattice Constant (a)	3.22	3.19 [1], 3.16 [13]	3.24	3.195 [1], 3.23 [6]
c/a ratio	3.84	3.86 [1], 3.89 [13]		

Table 3. Calculated band gap comparison with some reference

	MoS_2 Bulk		MoS_2 Monolayer	
	Result (eV)	Reference (eV)	Result (eV)	Reference (eV)
Bandgap	0.9	Theoritical Study 0.89 [1], 0.7 [4], 1.15 [8] Eksperimen 1.23 [7], 1.29 [2]	1.6	Theoritical Study 1.57 [1], 1.9 [5] Eksperimen 1.80 [7]

while other result seems pretty much different. But it show that experimental result show significant shift from theoretical result. Theoretical result gives lower magnitude of band gap than experiment. This shows that DFT can not really be a good estimation in calculation of exact band gap, yet still can be used as preliminary study of band gap and knows the shift trends of band gap while structure of material modified.
From electronic structure shown that 3R structure have larger band gap compared to 2H one. While in 2 dimensional state, band gap of MoS_2 become larger from the bulk one, since in dual layer has 1.26 eV band gap meanwhile it has 1.6 eV in monolayer state. This indicated that band gap of MoS_2 become shifted up while there is few layer only because of quantum confinement effect in c-axis direction. Interesting case in monolayer MoS_2, since the characteristic of indirect band gap that naturally occur in MoS_2 change to direct one in point K. This characteristic make MoS_2 more applicable in wide kind of devices. This also shows that band gap of MoS_2 can be tunable with variety of its number of layer. Resume of MoS_2 band gap properties show in table 4.

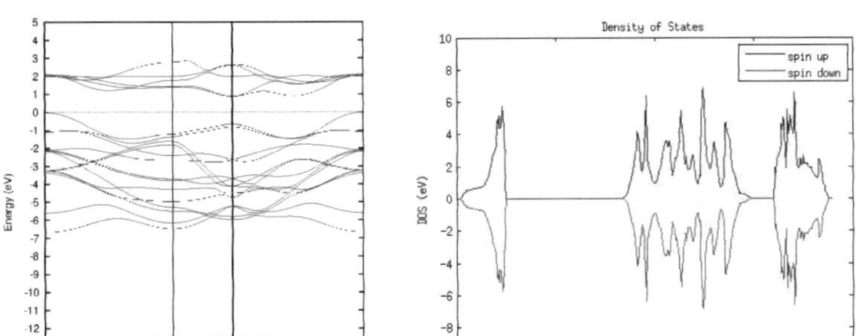

Figure 4. Electronic structure of 2H Bulk MoS_2 represented as band diagram (left) and density of states (right), this shown that 2H structure has 0.9 eV band gap lies between point Γ to $\Gamma - K$. Spin density shown to be very simmetry for spin up and spin down.

Table 4. Band gap of 2H, 3R, monolayer and dual layer MoS_2

	2H	**3R**	**monolayer**	**dual layer**
Band Gap (eV)	0.9	1.1	1.6	1.26
Band Gap Characteristic	indirect	indirect	direct	indirect

IOP Conf. Series: Journal of Physics: Conf. Series **877** (2017) 012026 doi:10.1088/1742-6596/877/1/012026

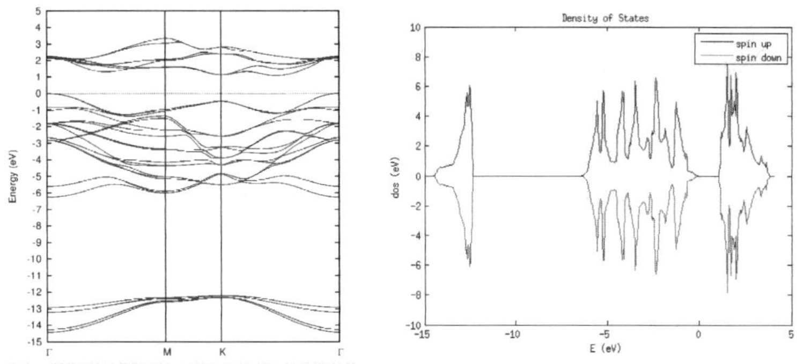

Figure 5. Electronic structure of 3R Bulk MoS_2 represented as band diagram (left) and density of states (right), this shown that 2H structure has 1.1 eV band gap lies between point Γ to $\Gamma - K$. Spin density shown to be very simmetry for spin up and spin down. This characteristic very similar with 2H, but with shifted up band gap.

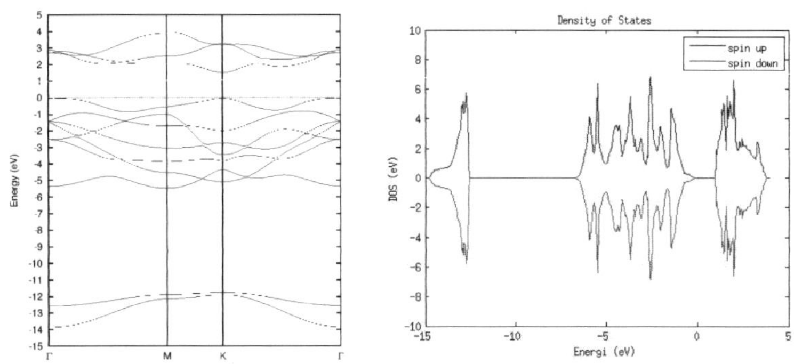

Figure 6. Electronic structure of monolayer MoS_2 represented as band diagram (left) and density of states (right), this shown that 2H structure has 1.6 eV band gap lies between point K to K. Spin density shown to be very simmetry for spin up and spin down. Interesting case within MoS_2 since in bulk phase, it has indirect band gap and yet shifted into direct one that lies in K point brillouin zone.

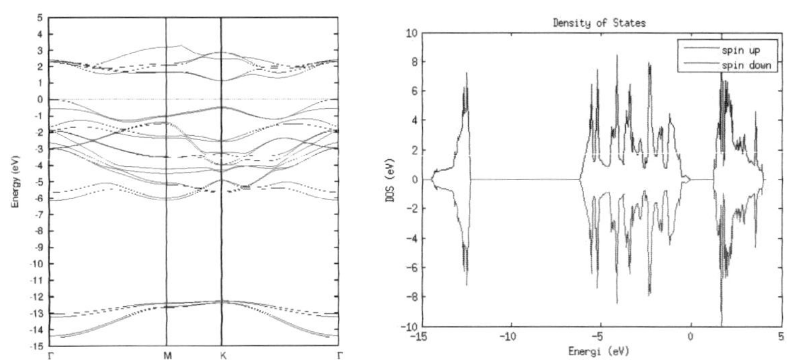

Figure 7. Electronic structure of dual MoS_2 represented as band diagram (left) and density of states (right), this shown that 2H structure has 1.26 eV band gap lies between point Γ to $\Gamma - K$. Spin density shown to be very simmetry for spin up and spin down.

From table 4, it shown that band gap of MoS_2 is varying around 0.9 eV to 1.6 eV. Also it properties is shifted from indirect band gap (bulk) into direct bandgap (monolayer). This unique properties of MoS_2 gives possibility on MoS_2 implementation on energy devices. Band gap with range 1-2 eV is suitable for application on semiconductor energy devices such as solar cell [11] and FET [10]. Also monolayer direct bandgap properties can lead into photoluminescence devices. This also shows that by shifting number of layer, also structure of MoS_2, can be tuned to suit more on energy devices design, which is, still open on future development.

Acknowledgments

The author(s) wish to acknowledge the contribution of LIPI high-performance computing facilities to the results of this research. URL http://grid.lipi.go.id.

References

[1] S. Ahmad and S. Mukherjee. A comparative study of electronic properties of bulk mos2 and its monolayer using dft technique: Application of mechanical strain on mos2 monolayer. *Graphene*, 3:52–59, 2014.

[2] Th. Böker, R. Severin, A. Müller, C. Janowitz, R. Manzke, D. Voß, P. Krüger, A. Mazur, and J. Pollmann. Band structure of mos_2, $mose_2$, and $\alpha - mote_2$: angle-resolved photoelectron spectroscopy and *ab initio* calculations. *Phys. Rev. B*, 64:235305, Nov 2001.

[3] K. Capelle. A bird's-eye view of density-functional theory. *eprint arXiv:cond-mat/0211443*, November 2002.

[4] Katsuyoshi Kobayashi and Jun Yamauchi. Electronic structure and scanning-tunneling-microscopy image of molybdenum dichalcogenide surfaces. *Phys. Rev. B*, 51:17085–17095, Jun 1995.

[5] A. Kuc, N. Zibouche, and T. Heine. Influence of quantum confinement on the electronic structure of the transition metal sulfide ts_2. *Phys. Rev. B*, 83:245213, Jun 2011.

[6] Ashok Kumar and P.K. Ahluwalia. A first principle comparative study of electronic and optical properties of 1h mos2 and 2h mos2. *Materials Chemistry and Physics*, 135(23):755 – 761, 2012.

[7] Kin Fai Mak, Changgu Lee, James Hone, Jie Shan, and Tony F. Heinz. Atomically thin mos_2: A new direct-gap semiconductor. *Phys. Rev. Lett.*, 105:136805, Sep 2010.

[8] L. F. Mattheiss. Energy bands for $2h - Nbse_2$ and $2h - Mos_2$. *Phys. Rev. Lett.*, 30:784–787, Apr 1973.

[9] K. S. Novoselov, A. K. Geim, S. V. Morozov, D. Jiang, Y. Zhang, S. V. Dubonos, I. V. Grigorieva, and A. A. Firsov. Electric field effect in atomically thin carbon films. *Science*, 306(5696):666–669, 2004.

[10] Xin Tong, Eric Ashalley, Feng Lin, Handong Li, and Zhiming M. Wang. Advances in mos2-based field effect transistors (fets). *Nano-Micro Letters*, 7(3):203–218, 2015.

[11] Meng-Lin Tsai, Sheng-Han Su, Jan-Kai Chang, Dung-Sheng Tsai, Chang-Hsiao Chen, Chih-I Wu, Lain-Jong Li, Lih-Juann Chen, and Jr-Hau He. Monolayer mos2 heterojunction solar cells. *ACS Nano*, 8(8):8317–8322, 2014. PMID: 25046764.

[12] J.A. Wilson and A.D. Yoffe. The transition metal dichalcogenides discussion and interpretation of the observed optical, electrical and structural properties. *Advances in Physics*, 18(73):193–335, 1969.

[13] J.A. Wilson and A.D. Yoffe. The transition metal dichalcogenides discussion and interpretation of the observed optical, electrical and structural properties. *Advances in Physics*, 18(73):193–335, 1969.

[14] Mingxiao Ye, Dustin Winslow, Dongyan Zhang, Ravindra Pandey, and Yoke Khin Yap. Recent advancement on the optical properties of two-dimensional molybdenum disulfide (mos2) thin films. *Photonics*, 2(1):288, 2015.

Electronic Properties of Rare-Earth Doped α-GaN

Muhammad Y H Widianto[1] and Acep Purqon[2]

[1]Department of Informatics, Del Institute of Technology, Jl. Sisingamangaraja Sitoluama Laguboti Toba Samosir Sumatera Utara 22381, Indonesia
[2]Physics of Earth and Complex Systems, Bandung Institute of Technology, Jl Ganesha 10 Bandung 40132, Indonesia

Email: [1] yusuf.widianto@del.ac.id and [2] acep@fi.itb.ac.id

Abstract. We carry out DFT calculation to study electronic properties of rare-earth (RE) doped α-GaN. GGA approximation and ultrasoft pseudopotential used in this calculation. The bond length and tetrahedral angle of pristine α-GaN are slightly changed. The direct band gap of α-GaN was detected at 2.56 eV, this is 1 eV lower from experiment. Electronic structure measurement is also performed for RE doped GaN (RE= Pr, Eu, Er, Gd and Tm), which the configuration is $Ga_{(1-x)}NRE_x$ (x=0.125). The bond length of Ga-RE is vary from 2.12-2.23 Å. The impurity energy is found relative between conduction bands and valence bands of host materials.

1. Introduction

In this recent decade, semiconductor III-V is used for electronic materials, magnetic and optical devices, particularly GaN [1][2]. Since Nakamura has found blue light emitting diodes (LED) from thin layer of GaN [3], these is possibly used in the optical devices operating ultraviolet until visible light. GaN has wide direct band gap 3.4 eV. This band gap is much stronger and produce more light emitting than indirect band gap. LED based on GaN starts to replace light bulbs or fluorescent lamp with more efficient and low cost energy.

Rare-earth (RE) doping GaN has gained more attention as an electroluminescent devices (ELDs). Modifying the active layer of GaN with activator RE can impact-excite RE dopants and produces emission of light. Visible spectrum is obtained by photoemission from excite RE states in GaN. GaN:Pr and GaN:Eu have been shown to produce light emission in red color (Pr at 650 nm and Eu at 621 nm)[4][5]. Green and blue are detected on the Tm (at 477 nm) and Er (at 537 nm) doped [6][7][8]. Mixed colors are created from combining primary colors [9]. RE doped GaN is good to generates high brightness LED.

In this paper, basic understanding of the electronic properties and optical transition RE doped GaN are needed to investigate photoemission. This requires first principle calculation based on density functional theory (DFT) to predict electronic structure from RE to host material GaN. First, optimize structure of pristine α-GaN. Then, optimize structure of GaN:RE and investigate the changes of bond length and tetrahedral angle[10]. Finally, analysis of electronic structure of GaN:RE will be conducted.

2. Model

PHASE0 code [11] is performed to calculate electronic properties of pristine α-GaN and GaN:RE. This calculation uses Generalized Gradient Approximation (GGA)-Ultrasoft Pseudopotential at zero Kelvin. The cut-off wave function is 25 Ry and cut-off charge density is 225 Ry. Geometry optimization of a force 1.0^{-3} Hartree/Bohr and the energy convergence is 1.0^{-3} Hartree/Bohr. In unit-cell structure of pristine α-GaN, two atoms Ga are placed at the (0,0,0) and (2/3,1/3,1/2) and two atoms N at the (0,0,u) and (2/3, 1/3, u+1/2), which u is internal parameter. The vector is $\mathbf{a}_1$=(a,0,0), $\mathbf{a}_2$=(a/2,a√3/2,0) and $\mathbf{a}_3$=(0,0,c), which a and c are lattice parameter[10]. For GaN:RE, the RE will be substituted in Ga site, see Figure 1. The RE will occupy cation site in the GaN. When Ga is substituted with RE, the configuration turns into $Ga_{(1-x)}NRE_x$ ($1 \leq x \leq 0$). In this calculation, x is concentration of RE in GaN structure which x=0.125. Pristine α-GaN and GaN:RE will be performed in k-point mesh (2x2x2) of hexagonal lattice on Brillouin zone (BZ) GaN.

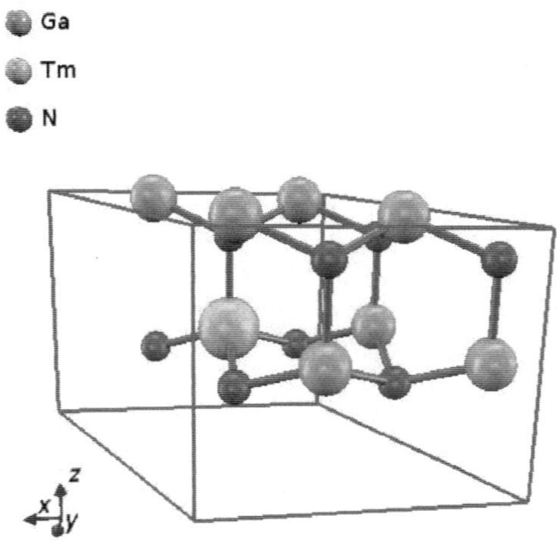

Figure 1. Model of supercell GaN:RE in 16 atoms, which one atom Ga substituted with one atom RE. For example the RE is Tm. Ga atom is symbolized with pink color, N with blue color and Tm with green color.

3. Result and Discussion

Optimization equilibrium geometry of the pristine α-GaN follows the procedure Stampfl et. al[12]. Lattice parameter of pristine α-GaN is found as a=3.1438 Å and c= 5.1254Å . This results is ~2% smaller than experimental data and from theoretical calculation [10][13]. Internal parameter of u=0.3768 and ratio of c/a=1.6303 have slightly changed. The bond length of Ga-N is 1.9266 Å and the tetrahedral angle of N-Ga-N is 109.16°, this tetrahedral angle is smaller from ideal wurtzite structure [14]. In general, the result of equilibrium lattice constant is slightly changed with experimental data and another theoretical calculations.

Figure 2 shows electronic properties of pristine α-GaN the band structure and total density of states (DOS). Outer orbital electron of *2p* N and *3d* Ga effect on lower valence band in the energy range between -12.5 eV until -15 eV. The band structure shows direct band gap, E_g, as it has detected 2.58 eV. This is 1 eV lower than experimental data. GGA calculation in electronic structure of GaN is still underestimated. Direct band occurs because of the minimum point of conduction band, C_b, and maximum point of valence band, V_b, are parallel in Γ-point.

The substitution of one atom RE to α-GaN made the structure changes. RE atoms is grater than Ga atom, this is effected to the bond length and tetrahedral angle [15]. Geometry optimization of GaN:RE was also implemented to get convergence of total energy. Four atoms of Nitrogen are the nearest neighboring atoms of RE with one atom bonded above and three atoms bonded in the lower. Table 1 shows the bond length and tetrahedral angle in several RE doped. The bond length of RE-N is varies with the closest bond is Ga-Tm (2.1277Å) and the longest is Ga-Pr (2.235Å). The bond length increases 12% from Ga-N, this is shorter from another calculation [6][10][16]. The bond length of RE-N impacts to the tetrahedral angle N-RE-N. This occurs because Tm has high density with electron $4f^{13}$ while Pr has $4f^{3}$.

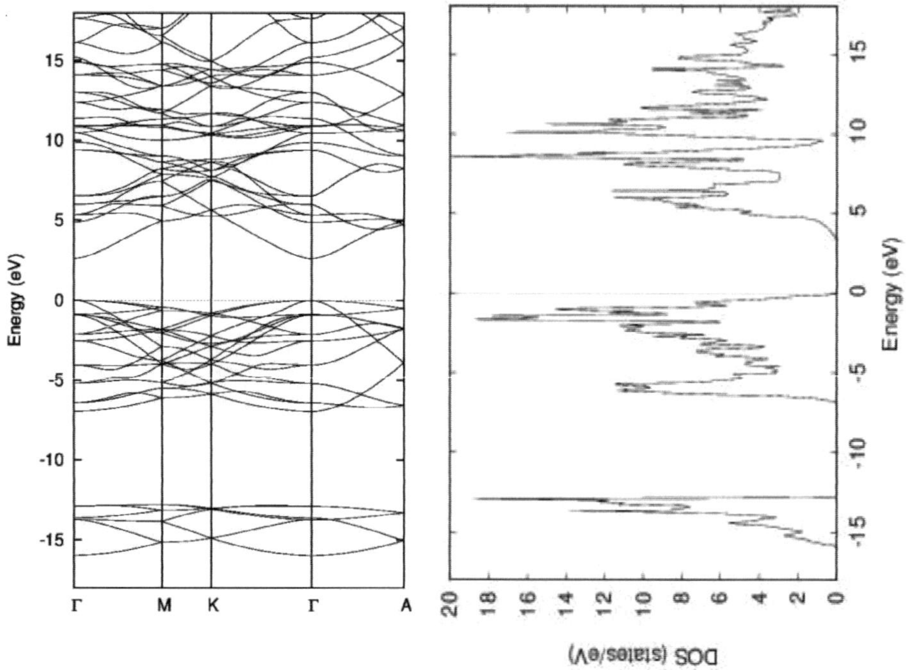

Figure 2. Electronic properties of pristine at 16 atoms supercell the band structure (left) and Density of States (DOS) (right). The position of Fermi energy, E_f, is denoted by vertical dotted line, which set to be 0.0 eV.

The band structure and DOS of GaN:RE are shown in the Fig. 3. Impurity energy is formed between conduction band (C_b) and valence band (V_b) of host material. On the band structure, the position energy level of C_b and V_b is decreases. The distance from minimum point C_b to maximum impurity energy level of GaN:Eu is 0.95 eV. The conduction level decreased with minimum point 1.24 eV. Maximum point of V_b is changed from Γ to K. The distance of minimum impurity energy level between V_b is 1.22 eV. It is predicted as the effect of Eu to crystal lattice where hexagonal lattice of BZ changed.

Table 1. Geometries optimization of GaN:RE (RE=Pr, Eu, Gd, Er, Tm). RE-N*** is three bonds and RE-N* is one bond.

Supercell	RE-N***(Å)	RE-N* (Å)	N-RE-N angles(°)
GaN:Pr	2.1672	2.2350	113.28
GaN:Eu	2.1427	2.2009	112.92
GaN:Gd	2.1346	2.1903	112.70
GaN:Er	2.1306	2.1887	111.93
GaN:Tm	2.1277	2.1278	111.50

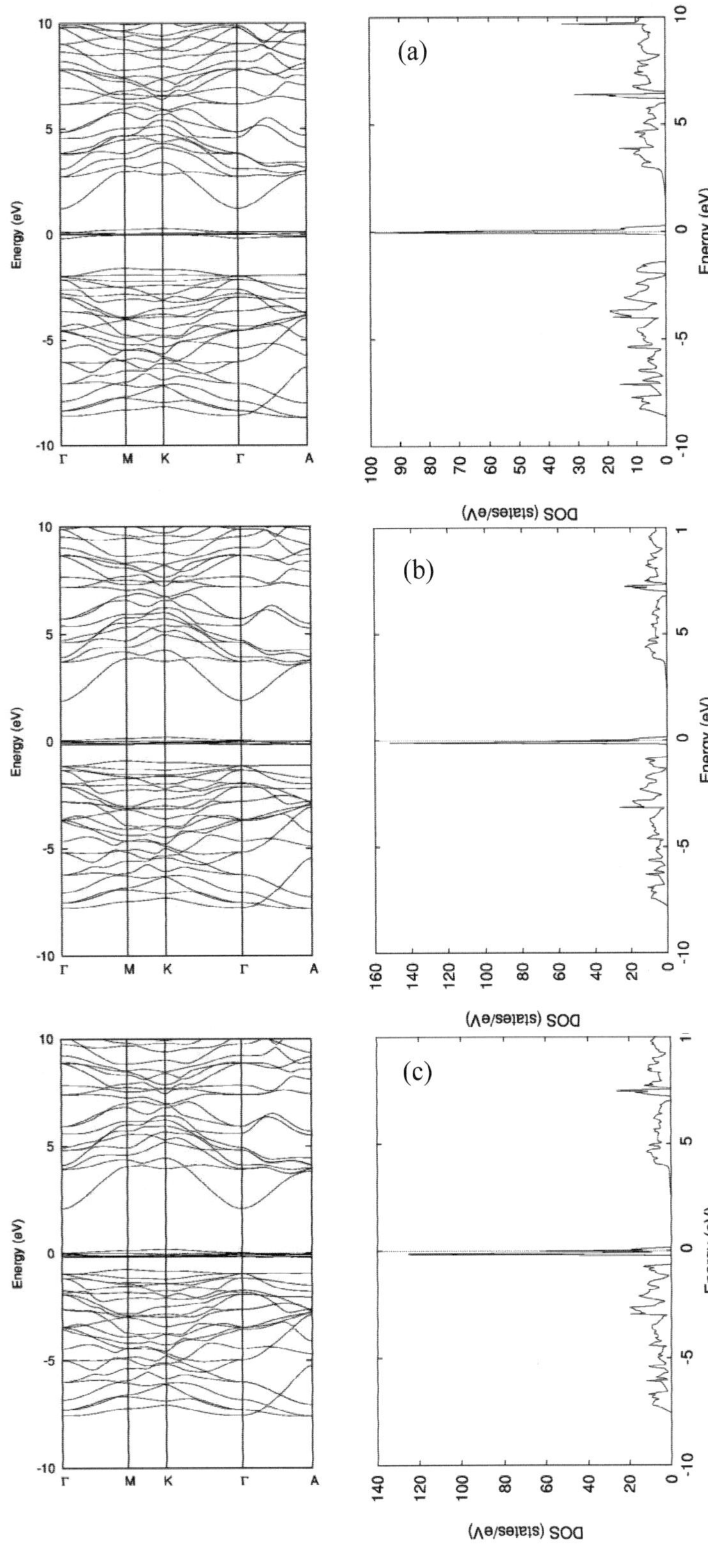

Figure 3. The band structure (left) and DOS (right) of (a) GaN:Eu, (b) GaN:Er and (c) GaN:Tm. All supercell formed 16 atoms. The position of Fermi energy, E_f, is denoted by vertical line, which has been set to be 0.0(eV).

The distance from C_b to maximum impurity of Gan:Er is 1.69 eV and GaN:Tm is 1.91 eV. The separation distances vary affecting to electron-hole excitation. It will produce different wavelength, three of them has red, green and blue colors [6][7]. Maximum valence band of GaN:Er and GaN:Tm are changed to M-point. The total energy DOS of the impurity energy RE has shown an increase then decrease. It shows the total DOS of GaN:Er (150 states/eV) is larger than GaN:Eu (100 states/eV) or GaN:Tm (120 states/eV). Impurity energy of RE in GaN has its own characteristics [17][18]. The energy level of C_b, V_b and impurity energy 4f are varied to host material for each RE doped. 4f electron of RE heavily influence to the band gap of host materials. Hybrid-DFT calculation is needed study in more details [17][18].

Summary

In this calculation, pristine α-GaN has direct band gap E_g=2.58 eV. The lattice constant has changed 24% from the experimental data. When RE atoms substitute Ga, impurity energy has detected between conduction band and valence band. This is affects the process of excite from RE to host materials. The energy level of conduction, valence and impurity RE have unique characteristics. Optical properties of GaN:RE can be predicted and can be used as optical devices.

Acknowledgments

MW thanks to HPC GRID LIPI for the computational calculation. This work is support by Thesis Grand FMIPA ITB. MW thanks to MS who has been supporting, discussing and advising.

References

[1] B. Monemar, Phys. Rev. B 10 (1974) 676–681.
[2] S. Denbaars, D. Feezell, K. Kelchner, S. Pimputkar, C. Pan, C. Yen, S. Tanaka, Y. Zhao, N. Pfaff, R. Farrell, M. Iza, S. Keller, U. Mishra, J. Speck, S. Nakamura, Acta Materialia 61 (3) (2013) 945–951.
[3] S. Nakamura, T. Mukai, M. Senoh, Applied Physics Letters 64 (13) (1994) 1687–1689.
[4] V. Kachkanov, M. J. Wallace, G. van der Laan, S. S. Dhesi, S. A. Cavill, Y. Fujiwara, K. P. O'Donnell, Scientific Reports 2 (2012) 969.
[5] J. Heikenfeld, M. Garter, D. S. Lee, R. Birkhahn, A. J. Steckl, Applied Physics Letters 75 (9) (1999) 1189– 1191.
[6] A. Steckl, J. Heikenfeld, D. Lee, M. Garter, Materials Science and Engineering: B 81 (13) (2001) 97 – 101.
[7] U. Hmmerich, J. Seo, C. Abernathy, A. Steckl, J. Zavada, Materials Science and Engineering: B 81 (13) (2001) 116-120.
[8] D. S. Lee, A. J. Steckl, Applied Physics Letters 82 (1) (2003) 55–57.
[9] A. J. Steckl, J. C. Heikenfeld, D.-S. Lee, M. J. Garter, C. C. Baker, Y. Wang, R. Jone, IEEE Journal of Selected Topics in Quantum Electronics 8 (4) (2002) 749–766.
[10] K. C. Mishra, V. Eyert, P. C. Schmidt, Zeitschrift fr Physikalische Chemie 221 (2009) 1663–1676.
[11] C. for Research on Innovative Simulation Software (CISS), the Institute of Industrial Science (IIS), the University of Tokyo, First-principles electronic structure calculation program phase/0.
[12] C. Stampfl, C. G. Van de Walle, Phys. Rev. B 59 (1999) 5521–5535.
[13] D. Andiwijayakusuma, M. Saito, A. Purqon, Journal of Physics: Conference Series 739 (1) (2016) 012027.
[14] X. L. CHEN, J. K. LIANG, Y. P. XU, T. XU, P. Z. JIANG, Y. D. YU, K. Q. LU, Modern Physics Letters B 13 (09n10) (1999) 285–290.
[15] S. Sanna, W. G. Schmidt, T. Frauenheim, U. Gerstmann, Phys. Rev. B 80 (2009) 104120.
[16] H. Bang, S. Morishima, Z. Li, K. Akimoto, M. Nomura, E. Yagi, Journal of Crystal Growth

237239, Part 2 (2002) 1027 – 1031.
[17] P. Dorenbos, E. van der Kolk, Applied Physics Letters 89 (6) (2006) 061122.
[18] G. Caroena, W. V. M. Machado, J. F. Justo, L. V. C. Assali, Applied Physics Letters 102 (6) (2013) 062101.

Study of Evaluation OSH Management System Policy Based On Safety Culture Dimensions in Construction Project

Yusuf Latief[1], Rossy Armyn Machfudiyanto[1], Rosmariani Arifuddin[1], R. Mahendra Fira Setiawan[1], Yoko Yogiswara[1]

[1]*Department Civil Engineering, Universitas Indonesia, Depok, Indonesia*

Abstract. Safety Culture in the construction industry is very influential on the socio economic conditions that resulted in the country's competitiveness. Based on the data, the accident rate of construction projects in Indonesia is very high. In the era of the Asian Economic Community (AEC) Indonesian contractor is required to improve competitiveness, one of which is the implementation of the project without zero accident. Research using primary and secondary data validated the results of the literature experts and questionnaire respondents were analyzed using methods SmartPLS, obtained pattern of relationships between dimensions of safety culture to improve the performance of Safety. The results showed that the behaviors and Cost of Safety into dimensions that significantly affect the performance of safety. an increase in visible policy-based on Regulation of Public Work and Housing No 5/PRT/M/2014 to improve to lower the accident rate.

Introduction

Based on the research International Labour Organization (ILO, 2003) mentioned that working accident in a country was mostly related to the rate of the competitiveness in that country. That research concluded that there was that points out the rate of working accident is inversely proportional with competitiveness in this case Indonesia is placed in the second lowest rate of the fatal working accident, in the 40 out of 100.000 workers (The Global Competitiveness Index, 2011). The research of ILO (2009) concluded that Indonesia had bad record on working accident with the record of number 152 of 153 Nations with the worst working accident aspect as showed in the Figure 1.

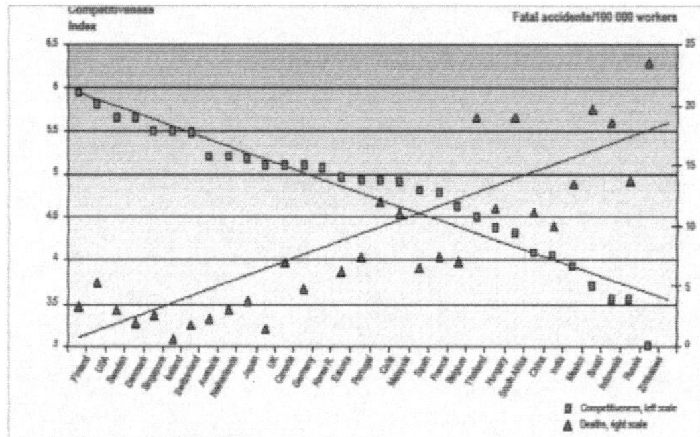

Figure 1. Relation Between The Competitiveness of the Country with The Working Accident Rate (Data IMD & ILO, 2003, page 34)

The working accident, particularly in the construction works has its impact either in micro, messo or macro level (Firdaus Ali, 2016). In the micro level, working accident may could make construction project hampered, over expenses, opportunity loss in the result in fatal injury of the workers. In the messo level, the impact of the working accident is influencing the construction service provides and the legal liability. In the macro level the working accident is influencing national product, competitiveness index, and the loss expenses (Firdaus Ali, 2006).

Based on data from *Badan Penyelenggara Jaminan Sosial* (BPJS) *Ketenagakerjaan,* from 2007 to 2015 indicated that the numbers fluctuate and an increasing trend as showed in the figure 2. Those figures show the highest working accident that compared to those of other South East Asian Nations (Anshori, 2008 in the Abduh's research, 2010). National Journal (2009) recorded that Indonesian construction industry contributed the highest working accident compared to that of other industry to the amount of 32% as showed in the Figure 3.

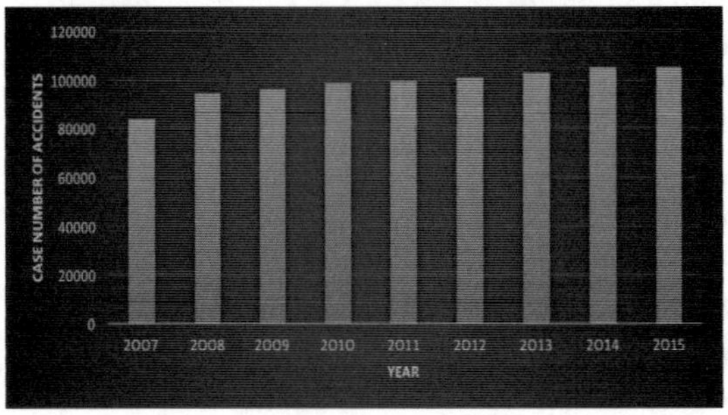

Figure 2. Working Accident Chart in Indonesia
Source: (BPJS Ketenagakerjaan), 2016

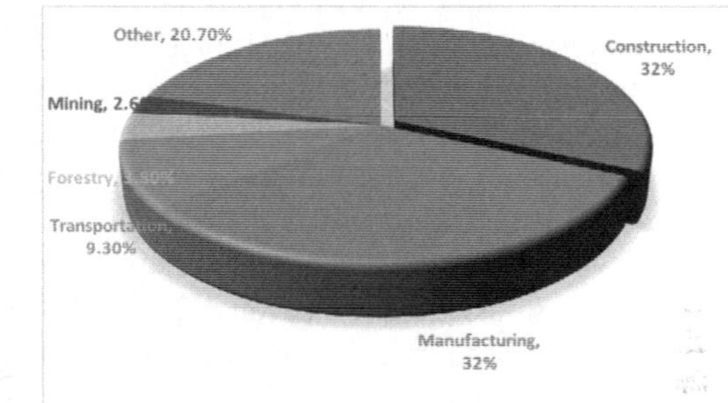

Figure 3. The Accident rate in a range of industries
Source: National Journal, 2009

The frequency of accidents in Malaysia has declined from 11.0 accidents per 1,000 workers in 2000 to 6.1 accidents per 1,000 workers in 2007. While the frequency of fatalities increased from 9.5 deaths per 100,000 workers in 2002 to 12 , 9 deaths per 100,000 workers in 2004, but has reduced the range of approximately 12.4 to 12.5 per 100,000 during the period 2004 to 2007. The Occupational Safety and Health Act 1994, enacted in February 1994, is the main law involving occupational safety and health in the manufacturing industry. The main purpose of this legislation would establish a health, safety, and welfare to health and safety hazards of work activities, protect people at work, adapting the work environment appropriate to meet the needs of physical and mental labor, and offers steps to develop regulations and code associated with this legislation to regulate and improve work health and safety standards Occupational Safety, Health, and Work Environment Act (2011) showed that the duty of companies to manage and maintain safe working conditions and healthy and the environment, including preventing employees from health and safety hazards in all forms. Employees also a duty to cooperate with the company to manage and maintain safe working conditions and healthy.

Based on the literature study, the rate of working accident which so high especially in construction industries, it is because of not having been formed of safety working culture yet. Based on the high working accident problems as well as yet on optimum safety culture, in this case, it is necessary be investigated by analysis. The analysis method which is used to observe the relation pattern and interaction or that is related to the various dimension to adapt the Research of Chinda et al (2007), Chinda (2011), dan Ismail et al (2012), is by being formed model and doing examination by using Structural Equation Modeling (SEM). This study aims to evaluate the safety management system of occupational health and safety based on cultural dimensions that are formed by the method of benchmarking

2. Theoretical Overview

2.1 Safety Culture

Practicing safety culture is the combination of attitudes, values, norms, and perception of the students or worker candidates in one organization that have connection.togheter with safety culture, safety behavior and its application to practical use in the production process (Clarke, 2000). Similar definition was done by The Advisory Committee on The Safety of Nuclear Installations (ACSNI, 1993) and then will be adapted, and clarifies that safety culture in one organization is the product of values, attitudes, perception, competence, and behavior pattern of individual and groups that have commitment to the safety culture. The basic culture of safety culture is attitude and perception to the safety culture (Gadd and Collins, 2002).

2.2 Safety Culture and Occupation Health in the Construction Industry

Some research study has been done in the department of safety culture of construction, Kartam et al (2000) such as studying problems, procedure, and construction safety problem in Kuwait and declared that upgrading the safety culture, especially in the department of management training and the commitment in the safety is required to prevent construction injury and accident. Little (2002) identify 14 elements and four topics (ownership and commitment, system and procedures, training and competence, and communication) for upgrading safety culture in UK construction. Ho and Zeta (2004), analyze safety culture in Hong Kong construction industry and make four factors of culture's key (surrounding, behavior, organization, and human) that affect safety culture.

Based on the previous research, can be concluded that dimension in the making of safety culture in the State-Owned Enterprises are 1) Leadership, 2) Policy, 3) Strategy, 4) Employee, 5) Process, 6) Behaviour, 7) Safety Culture Expenses, 8) Contract System, 9) Value System. Each dimension will be observed its influence towards safety culture, also its correlation between each dimension as well.

Based on the source which earned the author from journal, dissertation, and thesis, then obtained safety culture dimension. Following are safety cultural dimension that could potentially affect the performance of safety culture in the construction project.

3. Research Method

This research method using expert validation and benchmarking. expert validation is done to determine recommendations for the improvement of results-based policy pursued structuring dimensions of safety culture. method of benchmarking conducted to determine the right strategy compared with other countries in implementing measures already improved safety culture.

Benchmarking is the comparative method in which the company finds the best practices in a particular area and then tried to take his own performance in the region was in line with best practice. It is a reference point for measurement purposes and when implemented will produce superior results. Therefore, the company would have to be superior to that in the benchmark.

According to the American Productivity and Quality Center (APQC), benchmarking is the process of systematic and continuous measurement; the process of measuring and comparing sustainability on the business processes of an organization with figures of business processes anywhere in the world, to get information that will help the company's efforts to improve its performance (APQC, 1993).

Meanwhile, another definition of benchmarking is said by Westinghouse, benchmarking is the search and application practices are actually better continuously, which leads to superior competitive performance.

4. Research Results

4.1 Influential Correlation Pattern

Based on the result at SmartPLS 3.0 with Bootstrapping method (Path Coefficient) as big as 33 correlation between dimensions which meet significantly (T statistics > 1,96) is as big as 14 correlation between dimension, significantly, and interplayed each other, then obtained influential relationship pattern as follows:

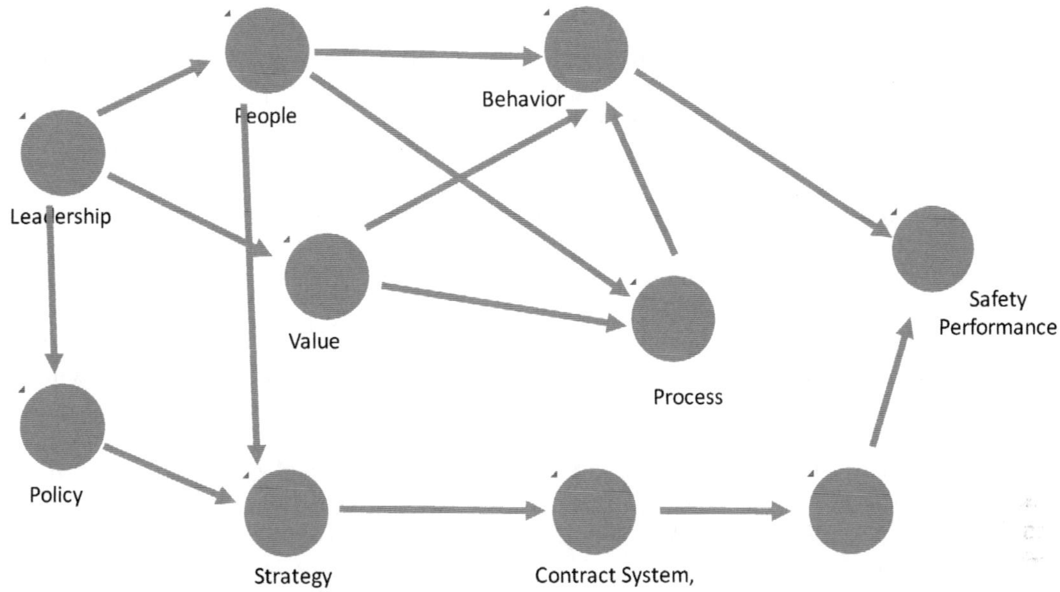

Figure 4. Correlation Pattern Between Dimension Significantly and Influence Each Other (Source: Latief, Machfudiyanto, Yogiswara, 2016)

Table 1. Correlation Pattern Between Dimension Significantly Interplayed

No.	Correlation	Explanation
1.	Leadership → Policy	Significant and interplayed
2.	Leadership → Employee	Significant and interplayed
3.	Leadership → Rate System	Significant and interplayed
4.	Policy → Strategy	Significant and interplayed
5.	Employee → Strategy	Significant and interplayed
6.	Employee → Process	Significant and interplayed
7.	Employee → Behavior	Significant and interplayed
8.	Rate System → Process	Significant and interplayed
9.	Rate System → Behavior	Significant and interplayed
10.	Process → Behavior	Significant and interplayed
11.	Strategy → Contract System	Significant and interplayed
12.	Contract System → Safety Culture Cost	Significant and interplayed
13.	Safety Culture Cost → Performance of Safety Culture	Significant and interplayed
14.	Behavior → Performance of Safety Culture	Significant and interplayed

With the result of the above research, so it is necessary to recommend additional duties, responsibilities, and authority on Regulation Ministry Public of Work and Housing No. 05/PRT/M/2014 as follows:

a. To develop good leadership in taking the job, responbility and authority on the level of the head coaching body construction, structural official echelon 1 of the technical working unit, structural official echelon 2 of the technical working unit, immediate supervisor working unit head, official commitment maker, and working unit ULP on the article 11,12,13,14,15,16, and 17 Permen PU No. 05/PRT/M/2014.

This leadership development could be met with the criteria:

1. To allocate safety responsibility on the entire organization effectively and efficiently
2. To support and motivate to the all organization member to create safety and healthy working
3. Give example safety and positively behavior and a role model in safety
4. Establish a good communication between organization member, supervisor and subordinates
5. Maintaining a commitment to the safety of the entire organization member, assets and environment

b. To motivate / support the worker engagement in safety and healthy work on the article 19 Permen PU No. 05/PRT/2014.

Work engagement could be met with the criteria:

1. To unite the same exception and awareness in the safety between organization and workers.
2. High involvement of workers creates positive effect on the safe working behavior.
3. To maintain good relationship and to develop habits to give each other advice / remaining about safety between workers.
4. Maintaining the level of pressure and a stable work and proportion in the climate of organization.
5. Working should have a proper knowledge about safety, skill, and competent that may minimize accident.
6. Provide safety equipment support as well as good working environment.

c. Developing strategy in implementing organization working safety on article 15 Regulation Ministry Public of Work and Housing 05/PRT/M/2014.

Developing strategy can be met with the criteria:

1. Safety is an integral part in formulating decision and goal of the organization.
2. Carrying out audit program implementing working safety as part of improvement policy of working safety.
3. To promote productivity harmoniously with the safety.
4. To reward incentive to the worker who has successfully implementing a good working safety culture.

d. Responsible for the certainty of safe behavior of worker in performing their duties, that is mentioned in article 19 Permen PU No. 05/PRT/M/2014.

Certainty of safe behavior worker can be met with the criteria:

1. Any individual on the project are obliged safety procedure
2. Applied the high discipline towards safety equipment storage
3. Implementing the high discipline towards safety equipment usage.
4. Ensuring workers have followed entire instruction from organization.
5. The creation of conditions in which the workers can freely report the accident and almost wretched.
6. The creation of conditions in which the workers remind each other about the danger and safety working.

e. Extra cost of implementing SMK3 in construction of PU department is allocated in the general cost in article 20 verse 1 of Regulation Ministry Public of Work and Housing No. 05/PRT/M/2014 are:

1. The cost of working accident compensation
2. Cost for maintaining safety equipment
3. Cost for external audit SMK3

f. Developing safety organization values which able to create the safe working climate on article 12 Regulation Ministry Public of Work and Housing No. 05/PRT/M/2014.
   Developing safety values can be met with the criteria:
   1. Make identification and analysis of the risk towards accident working potential as baseline start planning the safety.
   2. Developing psychological morality which determines how much a person's attention to safety work.
   3. Safety aspect as a topic that cannot be separated from the material discussed on any meeting agenda.
   4. Understanding that safety is a shared responsibility.

5. Discussion

Based on benchmark results, for the regulation of sanctions in force until today still needs to be improved because the penalties that apply are still relatively light and to enhance the deterrent effect of the company in case of accidents. As for the reward, need to be made regarding this policy regulations in the form of incentives if it reaches the target zero accident that motivates the company to continue to give priority to safety in the work environment

According Wirahadikusumah (2006) supervision by the regulatory Directorate of Occupational Health and Safety Norms (PNK3), the Ministry of Labor is still lacking due to a lack of supervisory personnel. According to the Secretary General of the Ministry of PUPR, intensive surveillance may facilitate the implementation of OSH because during this new surveillance carried out if there are events. In Malaysia, the government served as a policy maker, regulator and enforcer. In line with that principle, government agencies will be a role model for OSH. Not only government agencies tasked with monitoring the implementation of OSH, but also working with social agencies with roles and responsibilities clearly.

Recommendation for policy improvement associated with safety culture performance is as follows:

a. Responsible for certainty safe worker behavior in implementing duty on the chapter 19 Regulation Ministry Public of Work and Housing No. 05/PRT/M/2014. Certainty safe worker behavior can be met with the criteria:
   - Any individual in the project are obliged to follow the procedure of safety as follow.
   - Applying the high discipline towards the storage of safety equipment.
   - Applying the high discipline towards the usage of safety equipment.
   - To ensure the workers have followed the instruction from organization.
   - To create condition which the workers can freely report the accident and almost wretched.
   - The creation of condition which the workers remind each other about danger and safety working
b. Extra cost for implementation of SMK3 for construction in the department of PU that is allocated to the general cost on the chapter 20 article 1 Permen PU 05/PRT/M/2014 as follows:
   - The cost for working accident compensation
   - The cost for maintaining safety equipment
   - The cost for SMK3 external audit

Suggestion

As for the suggestion of the research are as follows:

1. The research can be used to arrange safety culture dimension especially in the construction industry that has high rate working accident.
2. Supervision of the central government especially the ministries that related to the construction works on the construction industry for tighter supervision in working accident and

collaboration of all parties (stakeholder) in construction industry to pay attention and organize safety culture continuously.
3. The cost of safety culture is not incorporated with the implementation of the project.
4. Further research needs to be done towards the effect on the safety culture cost towards safety culture performance and behavior that involving parties with government policy makes through focus group discussion Delphi round in order that this research recommendation result becomes a policy improvement which can be applied in Indonesia some day.
5. The next research needs to be done deeply associated with the review and complementing the existing safety culture regulation, like Government Regulation 50/2012, Regulation Ministry Public of Work and Housing. 05/PRT/M/2014, and others.

Acknowledment
The authors would like to acknowledge PITTA Grant of Univesitas Indonesia for the funding support of the research project

References
[1] Aksorn, T. and Hadikusumo, B.H.W., 2006. Critical success factors of safety programs implementation in Thai construction projects. *In*: D. Fang, R.M. Choudhry and J.W. Hinze, eds. *Proceedings of the CIB W99 2006 international conference on global unity for safety and health in construction, 28-30 June 2006, Beijing, China*. Beijing: Tsinghua University Press, 328-336.
[2] Andi et.al. 2005, '*Model Persamaan Struktural Pengaruh Budaya Keselamatan Kerja pada Perilaku Pekerja di Proyek Konstruksi'*, Jurnal Teknik Sipil, [online], Vol. 12 No. 3. Dari: http://www.ftsl.itb.ac.id/. [1 Mei 2014]
[3] Clarke, S. (2000). "*Safety Culture: Underspecified and Overrated?*". International Journal of management Reviews, 2(1), 65-90.
[4] Cooper, D., (2002), "*Safety Culture – A Model for Understanding & Quantifying Difficult Concept*", Professional Safety, 47(6), 3036.
[5] Gadd, S. and Collins, A.M., 2002. "*Safety culture: a review of the literature*". Report by the Health & Safety Laboratory, UK: Sheffield University.
[6] Ghozali, Prof. Dr. Imam, M.Com., Ak. 2005. Model Persamaan Struktural. Semarang: Badan Penerbit Universitas Dionegoro.
[7] Glendon, A.I. and Litherland, D.K., 2001. Safety climate factors, group differences and safety behaviour in road construction. *Safety science*, 39, 157-188
[8] Ho, J.K.L. and Zeta, K.C., 2004. *Cultural factors and their significance to the Hong Kong construction industry* [online]. Available from: http://www.ic.ployu.edu.hk/esh/kb/culture/Ho&Zeta.pdf [Accessed 20 September 2004].
[9] Kartam, N. A., Flood, I, and Koushki, P. (2000). "*Construction safety in Kuwait: issues, procedmes, problems and recommendations*". Safety Science. 36. 163-184.
[10] Lingard, H. and Blismas, N., 2006. "*Building a safety culture: the importance of shared mental models in the Australian construction industry*". In: D. Fang, R.M. Choudhry and J.W. Hinze, eds. Proceedings of the CIB W99 2006 international conference on global unity for safety and health in construction, 28-30 June 2006, Beijing, China. Beijing: Tsinghua University Press, 201-208.
[11] Little, A.D., 2002. "*Improving safety culture in the construction industry*". A workshop for senior management in construction contracting and client companies. Cambridge: University Press.
[12] Molenaar, K., Brown, H., Caile, S. and Smith, R., 2002. Corporate culture: a study of firms with outstanding construction safety. *Professional safety*, July, 18-27.
[13] Mohamed, S. and Chinda, T. (2011), *System dynamics model of construction safety culture*, J. of Engineering Construction and Architectural Management, 18(3), 266-281.

[14] Mohamed, S., 2002. Safety climate in construction site environments. *Journal of construction engineering and management*, 128(5), 375-384.

[15] Niskanen, T., 1994. Safety climate in the road administration. *Safety science*, 17, 237 255.

[16] Paparan Staff Ahli Kementerian Pekerjaan Umum dan perumahan Rakyat (PUPR), Firdaus Ali, pada Seminar Nasional Keselamatan dan Kesehatan Kerja (K3), di Jakarta, Kamis (11/2/2016).

[17] Peraturan Pemerintah Nomor 50 Tahun 2012 tentang Penerapan Sistem Manajemen Keselamatan dan Kesehatan Kerja

[18] Peraturan Menteri Pekerjaan Umum Nomor 5 Tahun 2014 tentang Pedoman Sistem Manajemen Keselamatan dan Kesehatan Kerja (SMK3) Konstruksi Bidang Pekerjaan Umum

[19] Surat Edaran Menteri Pekerjaan Umum dan Perumahan Rakyat Nomor 66 Tahun 2015 tentang Biaya Penyelenggaraan Sistem Manajemen Keselamatan dan Kesehatan Kerja (SMK3) Konstruksi Bidang Pekerjaan Umum

[20] Soeharto, Imam. (1995). Manajemen Proyek, dari Konseptual sampai Operasional, Penerbit Erlangga, Jakarta.

[21] Teo, E.A.L., Ling, F.Y.Y. and Chong, A.F.W., 2005. "*Framework for project managers to manage construction safety. International journal of project management*", 23(4), 329-341.

[22] Williamson, A.M., Feyer, A., Cairns, D. and Biancotti, D., 1997. The development of a measure of safety climate: the role of safety perceptions and attitudes. *Safety science*, 25(1-3), 15-27.

Scale Build-Up Prediction of FeS and $FeCO_3$ in Gas Production Pipes

R. K. Santoso[1,2], S. D. Rahmawati[1], A. Gadesa[1], D. Wahyuningrum[2]

[1]Petroleum Engineering Study Program, Institut Teknologi Bandung, Indonesia
[1]Chemistry Study Program, Institut Teknologi Bandung, Indonesia

[2]Corresponding author: ryan.kurniawan@students.itb.ac.id

Abstract. The existence of passive layer is important to control the corrosion-erosion rate in gas production pipes. However, passive layer can lead to two possible conditions: corrosion-erosion protection or scale build-up. Accurate prediction of scaling tendency is needed to plan treatment and operating condition during the production from gas field. In this study, we develop mathematical model to predict the scaling tendency in gas production pipes. The model consists of two basic equations: precipitation rate and erosion equation. Precipitation rate is calculated using semi-empirical correlation and erosion is calculated using Salama (2000) equation. Then, a modified parameter of scaling tendency (ST), which is the ratio between net precipitation rate and corrosion rate, is introduced to measure the scaling tendency in each segment of production pipe. From simulation, it was found that the interaction between pressure, temperature and fluid composition affected the scaling tendency at most. However, when sand was introduced in the pipe flow, scale formation occurred at low rate. Every segment of production tubing and pipeline gave different tendency condition.

1. Introduction

Passive layer is defined as corrosion product formed on the inner surface of pipes under the nature of corroding environment [21]. Passive layer could be advantageous as giving protection from further corrosion or disadvantageous as blocking the fluid flow. In protection mechanism, passive layer forms diffusion barrier for any species involved in corrosion process [27]. It covers and prevents the underlying steel from further dissolution. However, when the growth rate of passive layer is big, passive layer can block the fluid flow. It becomes scale that builds up and fills the internal pipe volume.

The common types of passive layer formed in the inner surface of the pipes are siderite for sweet corrosion and mackinawite for sour corrosion. When both CO_2 and H_2S exist in the gas stream, the formed passive layer is determined by the competiveness of mackinawite and siderite. The layering arrangement is also various, depends on the supersaturation of mackinawite and siderite and temperature [4] [33].

When moving fluid is introduced in the corrosion system, there will be erosion phenomenon that affects the growth of passive layer. Erosion can be caused by liquid particle or solid particle impingement [16]. In gas flow, liquid impingement is possibly triggered by the liquid condensation from the gas and solid impingement by produced sand. This erosion will reduce the thickness of passive layer or increase the material removal by direct damaging. Therefore, in order to predict the rate of material removal of the pipe accurately, the interaction between corrosion process, growth of passive layer and erosion process need to be further investigated.

In this paper, we present a mathematical model to predict the growth of mackinawite and siderite in CO_2/H_2S system and the tendency to form scale. The mathematical model relates the corrosion rate directly with the growth of passive layer using Faraday's Law and semi-empirical approach. Through the model's results, we will be able to see the effect of passive layer towards fluid flow and material removal in pipes.

2. Mathematical Model

The prediction model is built using Faraday's law and semi-empirical approach. Faraday's law is used to estimate the amount of Fe^{2+} ion in half-cell reaction. Several semi-empirical equations are utilized then as the main function to calculate the growth rate of passive layer. Scaling Tendency (ST) parameter, which is defined as the ratio of net growth rate and corrosion rate, is used to predict the tendency of scale to form. Therefore, the model is divided into three parts: calculation of Fe^{2+} ion, calculation of net growth rate of passive layer and calculation of scaling tendency. We also assume that

- corrosion and erosion occur uniformly in a specific segment of production pipe
- passive layer grows uniformly in the inner surface of the pipe
- reaction has reached its equilibrium condition at certain pressure and temperature

2.1. *Calculation of Fe^{2+} Ion*

Faraday's Law stated that the amount of dissolved cathode in liquid was proportional to the corrosion current [8]. Therefore, the amount of Fe^{2+} ion in solution is related to the amount of metal loss (corrosion rate). Thus, $[Fe^{2+}]$ can be estimated using

$$\left[Fe^{2+}\right] = \frac{CR \times A \times \rho_{metal} \times \Delta t}{MW_{metal} \times HU \times V_{pipe}} \tag{1}$$

where

CR : corrosion rate, L T^{-1}
A : inner pipe area, L^2
ρ_{metal} : metal density, M L^{-3}
MW_{metal} : molecular weight of metal
HU : liquid hold up, fraction
V_{pipe} : pipe inner volume, L^3
Δt : exposure time, T

Based on ASTM G102 [1], in half-cell reaction, molecular weight of a metal is determined using (equivalent molecular weight)

$$MW_{metal} = \frac{1}{\sum_{1}^{n} \frac{f_i}{\left(a_i / n_i\right)}} \tag{2}$$

where

f_i : mass fraction of an atom, fraction
a_i : molecular weight of an atom
n_i : valence

2.2. Calculation of Net Growth Rate of Passive Layer

In H_2S environment, the formation of mackinawite can be expressed by the following reaction

$$Fe^{2+}{}_{(aq)} + S^{2-}{}_{(aq)} \Leftrightarrow FeS_{(s)} \tag{3}$$

Therefore, the formation of mackinawite is related to its solubility product, which can be expressed mathematically by the following equation

$$Ksp_{FeS} = \left[Fe^{2+}\right]\left[S^{2-}\right] \tag{4}$$

There are also many researchers that introduce supersaturation parameter (ratio between concentration product of each ion that possibly contribute to form a solid and solubility product constant) as the contributed one in calculating the growth rate of mackinawite. Therefore, most of prediction equations relate solubility product, supersaturation and temperature to the growth rate of mackinawite. The growth of mackinawite can be calculated using Harmandas & Koutsoukos, 1996. In CO_2 environment, the formation of siderite can be expressed by the following reaction

$$Fe^{2+}{}_{(aq)} + CO_3^{\ 2-}{}_{(aq)} \Leftrightarrow FeCO_{3(s)} \tag{5}$$

Therefore, the solubility limit can be mathematically expressed as

$$Ksp_{FeCO_3} = \left[Fe^{2+}\right]\left[CO_3^{\ 2-}\right] \tag{6}$$

The semi-empirical equations of siderite formation rate also relate supersaturation, solubility product and temperature. Several correlations can be used: Tomson & Johnson, 1991; Greenberg & Tomson, 1992; Van Hunnik et al., 1998; Yean et al., 2008; and Gao et al., 2011.

When there is flowing fluid in the pipes, erosion occurs. The common equation used in petroleum industry for predicting erosion rate caused by either solid or liquid impingement is presented by Salama, 2000.

$$ER = \frac{M \times V_m^{\ 2} \times d_p}{S_m \times \rho_m \times D^2} \tag{7}$$

where

ER : erosion rate, L T^{-1}
S_m : a geometry dependent constant
M : amount of flowing sand, M T^{-1}
V_m : fluid mixture velocity, L T^{-1}
ρ_m : mixture density, M L^{-3}
d_p : sand particle diameter, L
D : pipe diameter, L

Finally, the net growth rate for each passive layer can be expressed mathematically by the following equation

$$NR_i = R_i - ER \tag{8}$$

where

NR_i : net formation rate of passive layer i, L T^{-1}
R_i : formation rate of passive layer i, L T^{-1}

When the value of erosion rate is bigger than formation rate, it can be concluded that net formation rate of passive layer is zero. Thus, there is no passive layer on the inner surface of the pipes and moving fluid is directly in touch with the pipes surface.

2.3. Calculation of Scaling Tendency

Scaling Tendency is used to determine whether the passive layer will protect the underlying surface from further corrosion or block the fluid flow. Nesic, Lee and Ruzic, 2002 have defined scaling tendency as the ratio of film growth rate and corrosion rate. However, when there is a fluid flow in the pipes, the definition should be changed into

$$ST_i = \frac{NR_i}{CR} \quad \textbf{(9)}$$

where

CR : corrosion rate, LT-1

3. Model Implementation

The execution step for the model follows the algorithm in **Figure 1**.

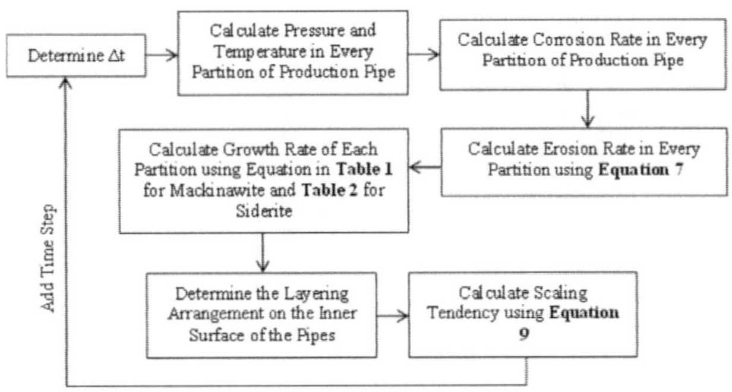

Figure 1. Algorithm for Execution

The input parameters for the model are pressure, temperature, fluid PVT and corrosion rate in every partition. One should measure them accurately to get good prediction of scaling tendency. Pressure and temperature in every partition can be measured directly using pressure and temperature log or calculated using empirical correlations. Corrosion rate can be measured directly using coupons or calculated using empirical correlations.

4. Case Study

To perform the model's simulation, we used an example of wet gas field composition. Its fluid is characterized by the significant amount of CO_2. The detail fluid's composition inside the case study can be seen at Table 1. The production data is shown at Table 2.

Table 1. Gas Composition Data

	Composition	Mole (%)
1	Hydrogen Sulfide (H_2S)	0.05
2	Carbon Dioxide (CO_2)	31.67
3	Nitrogen (N_2)	0.33
4	Methane (C_1)	56.92
5	Ethane (C_2)	0.11

IOP Conf. Series: Journal of Physics: Conf. Series **877** (2017) 012029 doi:10.1088/1742-6596/877/1/012029

6	Propane (C_3)	0.03
7	Water (H_2O)	10.90

Table 2. Production Data

	Parameter	Value	Unit
	Production Tubing		
1	Reservoir Pressure	1100	psig
2	Reservoir Temperature	325	F
3	Tubing Length	8152.9	ft
4	Tubing Internal Diameter	6.625	inch
5	Inclination (from Vertical)	0	degree
6	Wellhead Pressure	500	psig
7	Wellhead Temperature	255.3	F
8	Ambient Temperature	85	F
9	Gas Rate	57.13	MMSCFD
10	Tubing Material (Assumed)	API 5L-X 65	-
11	Sand Concentration (Assumed)	10	ppm
	Production Flowline		
1	Wellhead Pressure	500	psig
2	Wellhead Temperature	255.3	F
3	Ambient Temperature	85	F
4	Flowline Length	1875.4	ft
5	Flowline Outer Diameter	8	inch
6	Inclination (from Horizontal)	0	degree
7	Outlet Pressure	402.28	psig
8	Outlet Temperature	249.0	F
9	Gas Rate	57.13	MMSCFD
10	Tubing Material (Assumed)	API 5L-X 65	-
11	Sand Concentration (Assumed)	10	ppm

In this study, we assumed steady rate for two days period. We generated the pressure and temperature along tubing and flowline using commercial software with Beggs-Brill correlation for vertical and horizontal flow and black-oil model. The corrosion rate value was also assumed to be steady for two days, even there was passive layer on the surface of the metal. The corrosion rate was calculated using deWaard (1995) model. In the proposed model, concentration of S^{2-} and $CO_3{}^{2-}$ ion are determined using multiple step reaction. Mackinawite growth rate is calculated using Harmandas and Koutsoukos, 1996 and siderite growth rate is calculated using Greenberg and Tomson, 1992. Several constants in the equations are calculated using: Sander et al., 2011; Suleimenov & Seward, 1997; Kharaka et al., 1989; Sun et al., 2008; Prieto & Millero, 2002; Roy et al., 1993; and Benezeth et al., 2009.

5. Results and Discussions

Pressure and temperature calculation results using commercial software are shown at Figure 2. The corrosion rate profile is presented at Figure 3.

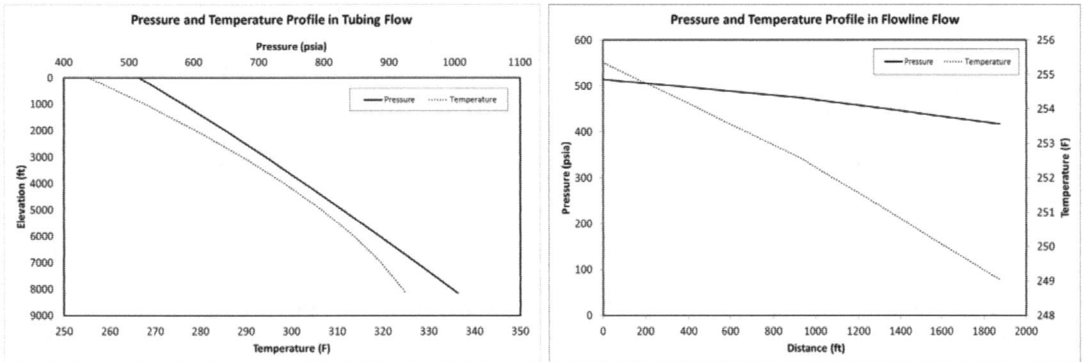

(a) (b)

Figure 2. Pressure and Temperature Profile in: Tubing Flow **(a)** and Flowline Flow **(b)**

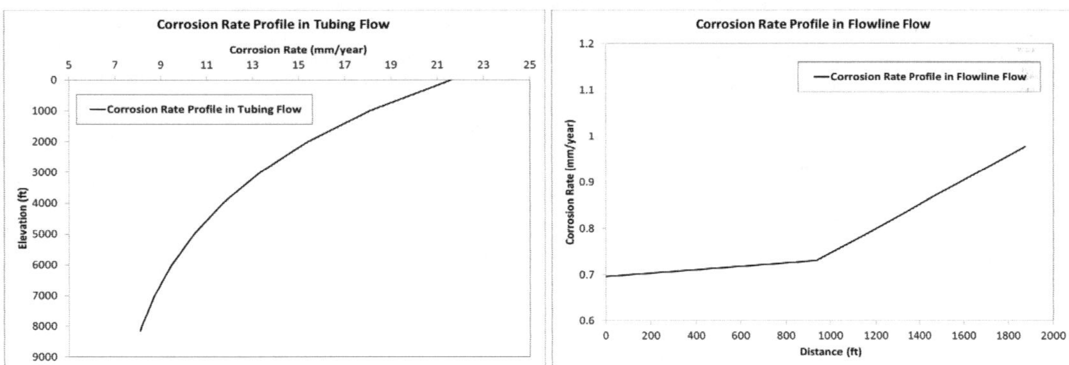

(a) (b)

Figure 3. Tubing's Corrosion Rate Profile **(a)** and Flowline's Corrosion Rate Profile **(b)**

Corrosion rate occurs worst at tubing because it has higher pressure and temperature than in flowline. Based on the deWaard prediction, corrosion rate has its biggest value at the near-outlet location of the pipes.

Because the small amount of sand concentration in the field, the calculation results using Salama (2000) equation show small value of erosion rate in both tubing and flowline. It is presented at Figure 4.

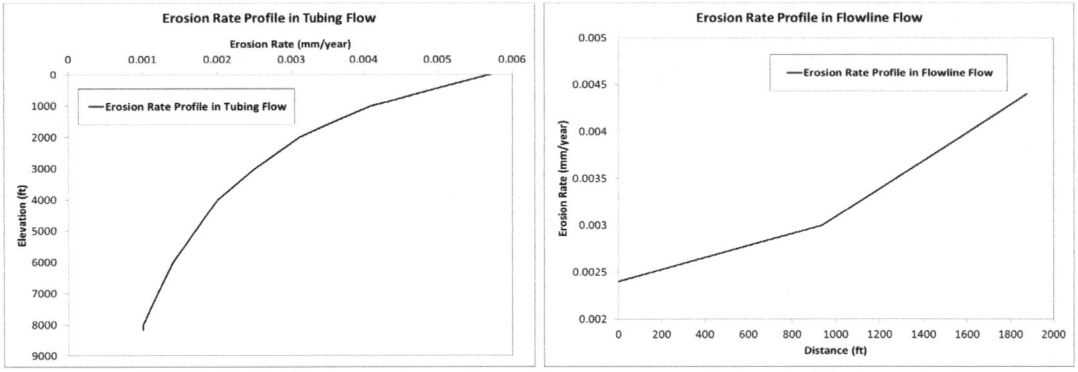

(a) **(b)**

Figure 4. Tubing's Erosion Rate Profile **(a)** and Flowline's Erosion Rate Profile **(b)**

The erosion rate values are the highest at the near-outlet of the pipes, compared to all partitions. It is caused by the expanding of gas volume. Increase in gas volume will increase the gas superficial velocity, thus, erosion rate value will grow higher.

The formation rate of mackinawite in tubing and flowline are presented at Figure 5.

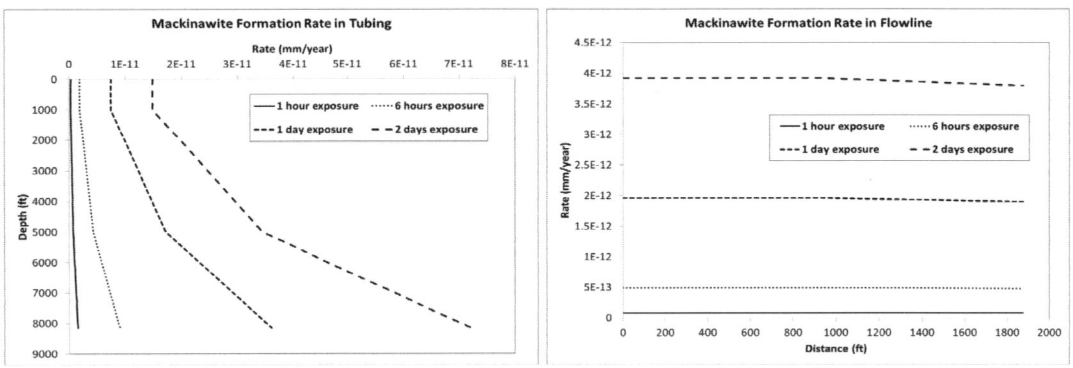

(a) **(b)**

Figure 5. Tubing's Mackinawite Formation Rate Profile **(a)** and Flowline's Mackinawite Formation Rate Profile **(b)**

Both graphs at Figure 5 depict the increasing mackinawite formation rate as increasing pressure and temperature. As going deeper from wellhead in tubing flow and going nearer to the wellhead in flowline flow, temperature and pressure tend to increase. Therefore, supersaturation value will be high. High value in supersaturation means that there are many mackinawite formed at specific section (high formation rate) and it also increases the possibility to deposit scale on the inner surface of pipes. Moreover, as increasing in time step, the supersaturation value will also grow high, thus, the formation rate will be increasing.

However, the formation rate of siderite is different as it is presented at Figure 6.

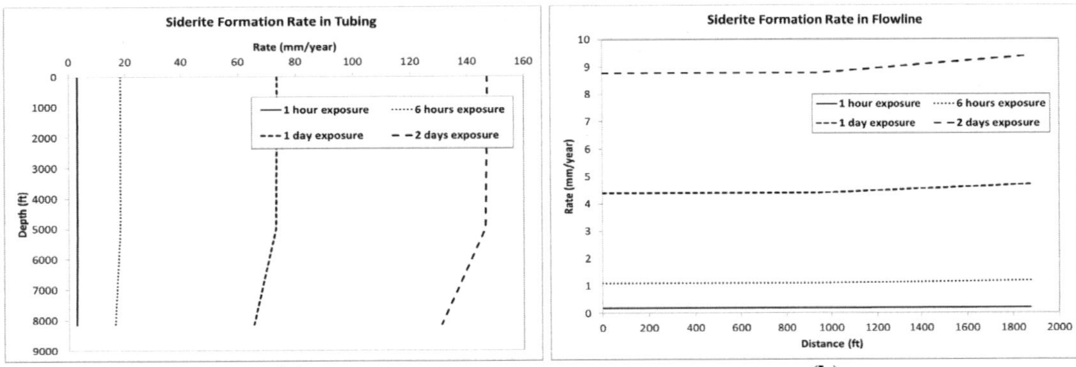

(a) (b)

Figure 6. Tubing's Siderite Formation Rate Profile (**a**) and Flowline's Siderite Formation Rate Profile (**b**)

The formation rate of siderite is strongly affected by the amount of Fe^{2+} ion in the solution. As going deeper from wellhead in tubing flow and going nearer to wellhead in flowline flow, the corrosion rate becomes smaller. Therefore, the concentration of Fe^{2+} ion will be also low. This condition will reduce the amount of siderite formed on the inner surface of the pipes. Moreover, the formation rate is increasing as the time is increasing. It is caused by the increasing amount of Fe^{2+} ion in the solution that can lead to high supersaturation value.

As carbon dioxide concentration in the system is higher than hydrogen sulfide and the formation rate of siderite is much higher than mackinawite, it can be said that siderite dominates the passive layer. Therefore, the scaling tendency is measured based on the siderite layer only. The scaling tendency profile in tubing and flowline are presented at **Figure 7**.

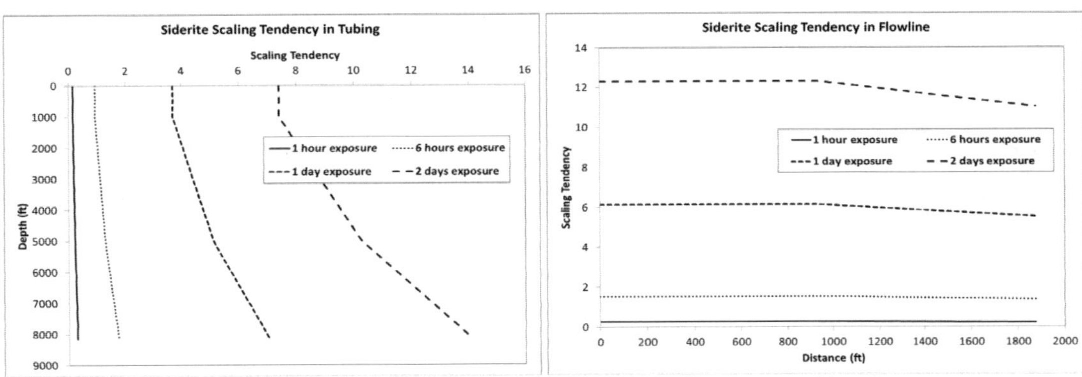

(a) (b)

Figure 7. Tubing's Scaling Tendency Profile (**a**) and Flowline's Scaling Tendency Profile (**b**)

As going deeper from the wellhead in tubing flow and going nearer to the wellhead in flowline flow, the scaling tendency increases. Thus, it explains why the corrosion rate is high at near the wellhead. Passive layer thickness is thin in near wellhead in tubing flow and far from wellhead in flowline flow. It makes corrosive agent easily attack the surface of the pipe.

Figure 7 also describes that the possibility of scale to block the fluid flow will only happens after 1 hour exposure because the scaling tendency value is greater than one. Corrosion will no longer exist anymore after 1 hour.

From all of the simulation results, it can be said that corrosion rate and formation rate of passive layer depend on each other. Corrosion rate determines the amount of Fe^{2+} ion in the solution, thus, it affect the supersaturation condition, then, the formation rate of passive layer. Formation rate of passive layer determines the thickness of passive layer that can reduce the value of corrosion rate. Thus, iteration process has to be involved to calculate the interaction between corrosion rate and passive layer formation rate in the future.

Brown and Nesic, 2005 has investigated the possibility of mixed siderite and mackinawite in the passive layer. However, the interaction of siderite and mackinawite when they exist at the same time is very complicated and hard to be predicted. Therefore, the proposed model has a limitation in predicting this condition. When using the proposed model, one has to assume type of passive layer that has high possibility to form on the inner surface of the pipe.

6. Conclusion

From the discussion above, it can be concluded that the proposed model successfully predicts the scaling tendency in tubing and flowline flow. However, the model is still limited in prediction when mackinawite and siderite exist at the same time.

Formation rate of passive layer is affected by three major parameters: pressure, temperature and supersaturation. Pressure and temperature determine the kinetic of reaction and the amount of S^{2-} and ${CO_3}^{2-}$ ion in the solution, then, affect the supersaturation value. Supersaturation value is also affected by the corrosion rate.

Scaling tendency has increasing trend as going deeper from the wellhead in tubing flow and going nearer to the wellhead in flowline flow. Therefore, corrosion in the tubing flow at near wellhead and in flowline flow far from the wellhead happens worst.

Acknowledgement
This research was funded by DIKTI 2016.

References

[1] 1989 ASTM G 102 (Philadelphia: American Society for Testing and Materials) p 416-422
[2] Andrzej Anderko and Robert D. Young 1999 *Corrosion 99* No. 31 (Houston: NACE International) p 1-19
[3] Bénézeth, P., Dandurand, J. L. and Harrichoury, J. C. Solubility product of siderite ($FeCO_3$) as a function of temperature (25–250 °C). *Chemical Geology Volume* **265**, Issues 1–2, 15 July 2009, Pages 3–12.
[4] B. Brown and S. Nesic 2005 *Corrosion 2005* No. 05625 (Houston: NACE International) p 1-29
[5] Carlos A. Palacios, Corpoven S. A., Valoy Chaudary 1996 *Fourth Latin American and Caribbean Petroleum Engineering Conference* No. SPE 36127 (Port of Spain: Society of Petroleum Engineers, Inc.) p 501-511
[6] Denny A. Jones 1996 *Principles and Prevention of Corrosion* 2nd edition (New Jersey: Prentice-Hall, Inc.) p 5
[7] E. W. J. van Hunnik, B. F. M. Pots and E. L. J. A. Hendriksen 1996 *Corrosion 96* No. 6 (Houston: NACE International) p 1-22
[8] F. C. Strong 1961 *J. Chem. Educ.* 1961, 38 (2), p 98
[9] Gao, M.; Pang, X.; and Gao, K. The growth mechanism of CO2 corrosion product films. Corrosion Science 53, 2011, 557–568.Greenberg, J. and Tomson, M. Precipitation and dissolution kinetics and equilibria of aqueous ferrous carbonate vs temperature. Appl. Geochem. 1992, **7**, 185–190.
[10] G. H. Al-Aithan, F. M. Al-Mutahar, J. R. Shadley, S. A. Shirazi, E. F. Rybicki and K. P. Roberts 2014 *Corrosion 2014* No. 3854 (San Antonio: NACE International) p 1-15
[11] G. H. Nancollas 1979 *Advances in Colloid and Interface Science* 10 (1979), p 215-252
[12] J. Greenberg and M. Tomson 1992 *Applied Geochemistry* 1992, Vol. 7, p 185-190
[13] Johnson, M. L.; and Tomson, M. B. Ferrous Carbonate Precipitation Kinetics and Its Impact on CO2 Corrosion. CORROSION/91, paper no. 268, (Houston, TX: NACE International, 1991).
[14] Kharaka, Y. K.; Perkins, E. H.; Gunter, W. D.; Debral, J. D.; and Bamford, C. H. Solmineq 88:

A Computer Program for Geochemical Modeling of Water Rock Interactions; Alberta Research Council: Menlo Park, CA, 1989.

[15] M. B. Tomson and M. L. Johnson 1991 *SPE International Symposium on Oilfield Chemistry* No. SPE 21025 (Anahelm: Society of Petroleum Engineers) p 1-8

[16] M. M. Salama 2000 *Corrosion 2000* No. 00085 (Houston: NACE International) p 1-18

[17] N. G. Harmandas and P. G. Koutsoukos 1996 *Journal of Crystal Growth* 167 (1996) p 719-724

[18] O. A. Nafday and S. Nesic 2005 *Corrosion 2005* No. 05295 (Houston: NACE International) p 1-27

[19] Oddo, J. E.; and Tomson, M.B. Simplified Calculation of $CaCO_3$ Saturation at High Temperatures and Pressure in Brine Solutions, *J. of Petroleum Technology*, 1982, pp.1583-1590.

[20] Prieto, F. J. M.; and Millero, F. J. The values of $pK_1 + pK_2$ for the dissociation of carbonic acid in seawater. *Geochimica et Cosmochimica Acta Vol. 66*, 2002, **No. 14**, pp. 2529–2540.

[21] Richard Woollam, Kavitha Tummala, Jose Vera and Sandra Hernandez 2011 *Corrosion 2011* No. 11076 (Houston: NACE International) p 1-14

[22] R. K. Santoso, I. Fauzi and S. D. Rahmawati 2015 *SPE Asia Pacific Oil and Gas Conference and Exhibition* No. SPE 176192 (Bali: Society of Petroleum Engineers, Inc.) p 1-17

[23] Roy, R. N.; Roy, L. N.; Vogel, K. M.; Porter-Moore, C.; Pearson, T.; Good, C. E.; Millero, F. J.; and Campbell, D. M. The dissociation constants of carbonic acid in seawater at salinities 5 to 45 and temperatures 0 to 45°C. *Mar. Chem.* 1993, **44**, 249–267.

[24] Sander, S. P.; Abbatt, J.; Barker, J. R.; Burkholder, J. B.; Friedl, R. R.; Golden, D. M.; Huie, R. E.; Kolb, C. E.; Kurylo, M. J.; Moortgat, G. K.; Orkin, V. L.; and Wine, P. H. Chemical Kinetics and Photochemical Data for Use in Atmospheric Studies, Evaluation No. 17, JPL Publication 10-6, 2011, Jet Propulsion Laboratory, Pasadena, available at: http://jpldataeval.jpl.nasa.gov (last access: 22 June 2016).

[25] Sander, R. Compilation of Henry's law constants (version 4.0) for water as solvent. Atmos. Chem. Phys., 2015, 15, 4399–4981.

[26] S. Nesic, S. Wang, H. Fang, W. Sun and K. K-L. Lee 2008 *Corrosion 2008 Conference and Expo* No. 08535 (Houston: NACE International) p 1-16

[27] S. Nesic, K. J. Lee and V. Ruzic 2002 *Corrosion 2002* No. 02237 (Houston: NACE International) p 1-35

[28] S. Nesic, M. Nordsveen, R. Nyborg and A. Stangeland 2001 *Corrosion 2001* No. 01040 (Houston: NACE International) p 1-28

[29] Suleimenov, O. M.; and Krupp, R. E. Solubility of hydrogen sulfide in pure water and in NaCl solutions, from 20 to 320 °C and at saturation pressures. Geochim. Cosmochim. Acta 1994, 58, 2433-2444.

[30] Suleimenov, O. M.; and Seward, T. M. A spectrophotometric study of hydrogen sulfide ionization in aqueous solutions to 350 °C. Geochim. Cosmochim. Acta 1997, 61, 5187-5198.

[31] Sun, W.; Nešić, S.; Young, D.; and Woollam, R. C. Equilibrium Expressions Related to the Solubility of the Sour Corrosion Product Mackinawite. *Ind. Eng. Chem. Res.* 2008, **47**, 1738-1742.

[32] Wei Sun and Srdjan Nesic 2007 *Corrosion 2007 Conference and Expo* No. 07655 (Houston: NACE International) p 1-26

[33] W. Sun, S. Nesic and S. Papavinasam 2006 *61ST Annual Conference & Exposition* No. 06644 (Houston: NACE International) p 1-26

[34] W. Sun, S. Nesic and S. Papavinasam 2006 *Corrosion* July 2008, Vol. 64, No. 7, p 586-599

[35] Yean, S.; Saiari, H.; Kan, A.; and Tomson, M. "Ferrous Carbonate Nucleation and Inhibition, in SPE International Oilfield Scale Conference," Society of Petroleum Engineers, Aberdeen, UK, 2008.

IOP Conf. Series: Journal of Physics: Conf. Series **877** (2017) 012030 doi:10.1088/1742-6596/877/1/012030

Modeling the Thermal Response of $Al_{0.3}Ga_{0.7}As$/GaAs/Ge Multijunction Solar Cells

T Sumaryada[1*], N E Damayanti[1], S Rohaeni[1], H Syafutra[1], Irzaman[1], A Maddu[1], H Hardhienata[1], H Alatas[1]

[1]Department of Physics, Bogor Agricultural University, Jalan Meranti Kampus IPB Dramaga Bogor 16680, Indonesia

E-mail : tsumaryada@ipb.ac.id

Abstract. The III-V group of semiconductor materials are well-known for their excellent performance as high efficient solar cells. This type of material is usually arranged in a multijunction structure which allows a continuous absorption of sun's radiation. The electronic transport properties of semiconductor materials are influenced by the temperature. In this paper, the effect of temperature to the performance of $Al_{0.3}Ga_{0.7}As$/GaAs/Ge multijunction solar cells was evaluated using a simulation approach. By varying the temperature of materials to 25 °C, 50 °C, 75 °C and 100 °C in PC1D solar cell simulation, we were able to determine the performance of each cell and the overall efficiency of $Al_{0.3}Ga_{0.7}As$/GaAs/Ge multijunction solar cells. The results have shown that the increasing temperature of the materials will reduce the V_{oc} of each cell by -1.40 mV/°C, -1.40 mV/°C and -1.30 mV/°C respectivelly, while the total efficiency of multijunction solar cells was reduced by -0.106%/°C.

1. Introduction

Solar energy is one of the most abundant source of clean energy in the world. One way to harvest this type of energy and convert it into electricity is by using a solar panel which consist of solar cells. Various types of solar cells and many different techniques had been invented as a result of intensive researchs all around the world for the last five decades. The most widely used solar cells so far are made from silicon based materials which is capable to gain an efficiency up to 25% [1]. Eventhough the efficiency of silicon based solar cells is quite low as compared to other types, but the abundant source of silicon in the world and the well-mastered fabrication's techniques has made the silicon based solar cells as the most available solar cells nowadays.

The III-V based solar cells like GaInP, AlGaAs, InP, and GaAs had been known to produce a higher efficiency of above 25% [2,3,4]. The combination of those semiconductor materials in the form of multijunction solar cells, recently hold the world record of efficiency of 44.7% [5]. In a multijunction solar cells, several p-n junction of semiconducting layers (subcells) were arranged from the top to the bottom following the order of their bandgap energies. The first layer has the highest bandgap energy which is purposed to absorb the solar radiation in the small wavelength region, while the next layers with smaller bandgap energies were set to absorb solar radiation in the longer wavelength regions [6]. Theoretically, a higher rate of solar cells's efficiency would be obtained by putting more junctions in the solar cells as more junctions would increase the amount of radiation absorbed and reduce the radiation loss [7].

One of the important factor that affecting the performance and the efficiency rate of semiconductor solar cells is the temperature. The increasing ambient temperature will reduce the performance of the solar cells [8-11].The thermal response of multijunction solar cells have been studied by some researchers [12,13]. This paper is mainly focused on the effect of temperature to the performance and efficiency of $Al_{0.3}Ga_{0.7}As/GaAs/Ge$ multijunction solar cells through a simulation approach. The simulation approach in this paper is limited by some assumptions, i.e: (a) for each subcell, an independent simulation were done in order to optimize the electricity power produced by each subcell, (b) the current flows in each subcell were not identical, (c) the boundary between two adjacent cells (tunnel junction) was not taken into account and assumed to play no role in this simulation.

2. Methods

There are two steps of simulation performed in this research as illustrated in Figure 1. The first step is simulating the incoming solar radiation arrives at each subcell. The amount of radiation intensity (in Watt.m^{-2}) received by the first subcell (I_0) is AM1.5G solar radiation (ASTM G-173), while the amount of radiation received by the next subcell depends on the amount of transmitted radiation from the previous subcell following this equation :

$$I_n = I_{n-1} e^{-\alpha_n (x_n)} \tag{1}$$

where n=1,2,3.. denotes the subcell index which depend on the number of subcell used in a multijunction solar cells, while α_n and x_n denote the absorption coefficient and the thickness of each subcell.

The second step is the current producing simulation using PC1D program [15], in which all parameters affecting the electrical output of solar cells including the temperature was set. Some of the relevant outputs from this simulation are the open circuit voltage V_{oc}, the short circuit current I_{sc}, the output power P_{out} and the optimum thickness of the subcell. The absorption coefficient of each subcell was modelled by using this equation [16] :

$$\alpha(\lambda) = 5.5\sqrt{E(\lambda) - E_g} + 1.5\sqrt{E(\lambda) - (E_g + 0.1)} \tag{2}$$

where $E(\lambda)$ is the photon energy of solar radiation at a particular wavelength and E_g the energy gap of the subcell. The power P is defined as the solar radiation intensity multiplied by the area of the solar cell. The total efficiency of the multijunction solar cells is calculated using :

$$\eta = \left(\frac{P_1 + P_2 + P_3}{P_0} \right) \tag{3}$$

where P_0 is the incoming power received by the first subcell, and P_1, P_2 and P_3 are the electric power produced by the first, second and the third subcells. Similar approach has been done in reference [17]. The temperature effect takes places in the single diode model (SDM) which governs the current producing simulation. Based on this model, for one sun radiation, the open circuit voltage V_{oc} is calculated through this equation :

$$V_{OC}(T) = \frac{nkT}{q} \ln(\frac{J_{SC}}{J_0(T)}) \tag{4}$$

where J_0 is the dark saturation current, J_{SC} is the short circuit current, k is the Boltzman constant, q is the electric charge and n is the number of flowing charge.

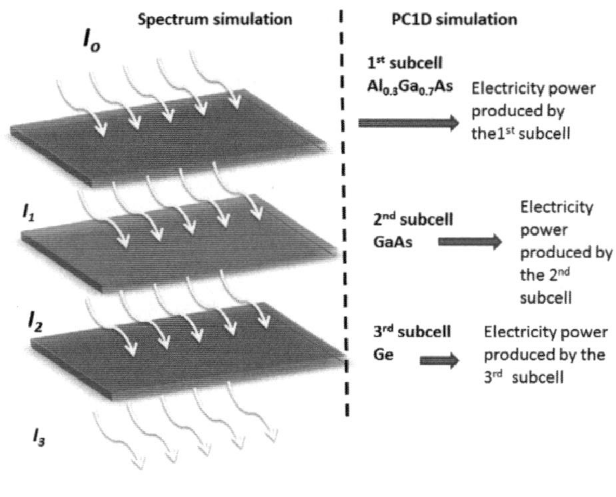

Figure 1. Ilustration of multijunction solar cells simulation.

3. Result and Discusions

The simulation results were shown in Table 1. The results of spectrum simulation for each subcell were shown in the incoming solar radiation intensity values. The amount of incoming AM1.5G solar radiation power per cm^2 received by the first subcell is 989.32×10^{-4} Watt, which is smaller than the solar constant 1353 $\times 10^{-4}$ Watt. The transmitted radiation to the second subcell is 538.40 $\times 10^{-4}$ Watt and to the third subcell was 335.80 $\times 10^{-4}$ Watt. As the temperature of the subcell is increased, the amount of electrical power produced by each subcell became smaller. The same behavior also found in the V_{oc} of each subcell. In Figure 2, the V_{oc} of each subcell was decreased as the temperature increased as seen on Figure 2(a). The decreasing rate of V_{oc} for $Al_{0.3}Ga_{0.7}As$ subcell and GaAs were similar -1.4 mV/°C. The I-V diagram in Figure 2(b) shows that the increasing temperature will reduced the overall electrical performance of multijunction solar cells.

Table 1. Simulation results

Subcell	Energy gap (eV)	Thickness (μm)	n doppin g	p dopping	Incoming solar radiation power ($\times 10^{-4}$ Watt)	Electric power produced in 1 cm^2 area at various temperatures ($\times 10^{-4}$ Watt)			
						25 °C	50 °C	75 °C	100 °C
$Al_{0.3}Ga_{0.7}As$	1.817	2.91	1.00×10^{20}	1.00×10^{16}	989.32	450	436	422	408
GaAs	1.42	3.00	1.00×10^{17}	1.00×10^{16}	538.40	198	190	182	174
Ge	0.667	5.00	1.80×10^{17}	1.00×10^{16}	335.80	37	33	28.8	24.7

The features of our simulations are consistent with the experimental results of other III-V based multijunction solar cells. Aho *et al* [13] found that the V_{oc} of GaInP/GaAs/GaInNAsSb decreased by -7.5 mV/°C and the total efficiency by -0.09%/°C, while our results have shown a performance degradation of about -1.40 mV/°C and-0.106%/°C for $Al_{0.3}Ga_{0.7}As$/GaAs/Ge multijunction solar cells. Eventhough the systems are different, but in general the performance degradation of III-V based multijunction solar cells due to the increasing temperature lies within the same range of about -1.0 to -10.0 mV/°C for V_{oc} and -0.10%/°C for the total efficiency. The degradation rate of V_{oc} and the subcell efficiency of each subcell in our simulation is similar as shown by proportional gradients in Figure 2(a) and Figure 3(a). The same behavior of III-V based solar cells due to the increasing temperature was also found in other reference [13,18]. Figure 3(b) shows the total efficiency degradation of $Al_{0.3}Ga_{0.7}As$/GaAs/Ge with the degradation rate of -0.106%/°C which is comparable to the result of others [13]. The lattice constant of $Al_{0.3}Ga_{0.7}As$, GaAs and Ge is 5.655 Å, 5.653 Å, 5.658 Å respectively [19]. With the average lattice mismatch of only 0.4%, the same monotonic behavior of each subcell is expected. The lattice matching arrangement of semiconductor materials in multijunction solar cells will minimize the dissipated power and allows a similar (monotonic) response in electrical performance and power efficiency of each subcell as shown in Figure 2(a) and 3(a).

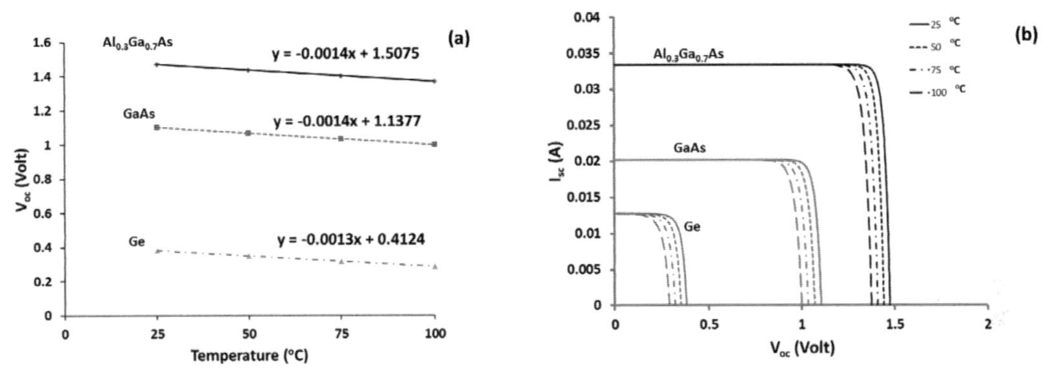

Figure 2. The thermal response of each subcell is shown in (a), while the performance of each subcell at various temperature (25 °C, 50 °C, 75 °C, 100 °C) is shown in (b).

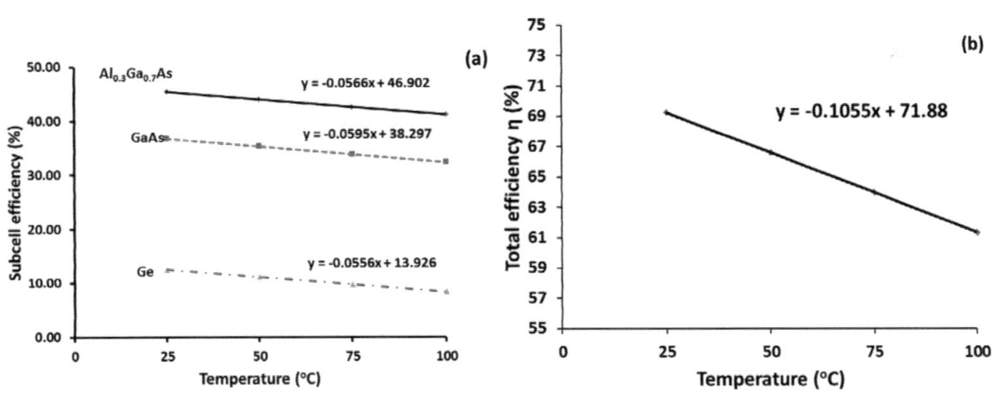

Figure 3. The efficiency degradation as a function of temperature for (a) each subcell, and for (b) the whole multijunction solar cells.

4. Conclusion

The study of temperature effect to the performance and the efficiency of $Al_{0.3}Ga_{0.7}As$/GaAs/Ge multijunction solar cells has been performed. The increasing temperature of multijunction solar cells will reduce the performance and the efficiency as shown by the decreasing V_{oc} and the efficiency of each subcell. Eventhough we use different system, but the general feature of our results is in a good agreement with the experimental results of other III-V based multijunction solar cells.

Acknowledgement
The Authors express their gratitude to the Directorate of Higher Education, Ministry of Research Technology and Higher Educations, Republic of Indonesia through Grant of Penelitian Institusi IPB under contract number 079/SP2H/LT/DRPM/II/2016.

References

[1] Masuko K, Shigematsu M, Hashiguchi T, Fujishima D, Kai M, Yoshimura N, Yamaguchi T, Ichihashi Y, Yamanishi T, Takahama T, Taguchi M, Maruyama E, Okamoto S 2014 *IEEE Journal of Photovoltaics*. **4** 1433
[2] Bertness K A, Friedman D J, Kurtz S R, Kibbler A E, Kramer C, Olson J M 1994 *IEEE Aerospace and Electronic Systems Magazine* **9** (12) 12
[3] Kayes B M, Nie H, Twist R, Spruytte S G, Reinhardt F, Kizilyalli I C, Higashi G S 2011 *Proceedings of the 37th IEEE Photovoltaic Specialists Conference* 000004
[4] King R R, Law D C, Edmonson K M, Fetzer C M, Kinsey G S, Yoon H, Sherif R A and Karam N 2007 *Applied Physics Letters* 90 (18) 182516
[5] Dimroth F, Grave M, Beutel P, Fiedeler U, Karcher C, Tibbits T N D, Oliva E, Siefer G, Schachtner M, Wekkeli A, Bett A W, Krause R, Piccin M, Blanc N, Drazek C, Guiot E, Ghyselen B, Salvetat T, Tauzin A, Signamarcheix T, Dobrich A, Hannappel T and Schwarzburg K 2014 *Progress in Photovoltaics: Research and Applications* **22**(3) 277
[6] Sumaryada T, Sobirin R, Syafutra H 2014 *AIP Conference Proceedings* **1554** 162
[7] Tobias I, Luque A 2002 *Progress in Photovoltaics: Research and Applications* **10** (5) 323
[8] Sze S M 1981 *Physics of Semiconductor Devices* (New York: John Wiley & Sons) 264
[9] Fan J C C 1986 *Solar Cells* **17** 309
[10] Singh P, Singh S N, Lai M, Husain M 2008 *Solar Energy Materials and Solar Cells* **92** 1611
[11] Guodan W, Kuen-Ting S, Noel C G, and Forest S R 2007 *Applied Physics Letters* **91** 223507
[12] Friedman D J 1996 *25th IEEE Photovoltaic Specialists Conference* Washington DC 89
[13] Aho A, Isoaho R, Tukiainen A, Polojärvi V, Aho T, Raappana M and Guina M 2016 *AIP Conference Proceedings* **1679** 050001
[14] Siefer G and Bett A W 2012 *Prog. Photovolt: Res. Appl* **22** 515
[15] Basore P A, Clugston D A 1997 *26th IEEE Photovoltaic Specialists Conference* 207
[16] Olson M., Friedman D J, Kurtz S 2003 *High-Efficiency III-V Multijunction Solar Cells, in Handbook of Photovoltaic Science and Engineering*, ed A. Luque (Chichester: John Wiley & Sons) 359
[17] Sumaryada T, Wahyuni E S, Syafutra H, Alatas H 2015 *Proc. SPIE **9444**, International Seminar on Photonics, Optics, and Its Applications (ISPhOA 2014)* 94440M
[18] Or A B, Appelbaum J 2014 *Solar Energy Materials & Solar Cells* **130** 234
[19] Heinzel T 2007 *Mesoscopic Electronics in Solid State Nanostructures*, 2nd Edition (Weinheim:John Wiley &Sons) 85

IOP Conf. Series: Journal of Physics: Conf. Series **877** (2017) 012031 doi:10.1088/1742-6596/877/1/012031

Comparison of Several Thermal Conductivity Constants for Thermal Hydraulic Calculation of Pebble Bed Reactor

Dwi Irwanto[1], Topan Setiadipura[2] and Asril Pramutadi[1]

[1]Nuclear Physics and Biophysics Research Group, Department of Physics, Institut Teknologi Bandung, Jl. Ganesha No 10, Bandung, 40132, Indonesia

[2]Center for Nuclear Reactor Technology and Safety, National Nuclear Energy Agency of Indonesia (BATAN), Puspiptek Complex OB. No. 80, Serpong, Tangerang Selatan 15310, Indonesia

*Email: dirwanto@fi.itb.ac.id

Abstract. There are two type of High Temperature Gas Reactor (HTGR), prismatic and pebble bed. Pebble Bed type has unique configuration because the fuels are randomly distributed inside the reactor core. In term of safety features, Pebble Bed Reactor (PBR) is one of the most promising reactor type in avoiding severe nuclear accidents. In order to analyze heat transfer and safety of this reactor type, a computer code is now under development. As a first step, calculation method proposed by Stroh [1] is adopted. An approach has been made to treat randomly distributed pebble balls contains fissile material inside the reactor core as a porous medium. Helium gas act as coolant on the reactor system are carrying heat flowing in the area between the pebble balls. Several parameters and constants are taken into account in the new developed code. Progress of the development of the code especially comparison of several thermal conductivity constants for a certain PBR-case are reported in the present study.

1. Introduction

Pebble Bed Reactor (PBR) is one of the most promising nuclear power reactor type. Neutronically, PBR has advantages in term of low excess reactivity during reactor operation, small absorption cross-section of moderator material and neutronically inert of helium gas as coolant. Based on how fuel loaded into the reactor core, there are three different fuel loading schemes named Once Through Then Out (OTTO), Peu a Peu (accumulative) and Multipass. Many studies have been conducted to analysed neutronic aspects of PBR. [2-4]

PBR has high output temperature that can be utilized for industrial processes such as sea water desalination, hydrogen production, district heating, oil recovery, etc. which increase its economical values. Pebble Bed Reactor also considered as a safety system due to low power density of the reactor, coating fuel particle which limits the release of radioactive material to environment and PBR also has high temperature limit, as high as 1600°C. Moreover, it is expected that natural mechanisms to remove the heat such as conductive and radiative heat transfer will works even if no active convective core cooling is provided. [5]

Indonesia is now preparing to build Experiment Power Reactor based on PBR technology. An in-house-computer code to calculate thermal hydraulic aspect of the system is needed to deeply analyze this reactor design. The objective of the present study is to develop a thermal hydraulic analysis tools for heat transfer on Pebble Bed Reactor with focused on comparison of several thermal conductivity constants.

2. Governing Equations and Calculation Procedures

In order to analyze the safety features of this reactor design, appropriate equations describing the condition of the reactor system were first established, including those related to the safety parameters of the reactor. Heat production is one of the important parameters in the present analysis; it was calculated using power density information obtained by neutronic analysis.

As a first step to develop thermal hydraulic code for PBR is to implement calculation procedures proposed by Kenneth Stroh [1]. An approach has been made to treat randomly distributed pebble balls contains fissile material inside the reactor core as a porous medium. Helium gas as reactor coolant are carrying heat and flowing in the area between the pebble balls. The following equations are implemented in the computer code:

Forcheimer equation:

$$\frac{\Delta P}{L} = a_1 v + a_2 v^2 \tag{1}$$

$v: speed\ of\ fluid$
$P: pressure$
$L: length\ of\ the\ medium$
$a_1\ and\ a_2: constant\ that\ depend\ on\ fluid\ and\ porous\ medium\ characteristics$

Ergun equation:

$$\frac{\Delta P}{L} = \frac{1-\varepsilon}{\varepsilon^3}\left[150(1-\varepsilon)\frac{\mu v}{{d_p}^2} + 1.75\frac{\rho}{d_p}v^2\right] \tag{2}$$

$v: speed\ of\ fluid$
$\varepsilon: fuel\ void\ fraction$
$P: pressure$
$L: length\ of\ the\ medium$
$d_p: pebble\ ball\ diameter$

Implementation on axisymmetric cylinder 2D (R-Z)

$$v_r\frac{\partial v_r}{\partial r} + v_z\frac{\partial v_r}{\partial z} = -\frac{1}{\rho}\frac{\partial \rho}{\partial r} - \frac{k_1}{\rho}v_r - \frac{k_2}{\rho}v_r v \tag{3}$$

$$v_r\frac{\partial v_z}{\partial r} + v_z\frac{\partial v_z}{\partial z} = -\frac{1}{\rho}\frac{\partial \rho}{\partial z} - \frac{k_1}{\rho}v_z - \frac{k_2}{\rho}v_z v \tag{4}$$

The heat transfer within pebbles in the pebble bed involves both heat conduction through pebbles and heat radiation through the voids between pebbles. The helium conduction in the voids between pebbles is negligible in the present study. The effective thermal conductivity for this system using the German equation obtained by experimentation can be written as follows [6]:

$$\lambda_{\text{eff}} = \lambda_{\text{ruhe}} + \lambda_{\text{konv}} \tag{5}$$

where

$$\lambda_{\text{ruhe}} = 0.3 + 0.002\ (\mathrm{T_s} - 135.0)^{1.29}\ \ \frac{\mathrm{W}}{\mathrm{m}}\cdot\mathrm{K} \tag{6}$$

$$\lambda_{konv} = \frac{\rho_G \; c_p \, v \, d_k}{K_r} \qquad (7)$$

where $K_r = 8\left(2 - \left(1 - 2\,\frac{d_k}{D}\right)^2\right)$

ρ_G : Gas (Helium)density [kg/m^3]
c_p : Heat capacity [J/kg K]
v : Gas (Helium) velocity [m/s]
d_k : Fuel pebble diameter [m]
D : Core diameter [m]

However, General Electric also developed a correlation for the pebble bed thermal conductivity as follows: [7]

$$\lambda = 1.1538\times10^{-6}\,(T_s + 100)^{1.6622} \qquad (8)$$

These effective thermal conductivity constants are implemented in the code and the effect are studied. This new developed code is written with C++ language.

3. Calculation Results and Discussion

Calculation was performed for KFA (Kernforschunganlage) reactor design with 3000MWt and 4 MPa pressure. There are three cases; Case A: calculation results obtained from reference; Case B: calculation using the developed code with thermal conductivity equation used is proposed by German researchers: Case C: calculation using the developed code with thermal conductivity equation used is proposed by GE. Comparison of calculation results for these three cases for coolant outlet temperatures and coolant bulk temperatures can be seen in Fig. 1 and 2 as follows:

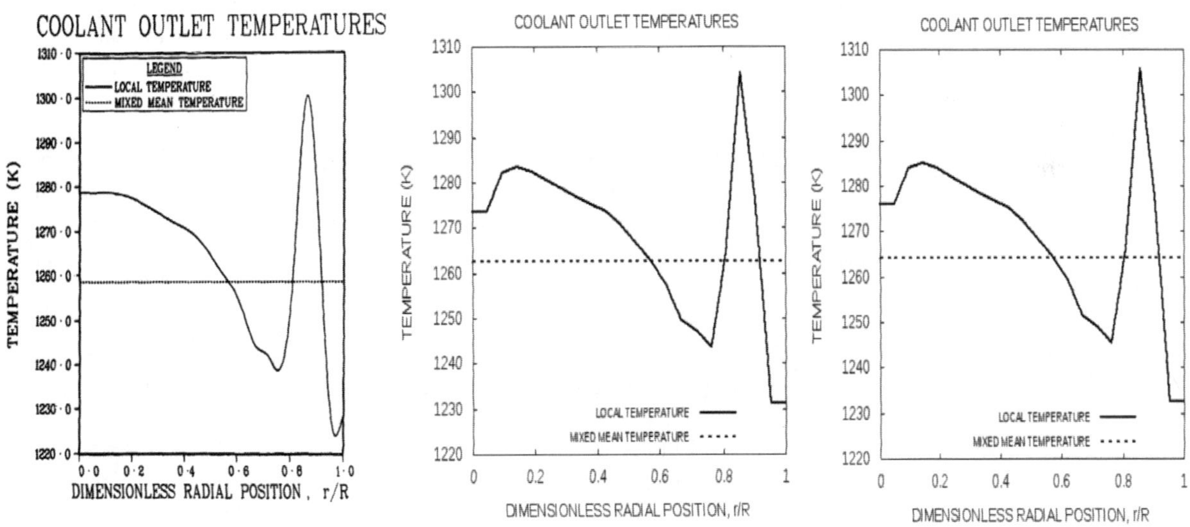

(a) (b) (c)

Figure 1. Comparison of coolant outlet temperatures between (a) reference (b) with German thermal conductivity constant and (c) GE thermal conductivity constant

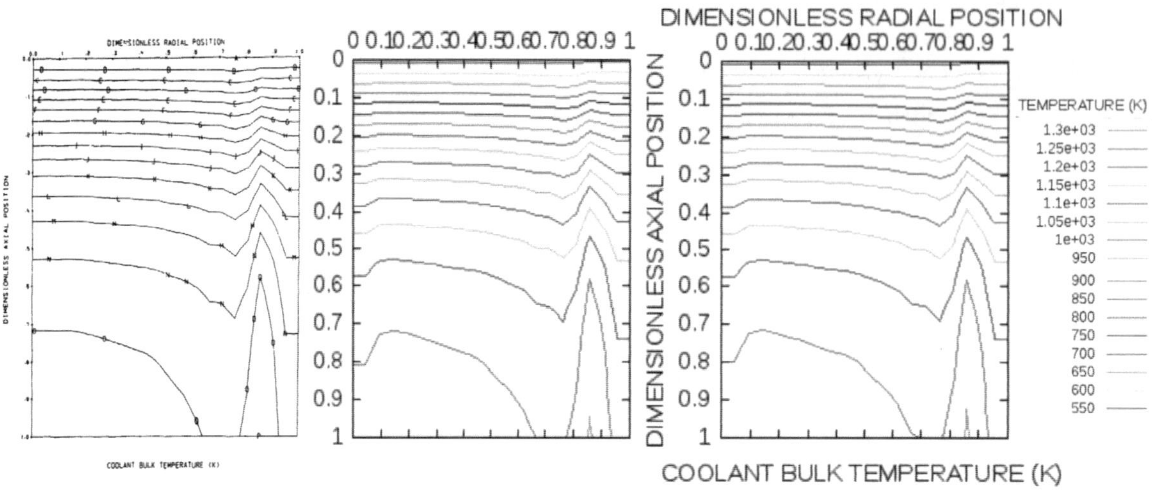

(a) (b) (c)

Figure 2. Comparison of coolant bulk temperatures between (a) reference (b) with German thermal conductivity constant and (c) GE thermal conductivity constant

Figure 1-2 shows comparison of coolant outlet and bulk temperatures calculated using different thermal conductivity constants. Calculation results is performed by the developed code generally show fairly good agreements with the reference. As shown in Figure 1.a-1.c, compared to the reference, calculation result for coolant outlet temperatures in normalized radial direction using German thermal conductivity constants generated almost similar results except that at the 0.8 radial position. At this area, temperatures are overestimated for around 5 K. Meanwhile, calculation results in case C in general have slightly overestimated value compared to the reference and also to the case B. This could be explained as the GE thermal conductivity correlation produced slightly higher value than the German equation.

Calculation shows that case B and C on average have several degree higher temperatures than the reference. This also could be observed for the peak temperature at around 0.8 radial position. The use of thermal conductivity in case B and C generally will produced higher temperature than the reference. Some different result at the first parts of the normalized radial position need to be fixed and improved in the future.

4. Conclusion

A thermal hydraulic computer code for Pebble Bed Reactor are now under development. At the first step, methods proposed by K.Stroh is adapted and several parameters are updated. At the present study, several thermal conductivity constants are used in the code and the effects are analysed. Generally, both thermal conductivity proposed by German and GE have fairly good agreement with the reference. Calculation using GE's constants produced the highest temperatures. For a conservative analysis on the operation of PBR, this thermal conductivity constant could be used in the calculation.

Acknowledgment

The authors express their gratitude for the support from The Ministry of Research Technology and Higher Education of The Republic of Indonesia through the scheme of Penelitian Unggulan Perguruan Tinggi (PUPT) RISTEKDIKTI 2016.

References

[1] K.R. Stroh, 1979, *Thesis Colorado State University*
[2] D. Irwanto, T. Obara, K. Yukitaka and I. Yamanaka, 2010 *Trans Am Nucl Soc* 103 p.791
[3] D. Irwanto and T.Obara, 2011 *J. Nucl. Sci. Technol*, vol. 48, No.11, p. 1385-1395
[4] D. Irwanto and T.Obara, 2012 *J. Nucl. Sci. Technol*, vol. 49, No.2, p.222-229
[5] D. Irwanto and T. Obara, 2013 *Ann. Nucl. Energy*, vol. 60, p. 383-395
[6] K. Verfondern, 1983, *Jul*-1826
[7] M.G. Savage, 1984 *ORNL/TM*-9215

Probabilistic Assessment of Severe Accident Consequence in West Bangka

Sunarko[1], **Zaki Su'ud**[2]

[1]Center for Energy System Assessment-National Nuclear Energy Agency, Jakarta, Indonesia
[2]Physics Department, Institut Teknologi Bandung, Bandung, Indonesia

[1]e-mail: sunarko@students.itb.ac.id
[2]e-mail: szaki@fi.itb.ac.id

Abstract. Probabilistic dose assessment for severe accident condition is performed for West Bangka area. Source-term from WASH-1400 reactor analysis is used as a conservative release scenario for 1000 MWe PWR. Seven groups of isotopes are used in the simulation based on core inventory and release fraction. Population distribution for Muntok district and the area within a 100 km radius is obtained from 2014 data. Meteorological data is provided through cyclic sampling from a database containing two-year site-specific hourly records in 2014-2015 periods. PC-COSYMA segmented plume dispersion code is used to investigate the assumed the consequence of the accident scenario. The result indicates that early or deterministic effect is important for areas close the release point while long-term or stochastic effect is related to population distribution and covers area of up to 100 km from the release point. The mean annual expected values for early mortality and late mortality for the population within 100 km radius from Muntok site are $2.38\times10^{-4}\ yr^{-1}$ and $1.33\times10^{-3}\ yr^{-1}$ respectively.

1. Introduction

Spatial distribution of radioactive pollutant concentration and its probability is predicted using a single source-term and a set of meteorological data. The resulted distribution can aid emergency response officials to prepare evacuation route and the areas to be evacuated during an accident. The recent example from Fukushima Daiichi nuclear accident has demonstrated that timely evacuation has prevented catastrophic radioactive exposure to the public.

Dose analysis can be performed using both deterministic and probabilistic methods. Probabilistic Risk Assessment (PRA) or Accident Consequences Assessment (ACA) is a process where potential consequence of radionuclide release is analyzed, taking into consideration a number of conditions that may persist during the sequence of the accident. Consequence is analyzed for a range of conditions. A pair of condition under consideration is a probable occurrence and this pair is related to a certain probability.

Generally, it is not reasonable to adopt any single specific accident scenario as the basis for the establishment of nuclear emergency planning. The spectrum of the source terms of the nuclear facility in question and the whole scope of local weather conditions should all be taken into consideration. The consequences of most common nuclear accidents are negligible, including the rare occurrence accidents included in the Design Basis Accidents (DBA). It is only in the case of Beyond Design Basis

Accidents (BDBA), which occurs with extremely low probabilities that any problem of relocation can appear[1]. In this paper, the postulated accident is a BDBA using core inventory and release fraction from WASH-1400 data, which were obtained from probabilistic safety analyses based on the operation experience of the PWRs in USA[2][3]. The source terms were more conservative and the calculated doses and risks are relatively large.

Site feasibility study performed in Bangka Island of Bangka Belitung Islands Province had identified several possible locations for future nuclear power plants. One of the candidate sites is located in Muntok district in western tip of Bangka Island. The candidate site is located relatively far from population centers however it is necessary to prepare for emergency planning actions in case of an accident. This paper aims to evaluate health effects and risks based on a hypothetical and conservative source-term and on site-specific meteorological database providing a range of possible weather condition.

2. PC-COSYMA accident analysis code

The description provided in this section is referenced from Jones et al. (1996) and PC-COSYMA User's Manual[4][5] unless otherwise stated.

PC-COSYMA was developed by NRPB and FZJ to evaluate radioactive dispersion across Europe in the framework of MARIA program. COSYMA (Code SYstem from MAria) is a software package for assessing the offsite consequences of accidental releases of radioactive material to the atmosphere. It was developed as part of the European Communities program Methods for Assessing the Radiological Impact of Accidents (MARIA). The code contains three different accident consequence codes, designed for application in different time periods and distance regimes. The Near Early (NE) program calculates early health effects and the influence of emergency actions to mitigate the consequences, and is used in the near-field. The Near Late (NL) program considers only late health effects and the associated mitigating countermeasures, and is also used in the near-field. The Far Late (FL) subsystem is concerned with late health effects at distances far from the site, together with the appropriate countermeasures. The NE and NL subsystems use the MUSEMET program, a segmented Gaussian plume model, allowing incorporation of temporal changes in the meteorological conditions. The FL subsystem uses the MESOS model intended to address transport and dispersion over large distances. The NL subsystem can also be used with ISOLA, a model for very long duration releases which are sufficiently small that no countermeasures will be expected. External irradiation pathways are considered with a variety of countermeasures to be considered. Doses, early and latent health effects are calculated based on models from United Kingdom, United States, and Federal Republic of Germany.

The input for the code includes user-specified deposition velocities for five different isotope classes: noble gases, vapor, aerosols as well as iodine in elemental, organic or aerosol-type phase. Meteorological data input includes hourly wind speed, wind direction, rainfall rate and atmospheric stability class. The output of the code includes doses, health effects and risks at specific grid points and associated probability distributions.

The wind speed at the height z is calculated using $U(z) = U_0(z/z_0)^p$ where z_0 is wind measurement height and p is the wind profile index. Surface roughness of ≥ 1m is assumed and the values for p in Table 1 is used.

Table 1. Values of p for different atmospheric stability categories

Atmospheric stability category	A	B	C	D	E	F
p	0.07	0.13	0.21	0.34	0.44	0.44

Health effects in PC COSYMA include early morbidity for lung function impairment, hyperthyroidism, skin burns, cataracts and mental retardation. The early mortality effects include pulmonary syndrome, hematopoietic syndrome, gastrointestinal syndrome, pre- and neonatal death and death from skin burns. The risk R of early effects to an individual is calculated using the hazard function H:

$$R = 1 - e^{-H}$$

where

$$H = \ln 2 \times \left(\frac{D}{D_{50}}\right)^{S}$$

D (Gy) is the dose received by a certain organ in a certain period, D_{50} (Gy) is the dose causing the effect in 50% of the exposed population, and S is the shape parameter characterizing the slope of the dose-risk relationship. For the same dose, a higher dose rate is more effective in causing early health effects. PC COSYMA divided the irradiation into several periods and replaced the ratio (D/D_{50}) by:

$$\frac{D}{D_{50}} = \sum_{i} \frac{D^{i}}{D_{50}^{i}}$$

where i specified the ith period and:

$$D_{50}^{i} = D_{\infty} + \frac{D_{0}}{X^{i}}$$

Where X^{i} is the average dose rate in the ith period, D_{∞} is the value for D_{50} at high dose rate and D_{0} is a parameter. There is no threshold for early effects but when the value for R is less than 1%, the corresponding dose was assumed to be below the threshold and R is reset to zero.

The late effects considered in PC COSYMA include 11 types of cancers and hereditary effects. The cancers are leukemia and cancers to the bone surface, breast, lung, stomach, colon, liver, pancreas, thyroid, skin and the remainder of human body. The absolute risk (additive model) is applied for leukemia and bone surface cancer and the relative risk (multiplicative model) is applied for the rest. Additive risk assumes that the excess mortality from cancer caused by a given dose of radiation is expressed as a constant number of extra cancer deaths per year, irrespective of the underlying spontaneous rate in the population while multiplicative risk assumes that excess mortality is expressed as a constant percentage of the underlying rate. The incidence of late effects is calculated in PC COSYMA using no-threshold linear dose-risk relationship. As an example, the risk of a certain late health effect caused by external irradiation from material deposited on the ground could simply be expressed as:

$$r = SF \times \sum_{k} (AG(k) \times ARC(k))$$

Where k specifies a particular radionuclide, SF is the mean shielding factor for external irradiation from materials deposited on the ground, AG is the radionuclide concentration deposited on the ground ($Bq\ m^{-2}$). The risks of other late health effects caused by other exposure pathways are calculated with formulas similar to the above equation. The coefficients used in this paper are given in ICRP-60[6].

3. Methodology

The meteorological data for the modeling is obtained from an on-site monitoring in west Bangka area. The duration of the data is approximately 2 years (2014-2015) with a completeness of more than 90% in each year[7]. The required parameters by PC-COSYMA are wind speed and direction, atmospheric stability condition, rainfall rate and boundary layer height. Atmospheric stability condition is determined by lapse-rate method using temperature difference at the heights of 80 and 10 meters.

Probabilistic dispersion modeling is performed using cyclic sampling for meteorology data. In the cyclic sampling, a period of 61 hours is selected as the interval and therefore for one year data of about 8760 hours there are 144 sample hours.

Simulation is performed for west Bangka area, centered at the meteorological station location. This is located at the westernmost tip of Bangka Island in Bangka-Belitung Province. Flat topography with rough surface is assumed considering various palm oil and rubber plantations exist in the neighboring areas. The reactor is assumed to be a 40×40 m building and the stack height is 25 m. The actual duration of the release is 0.5 hours but because of the requirement in the model it is changed to 1 hour as the minimum duration of release.

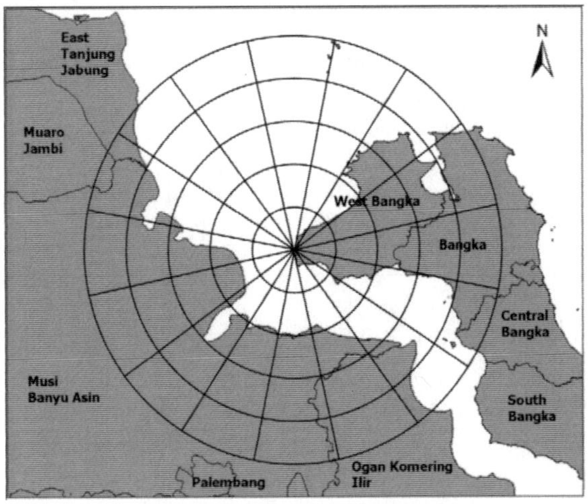

Figure 1. Division of sector and radial distance from release point (20 km radius intervals are shown)

The investigated area in Muntok area, West Bangka is included in a 100 km radius and is being divided into 20 concentric stripes and 16 sectors each having 22.5 degree angle provided in Figure 1. The assumed release point is located at the center of the concentric circles. The distances of the circles are 0.8, 2, 3, 4, 5, 6, 7, 8, 9, 10, 15, 20, 30, 40, 50, 60, 70, 80, 90 and 100 km. the sectors are numbered in clockwise direction starting from the north direction in the middle of the 1st sector.

Deposition velocities are set as provided in PC COSYMA as follows; aerosols: 0.001 ms^{-1}; elemental iodine: 0.01 ms^{-1}; organic iodine: 5×10^{-4} ms^{-1}. Washout coefficient is provided as $\lambda = a \times i^{b}$ where i is the rainfall intensity in mm h^{-1}. The coefficient a and index b is provided as the default by PC COSYMA as in Table 2.

Table 2. Deposition coefficients

Category	Aerosol	Elemental iodine	Organic iodine
a	8×10^{-5}	8×10^{-5}	8×10^{-7}
b	0.8	0.6	0.6

The source term data of WASH-1400 were obtained from probabilistic safety analyses based on the operation experience of the PWRs in USA. The source terms were more conservative and the calculated doses relatively large. The model itself contains nine release scenarios for Pressurized Water Reactor (PWR) and seven radionuclide groups. Scenario PWR1, with annual probability of 9×10^{-9}, is selected since it offers the most conservative amount of release. The details on radionuclides, their groups and release fractions can be seen in detail in the WASH-1400 report. PWR1 accident can be described as a condition after an accident involving core uncovery and followed by core melt. In the next phase, early total containment failure occurred with early major containment leakage and later containment melt-through. The spray or suppression pool failure occurred and radioactive materials are released to the environment.

Source term is one of the determining factors affecting accident consequences. It is defined as the amount and form of radioactive release, duration of release, release energy, release height, etc. At the moment, the reactor type and power have not been decided yet. In this paper, a pressurized water reactor of 1000 MWe class is assumed and correspondingly WASH-1400 data for PWR1 type of accident is used as the most conservative scenario.

PWR1 scenario is further divided into PWR1A and PWR1B each having annual probability and released energy of 4×10^{-7} yr^{-1} and 5.86 MW for PWR1A, respectively, and 5×10^{-7} yr^{-1} and 152 MW for PWR1B, respectively. The release fractions relative to core inventory for 7 nuclide groups are provided in the following Table. The fraction of organic iodine is 1% of all types of iodine.

No further retention or reduction is performed and the fraction is directly released to the environment. No countermeasures are taken.

Table 3. Release fractions of various nuclide groups in the core for PWR1A

Category	Group 1 Xe-Kr	Group 2 iodine	Group 3 Cs-Rb	Group 4 Te-Sb	Group 5 Ba-Sr	Group 6 Ru(a)	Group 7 La(b)
PWR1	0.9	0.7	0.4	0.4	0.05	0.4	3.00E-03

Notes: (a) includes Mo, Rh, Tc, Co, Ru
(b) includes Nd, Y, Ce, Pr, La, Nb, Am, Cm, Pu, Np, Zr

In this model, the stack height was taken to be 25.0 m, the height and width of the reactor building were set at 40 and 40 m, respectively, to include consideration of building wake effect. The calculations assumed no countermeasures. The other input parameters were based on those from Cao (2000)[8] or default values in the COSYMA code. Other adopted parameters included the following: fraction of contaminated skin: 10%; integration time for late skin dose: 10,000 d; integration time for early individual dose: 7 d; half life of radioactivity deposited on skin: 30 d; radioactivity concentration deposited on skin is equal to that deposited on the ground.

The population distribution around the release point is obtained from the field survey during the site study using the 2014 data in Bangka Island and South Sumatra province located to the south of Bangka Island on the other side of Bangka Strait. The data is also provided 20 concentric stripes and 16 sectors as shown in Table 4. The calculated population density is provided in Figure 2.

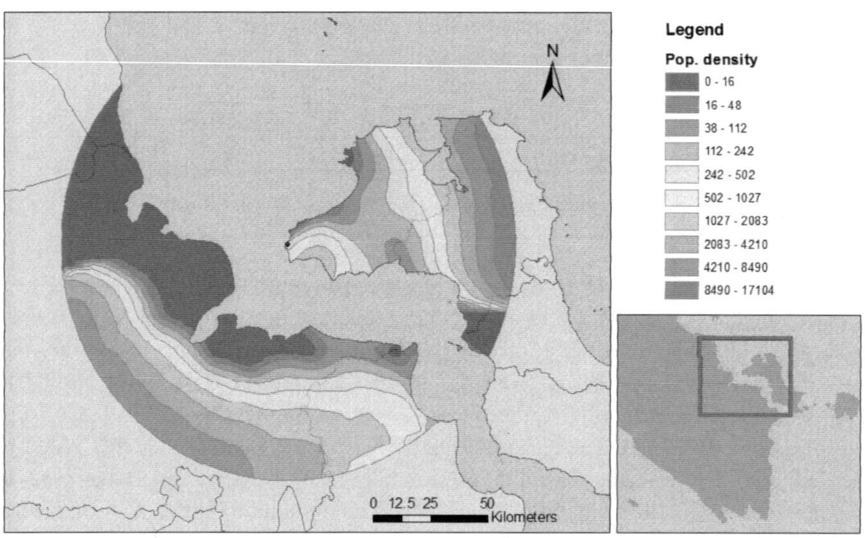

Figure 2. Population density witihin 100 km radius.

Table 4. Population distribution within 100 km radius from release point[7].

Sector		Outer radius									
Num.	Direction	0.8	2	3	4	5	6	7	8	9	10
1	N	0	0	0	0	0	0	0	0	0	0
2	NNE	0	0	0	0	0	0	0	0	0	0
3	NE	0	0	0	0	0	0	0	0	0	0
4	ENE	0	0	0	0	0	0	0	0	0	0
5	E	0	0	0	0	0	0	0	0	0	0
6	ESE	0	0	0	0	0	0	0	982	699	1119
7	SE	0	0	0	0	0	0	0	406	1119	2118
8	SSW	0	0	0	0	0	0	0	0	1439	2159
9	S	0	0	0	0	0	0	0	0	0	406
10	SSW	0	0	0	0	0	0	0	0	0	0
11	SW	0	0	0	0	0	0	0	0	0	0
12	WSW	0	0	0	0	0	0	0	0	0	0
13	W	0	0	0	0	0	0	0	0	0	0
14	WNW	0	0	0	0	0	0	0	0	0	0
15	NW	0	0	0	0	0	0	0	0	0	0
16	NNW	0	0	0	0	0	0	0	0	0	0

Sector		Outer radius									
Num.	Direction	15	20	30	40	50	60	70	80	90	100
1	N	0	0	0	0	0	0	0	0	0	0
2	NNE	0	0	0	0	0	0	0	0	0	0
3	NE	0	0	0	0	0	334	14055	53413	56038	20378
4	ENE	0	0	0	5610	7155	5090	22230	33500	89601	283749
5	E	0	0	0	2541	0	0	11983	41700	109517	335845
6	ESE	3718	11415	3586	9508	0	816	0	0	0	0
7	SE	5213	3151	5342	13687	0	0	0	0	9607	1931
8	SSW	1625	0	0	0	0	0	14246	21464	32774	11582
9	S	406	0	0	0	0	0	23821	32893	62470	50650
10	SSW	0	0	0	0	0	0	0	48726	79079	154802
11	SW	0	0	0	0	0	0	0	40733	88762	154802
12	WSW	0	0	0	0	0	0	0	4202	78285	129161
13	W	0	0	0	0	0	0	0	0	0	0
14	WNW	0	0	0	0	0	0	0	0	0	0
15	NW	0	0	0	0	0	0	0	0	0	0
16	NNW	0	0	0	0	0	0	0	0	0	0

4. Result and Discussion

Meteorological data record from monitoring height of 10 meter shows that wind is distributed into all 16 sectors with the highest frequency of 14.7% from the South-Southeast direction for 2014 and 20.9% from the Southeast direction for 2015. Wind speed is dominated by 2.1-3.6 m/s class with a frequency of 58% and 62.8% for 2014 and 2015 respectively. There is no occurrence of calm wind condition for both years. Average wind speed2 at 10 meter level are 2.43 m/s and 2.35 m/s for 2014 and 2015 respectively. Wind condition can be summarized by the windrose plot for the monitoring period for 2014 and 2015 in Figure 3. The colors in the windrose plot signify the classes of wind speed while the bars represent the direction from which the wind is blowing. Zero degree is to the North direction and rotates clockwise in 16 directions classes.

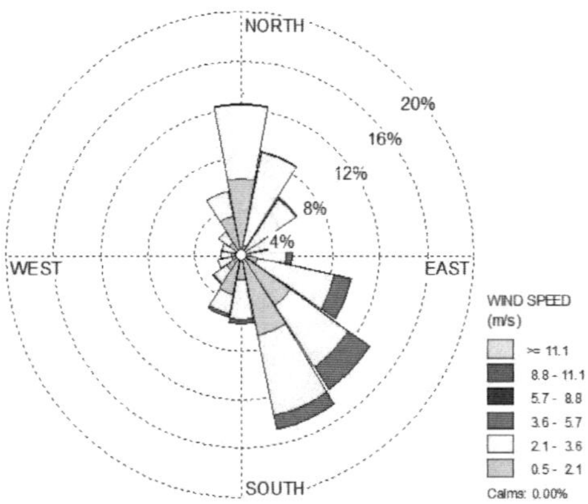

(a)

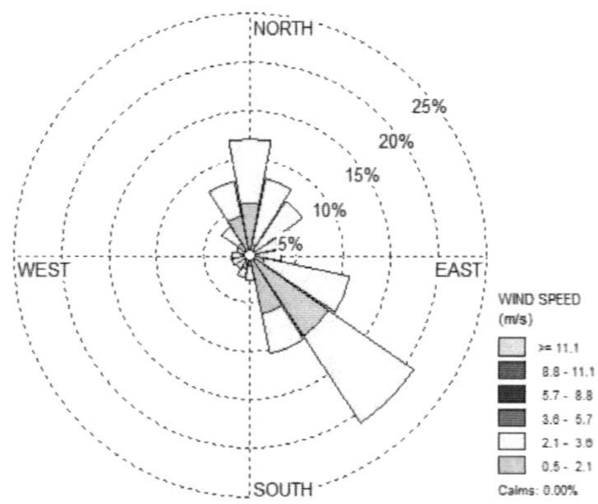

(b)

Figure 3. Windrose in Bangka station at 10 m height

The average individual effective dose at a certain radius from release point is presented in Figure 4. The maximum and minimum doses for the first 24 hours after the release was initiated (early effect) are 4.1 Sv at radius 0.4 km and 0.64 mSv at radius 95 km respectively. The late effect maximum and minimum doses are 37.7 Sv at radius 0.4 km and 14.1 mSv at radius 95 km respectively.

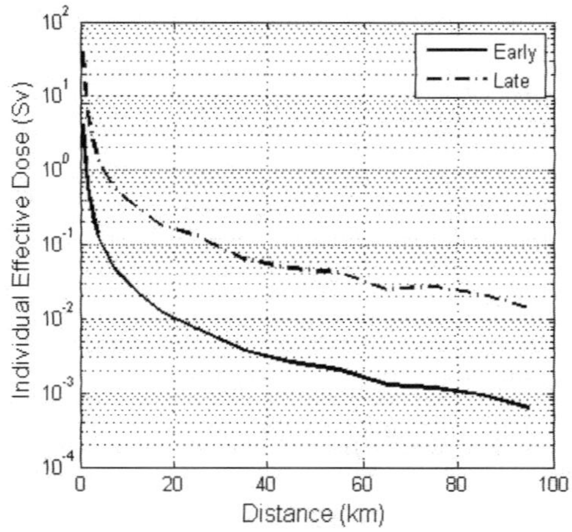

Figure 4. Early and late average individual effective dose by distance

The predicted numbers of early morbidity, early mortality and fatal cancers for general public within 80 km radius from release point for PWR1A accident scenario is provided in Figure 5. It can be observed that the early (deterministic) effect is different from late (stochastic) effect. Most of the early effect occurred in closer proximity of about 10 km from the site which includes population center in the city of Muntok and surrounding area. The effect does not transgress to over 40 km radius. On the other hand, late mortality affect mostly areas far from the site causing almost 500 deaths per year at a distance of 100 km, almost five times the number for the population within 40 km radius. Cao et al. (2000)[8] stated that this is caused by the threshold value effect for close distance which cannot be exceeded at larger distances. Late effect is calculated using linear relationship between dose and risk and there are finite probabilities at larger distances although the actual dose values are small. The number of cancer fatality is obtained from the product of cancer probabilities and the number of population in the area and therefore it is in close relation with population density. Taking into consideration the dominant wind direction in the area, most of the early effect will be incurred by the population in the South East to South sector within the 40 km radius around the city of Muntok.

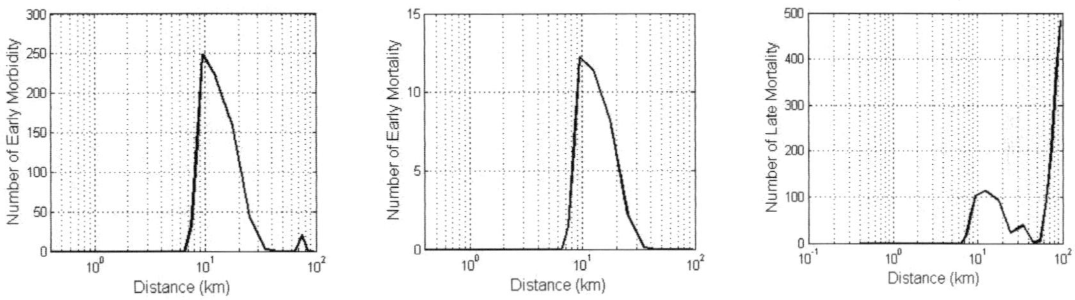

Figure 5. Total number of early morbidity, early mortality and late mortality (fatal cancers) within 80 km radius (yr^{-1}) for PWR1A accident

Figure 6 provides the complementary cumulative distribution functions (CCDFs) for early morbidity and early mortality. Each point the graphs can be interpreted as the probability of having results larger than or equal to the value in the horizontal axis. The number of early morbidity is dominated by skin burns, followed by hematopoietic and gastrointestinal (GI) syndromes. The frequencies for other types of morbidity are very small in this case. Early mortalities are caused mostly by pre- and neo-natal syndromes while other causes have lower frequencies of occurrence.

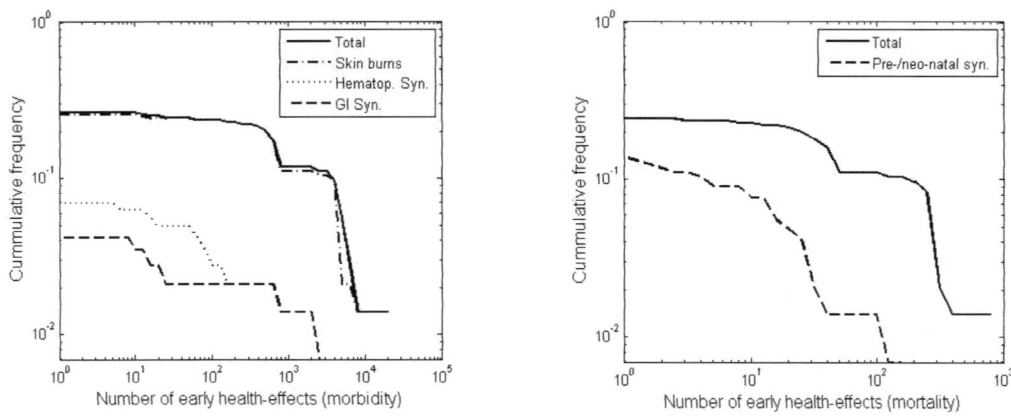

Figure 6. Complementary Cumulative Distribution Function (CCDF) for early morbidity and early mortality (yr^{-1}) within 80 km radius for PWR1A accident

Figure 7 provides the CCDFs for fatal cancers (late effect) for PWR1A release scenario. The effect is dominated by lung cancer from inhalation of radioactive substance in extended period of time, followed by remaining internal organs. Smallest effects are incurred by the skin and bone surface. In comparison to the early health effects, where skin effect is dominant, in the late effect skin effect is the least dominant because of larger distance and concentration factor due to dispersion effect. Most of the effect will be felt by the South Sumatran population across the Bangka Strait in the Southeast to West Southwest sectors at radius 60 km and more where there are densely populated areas. Although there are increasingly heavier populations to Northeastern to Eastern sectors, they are less likely to be affected due to the small probability of wind blow to this direction.

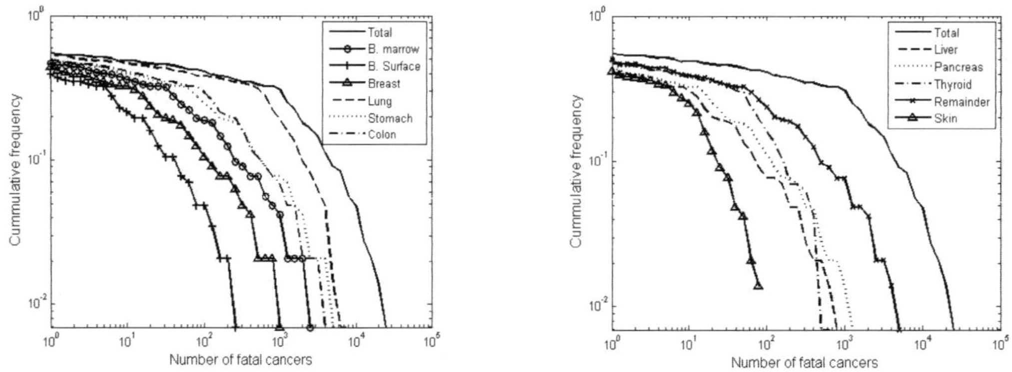

Figure 7. CCDF for fatal cancers (yr^{-1}) for PWR1A release

Using the CCDFs for early and late effects as well as the annual probability of PWR1A of $4{\times}10^{-7}\,yr^{-1}$ the annual averaged expected values for early morbidity, early mortality and late mortality for population within 100 km radius from the release point are $4.51{\times}10^{-3}\,yr^{-1}$, $2.38{\times}10^{-4}\,yr^{-1}$ and $1.33{\times}10^{-3}\,yr^{-1}$ respectively. In comparison to traffic accidents deaths of 13 people in 2014[9], the expected values for early and late mortality from PWR1A severe accident is about four to five degrees of magnitude smaller.

The Nuclear Energy Regulatory Agency of Indonesia (BAPETEN) Chairman Decree No. 1 (2010)[10] stipulates that reactor with a capacity of more than 100 MWt and installation or facility with very high potential risk able to generate radioactive releases causing severe deterministic effects offsite is included in the Radiological threat Category 1. Emergency Zones and radius for installation/facility with radiological threat Category 1 are defined as follows: 3 – 5 km radius for Precautionary Action (PAZ), 25 km radius for Urgent Protective Zone (UPZ), and 300 km radius for Food Restriction Planning. In UPZ, preparations are made to promptly take urgent protective measures based on environmental monitoring. In this zone, sheltering is recommended when the effective dose exceed 10 mSv. However, considering the early effect which includes areas within 40 km boundary, it is recommended that UPZ is extended to 40 km for West Bangka site.

5. Conclusion

Probabilistic analysis for severe accident consequence in West Bangka NPP candidate site has shown that early health effect is an important consideration for areas surrounding the site to a radius of 40 km. late heath effect affects farther areas due to the linear relationship between dose and risk and also higher population density beyond 40 km up to 100 km to the South of the site in South Sumatra Province. Average yearly probability for early and late mortalities are $2.38{\times}10^{-4}\,\mathrm{yr}^{-1}$ and $1.33{\times}10^{-3}\,\mathrm{yr}^{-1}$ respectively. In comparison to the fatalities from traffic accident in West Bangka in 2015, the late and early mortality risks from a severe nuclear accident are between 4 and 5 order of magnitude smaller respectively. In relation to emergency planning zones, it is suggested that the

Urgent Protective Zone being extended from 25 km to 40 km from the proposed plant considering the early effects due to existing population and meteorological factors.

Acknowledgement
The authors acknowledge the Ministry of Research and Technology for funding this research and Nuclear Physics and Biophysics Laboratory in Institute Teknologi Bandung for the computation facility.

References

[1] Qu J 2003 The determination of the area affected by relocation after nuclear accidents, *Nucl. Eng. and Des.* **223** 41.
[2] Rasmussen Professor Norman C et al 1975 Reactor safety study. An assessment of accident risks in U. S. commercial nuclear power plants. Executive Summary WASH-1400 (NUREG-75/014) (Rockville, MD, USA: Federal Government of the United States, U.S. Nuclear Regulatory Commission).
[3] McKenna T J, Giitter J G 1988 NUREG 1228 Source term estimation during incident response to nuclear power plant accidents (Rockville, MD, USA: Federal Government of the United States, U.S. Nuclear Regulatory Commission).
[4] J A Jones et al 2001 Probabilistic Accident Consequence Uncertainty Assessment Using COSYMA: Methodology and Processing Techniques, EC 2001
[5] Haywood S M, Jones J A, Ehrhardt J 1993 PC COSYMA Version 1.0 User Guide (Brussels-Luxembourg: National Radiological Protection Board and Kernforschungszentrum Karlsruhe GmbH, ECSC-EEC-EAEC).
[6] ICRP 1991 1990 Recommendations of the International Commission on Radiological Protection. ICRP Publication 60. Ann. ICRP 21.
[7] Anonymous 2014 West Bangka, Dispersion and dose assessment Final Topical Report Year-2 2012 (Jakarta: PT. Surveyor Indonesia).
[8] Cao, J. Z. et al.2000 Adaptation of COSYMA and assessment of accident consequences for Daya Bay nuclear power plant in China *J. Environ. Radioactivity* **48** 265.
[9] Anonymous 2015 West Bangka in Figures (Muntok: BPS-Statistics of Bangka Barat Regency).
[10] Chairman Regulatio Number 1 2010 Nuclear emergency preparedness and mitigation (Jakarta: Indonesian Nuclear Energy Regulatory Agency - BAPETEN).

Looking at groundwater research landscape of Jakarta Basin for better water management

Dasapta Erwin Irawan[1*], Adhi Priyambodho[2], Cut Novianti Rachmi[3], and Dimas Maulana Wibowo[1]

[1*] Faculty of Earth Sciences and Technology, Institut Teknologi Bandung;
[2] Water Resources Program, Delft University, The Netherlands;
[3] Discipline of Child and Adolescent Health at The Childrens Hospital at Westmead Clinical School, The University of Sydney

dasaptaerwin@outlook.co.id

Abstract. Based on our experience, defining the gap between what we know and what we don't know is the hardest part in proposing water management strategy. Many techniques have been introduced to make this stage easier, and one of them is bibliometric analysis. The following paper is the second part of our bibliometric project in the search for a gap in the water resources research in Jakarta. This paper starts to analyse the visualisations that had been extracted from the previous paper based on our database. Using the keyword "groundwater Jakarta", we managed to get 70 relevant papers. Several visualisations have been built using open source applications. Word cloud analysis shows that the trend to discuss groundwater in scientific sense had only been started in the early 2000's. This is presumably due to the emerging regional autonomy in which forcing regions to understand their groundwater setting before creating a management strategy. More papers in the later time has been induced by more geo-hazards (land subsidence and floods) resulted in the vast groundwater pumping. More and more resources have been utilized to get more groundwater data. Water scientists by then understood that these hazards had been started long before the 2000's. This had become the starting point of data era later on. The next era will be the era of water management. Hydrologists had been proposing integrated water management Jakarta and its nearby groundwater basins. Most of them have been strongly suggested to manage all water bodies, rainfall, surface water, and groundwater as one system. In the 2010's we identify more papers are discussing in water quality following the vast discussion in water quantity in the previous era. People have been more aware the importance of quality in providing water system for the citizen. Then five years later, we believe that water researchers have also put their mind in the interactions between surface water and groundwater, especially in the riverbank, where most of the slums are located. Based on the results, we believe that more researches to understand interactions between groundwater and surface water would fill the gap to come up with better water management system in Jakarta.

1. Introduction

Topic modelling and network visualization is a very important for researchers to choose the least discussed topic and the gap in between research topics. To solve the problem, we need to have a systematic extraction of topics and visualize it for easy analysis[1,2]. This paper analyses the results from topic modelling and its visualizations using open source applications: VosViewer[3] and Zotero and PaperMachine plug in[4,5]. We brought a case in hydrogeological science, which presents research landscape in Jakarta area. The work discussed in this paper was mainly derived from our previous paper[6], which show case our topic modelling technique.

2. Methods

This paper is the second part of our bibliometric work in which the first paper had been published in RioJournal[6]. We're then try to give more analysis based on the word cloud plots presented in the paper. We extract a reference database in RIS format[8] using VosViewer software[9,10]. The original dataset and plots are available on our Figshare Repository[7].

3. Word cloud

The basis of this analysis is the following word cloud plot using VosViewer software (Figure 1). The trend to discuss groundwater in scientific sense had only been started in the early 2000's. This is presumably due to the emerging regional autonomy in which forcing regions to understand their groundwater setting before creating a management strategy. Before 2000's, groundwater is placed as non-economical goods.

More papers in the later time has been induced by more geo-hazards (land subsidence and floods) resulted in the vast groundwater pumping. More and more resources have been utilized to get more groundwater data. Water scientists by then understood that these hazards had been started long before the 2000's. This had become the starting point of data era later on.

The next era will be the era of water management. Hydrologists had been proposing integrated water management Jakarta and its nearby groundwater basins. Most of them have been strongly suggested to manage all water bodies, rainfall, surface water, and groundwater as one system.

In the 2010's we identify more papers are discussing in water quality following the vast discussion in water quantity in the previous era. People have been more aware the importance of quality in providing water system for the citizen. Then five years later, we believe that water researchers have also put their mind in the interactions between surface water and groundwater, especially in the riverbank, where most of the slums are located.

4. Word connection

Using Zotero reference manager, we managed to build word connections in the paper collections (Figure 2). We thought this tool is very useful to understand the nature of papers when they were written, instead of only relying on the keywords. We can see that the word urban is highly connected to floods, ecosystem, development, river, and river-rehabilitation. This, in a way, is referring the impacts of urban development to the change of ecosystem, the modification of river environment, and the rising frequency of flood hazards.

Another word connection, "land – subsidence", would be another way to look into the impact of excessive groundwater pumping, as has been in the perception of many since the early 2000's. We don't argue that this perception is vague, but this is likely due to the lack of researches in this subject. However more maps have been generated to identify the distribution of land subsidence, and try to connect the findings with the distribution of groundwater pumping.

The next observed word connection is "groundwater - interactions, management, age, excessive (pumping)". This indicates the rising awareness of groundwater as part of public goods, which needs to be managed. Hence defining its interaction with physical and social environment is very important. More problems in "water – supply" and "water – quality" are also important since more people means more water demand in high quality. Another word pair "subsurface – environments" is also notable in the plot. This is very interesting to us that the rising environmental problems before 2000's have never been connected with subsurface environment. Regional planning documents released before 2000's have been only occupied mainly by surface environment.

Figure 1. Word cloud from the year 2004 to 2015. It shows the setting of groundwater research in each interval.

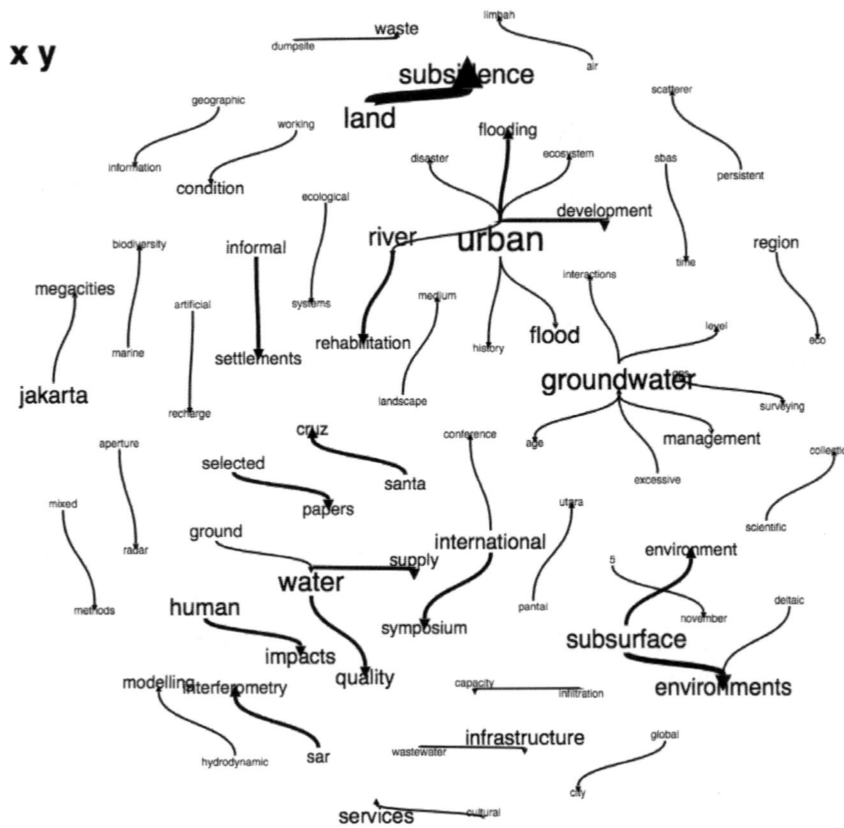

Figure 2. Word connection analysis.

5. Concluding remarks

We believe that the year 2000 is the major milestone of groundwater research in Jakarta area, with the rise of regional autonomy. Our visualizations of bibliometric data shows that groundwater before 2000's has only been placed as non-economical goods, which then shifted as the major spot of economical goods after the 2000's. Such shifting has resulted to many geo-hazards, as perceived by the community. More and more unique keywords and word connections can be extracted since then with more environmental complication and social interaction. We also believe that more research should be directed to groundwater and river/surface water interaction to understand the nature of water quality in such megacity like Jakarta.

Another take home note that we want to deliver is the importance of open source resources in academic life. More open source applications have been built to support researchers with lack of funding and sponsor. These open source applications as well as open source movement also embraces the power creativity and networking among academia.

Acknowledgment

The authors would express their gratitude for ITB for the partial funding for this research and anonymous reviewers who had enrich our analysis and chosen this paper to be presented oral in this event.

References

[1] Q. Mei, D. Cai, D. Zhang, and C. Zhai, "Topic modelling with network regularization" in *Proceedings of the 17th international conference on World Wide Web*, 2008, pp. 101–110. [Online]. Available at: www.nvc.cs.vt.edu/~ctlu/Seminar/2009/2009-SDM-Summer/Topic-Modeling.pdf [Accessed: 25-Dec-2016].

[2] C. Wang and D. M. Blei, "Collaborative topic modelling for recommending scientific articles," in *Proceedings of the 17th ACM SIGKDD international conference on Knowledge discovery and data mining*, 2011, pp. 448–456. [Online]. Available at: https://www.cs.princeton.edu/~blei/papers/WangBlei2011.pdf [Accessed: 25-Dec-2016].

[3] N. J. P. Eck and L. Waltman, "VOSviewer: A computer program for bibliometric mapping," ERIM Report Series Research in Management, 2009. [Online]. Available at: http://repub.eur.nl/pub/14841/ERS-2009-005-LIS.pdf [Accessed: 25-Dec-2016].

[4] "Zotero: An open source reference manager." [Online]. Available at: https://www.zotero.org/ [Accessed: 25-Dec-2016].

[5] J. Guldi and C. J. Roberson, *Papermachines: a text mining plugin for Zotero*. 2011. [Online]. Available at: http://papermachines.org [Accessed: 25-Dec-2016].

[6] D. Irawan, A. Priyambodho, C. Rachmi, D. Wibowo, and A. Fahmi, "Bibliometric study to assist research topic selection: a case from research design on Jakarta's groundwater (part 1)," *Res. Ideas Outcomes*, vol. 2, p. e9841, Jul. 2016. [Online]. Available at: http://riojournal.com/articles.php?id=9841 [Accessed: 25-Dec-2016].

[7] D. E. Irawan, A. Priyambodho, D. M. Wibowo, and A. K. Fahmi, "Literature visualisation: Bibliometric of groundwater research in Jakarta," 2016. [Online]. Available at: https://doi.org/10.6084/m9.figshare.3405685.v3 [Accessed: 25-Dec-2016].

[8] "RIS (file format)," *Wikipedia, the free encyclopedia*. 06-Jul-2016.

[9] N. J. Van Eck and L. Waltman, "Text mining and visualization using VOSviewer," *ArXiv Repository. ArXiv: 11092058*, 2011 [Online]. Available at: https://arxiv.org/abs/1109.2058 [Accessed: 25-Dec-2016].

[10] Z. Zahedi and N. J. Van Eck, "Visualizing readership activity of Mendeley users using VOSviewer," in *altmetrics14: Expanding impacts and metrics, Workshop at Web Science Conference*, 2014. [Online]. Available at: https://figshare.com/articles/Visualizing_readership_activity_of_Mendeley_users_using_VOSviewer/1041819 [Accessed: 25-Dec-2016].

IOP Conf. Series: Journal of Physics: Conf. Series **877** (2017) 012034 doi:10.1088/1742-6596/877/1/012034

The Effects of Hydration on Protein of Azurin using Coarse-Grained Method and The Free-Energy Analysis

Dian Fitrasari and Acep Purqon

Physics of Earth and Complex Systems, Institute of Technology Bandung, Indonesia

email: dianfitrasari30@gmail.com

Abstract. Proteins play important roles in body metabolism. However, to reveal hydration effects, it is cost computing especially for all-atom calculation. Coarse-grained method is one of potential solution to reduce the calculation and computable in longer timescale. Furthermore, the protein of Azurin is interesting protein and potentially applicable to cancer medicine for the stability property reason. We investigate the effects of hydration on Azurin, the conformation and the stabilities. Furthermore, we analyze the free-energy of the conformation system to find the favorable structure using free energy perturbation (FEP) calculation. Our calculation results show that free energy value of azurin is -136.9 kJ/mol. It shows a good agreement with experimental results with relative error index remained at 0.07%.

1. Introduction

The hydration process is important for physical and biological properties of protein [1]. We investigate hydration effects and protein stability. Basically, the stability of protein can be revealed by the amount free energy of hydration such as the negative free energy value indicates the system is favorable. Azurin is interesting protein for potentially applied as anticancer therapeutic agents. It causes apoptosis without much negative effect in cancer patient when enter the human breast cancer cells [2]. Moreover, it can be found in bacterial species *Pseudomonas aeruginosa* and it is known as donor at electron transfer process [3]. Furthermore, blue copper ions in polypeptide chains contribute the stability of Azurin protein [3-5].

In this study we perform molecular dynamic (MD) simulation with various parameters consist of simulation parameters (force field), integrator parameters (time step, rigid bonds), constant temperature control, constant pressure control and coordinate structure. The characterization of Azurin is calculated by coarse-grained methods and the free energy is calculated by using NAMD 2.12-multicore software with MARTINI force field and pseudo atom.

2. Method

A coarse-grained (CG) describes model on smaller scale (fine details) for smoothed or averaged. It observes molecular level systems with longer time scale, while maintaining physical properties similar to more detailed scale (small) in the all-atom system. The group of atoms is mapping in cluster to CG site called beads.

A Hamiltonian of system can be described by the function of the general extent parameter, λ, that connect between state a to state b. The connection can be achieved by linear combination of Hamiltonian as can be found in reference [6]:

$$H(x,p_x;\lambda)=H_0(x,p_x)+\lambda H_b(x,p_x)+(\lambda-1)H_a(x,p_x) \tag{1}$$

where $H_a(x,p_x)$ describes the interaction of the group of atoms representing reference state, a. $H_b(x,p_x)$ characterizes interaction of the final state, b. $H_0(x,p_x)$ is the Hamiltonian describing those atoms that is not transforming during the MD simulation. λ is the initial Hamiltonian parameter and $\lambda-1$ is the final Hamiltonian parameter, this parameter describes the function of energy and forces.

The coupling parameter λ_{LJ} scales the Lennard-Jones interactions and λ_{elec} scales electrostatic interactions, so the actual value of λ can be provided by:

$$V_{NB}(r_{ij})=\lambda_{LJ}\varepsilon_{ij}\left[\left(\frac{R_{ij}^{\min 2}}{r_{ij}^2+\delta(1-\lambda_{LJ})}\right)^6-\left(\frac{R_{ij}^{\min 2}}{r_{ij}^2+\delta(1-\lambda_{LJ})}\right)^3\right]+\lambda_{elec}\frac{q_iq_j}{\varepsilon_1 r_{ij}} \tag{2}$$

The free energy calculation does not change intermolecular bonded potentials. The interactions within the perturbed atoms in vacuum are scaled but in hydration state only non-bonded interactions are scaled. The free energy difference between initial and final state is expressed by:

$$\Delta A_{a\to b}=-\frac{1}{\beta}\ln\left\langle\exp\{-\beta[H_b(x,p_x)-H_a(x,p_x)]\}\right\rangle_a \tag{3}$$

Here, $\beta^{-1}\equiv k_BT$, where k_B is the Boltzmann constant, T is the temperature. $H_a(x,p_x)$ and $H_b(x,p_x)$ are the Hamiltonians describing states a and b, respectively. $\langle\ldots\rangle_a$ denotes an ensemble average over configurations representative of the initial, reference state, a. Transformation between two thermodynamic states is replaced by a series of transformations between non-physical, intermediate states along a well-delineated pathway that connects a to b. This pathway is characterized by the general extent parameter, λ, it cause the Hamiltonian and, hence, the free energy, a continuous function of this parameter between a and b:

$$\Delta A_{a\to b}=-\frac{1}{\beta}\sum_{i=1}^{N}\ln\left\langle\exp\{-\beta[H(x,p_x;\lambda_{i+1})-H(x,p_x;\lambda_i)]\}\right\rangle \tag{4}$$

Here, N stands for the number of intermediate stages.

In this study, we use NAMD 2.12-multicore program package for calculating the free energy hydration in coarse-grained scale, to determine the native state of protein. The hydration system of azurin is visualized by VMD 1.9.3. The initial structure coordinates of azurin obtain from protein data bank prepare by VMD 1.9.3 through coarse-grained builder menu with Residue-Based Coarse Graining (RBCG) method by NAMD. The schematic diagram of molecular dynamic simulation for calculating the free energy hydration of Azurin using coarse-grained can be shown in Figure 1.

X-ray crystal structure of oxidized AZ derived from the Protein Data Bank (PDB) (4AZU). AZ structure has 128 residues and is composed of two layers of sheet-β with eight β-strands arranged in topology. Copper ions are bound between two β-sheets bi-pyramidal coordinated trigonal geometry to five atoms, three equators (Nd from His46, N of His117, and SC of Cys112) and two axial (S of Met121 and carbonyl groups of Gly45) atom. Amino acids copper protrudes at His117 through the surface of the protein and is surrounded by a group of hydrophobic residues (Met13, Leu49, Val43, Met44, Met64, Phe114, Pro115, Gly116, Ala119 and Leu120), and this area is known as hydrophobic patch.

The initial state of the system is prepared by using X-ray structure at 2.7 Å resolution. In this study, three subunit proteins, moisture and nitrate removed to get just one protein molecule (930 protein molecules with one molecule blue copper ion), this structure obtained from protein data banks (1AZU) as the initial configuration. Azurin then modeled in CG structure to reduce the number of molecules to

269 molecules with the addition of 1746 water molecules (water sphere) and neutralized by the addition of 5 molecules of Na^+ and 2 molecules of Cl^-. The number of atoms changes to 2022 atoms. At the end of the addition of water put anti-freeze molecules (AF) in the form BP_4 as much as 10% of the total water molecules. This is because when the all-atom models turned into a coarse-grained model the freezing point of water increases. This model represents a system with a salt concentration of 0.1 mol / L.

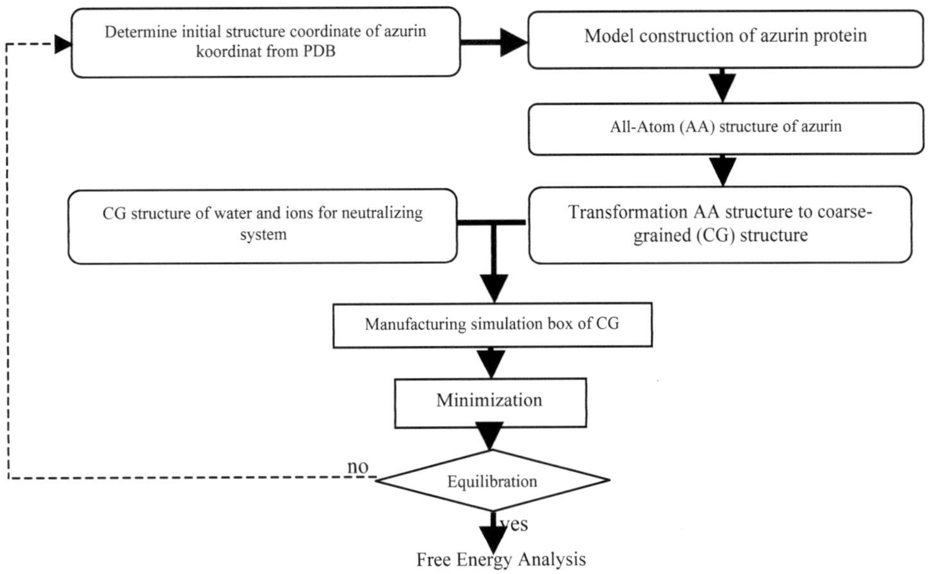

Figure 1. Diagram of MD simulation using CG method.

The initial state of the copper ion surrounded by three amino acids azurin planar (His46, His117, Cys112) and two axial amino acids (Met121 and Gly45). Coulomb potential with cutoff 9 Å and 15 Å. Time step integration time of minimization is 40 fs and time step integration time of equilibration is 20 fs with NPT ensemble (no change in temperature and pressure, wherein the temperature of 310 K to follow the body temperature and standard pressure 1 atm).

The final state azurin is modeled in CG structure so as to reduce the number of molecules to 269 molecules with the addition of as many as 1746 molecules of water molecules (water sphere) and neutralized by the addition of 5 molecules of Na + and Cl 2 molecules. The number of atoms changes to 2022 atoms. This model represents a system with a salt concentration of 0.1 mol / L.

3. Results and Discussion

Figure 2.a. illustrates the structure of the all-atom 1 azurin protein molecule composed of 930 protein atoms and one atom from blue copper ion. Then, in figure 2.b., all-atom structure of azurin reduced to coarse-grained structure. Figure 2.c. show every four heavy atom represent an interaction center.

Figure 3 shows the beads light blue indicates water molecules, with a dark blue bead as anti-freeze molecules (AF) in the form BP4. Anti-freeze molecules included in the simulation box because the current structure was changed into structure AA, CG freezing point of water increases water can freeze at room temperature 310 K, which is the temperature used in the simulation. Red bead is Na^+ ions that amount five atoms and green beads is Cl^- ions that amount two atoms. The additional function of ions completes the lack of electrons in each of amino acid residues so the system is in neutral. Neutral state reduces interaction outside of the protein and the water caused the conformation of the protein in water is not affected by other interactions.

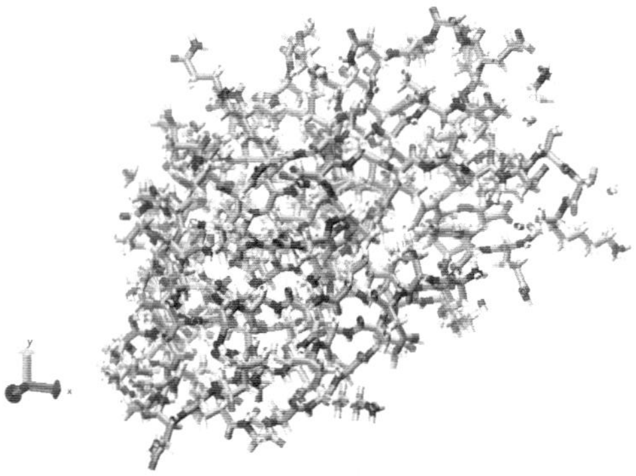

(a) All-Atom Structure of Azurin.

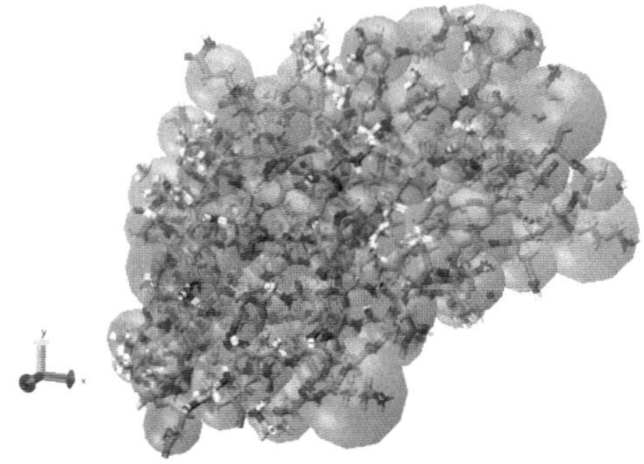

(b) Coarse-Grained (CG) Structure of Azurin.

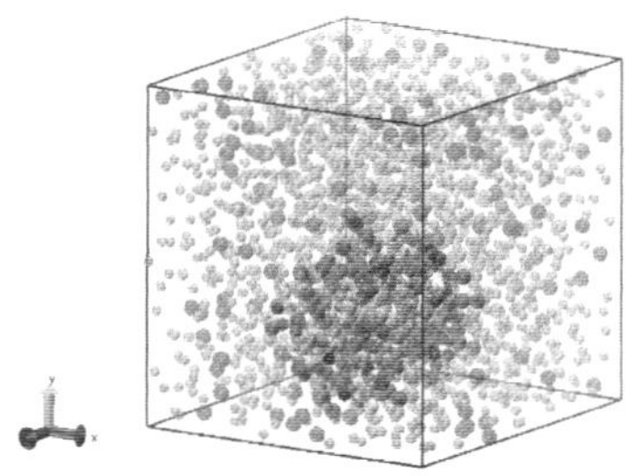

(c) CG Structure of Neutralize Azurin in water box simulation.

Figure 2. (a) Before the structure change into CG model (b) Transparent beads show four heavy atoms that represent one interaction center (c) Light blue bead show water atom; ice blue bead show anti freeze molecule (BP_4); red beads show Na^+ ions; green beads show Cl^- ions; protein of azurin show in the middle of water box simulation, every residue of amino acid show in different color.

The simulation show that there are some amino acids that actually is not hydrophobic out from the water such as asparagine, aspartic acid, lysine, and threonine. This is due to the scheme of conformational azurin of 10 ns still move to find the stable native state. The interaction between the charge of water, BP_4 and NaCl can also cause it. Some of the amino acids of azurin also out from the water in this state such as histidine and methionine. Methionine is known to have hydrophobic properties while histidine residue with the mutant has a positive. It is different with other amino acids (glycine and cysteine), which is an amino acid uncharged polar. The conformation processes of two amino acids are in water simulation box.

At the end of the state 1000 ns simulation show azurin protein forming a favorable conformation and overall state of the protein in the water. This indicates azurin achieve a more stable state conformation in water, it is shown by the end position conformation.

IOP Conf. Series: Journal of Physics: Conf. Series **877** (2017) 012034 doi:10.1088/1742-6596/877/1/012034

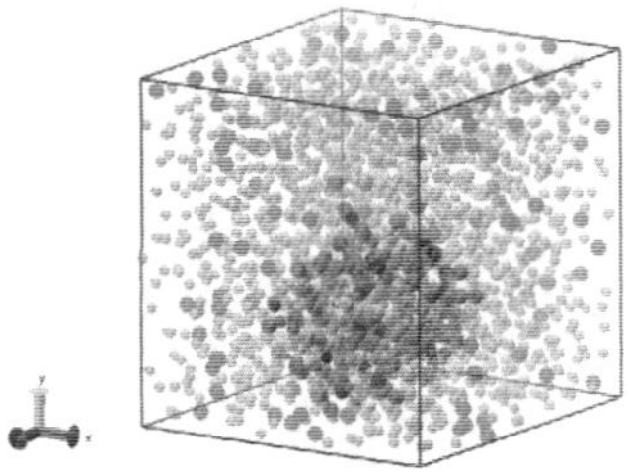

(a) 0 ns

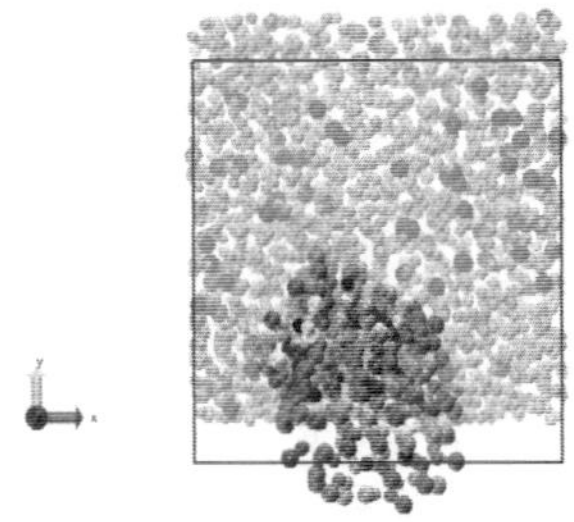

(b) 10 ns

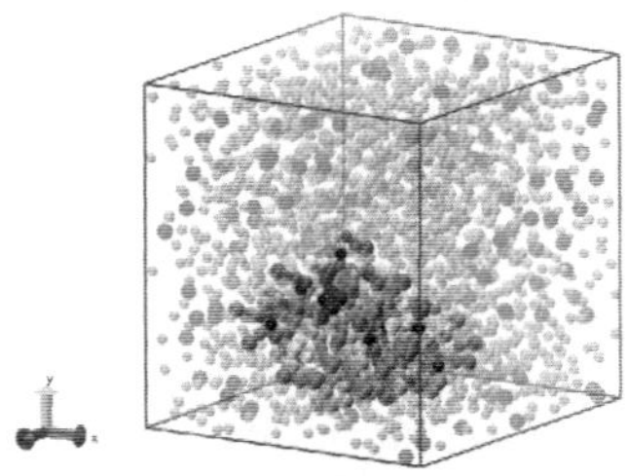

(c) 100 ns

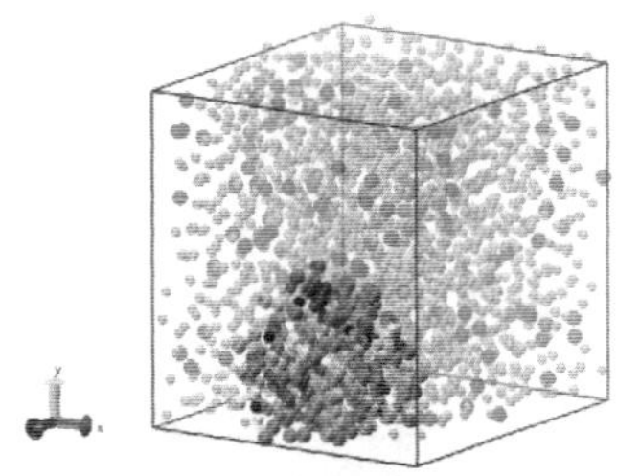

(d) 1000 ns

Figure 3. (a) Equilibration simulation of azurin protein in simulation time 0 ns (initial state); (b) Equilibration simulation of azurin protein in simulation time 10 ns; (c) Equilibration simulation of azurin protein in simulation time 100 ns; (d) Equilibration simulation of azurin protein in simulation time 1000 ns.

In this study, we perform free energy perturbation simulations with molecular dynamics using MARTINI CG force field [8]. Figure 4 shows the free energy value of azurin. The straight line shows free energy of hydration and dotted line show free energy of vacuum. The net solvation free energy change *ΔΔG* for protein of azurin is the difference between free energy of hydration value and free energy of vacuum value. We find the net solvation free energy change of azurin, which shows in Figure 4, is -136.9 kJ/mol.

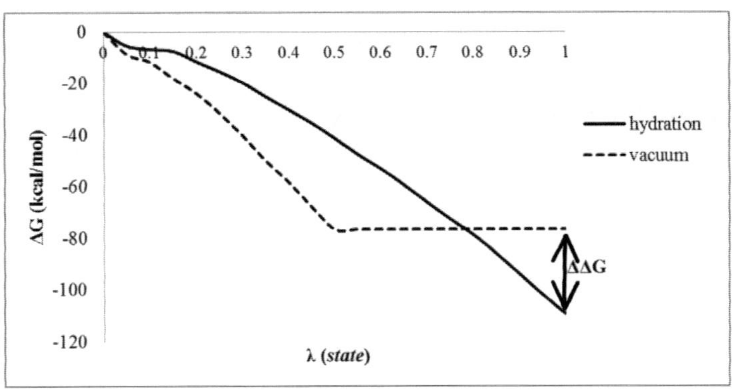

Figure 4. The free energy of azurin, straight line shows free energy of hydration and dotted line show free energy of vacuum.

The validity of FEP calculation compared with the value from a previous study, which is using three different force fields, AMBER, CVFF and CFF for validating the result. The difference between FEP calculation and another method to find free energy protein of azurin is shown in table 1.

Table.1 The comparison of free energy value using FEP method and another method.

	$\Delta G_{binding}\left(kJ\ mol^{-1}\right)$					Relative error index (This research)
Azurin	AMBER	CVFF	CFF	Experiment	FEP (This research)	
Native	-200.0 ± 1.6	-278 ± 12	-127.0 ± 1	-137.0 ± 3	-136.9	0.07%
Unfolded	-	-	-91.0 ± 1.2	-99.3 ± 2.0	-	-

4. Conclusion

Coarse-grained method provides alternative methods to ensure all-atom simulation calculations for larger systems such as proteins and can reduce the calculation. We have investigated the effects of hydration on Azurin, the conformation and the stabilities and we have analyzed the free energy of the conformation system to find the favorable structure using free energy perturbation (FEP) calculation. Our calculation results show that the calculation of free energy perturbation values obtained free energy of -136.9 kJ / mol. It shows a good agreement with experimental result with relative error index remained at 0.07%.

References

[1] Frauenfelder H et al *Proc. Nat'l Acad Sci* (USA)
[2] Yamada T et al 2002 *Proc. of the Nat. Academy of Sciences (USA)* vol 99 (Chicago: University of Illinois National Academy of Sciences) pp 14098-14103
[3] Pozdnyakova I et al 2001 *Archives of Biochemistry and Biophysics* vol 390 (New Orleans: Tulane University Academic Press) pp 146–148
[4] Pozdnyakova I et al 2002 *Biophysical Journal* vol 82 (New Orleans: Tulane University Biophysical Society) pp 2645-2651
[5] Pozdnyakova I and Wittung-Stafshede P 2001 *Biochemistry* vol 40 (New Orleans: Tulane University American Chemical Society) pp 13728-13733
[6] Beveridge D L and DiCapua F M 1989 *Ann. Rev. Biophys. Biophys. Chem.* vol 18 (Birmingham: Annual Reviews) pp 431-492
[7] Mark A E 1998 *Free energy perturbation calculations* vol 2 Schleyer P (New York: John Wiley and Sons) pp 1070-1083
[8] Marrink S J, et al 2007 *J. Phys. Chem. B Chemistry* (American Chemical Society) vol 111 pp 7812-7824

IOP Conf. Series: Journal of Physics: Conf. Series **877** (2017) 012035 doi:10.1088/1742-6596/877/1/012035

Accuracy and Numerical Stabilty Analysis of Lattice Boltzmann Method with Multiple Relaxation Time for Incompressible Flows

Pradipto[1], **Acep Purqon**[1]

[1]Department of Physics, Bandung Institute of Technology

E-mail: pradipto711@gmail.com

Abstract. Lattice Boltzmann Method (LBM) is the novel method for simulating fluid dynamics. Nowadays, the application of LBM ranges from the incompressible flow, flow in the porous medium, until microflows. The common collision model of LBM is the BGK with a constant single relaxation time τ. However, BGK suffers from numerical instabilities. These instabilities could be eliminated by implementing LBM with multiple relaxation time. Both of those scheme have implemented for incompressible 2 dimensions lid-driven cavity. The stability analysis has done by finding the maximum Reynolds number and velocity for converged simulations. The accuracy analysis is done by comparing the velocity profile with the benchmark results from Ghia, et al and calculating the net velocity flux. The tests concluded that LBM with MRT are more stable than BGK, and have a similar accuracy. The maximum Reynolds number that converges for BGK is 3200 and 7500 for MRT respectively.

1. Introduction

The Lattice Boltzmann Method (LBM) was an extension from the Lattice Gas Automata (LGA) [1]. LBM has developed in significant wey and already applied to several kinds of fluid dynamics such as flow in porous media and multiphase flow [2, 3]. The common collision process is the Bhatnagar-Gross-Krook (BGK) with a single relaxation time τ.

Despite its simplicity, BGK suffers numerical instability. This problem has already solved by implementing the multiple relaxation time (MRT)[7, 8]. This generalized view has an agreement with the kinetic theory of gases since the collision according to the kinetic theory of gases has various relaxation time and is related to several physical parameters [6]. Beside the numerical stability, another important aspect to analyze from Computational Fluid Dynamics(CFD) is the accuracy. The test-case for the accuracy and numerical stabilty analysis is the incompressible 2 dimensional Lid-Driven Cavity. Numerical stability and accuracy analysis in 2 dimensional lid-driven cavity has been done for low Reynolds number by Luo, et. al.[14].

This research will continue the works by Luo, et. al [14] for higher Reynolds number. Moreover, we will do the continuity analysis by Aydin [13] and further accuracy analysis by comparing the benchmark results from Ghia, et. al.

In general, this paper begins with the discussion about LBM with BGK and MRT for D2Q9 lattices, the initial and boundary condition, and the lid-driven cavity itself. Last, we will discuss the accuracy and numerical stability.

2. The Lattice Boltzmann Method.

2.1. Bhatnagar-Gross-Krook Approximation,

The evolving equation for the distribution function with respect to canonical coordinate $f(\boldsymbol{r}, \boldsymbol{p}, t)$ in phase space $d^3\boldsymbol{r}d^3\boldsymbol{p}$ is the Boltzmann's equation,

$$\left(\frac{\partial}{\partial t} + \frac{\boldsymbol{u}}{m} \cdot \nabla_{\boldsymbol{r}} + \boldsymbol{F} \cdot \nabla_{\boldsymbol{p}}\right) f(\boldsymbol{r}, \boldsymbol{p}, t) = \left(\frac{\partial f}{\partial t}\right)_{coll} \tag{1}$$

which stated that the spatial and temporal evolution depend on the changes from collision $\left(\frac{\partial f}{\partial t}\right)_{coll}$. The macroscopic parameters can be calculated from the moment integral.

$$\rho = \int f(\boldsymbol{r}, \boldsymbol{p}, t) d\boldsymbol{u} \qquad \rho\boldsymbol{v} = \int \boldsymbol{u} f(\boldsymbol{r}, \boldsymbol{p}, t) d\boldsymbol{u} \tag{2}$$

With Bhatnagar-Gross-Krook approximation, $(\frac{\partial f}{\partial t})_{coll}$, is affected by single relaxation time τ,

$$\frac{\partial f(\boldsymbol{r}, \boldsymbol{p}, t)}{\partial t} + \frac{\boldsymbol{u}}{m} \cdot \nabla_r f(\boldsymbol{r}, \boldsymbol{p}, t) = -\frac{1}{\tau}[f(\boldsymbol{r}, \boldsymbol{p}, t) - f^{eq}(\boldsymbol{r}, \boldsymbol{p}, t)] \tag{3}$$

Furthermore, the integral on equation (2) is discretized with quadrature,

$$\rho = \sum_i f_i \qquad \rho\boldsymbol{v} = \sum_i f_i \boldsymbol{c}_i \tag{4}$$

f_i is discrete distribution function and $\boldsymbol{c}_i$ is discrete velocity. Hence we have the discrete form of Boltzmann's equation,

$$f_i(\boldsymbol{r} + c_i \Delta t, t + \Delta t) - f_i(\boldsymbol{r}, t) = -\frac{1}{\tau}[f_i(\boldsymbol{r}, t) - f_i^{eq}(\boldsymbol{r}, t)] \tag{5}$$

The l.h.s is the streaming process and the r.h.s is the collision process. With Chapman-Enskog expansion [2, 3], we obtained the exact relation between LBM and Navier-Stokes for fluid dynamics namely the kinematic viscosity,

$$\nu = c_s^2\left(\tau - \frac{1}{2}\right) \tag{6}$$

The solution of equation (5) is executed in lattice domain. We will use D2Q9 lattice with 9 discrete velocities,

$$c_i = \begin{cases} (0,0), & i = 0 \\ (0, \pm 1)(\pm 1, 0), & i = 1,2,3,4 \\ (\pm 1, \pm 1), & i = 5,6,7,8 \end{cases} \tag{7}$$

In D2Q9 the lattice speed will be $c = \frac{\Delta x}{\Delta t} = 1$ and sound speed $c_s^2 = RT = c^2/3$. As a result, the equilibrium distribution function f^{eq},

$$f_i^{(eq)} = w_i \rho\left[1 + \frac{3(c_i \cdot \boldsymbol{v})}{c^2} + \frac{9(c_i \cdot \boldsymbol{v})^2}{2c^4} - \frac{3v^2}{2c^2}\right] \tag{8}$$

f^{eq} is obtained by second order Taylor expansion for velocity from the equilibrium Maxwell-Boltzmann distribution. Furthermore, the moment integral is evaluated with third-order Gauss-Hermite quadrature [5], the value of the weight w_i then become,

$$w_i = \begin{cases} 4/9, & i = 0 \\ 1/9, & i = 1,2,3,4 \\ 1/36, & i = 5,6,7,8 \end{cases} \tag{9}$$

2.2. Multiple Relaxation Time

The collision process on MRT is solved on moment space $\mathbb{M}$, while the streaming process still on the velocity space $\mathbb{V}$, the transformation to moment space is done with a group of base vectors $\{|\phi_k\rangle | k = 0, 1, \ldots, N\}$ which are the results of the Gram-Schmidt orthogonalisation for polynomials of the velocity. Hence, the base vectors for 9 discrete velocities [8],

$$|\phi_1\rangle = |\boldsymbol{c_i}|^0 \tag{10}$$

$$|\phi_2\rangle = -4|\boldsymbol{c_i}|^0 + 3(c_{ix}^2 + c_{iy}^2) \tag{11}$$

$$|\phi_3\rangle = 4|\boldsymbol{c_i}|^0 - \tfrac{21}{2}(c_{ix}^2 + c_{iy}^2) + \tfrac{9}{2}(c_{ix}^2 + c_{iy}^2)^2 \tag{12}$$

$$|\phi_4\rangle = c_{ix} \tag{13}$$

$$|\phi_5\rangle = [-5|\boldsymbol{c_i}|^0 + 3(c_{ix}^2 + c_{iy}^2)]c_{ix} \tag{14}$$

$$|\phi_6\rangle = c_{iy} \tag{15}$$

$$|\phi_7\rangle = [-5|\boldsymbol{c_i}|^0 + 3(c_{ix}^2 + c_{iy}^2)]c_{iy} \tag{16}$$

$$|\phi_8\rangle = c_{ix}^2 + c_{iy}^2 \tag{17}$$

$$|\phi_9\rangle = c_{ix}c_{iy} \tag{18}$$

One can state the base vectors with the transformation matrix M,

$$M = \begin{pmatrix} 1 & 1 & 1 & 1 & 1 & 1 & 1 & 1 & 1 \\ -4 & -1 & -1 & -1 & -1 & 2 & 2 & 2 & 2 \\ 4 & -2 & -2 & -2 & -2 & 1 & 1 & 1 & 1 \\ 0 & 1 & 0 & -1 & 0 & 1 & -1 & -1 & 1 \\ 0 & -2 & 0 & 2 & 0 & 1 & -1 & -1 & 1 \\ 0 & 0 & 1 & 0 & -1 & 1 & 1 & -1 & -1 \\ 0 & 0 & -2 & 0 & 2 & 1 & 1 & -1 & -1 \\ 0 & 1 & -1 & 1 & -1 & 0 & 0 & 0 & 0 \\ 0 & 0 & 0 & 0 & 0 & 1 & -1 & 1 & -1 \end{pmatrix} \tag{19}$$

Hence, the moment distribution function,

$$\boldsymbol{m} = \langle f|\phi_k\rangle = (\rho, e, \epsilon, \boldsymbol{j}_x, \boldsymbol{q}_x, \boldsymbol{j}_y, \boldsymbol{q}_y, \boldsymbol{p}_{xx}, \boldsymbol{p}_{xy})^T \tag{20}$$

The first three elements (ρ, e, ϵ) are rank-0 tensor or scalar, ρ is density, e stands for kinetic energy and ϵ is the energy square. The next 4 elements $(\boldsymbol{j}_x, \boldsymbol{q}_x, \boldsymbol{j}_y, \boldsymbol{q}_y)$ are rank-1 tensor which denotes the x-axes momentum, energy flux on x-axes, y-axes momentum, and energy flux on y-axes, respectively. The last 2 elements $(\boldsymbol{p}_{xx}, \boldsymbol{p}_{xy})$ are rank-2 tensors that denotes the diagonal and off-diagonal off stress tensor.

Then, the equilibrium distribution function on moment space [3].

$$\boldsymbol{m}^{eq} = \rho(1, -2 + 3v^2, 1 - 3v^2, v_x, -v_x, v_y, -v_y, v_x^2 - v_y^2, v_x v_y)^T \tag{21}$$

The relaxation rate can be set to each physical process related to equation (20). As a result, there are 9 relaxation rate for different processes.

$$S = (0, s_e, s_\epsilon, 0, s_q, 0, s_q, s_\nu, s_\nu) \tag{22}$$

Hence, the evolving equation for LB-MRT,

$$|f(\boldsymbol{r} + c_i\Delta t, t + \Delta t)\rangle - |f(\boldsymbol{r}, t)\rangle = -M^{-1}S[|m(\boldsymbol{r}, t)\rangle - |m^{eq}(\boldsymbol{r}, t)\rangle] \tag{23}$$

Furthermore, the relaxation rate for density and momentum is equal to zero because they conserved against collision. Other relaxation parameters are set with such values,

$$s_e = 1.64 \qquad s_\epsilon = 1.54$$

$$s_q = 1.92 \qquad s_\nu = 1/\tau$$

3. Lid-Driven Cavity

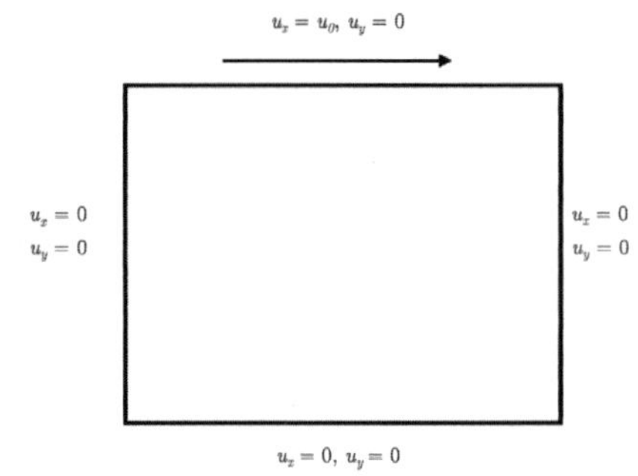

Figure 1: 2 dimensional lid-driven cavity.

In 2 dimensional lid-driven cavity, the fluid is in a $L_x \times L_y$ square cavity. The top wall of the cavity is moving with constant velocity v_0. The benchmark velocity profile for 2 dimensional lid-driven cavity from Ghia, et. al [12] is used for accuracy comparison. The characterization parameter for incompressible flow in lid-driven cavity is the Reynolds number (Re),

$$Re = \frac{Lv_0}{\nu} \tag{24}$$

The total horizontal and vertical velocity flux must equal to zero since the flow is closed. One can determine the total flux by calculating the volumetric integrals [13],

$$Q_{hor} = \int_0^{L_x} v_y dx \tag{25}$$

$$Q_{ver} = \int_0^{L_y} v_x dy \tag{26}$$

The first integral is evaluated at $L_y/2$ and the latter is evaluated at $L_x/2$.

4. Accuracy and Numerical Stability Analysis

The simulation is done with various Reynolds number from 100 to 7500, wall velocity $v_0 = 0.1$, and two lattice sizes 128×128 and 256×256. First, the evaluation of streamline plot for $N = 128 \times 128$

There are differences between the streamline plot of the BGK and MRT implementation. BGK cannot capture the vortex in the bottom left of the cavity on Re=400 and Re=1000. We get different outcome when using $N = 256 \times 256$ where the results are identical. Here, we could indicate that the implementation of MRT have greater accuracy for smaller lattice size and the MRT implementation is more stable against the lattice size. Further accuracy test is comparing the velocity profiles with the benchmark results from Ghia, et. al.[12].

The velocity profile is slightly deviates from the references for N=128. Most of the errors are occured near the boundary. Surprisingly, the results for BGK and MRT are again identical with N=256. Here it indicates once more that implementation of MRT is more stable against the variation of lattice size.

Moreover, the pressure contour will also be evaluated. The value of pressure is obtained from the equation of ideal gas $p = \rho c_s^2$ and normalized with v_0^2. Here we found that the pressure contour plot is oscilating at top corner of the cavity and increased as the Reynolds number increased. The MRT implementation cannot eliminate this effect but it can damp the oscilation. This oscilation is caused by the singular velocity on top corner of the cavity since the top wall is

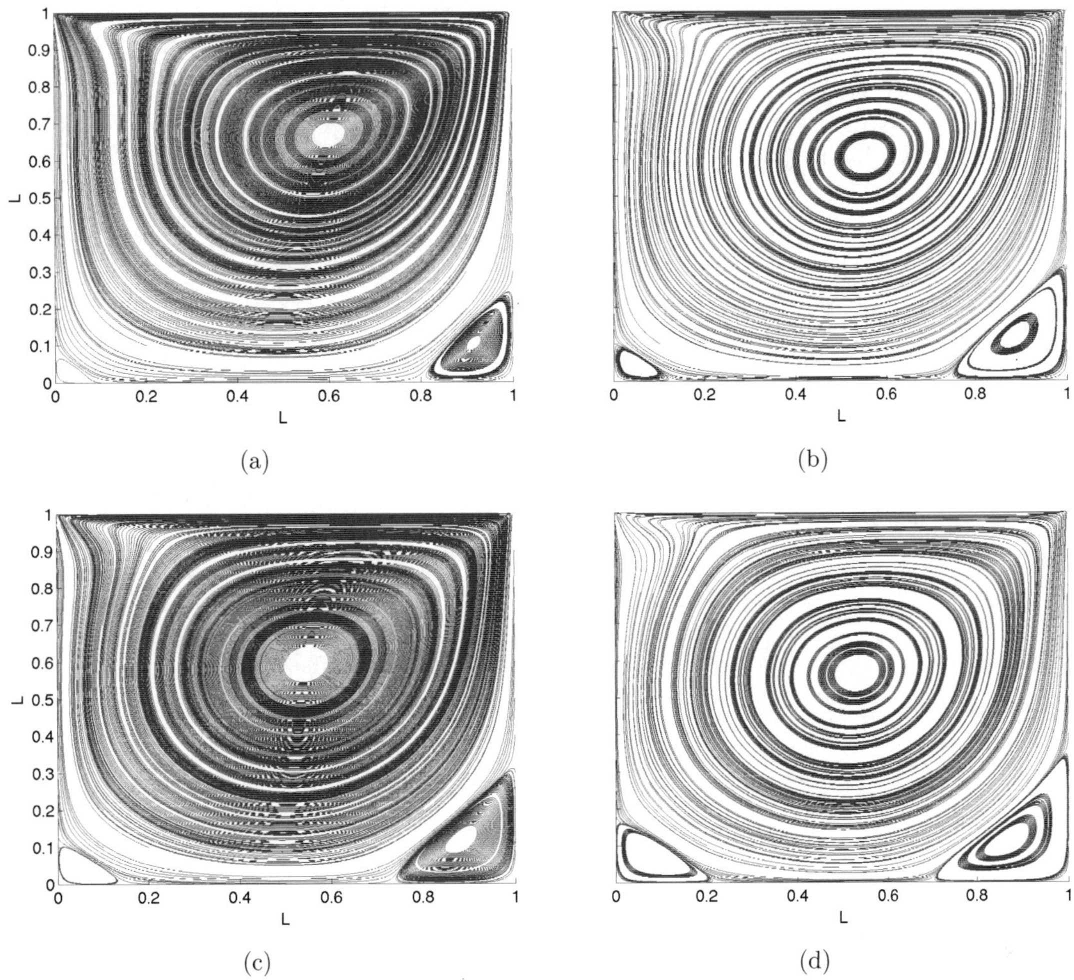

Figure 2: Streamline for Re=400 with N=128: (a)BGK, (b)LB-MRT, Streamline for Re=1000 with N=128: (c)BGK, (d)LB-MRT

moving and the side wall is stationary. This is called the compressibility effect which is the main source of the density fluctuation in the corner that caused the oscilation. This effect is intrinsic within LBM since it is used to simulate small velocity but cannot fully incompressible. These effect can be reduced by setting the dissipation of the density fluctuation. This can be done only if we set the relaxation rate with respect to the energy s_e. Hence, it is obvious that only the MRT implementation could done that since the relaxation rate for BGK only depends on τ.

In order to analyze the numerical stability, we tried to found the maximum value of Reynolds number that still converges with each implementation. The BGK can converges until Re=3200 and the MRT can converges until Re=5000 with N=128 and Re=7500 with N=256.

The final test for the numerical stability analysis is to determine the maximum wall velocity that still converges within 1000 iterations.

The test showed the MRT implementation is more stable than BGK. On BGK, when $1/\tau$ approach 2, then the viscosity ν approach 0, and when $1/\tau > 2$, we have a negative viscosity which is certainly unphysical. From the graph we can see that the highest velocity is reached when $1/\tau = 1.92$ which is the exact value of s_q. Furthermore, the numerical stability could be

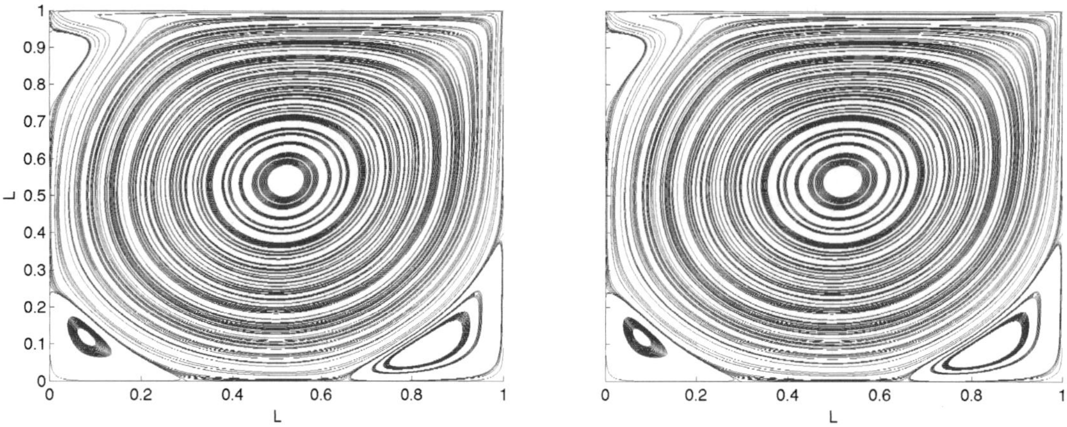

(a) (b)

Figure 3: Streamline for Re=3200 with N=256: (a)BGK, (b)LB-MRT.

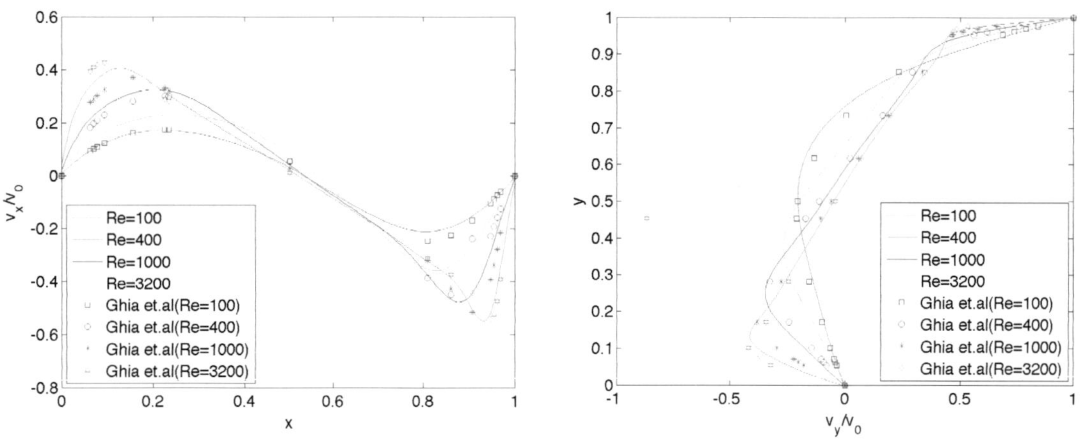

(a) (b)

Figure 4: Velocity Profiles of BGK compared with Ghia,et. al results [12], N=128: (a)Horizontal, (b)Vertical.

enhanced if we set the value of s_q equal to s_ν, to realized the concept of no-slip at bounce-back boundary condition [14].

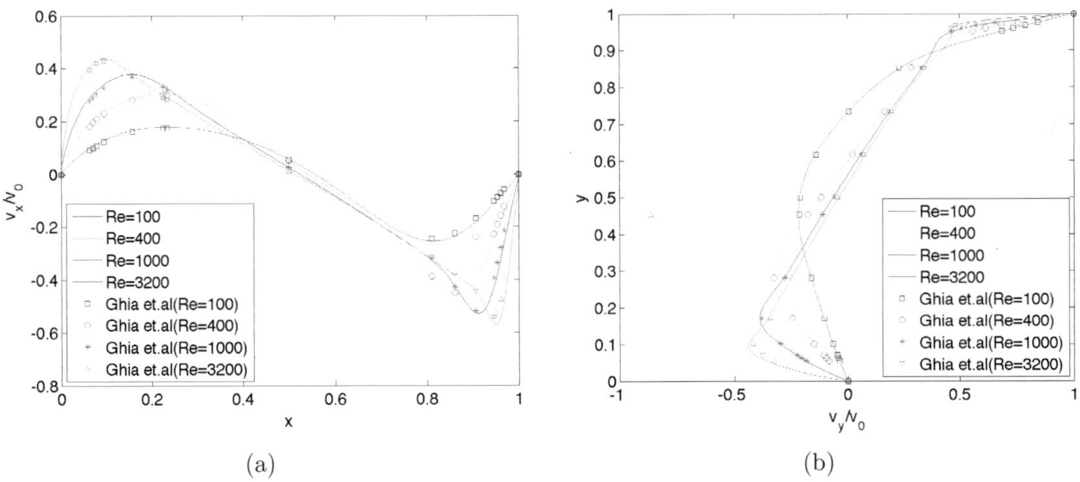

Figure 5: Velocity Profiles of MRT compared with Ghia,et. al results [12], N=128: (a)Horizontal, (b)Vertical.

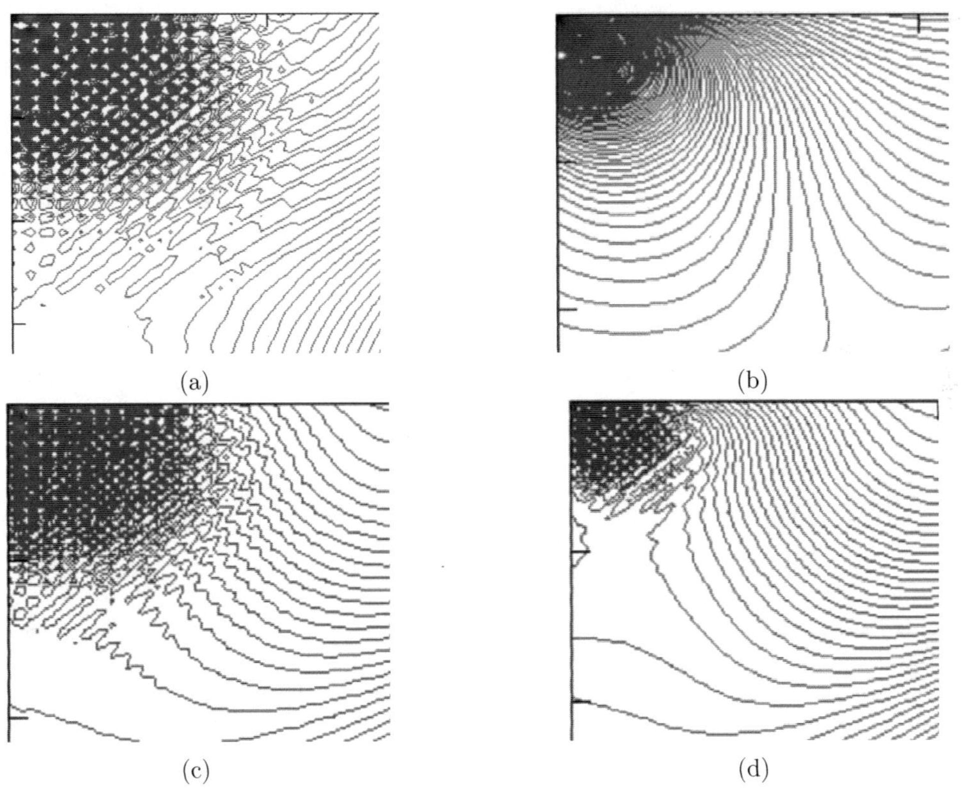

Figure 6: Pressure contour on top-left of the cavity, (a)Re=400,BGK (b)Re=400,MRT, (c)Re=1000,BGK, (d)Re=1000,MRT

5. Conclusions

The implementation of multiple relaxation time (MRT) for collision process of the lattice Boltzmann method for 128×128 lattices has better accuracy compared to the implementation

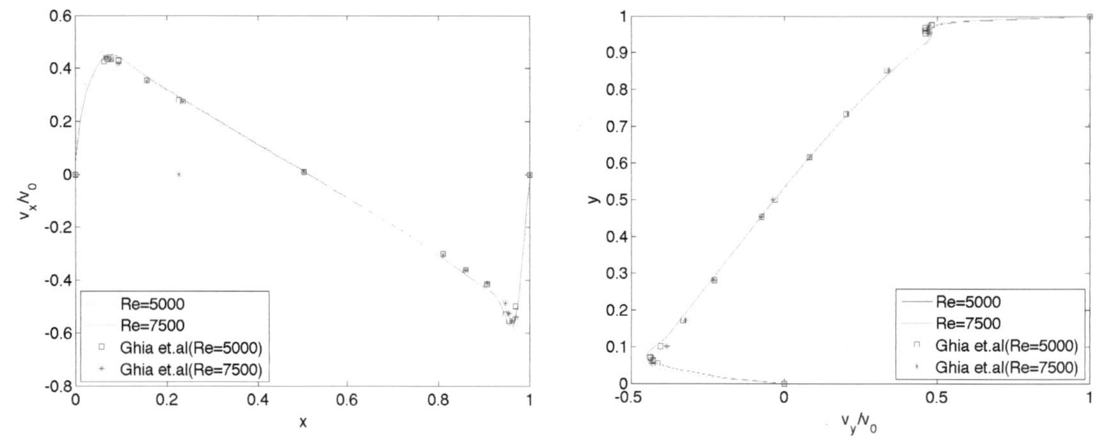

(a) (b)

Figure 7: Velocity profiles for Re=5000 and Re=7500.

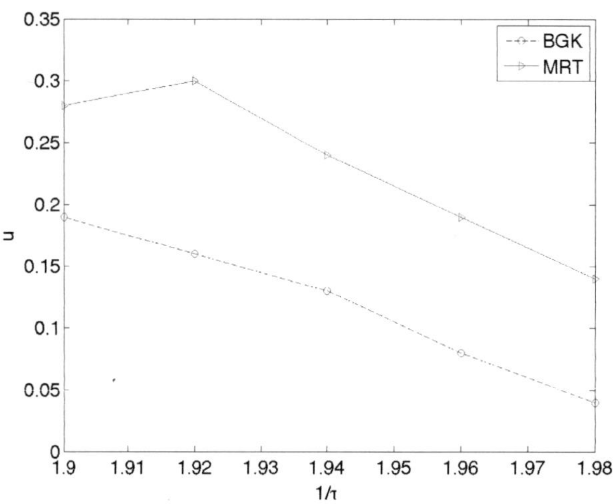

Figure 8: Numerical stability test.

of Bhatnagar-Gross-Krook (BGK) approximation, both of them have identical accuracy on 256×256 lattices.

The maximum Reynolds number that still converges is 3200 for BGK and 5000 and 7500 for MRT with N=128 and N=256 respectively. Hence, we can conclude that the implementation of MRT is more stable against variation of Reynolds number and lattice size.

Moreover, the MRT implementation can reduce the compressibility effect by dissipating the density fluctuation, this can be done by setting the relaxation rate of energy s_e. The numerical stability test also showed that MRT is more stable than BGK, and can enhance the numerical stabilty by setting the s_q equals to s_ν. The test showed that it reach its maximum velocity when $1/\tau = 1.92$ which is the value of s_q.

References

[1] McNamara .G, Zanetti .G, 1988, *Use of The Boltzmann Equation to Simulate Lattice Gas Automata*, (Physical Review Letter) 61, 2332,

[2] Succi .S, 2001, *The Lattice Boltzmann Equation for Fluid Dynamics and Beyond*, (Oxford University Press).

[3] Guo .Z, Shu .C, 2013, *Lattice Boltzmann Method and its Applications in Engineering*, (Advances in Computational Fluid Dynamics Vol 3,World Scientific)

[4] A.A. Mohammad, 2011, *Lattice Boltzmann Method : Fundamentals and Engineering Applications with Computer Codes*, (Springer).

[5] He. X, Luo. L, 1997, *Theory of the Lattice Boltzmann Method:From the Boltzmann Equation to the Lattice Boltzmann Equation*, (Physical Review E)56,6.

[6] Huang, Kerson, 1987, *Statistical Mechanics 2nd Edition*, (John Wiley and Sons)

[7] d'Humieres, 1992, *Generalized lattice Boltzmann equations*, (Progress in Aeronautics and Astronautics) 159, 450-458,

[8] Lallemand. P , Luo L. S, 2000. *Theory of the Lattice Boltzmann Method:Dispersion, dissipation, isotropy, Galilean invariance, and stability*, (Physical Review E)61,6.

[9] McNamara. G.R, Garcia.A.L, Adler.B.J, , *Stabilization of Thermal Lattice Boltzmann Models*, (Journal of Statistical Physics)81, 395.

[10] Chikatamarla .S.S, Karlin .I.V, 2009 *Lattices for the lattice Boltzmann method*, (Physical Review E) 79, 046701.

[11] Ladd. A. J. C, 1994 *Numerical simulations of particulate suspensions via a discretized Boltzmann equation Part II. Numerical results*, (Journal of Fluid Mechanics) 271, 311.

[12] Ghia. U, Ghia. K. N, Shin. C. T, 1982 *High-Re Solutions for Incompressible Flow Using the Navier-Stokes Equations and a Multigrid Method*, (Journal of Computational Physics) 48, 387-411.

[13] Aydin. M, Fenner. R. T, 2001 *Boundary element analysis of driven cavity flow for low and moderate Reynolds numbers*, (International Journal of Numerical Method of Fluids 2001)37 pp: 4564,

[14] Luo. L. S, Liao. W, Chen. X. W, et al, 2011, *Numerics of the latticd Boltzmann method: Effects of collision models on the lattice Boltzmann simulations*, (Physical Review E) 83, 056710.

[15] Erturk. E, 2009, *Discussions On Driven Cavity Flow*, (International Journal of Numerical Methods of Fluids) 60:275-294,

[16] Prasad. A.K, Kossef. J.R, 1989, *Reynolds Number and End-Wall Effects on a Lid-Driven Cavity Flow*, (Physics of Fluids A 1) 208-218

Structural and optical properties analysis of MoS_2 nanoflakes on quartz substrate as prepared by mechanical exfoliation

Sandy Jaka Adilla, Eka Nurfani, Robi Kurniawan, Christoforus Dimas Satrya and Yudi Darma[a)]

Quantum Semiconductor and Devices Laboratory, Department of Physics, Bandung Institute of Technology, Bandung 40132 Indonesia.

Email: [a)]yudi@fi.itb.ac.id

Abstract. We study the structural and optical properties of MoS_2 nanoflakes on quartz substrate as prepared by mechanical exfoliation method. The structural and morphological properties of MoS_2 nanoflakes were characterized by SEM, EDS, and XRD, while the high-resolution spectroscopic ellipsometry (SE) with the photon energy of 1.27 to 6.53 eV is used to study its optical characteristics. As the results, SEM data shows that MoS_2 appears to be nanoflakes covering around 25-35% on the surface of quartz substrate, and XRD spectra shows the dominant orientation along c-axis (002). Based on spectroscopic ellipsometry analysis, the average thickness of MoS_2 nanoflakes is around 12 nm or about 6-8 layers. By using Tauc plot method, we confirm that MoS2 nanoflakes have a semiconductor characteristic with the optical bandgap as high as 1.68 eV. Our study shows the important of structural and optical properties of MoS_2 nanoflakes that can be utilized for future optoelectronic devices and energy-harvesting purposes.

1. Introduction

Since 1859 scientists always failed to synthesize the two-dimensional materials, until finally graphene was found in 2004 with scotch tape by Novoselov and Constantine and it triggered the development of two-dimensional material [1–4]. Since that time, many potential materials such as MoS_2 were investigated further due to their potential merits in the future. MoS_2 is one of the TMDC (transition metal dichalcogenides) material that forms layered structure [1-3] with strong bonding (covalent bond) for each atom S-Mo-S in one layer or in one plane. MoS_2 has weak bonding (Van Der Waals bonding) for each layer which then can be detached into an individual plane or thin layers [4-6]. In the past, MoS_2 was usually used for lubricants in bulk size and its properties become interesting when it reaches nanosize, especially in two-dimensional material like graphene. The mobility reaches up to 410 $cm^2\ V^{-1}\ s^{-1}$ at room temperature, while its high on/off ratio is about 10^8. On the other hand, graphene has conductive properties low on/off ratio making less suitable for electronic devices. Additionally, MoS_2 has a sizeable bandgap that depends on the thickness. At monolayer thickness, it exhibits direct energy bandgap around 1.3 eV [7,8] resulted in the photoluminescence properties [9]. Therefore, it is very applicable for optoelectronic devices, namely solar cell and photodetector. At bulk structure, the band gap will turn into indirect around 1.8 eV [10] and it can be applied effectively for devices such as FET, memory, and battery.

Nowadays, the knowledge of optical properties and its correlated nanostructure of MoS_2 is still rarely investigated, especially in the broad energy spectrum. In this paper, we investigate the structural and optical properties of MoS_2 nanoflakes in the energy range of 1.27 to 6.53 eV. Spectroscopic ellipsometry (SE) was used to determine the dielectric function of MoS_2. Dielectric function is the optical properties that can describe light-matter interactions and explain the excited-state properties of material. The dielectric function can also explain electronic structure of MoS_2 and its potential application. Hence, it is crucial to determine the dielectric function to profoundly understand the potential of MoS_2.

2. Method

Mechanical exfoliation is one of the top-down methods that is very handy and simple compared to any epitaxial synthesis, and this method is still reliable enough to represent the characteristic of the attached material. MoS_2 bulk was peeled into nanoscale by adhesive tape [11-13]. Equalization or exfoliation process was repeated for 60-70 times or ten minutes every sample. Then it is pasted above quartz substrate as many as 10 times. For the comparison, we also attached bulk MoS_2 upon silicon. The structural and morphological properties of MoS_2 nanoflakes were characterized by SEM (scanning electron microscopy), EDS (Energy Dispersive X-Ray Spectroscopy), and XRD (x-ray diffraction) spectra.The high-resolution Spectroscopic ellipsometry (SE) with the photon energy of 1.27 - 6.53 eV was used to study its optical characteristics. SE measurements were carried out using an incident angle of 70° at room temperature. Amplitude ratio ψ and phase difference Δ of the polarized light were directly obtained from this measurement. We used RefFit software to analyze the dielectric function of MoS_2 by fitting both parameters using Drude-Lorentz model, as shown in equation 1. This model describes the free carrier absorption and electrical polarization of oscillated electron. The model is represented as

$$\varepsilon(\omega) = \varepsilon_\infty + \sum_i \frac{\omega_{pi}^2}{\omega_{oi}^2 - \omega^2 - i\gamma_i\omega} \quad (1)$$

where ε_∞ describes the infinite frequency, ω_p describes plasma frequency, ω_0 describes the angular frequency and γ describes scattering rate [14]. From the generated model, it will produce the complex dielectric function that are mutually connected by Kramers-Kronig relation, as shown in equation 2 and 3. So, once we find the real part of the dielectric function, we can get the imaginary part simultaneously [15].

$$\varepsilon_1(\omega) = 1 + \frac{2}{\pi} P \int_0^\infty \frac{\omega' \varepsilon_2(\omega')}{\omega'^2 - \omega^2} d\omega' \quad (2)$$

$$\varepsilon_2(\omega) = -\frac{2\omega}{\pi} P \int_0^\infty \frac{\varepsilon_1(\omega') - 1}{\omega'^2 - \omega^2} d\omega' \quad (3)$$

3. Results and Discussion

Based on the SEM image shown in figure 1, MoS_2 is attached well with the coverage area around 25-35% on quartz substrate. The thickness of MoS_2 is varied, which depends on its position. Figure 1(d) shows that the number of layers are around 2 layers. As we can see in figure 1, MoS_2 is formed by multiple layers which is bounded by Van Der Waals force for each layer, like a stacking structure. The most important information from this microscopy images is that the MoS_2 was not grown in the homogenous area, however it form a very small flake called as the nanoflakes.

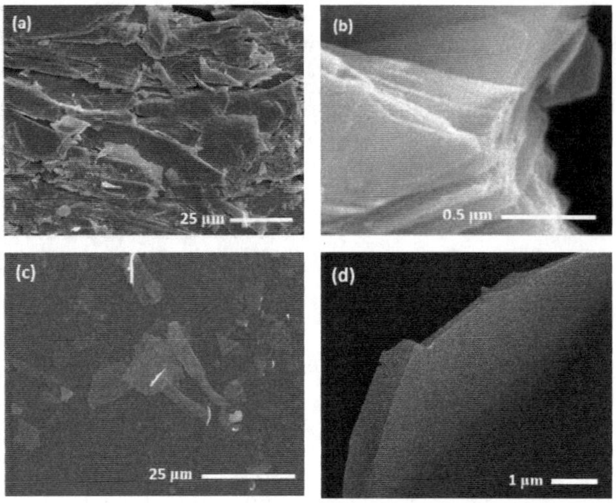

Figure 1. SEM images of MoS_2 bulk on Si with magnification (a) 500 times and (b) 40.000 times. Nanoflakes MoS_2 shows the coverage area 25-35% on top of quartz with magnification 500 timesand (d) 40.000 times.

EDS was used to determine the composition of the sample. If we compared between figure 2(a) and 2(b), there are still remain the molybdenum and sulfur characteristic in the different quantity. It is obvious that the elemental composition of bulk MoS_2 is more visible rather than the composition of MoS_2 nanoflakes, and confirm that the attached MoS_2 is not fully cover quartz substrate and consist of a small pieces of nanoflakes. Besides Mo and S, there are no others atomic element are identified, so we conclude that the attached material on quartz is pure MoS_2.

Figure 3 shows the XRD spectra of MoS_2 in bulk and multiple layer form. Bulk MoS_2 have many peak or miller indices. This indicates that MoS_2 in bulk form tend to be polycrystalline than in the nanoflakes form. The synthesized nanoflakes just have one peak (002) along c-axis. As the decrease of the composition, the miller indices will be decreased as well. By using equation 4, lattice parameters of hexagonal MoS_2 is obtained (a=b=3.160 Angstrom and c = 12.30 Angstrom). As compared to the reference [16], we find the error of a = b = 0.032% and c = 0.040%.

$$\frac{1}{d_{hkl}^2} = \left[\frac{4}{3}(h^2 + k^2 + hk) + l^2\left(\frac{a}{c}\right)^2\right]\frac{1}{a^2} \quad (4)$$

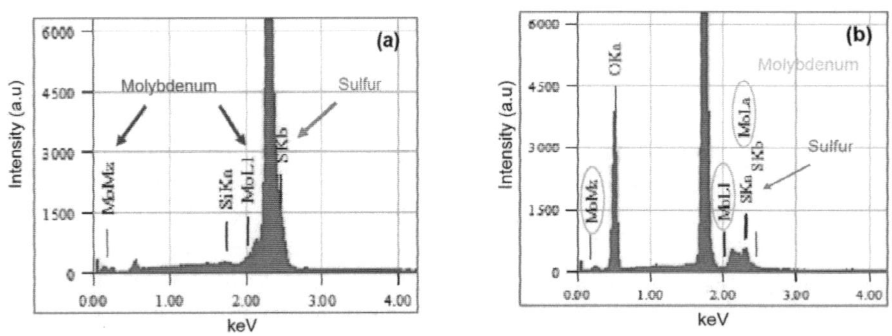

Figure 2. EDS analysis (a) MoS_2 bulk (b) MoS_2 nanoflakes.

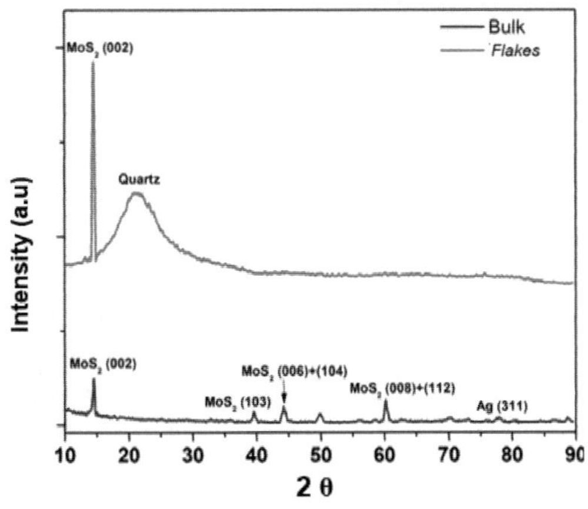

Figure 3. XRD data of bulk MoS_2 on Si (blue line) and nanoflakes MoS_2 on quartz (red line). The obtained lattice parameters of hexagonal structure are a=3.16 Å and b=12.3 Å.

Because of Spectroscopic Ellipsometry is one of the indirect measurement, fitting ψ and Δ data should be carried out to analyse the optical properties and its derivative. Drude-Lorentz model was chosen in the fitting process. In figure 4 (a) and (b), ψ and Δ data of quartz substrate are almost reach 0° until having a peak around 6.2 eV. This peak indicates that substrate quartz is one of the isolator material. In Figure 4 (c) and (d), we can observe clearly the optical characteristic of MoS_2 nanoflakes. We assume that the resulted thickness is determined by the average of each nanoflakes.

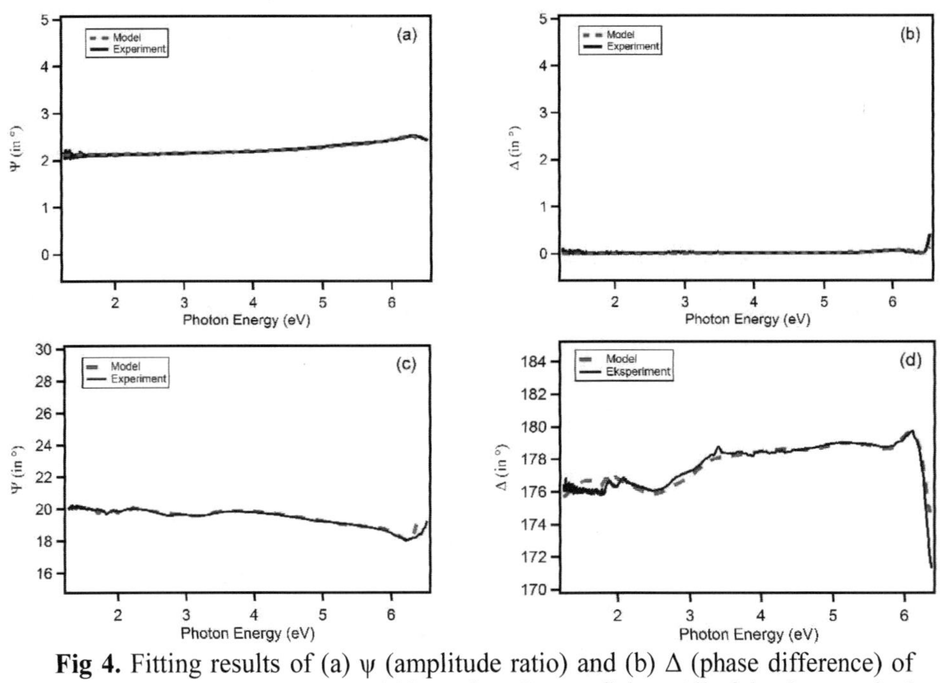

Fig 4. Fitting results of (a) ψ (amplitude ratio) and (b) Δ (phase difference) of quartz substrate, and (c) ψ and (d) Δ of MoS_2 nanoflakes. All of the data resulted by the incident angle of 70^0 from SE equipment.

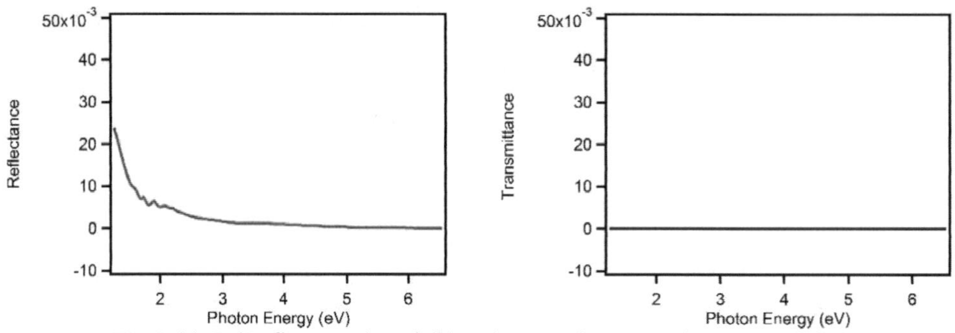

Fig 5.(a) R (Reflectance) and (b) T (Transmittance) of MoS_2 nanoflakes.

From the best fit of experimenta data, we can extract optical constants of nanoflakes MoS_2, such as reflectance (R) and transmittance (T), as shown in fig. 5. Reflectance spectra represents the process of electromagnetic waves returned at the boundary of material, whereas transmittance represents the passage of electromagnetic waves through a material. Those R and T represent the specific character of MoS_2 nanoflakes attached above quartz substrate. From reflectance curve, MoS_2 have responses in visible region area. The unusual shape of the curve is caused by the absorbance responses of the excitation process in MoS_2 [17,18], till the shape become constant at the high photon energy. Because the small intensity of reflectance and transmittance, MoS_2 nanoflakes is indicated to be one of the absorbing material and show the opaque characteristic material. In addition, the thickness of MoS_2 is around 15 nm or around 6-8 layers. From its layer, this MoS_2 nanoflakes must be an indirect type bandgap.

From the penetration depth parameters, Tauc-plot method is used to determine the optical bandgap.The obtained optical band gap is ~1.68 eV. It was matched with the thickness of MoS_2 around 6-8 layers. This bandgap is suitable for the future devices, like electronic devices or optoelectronic devices.

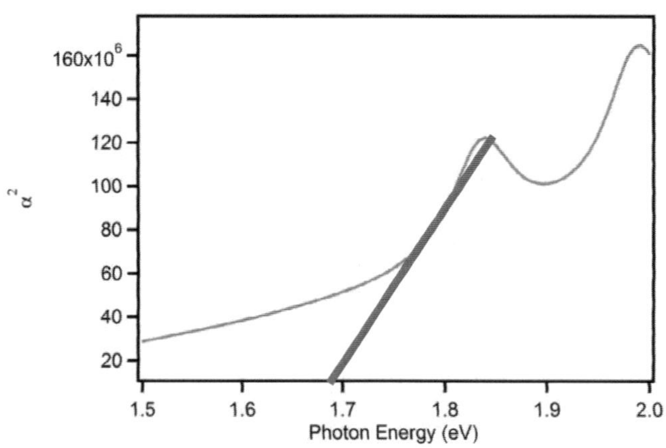

Fig 6. Extrapolation of Tauc-plot method to determine the bandgap of MoS_2 nanoflakes.

4. Conclusion

EDS spectra and SEM images confirms that the nanoflakes MoS_2 is properly attached on the quartz substrate, and XRD spectra show that the crystallinity of MoS_2 is hexagonal with the lattice parameter $a = b = 3.160$ Angstrom and $c = 12.30$ Angstrom. More interestingly, R and T spectra of MoS_2 nanoflakes are small that indicates an opaque characteristic material and might be useful as an absorbing material. Based on the Tauc-plot method, the optical bandgap is ~1.68 eV that come from the valence band to the conduction band transition. This study reveal the fundamental properties of MoS_2 nanoflakes that will be promising candidates for future optoelectronic devices and renewable energy application.

Acknowledgement

This work is partly supported by Riset Kompetensi (586n/I1.C01/PL/2016) by Kemenristek DIKTI 2016 program and RUPT 2016 from Indonesian government. We also thanks to SSLS-NUS Singapore for SE measurement facility.

References

[1] Wilson J A and Yoffe A D 1969 Transition metal dichalcogenides: discussion and interpretation of observed optical, electrical and structural properties. *Adv. Phys.* **18** 193-335.
[2] Yoffe A D 1993 Layer compounds *Annu. Rev. Matter. Sci,* **3** 147-170.
[3] Yoffe A D 1993 Low-dimensional systems : quantum size effects and electronic properties of semiconductor microcrystallites *Adv. Phys.* **42** 173-266.
[4] Sipos B 2008 From Mott state to superconductiity in 1T-TaS_2*Nature Mater* **7** 960-965.
[5] Wilson J A Disalvo F J and Mahajan S 1975 Charge-density waves and superlattices in metallic layered transition-metal dichalcogenides *Adv Phys***24** 117-201.
[6] Castro Neto A H 2001 Charge density wave, superconductivity, and anomalous metallic behavior in 2D transition metal dichalcogenides *Phys. Rev. Lett.* **86** 4382-4385.
[7] Yang C L 2010 Determination of the sign of g factors for conduction electrons using time-resolved Kerr rotation *Appl. Phys. Lett***96** 152109.
[8] Molina S W 2012 Phonons in single-layer and few-layer $MoS_2$and WS_2*Phys. Lett.* **101**(22) 221907.
[9] Qing H W and Michael S S 2012 Electronics and optoelectronics of two-dimensional transition metal dichalcogenides *Nat. Nanotech.***193** 699-712
[10] Zibouche K A and Heine T 2011 Influence of quantum confinement on the electronic structure of the transition metal sulfide TS_2*Phys. Rev. B.* **83** 245213.
[11] Alem N 2009 Atomically thin hexagonal boron nitride probed by ultrahigh resolution transmission electron microscopy. *Phys. Rev. B.* **80** 155425.
[12] Lee C 2010 Anomalous lattice vibration of single and few layer MoS_2*ACS Nano* **4** 2695-2700.
[13] Mak K F Lee C Hone J Shan J and Heinzz 2010 Atomically thin MoS_2a new direct-gap semiconductor *Phys. Rev. Lett.* **105** 1366805.
[14] Fujiwara H 2007 *Spectroscopic ellipsometry: Principles and applications* (West Sussex: John Wiley and Sons).
[15] Tipler P A and Ralph A 2003 Modern Physics Freeman and Company 222.
[16] Schonfeld B 1983 Acta Crustallographica **B39** 404-407.
[17] Qiu D Y Jornada F H and Louie S G 2013 Optical Spectrum MoS_2: Many-Body Effects and Diversity of Exciton States *Phys. Rev. Lett.* **111** 216805.
[18] Zhu Z Y Cheng Y C and Schwingenschlogl U 2011 Giant spin-orbit-induced spin splitting in two-dimensional transition-metal dichalcogenide semiconductors *Phys. Rev. B.***84** 153402.

Biodiesel Production by Using CaO Catalyst and Ultrasonic Assisted

W Widayat[1,2,*], T Darmawan[2], R Ar Rosyid[2] and H Hadiyanto[2]

[1]Laboratory of Advanced Material Center of Research and Service Diponegoro University
[2]Department of Chemical Engineering Diponegoro University
Jl. Prof. Soedarto SH, Tembalang, Semarang, 50275

[*]Email. widayat@live.undip.ac.id

Abstract. Biodiesel is an alternative energy and sustainable products. Biodiesel can be produced from vegetable oil by using homogenous and heterogenous catalyst. In this research, biodiesel was produced by using CaO catalyst from limestone and assisted by ultrasonic wave. CaO catalyst was used 1- 4.4% of oil weight. Ultrasonic wave was generated by using ultrasonic cleaner at 28 and 42 kHz of frequency. Cooking oil was preparated by esterification process by using sulphuric acid as catalyst until amount of free fatty acid (FFA) is under 0.5%. The results shown yield of methyl ester obtained highest 89.98% on the reaction carried on mole ratio methanol to oil, 9: 1 and frequency of 40 kHz. Increasing frequency of ultrasonic wave cause increasing of yield of biodiesel.

1. Introduction

Biodiesel is a diesel engine fuel made from biological sources or biomass. Meanwhile, according to today's understanding of the industry, biodiesel is a fuel for diesel engines consisting of alkyl esters of fatty acids **Error! Reference source not found.**. Ester is a chemistry term that means a compound formed from condensation of alcohol with acid. The process of making biodiesel by reacting alcohol with fatty acids can be done in two ways, namely esterification and transesterification [5]. In making the ester biodiesel can be made from vegetable oils by esterification or transesterification reaction or a combination of both reactions. The esterification reaction is a reaction between free fatty acid with an alcohol to form an ester and water. The reaction that occurs is an endothermic reaction and temperature reaction at 55-60°C [1-2].

Commonly, biodiesel was produced with homogenous catalyst[1-2,5], like sulphuric acid for esterification and potasium hydroxide for transesterification reaction. Heterogenous catalyst was also used in biodiesel production like as KI/KIO_3/H-zeolite[6], calcium oxide[7-12]. Widayat et.al (2016) were used KI or KIO_3 that loading on H-zeolite for catalyst on biodiesel production from palm oil. H-zeolite was prepared from natural zeolite with chemical and physical treatment[6]. Hadiyanto et.al (2016) were used calcium oxide as catalyst on biodiesel production. Calcium oxide catalyst was prepared from mollusk shell and combining with fly ash from coal combustion[7,8]. Waste of animal bone was also used as catalyst in biodiesel production and physical treatment was used for preparation of CaO catalyst[9]. Calcium oxide has potential catalyst in biodiesel production[10- 12].

Catalysts are compounds that speed up reactions by lowering the activation energy without participate in reaction[13]. The advantages of heterogenous catalyst are separation from products were quite simple, namely by using filtering. One sources of CaO in natural resources is limestone. Limestone is a sedimentary rock composed largely of the minerals calcite and aragonite, which are different crystal forms of calcium carbonate ($CaCO_3$). CaO catalyst was used in transesterification reaction .

Biodiesel or Fatty Acid Methyl Ester (FAME) can produced with esterification and transesterification[1-2,14].The esterification reaction is done before the transesterification reaction if the used oil containing high free fatty acid (> 0.5%)[1,14]. In esterification reaction, free fatty acid convert to FAME with adding of methanol and acid catalyst. Transesterification reaction also called alcoholysis reaction, reaction between triglycerides with alcohol and alkaline catalyst to produce an methyl ester /FAME and glycerol. Alcohols that used transesterification reaction were methanol, ethanol and isopropanol. FAME and glycerol are easily separate with decantation processing [13].

The objective of this research is to study effect of mole ration methanol to triglycerides and CaO catalyst concentration on transesterification reaction.

2. Methods

2.1. *Material*

Palm oil was purchased from local market. Limestone was bought from material store. Methanol has industrial specification and sulphuric acid obtained from Merck Ltd. In biodiesel production was used reactor that require with mechanical mixing and ultrasonic assisted. Ultrasonic wave was generated in ultrasonic cleaner with 200 watt of power and 28, 40 kHz of frequency. This equipment like presented in Figure 1.

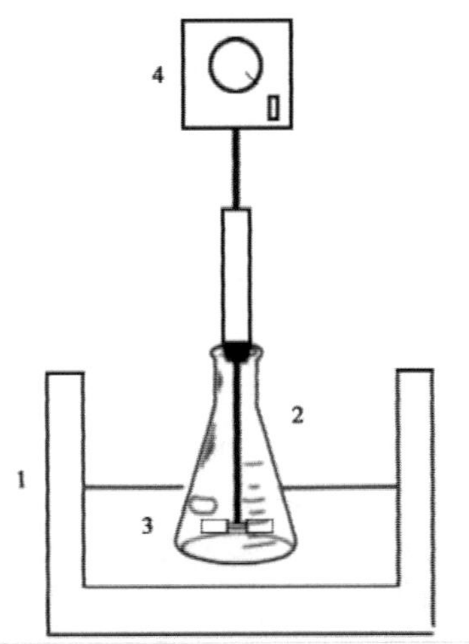

Figure 1. Apparatus for biodiesel production (1.ultrasonic cleaner; 2. Erlenmeyer as reactor; 3. Water as transmition medium; 4. Mechanical stirer)

2.2. *Catalyst preparation*

100 grams of limestone were washed by deionized water to remove dust and impurities, and then dried overnight in an oven at 105°C. The acquired limestone as well as commercial $CaCO_3$ were crushed and sieved, and then calcined at 900°C for 1 ½ hours. The characterization of the limestone and catalyst were analyzed by The X-ray diffractometer (XRD) for knowing the composition and Scanning Electron Microscopy (SEM) to discover the size of crystals formed.

IOP Conf. Series: Journal of Physics: Conf. Series **877** (2017) 012037 doi:10.1088/1742-6596/877/1/012037

2.3. Cooking oil preparation

Cooking oil first is tested free fatty acid (FFA) composition. The FFA value must be under 0.5%. If the certain condition did not achieve, the esterification had to be done in order to decrease the FFA value until reached under 0.5%[1,14]. The esterification reaction was carried out in a three-neck glass batch reactor equipped with a condenser and a thermometer. 800 ml cooking oil was mixed with 10 ml sulfuric acid and methanol by mole ratio to FFA 40:1, which all of the material had already been heated to 60°C. The reactor was placed in a water bath and heated on a hotplate. The agitating speed was kept at 600 rpm to ensure the efficient mixing. The reaction kept going for 2 hours.

2.4 *Biodiesel production*

The transesterification reaction was carried out in a batch reactor equipped with a propeller. The reactions were performed using Branson (USA) ultrasonic processor (28 kHz and 42 kHz) with a full power of 200 W. 100 ml of triglycerides were filled in reactor. Methanol was added in reactor at mole ratio 3:1 -15:1. Amount of CaO catalyst was loaded in reactor with concentation 1 -4.4% (w/w). A mixture was heated until 60°C. Ultrasonic radiation was setted in reaction time 45 minutes. After the transesterification reaction, the catalyst was removed from the mixture by filtration. The filtrate was kept in the filter paper for 24 h or more in 70°C oven. The filtrate was decanted until glycerol (lower layer) and methyl ester (upper layer) by using separation column. FAME product was analyzed with Gas chromatography (GCMS) for composition of methyl ester.

2.5. Catalyst characteristic

The CaO catalyst products were analyzed of characterization of morphology and crystallite. The analysis of crystallography was used x-ray diffractometer, XRD-7000S model Shimadzu brand with X-ray tube target Cu, voltage 30 kV, current 30 mA, and Kα radiation. XRD data was analyzed with PCXRD software. Morphology of catalyst was analyzed with JEOL PC Scanning Electron Microscope (PCSEM) model JSM-6510LA with magnification x5000. The analysis process was conducted in Center of Research and Service Diponegoro University (CORES DU).

3. Result and discussions

3.1 *Characterization of CaO catalysts*

The characterization of CaO catalysts and limestone about morphology like presented in Figure 2. Figure 2.a. is a photograph a limestone that a fragile material. Limestone was calcinied in 900°C temperature at 1.5 hours. This process affected in structure and morphology, because compounds was convert from $CaCO_3$ to CaO and $Ca(OH)_2$ convert to CaO and H_2O[11-12]. Morphology of CaO like presented in Figure 2.b. CaO catalyst has more crystaline than limestone. Cao catalyst was not rigid and had surface area more than limestone.

a. Limestone

b. CaO catalyst

Figure.2. Photographs of limestone and CaO catalyst through Scanning Microscope

XRD results of limestone and CaO catalyst was presented Figure 3. The color of catalyst products commonly has white. The calcination processing don't cause change of color Figure 3. also indicate that physical treatment influence in preparation of CaO catalyst. XRD patterns have peaks as similar in 2 theta degree 18. Intensity of limestone pattern has peak lower than CaO catalyst. This indicates Caocatalyst has crystallinity more than limestone. Physical treatment with calcination process affected appearance of peak in 2 theta degree 32,38, 68 and 80. This peak shwon an new compunds likes calcium oxide (CaO). CaO was form calcination calcium carbonate and calcium hydroxide[7,11-12]. The circle signs in Figure 3 as indicate CaO.

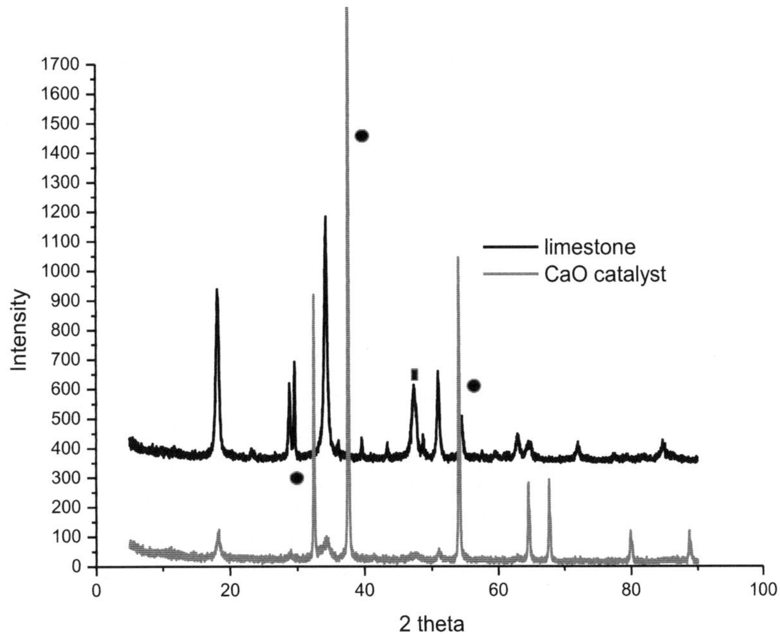

Figure 3. XRD pattern of limestone and catalyst product

3.2. Effect of ultrasonic frequency

The results of the frequency effect like presented in Fgure 4. Ultrasonic frequency has a significant effect in biodiesel forming. Biodiesel processing that use of a greater frequency of 42 kHz produces methyl ester conversion of 89.98 % and 28 kHz frequency jusat 64.68 %. This is due to the influence of ultrasound on the reaction can be described by the Arrhenius equation. Frequency influnce in power that generate ultasonic wave. So, power can affected in activation energy on transesterification reaction. Widayat et.al (2015) describe that activation energy as ultrasonic power function (P). Effect of ultrasonic can be described by quadratic equations [5] as follows:

$$Ea = a.P^{0.9} + c \quad (1)$$

In this equation Ea is activation energy, P as ultrasonic power and a,c as constante

The activation energy is the minimum energy needed for certain chemical reactions can occur. Increasing the ultrasonic frequency is proportional to the increase in power , where the greater power of ultrasonic cavitation bubbles can form more. The existence of cavitation bubbles may affect the particles that damage the walls will decrease the value of the activation energy [5].

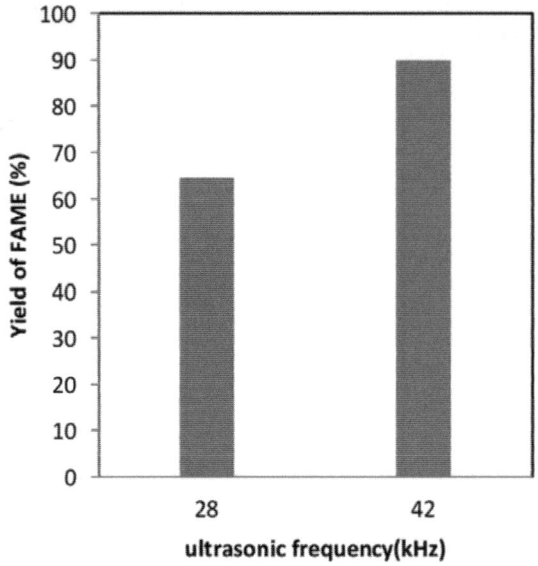

Figure 4. effect of difference frequency on transesterification on cooking oil

3.3. Effect of mole ratio methanol to triglyceride

The influence of mole ratio methanol to triglyceride in biodiesel production with resphonse variable yield of biodiesel was presented in Figure 5. Increasing moles of methanol occurs on the ratio 3: 1 to 9 : 1, and the higher obtained yield of biodiesel 89.98 %. After that, yield of biodiesel was decrease. However, the addition of further methanol does not increase yield biodiesel, Yield of biodiesel has tends to decrease when mole ratio of methanol to oil change to 12: 1 and 15: 1. This is can be occur because by-product of glycerol formed largely will react with methanol excess. Furthermore, it hinder reaction of methanol with catalyst. In mechanism of biodiesel production methanol was react with catalyst and formed methoxyde compounds (catalyst). In this research methanol react with CaO catalyst form Camethoxide ($Ca(CH_3O)_2$) [7-8]. Thereby disrupting separation of glycerine is then lowered conversion to shift the equilibrium towards the opposite (reactants) [9]. In this research We obtain the best ratio of methanol to oil is 9:1. Widayat et.al reported biodiesel production can obtained maximum at ratio methanol to oil 10:1[1;2]. So this research more better than previous research, because can save methanol as reactant.

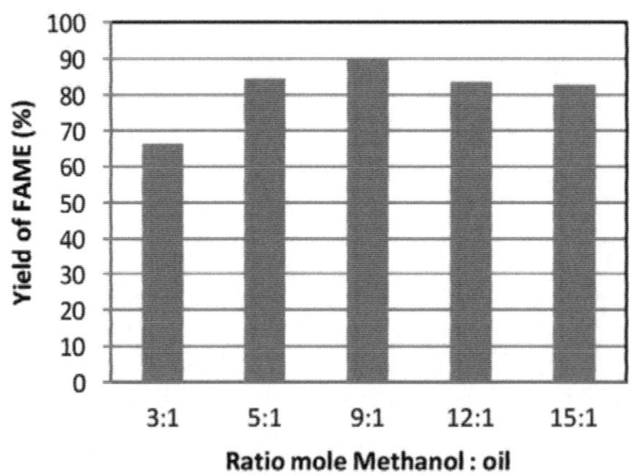

Figure 5. effect of mole ratio on transesterification on cooking oil

3.4. *Effect of catalyst concentration*

The effect of the catalyst concentration presented in Figure 6. Yield of FAME could be analysed by varying catalyst concentration of 1- 4.4% wt. The study was conducted with the help of ultrasonic systems while stirring using a propeller. Figure 6 shown yield of FAME was increase along with the increased amount of catalyst loading. This experiments similar with Chen et.al (2014) [15]. Concentration of catalyst was increased from 1% wt. to 2.6% wt., yield of FAME produced gradually increased and reached the best results at 2.6% wt with yield of 89.98%. However, excess addition of a catalyst in the sample (above 2.6 wt%) will reduce the yield produced. This is due to the addition of the catalyst will occur the saponification reaction. Saponification reaction is caused by $Ca(OH)_2$ which has base properties[7-8].The following is the reaction of the saponification:

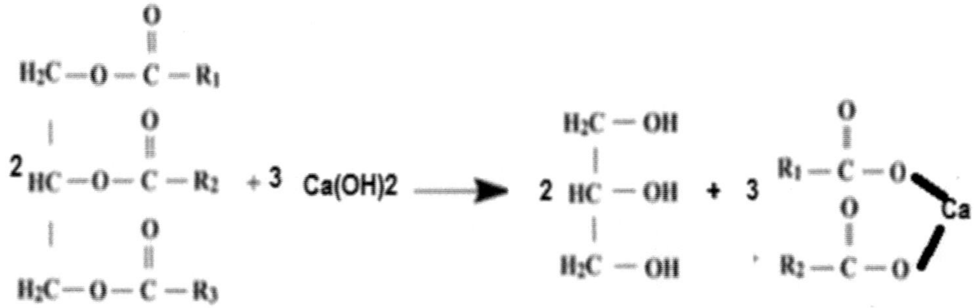

Due to the formation of soap (saponification), it made the solution too viscous so that it would be difficult to dissolve and reacting [15].

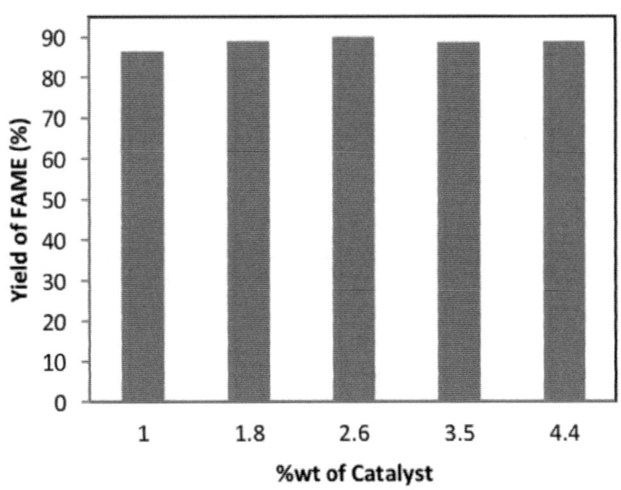

Figure 6. Yield amount of FAME with variation of catalyst concentration (ratio of metanol : oil = 9 : 1)

3.5. *Catalyst reusability in biodiesel production*

The reusability of CaO catalyst was examined by varying the reaction cycles. Before reuse, the spent catalyst was washed with n-hexane to remove the absorbed materials and dried overnight in an oven at 105°C. Figure 7 shows yield of FAME after reuse of the catalysts in the magnetically stirred system combined ultrasonic system. After the 3th cycle of transesterification, yield of FAME was already 66%. This results of reuseability was diference with Chen et.al (2014) that has 8th cylce of transesterification reaction[16]. The first cycle, yield of FAME reach 89.98%, second cycle yield of FAME is 68.96%, and third cycle yield of FAME is 66.0%. Each experiment generates decreases result in methyl esters composition. This is caused by the formation of $Ca(OH)_2$ and diglyceroxide.

$Ca(OH)_2$ formed by the hydration reaction of CaO that can lower the catalytic activity. Meanwhile diglyceroxide also react with calcium by forming Ca-glyceroxide for their incorporation byproducts are glycerol and CaO catalyst[9]. Ca-glyceroxide can still work as a catalyst for the transesterification, but has a lower catalytic activity than CaO [10]. The decreasing of catalyst activity can also be caused due to sensitivity CaO to CO_2 in any trial process [11] as well as leaching of CaO for reuse [9]

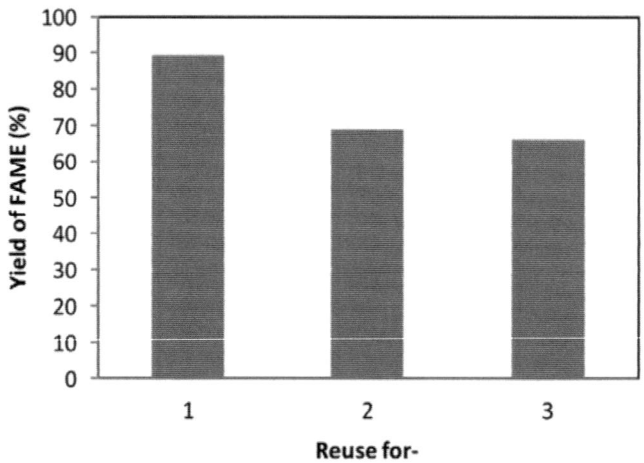

Figure 7. Reuseability of CaO catalyst

4. Conclusions

CaO catalyst was prepared from limestone with calcination process at 900°C and 1.5 hours. Yield of biodiesel /FAME has 98.89% that obtained at mole ratio methanol to oil 9 : 1 and 2.6%w catalyst concentration. The addition of methanol would cause excess methanol reacts with glycerol to inhibit the reaction of methanol and reactants, as well as reversing the direction of the reaction. the best frequency of ultrasonic frequency is 42 kHz.

References

[1] Widayat W, Satriadi H, Nafiega FN, Alimin AJ and Ali MFM 2015 *AIP Conference Proceedings*, **1699, 030020**

[2] Widayat W, Satriadi H, Yuariski O and Murwono D. 2013 *Research Journal of Applied Sciences, Engineering and Technology*, **6 (10),** 1732-1739

[3] Widayat W, Wibowo ADK, Hadiyanto and Jos B 2012 *International Review of Mechanical Engineering (IREME)* **6 (7),** 1601-1608

[4] Widayat W, Wibowo ADK and Hadiyanto 2013 *Energy Procedia* **72,** 64-73

[5] Widayat W, Aulia H N, Hadiyanto and Sasongko S B 2015 *Journal of Engineering and Technological Sciences* **47** pp. 374-388.

[6] Widayat W, Wicaksono AR, Firdaus LH and Okvitarini N 2016 *IOP Conference Series: Materials Science and Engineering*

[7] Hadiyanto H, Lestari SP and Widayat W 2016 *Bulletin of Chemical Reaction Engineering and Catalysis* **11(1)** pp. 21-26

[8] Hadiyanto H, Lestari SP, Abdullah A, Widayat W and Sutanto H 2016 *International Journal of Energy and Environmental Engineering*

[9] Obadiah A, Swaroopa G A, Kumar S V and Jeganathan K R 2011 *Bioresource Technology* **116** pp. 512-516

[10] Granadosa M L, Alonsoa D M, Sádabaa I, Mariscala R and Ocón P 2009 *Applied Catalysis B: Environmental* **89** pp. 265-272.

[11] Kouzua M, Kasuno T, Tajika M, Sugimoto Y, Yamanaka S and Hidaka J 2008 *Fuel* **87,** pp. 2798-2806.
[12] Tanga Y, Menga M, Zhanga J and Lub Y 2011 *Applied energy* **88** pp. 2735-2739.
[13] Leach, B E, 1983. Applied industrial catalysis, Volume 2. Michigan: Academic Press.
[14] Ma F and Hanna MA 1999 *Bioresource Technology* **70** pp.1–15
[15] Salamatinia B, Mootabadi H, Bhatia S and Abdullah A Z 2009 *Fuel Processing Technology* **91** pp. 441–448
[16] Chen G, Shan R, Shi J and Yan B 2014. *Bioresource Technology* **171** pp. 428-432

Heat exchange studies on coconut oil cells as thermal energy storage for room thermal conditioning

I M Sutjahja[1], Widya A Putri[1], Z Fahmi[1], S Wonorahardjo[2], D Kurnia[1]

[1] Dept. of Physics, Faculty of Mathematics and Natural Sciences, Institut Teknologi Bandung
[2] Building Technology Research Division, School of Architecture, Planning and Policy Development, Institut Teknologi Bandung

inge@fi.itb.ac.id

Abstract. As reported by many thermal environment experts, room air conditioning might be controlled by thermal mass system. In this paper we discuss the performance of coconut oil cells as room thermal energy storage. The heat exchange mechanism of coconut oil (CO) which is one of potential organic Phase Change Material (PCM) is studied based on the results of temperature measurements in the perimeter and core parts of cells. We found that the heat exchange performance, i.e. heat absorption and heat release processes of CO cells are dominated by heat conduction in the sensible solid from the higher temperature perimeter part to the lower temperature core part and heat convection during the solid-liquid phase transition and sensible liquid phase. The capability of heat absorption as measured by the reduction of air temperature is not influenced by CO cell size. Besides that, the application of CO as the thermal mass has to be accompanied by air circulation to get the cool sensation of the room's occupants.

1. Introduction

In big cities in Indonesia, the highest electrical energy consumption, i.e. above 50% from the total electrical energy consumption, are in the household and building sectors, and in particular it is used for air conditioning (AC) system [1]. This is related to the fact that Indonesia is a tropical country located around the equator, and thus making relatively high the average air-temperature throughout the year [2]. However, it is already commonly known that the use of AC system have negative impacts both directly to the human and also indirectly to the environment due to production of CO_2 gas that can cause local and global environment warming effects [3].

As a passive strategy to reduce the use of AC system in order to achieve the human comfort temperature in the room, one may use thermal energy storage (TES) system. Based on the change in the medium temperature TES system work based on sensible and latent heat concepts, where in the sensible TES there is a change in medium temperature during heat absorption or heat release while latent TES work at constant temperature [4]. The conventional thermal mass system such as wood, brick and concrete wall that work based on sensible heat storage have relatively low effectiveness related to its heat storage capacity and temperature change during heat absorption or heat release. For that reason, thermal mass system based on phase change material (PCM) offer more advantages due to the fact that PCM show superior heat storage density at temperatures around its phase transition without significant change in its temperature [5]. Among many potential PCM, organic PCM from coconut oil (CO) is very comfort and suitable for application in Indonesia due to its characteristic of human and environmental friendly and its abundant amount. In previous references it have been reported the potential application of CO as the medium of latent thermal mass system, namely assessed from physical parameters [6] and capability to lowering the air temperature [7] or water temperature [8] depending on the environment where CO was placed.

In this paper we study the heat exchange of CO in the solid, solid-liquid, and liquid phases by means of temperature measurements arranged in the core and perimeter parts of the CO, by taking the case of heat absorption process. We argue that since CO has relatively low thermal conductivity, its optimum capability to absorb and release the heat might depend on the cell size. The end part of this paper will show how is the cell size of CO might have the role to the capability for the heat absorption of CO to reduce the air temperature in the thermal chamber.

2. Experiment

The coconut oil (CO) was purchased from local supermarket in Bandung, Indonesia. The measurements were divided into two parts, namely the heat exchange study of CO, and how the effectiveness of CO to absorb the heat is vary by the cell size. For the heat exchange study, large amount of CO was placed in a large cylinder plastic container with diameter of 19.5cm and height of 13cm. The system was equipped with 3 termometers placed in some parts of CO, namely 1 termometer at core and 2 termometer at perimeter, at distance one-third of its diameter. The vertical positions arrangement of termometers are described in table 1.

Table 1. The horizontal and vertical positions of thermometer in the cylinder container.

Thermometer Number	Horizontal position	Distance from bottom (cm)
I	perimeter	11.5
II	core	6.2
III	perimeter	8.8

The overall system was placed in a thermal chamber with its temperature can be regulated for cooling and heating, and that the temperatures of the system are directly monitored during the solidification and melting process. The measurements were repeated for two times and the results show good agreement of the data. Before the measurement, the thermometers were calibrated with the procedure described previously [9].

For the effectiveness study, the CO specimen was placed in some cylindrical containers with different cell size that has the same diameter and height, namely 8×8 cm, 2×12cm and 1×16cm for the same total mass of about 3kgs (table 2). The CO' temperature during heat absorption was monitored by a thermometer placed at the core part of each CO cell. The CO was initially cooled so that its temperature is about 10°C. For this measurement, the heater system in the thermal chamber was turned on so that the air temperature reaches the value higher than 45°C before turned off. The effect of CO was studied by two kinds of air temperature measurement: dry bulb thermometer (DBT) and wet bulb thermometer (WBT). While the DBT represent the normal air temperature, the difference between them is used to represent the relative humidity of the environment.

Table 2. The number of CO cell, cell size, and mass of each CO cell for the effectiveness study.

Cell size (cm)	The number of CO cell	Mass of each CO cell (kg)
8	8	0.37
12	2	1.46
16	1	2.96

3. Results and Discussion

The result for heat absorption of CO or equivalently the melting process is described in figure 1. Start from low temperature solid phase, the heat absorption is dominated by sensible heat (phase I) as shown by rather large temperature change. In this region, the temperature value shown by thermometer I show the highest value due to the fact that it is closer to the air environment as the heat source, followed by thermometer III and thermometer II. As the fact, the temperature value shown by thermometer II which is deepest in vertical position initially constant before rising with the time. We argue that the heat exchange is dominated by heat conduction directed vertically rather than horizontally, due to the fact that the plastic container might block the heat from environment to PCM CO.

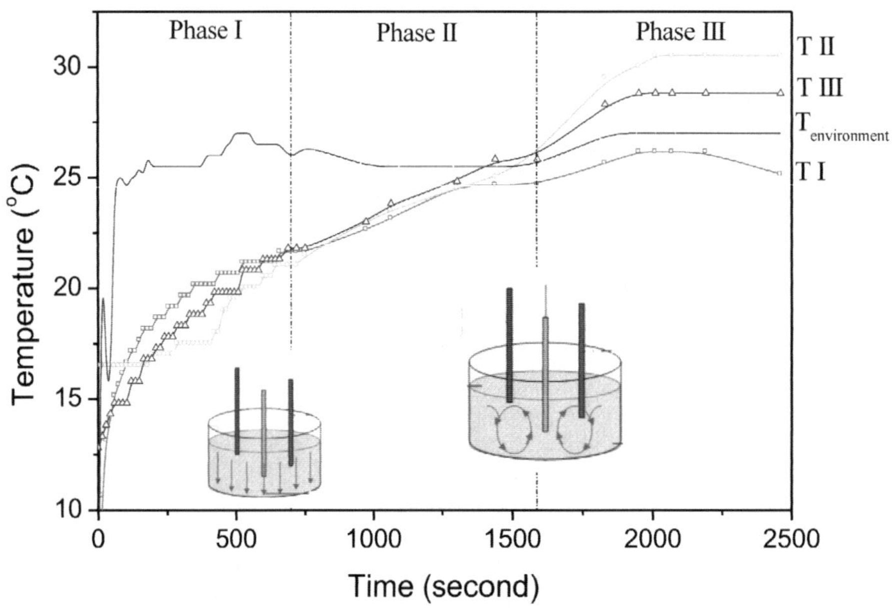

Figure 1. The temperature-time dependent of heat absorption of PCM coconut oil (CO) during melting process, with the position of thermometer I, II and III are described in table 1. The phase regions are described based on the temperature profiles from each thermometer. The inset show the heat exchange process at each phase.

The next region is the latent heat related to solid-liquid phase transition (phase II). Theoretically, this process occur at a constant temperature. However, due to the fact that CO consist of many kinds of fatty acids with more than 50% lauric acid, the solid-liquid phase transition occur with the relatively small temperature change. During this process, the temperature values shown by all thermometers almost the same. Following latent heat absorption is sensible liquid (phase III) characterized with temperature increase upto higher than air temperature for termometer II and thermometer III that are far from the CO'surface as the border between air environment and PCM CO. As the matter of fact, at this two phases, the thermometer I always show the lowest value compared with those of thermometer II and III, which have the values alternately. It seems that CO with characteristic of low thermal conductivity have the property of storing the heat, and once stored the accumulated heat will concentrated at the core and deepest part of CO (part around thermometer II). The heat exchange for this two phases is convective heat (as shown in the lower part of figure 1) with the horizontal component is contribution from the heat exchange through the plastic container.

The results for effectiveness study of CO is described in figure 2 and figure 3. From figure 2 one can see that the air temperature as measured by dry bulb thermometer (DBT) is almost the same for the CO with different cell size, as long as the total amounts of CO are the same. Moreover, by adding the CO, the air temperature decrease by the value of about 3°C compared with those without CO. An interesting fact is that although the effectiveness of CO is not depend on the cell size, the melting process of CO are not the same for different CO cell size. Qualitatively, we found that for CO with smallest cell size, after 4 hours heat absorption process, all part of CO have become liquid, while larger amount of CO mass can sustain up to 7 hours heat absorption, with medium cell size become melted up to 70% and largest cell size become melted up to 20%.

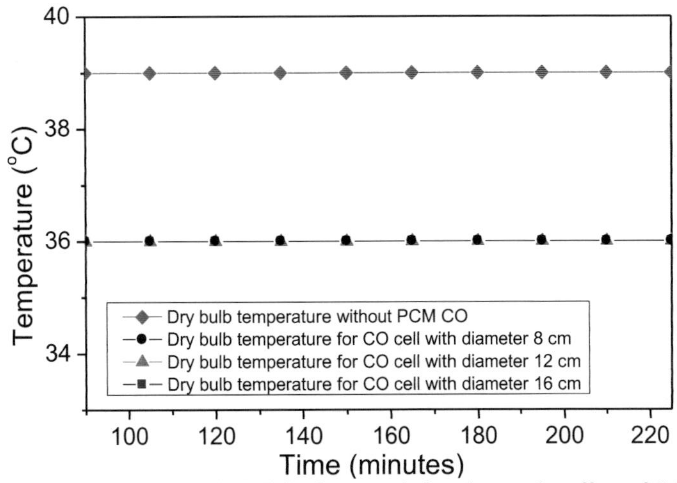

Figure 2. The decrease of air temperature (DBT) in the termal chamber as the effect of CO with varying cell size of cylinder with 8cm, 12cm and 16cm, and the same total mass of about 3kgs.

In difference with the values for dry bulb thermometer (DBT), the values of wet bulb thermometer (WBT) as shown in figure 3 show almost the same values for thermal chamber filled by PCM CO with different cells and for the thermal chamber without any PCM CO. This means that there is no change of the relative humidity of the stable air temperature condition both before and after CO work working on a thermal chamber that is not ventilated. This is in difference with aircond (AC) system that work based on air exchange that cause significant change in the relative air humidity due to that system.

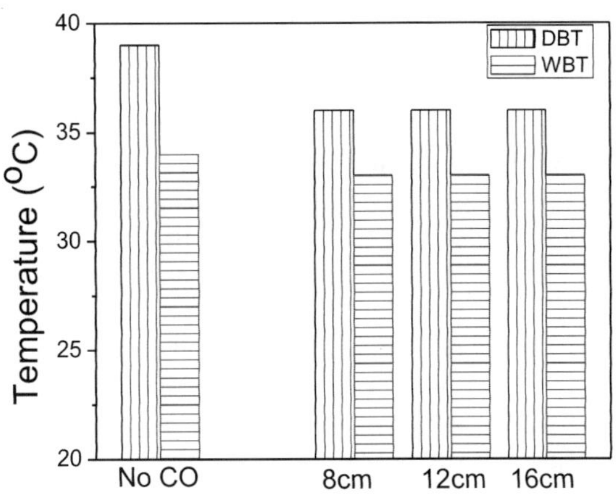

Figure 3. The air temperature as measured by dry bulb thermometer (DBT) and wet bulb thermometer (WBT) in the thermal chamber, without PCM CO and with PCM CO for different cylinder cell size of 8cm, 12cm, and 16cm.

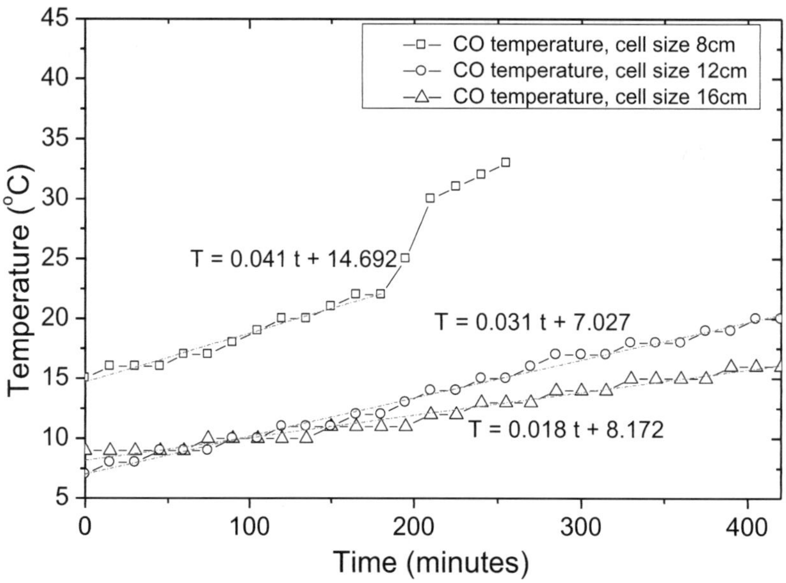

Figure 4. Temperature profiles of PCM CO at various sizes cell during the process of heat absorption in the thermal chamber. The equations show the linear fitting of the data for each CO cell.

The rate for heat absorption of CO at various sizes of cell is shown in figure 4. For 8cm CO cell, the data was taken during the time duration shorter than the others (4 hours) due to the fact that during this time interval all part of CO was melted. No clear signature for the solid-liquid phase transition might be due to high power radiation of heating for this small CO cell. For larger CO cells, namely 12cm and 16cm, for total time duration of 7hours the temperature respons of CO show linier trend with respect to the time. Fitting the linear part of the temperature rising by the time one can see that smaller CO cell have resulted in the larger slope, which might be related to the larger heat absorption rate.

4. Conclusion

We have described in this paper the heat exchange process of organic phase change material (PCM) coconut oil (CO) from the measurement of temperatures in several parts of CO during the heat absorption (melting) process. In the sensible solid, the heat exchange is dominated by heat conduction from high temperature to low temperature. During the solid-liquid phase transition and in the sensible liquid, the heat exchange is convective cooling. The solid-liquid phase transition of CO does not occur at a constant temperature, which might be due to the fact that CO consist of many kinds of fatty acid, with the largest amount of lauric acid (about 50%). With this property, CO behaves as sensible as well as latent heat energy storages. Intensive study to increase the effectiveness of heat exchange between CO and air environment through the cell size variation showed almost no significant results although it is known that CO has relatively low thermal conductivity. This suggests that the role of thermal conductivity in CO may be replaced by heat propagation through convection in the solid-liquid and liquid phases. On the application of air conditioning using thermal mass, there is no decrease in the air humidity, and therefore to get a cool sensation of the occupants, the use of thermal mass have to be followed by the air circulation degenerated from fan or ventilation opening.

Acknowledgments
The results in this research are funded by Desentralisasi DIKTI 2016 research program under contract number: 506q/I1.C10/KU/2016.

References

[1] JICA (Project Research) August 2015 *Development of Evaluation Method on Energy Efficiency and Conservation Policy (MACC)*

[2] S Aisyah, S Ahmad, H Bahadur, and S Wonorahardjo 2016 Field study on adaptive thermal comfort in office buildings in Malaysia, Indonesia, Singapore, and Japan during hot and humid season, *Build. Environ.* vol 109 p 208–223

[3] S Wonorahardjo 2012 New concepts in districts planning, based on heat island investigation *Procedia - Soc. Behav. Sci.* vol 36 p 235–242

[4] I Dincer and M. Rosen 2010 *Energy Storage System and Aplications* (Ontario Canada: A John Wiley and Sons, Ltd., Publication)

[5] Mehling, Harald and L F Cabeza 2008 *Heat and Cold Storage with PCM An Up to Date Introduction Into Basics and Applications* (Berlin, Germany: Springer)

[6] Widya A. Putri, Zulfikar Fahmi, I. M. Sutjahja, D. Kurnia 2016 Thermophysical parameters of coconut oil and its potential, *J. Physics Conf. Ser.* vol 739 p 012065

[7] E. S. Mettawee and A. Ead 2013 Energy saving in building with latent heat storage, *Int. J. Therm. Environ. Eng.* vol 5 no. 1 p 21–30

[8] Sri Rahayu A. U., Widya A. Putri, I. M. Sutjahja, D. Kurnia 2016 The effectiveness of organic PCM based on lauric acid from coconut oil and inorganic PCM based on salt hydrate $CaCl_2.6H_2O$ as latent heat energy storage system in Indonesia, *J. Phys. Conf. Ser.* vol 739 p 012119

[9] Widya A Putri, I M Sutjahja, D Kurnia, and S Wonorahardjo 2015 Potensi minyak kelapa sebagai media penyimpan kalor laten (studi kasus: analisa lepasan kalor pada proses solidifikasi), *Prosiding Simposium Nasional Inovasi dan Pembelajaran Sains 2015* p 53-56

Second Vertical Derivative Using 3-D Gravity Data for Fault Structure Interpretation

E J Wahyudi[1], Y Kynantoro[1], and S Alawiyah[1]
[1]Geophysical Engineering, Institut Teknologi Bandung, Jl. Ganesha 10 BSC-B Building, Bandung, West Java, Indonesia

Abstract. Derivative calculation of potential field for signal enhancement is common to be applied in order to provide geologic interpretation. Second Vertical Derivative (SVD) can be used to estimate fault structure and also tends to emphasize shallower geologic anomalies. In this work, we evaluate SVD results using 2-D and 3-D gravity data from two synthetic models (combination of two faults at different depth). Horizontal grid stations designed in the flat surface with 100 m interval for 2-D synthetic data calculation, while 3-D synthetic data calculation designed with addition of 50 and 100 cm in vertical levels (above each surface grid stations). Comparison SVD calculation using 2-D and 3-D synthetic gravity data conducted to perform the possibility of combined gradient interpretation proceedure. SVD from real data aplication identify 2 normal fault with (NE-SW trend) and 1 reverse fault (with NW-SE trend).

1. Introduction

Gravity method is common to be conducted in early stage of oil and gas exploration to provide regional geology information using Bouguer anomaly map. In order to delineate geology subsurface structure, derivative techniques can be applied to gravity data. The importance of the second derivative for gravity interpretation arises from the fact that the double differentiation with respect to depth tends to emphasize the smaller, shallower geologic anomalies at the expense of larger, regional features [1]. Microgravimetric and gravity gradient surveying techniques are applicable to the detection and delineation of shallow subsurface cavities and tunnels [2].

The second derivative map is also often provide better image to interprete anomalies type which are important in oil, mineral, or geothermal exploration. Some publication ([3] and [4]), described gradient interpretation techniques for gravity data in geothermal field. An example of gradient gravity application to identify fault structure and hydrocarbon prospect in oil field area was also conducted in Indonesia [5].

Based on the study of integrated gradient interpretation techniques for 2D and 3-D gravity data (Saibi et al., 2006) results, the second vertical derivative (SVD) enhances near surface effects at the expense of deeper anomalies. So, in this paper we try to understand the advantage of using 3-D gravity data from simple analysis of SVD using synthetic model. Simple fault structure model will be used to provide synthetic data and then calculated using horizontal and vertical gradient. SVD may amplifies noise and SVD can produce many second derivative anomalies that could be artificial. Using SVD analysis from synthetic study, we will describe the possibility of combined gradient interpretation proceedure in real data. More discussion of SVD application will arise in real data application using 3-D gravity data, especially in the interpretation stage.

2. Methodology

In this work, gravity data was analyzed and interpreted using horizontal and vertical derivative methods. Performance of horizontal and vertical derivative were evaluated using synthetic data. SVD calculation using 2-D and 3-D real gravity data were applied to identify fault structure. Horizontal grid stations designed in the flat surface with 100 m interval for 2-D synthetic data calculation. For 3-D synthetic data calculation designed with addition of 50 and 100 cm in vertical levels (above each surface grid stations). The ilustration of 2-D and 3-D grid station for synthetic data calculation shown in figure 1.

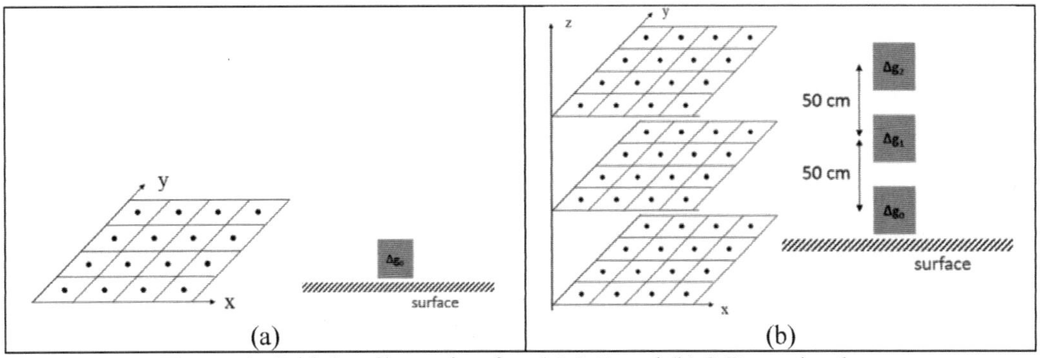

Figure 1. Grid data ilustration for: (a) 2-D and (b) 3-D gravity data.

SVD calculation using 2-D gravity data (with example of grid data ilustration in figure 1a) can be calculated using Laplace's equation. The formula for SVD calculation [1] using horizontal gradient as follow:

$$\frac{\partial^2 \Delta g_0}{\partial x^2} + \frac{\partial^2 \Delta g_0}{\partial y^2} + \frac{\partial^2 \Delta g_0}{\partial z^2} = 0 \tag{1}$$

$$\frac{\partial^2 \Delta g_0}{\partial z^2} = -\left(\frac{\partial^2 \Delta g_0}{\partial x^2} + \frac{\partial^2 \Delta g_0}{\partial y^2}\right) \tag{2}$$

SVD calculation using 3-D gravity data (with example of grid data ilustration in figure 1b, $\Delta z = 50$ cm) can be calculated using partial differential equation [6] as follow:

$$\frac{\partial^2 \Delta g_0}{\partial z^2} = \frac{\Delta g_0 - 2\Delta g_1 + \Delta g_2}{\Delta z^2} \tag{3}$$

In synthetic data analysis, we show the comparation of SVD maps between equation (2) and (3). The advantage from combination of 2 maps interpretation will be identify before we perform in real data aplication.

3. Synthetic Data Analysis

Three layers density (-0.2 g/cc, 0.0 g/cc, and +0.2 g/cc) used in synthetic model. Synthetic data calculated over (13x13) km^2 with 100 m horizontal grid interval and 50 cm vertical grid interval. Shallow fault located in 100 – 500 m depth. Deep fault located in 800 – 1500 m depth. Synthetic data calculated using forward calculation of polygon [7].

We used two synthetic models with combination of two faults at different depth (ilustration shown in figure 2). Model A as shown in figure 2a has shallow fault in Southern part (normal fault) and deep fault in Northern part (also normal fault). Model B as shown in figure 2b has shallow fault in Southern part (reverse fault) and deep fault in Northern part (normal fault). Figure 3 and 4 show the illustration of forward calculation for model A and B respectively. After synthetic data calculated as shown in figure 3 and 4, SVD of 2-D and 3-D grid data calculated. SVD maps of model A and B shown in figure 5.

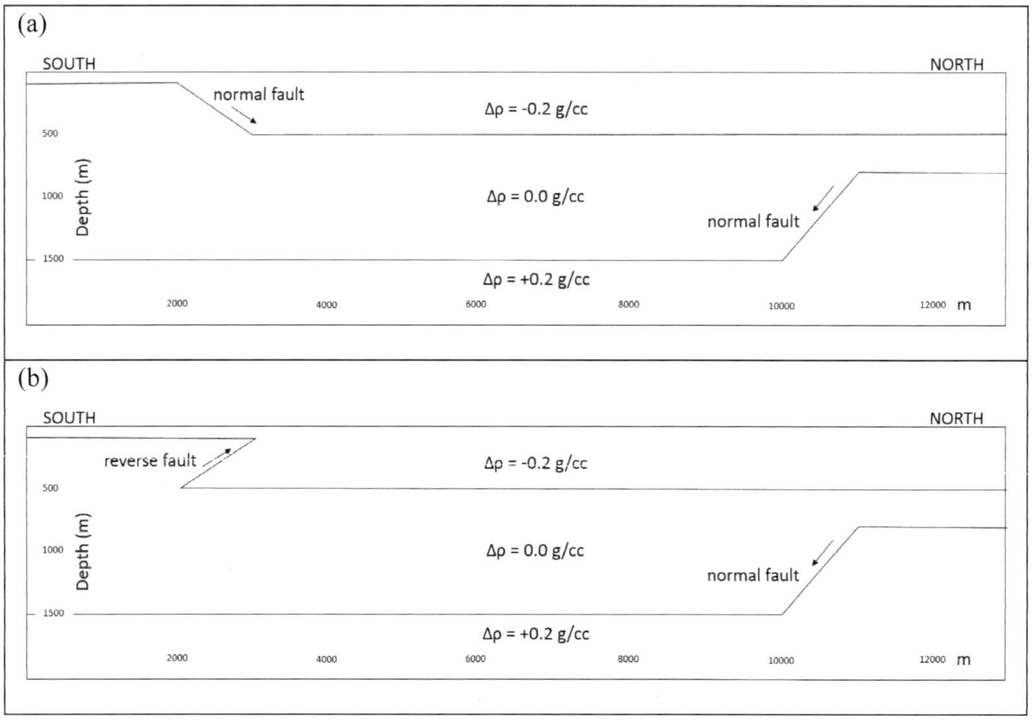

Figure 2. Synthetic model: (a) model A and (b) model B.

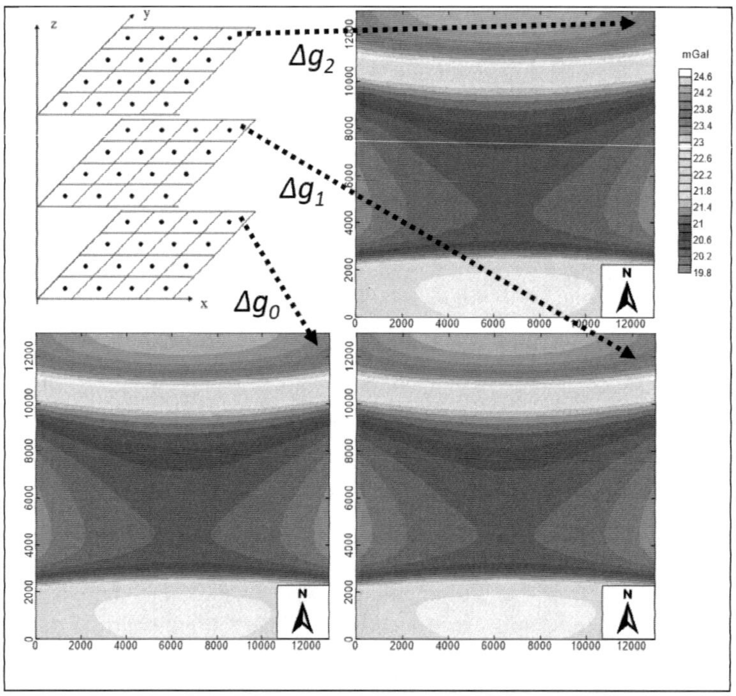

Figure 3. Ilustration for synthetic data calculation using model A.

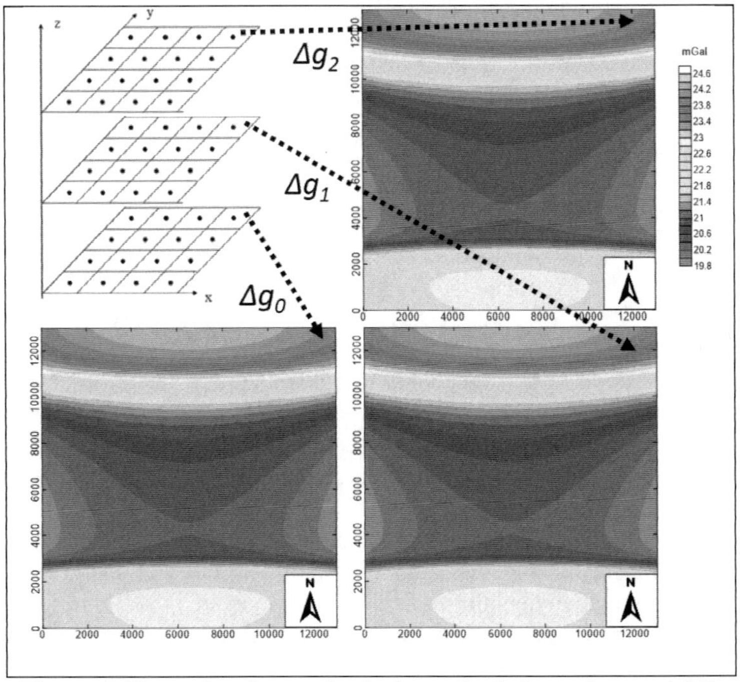

Figure 4. Ilustration for synthetic data calculation using model B.

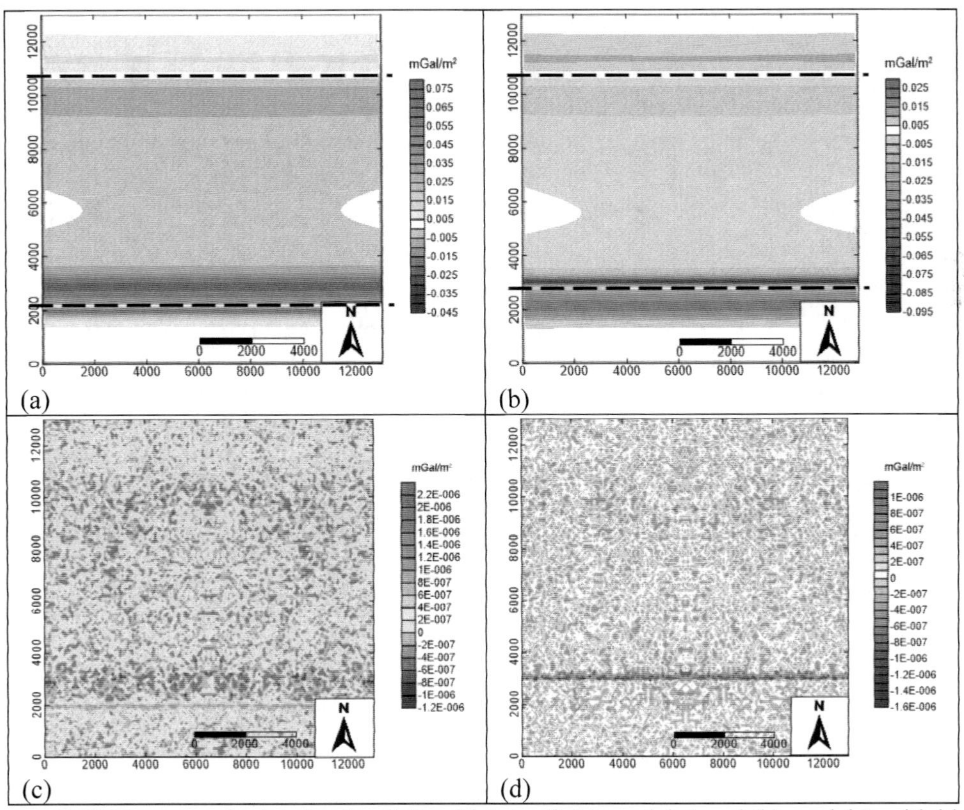

Figure 5. SVD maps calculated using 2-D grid data of: (a) model A and (b) model B with black dashed lines are interpretation of fault location. SVD maps calculated using 3-D grid data of: (c) model A and (d) model D.

IOP Conf. Series: Journal of Physics: Conf. Series **877** (2017) 012039 doi:10.1088/1742-6596/877/1/012039

Based on SVD maps calculated using 2-D grid data able to detect shallow fault and deep fault location. Fault structure location interpreted in zero SVD value that separate pairs of positive and negative SVD value. The location of fault structure drew with black dashed lines in figure 5a and 5b. From the colour bar, we can see that normal fault has |+SVD| > |-SVD|, while the reverse fault has |+SVD| < |-SVD|. We can more easy to see the difference between |+SVD| and |-SVD| in shallow structure, while in deep structure the difference between|+SVD| and |-SVD| become smaller (and still we can abserve zero value of SVD that separate pairs of positive and negative SVD value). We also observe there are zero value that artificially constructed in the Eastern and Western part of the SVD maps figure 5a and 5b. It gives us some consideration in the interpretation stage that zero SVD value not always fault structure.

Based on SVD maps calculated using 3-D grid data, we observe more chaotic patern and smaller value in colour bar of SVD maps figure 5c and 5d. In Southern part of SVD maps figure 5c and 5d, we see maksimum and minimum global respectively and align very close to the shallow fault location. Maksimum global of SVD maps in figure 5c indicate normal fault structure, and minimum global of SVD maps in figure 5d indicate reverse fault structure. Deep faults in SVD maps (figure 5c and 5d) are not visible to identify, so we understand that SVD calculation using 3-D grid data has depth limitation to detect fault structure.

Ilustration in figure 6 provide us closer look of the minimum global of SVD value. Figure 6a is ilustration of slicing in SVD map (figure 5d as this example) with AB section. SVD value in AB section then shown in figure 6d, and we notice minimum global high-lighted to help our optimization of fault response. Figure 6c shows the ilustration if we mute several colour scale to emphasize fault structure location.

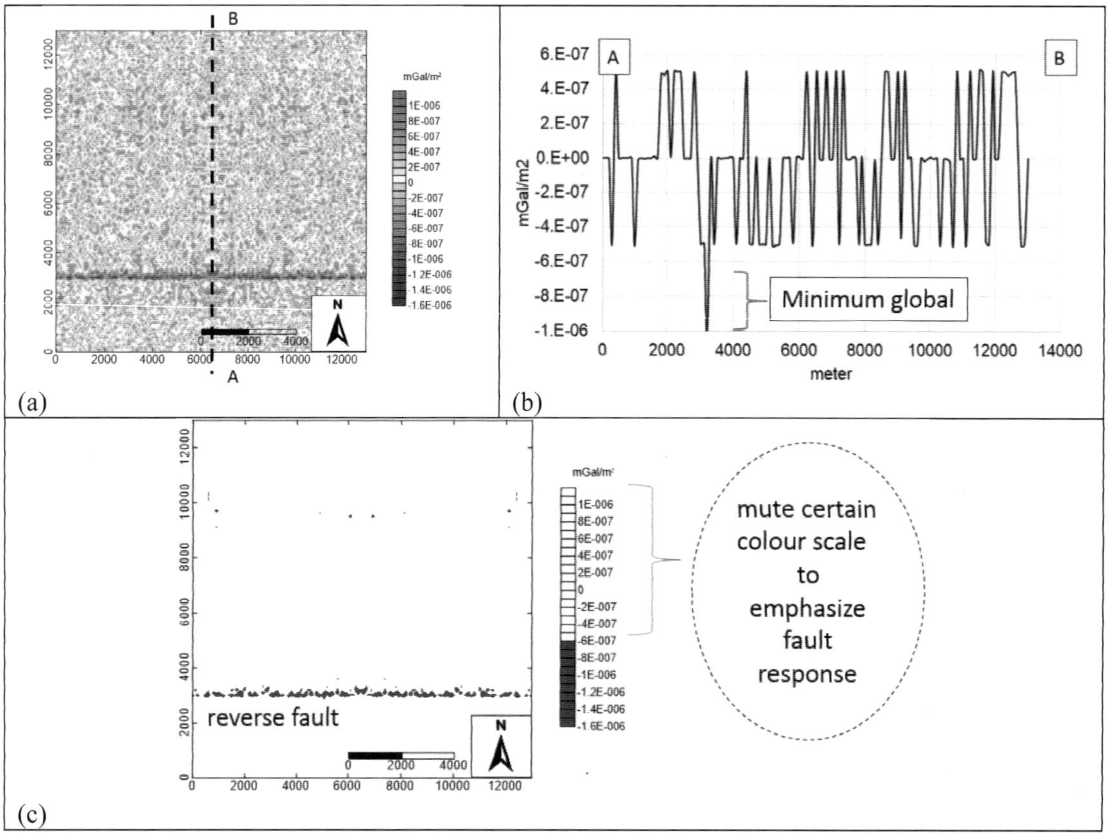

Figure 6. Ilustration of step-by-step to emphasize shallow fault structure location of SVD map (from 3-D grid data): (a) slice SVD map, (b) identify range of minimum SVD value, and then (c) we mute other SVD value.

4. Real Data Application

Study area located in East Java Province, Indonesia (figure 7a). Gravity data collected during 2014 using Scintrex CG5 with average interval grid in x-, y-, and z-axis respectively are 326, 330, and 0.476 meter. A density of 2.2 g/cm^3 (derived from Nethleton's density estimation methods) was used to produce the Bouguer anomaly map of the study area (figure 7b). Bouguer anomaly map calculated with ground surface data has range value from 77 – 105 mGal. Spectral analysis (figure 7c) provide the information of estimated shallow source ranged in the depth of 39 - 207 meters. Residual anomaly (figure 7d) map derived using moving average filtering to separate regional anomaly from Bouguer anomaly.

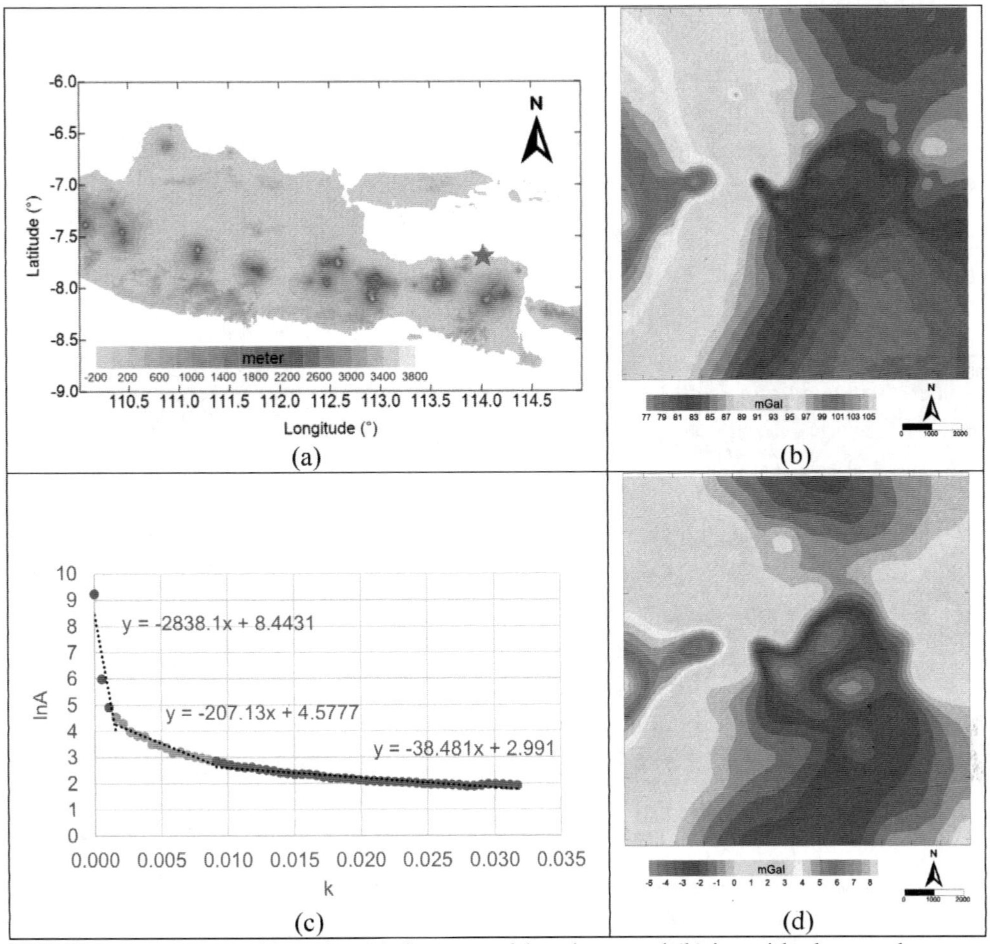

Figure 7. Study area shown in: (a) the map with red star and (b) is residual anomaly map.

The residual anomaly map, that has range value from -5 to 8 mGal, then become an input for 2-D grid data SVD calculation. From synthetic data analysis we can identify deep and shallow structure from SVD map using 2-D grid data. SVD map shown in figure 8a interpreted with six trends (drew with black lines). Three NW-SE trends are: N34°W, N37°W, and N44°W. Three NE-SW trends are: N32°E, N32°E, and N57°E.

SVD map resulted with 3-D grid data also calculated as shown in figure 8b. After we mute certain colour bar, we can observe similar trend considering maksimum and minimum global SVD value. The fault trends are easier to be conducted (as shown in figure 8b). SVD map shown in figure 8b interpreted with six trends (drew with black lines). Three NW-SE trends: N29°W, N36°W, and N38°W. Three NE-SW trends are: N35°E, N48°E, and N49°E. Integrated interpretation from figure 8a and figure 8b shown in figure 9.

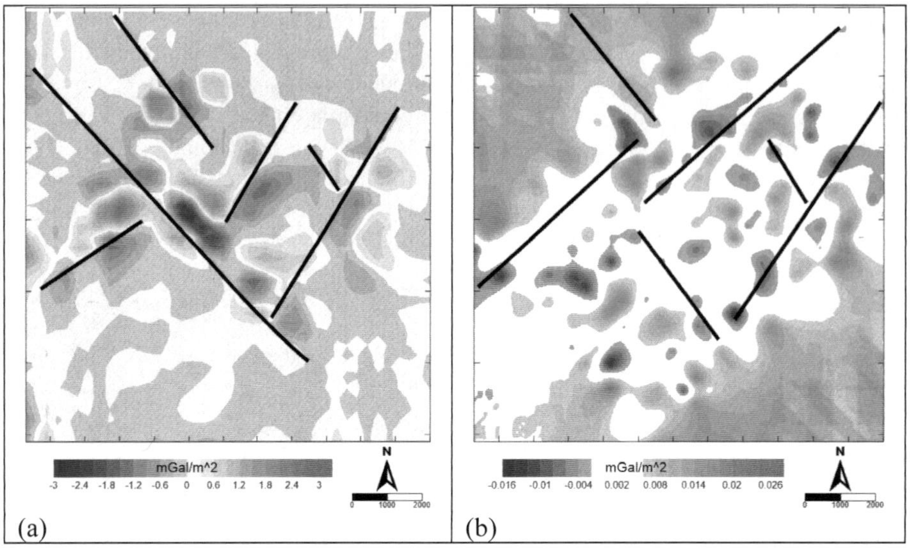

Figure 8. SVD map: (a) calculated using 2-D grid data, (b) calculated using 3-D grid data. Each SVD map interpreted with six trends.

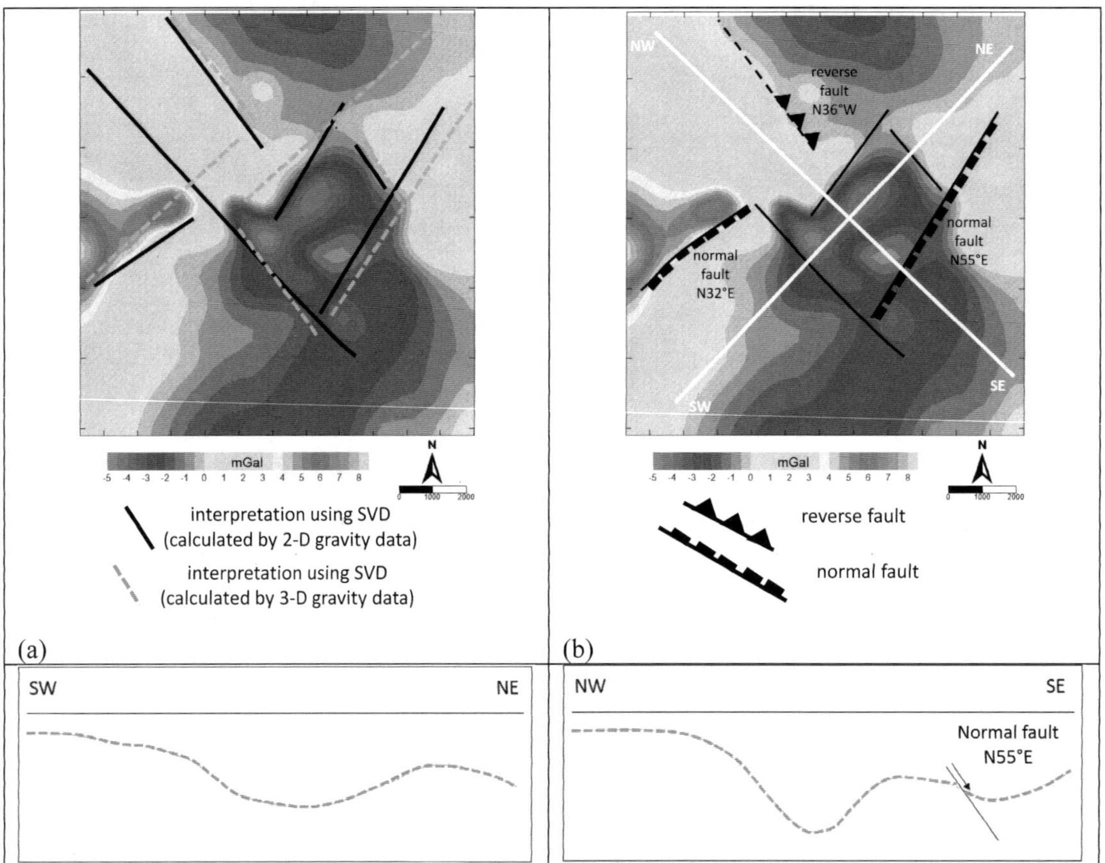

Figure 9. Integrated results from SVD calculation using 2-D grid 3-D grid data: (a) trends comparison between figure 8a and figure 8b overlied with residual anomaly map, (b) combine interpretation overlied with residual anomaly map, (c) schematic model section SW-NE, and (d) schematic model section NW-SE.

Figure 9a shows trends comparison derived from SVD map figure 8a (black lines) and figure 8b (gray dashed lines). As we can see, six of the trends relatively close to each other and we confirm the fault structure with contour patern of residual anomaly map. We optimize the interpretation for shallow structure as shown in figure 9b. We identify two normal faults with NE-SW trend in the Western part and Eastern part of study area. We also identify weaker SVD signature in the Northern part of study area as reverse fault. In center part of study area, three trends seems like circling significance low anomaly. Considering geology information in the study area, three trends in center part study area seems like to show syncline or small bawl shape. Subsurface model based on residual anomaly shown in the schematic model in SW-NE section (figure 9c) and in NW-SE section.

5. Conclusion

SVD from synthetic data calculation using 2D gravity data can be used to detect shallow fault and deep fault. SVD from synthetic data calculation using 3D gravity data can be used to optimize shallow fault location. SVD from real data aplication identify 2 normal fault with (NE-SW trend) and 1 reverse fault (with NW-SE trend).

Acknowledgements

Authors acknowledge support and discussion from Prof. Wawan Gunawan A. Kadir and FTTM ITB.

References

[1] Elkins T A 1951 The Second Derivative Method of Gravity Interpretation *Geophysics* **16** 29
[2] Butler D K 1984 Microgravimetric and Gravity Gradient Techniques for Detection of Subsurface Cavities *Geophysics* **49** 1084
[3] Saibi H, Nishijima J, Ehara S and Aboud E 2006 Integrated gradient interpretation techniques for 2D and 3D gravity data interpretation *Earth Planets Space* **58** 815
[4] Alawiyah S, Kadir W G A, Setianingsih, and Wahyudi E J 2013 *Proceeding of International Symposium Hanoi Geoengineering (Hanoi)* Well Targeting for Geothermal Field Development based on Integration of Gravity Study, Geology, and Reservoir Analysis in the TMS Field, Sulawesi, Indonesia
[5] Kadir W G A, Alawiyah S, Dahrin D, Santoso D, Widianto E, Setianingsih, Wahyudi E J and Adabi A 2013 *Proceeding of the 11th SEGJ International Symposium (Yokohama)* Identification of Fault Structure and Hydrocarbon Prospect Areas based on Integration of Gravity Anomalies, Geology and Production Well Analysis in the 'X' Oil Field, East Kalimantan
[6] Chapra S C and Canale R P 2010 *Numerical Methods for Engineer: Sixth Edition* (New York: McGraw-Hill)
[7] Talwani M, Worzel J L and Landisman M 1959 Rapid Gravity Computations for Two-Dimensional Bodies with Application to the Mendocino Submarine Fracture Zone, *Journal of Geophysical Research* **64** 49

IOP Conf. Series: Journal of Physics: Conf. Series **877** (2017) 012040 doi:10.1088/1742-6596/877/1/012040

Dielectric-function analysis of metals for plasmonic-device application

Christoforus Dimas Satrya and Yudi Darma[a]

Quantum Semiconductor and Device Laboratory, Department of Physics, Institut Teknologi Bandung, Bandung 40132, Indonesia.

Email: [a]yudi@fi.itb.ac.id

Abstract. We study the potential of various metals (Pt, Al, Cu and Ni) as plasmonic material used for devices by analyzing their complex permittivity and comparing with other metals. Metals were characterized by using high-resolution spectroscopic ellipsometry covering energy range of 0.5 to 6.5 eV. In fitting process, instead using Drude model, we used the combination of Lorentz model to describe optical properties of metals. The results show that each metal has unique different features of ε_1 and ε_2 in range of far-infrared to vacuum-ultraviolet. Also, the loss by interband transition is observable for some metals. Furthermore, the plasmonic quality-factor, which are related to electric-field enhancement and heat production generated by surface plasmon, of metal nanoparticle have been calculated and we found the optimum region of device application for each metal. From this study, Cu is promising metal working in near-infrared to visible area potentially to substitute noble Ag and Au. On the other hand, Al is the best metal to be applied as plasmonic device working in ultraviolet region. Moreover, enhancement of plasmonic quality-factor by changing geometry and environment of metal is also discussed. Our studies give an alternative of fundamental perspective for plasmonic-device development especially for energy-harvesting purposes.

1. Introduction

Plasmonic materials are recently used as complementing materials for various and advanced devices for many field application [1-2]. By definition, plasmonic materials are substances that have negative values of real-complex permittivity ε_1 within certain energy range. Metals are still prime in practice as plasmonic materials because of their high electron mobility in broad energy spectrum. They can be applied for optoelectronic devices as part of sensors, data storage, imaging, and waveguide systems [3-6]. Specially, plasmonic materials can be utilized as part of heterostructure system for energy-harvesting purposes, for solar-cell and water-splitting devices [7-11]. Devices based on plasmonic material work relied on surface-plasmon characteristic, which is an electron oscillation, at metal-insulator boundary. The electron oscillation is caused by interaction between collective electrons and incident electromagnetic (EM) wave, which produces intense heat and electric field around metal's surface [12]. Surface-plasmon behavior is divided into two-type oscillation based on its host medium in metal's structure. Surface plasmon emerges on metal slab that borders with insulator material and creates evanescent wave called surface plasmon polariton (SPP) along metal-insulator interface. This SPP, which is decay quickly, enhances considerably heat and electric field alongside metal's surface [12]. Otherwise, localized surface plasmon resonance (LSPR) is aroused when subwavelength EM interacts free electrons in metal nanoparticle structure. The collective electrons oscillate with the same frequency as that of the incident EM wave and it creates heat inside nanoparticle and electric field outside nanoparticle. During resonance condition, LSPR can produce tremendous heat and electric field. In this LSPR mode, there is no SPP wave due to existence of an effective restoring force produced by the small curved nanoparticle against driven electrons [12]. The properties of two-type surface-plasmon modes that are the energy resonance and the level of enhanced heat and electric field extremely depend on material type, geometry, size and surrounding medium of metals, and all of these factors are related with metal's dielectric function [13-19].

1.1 Plasmonic materials for energy-harvesting devices
Energy-harvesting devices based on plasmonic material exploit SPP or LSPR features coupled with other materials (e.g. semiconductor thin film) to enhance the photocatalytic process of entire system to improve its energy-conversion efficiency [7-11, 20]. For solar-cell and water-splitting devices, various metal nanoparticles that utilize LSPR's attributes are commonly used to complement oxide-metal semiconductor in order to improve the efficiency of devices. Metal nanostructures are located on top of metal-oxide semiconductor and it arranges Schottky-barrier junction (figure 1) [11, 21-22]. Gold (Au) and silver (Ag) are plasmonic materials which have been usually employed by many researchers in solar-energy-converter devices on account of their high plasmonic response in visible energy spectrum [11, 21-22]. Unfortunately, Au and Ag are high-cost materials and as noble metals they can be oxidized easily which decreases the capability of plasmonic devices, and these problems make the system inefficient for energy conversion. The enhancement of producing efficient energy in devices based on plasmonic materials is relied on plasmon-semiconductor-coupling mechanism which utilizes LSPR's properties to produce electron-hole pairs (charge separation) [23-25].

There are two-type mechanisms of plasmon-semiconductor coupling in solar-cell and water-splitting devices: first, plasmon-induced hot-electron generation (PIHE), second, plasmon-induced-energy transfer (PIET) [23-25]. As mentioned before, when EM wave interacts electrons in nanoparticle, LSPR is produced. After that, scattering and damping process of electron oscillation in metal nanoparticle produce heat inside and near-electric-field enhancement outside metal. PIHE mechanism performs when heat produced by surface plasmon transfers its energy to electrons in metal, which it generates hot electrons to surpass Schottky-barrier energy and migrate through conduction band of semiconductor. Then, hole is existing in metal and charge separation is emerged in metal-semiconductor system resulting an electric current. Another aspect that is crucial for this mechanism is the existence of electron-donor solution or hole-transporting material (HTM) surrounding around metal. HTM solution gives its electrons to metal to keeping the charge balance and sustaining the electric current (figure 1) [24].

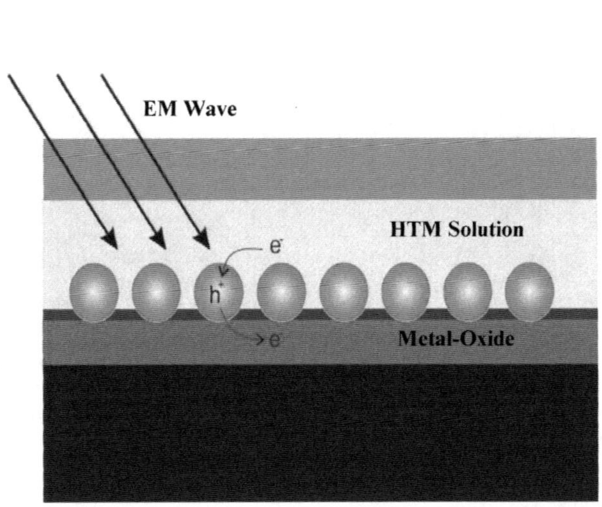

Figure 1. Solar-cell heterostructure system using metal nanoparticles on top of metal-oxide semiconductor.

For PIET mechanism, the energy from near-electric-field enhancement outside metal is harnessed to excite electrons of semiconductor in valence band moving through conduction band, and the energy transferred by PIET to electrons have to be greater than the energy bandgap of semiconductor. As result of this, charge separation exists in semiconductor and creates an electric current. To quantifying the goodness ability of metal to produce heat and electric field, Andrien Lalisse *et al.* proposed the new way to calculate accurately plasmonic quality-factor of nanoparticle; Faraday number and Joule number, which will be explained later [19].

1.2 Plasmonic quality-factor of nanoparticle

The efficiency of energy-harvesting devices or for other devices that work in plasmon-semiconductor mechanism is directly associated to plasmonic quality-factor (PQF), which measures the goodness of plasmonic response when EM wave interacts materials; devices with higher PQF will produce more efficient energy. PQF is expressed with the same frequency as that of the incident EM wave within certain energy. Andrien Lalisse *et al.* suggested two dimensionless PQF parameters as function of plasmonic material's dielectric function, which are Faraday number (Fa) and Joule number (Jo), in order to quantify the ability of a plasmonic nanoparticle to enhance the electric-field intensity and produce heat. Fa and Jo describe the properties caused by LSPR [19], while Fa describing near-field enhancement relates with PIET mechanism and heat generation is expressed by Jo which corresponds to PIHE mechanism. For nanospherical structure, Fa and Jo can be calculated by equation 1 and 2; $\boldsymbol{\varepsilon}$ is complex permittivity consisting real ε_1 and imaginary ε_2 part, n_s is refractive index of surrounding medium and E is the energy of EM wave.

$$Fa = 9.\left|\frac{\boldsymbol{\varepsilon}}{\boldsymbol{\varepsilon} + 2.{n_s}^2}\right|^2 \quad (1)$$

$$Jo = 9.\frac{E.\varepsilon_2}{n_s}\left|\frac{{n_s}^2}{\boldsymbol{\varepsilon} + 2.{n_s}^2}\right|^2 \quad (2)$$

To study the alteration of PQF by different geometry apart spherical nanoparticle, we also discuss another nanoparticle structure that is spheroid, which is divided into prolate and oblate structure (figure 2). These structures are elongated sphere with $R_x = R_y$, stretched in z-axes for prolate and in x- and y-axes for oblate, and we define r as aspect ratio that is R_z/R_x. For prolate, r is always higher than 1, otherwise, the value of r is less than 1 for oblate structure.

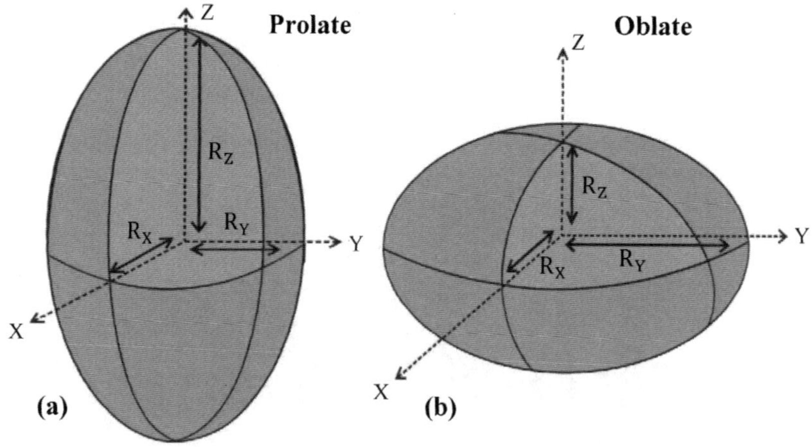

Figure 2. Spheroid nanoparticle: (a) prolate and (b) oblate structure.

Equation 3 and 4 are modified formula for Fa and Jo in arbitrary nanoparticle structures [19]. L_j (equation 5) is depolarization factor, which describes polarization of EM in *j*-axes of structure fulfilling restriction $L_X + L_Y + L_Z = 1$, and it has different equation's solution and value for specific type structure, for sphere $L_j = 1/3$. Deriving analytical solution of L_j in other particular geometries is the challenge. Fa_j and Jo_j explicate the PQF that is aroused at around nanoparticle's boundary which is parallel with *j*-axes.

$$Fa_j = \left| \frac{\boldsymbol{\varepsilon}}{n_s{}^2 + L_j(\boldsymbol{\varepsilon} - n_s{}^2)} \right|^2 \quad (3)$$

$$Jo_j = \frac{E.\varepsilon_2}{n_s} \left| \frac{n_s{}^2}{n_s{}^2 + L_j.(\boldsymbol{\varepsilon} - n_s{}^2)} \right|^2 \quad (4)$$

$$L_j = \frac{R_x.R_y.R_z}{2}.\int_0^\infty \frac{du}{(u + R_j^2).\sqrt{(u + R_x^2).(u + R_y^2).(u + R_z^2)}} \quad (5)$$

In spheroid case, because radius of x and y are same, in both axes will have the same depolarization factor and the same value of Fa and Jo, so that, only L_Z have to be calculated, then $L_X = L_Y = (1 - L_Z)/2$. The analytical approach of the depolarization factor of prolate ($L_{p,z}$) and oblate ($L_{o,z}$) in z-axes can be calculated differently by equation 6 and 7 as function of aspect ratio of spheroid [26].

$$L_{p,z} = \frac{1}{1 - r^2} + \frac{r}{\sqrt{(r^2 - 1)^3}}.\ln(r + \sqrt{r^2 - 1}) \quad (6)$$

$$L_{o,z} = \frac{1}{1 - r^2} - \frac{r}{\sqrt{(1 - r^2)^3}}.\arccos(r) \quad (7)$$

1.3 Dielectric function of metals

Determination of certain type, size and geometry of metals to be applied as plasmonic material is pivotal so as to yield the highest efficiency of energy conversion. All of these important aspects are related strictly to the dielectric function of metals. It is important to get precisely the dielectric-function data in broad spectrum from far-infrared to vacuum-ultraviolet of metals in order to understand the metals to be applied as plasmonic material in extensive application of broad energy range. Visible range is only special case for solar-energy converter devices (e.g. solar cell and water splitting). Many studies by researchers characterized various metals only in limited energy spectra while some works obtained the data in very extensive spectrum using electron-scattering spectroscopy, which was defective method for materials due to changing the structure and morphology of samples, so that, the obtained data will be slightly different compared to the real data [27-28]. Therefore, other apparatus that is non destructive and working in broad spectrum is needed.

Spectroscopic ellipsometry (SE) is a non-invasive tool to characterize materials and can be operated in energy ranging from far-infrared to vacuum-ultraviolet. By fitting process in SE-data analysis, we can obtain the both of real and imaginary part of complex permittivity which are mutually Kramers-Kronig consistent. Our group showed that SE was the best method to obtain the optical properties of materials in broad spectrum of energy and could be used precisely to determine the state of plasmon and its interaction to exciton in ZnO film [29-36]. In this work, we study the potential of various metals: platinum (Pt), aluminum (Al), copper (Cu) and nickel (Ni), as an alternative plasmonic materials compared to gold (Au) and silver (Ag) by analyzing their complex dielectric function. Metals were characterized by SE apparatus in energy from 0.5 to 6.5 eV. The obtained dielectric-function of metals was used in calculating plasmonic quality-factor for each metal in certain structure.

2. Experimental method

In order to acquire the complex permittivity of metals in vast spectrum from far-infrared to vacuum-ultraviolet, a SENTECH SE850 ellipsometer was equipped by three different light sources: UV (deuterium), UV-VIS source (Xe-lamp) and NIR source (halogen lamp from FTIR spectrometer). Attached by polarizer, this instrument provides a linear-polarized EM wave with energy ranging from 0.5 to 6.5 eV. The EM incident wave illuminated metal's surface plane at angle 50°, 60° and 70° to get information about amplitude ratio (ψ) and phase difference (Δ) between p- and s-polarized light which are detected by detector. The dispersive SE's data experiment is attributed in simple complex number ρ associating with complex reflection amplitude of p-light (r_p) and s-light (r_s) when the EM wave is reflected by metals' surface (equation 8) [37].

$$\rho = \tan(\psi).\exp(\mathrm{i}.\Delta) = \frac{r_p}{r_s} \tag{8}$$

In SE data analysis, fitting process is required to extract the complex dielectric function of metals. In this process, to match the SE's data experiment (equation 8) to calculated model, non-linear regression algorithm Levenberg-Marquardt was performed in order to get the best fitting. The goodness of fitting is judged intuitively by looking in graph how match ψ and Δ data between experiment and generated model, and evaluated numerically by MSE (mean squared error). The best value of MSE is approximately 1, and it may exhibit much larger value (>10) depending on sample's structure; thick sample and multilayer sample would result higher MSE and it is acceptable as long as the functions (data between measurement and model) are fit [38]. The MSE value can be calculated by equation 9; n is a number of EM-wave data, m is a number of model's parameters, $N = \cos(2\psi), C = \sin(2\psi).\cos(\Delta)$ and $S = \sin(2\psi).\sin(\Delta)$, while E and M indicate experimental and model data respectively [38]. CompleteEASE software was used in our fitting process [38]. Because of our simple air/bulk architectures with indifference of roughness, the complex pseudo dielectric function can be used in fitting and be computed by equation 10, θ_0 is angle of the incident wave to the line which is perpendicular to the sample's plane [37]. Then, we have to build our dielectric function model to fit with pseudo dielectric function by using equation 8 and 10.

$$MSE = \sqrt{\frac{1}{3.n-m}\sum_{\mathrm{i}=1}^{n}[(N_{\mathrm{E}i} - N_{\mathrm{M}i})^2 + (C_{\mathrm{E}i} - C_{\mathrm{M}i})^2 + (S_{\mathrm{E}i} - S_{\mathrm{M}i})^2] \times 1000} \tag{9}$$

$$< \boldsymbol{\varepsilon} > = < \varepsilon_1 > + \mathrm{i}. < \varepsilon_2 > = \sin^2(\theta_0).\left(1 + \tan^2(\theta_0).\frac{(1-\rho)^2}{(1+\rho)^2}\right) \tag{10}$$

$$\boldsymbol{\varepsilon}(E) = \varepsilon_\infty + \sum_{\mathrm{j}=1}^{n}\frac{A_j.\gamma_j.E_{0j}}{E_{0j}^2 - E^2 - \mathrm{i}.\gamma_j.E} + \sum_{\mathrm{j}=1}^{n}\frac{B_j}{E_{1j}^2 - E^2} \tag{11}$$

Instead of using Drude model, which cannot take into account absorption by interband transition, we used combination of Lorentz model to represent the optical properties of metals (equation 11). The first term is constants to describe real permittivity at infinite energy, second term is a Lorentz model and third term is a pole function which only delineates real permittivity in infrared and UV region outside data measurement [38].

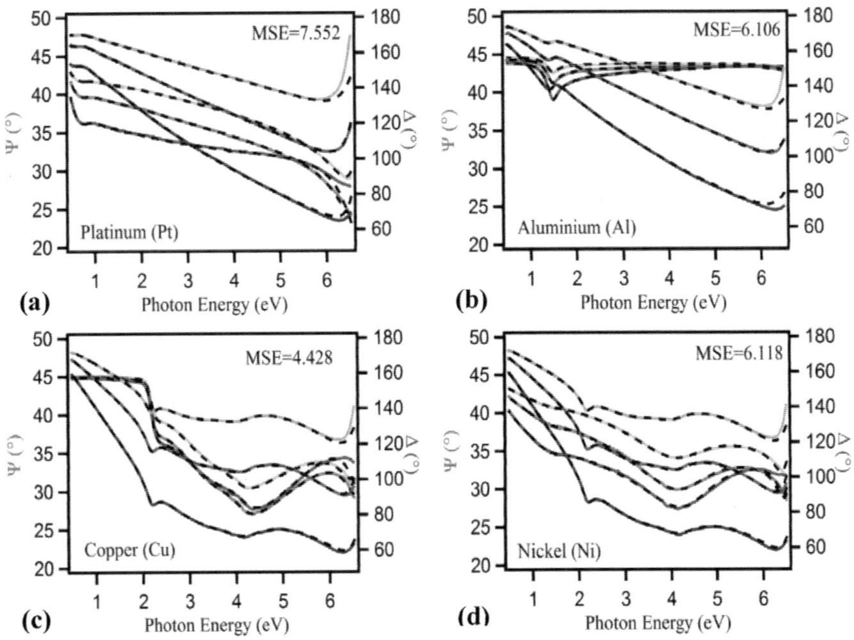

Figure 3. The matched optical dielectric model to ψ and Δ data measurement of (a) Pt, (b) Al, (c) Cu and (d) Ni.

3. Result and discussion

Figure 3 is the results of fitting process of Pt, Al, Cu and Ni with their each error (MSE). It is shown in figure 3 that our Lorentz model (black line) is fit with ψ (red) and Δ (blue) data measurement for 70°, 60° and 50° respectively in brighter gradation for almost all spectrum with MSE less than 8, and this optical model is admissible to be extracted as permittivity of metals as we can see that the model is fit enough to the measured data. The complex permittivity of all metals is depicted in figure 4. For comparison, also Au and Ag from Palik's optical constants handbook are inserted in energy range 1.2-5 eV [28].

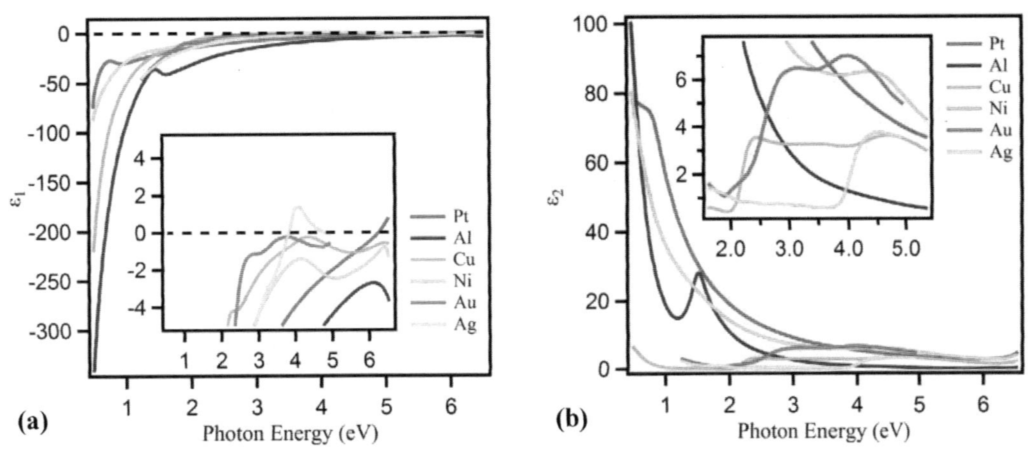

Figure 4. The extracted complex-permittivity of various metals: (a) real permittivity and (b) imaginary permittivity.

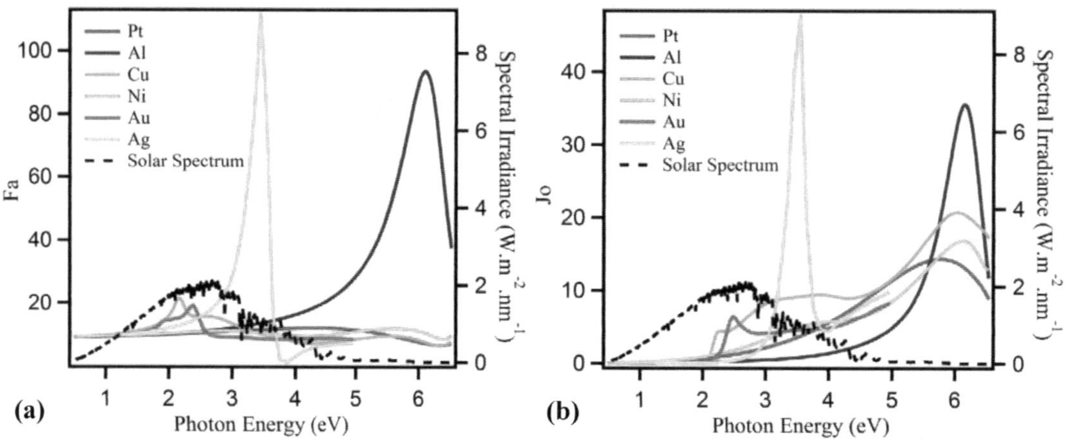

Figure 5. PQF of various spherical metal nanoparticles: (a) Faraday number and (b) Joule number.

Metals have negative ε_1 in broad energy spectrum and the patterns are unique for each metal in around $\varepsilon_1 \sim 0$, which is the key state of existence of surface plasmon that only will come up in negative ε_1 and it's value have to be small closer to zero. Cu, Ag and Au approach this condition in visible area whereas Pt, Al and Ni in UV region. In absorption ε_2 spectra, Drude response is highly observed for Pt, Al and Ni in infrared energy and is low for Cu. The low absorption in plasmonic material is critical to keep the existence of surface plasmon in condition low loss. Also, for Al, Cu and Ni, the loss by interband transition are observed.

To numerically quantifying the response of surface plasmon in nanoparticle metals, PQF for Fa and Jo in spherical geometry surrounded by vacuum ($n_s = 1$) was calculated, and it is showed in figure 5 and is superimposed with solar spectrum to understand the possibility of metals applied for solar-energy-harvesting devices. It can be seen clearly that in both for electric enhancement (Fa) and producing heat (Jo), Cu is promising metal working well in all visible-light area comparable to superiority of Ag and Au. Al is the best plasmonic material for devices working in blue-visible to vacuum-ultraviolet. In enhancing electric field, Ni and Pt seem ineffective, however they are promising as heat producer in UV region. The highest peak of each metal is a condition during LSPR resonance. Because of the high ability of Cu to collect solar energy in almost all visible spectrums, for further analysis to comprehend the alteration of PQF by changing geometry and surrounding medium, only Cu will be investigated.

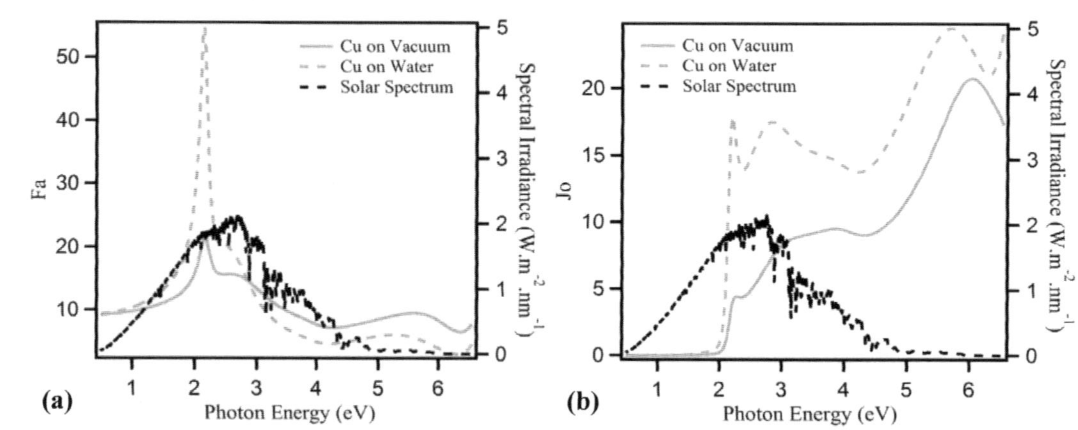

Figure 6. The improvement of (a) Fa and (b) Jo in higher refractive index of surrounding medium

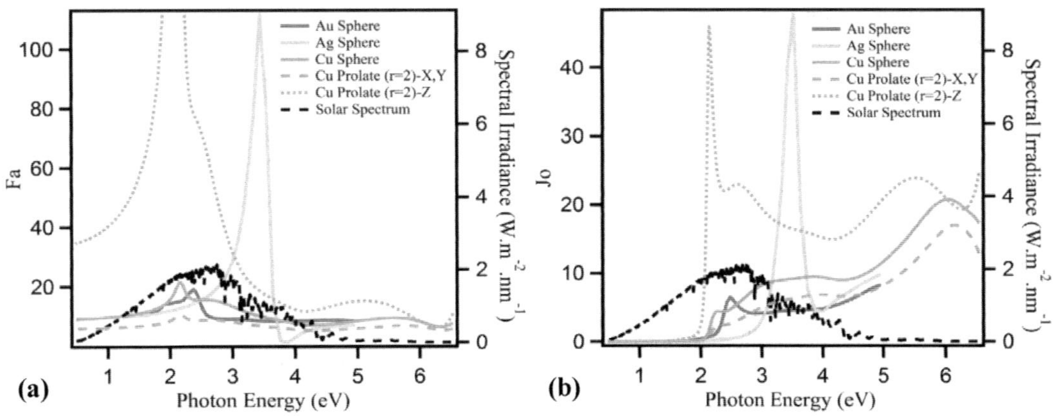

Figure 7. The enhancement of (a) Fa and (b) Jo for Cu prolate structure.

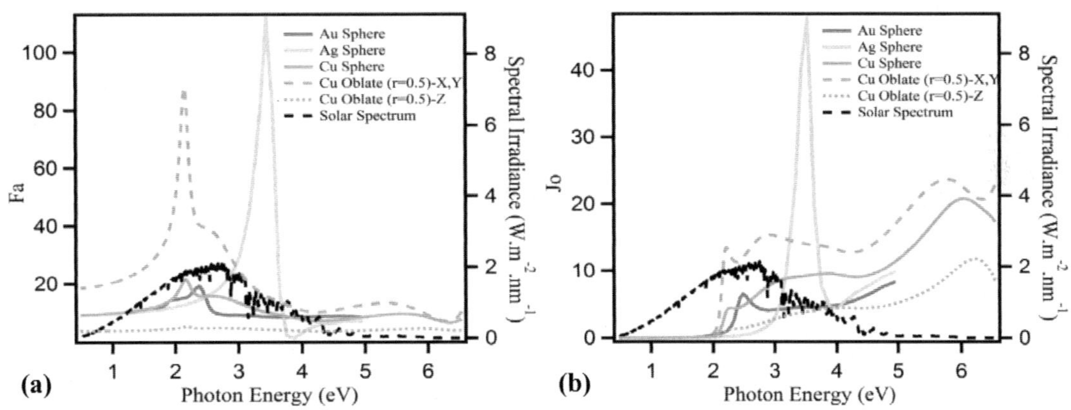

Figure 8. The enhancement of (a) Fa and (b) Jo for Cu oblate structure.

In solar-energy-harvesting devices, using plasmonic material that is responsive in visible region is important. As calculated before in vacuum condition, Cu is a great material for energy application. In fact, as we explained above, in solar-cell and water-splitting devices, the metals are surrounded by HTM solution contributing its electrons to metal to sustain electric current through the system, so that, n_s should be higher than 1. Nowadays, there is still lack of characterization of various types of HTM solution used in devices and the value of refractive index n_s for various HTM solutions is still unrevealed. To know this HTM effect to Fa and Jo, we used water (n_s=1.33) to represent refractive index of HTM. Figure 6 displays the improvement of PQF in terms of increasing refractive index from 1 to 1.33. This result is substantial for further development of plasmonic-device system which is not only to discover effective HTM solution to donor electrons but also to fabricate HTM with higher values of refractive index to increase efficiency of devices.

There are many ways to elevate PQF of metal nanoparticle, which one of them is to tune geometry of nanoparticle. Spheroid (prolate and oblate) architectures have been proposed for enhancing PQF than that of sphere. We recalculated PQF in Cu prolate structure with aspect ratio 2, and figure 7 shows considerably enhancement of Fa and Jo in z-axes of metal while it decreases in x- and y-axes. By changing Cu prolate to aspect ratio 2, we can produce PQF greater than that of Au and Ag in spherical geometry, and higher aspect ratio will improve extremely PQF of Cu. Also, the energy resonance of LSPR is shifted to red frequency. On the other hand, Cu oblate with aspect ratio 0.5 creates rising of PQF and results blue-shifted LSPR of resonance energy in x- and y-axes of metal, while lower aspect ratio of oblate will enhance greatly plasmonic quality-factor (figure 8). In device application, it is necessary to consider attached position of prolate and oblate along semiconductor to generate optimum efficiency. These results of prolate and oblate contexture give us another perspective that it is not need to fabricate high-cost plasmonic materials when we can manipulate the geometry of low-cost materials to have greater PQF.

4. Conclusion

Cooper (Cu) is promising plasmonic material with optimum region in near infrared to visible energy potentially to substitute gold (Au) and silver (Ag). Aluminium (Al) is the best metal to be applied as plasmonic devices performing in UV region. Nickel (Ni) and platinum (Pt) are impractical for plasmonic devices utilizing electric-near-field enhancement but are good metals to produce heat in UV region. The enhancement of plasmonic quality-factor of Fa (electric-field enhancement) and Jo (heat generation) of metal nanoparticle can be achieved by changing the geometry to spheorid (prolate and oblate) instead of sphere. For prolate, increasing aspect ratio will improve PQF of metal, otherwise, by decreasing aspect ratio in oblate will raise efficiency of PQF. Also, alteration of refractive index of surrounding medium around metal to higher value will lift up plasmonic response of metal. Our studies give an alternative of fundamental perspective for plasmonic-device development especially for energy purposes.

Acknowledgment

This work is supported by Riset Kompetensi 2016, Desentralisasi kemenristekDIKTI 2016 research program and RUPT 2016 from the Indonesian government. Thanks to SSLS-NUS Singapore for SE measurement facility. We thank to quantum semiconductor and devices laboratory members for supporting and to Luthfi Naufal for nice pictures and graphs.

References

[1] Murray W A and Barnes W L 2007 *Adv. Mater.* **19** 3771–3782.
[2] Dionne J A, Baldi A, Baum B, Ho C-S, Janković V, Naik G V, Narayan T Scholl J A and Zhao Y 2015 *MRS Buletin* **40** 1138-1145.
[3] Barnes W L, Dereux A and Ebbesen T W 2003 *Nature* **424** 824-830.
[4] Oulton R F, Sorger V J, Genov D A, Pile D F P, and Zhang X 2008 *Nat. Photonics* **2** 496-500.
[5] Bolduc O R and Masson J-F 2011 *Anal. Chem.* **83** 8057–8062.
[6] Wong C L and Olivio M 2014 *Plasmonics* **9** 809-824.
[7] Boriskina S V,Ghasemi H and Chen G 2013 *Mat. Today* **16** 375-386.
[8] Smith J G, Faucheaux J A and Jain P K 2015 *Nano Today* **10** 67-80.
[9] Li X, Choy W C H, Lu H, Sha W E I and Ho A H P 2013 *Adv. Funct. Mater.* **23** 2728–2735.
[10] Warren S C and Thimsen E 2012 *Energy Environ. Sci.* **5** 5133-5146.
[11] Liu Z, Hou W, Pavaskar P, Aykol M and Cronin S B 2011 *Nano Lett.* **11** 1111–1116.
[12] Maier S A 2007 *Plasmonic: Fundamentals and applications* (New York: Springer).
[13] Li X D, Chen T P, Liu Y and Leong K C 2014 *Opt. Express* **22** 5124-5132.
[14] Li X D, Chen T P, Liu Y and Leong K C 2015 *J. Nanopart. Res.* **17** 67.
[15] Gong J, Dai R, Wang Z and Zhang Z 2015 *Sci. Rep.* **5** 9279.
[16] Langhammer C, Yuan Z, Zoric I and Kasemo B 2006 *Nano Lett.* **6** 833-838.
[17] Haynes C L and Van Duyne R P *J. Phys. Chem. B* **105** 5599–5611.
[18] Zhang Q, Large N, Nordlander P, and Wang H 2014 *J. Phys. Chem. Lett.* **5** 370–374.
[19] Lalisse A, Tessier G, Plain J, and Baffou G 2015 *J. Phys. Chem. C* **119** 25518–25528.
[20] Wang F and Melosh N A *Nano Lett.* **11** 5426–5430.
[21] Nishijima Y, Ueno K, Yokota Y, Murakoshi K and Misawa H 2010 *J. Phys. Chem. Lett.* **1** 2031–2036.
[22] Wu F, Hu X, Fan J, Liu E, Sun T, Kang L, Hou W, Zhu C and Liu H 2013 *Plasmonics* **8** 501–508.
[23] Cushing S K and Wu N 2013 *Interface* **22** 63-67.

[24] César Clavero 2014 *Nat. Photonics* **8** 95–103.
[25] Ma X-C, Dai Y, Yu L and Huang B-B 2016 *Light: Science & Applications* **5** e16017.
[26] Doyle W T and Jacobs I S 1992 *J. Appl. Phys*. **71** 3926-3936.
[27] Werner W S M, Glantschnig K and Ambrosch-Draxl C 2009 *J. Phys. Chem. Ref. Data* **38** 1013-1092.
[28] Palik E D 1985 *Handbook of optical constants of solids vol 1* (New York: Academic).
[29] Darma Y, Herng T S, Marlina R, Fauziah R, Ding J, Rusydi A *Appl. Phys. Lett*. **104** 081922.
[30] Kurniawan R, Sutjahja I M, Winata T, Rusydi A and Darma Y 2015 *AIP Conf. Proc*. **1677** 070002.
[31] Nurfani E, Sutjahja I M, Winata T, Rusydi A and Darma Y 2015 *AIP Conf. Proc*. **1677** 070008.
[32] Darma Y and Rusydi A 2015 *Adv. Mat. Res*. **1112** 3-6.
[33] Marlina R, Rusydi A and Darma Y 2015 *Adv. Mat. Res*. **1112** 124-127.
[34] Nurfani E, Kurniawan R, Muhammady S, Marlina R, Sutjahja I M, Winata T, Rusydi A and Darma Y 2016 *AIP Conf. Proc*. **1725** 020057.
[35] Darma Y, Marlina R, Herng T S, Ding J and Rusydi A 2016 *IEEE Photonics Journal* **8** 3.
[36] Darma Y, Satrya C D, Marlina R, Kurniawan R, Herng T S, Ding J and Rusydi A 2017 *Jpn. J. Appl. Phys*. **56** 01AD06.
[37] Fujiwara H 2007 *Spectroscopic ellipsometry: Principles and applications* (West Sussex: John Wiley and Sons).
[38] Woollam J A 2011 *CompleteEASE^{TM} data analysis manual* (Nebraska: J. A. Woollam Co. Inc.).

Construction Performance Optimization toward Green Building Premium Cost Based on Greenship Rating Tools Assessment with Value Engineering Method

Yusuf Latief, Mohammed Ali Berawi, Van Basten*, Riswanto, Rachmat Budiman

Civil Engineering Department, Faculty of Engineering, Universitas Indonesia, Kampus Baru UI, Depok, 16424, Indonesia

*basten89@yahoo.com

Abstract. Green building concept becomes important in current building life cycle to mitigate environment issues. The purpose of this paper is to optimize building construction performance towards green building premium cost, achieving green building rating tools with optimizing life cycle cost. Therefore, this study helps building stakeholder determining building fixture to achieve green building certification target. Empirically the paper collects data of green building in the Indonesian construction industry such as green building fixture, initial cost, operational and maintenance cost, and certification score achievement. After that, using value engineering method optimized green building fixture based on building function and cost aspects. Findings indicate that construction performance optimization affected green building achievement with increasing energy and water efficiency factors and life cycle cost effectively especially chosen green building fixture.

1. Introduction

Building facility has been a human need for some activities platform such as social, economic, and environment. In the other hand, building construction has both positive and negative impacts that there are not only on construction phase but also on operational, maintenance, and demolition phase. The negative impacts of conventional building are high energy consumption which up to 40% of total global energy production, consuming 12% of the total clean water supply in the world, and spending 30% of the world resources for building construction phase. In addition, indirect negative impact of building operational and maintenance is greenhouse gas emissions production that up to 40% of total emission in the world. In 2030, this emission was predicted 30% come from countries in Asia. Therefore, these conditions should be prevented to decrease greenhouse gas emissions with some concrete steps such as sustainable environmentally development concept [1].

Building concept that concerned on sustainability environment in several countries is known as green building concept. Green building concept is the concept with friendly environment buildings that accommodated the interests of environmental, social, and economic integration aspects of buildings according to standards that it has been determined from the building life cycle plan. Some proponents of green building concepts walk with the owners needs of their building, rating tools and knowledge, government aid, and government regulations [2]. The Green Building Concept in Indonesia has been formulated in several laws and provisions such as Greenship. It was published by Green Building Council Indonesia (GBCI). Some of Government regulations that supporting green building such as Presidential Regulation Number 38/ 2015 on public-private partnership on energy conservation infrastructure, Ministry of Labor Regulation Number 80/ 2015 on national competence in energy management for industry and buildings, and Ministry of Public Works and Housing Regulation Number 02/ PRT/ M/ 2015 about green buildings[3].

The main problem of conventional buildings as the previous discussion is the greenhouse gas emissions that harm directly. The main cause of the greenhouse gas emissions is come from energy production and consumption, so that the necessary action for energy efficiency and conservation. Based on Energy Conservation Partnership Program and JICA Study (2010), construction of commercial buildings in Indonesia is still in poor pursuing energy efficiency. The largest energy is used on office buildings, malls, hotels, and hospitals (Figure 1). It is strengthened further by the low number of green building in Indonesia when compared with countries which they have implemented the concept of Green Building [4]. Based on Green Building Council Indonesia record, there are only eighteen green buildings in Indonesia (Table 1). The main reasons of the low interesting in green building implementation based on Green Building Market Report South East Asia (2014) were higher cost of application green building materials, additional time and cost for research material, training, documentation, and design stage. (Figure 2).

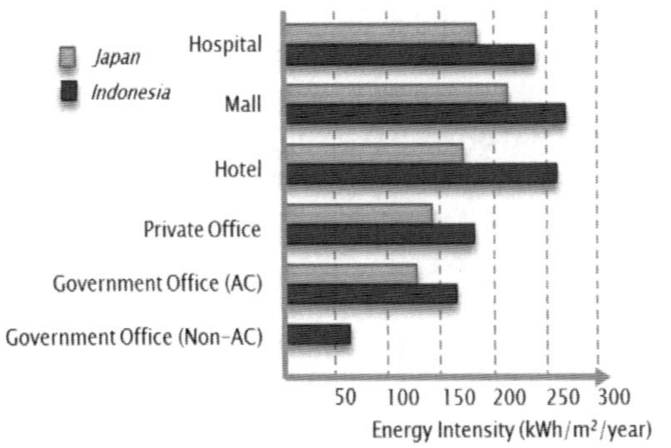

Figure 1. Building Energy Performance Compilation between Indonesia and Japan [5]

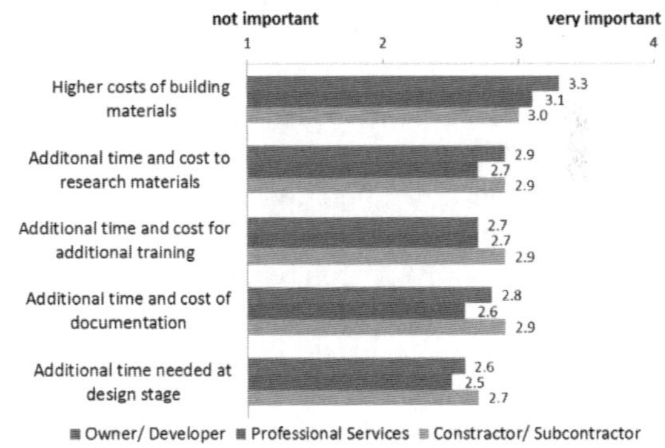

Figure 2. Reason of Low Interest on Green Concept in Indonesia [6]

Table 1. Number of Green Building in Some Countries [7-10]

Country Name	Year	Number of Green Building
US	1998-2012	13,000
Hongkong	2005-2014	1,500
Singapore	2005-2014	509
Germany	2009-2016	490
Malaysia	2009-2013	61
South Korea	2008-2012	53
Israel	2005-2013	45
Indonesia	2009-2015	18

This research was carried out to solve two major problems; there are optimum cycle cost of green building in Indonesia and performance optimization key factors on green building construction premium cost to achieve green building operation and maintenance goals. Therefore, this research was observed green building from design recognition data up to operational and the maintenance phase data as an assessment on final assessment for new and existing building. Observations were constrained by building function as office buildings in Jakarta with Value Engineering Method.

2. Green Building Construction

Building life cycle generally consists of design and material production phase, transport and construction phase, and operation and maintenance phase. Greenhouse gases and other building emissions is not only produced in the operation and maintenance phase, but also prior to this, such as in the design and material production phase, and the material transport and construction phase, which are considered to minimize the negative effect for environment [11]. Green building that minimized the negative impact of building construction has been one of the ways to get the sustainable infrastructure because it decided building operational and maintenance in the beginning of building design. Beside that, Green building concept knowledge that makes simply in monitoring, controlling, and improving of building cycle stage especially in building construction phase, could accelerate design process. After planning phase was finished, the preliminary assessment valued building plan based on green building rating tools to control construction phase sustainability [12].

3. Greenship

According to Greenship workflow, green building development has been started from initiation stage. Therefore, understanding green building concept is already required for all of the green building initiator. Green building assessment will be decided on the green building expert court to bring design recognition (DR) certification. In construction phase, building will be built according to design recognition that it needs fully integrated part of building stakeholder such as owner, consultant, and contractor. Each phase of the work will be supervised by a professional on greenship association. At the end of the construction, Green Building Council will make the final assessment based on expert court decision to determine building achievement on green building rating tools to enter operational and maintenance phase. This phase certification is named as Final Assessment (FA) certification for New Building (NB). Green building monitoring walks on the operation and maintenance phase by the building management and evaluate by green building council team yearly [13]. The final result of green building council team monitoring will be used as evidence for building certification on operational and maintenance phase that named Final Assessment (FA) certification for Existing Building (EB). Green building assessment process flow based on Greenship Assessment Tool can be drawn as Figure 3.

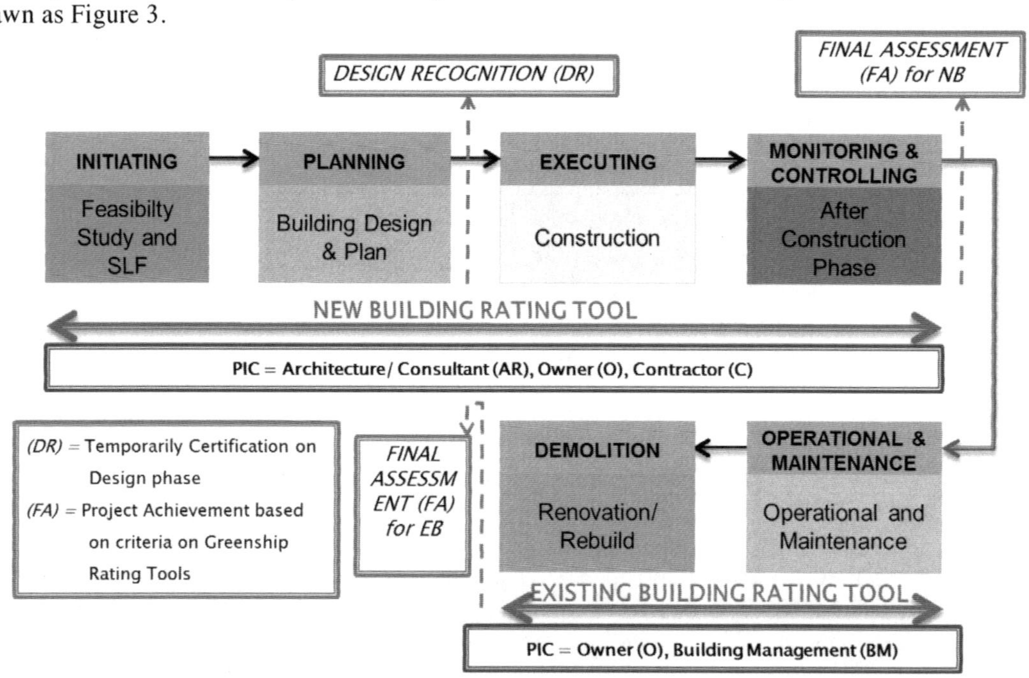

Figure 3. Flow Process on Green Building Assessment on Greenship Rating Tools [7]

IOP Conf. Series: Journal of Physics: Conf. Series **877** (2017) 012041 doi:10.1088/1742-6596/877/1/012041

4. Research Methodology

This research was conducted by observation of green building design innovations that it already has obtained Final Assessment Certificate. After that, value engineering method process was begun with information and function analysis phase. Value Engineering (VE) is an effort with systematic and organized system function, production, and service to reach essential function of life cycle cost [14]. The essential function is the optimization result of consistent performance, reliability, quality, and safety implied [15, 16]. This study begins with data collection from green building in Indonesia to benchmark achievement of aspects that was valued with greenship (green building rating tools in Indonesia). After that, creative phase began the new model that was illustrated by Function Analysis System Technique (FAST) diagram. It will be evaluated to get the result in the workshop step, analyzed the project models understand what the required functions was needed.

Generate ideas on all the possible ways accomplished the required functions. Synthesize ideas and concepts that were selected, were feasible for development into specific value improvements. Selected the 'best' alternative improved the building performance; it was continued with life cycle cost analysis. Life cycle cost analysis is a part of value engineering process, it will value for money in the owner of physical assets during all of building life cycle [17]. The best results are achieved by the optimum total costs of asset ownership.

5. Result and Discussion

This research is developed by four phase, there are information phase, function analysis phase, creative phase, and evaluation phase. In the first phase, building benchmark was based on data collection of two green office buildings in Indonesia, there are Indonesian Ministry of Public Works Building and BCA Tower. According to data survey, Both Green Building of Public Works and Housing and Green Building of BCA Tower, have premium costs for green building fixture application. Cost components of green building fixtures come from structure and architecture works and mechanical and electrical works. Initial cost breakdown structure of green building can see on Table 2.

Table 2. An Example of Green Building Initial Cost Modelling

Nu.	Building Component	Price	Unit
A	*INITIAL COST for GREEN FEATURE*		IDR
1	**Structure & Architecture**	**10,119,889,115.07**	**IDR**
	- Windows & Frame Works	862,752,000.00	IDR
	Glass Volume (Facades)	15,136.00	m^2
	Conventional Glass (Clear Glass)	785,000.00	IDR/ m^2
	Green Glass (Blue Green)	842,000.00	IDR/ m^2
	- Sun Shading	4,597,560,000.00	IDR
	Sun Shading Volume	7,568.00	m^2
	Sun Shading Prices	607,500.00	IDR/ m^2
	- Insulation	85,057,474.60	IDR
	- Ceiling	443,525,119.21	IDR
	- Floor + Finishing	944,782,238.75	IDR
	- Façade Work (Wall Envelope Structure)	3,186,212,282.51	IDR
2	**Mechanical & Electrical installations**	**6,687,828,935.64**	**IDR**
	- HVAC	2,405,008,390.80	IDR
	- Electrical System	516,554,224.27	IDR
	- Lighting System	121,291,072.28	IDR
	- Security System	42,224,017.67	IDR
	- Fire Alarm System	179,673,488.75	IDR
	- Internal Feature Work	2,887,012,337.45	IDR
	- Water Supply System	536,065,404.44	IDR

Green building fixtures may not only give affect in initial costs of building but also in operational and maintenance costs of building. Therefore, green building operational and maintenance cost data breakdown can see on Table 3. There are benefits from green building operational, although there are some increasing maintenance costs. Benefits of green building in operational phase come from energy efficiency and water conservation. In addition, green building energy efficiency makes impact to reduce carbon emission (Table 4). This research was limited with some assumptions such as inflation, discounted rate, and salvage value (Table 5).

Table 3. An Example of Green Building Maintenance and Operational Cost Modelling

B.	***OPERATIONAL AND MAINTENANCE COST***		
	Building Power		
1	Energy Consumption Intensity	250	kWh/m^2/yr
2	Power needs	5,676,000.00	kWh/ year
		473,000.00	kWh/ month
3	Energy saving average	54%	per year
	(Based conventional building)	3,065,040.00	kWh/ year
4	Power Prices	1,200	IDR/ kWh
5	**Power Cost (PLN) - Conventional**	**6,811,200,000.00**	**IDR/ year**
	Water		
1	Water consumpsion intensity	50	liter/prs.dy
2	Water Needs	34,050.00	m^3/ year
		2,837.50	m^3/ month
3	Water saving average	35%	per year
	(Based conventional building)	11,917.50	m^3/ year
4	Water price	12,000	IDR/ m3
5	**Water Cost (PLN) - Conventional**	**408,600,000.00**	IDR/ year
	Maintenance/ Renewal	**603,613,154.44**	IDR/ year
1	Air conditioner	44,794,802.87	IDR/ year
2	Wall painting	24,634,123.52	IDR/ year
3	Cleaning service and indoor quality test	333,623,697.27	IDR/ year
4	Building structure maintenance	22,400,754.84	IDR/ year
5	Lift	18,115,101.45	IDR/ year
6	Electical	81,434,121.35	IDR/ year
7	Plumbing	30,851,339.01	IDR/ year
8	Safety	31,877,481.37	IDR/ year
9	Sanitair	15,881,732.78	IDR/ year

Table 4. Carbon Emission Data

C	**Carbon Emission Reduction**		
1	CO_2 convertion	0.891	$KgCO_2$/ kWh
2	Carbon emission reduction	2,731	Ton/ year

Table 5. Assumption in Analysis

	Assumption		**Source**
1	Current year	2014-2015	BI Rate
2	Inflation (Average)	6.81%	BI Rate
3	Discount rate (Average)	12.75%	BI Rate
4	Salvage Value	25.00%	Trixy, 2012

In function analysis phase, this research used Function Analysis System Technique (FAST) Diagram to analyze green building construction component which could optimize building construction cost by stimulate innovation. Based on the literature, there are four main design elements which influence green building construction such as prerequisite design, given design, created design, and passive design (Figure 4). For prerequisite design, the owner has to construct prerequisite elements before getting rating point from rating criteria. In the other hand, given design was the building element that was given by the area development such as road network, bank, terminal, post office, fire department office, library, government office, medical facilities, etc.

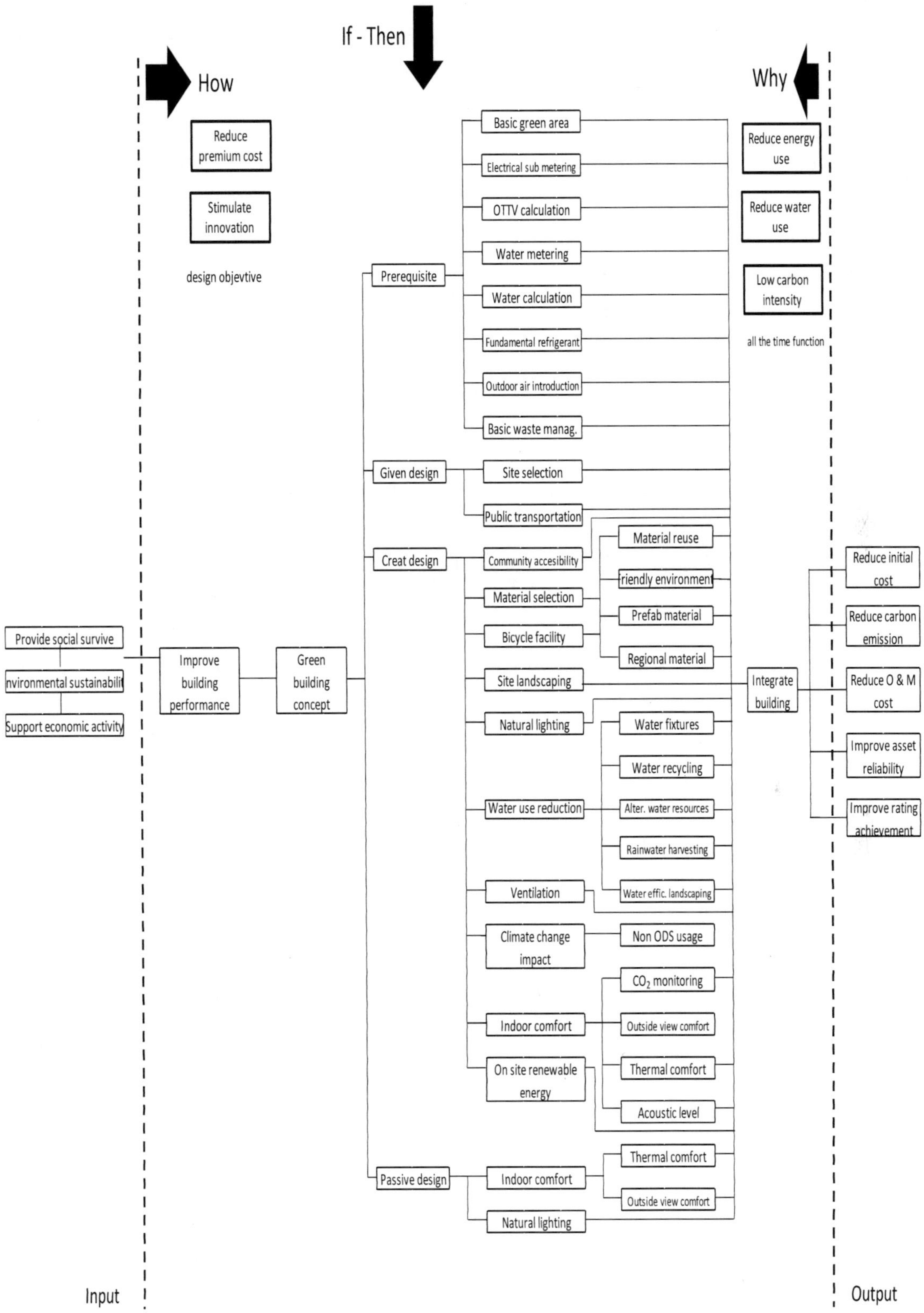

Figure 4. Function Analysis System Technique Diagram

Creating building design in rating tools, ways to get the green building criteria point depend on the owner needs. In generally, creating design is decided by how many points that the building owner wants to get certification achievement. From three design factors which influences green building concept achievement, have affect to green building passive design. The example of green building passive design used efficiency on ventilation energy which can value the indoor comfort such as CO_2 monitoring and outdoor air introduction controlling. Based on this condition, using efficiency energy on ventilation system would get credit point from energy efficiency and conservation criteria. In the other hand, increasing indoor comfort by using this ventilation would make building get extra point indirect from indoor health and comfort criteria. The output of green building design that influence construction phase, is design integration of reduce initial cost, carbon emission, operational and maintenance cost, improve asset reliability, and rating achievement.

In creative phase, this research was collecting results of green building operational and maintenance. The green building managements were focus on water conservation and energy efficiency and conservation optimization. Energy saving analysis is shown in Table 6 and Table 7. Both of them describe the average calculation of annual energy efficiency. The efficiency of energy is compared to average office building standard in Indonesia and design recognition phase. Indonesia Ministry of Public Works gets platinum achievement certificate from Green Building Council Indonesia in 2012. Energy efficiency in Indonesian Ministry of Public Works reach up to 53% if it is compared to conventional office buildings energy standard value and 19% if it is compared to its design recognition. Similarly, the efficiency value in 2013 and 2014 are presented in Table 6. BCA Tower is existing building which it was retrofitted to reach Platinum rating achievement. Energy efficiency value is 18% in 2010 if it is compared to average office building standard. All of the building value by the year could be seen on Table 7. BCA Tower can't be compared with the design because it was existing green building without energy target value in the beginning.

Table 6. Analysis of Energy Consumption Intensity in Indonesian Ministry of Public Works

Year	Energy Consumption Intensity (kWh/m^2.tahun)			Energy Efficiency	
	ASEAN	Design	Actual	Toward ASEAN Office Building	Toward Green Building Design
2012	240	140	112.88	53%	19%
2013	240	140	118.08	51%	16%
2014	240	140	111.20	54%	21%

Table 7. Analysis of Energy Consumption Intensity in BCA Tower

Year	Energy Consumption Intensity (kWh/m^2.tahun)		Energy Efficiency
	ASEAN	Actual	Toward ASEAN Office Building
2010	240	196.45	18%
2011	240	161.29	33%
2012	240	158.30	34%
2013	240	136.16	43%

Both of green building which was benchmarking in this research is focused in energy and water efficiency for cost saving. Based on Indonesian standard, water usage target is 50 liters per user per day, but with green building design target is expected to achieve maximum 18.5 liters of water usage per user per day. The data collected in Table 8 and Table 9 show the values of efficiency that there was achieved by two green building in Indonesia. In the Year 2012, The Indonesian Ministry of Public Works Building saves water reach up to 35% compared to Indonesian standard but still beyond expectations green design target is -76%. However, the BCA tower building there is only reference value data efficiency because the building is the existing building (not planned from the beginning as a green building).

Table 8. Analysis of Water Consumption Intensity in Indonesian Ministry of Public Works

Year	Water Consumption Intensity liter/person.day			Water Efficiency	
	SNI	Design	Actual	Toward SNI Office Building	Toward Green Building Design
2012	50	18.5	32.50	35%	-76%
2013	50	18.5	36.50	27%	-97%
2014	50	18.5	27.80	44%	-50%

Table 9. Analysis of Water Consumption Intensity in BCA Tower

Year	Water Consumption Intensity liter/person.day		Water Efficiency
	SNI	Actual	Toward SNI Office Building
2010	50	8.63	83%
2011	50	36.72	27%
2012	50	49.67	1%
2013	50	53.59	-7%

Based on information phase, function analysis phase, and creative phase, is created an evaluation to get the optimum of construction performance of green building. Evaluation phase is synthesizing ideas and concepts and select those that are feasible for development into specific value improvements. According to green building observation, operational cost could be reduced by energy efficiency and water efficiency. Green building futures is modelled from initial investment looked like on Table 2 and the energy efficiency and water efficiency results for operational phase can see on Table 10. Actual Ministry of Public Works (MoPW) application for green building futures make 54% energy efficiency and 35% because there are some different system of operational. Design of MoPW begins with 44% energy efficiency and 63% water efficiency. For energy design application is good, but in water efficiency is not achieved. Therefore, this research is evaluation the cause of water efficiency problem in MoPW such as increasing building tenants and water system leakage that is not treated immediately. Green Building BCA Tower is an existing building which got retrofitting system. At the operational phase, BCA Tower can reach 35% energy efficiency and 81% water efficiency. The last system that was threated in this research was renewable energy which never used in Indonesia.

Table 10. Optional of Green Building Efficiency Results for Energy and Water Used

Design Name	Optional	Energy Efficiency	Water Efficiency
Actual MoPW	1	54%	35%
Design MoPW	2	44%	63%
Comb. of Opt. 1 & 2	3	54%	63%
Actual BCA Tower	4	35%	81%
Comb. Opt. 3 & 4	5	54%	90%
Comb. Ops. 5 + Ren. E	6	100%	90%

This case study was formatted for combination of green building fixtures to develop the optimum green building model based on construction result until operational and maintenance phase. Indonesia Passive design in Ministry of Public Works, concerns on indoor comfort, water usage reduction, and natural lighting system. Indoor comfort can be accommodated with structure and architecture design such as window fixture, roof insulation, ceiling, and facade work. Beside that, there is mechanical and electrical contribution such as HVAC, electrical, lighting, and security/ alarm/ sensor system. Water usage reduction is worked by water supply system to detect the water demand. Natural lighting can be controlled by sun shading, beside that it can reduce UV radiation.

For building maintenance, there are some fixtures will get maintenance and reduce operational cost. Maintenance cost elements were HVAC system, structural maintenance, mechanical and electrical system, sanitaria, and safety. In building operational, this research only considered water and energy cost. Based on deep interview, operational and maintenance aspects of green building can be optimized by enforcing smoke-free laws, commitment of top management to carry out a performance audit program building, energy/ water/ guidance of building usage campaign, convenience building used survey, pest control, installing safety equipment in the building, energy/ water metering calibration, and using automatic features/ sensor in public areas.

Green building development results presented life cycle cost analysis. First of all, Figure 5 describes relation of energy and water efficiency toward building internal rate of return. The bigger option value of energy and water efficiency will increase the internal rate of return. The second, Figure 6 describes the relation NPV of initial cost, efficiency, and salvage value. The highest initial cost for green building fixtures infestation is optional 6 because there is additional cost for renewable energy application. It makes NPV of efficiency and salvage value is higher than the others. The third, Figure 7 describes payback period of green building infestation. Optional 6 has the fastest payback period time of investment on green building (Figure 7). But, the optimum model is optional 4 because with lowest investment cost can get the moderate energy and water efficiency (Figure 8).

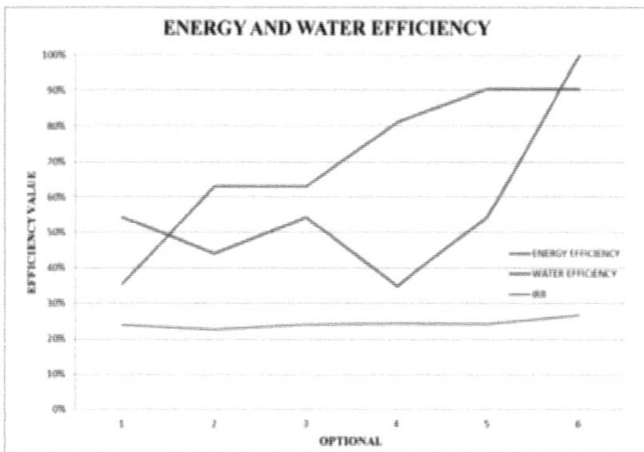

Figure 5. Value of Energy, Water Efficiency, and IRR

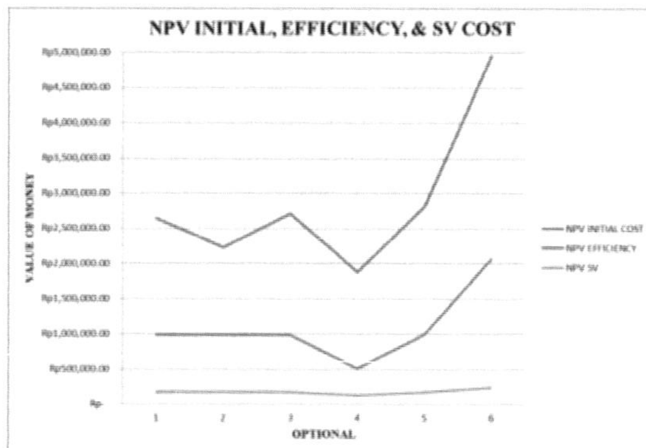

Figure 6. NPV of Initial cost, Efficiency Total, and Salvage Value

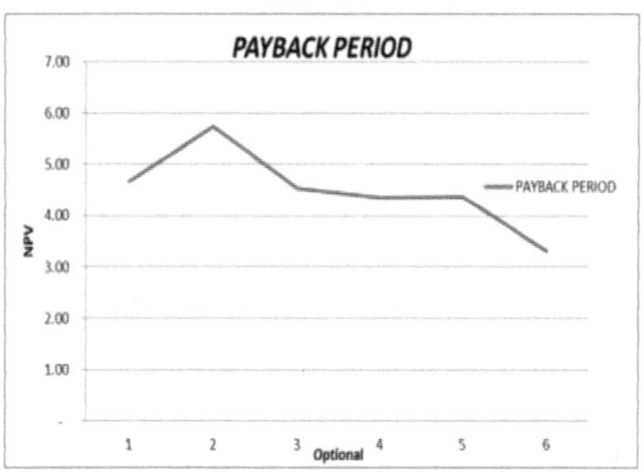

Figure 7. Payback Period of Green Future

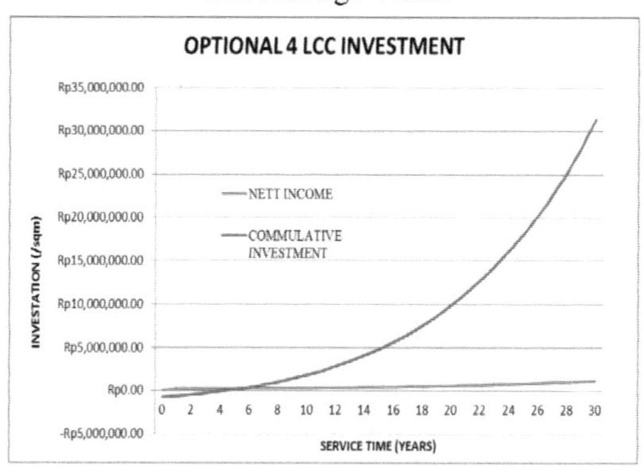

Figure 8. Optimum Cost of Building Life Cycle

6. Conclusion

According to data survey, actual life cycle cost of green building in Indonesia with platinum certificate achievement needs 13.72% additional cost. Based on Y. Lin et al. research, USA building only needs 6.50% additional cost to reach platinum achievement certification. So, green building investment in Indonesia is still in high cost. The optimum optional of green building in Indonesia was optional 4 with 7.34 % for additional cost, optimum energy and water efficiency, and more feasible infestation in Green Building Concept

Performance optimization key factors on green building construction premium cost to achieve green building operation and maintenance goals were building envelope design, kind of material (glasses, shading, and insulation), and construction method. Almost all of the creating design influences the passive design for building operational and maintenance. So, the accuracy of construction has to decide to get the best performance in operational phase. Combination of green fixtures has to measures to get the optimum in energy and water efficiency. All of the building cycle costs components have affect one to the others.

Acknowledgement

The authors wish to thank for funders of this research that is International Indexed Publication Grants for Final Project Students in Universitas Indonesia (Hibah PITTA UI). Beside that, we thank to all the organizations and individuals who kindly participated in the research. Their time and input in the interview and survey are much appreciated. The authors would also like to thank and acknowledge Project Management Study Program, Faculty of Engineering, Universitas Indonesia, Indonesia, for generous support and encouragement of this research.

References

[1] L. Bernstein, P. Bosch, O. Canziani, Z. Chen, R. Christ, K. Riahi, IPCC, 2007: climate change 2007: synthesis report, IPCC2008.

[2] O. Jasudaite, I. Associated Consultant, E. Kwast, M. Huse, A. Lüders, A. Valge, I. Associated Consultant, I. Andersone, I. Associated Consultant, F. Jaren, Green Building Market Report, DOI (2014).

[3] M. Rimmer, The Greenwashing of the Trans-Pacific Partnership, Submission to the Productivity Commission, the Joint Standing Committee on Treaties, and the Senate Foreign Affairs, Trade, and References Committee, DOI (2016).

[4] R.D. Miller, J.M. Ford, Shared Savings in the Residential Market: Public/private Partnership for Energy Conservation, The Office1985.

[5] J. Marquardt, A struggle of multi-level governance: promoting renewable energy in Indonesia, Energy Procedia, 58 (2014) 87-94.

[6] H.-T. Nguyen, M. Gray, M. Skitmore, Comparative study on green building supportive policies of Pacific-Rim countries most vulnerable to climate change, DOI (2016).

[7] N.S. Adiwoso, S.P. Prasetyoadi, Towards Indonesia Sustainable Future through Sustainable Building and Construction, Green Building Council Indonesia, DOI (2010).

[8] C.F. Yiing, N.M. Yaacob, H. Hussein, Achieving sustainable development: accessibility of green buildings in Malaysia, Procedia-Social and Behavioral Sciences, 101 (2013) 120-129.

[9] I. Mayasari, C. Utomo, KONSEP ANALISA PENGARUH KRITERIA GREEN BUILDING TERHADAP KEPUTUSAN INVESTASI PADA PENGEMBANG PROPERTI DI SURABAYA, DOI.

[10] J. Zuo, Z.-Y. Zhao, Green building research–current status and future agenda: A review, Renewable and Sustainable Energy Reviews, 30 (2014) 271-281.

[11] B.K. Oh, J.S. Park, S.W. Choi, H.S. Park, Design model for analysis of relationships among CO 2 emissions, cost, and structural parameters in green building construction with composite columns, Energy and Buildings, 118 (2016) 301-315.

[12] A.Y. Bahaudin, E.M. Elias, A.M. Saifudin, A comparison of the green building's criteria, E3S Web of Conferences, EDP Sciences, 2014.
[13] V. Soebarto, D. Ness, Rethinking the adoption of green building rating systems in developing countries, SENVAR, 2010.
[14] A.K. Mukhopadhyaya, Value engineering mastermind: from concept to value engineering certification, SAGE Publications India2009.
[15] A. Dell'Isola, Value engineering, Practical Applications... for Design, Construction, Maintenance & Operations. RS Means Company Inc. Kingston MA, DOI (1997).
[16] L.W. Zimmerman, G.D. Hart, Value Engineering a Practical approach for owners, Designers and Contractors, DOI (1982).
[17] D.G. Woodward, Life cycle costing—theory, information acquisition and application, International journal of project management, 15 (1997) 335-344.

Modelling of Mechanical Coupling for Piezoelectric Energy Harvester Adapted to Low-Frequency Vibration

T Untoro1, S Viridi[2], Suprijanto[3], E Ekawati[4]

[1]Master Program of Instrumentation and Control, Faculty of Industrial, Institut Teknologi Bandung, Bandung, Indonesia
[2]Nuclear Physics and Biophysics Research Division, Faculty of Mathematics and Natural Sciences, Institut Teknologi Bandung, Bandung, Indonesia
[3, 4]Instrumentation and Control Research Group, Faculty of Industrial, Institut Teknologi Bandung, Bandung, Indonesia

[1]triu@fi.itb.ac.id, [2]dudung@fi.itb.ac.id, [3]supri@tf.itb.ac.id, [4]esti@tf.itb.ac.id

Abstract. In our previous work, we have developed a mechanical coupling for energy harvester from vibration source. This energy harvester uses piezoelectric with additional cantilever beam and permanent magnets. Our work proposed alternative scheme of mechanical coupling for tune the vibration input into resonant frequency of piezoelectric. Based on the experiment, correlation between the length of cantilever beam and the output power also evaluated. In this paper, we try to modelling our work into mathematical model and apply it to some case study. For example application, we apply our energy harvester system to generate electrical energy to enlighten the street. The human footsteps can be used as vibration source to generate electrical energy.

1. Introduction

Harvesting vibration energy in several recent years has become a trend. The main purpose is how to take advantage of daily life activity as sources of electrical energy. The example of sources are vibration on some surface when stepped by human footstep, machine vibration, bridge vibration, or ship vibration caused by sea wave. Many ways are used to convert those vibrations into electrical energy. This conversion mechanism is commonly known as transduction. There are three famous transduction mechanisms to pursue energy scavenging, i.e. electrostatic, electromagnetic, and piezoelectric [1]. Electrostatic transduction is electrical energy that generate by the movement of two capacitive plates which separated by air medium, vacuum, or dielectrics. Principle of electromagnetic transduction is using Faraday Law by moving coil relatively to or from permanent magnetor or otherwise [2, 3]. Different from the two of transduction mechanism, piezoelectric transduction is unique. Piezoelectric can generate electrical energy by changing strain on piezoelectric material. Vibration on the surface of piezoelectric caused strain alteration. Piezoelectric transduction has advantages compared first two transduction methods. There are large power densities and ease of application [4].

Every piezoelectric have resonance frequency that can optimize power output. Therefore, recent research is focused to wider resonance bandwidth. Material, mass, dimension, or the thickness of the piezoelectric structure is main factor that determine of resonant frequency. Not all researchers have to dig some new from all main factors to determine suitably resonant frequency. Some researcher just have utilize some product that produced by some piezoelectric vendor. The main problem for this kind of researcher is how to tune frequency of the vibration source with resonant frequency of the piezoelectric [5]. When the piezoelectric reaches the resonant frequency, it can produce maximum output power. Variety techniques are determined, such as integrate piezoelectric with cantilever in many configurations, combine two piezoelectric with blast technique, combine some piezoelectric into array that have different resonant frequencies, or integrate piezoelectric with combination of cantilever-permanent magnet [1, 5-8].

The last technique was developed with many efforts. Our previous work is included with that last technique. Additional cantilever and pairs of permanent magnet was utilized to reach maximum power from piezoelectric. Based on empirical data, the maximum power could be reached when the length of additional cantilever beam is twice of the length of the piezoelectric dimension, although there was use of some constraint on that case. When it gives impulse input, the power reaches 12.54 mW [5]. The next project is how to apply the previous result to the real application. Before reaching that goal, numerical simulation could be an important guidance to determine the suitable output power. Vibration source in this simulation comes from surface that is trampled by human walking activity.

2. Theoretical background

2.1. Introduction

Device of vibration energy harvester will harvest energy from low-level ambient energy, below 10 Hz for the human movement and above 30 Hz for vibration on the engine [10]. On the market, several types of piezoelectric for vibration energy harvester have a resonant frequency range between 30-1500 Hz [9, 10]. There is no problem when the vibration source has same frequency with piezoelectric. But, when it below the resonant frequency piezoelectric, it will cause piezoelectric can not generate output power optimally. Therefore, it requires some way to solve that problem.

Mechanical coupling is a system that used to amplify low vibration frequency in order to tune with resonant frequency of piezoelectric. Many techniques used in mechanical coupling. But one of the easier ways is to utilize permanent magnet and additional cantilever [1, 5]. Attractive magnetic force and repulsive force will shift natural frequency of cantilever compared with unapplied additional permanents magnets. The permanent magnets also alter the effective stiffness of the cantilever [1]. The permanent magnets and additional cantilever also would generate vibrations a lot more and longer than without it [5].

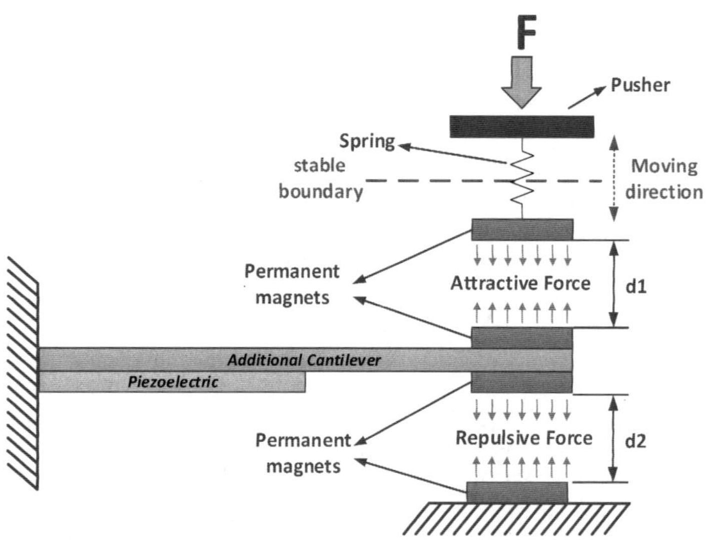

Figure 1. Schematic of mechanical coupling for energy harvesting device.

Four permanent magnets are used. Two magnets are fixed at the free end of the cantilever beam, while the other one magnet is fixed to the enclosure of the device at the bottom, and the last one placed at the top that can move with vertical direction as depicted in figure 1. The magnets are placed such that attractive and repulsive magnetic forces can be applied on each side of the beam. The function of spring is to make top magnet movable. When top magnet is pushed up, it can force to move the magnets on the beam down. The oscillations of the spring can makes vibrations on the cantilever a lot more and longer than without use of the spring. Figure 2 depicted the design of the mechanical coupling for energy harvesting device. When the pusher is pushed by an applied force, the cantilever beam has to move-down. After the force removed, i.e. no longer stepped by human, the cantilever beam will oscillate and the system will produce electricity.

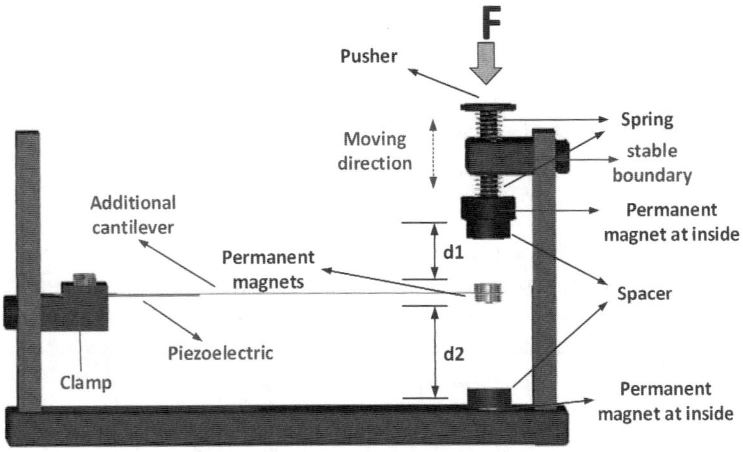

Figure 2. Design of the mechanical coupling for energy harvesting device.

Spacers at top and bottom side are preventing the cantilever beam adheres the permanent magnets. The cantilever and piezoelectric are fixed on a clamp that can be vertically displaced using a screw mechanism.

2.2. Magnetic force impact

Cylindrical magnets are used to apply attractive magnetic force and repulsive force on the mechanical coupling system. The magnetic force between any two cylindrical magnets is given as

$$F_{\text{mag}}(d)=\left[\frac{B_r^2 A_m^2 (l+r)^2}{\pi\mu_0 l^2}\right]\left[\frac{1}{d^2}+\frac{1}{(d+2l)^2}+\frac{1}{(d+l)^2}\right], \quad (1)$$

where B_r is the residual flux density of the magnet, A_m is the common area between the magnets, l is the length of the magnet, r is the radius of the magnet, d is the distance between the magnets, and μ_0 is the permeability of the intervening medium. Applying magnetic force will alter the effective stiffness of the beam and it will be changing on resonant frequency of the beam [1].

2.3. Equivalent model for a mechanical coupling

Designed mechanical coupling consists of cantilever, permanent magnets and piezoelectric. Piezoelectric is mounted under cantilever. The setup description is depicted in figure 2. From the information in figures 1 and 2, the mathematical model can be developed as depicted in figure 3.

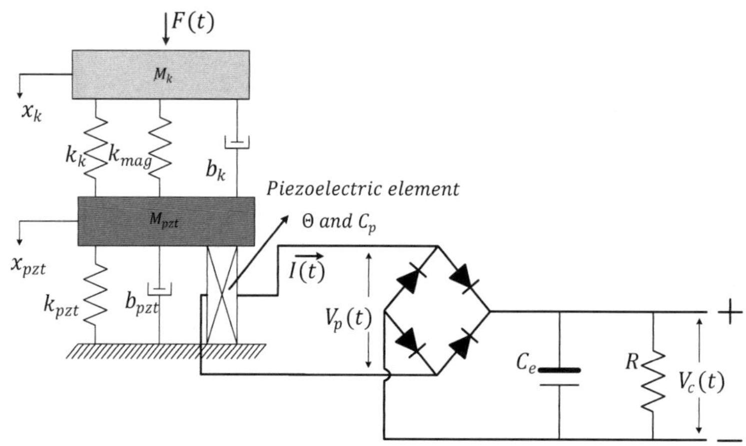

Figure 3. Equivalent model of mechanical coupling for energy harvester.

where $F(t)$ is force function that applied to the system, M_k is mass of cantilever, k_k is elasticity coefficient of cantilever, b_k is damping coefficient of cantilever, k_{mag} is elasticity coefficient of permanent magnet, x_k is the displacement position of the cantilever, M_{pzt} is mass of piezoelectric, k_{pzt} is elasticity coefficient of piezoelectric, b_{pzt} damping coefficient of piezoelectric, and x_{pzt} is the displacement position of the piezoelectric, Θ is effective piezoelectric coefficient, C_p capacitance of piezoelectric, $I(t)$ is current function through the circuit, $V_{\mathrm{p}}(t)$ is the voltage across the piezoelectric element, and $V_c(t)$ is the voltage across the rectifier and filter circuit.

The governing equations of the piezoelectric vibrator can be described by following equations

$$F(t)-k_{\mathrm{eff}}\left(x_k+x_{\mathrm{pzt}}\right)-b_k\left(\dot{x}_k+\dot{x}_{\mathrm{pzt}}\right)=M_k\ddot{x}_k\,, \tag{2}$$

$$k_{\mathrm{eff}}\left(x_k+x_{\mathrm{pzt}}\right)-b_k\left(\dot{x}_k+\dot{x}_{\mathrm{pzt}}\right)-k_{\mathrm{pzt}}x_{\mathrm{pzt}}-b_{\mathrm{pzt}}\dot{x}_{\mathrm{pzt}}=M_{\mathrm{pzt}}\ddot{x}_{\mathrm{pzt}}+\Theta V_p(t)\,, \tag{3}$$

$$\Theta\dot{x}_{\mathrm{pzt}}+C_p\dot{V}_p(t)=-I(t)\,, \tag{4}$$

where k_{eff} is $k_{\mathrm{mag}}+k_k$. A mechanical coupling system is part of a harvesting energy system from vibration energy. Ideally, the system has no damping factor, so when given the instantaneous vibration input, a mechanical coupling system will continually generate vibration for piezoelectric. But in fact, damping factor will always appear in any system, including mechanical coupling. This can be caused by friction between materials, material with the air, properties of materials used, and theoretically there is applied the law of conservation of energy. Although there is no ideal condition, but in this research will be determine how to maintain the vibrations as long as possible when no more vibration input.

Influencing parameters of the equation of mechanical coupling is M_k k_k, b_k, k_{mag}, M_{pzt}, b_{pzt}, k_{pzt}, Θ, dan C_p. For M_{pzt}, k_{pzt}, b_{pzt}, Θ, dan C_p are default parameters from piezoelectric, so those parameters can not be changed nor sought for optimal value. This is because we only use one type of piezoelectric that is Vulture V21BL. So that the parameters that can be searched or altered for the purposes of the experiment is M_k, k_k, b_k, k_{mag}. Thus the material type, length, width and thickness of the cantilever will change the value of M_k, k_k, b_k. In this research, the type of material, width, and thickness of the cantilever are constant, so only a cantilever length which will be altered for the purpose of experimentation and data collection. kmag from magnet position is also constant.

The associated values of the variables as used in the prototype described later in this paper are listed in table 1 [12].

Table 1. Variable descriptions of mechanical coupling.

Symbol	Description	Value	Unit
$F(t)$	Forcing function that applied to the system		N
M_k	Mass of cantilever	225	g
k_k	Effective stiffness of cantilever	5623.06	N/mm
b_k	Damping coefficient of cantilever	0.015	N/mm
x_k	Displacement position of the cantilever		mm
k_{mag}	Effective stiffness of magnet	4234.08	N/mm
M_{pzt}	Mass of piezoelectric	2.92	g
b_{pzt}	Damping coefficient of piezoelectric	0.09	N/mm
k_{pzt}	Effective stiffness of piezoelectric	18013.53	N/mm
x_{pzt}	Displacement position of the piezoelectric		mm
Θ	Effective piezoelectric coefficient	390×10^{-12}	C/N
k_{eff}	Effective stiffness of beam	12390.47	N/mm
C_p	Capacitance	43×10^{-9}	F
$I(t)$	The current flowing into the specified circuit		A
$V_{p(t)}$	The voltage across the piezoelectric element		V
$V_c(t)$	The voltage across the rectifier and filter circuit		V

In this simulations $F(t)$, x_k, x_{pzt}, $I(t)$, $V_p(t)$, and $V_c(t)$ are functions of time t, which could be obtained by solving equations (2) – (4).

2.4. Changing the value of natural frequency

In our prototype, changing the M_k, k_k, b_k will affect the value of the natural frequency of the cantilever ω_n and damping factor of the cantilever ς. If $\omega_n = \sqrt{k/m}$, $\varsigma = b/2\sqrt{km}$, and $k = 3EI/l^3$, then

$$k \approx \frac{3Em}{l}, \tag{5}$$

$$\omega_n = \sqrt{\frac{3E}{l}}, \tag{6}$$

$$\varsigma = \frac{b\sqrt{l}}{2\sqrt{3Em^2}}. \tag{7}$$

A critical factor when designing the mechanical coupling for energy harvester is how to keep the natural frequency of the cantilever can match the natural frequency of the piezoelectric. And also has a low damping coefficient, so the cantilever has a high total deviation. From the equation 6 and 7 show that to get ω_n or natural frequency of the cantilever can match the natural frequency of the piezoelectric is by changing l or the length of cantilever. And to get low value of ς or damping factor in order to cantilever has continually oscillate is by increasing the value of l. The role of the permanent magnet in this system is to reduce the damping.

Table 2. Variable descriptions of changing the value of the natural frequency.

Symbol	Description	Value	Unit
ω_n	Natural frequency of the cantilever	299	rad/sec
ς	Damping factor	4,42 x 10e-8	
E	Young's modulus [13]	189 x 10e9	N/mm^2
l	The length of cantilever	100	mm

2.5. Output power
Power output from rectifier and filter circuit can obtained from

$$P(t) = V_c(t)\, I(t), \tag{8}$$

where $I(t)$ and $V_c(t)$ are to be found.

3. Simulation with human footsteps as vibration source

3.1. Vibration source
Vibration on a surface that caused by human footsteps is used in our simulation. The purpose is for calculate how many foot step or mechanical coupling needed to produce desired electrical energy. When a footstep contacts a surface, it will trigger the mechanical coupling to generate vibrations that needed. The explanation of the process of collecting human footsteps is depict in figure 4

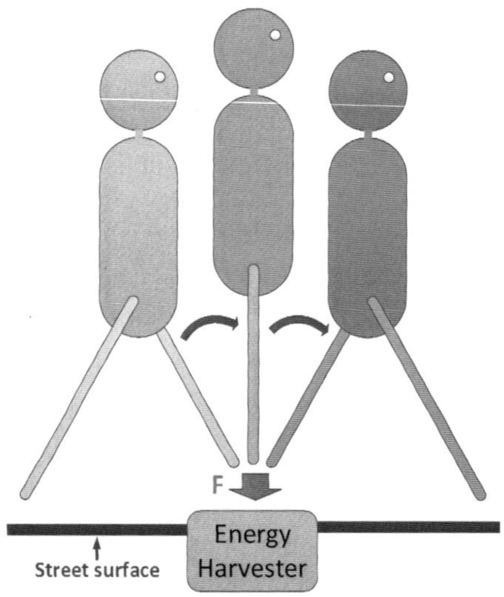

Figure 4. Collecting process of human footsteps as vibration source.

Supposed that a human has mass M_h dan he start stepping on the pusher at time t_i and leave it at time t_f then $F(t)$ can be proposed as

$$F(t)=u(t-t_i)\left[1-u(t-t_f)\right]\ M_h g\,, \tag{9}$$

where $u(t-a)$ is a step function

$$u(t-a)=\begin{cases}0, & t<a,\\ 1, & t\ge a.\end{cases} \tag{10}$$

If there are some humans stepping on the pusher, that series of equation (9) are required. Each human will have its own t_i and t_f.

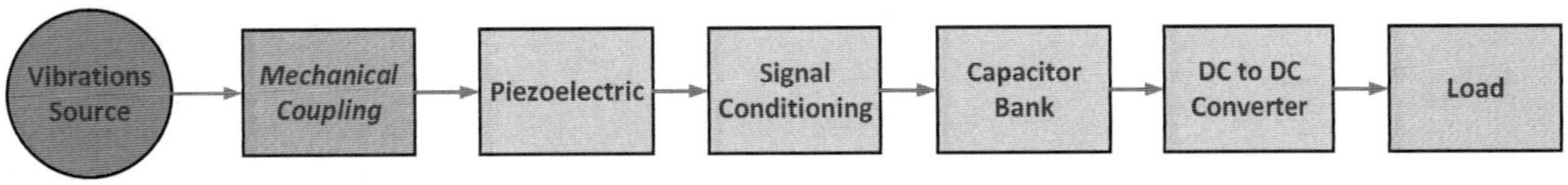

Figure 5. Single mechanical coupling application.

Single unit of energy harvesting system has schematic is shown in figure 5, while multiple units of the same system are shown in figure 6. Details how the schematic will be implemented in the real system are not in the scope of this work, which are subject to be discussed in future work.

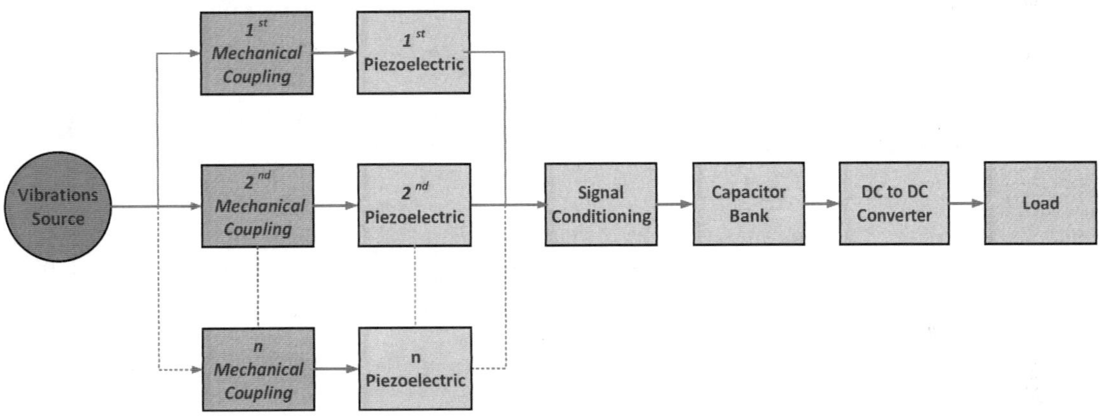

Figure 6. Multi-mechanical coupling application.

3.2. Finite difference method
In order to solve equations (2) – (4) following finite difference relations

$$\dot{x}=\frac{x(t+\Delta t)-x(t)}{\Delta t} \tag{11}$$

and

$$\ddot{x}=\frac{x(t+\Delta t)-2x(t)+x(t-\Delta t)}{(\Delta x)^2} \tag{12}$$

are required.

4. Results and discussion

Implementation of equations (11) and (12) in equations (2) – (4) will lead to following numerical equations

$$\begin{aligned}&(M_k+\Delta t\,b_k)x_k(t+\Delta t)+\Delta t\,b_k\,x_{pzt}(t+\Delta t)=\\&(\Delta t)^2F(t)+\left[\Delta t\,b_k-(\Delta t)^2k_{eff}\right]x_{pzt}(t)\\&+\left[2M_k+\Delta t\,b_k-(\Delta t)^2k_{eff}\right]x_k(t)-M_kx_k(t-\Delta t),\end{aligned}\tag{13}$$

$$\begin{aligned}&\Delta t\,b_kx_k(t+\Delta t)+\left[M_{pzt}+\Delta t\left(b_k+b_{pzt}\right)\right]x_{pzt}(t+\Delta t)=\\&\left[2M_{pzt}+\Delta t\left(b_{pzt}+b_k\right)+(\Delta t)^2\left(k_{eff}-k_{pzt}\right)\right]x_{pzt}(t)-M_{pzt}x_{pzt}(t-\Delta t),\\&\left[\Delta t\,b_k+(\Delta t)^2k_{eff}\right]x_k(t)-(\Delta t)^2\Theta V_p(t),\end{aligned}\tag{14}$$

$$\Theta\dot{x}_{pzt}+C_p\dot{V}_p(t)=-I(t),\tag{15}$$

$$\begin{aligned}&(\Delta t)^2F(t)-\Delta t\,k_{eff}\left[x_k(t)+x_{pzt}(t)\right]-b_k\left[x_k(t+\Delta t)-x_k(t)+x_{pzt}(t+\Delta t)-x_{pzt}(t)\right]\\&=M_k\left[x_k(t+\Delta t)-2x_k(t)+x_k(t-\Delta t)\right],\end{aligned}\tag{16}$$

$$\begin{aligned}&\Delta t\,k_{eff}\left[x_k(t)+x_{pzt}(t)\right]-b_k\left[x_k(t+\Delta t)-x_k(t)+x_{pzt}(t+\Delta t)-x_{pzt}(t)\right]\\&-\Delta t\,k_{pzt}x_{pzt}-\Delta t\,b_{pzt}\left[x_{pzt}(t+\Delta t)-x_{pzt}(t)\right]\\&=M_{pzt}\left[x_{pzt}(t+\Delta t)-2x_{pzt}(t)+x_{pzt}(t-\Delta t)\right]\ddot{x}_{pzt}+(\Delta t)^2\Theta V_p(t),\end{aligned}\tag{17}$$

$$\Theta\left[x_{pzt}(t+\Delta t)-x_{pzt}(t)\right]+C_p\left[V_p(t+\Delta t)-V_p(t)\right]=-I(t).\tag{18}$$

5. Summary
Numerical model of mechanical coupling for harvesting energy using piezoelectric at low frequency vibration has been formulated. The numerical scheme has not been tested.

Acknowledgements
This research previously was supported by LPDP (Lembaga Pengelolaan Dana Pendidikan) from Ministry of Finance Republic of Indonesia. The authors gratefully acknowledge the supports from Instrumentation and Control Research Group (KINK) and Nuclear Physics and Biophysics Research Division (KKNB) in Institut Teknologi Bandung, Indonesia.

References

[1] Vinod R Challa, M G Prasad, 2008, Yong Shi and Frank T Fisher, "A vibration energy harvesting device with bidirectional resonance frequency tunability", Smart Material Structure 17, IOP Publishing Ltd
[2] S. Boisseau, G. Despesse, and B. Ahmed Seddik, 2012, "Electrostatic Conversion for Vibration Energy Harvesting", Small-Scale Energy Harvesting, Intech
[3] Qian Zhang and Eun Sok Kim, 2014, November 2-3, 2015, "Micromachined Energy-Harvester Stack With Enhanced Electromagnetic Induction Through Vertical Integration of Magnets", Journal of Microelectromechanical System, IEEE
[4] Alper Erturk and Daniel J. Inman, 2011, "Piezoelectric Energy Harvesting", First Edition, John Wiley & Sons
[5] Tri Untoro, Suprijanto, and Estiyanti Ekawati, "Development of mechanical coupling for piezoelectric energy harvester adapted to low-frequency vibration", 2015 4th International Conference on Instrumentation, Communications, Information Technology, and Biomedical Engineering (ICICI-BME), Bandung
[6] David F. Berdy et al, Mei 2012, "Low-Frequency Meandering Piezoelectric Vibration Energy Harvester", "IEEE Transactions on Ultrasonics, Ferroelectrics, and Frequency Control", vol. 59, no. 5
[7] M.Renaud et al, 2009, "An Impact Based Piezoelectric Harvester Adapted to Low Frequency

Environmental Vibrations", IEEE Transducers 2009

[8] Huan Xue et al, "Broadband Piezoelectric Energy Harvesting Devices Using Multiple Bimorphs with Different Operating Frequencies", "IEEE Transactions on Ultrasonics, Ferroelectrics, and Frequency Control", vol. 55, no. 9, September 2008

[9] Volture™ vibration energy harvester, accessed on 20/02/2015 - 15.58, Mide, "http://www.mide.com/products/volture/piezoelectric-vibration-energy-harvesters.php"

[10] BOLT ™ Energy Harvesting Products, MicroGen Systems, Inc, ", accessed on 20/02/2015 - 15.55 "http://www.microgensystems.co/content/MicroGen_BOLT-INDUSTRIAL_Jun2013.pdf

[11] Tom J. Kazmierski and Steve Beeby, 2011, "Energy Harvesting Systems: Principles, Modeling and Applications", Springer Science+Business Media LLC, DOI 10.1007/978-1-4419-7566-9, chapter-1

[12] Tri Untoro, 2015, "Pengembangan Kopling Mekanik Untuk Pemanenan Energi dengan Piezoelektrik yang Memanfaatkan Vibrasi Frekuensi Rendah", Program Studi Instrumentasi dan Kontrol, Fakultas Teknologi Industri, Institut Teknologi Bandung, chapter-IV

[13] Cambridge University Engineering Department, 2003, "Materials Data Book", Page 11

IOP Conf. Series: Journal of Physics: Conf. Series **877** (2017) 012043 doi:10.1088/1742-6596/877/1/012043

The Temperature Effect on the Working Characteristics of Solar Cells Based on Organometal Halide Perovskite Crystals

Tanti Dewinggih[1], Shobih[2], Lia Muliani[2], Herman[1], Rahmat Hidayat[1]

[1] Physics of Photonics and Magnetism Research Division, Physics Program Study, Faculty of Mathematics and Natural Sciences, Institut Teknologi Bandung, Jl. Ganesha 10, 40132, Bandung, Indonesia
[2] Electronics and Telecommunication Research Centre LIPI, Jl. Sangkuriang Komp.LIPI Gd.20 Bandung, Indonesia

e-mail: rahmat@fi.itb.ac.id

Abstract. Organometal halide perovskites have been much studied as an active material in a new generation of solar cell with high power conversion efficiency. The chemical reactions involved in their crystallization process are simple but the crystallization process and the formed crystal are very sensitive to temperature and humidity. In general, if the electronic structure of this active material is easily affected by temperature, the working performance of its solar cell will be also easily affected by temperature. In this work, we investigated the temperature effect on the working performance, namely the *J-V* characteristics, of $CH_3NH_3PbI_3$ perovskite based solar cell. The measurement result show that the *J-V* characteristic significantly changed with temperature. The *J-V* curve shows a diode characteristic at room temperature but it changes to an Ohmic characteristic at high temperature. This characteristics change may be due to the degradation of the perovskite crystals, which may be caused by separation and recrystallization PbI_2 inside the perovskite layer.

1. Introduction

Two research groups led by Snaith and Grätzel have independently reported the use of organometal halide perovskite as active material for a new generation of highly efficient solar cells. Since then the study of this material as an active layer for solar cells has attracted much attention and rapidly developed. [1-4] This perovskite crystal contains organic group, which is different from conventional inorganic perovskite crystals that have been known for long time. In general, these crystals have a similar structure as calcium titanium oxide ($CaTiO_3$). The general formula to represent perovskite structure is AMX_3, where 'A' and 'M' are two cations with different size and X is an anion. The ideal cubic-symmetry structure has the M cation in 6-fold coordination that is surrounded by an octahedron of anions (MX_6), while the A cation in 12-fold cuboctahedral coordination, as illustrated in Figure 1a and 1b. The organometal halide perovskite, such as *methylammonium lead tri-iodide* ($MAPbI_3$), has the same the AMX_3 formula, where $CH_3NH_3^+$ (MA^+) is as the A cation, Pb^{2+} is as the M cation, and I is as the X anion.

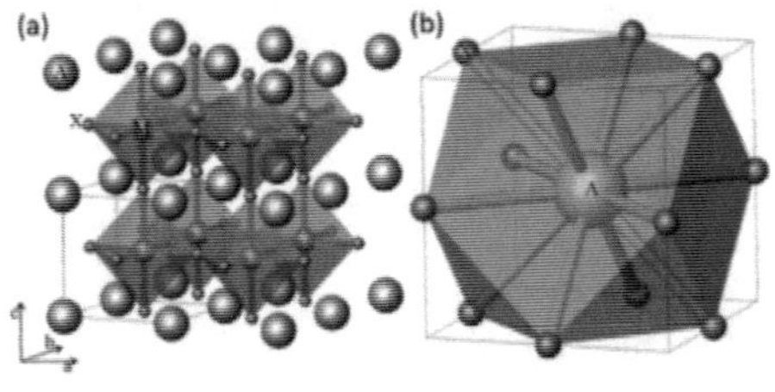

Figure 1. (a) Crystal structure of cubic perovskite with general formula of AMX_3; (b) Twelve-fold coordination of the A-site cation. [2]

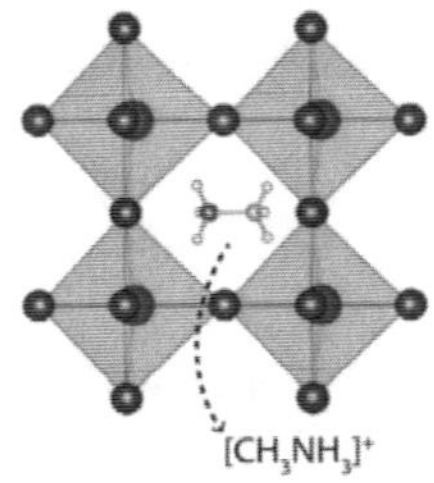

Figure 2. The crystal structure of organometal halide perovskite with MA^+ as the A cation.

Conventional inorganic perovskite crystals have been much studied in relation to their magnetic and electronic properties including ferroelectricity and the superconductivity properties. Most of inorganic perovskite crystals are stable. In contrast to those perovskite crystal, organometal halide perovskite crystals exhibits photovoltaic and other photoresponsive properties. Organometal halide perovskite, such as $CH_3NH_3PbI_3$, can be prepared by a simple chemical reaction, namely:

$$PbI_2\ (s) + CH_3NH_3I\ (aq) \leftrightarrow CH_3NH_3PbI_3\ (s) \quad (1)$$

However, the chemical stability during the crystal formation and fabrication processes should be always taken into consideration. [5-6] Oxygen and moisture in the atmosphere have crucial effects leading to the degradation of perovskite crystal, which may occur through the following reactions.

$$CH_3NH_3PbI_{3(s)} \leftrightarrow PbI_{2(s)} + CH_3NH_3I_{(aq)} \quad (2)$$

$$CH_3NH_3I_{(aq)} \leftrightarrow CH_3NH_{2(aq)} + HI_{(aq)} \quad (3)$$

$$4HI_{(aq)} + O_{2(g)} \leftrightarrow 2I_{2(s)} + 2H_2O_{(l)} \quad (4)$$

$$2HI_{(aq)} \leftrightarrow H_{2(g)} + I_{2(s)} \quad (5)$$

It is also evident that the formed crystal structure significantly affected by temperature. The perovskite crystals shows three phases, namely α, β, and γ phases, depending on the temperature. [7-9] This paper presents a report concerning the temperature effect on the *J-V* characteristics of this perovskite solar cell.

2. Experiments

The structure of the solar cell investigated here is a multilayer structure of FTO/c-TiO_2/mp-TiO_2/Perovskite/*spiro* OMeTAD/Au, as illustrated in Figure 3. Compact TiO_2 (c-TiO_2) layer was used as the blocking layer and the mesoporous TiO_2 (mp-TiO_2) was used as the scaffold mesopori layer. This layer was deposited on the top of FTO substrate by using a spincoater at 5000 rpm for 30 s with acceleration 2500 rpm/s. The layer was sintered up to 550°C for the c-TiO_2 and 475°C for the mp-TiO_2 layer.

In order to prepare the perovskite crystal in TiO_2 mesoporous layer, a precursor solution of perovskite was deposited by one step precursor deposition method at 3000 rpm for 30 s. The precursor solution have 1:1 mol ratio of the PbI_2 and MAI in the DMF. This solution was previously heated at 120°C for 20 minutes to promote complete dissolution and is preserved at 70°C before spin coating process.

The spiro-OMeTAD as the hole transport material (HTM) was deposited on the top of perovskite layer by using a spincoater at 5000 rpm for 30 s. Spiro-OMeTAD solution was prepared from a chlorobenzene solution containing 72,3 mg/mL spiro-OMeTAD, 17,5 μL/mL of LiTFSI stock solution (stock: 520 mg/mL LiTFSI in acetonitrile), and 26,3 μL/mL of FK 102 Co(III) TFSI stock solution (stock: 300 mg/mL FK102 in acetonitrile). Finally, Au layer was deposited by using physical vapour deposition as the electrode.

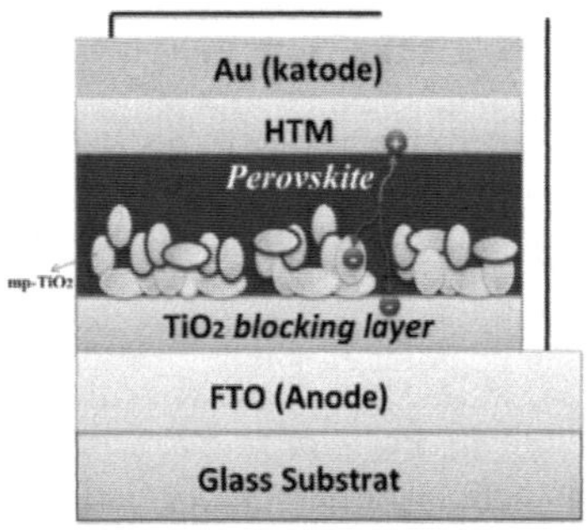

Figure 3. The structure of the perovskite solar cell investigated here.

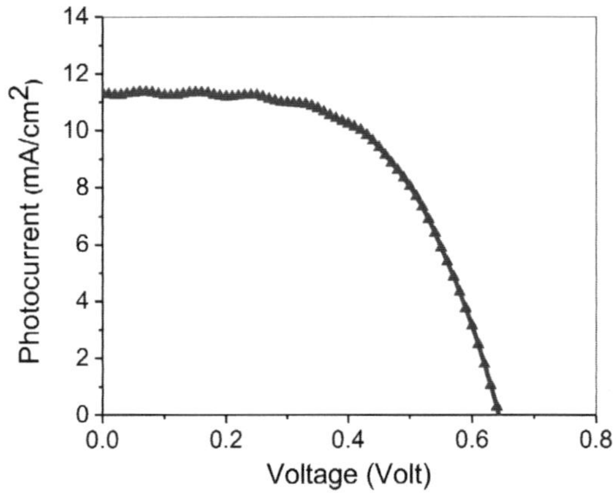

Figure 4. The measured *J–V* characteristic of a perovskite solar cell sample fabricated in this work under light illumination (1 sun).

3. Results and Discussion

Figure 4 shows typical characteristic of J-V curve measured from perovskite solar cells fabricated in this work. This sample shows the efficiency, short circuit photocurrent (J_{sc}), open voltage bias (V_{oc}), and Filling Factor (FF) of about 4.3%, 11.5 mA/cm^2, 0.63V and 0.58, respectively. In order to study the effect of temperature on the solar cells, the *J-V* characteristics of the cells were measured under dark condition. Figure 5 (a) and (b) clearly shows that the *J-V* characteristic is significantly affected by temperature. At room temperature, the *J-V* curve exhibits a diode characteristic. However, it much deviates from a diode characteristic at high temperature.

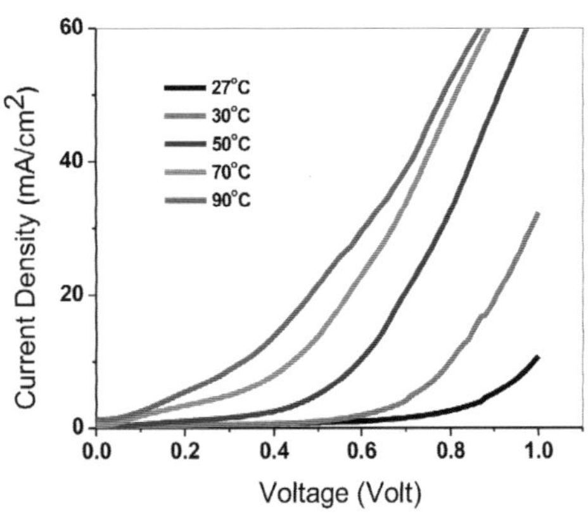

(a)

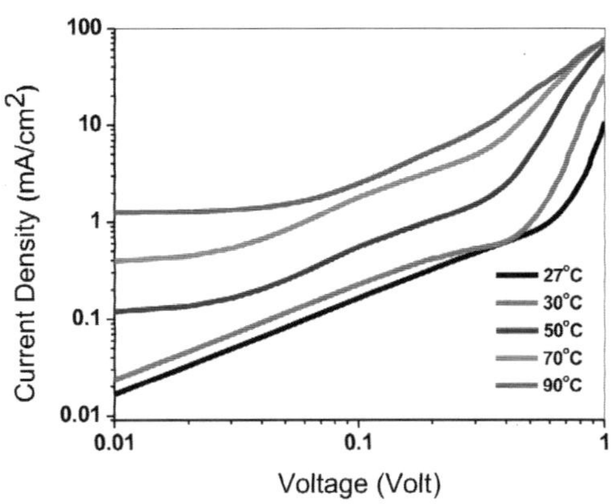

(b)

Figure 5. The measured *J–V* characteristic of this perovskite solar cell at various temperatures presented in (a) in normal scale and (b) in logarithmic scale for both x-axis and y-axis.

For understanding the change on these *J-V* characteristics, the curves were fitted by the Murgatroyd equation. [10] However, the fitting for temperature larger room temperature were failed if we did not add an additional parameter that is related to an internal resistance *r*. The fitting equation then becomes

$$J\,SCLC = \frac{9}{8}\varepsilon\mu\frac{\left(V - V_{bi}\right)^2}{d^3}\exp\left(\frac{0{,}98\beta\sqrt{V - V_{bi}}}{\sqrt{d}}\right) + \frac{V}{r} \tag{6}$$

where μ is the mobility, ε is the permitivity, β is the Poole–Frenkel parameter for taking into account the electric field-activation factor and the degree of disorder, V_{bi} is built-in potential, and d is the thickness of perovskite layer. The second term in eq. (6) is introduced to represent Ohmic medium governed by PbI_2 crystals inside the perovskite layer, where the amount of this PbI_2 crystals may increase when the cell degrades. The fitting results are shown in Figure 6, while the fitting parameters are tabulated in Table 1.

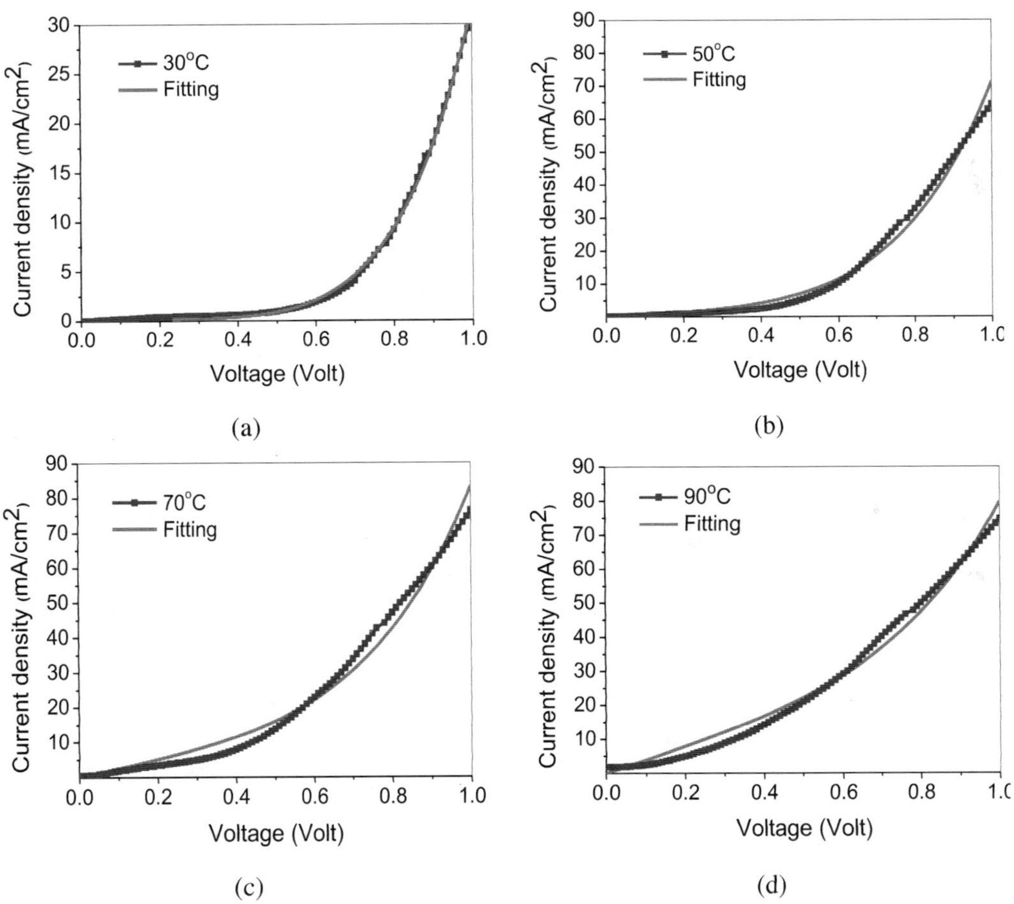

Figure 6. Fitting results of the *J-V* curves using eq. (6) for measurements at (a) 30°C, (b) 60°C, (c) 70°C and (d) 90°C.

In a solar cell, Schottky barrier (SB) may be formed at the interface between two semiconductors or between semiconductor and metal. The Schottky barrier causes the formation of built-in potential (V_{bi}) because the Fermi energy alignment causes the energy bending of valence band and conduction band. The observation of a diode characteristic measured at room temperature here may suggest the formation of built-in potential. As indicated in Table 1, the built-in potential values are zero for

Table 1. Parameter values obtained fitting using eq. (6).

	μ_0 (10^{-10} cm^2/Vs)	V_{bi} (eV)	r (Ω)
30 °C	2,6394	0,23	0,1451
50 °C	2,36804	0	0,01535
70 °C	2,86656	0	0,00405
90 °C	2,49048	0	0,00257

measurements above the room temperature. On the other hand, the internal resistance associated with *r* parameter decreases significantly with the increase of temperature. This may indicates the change of perovskite layer from semiconductor to Ohmic materials. Such situation may occur due to the degradation of perovskite layer, which possibly by separation of the inorganic and organic entities producing separate phases of PbI_2 and MAI again.

4. Conclusion

The measurement results show that the *J-V* characteristic of this perovskite solar cell is significantly affected by temperature. The *J-V* curve measured at room temperature has a diode characteristic. However, the fitting results indicate that the solar cell loses the diode characteristic and turns into Ohmic characteristic at higher temperature. This may be due to the degradation of the perovskite crystal at temperature, which may be caused by separation and recrystallization PbI_2. Special attention to overcome this problem is required for improving the stability of this solar cell performance.

Acknowledgment

This work was supported by Ministry of Research and Technology of Indonesia through National Innovation System Research Incentive 2015.

References

[1] Niu G, Guo X and Wang L 2014 Review of recent progress in chemical stability of perovskite solar cells *J. Mater. Chem.* **3** 8970-80

[2] Peng G, Xu X and Xu G 2014 Hybrid organic-inorganic perovskites open a new era for low-cost, high efficiency solar cells *Nanomaterials*

[3] Gratzel M 2014 The light and shade of perovskite solar cells *Nature Materials* **13** 838-42

[4] Kim H S *et al* 2012 Lead iodide perovskite sensitized all-solid-state submicron thin film mesoscopic solar cell with efficiency exceeding 9% *Scientific Reports* **2** 1-7

[5] Jeng J Y, Chen K C, Chiang T Y, Lin P Y, Tsai T D , Chang Y C , Guo T F, Chen P , Wen T C and Hsu Y J 2014 Nickel oxide electrode interlayer in $CH_3NH_3PbI_3$ perovskite/PCBM planar-heterojunction hybrid solar cells *Adv. Mater.* **26** 4107-13

[6] Bi D, El-Zohry A M, Hagfedlt A and Boscloo G 2015 Unraveling the effect of PbI_2 concentration on charge recombination kinetics in perovskite solar cells *ACS Photonics* **2** 589-94

[7] Butler K B, Frost JM and Walsh A 2015 Ferroelectric materials for solar energy conversion: photoferroics revisited *The Royal Society of Chemistry*. **8** 838-48

[8] Gao P, Nazeeruddin M K 2013 Organohalide lead perovskites for photovoltaic lead perovskites for photovoltaic *The Royal Society of Chemistry* **7** 2448-63

[9] Bretshneider S A, Weickert J, Dorman J A and Mende L S 2014 Physical and electrical characteristics of lead halide perovskites for solar cell applications *APL Materials.* **2** 040701

[10] Kirchartz T 2013 Influence of diffusion on space charge limited current measurements in organic semiconductors *Beilstein J. Nanotechnology* **4** 180–88

IOP Conf. Series: Journal of Physics: Conf. Series **877** (2017) 012044 doi:10.1088/1742-6596/877/1/012044

Construction of Maximum Tortuosity of Single Fluid Path in Grid-based 2-d System (5×20) and 3-d system (5×20×3) for Certain Value of Porosity

S Viridi[1], FDE Latief[2], SN Khotimah[3]

Physics Department, Faculty of Mathematics and Natural Sciences, Institut Teknologi Bandung, Jalan Ganesha 10, Bandung 40132, Indonesia

[1]dudung@fi.itb.ac.id, [2]fourier@fi.itb.ac.id, [3]nurul@fi.itb.ac.id

Abstract. Tortuosity is an important physical property in porous materials since it describes path length of fluid through the materials, which means how much the loss of kinetic energy. The simplest definition of tortuosity T is λ/L with λ is length of the path and L is the distance between two ends of the path. System is discretized using grid to limit number of possible paths, which could be infinity for continuous system. Porosity of the system is also constrained. Variable T is chosen as macrostates in statistical physics point of view, while all possible paths within this T are the microstates. It is found that some macrostates (more appropriately is its maximum value) have larger thermodynamics probability than the others. It should be a relation between this probability and reported tortuosity.

1. Introduction

Tortuosity is still an interesting parameter in recent years, since it can be implemented in various fields, such in botany for defining root path formation [1], in cardiac cell for studying cytoplasmic diffusivity [2], in battery for investigating ohmic losses [3], and in soil for predicting its value from hydraulic conductivity instead of porosity [4]. Its relations to porosity have been reported in various forms of relation [5, 6]. In this work instead of tortuosity another related term, average tortuosity, is introduced, which is averaged though all possible values within certain value of porosity.

2. Model

Porosity ϕ is fraction of void space V_V in material with total volume V_T [7]

$$\phi = \frac{V_V}{V_T}, \tag{1}$$

hydraulic tortuosity T is [8]

$$T = \frac{\lambda_{avg}}{L}. \tag{2}$$

where λ_{avg} is average length of fluid path and L is the geometrical length of the sample, and specific surface area S is [9]

$$S = \frac{N l_{\text{avg}}}{1 - \phi}, \tag{3}$$

with N is number of channels per cross-sectional area and l_{avg} is average peripheral length of each pore, which could be simply circumference of a pore.

A two dimensional model of porous medium is considered, which is relatively similar to [10], where upper and lower plate are assumed to have $\phi = 0$.

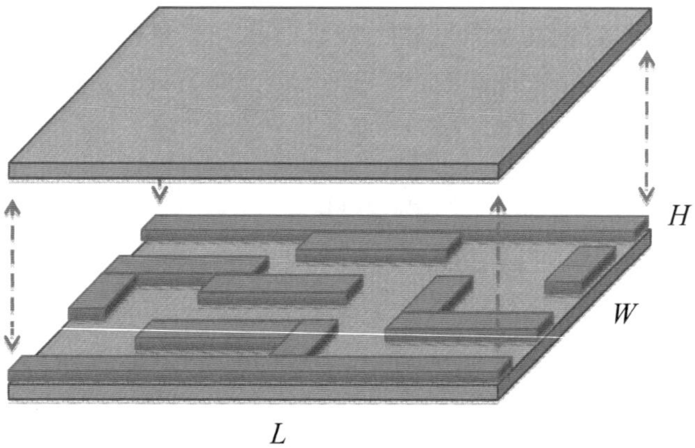

Figure 1. Model of two-dimensional porous medium with length of L and width W.

Part with blue plates in figure 1 represent the material, while the empty spaces represent the void in the porous medium. It can also be shown in two-dimension as given in figure 2. The system is simplified into rectangular grids with size in order of H. Number of grid in L direction is N_L and in W direction is N_W. Total number of grids in the system is

$$N = N_L N_W \tag{4}$$

with

$$N_L = \frac{L}{H}, \tag{5}$$

$$N_W = \frac{W}{H}. \tag{6}$$

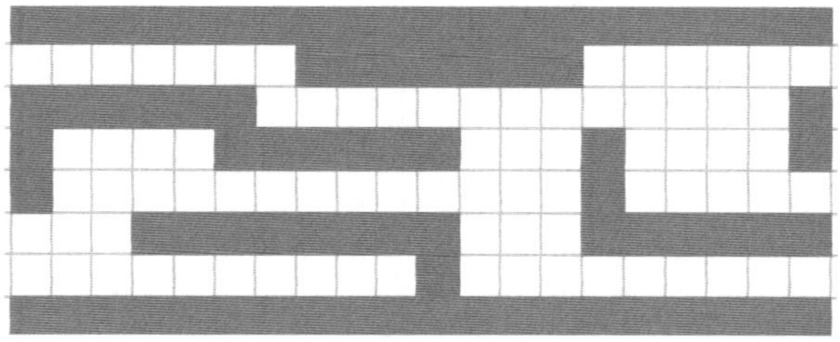

Figure 2. Two-dimension representation of porous medum between upper and lower plates.

Indexing of grids (i, j) begins at the lower left corner (0, 0) and ends at upper right corner (N_L, N_W). Position of grid (i, j) is determined through

$$\left(i\frac{H}{L}, j\frac{H}{W} \right). \tag{7}$$

Grid (i, j) has value 1 for material and 0 for void, which can be represented in the form of

$$G_{ij} = \begin{cases} 0, & \text{for void,} \\ 1, & \text{for matrial.} \end{cases} \tag{8}$$

Using equation (8) porosity of the system is simply

$$\phi = \frac{1}{N} \sum_{j=0}^{N_W} \sum_{i=0}^{N_L} G_{ij} \;. \tag{9}$$

Minimum porosity will be 0, while the maximum will be

$$\phi_{\max} = 1 - \frac{2}{N_W}, \tag{10}$$

since the system must have two side walls with length of N_L and width one grid (or H).

Tabel 1. Configurations in system with size 5×20 and $\phi_{\max} = 0.6$.

Configuration of void and material grids	ϕ
1 1 1 1 1 1 1 1 1 1 1 1 1 1 1 1 1 1 1 1 1 1 1 1 1 1 1 1 1 1 1 1 1 1 1 1 1 1 1 1 1 1 1 1 1 1 1 1 1 1 1 1 1 1 1 1 1 1 1 1 1 1 1 1 1 1 1 1 1 1 1 1 1 1 1 1 1 1 1 1 1 1 1 1 1 1 1 1 1 1 1 1 1 1 1 1 1 1 1 1	0.0
1 1 1 1 1 1 1 1 1 1 1 1 1 1 1 1 1 1 1 1 1 1 1 0 1 1 0 1 0 1 0 1 1 0 0 0 0 1 1 1 1 1 1 1 1 0 1 1 1 1 0 0 1 1 1 1 1 1 1 1 0 0 1 1 1 0 1 1 1 0 1 1 0 0 0 1 1 0 1 0 1 1 1 1 1 1 1 1 1 1 1 1 1 1 1 1 1 1 1 1	0.2
1 1 1 1 1 1 1 1 1 1 1 1 1 1 1 1 1 1 1 1 1 0 1 0 1 0 0 0 0 0 0 0 0 1 0 0 0 0 0 0 0 1 0 1 1 0 1 0 0 1 0 0 0 0 0 0 1 0 1 0 1 0 1 0 1 0 1 0 0 0 1 0 0 1 1 0 0 0 1 1 1 1 1 1 1 1 1 1 1 1 1 1 1 1 1 1 1 1 1 1	0.4
1 1 1 1 1 1 1 1 1 1 1 1 1 1 1 1 1 1 1 1 0 0 0 0 0 0 0 0 0 0 0 0 0 0 0 0 0 0 0 0 0 0 0 0 0 0 0 0 0 0 0 0 0 0 0 0 0 0 0 0 0 0 0 0 0 0 0 0 0 0 0 0 0 0 0 0 0 0 0 0 1 1 1 1 1 1 1 1 1 1 1 1 1 1 1 1 1 1 1 1	≥ 0.6

In generating certain porosity number of grids representing material is start with value $2N_L$ until N, which corresponds to porosity $2/N_W$ and 1. A value of certain porosity ϕ will be considered as a macrostate, while all possibilities to construct the state are the microstates. If there are $(N - 2N_L)$ grids and porosity ϕ is the target, then there will be

$$W_\phi = \frac{(N - 2N_L)!}{(N\varphi)!(N - 2N_L - N\varphi)!} \tag{11}$$

microstates for it. Random process is used to generate the microstates, which is related to values of i and j, but not for the $2N_L$ grids. If f_U is a function to generate number, where $0 \leq f_U \leq 1$ then

$$k = 1 + \lfloor f_U (I_u - I_l) \rfloor + I_l, , \tag{12}$$

will generate $k = I_l, I_l + 1, .., I_u - 1, I_u$, where I_l and I_u is lower and upper values of desired integer. In the system k will represent i and j. Symbol of floor operation is used in equation (9)

$$\lfloor x \rfloor = \max \{ m \in Z \mid m \leq x \}, \tag{13}$$

where m is integer. Beside the $2N_L$ grid for materials in order to get a particular value of ϕ

$$N_\phi = \max(0,\ N - N\phi - 2N_L) \tag{14}$$

random numbers should be generated. Maximum porosity $\phi_{\max}$ is obtained when $N_\phi = 0$ (no random numbers generated) and it will turn equation (14) into equation (11) with help of equation (4). Illustration of system 5×20 is given in table 1, where $\phi_{\max} = 0.6$ in this case. Increasing system size can reduce this value, but it is also increasing computation time. If $N \rightarrow \infty$ then $\phi_{\max} \rightarrow 1$.

From table 1 it can be seen that a channel can be constructed if $0.2 \leq \phi \leq \phi_{\max}$, but not always, since for $\phi = 0.2$ it must has $T = 1$ with only one channel. This possibility is not favorable. In this work only perpendicular turns are considered for simplifying the calculation. This will give that the next channel length will be more one grid in length than the channel with $T = 1$, which is given in figure 3.

```
1 1 1 1 1 1 1 1 1 1 1 1 1 1 1 1 1 1 1 1
1 1 1 1 1 1 1 1 1 1 1 1 1 1 1 1 1 1 1 1
0 0 0 0 0 0 0 0 0 0 0 0 0 0 0 0 0 0 0 0
1 1 1 1 1 1 1 1 1 1 1 1 1 1 1 1 1 1 1 1
1 1 1 1 1 1 1 1 1 1 1 1 1 1 1 1 1 1 1 1
                  (a)

1 1 1 1 1 1 1 1 1 1 1 1 1 1 1 1 1 1 1 1
0 0 0 0 0 0 0 0 0 0 1 1 1 1 1 1 1 1 1 1
1 1 1 1 1 1 1 1 1 0 0 0 0 0 0 0 0 0 0 0
1 1 1 1 1 1 1 1 1 1 1 1 1 1 1 1 1 1 1 1
1 1 1 1 1 1 1 1 1 1 1 1 1 1 1 1 1 1 1 1
                  (b)

1 1 1 1 1 1 1 1 1 1 1 1 1 1 1 1 1 1 1 1
0 0 0 0 0 0 0 0 0 0 1 1 1 1 1 1 1 1 1 1
1 1 1 1 1 1 1 1 1 0 1 1 1 1 1 1 1 1 1 1
1 1 1 1 1 1 1 1 1 0 0 0 0 0 0 0 0 0 0 0
1 1 1 1 1 1 1 1 1 1 1 1 1 1 1 1 1 1 1 1
                  (c)

1 1 1 1 1 1 1 1 1 1 1 1 1 1 1 1 1 1 1 1
0 0 0 0 0 0 0 1 1 1 1 1 1 1 1 1 1 1 1 1
1 1 1 1 1 1 0 0 0 0 0 0 0 0 1 1 1 1 1 1
1 1 1 1 1 1 1 1 1 1 1 1 1 0 0 0 0 0 0 0
1 1 1 1 1 1 1 1 1 1 1 1 1 1 1 1 1 1 1 1
                  (d)
```

Figure 3. Some examples of single channel with different tortuosity T: (a) 1, (b) 1.05, (c) 1.1, (d) 1.1.

Zeros with red color in figure 3 are the perpendicular turns, where fluid flows could exhibit different fluid friction compare to the straight paths. These types of friction will be adapted from the previous work that in straight line friction will be proportional to velocity, while in the perpendicular turn it will be proportional to square of velocity due to change of coordinate system [11].

3. Results and discussion

For a particular value of porosity ϕ there will be some number of possible configuration or microstates W_ϕ, which is given by equation (11) and number of tortuosity may be varied. What easier to determine is maximum tortuosity $T_{\max}$ for a particular value of porosity ϕ as shown in figure 4.

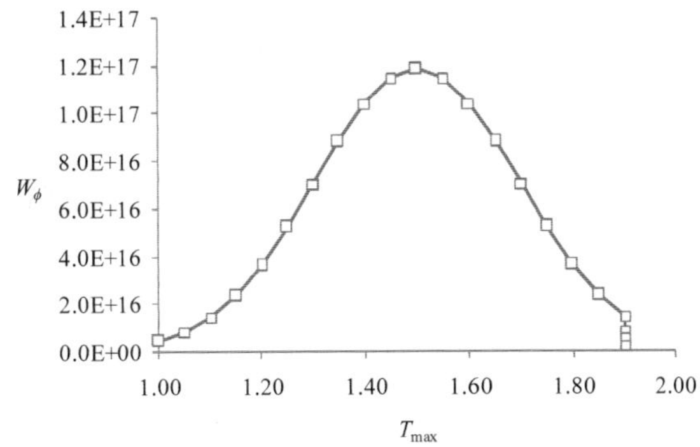

Figure 4. Number of microstates W_ϕ for particular value of maximum tortuosity T_{max}.

It seem that T_{max} = 1.5 (or ϕ = 0.3) gives maximum value of microstates W_ϕ, which can be seen in figure 4 as the peak in the curve. For $\phi \geq 0.38$ value of T_{max} remains at 1.90, where increasing of porosity does not increase value of maximum tortuosity. Several examples of microstates for several value of ϕ, T_{max}, and W_ϕ are given in figure 5. Path with more perpendicular turns give higher value of ϕ and T_{max}, but not always value of W_ϕ.

1	1	1	1	1	1	1	1	1	1	1	1	1	1	1	1	1	1	1	1
1	0	0	0	1	0	0	0	1	0	0	0	1	0	0	0	1	0	0	0
1	0	1	0	1	0	1	0	1	0	1	0	1	0	1	0	1	0	1	1
0	0	1	0	0	0	1	0	0	0	1	0	0	0	1	0	0	0	1	1
1	1	1	1	1	1	1	1	1	1	1	1	1	1	1	1	1	1	1	1

1	1	1	1	1	1	1	1	1	1	1	1	1	1	1	1	1	1	1	1
0	0	0	0	1	0	0	0	1	1	1	1	1	0	0	0	1	1	1	1
1	1	1	0	1	0	1	0	1	1	1	1	1	0	1	0	1	1	1	1
1	1	1	0	0	0	1	0	0	0	0	0	0	0	1	0	0	0	0	0
1	1	1	1	1	1	1	1	1	1	1	1	1	1	1	1	1	1	1	1

1	1	1	1	1	1	1	1	1	1	1	1	1	1	1	1	1	1	1	1
0	0	0	0	0	0	0	0	1	1	1	1	1	0	0	0	0	0	0	0
1	1	1	1	1	1	1	0	1	1	1	1	1	0	1	1	1	1	1	1
1	1	1	1	1	1	1	0	0	0	0	0	0	0	1	1	1	1	1	1
1	1	1	1	1	1	1	1	1	1	1	1	1	1	1	1	1	1	1	1

Figure 5. Examples of configuration with: ϕ = 0.38, T_{max} = 1.9, W_ϕ = 1.41543E+16 (top), ϕ = 0.30, T_{max} = 1.5, W_ϕ = 1.18265E+17 (center), and ϕ = 0.24, T_{max} = 1.2, W_ϕ = 13.60524E+16 (bottom).

Average tortuosity can be calculated from

$$T_{avg} = \frac{1}{W_\phi} \int_1^{T_{max}} m(\mathrm{T}) dT , \tag{15}$$

where $m(T)$ is number of microstates for every tortuosity T on a certain value of porosity ϕ. Lower bound of the integral is due to shortest path which gives tortuosity of one, while the upper bound is due to the possible maximum value of tortuosity.

If it is assumed that every value of tortuosity T less than T_{max} has only one microstate on a certain value of porosity ϕ or $m(T) = 1$ then equation (15) will be simplified into

$$T_{avg} \approx \frac{1}{W_\phi} \int_1^{T_{max}} dT = \frac{T_{max} - 1}{W_\phi} . \tag{16}$$

IOP Conf. Series: Journal of Physics: Conf. Series **877** (2017) 012044 doi:10.1088/1742-6596/877/1/012044

Results of equation (16) is presented in figure 6. Values of T_{avg} are quite small since for a certain value of porosity ϕ there is a lot of microstates which do not construct a channel through the system. Minimum value of T_{avg} is found at ϕ about 0.28.

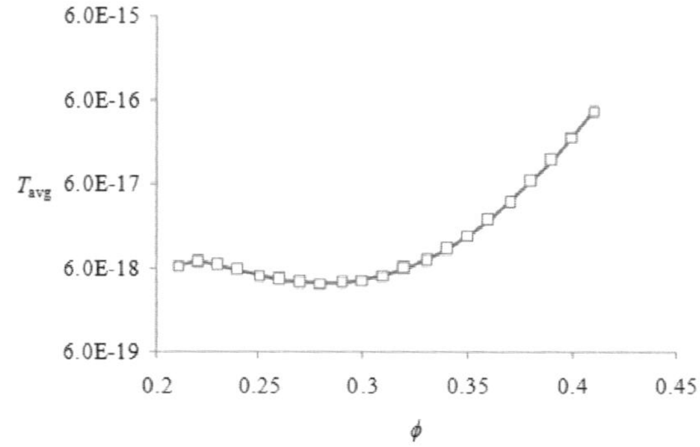

Figure 6. Average tortuosity T_{avg} of certain value of porosity ϕ.

From this point forward let minimum value of porosity $\phi = 0.28$ be the focus of this work. For $\phi = 0.28$ two u-valleys are required, which is shown in figure 7 with $w_1 = w_2 = 1$, where w_n is width of the n-th u-valley. Another possibility is given in figure 8, where $w_1 = 1$ and $w_2 = 1, .., 12$.

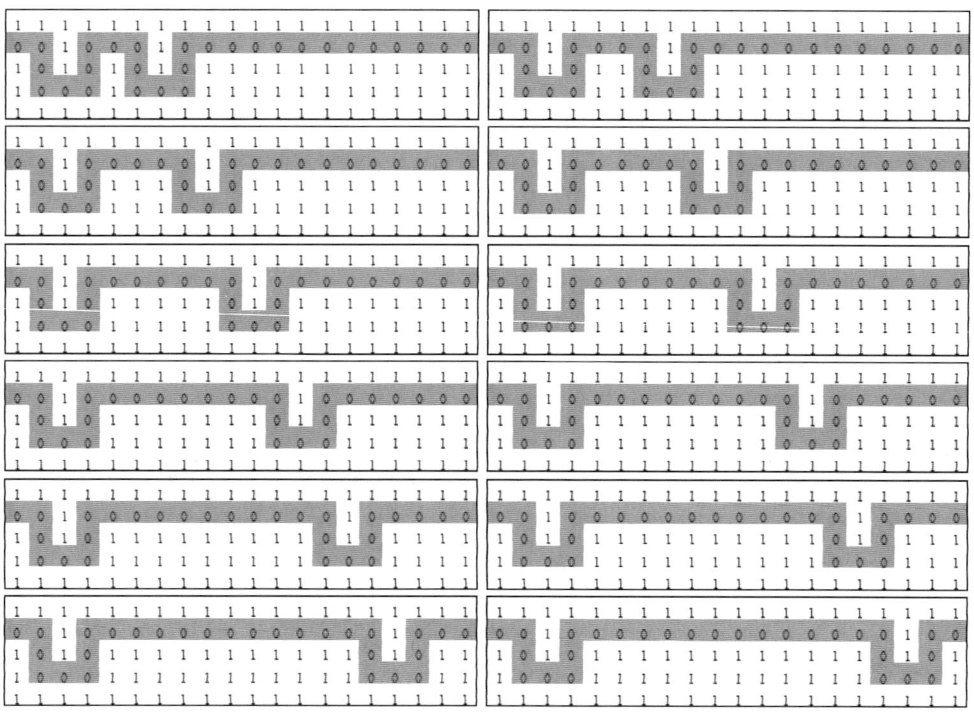

Figure 7. Two u-valleys ($\phi = 0.28$ $w_1 = 1$, $w_2 = 1$, $l_{12} = 1$ .. 12) can make 78 microstates by changing position of first u-valley from 1 to 12 (only first 12 is shown).

Distance between the two u-valleys i and j is defined as l_{ij} as shown in figure 7 for l_{12}. For the system value of l_{12} is between 1 and 12, while for longer system with the same width it will be

$$l_{12} \in [1, N_L - 8]. \tag{17}$$

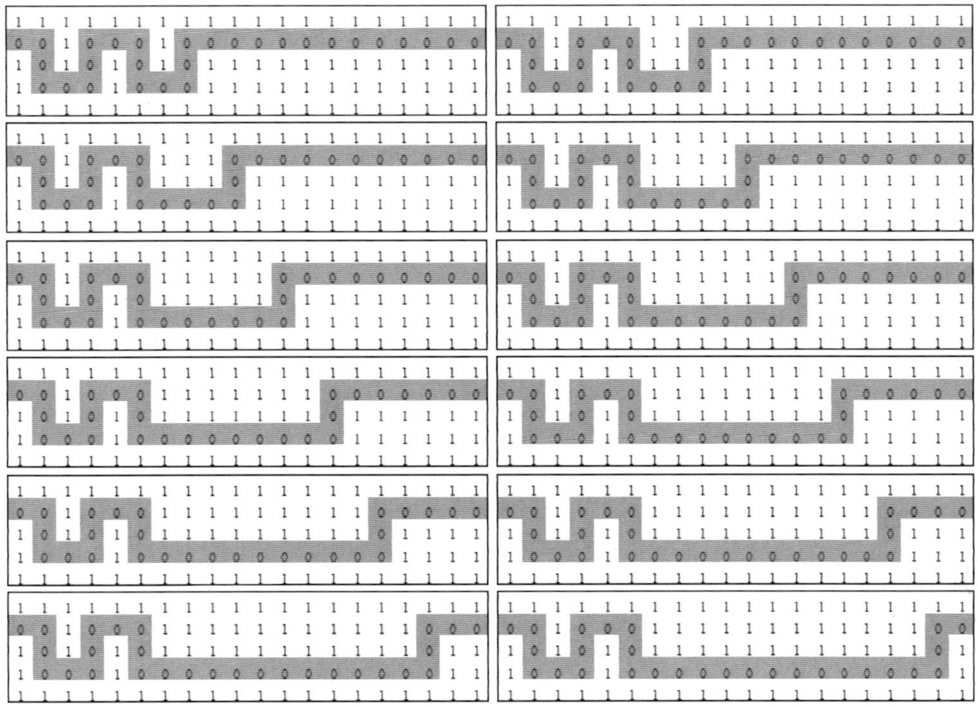

Figure 8. Two u-valleys ($\phi = 0.28$ $w_1 = 1$, $w_2 = 1..12$, $l_{12} = 1$) can make 364 microstates by changing position of first u-valley from 1 to 12 (only first 12 is shown).

As in equation (17) for system with longer size but with same width value of w_2 can be summarized as

$$w_{12} \in [1, N_L - 8]. \quad (18)$$

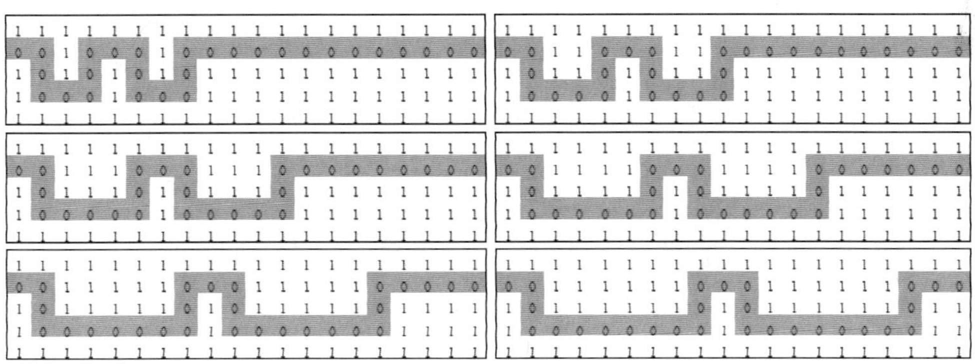

Figure 9. Two u-valleys ($\phi = 0.28$ $w_1 = 1.. 6$, $w_2 = 1..6$, $l_{12} = 1$) can make 40 microstates by changing position of second u-valley (only first 6 is shown).

For l_{12} is fixed at one as shown in figure 9 and $w_1 = w_2$ then

$$w_1 = w_2 \in [1, \tfrac{1}{2} N_L - 4], \quad (19)$$

where N_L should be even. A more compact relation that can merge equations (17) – (19) is one of the homework for future investigation, also when $N_W > 5$.

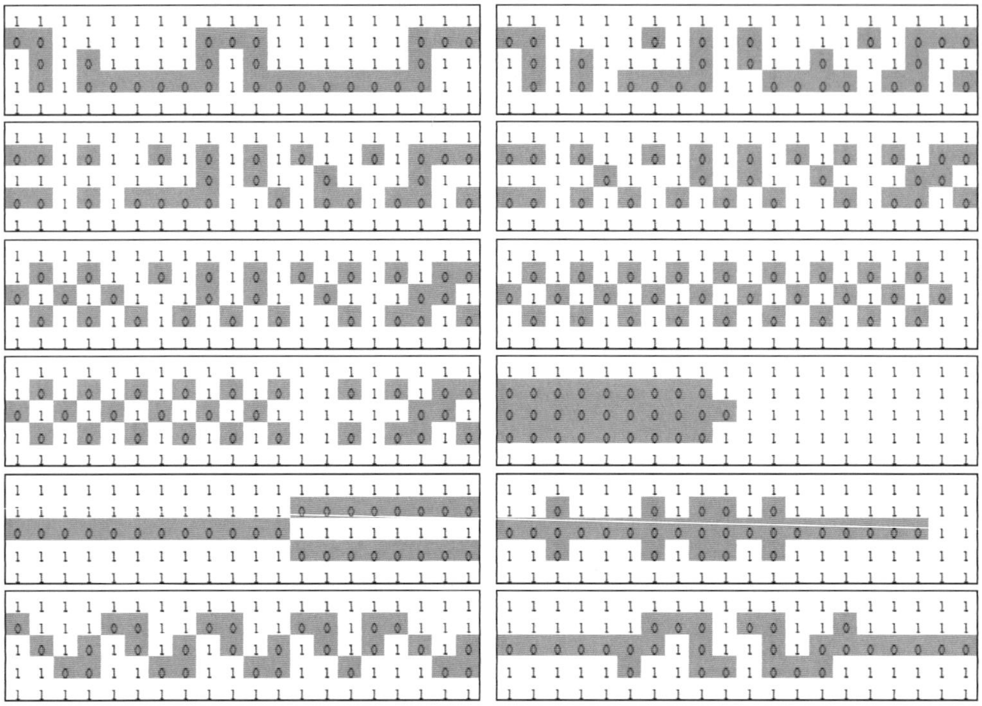

Figure 10. Some microstates ($\phi = 0.28$) which do not have tortuosity.

For now it is assumed impossible to figure out how many configurations which can construct a single tortuosity value. It is the reason why equation (16) considers only one microstate. Figures 7-9 shows only some possibilities, while there are still a lot of others to be found.

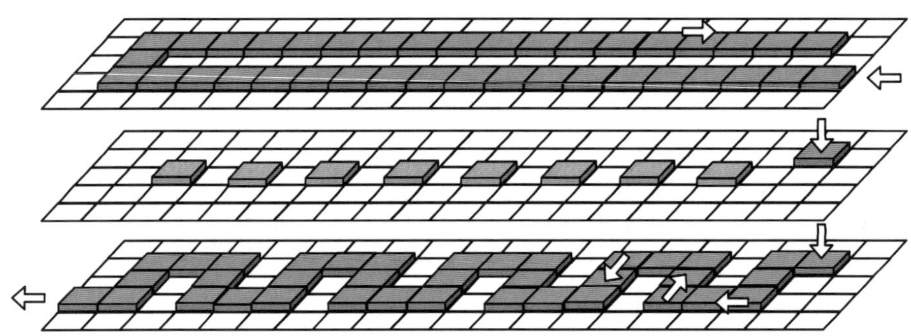

Figure 11. Maximum tortuosity T_{max} in system 5×20×3 for porosity ϕ = 0.28 with 8 unconnected voids.

Supposed that the system is increased in H direction from 1 to 3, which means that current 5×20 (2-d) system is changed into 5×20×3 (3-d) system as shown in figure 11. Number of additional materials grids will be

$$N_{\varphi} = \max(0, N - N\varphi - 6N_L) \tag{20}$$

from equation (14). Then $N = 300$ and similar porosity ($\phi = 0.28$) will required $N_{\varphi} = 96$, number of random number to be generated or simply 84 voids inside the 5×20×3 grids. Figure 11 gives the maximum possible tortuosity T_{max} = 3.8 for the system. It can be understood that porosity between

0.253 and 0.280 will have the same value of T_{max}, since the unconnected voids do not give any contribution to T_{max}. In future work permeability must also be investigated related to tortuosity instead only related to porosity [12], since tortuosity determines permeability more than porosity. Current results are still not sufficient to show relation between tortuosity to porosity in the forms as shown in [5, 6], which are very handy for further use. More microstates are required in order to obtain such formulation.

4. Summary

System in two-dimension for modeling tortuosity of single channel fluid path has been developed. Based on statistical physics postulate, thermodynamics probability of a certain porosity value to produce a tortuous channel is very small, in order of 10^{-17}. This means that the postulate only is not sufficient to predict how such channel can be formed in nature.

Acknowledgements

This work and the presentation are supported by research scheme RIK ITB with contract number 006q/I1.C01/PL/2016 and ICES committee – Physics Department, FMIPA, ITB, respectively.

References

[1] Popova L, van Dusschoten D, Nagel KA, Fiorani F, Mazzolai B 2016 *Ann. Bot.* **118** xx
[2] Richards M, Lomas O, Jalink K, Ford KL, Vaughan-Jones RD, Lefkimmiatis K, Swietach P 2016 *Cardiovasc. Res.* **110** 395
[3] Landesfeind J, Hattendorff J, Ehrl A, Wall WA, Gasteiger HA 2016 *J. Electrochem. Soc.* **163** A1373
[4] Carey GR, McBean EA, Feenstra S 2016 *Groundwater* **54** 476
[5] Archie G 1942 Trans. Am. Inst. Min., Metall. Pet. Eng. 146 54
[6] Ahmadi MM, Mohammadi S, Hayati AN 2011 Phys. Rev. E 83 026312
[7] Hilfer R 1992 *Phys. Rev. B* **45** 7115
[8] Matyka M, Khalili A, Koza Z 2008 *Phys. Rev E* **78** 026306
[9] Schwertz FA 1949 *J. Am. Ceram. Soc.* **32** 390.
[10] Hemmatifar A, Stadermann M, Santiago JG 2015 *J. Phys. Chem. C* **119** 24681
[11] Viridi S, Novitrian, Nurhayati, Hidayat W, Latief FDE, Zen FP 2014 *AIP Conf. Proc.* **1615** 199
[12] Khotimah SN, Branarto BB, Viridi S, Arif I 2012 *Indonesian J. Phys.* **23** 5

First Principle Calculation : Investigation on interaction of Pt/Graphene as Catalyst

Nuning Anugrah Putri Namari, Suprijadi

Departement of Physics, Institut Teknologi Bandung

Email : nuningapn@gmail.com, supri@fi.itb.ac.id

Abstract. The increasing in energy needs and the lack of non-renewable energy sources becomes a challenge for the human being to be able to use renewable energy sources. One of the devices to process renewable energy is Polymer Electrolyte Membrane Fuel Cell (PEMFC) . PEMFC use hydrogen and Oxygen as an energy sources . The most important reaction in fuel cell is Oxidation and reduction process. Therefore, a catalyst is needed to help the OR process. Study of catalyst shows that the most effective fuel cell for now is Platinum. Many fuel cell have use platinum as the catalyst. However, Platinum is a rare and expensive element. Therefore, to reduce the cost of fuel cell fabrication, we need to increase the activity of platinum. In this research, we use graphene as a support material. Then, we will study about the interaction of platinum on graphene and analyze its morphological change and electronic properties.The research conduct using Density Functional Theory (DFT). The calculation result shows that Pt/graphene can break H_2 into H^+ and the binding between Pt cluster is stronger than binding with the substrate.

1. Introduction

Human population is increasing everyday. Therefore, energy demands is also increasing. There are two types of energy : non-renewable and renewable. Non-renewable energy is an energy which comes from the sources that can't be replenished in a short period of time, such as fossil fuel. Meanwhile, renewable energy is an energy which comes from a renewable sources that can be replenished in a short time such as solar and wind. Until now, many country such as Indonesia and other develop country still depend on non-renewable energy. If this condition is continue, petroluem is predicted to run out in the next 14 years . Because of that, we need to consider to use renewable energy. However, renewable energy has some disadvantages, the cost of technology is still expensive.

One of the technology that we use for renewable energy processing is fuel cell. Fuel cell produces electricity through chemical reaction. Some of the advantages of fuel cell are eco-friendly and has higher efficiency than diesel or gas engines. One of the most efficient type of fuel cell is Polymer Electrolyte Membrane Fuel Cell (PEMFC). This fuel cell use Platinum as a catalyst. The main fuel of this fuel cell is just Hydrogen and Oxygen. Therefore, it is an eco-friendly technology. The most important part of the fuel cell is catalyst, it determines how fast the reaction works. But, as we know, platinum as the catalyst is an expensive material . That is why the cost of the fuel cell is still high. Because of that, to lower the cost, some researcher try to add support material into the catalyst to get higher activity of catalyst without using much platinum. Nowadays, researcher try to use graphene as a support material because it has high surface area (~2630 m^2/g) which is higher than graphite (~10m^2/g) and CNT (1300 m^2/g). The experimental work also shows that using graphene as a support material will give *Electrochemical Active Surface Area* (ECSA) about 53 m^2/g which is higher than using carbon black (26 m^2/g). [1]

Better understanding of the interaction between platinum and graphene is needed to achieve the highest activity of platinum and to be able to do the optimation of platinum use. In this research, we will do the investigation on Pt/graphene interaction.

2. Computational Details

All of the calculation in this work is performed using Vienna *ab initio* simulation package VASP that based on DFT. We use the generalized-gradient approximation (GGA) as an exchange correlation. The K-points that we use is 5x5x1 Monkhorst-Pack grid with energy cut-off 400.00 eV. We also use 5x5 graphene on this work.

3. Results and Discussion

In this work, we will calculate energy adsorption. For a single platinum on graphene (Pt/graphene) system, we will calculate the energy using this equation [2]

$$E_{ad} = \left(E^{Pt} + E^{graph}\right) - E^{Pt/graph} \qquad (1)$$

$E^{Pt/graph}$ is the total energy of the Pt-graphene layer, E^{graph} is the total energy of clean graphene layer, and E^{Pt} is the total energy of an isolated Pt-atom. For the other Pt cluster, we will calculate 2 types of adsorption energy taken from Pitor Błoński,et all work. [3] First, we will calculate interaction energy of Pt-cluster of a certain size and shape with graphene sheet, defined by

$$E_{ad}^{int} = \frac{1}{n}\left(E^{Pt_n/graph} - E^{graph} - E^{Pt_n}\right) \qquad (2)$$

$E^{Pt_n/graph}$ is the total energy of the Pt_n-graphene system, E^{Pt_n} is the total ground-state energy of the Pt_n cluster in the gas phase. Next, we wil calculate cohesive energy of the adsorption, defined by

$$E_{ad}^{coh} = \frac{1}{n}\left(E^{Pt_n/graph} - E^{graph} - nE^{Pt}\right) \qquad (3)$$

The Equation (3) measures the binding between Pt in the cluster.

3.1 Adsorption of H_2

A catalyst in fuel cell have a function to break H_2 into ion H^+. In this section, we will prove that Pt/graphene can be use as a catalyst.

(a) (b)

Figure 1. H_2 on Pt/graphene (a) initial position (b) after relaxation

Figure 1 (a) shows the initial position of Pt/graphene. Figure 1 (b) shows the final position after the relaxation. From the Figure 1 above , we can see that H_2 is break into H^+ after interact with Pt/graphene.

3.2 Pt/graphene

Before we calculate the energy adsorption, first of all, we have to find the most stable position of Pt on graphene. In this work, we placed the graphene in 3 site : Top, Bridge, and Hollow. Top is the position on top of C atom, Bridge is the place between C-C bonds, and Hollow is the place between 6 atom carbon (see figure below)

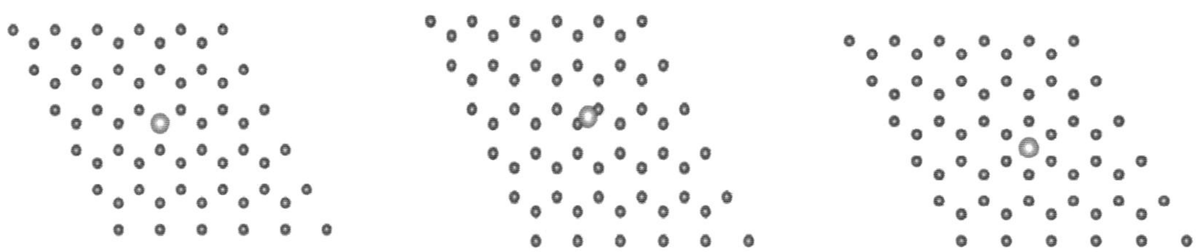

(a) (b) (c)

Figure 2. Adsorption site of Pt (a) Top (b) Bridge (c) Hollow

Static calculation was performed to get the most stable position

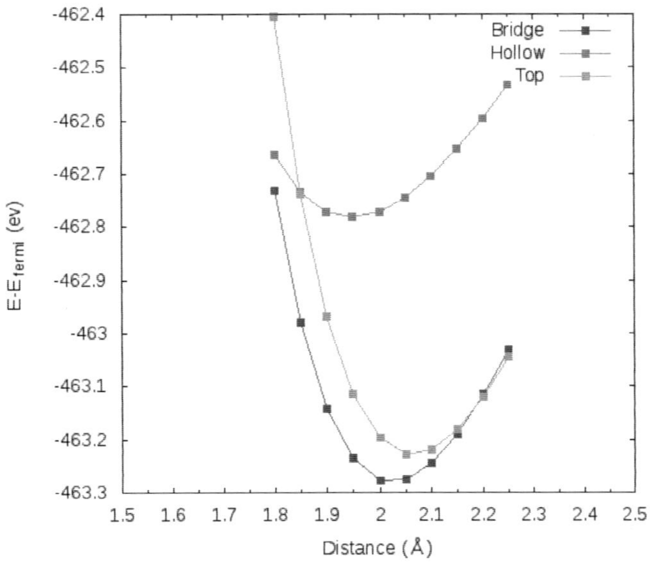

Figure 3. Total energy on every adsorption site

Figure 3, shows that the most stable position is Bridge with distance 2Å from the graphene layer. After that, relaxation was performed on the structure to get the most stable structure (Figure 4) . We can see from the figure 4 that the structure of graphene is change and there is some transfer charge between platinum and graphene.

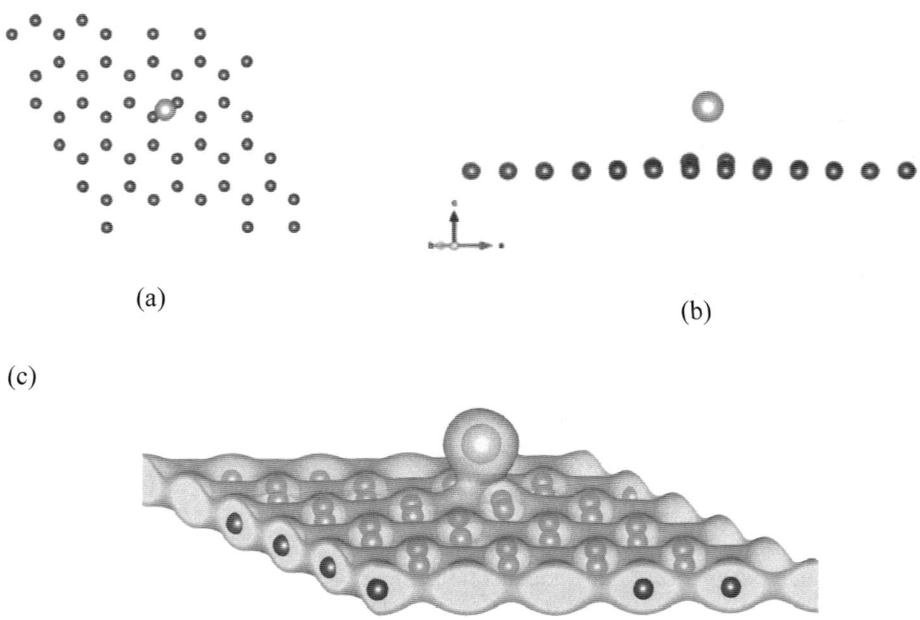

Figure 4. Structure of Pt/graphene after relaxation (a) Top view (b) Side view (c) Transfer charge

Next, we will investigate the reversibility of Pt/graphene. Figure 5 shows Pt/graphene structures before react with H_2, when react with H_2, and after reacted with H_2. The figure shows that there are some changes in the height of graphene structure (Δx) when react with H_2.

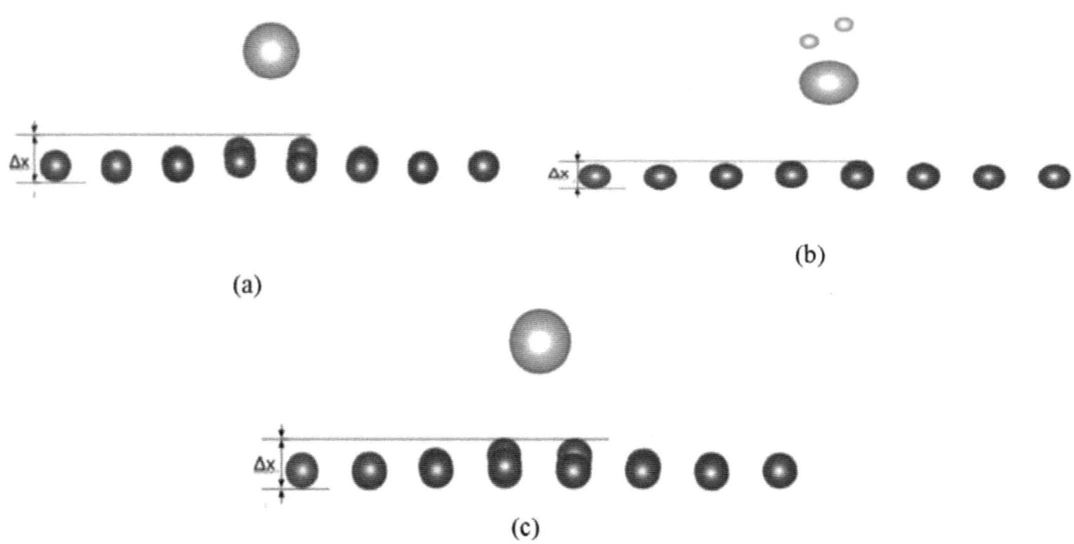

Figure 5 Structure of Pt/graphene (a) without H_2 ,(b) react with H_2, (c) after H_2 removed

If we calculate Δx on each condition we will get the graph on figure 6. On the graph below, we can see that Pt/graphene structure is 95% reversible. Therefore, it can be use for many times.

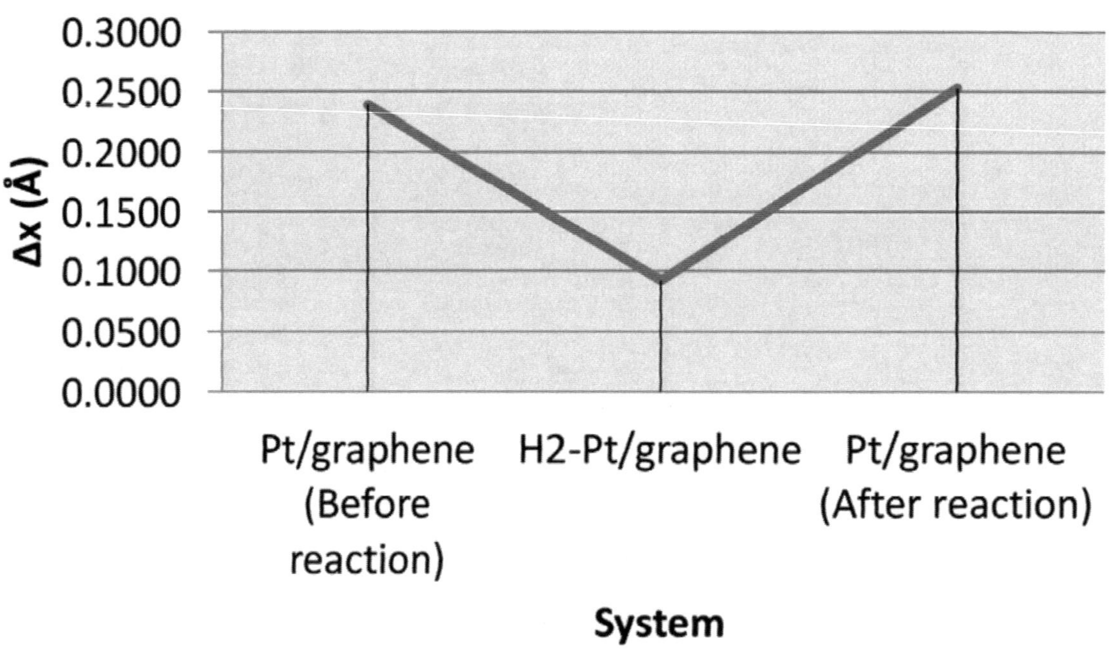

Figure 6. Graph that shows the reversibility of Pt/graphene

3.3 Pt_2/graphene

Now, we will investigate the structure of Pt_2 cluster on graphene. In this work, we use parallel position of Pt_2 cluster.

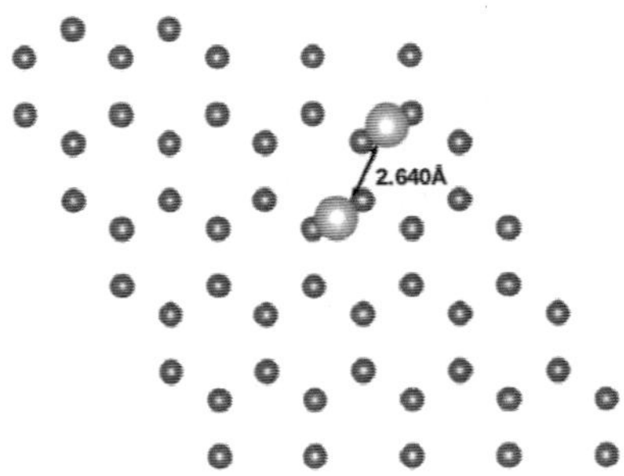

(a)

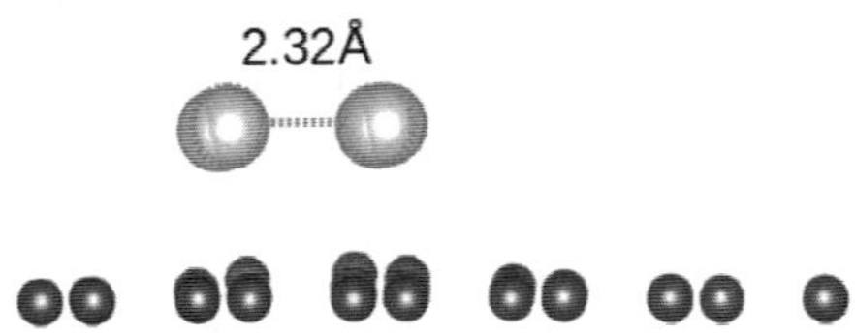

(b)

Figure 7. Structure of Pt_2/graphene after relaxation (a) Top view (b) Side view

The red circles in Figure 7(b) shows the distance between Pt before relaxation. If we compare the result with Figure 7(a) which shows the Pt distance after relaxation, we can see that the distance between Pt is become longer.

3.4 Pt_3/graphene

For Pt_3 cluster, we will use Pt_3 cluster with flat triangular configuration where every Pt is in the bridge site.

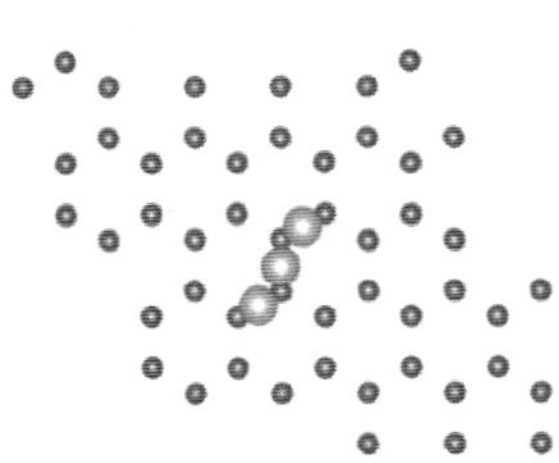

(a)

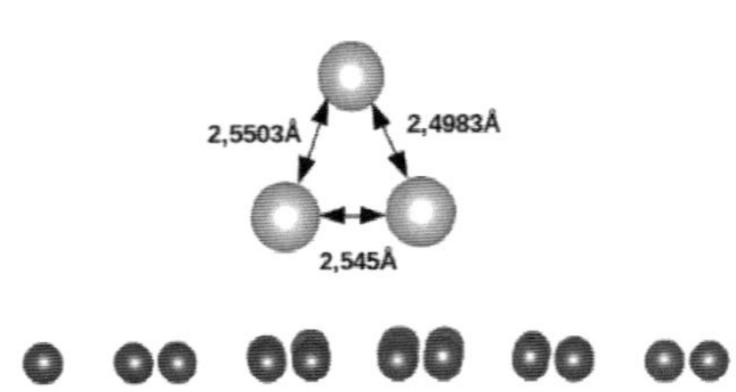

(b)

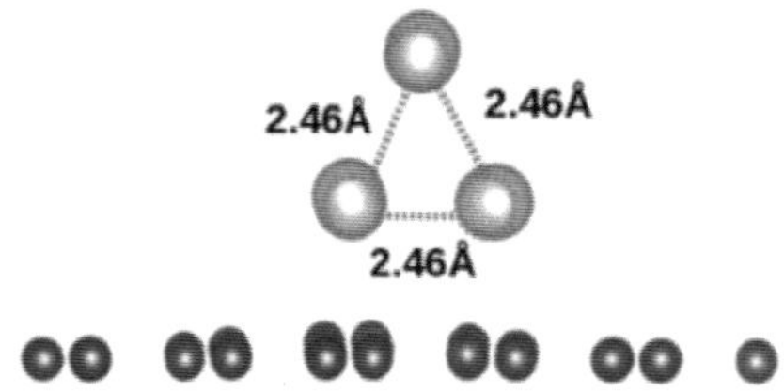

(c)

Figure 8. Structure of Pt_3/graphene after relaxation (a) Top view (b) Side view(c) Side view with Pt position before relaxation

Figure 8 shows that the distance between the platinum is larger after the relaxation, it means that they encounter repulsive force with each other

3.5 Energy adsorption

Using equation (1), we get that the adsorption energy of Pt/graphene system 1.97 eV and this is compareable with previous work [2] .

For the Pt_2/graphene and Pt_3/graphene, we calculate the adsorption energy using equation (2) and (3)

Table 1. Adsorption energy of Pt-cluster on graphene

System	**E^{int} (eV)**	**E^{coh} (eV)**
Pt_2/graphene	-0.595	-2.483
Pt_3/graphene	-0.38	-3.188

From Table 1, we can see that when we add more platinum, the energy adsorption between cluster and substrate is getting lower while the binding between cluster is getting stronger.

4. Conclusion

Structural and electronic properties of Pt cluster on graphene has been investigated using DFT method. The Calculation of adsorption energy shows that the binding between the cluster is much stronger than the binding with the substrate. From the calculation above, we can see that Pt/graphene can break H_2 into H^+, it means that it can be use as catalyst. The calculation also shows that Pt/graphene structure is 95% reversible. Therefore, we can use it many times .

References

[1] H. J. Choi, S. M. Jung, J. M. Seo, D. W. Chang, L. Dai dan J. B. Baek, "Graphene for energy conversion and storage in fuel cells and supercapacitors," *Nano Energy,* pp. 534-551, 2012.

[2] O. M. K., M. Yoshitada, T. S dan K. M, "Structures of Pt clusters on graphene by first-principles calculations," *Surface Science,* vol. 604, pp. 144-154, 2010.

[3] P. Blonski dan J. Hafner, "Geometric ad magnetic properties of Pt cluster supported on graphene : Relativistic density-functional calculations," *The Journal of Chemical Physics,* p. 134, 2011.

IOP Conf. Series: Journal of Physics: Conf. Series **877** (2017) 012046 doi:10.1088/1742-6596/877/1/012046

Influences of Precursor Solution Concentration and Temperature on $CH_3NH_3PbI_3$ Perovskite Layer Morphology and the Unconverted PbI_2 Proportion to their Perovskite Solar Cell Characteristics

Mhd. Ikhsan Alturisa[1], Joe Wira[1], Mardiyati[1], Herman[2] and Rahmat Hidayat[2]

[1] Materials Engineering Program Study, Faculty of Mechanical and Aerospace Engineering, Institut Teknologi Bandung, Indonesia
[2] Physics of Magnetism and Photonics Research Division, Physics Program Study, Faculty of Mathematics and Natural Sciences, Institut Teknologi Bandung, Indonesia

e-mail: rahmat@fi.itb.ac.id

Abstract. The formation of organometal-halide perovskite crystal is simply based on stoichiometry reaction. However, the formation process seems to be sensitively affected by some process parameters such as concentration and temperature of precursor solution. In this work, we studied the formation of $CH_3NH_3PbI_3$ ($MAPbI_3$) crystal in mesoporous TiO_2 layer deposited by spin coating technique. Perovskite layers were characterized by X-Ray Diffraction (XRD) and its surface morphology were investigated by Scanning Electron Microscope (SEM). Perovskite layers obtained from precursor solution at 40°C and 60°C show a mesh-like structure. Meanwhile, for higher precursor solution temperature (80°C), it shows a flake-like structure. However, precursor solution at 60°C produced the best conversion of precursor solution with the smallest percentage of the remaining unconverted PbI_2 inside the produced crystal layer. These experimental results also showed that the best crystallization degree is obtained from 1.0 M precursor solution. The photovoltaic characteristics of their solar cells seem to have a correlation with the quantity of unconverted PbI_2. High PbI_2 content significantly decreased the Filling Factor of the solar cell.

1. Introduction

Research in the application of organometal-halide perovskite (hybrid perovskite) as an active layer in solar cells has significantly increased in the past 5 years [1,2]. This material exhibits photovoltaic characteristics with relatively small energy gap [3], and therefore becomes a good candidate for active layer in solar cell. Despite of its simplicity in its preparation, this hybrid perovskite material produced remarkable high solar cell efficiency. The key feature of this material is that it can be synthesized with wet-chemical process at low temperature. This low temperature process feature could reduce the cost of cell fabrication which leads to open a possibility of much cheaper solar cell among the third generation of solar cells.

Research works on hybrid perovskite material as an active layer in a solar cell were pioneered by Miyasaka and co-workers in 2009 [4]. The best efficiency of this perovskite solar cell (PSC) was just 3.81% at that time. However, in just 5 years since that report, the PSCs efficiency records increased significantly up to 20.1% [5]. The most studied hybrid perovskite material in PSC is methylammonium lead iodide ($CH_3NH_3PbI_3$ or $MAPbI_3$). Thin film perovskite layer is commonly deposited onto a substrate using spin coating technique from precursor solution, which is prepared by dissolving lead halide salt and organic halide salt powders at a specific mole ratio in a high boiling point aprotic polar solvent [6]. Commonly, a mesoporous substrate, for instance from TiO_2, is used as the substrate.

The formation of $MAPbI_3$ crystal is given by the following reaction scheme:

$$PbI_2 + MAI \rightarrow MAPbI_3, \quad (1)$$

which is a reaction with simple reaction stoichiometry. The formed crystal is believed to be formed by ionic bond. However, weak interaction such as a Coulomb force, which works in two opposite charges, may also contribute in this crystal formation. The reaction rate is very fast but the reaction seems to be highly sensitive with some reaction parameters such as precursor concentration, air humidity and processing temperature. Those parameters seem to affect the conversion degree from precursor solution to crystal perovskite layer. Therefore, the formed perovskite crystal layer from this reaction may also include the unreacted PbI_2, which affect the solar cell performance [7]. PbI_2 has poor electrical conductivity and higher conduction band than perovskite, which may prevent electron injection to TiO_2 and provide charge carrier transport channel with Ohmic characteristics. In the research work reported here, we investigated the effect of preparation parameters from precursor solution (i.e. temperature and concentration) in one-step solution process on the formed perovskite crystal layer characteristics and their solar cell performances.

2. Experimental

2.1. Substrate preparations and $MAPbI_3$ layer formation

Flourine-doped Tin Oxide (FTO)-coated glass (TCO22-7, Solaronix) was used as the solar cell substrate. This FTO substrate was cut into small pieces with size around 2.5 × 2.5 cm^2. Each substrate was etched with HCl and zinc powder to form a strip-like electrode pattern. The patterned substrates were then cleaned by sonication in washing liquids consecutively, namely Teepol solution, aqua DM, acetone and ethanol, for 5 minutes in each liquid. After being cleaned, the substrates were dried with air gun. A compact-TiO_2 (c-TiO_2) precursor solution (Ti-nanoxide BL/SC, Solaronix) was spin-coated onto the substrate at 5000 rpm for 30 s and then followed by sintering at 550°C for 1 hour. After c-TiO_2 deposition, the substrates were then coated with mesoporous-TiO_2 (m-TiO_2) by spin coating technique at 5000 rpm (acc. 2000 rpm/s) for 30 s from diluted m-TiO_2 precursor solution (Ti-nanoxide T/600SC, Solaronix) with ethanol (Sigma-Aldrich) at volume ratio 1:1. After spin coating, those TiO_2 coated substrates were sintered at 475°C for 30 minutes.

Perovskite precursor solutions were made by dissolving equimolar mixture of PbI_2 (Sigma-Aldrich) and MAI (Luminescence Tech.) in DMF (Sigma-Aldrich). Precursors were made into two concentrations: 0.5 M (PSX-C05 sample) and 1.0 M (PSX-C10 sample). Precursor solutions were heated up to 120°C for 20 minutes to promote complete dissolution of PbI_2 [8]. Perovskite solution with 1.0 M concentration was also used for producing samples for studying the temperature effect. The precursor solution temperature prior to spin coating were set to 40°C (PSX-T40 sample), 60°C (PSX-T60 sample), and 80°C (PSX-T80 sample). The precursor solutions were then spin-coated at 3000 rpm for 30 s and heated at 90°C for 90 minutes. For producing samples for studying the effect of precursor solution concentration, all precursor solutions prior to spin coating were set at 60°C. These perovskite layers were made inside a glove-box with nitrogen atmosphere to minimize the effect of moisture and oxygen.

2.2. Perovskite solar cell fabrication

After the deposition of active layer, the next process is the deposition of hole transport material (HTM) and gold (Au) electrode. HTM solution was deposited by spin coating at 5000 rpm for 30 s, which was prepared by dissolving 72.3 mg of Spiro-OMeTAD (Luminescence Tech.) in 1 ml chlorobenzene. Additional substances were added to this HTM solution to improve the conductivity of the formed layer, namely 27 μl of FK 102 Co (III) TFSI ($Co[PyPz]_3[TFSI]_3$) solution (with concentration of 300 mg/ml acetonitrile) and 20 μl of Lithium bistrifluoromethanesulfonimidate (LiTFSI) solution (with concentration of 455 mg/ml acetonitrile). After the HTM layer deposition, the substrates were left overnight inside the glovebox to let oxidization process in the HTM layer. Finally, 80 nm of gold (Au) layer was thermally evaporated onto the HTM layer to form the back electrode contact. In order to prevent direct contact with moisture and oxygen in air atmosphere during the solar cell characterizations, each cell was covered with sodalime glass and sealed with epoxy adhesive. Figure 1 shows a PSC sample and its structure. PSC samples for studying the effect of precursor solution concentration were named as PSC-05 (for 0.5 M precursor concentration) and PSC-10 (for 1.0 M precursor concentration).

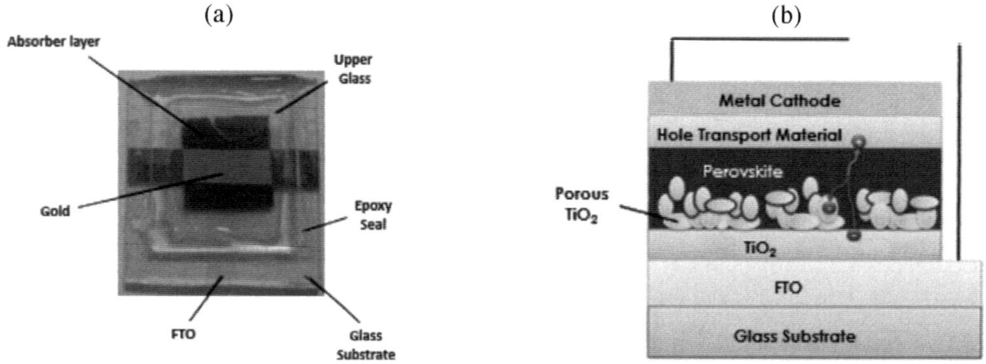

Figure 1. Perovskite solar cell (a) and its structure (b).

2.3. $MAPbI_3$ thin films characterizations

Each perovskite layer was characterized by scanning electron microscopy (SEM JEOL 6030 LA) to see the morphology of the formed crystal layer. In order to evaluate the perovskite crystal formation, XRD measurements were conducted by using a Bruker D8 Advanced X-ray Diffractometer with Kα Cu as the X-Ray source. Step size of 0.02 degree and an acquisition time of 2 s/degree were chosen in the measurement setup. The measured XRD patterns were analyzed using Match! ver. 3.0 software and Crystallography Open Database (COD) ver. 20150706 as reference library.

2.4. Perovskite solar cell (PSC) characterization

The measurements of J-V characteristic of PSC samples were performed by using Oriel Sol 3A™ solar simulator (Newport) that was connected with a digital sourcemeter Keithley 2400. The cell was illuminated under 1 sun (intensity: 100mW/cm^2) with active area of 0.25 cm^2. Forward bias voltage from -0.2 V to 1.0 V was applied to the cell during the measurement.

3. Results and discussion

3.1. Influence of precursor solution temperature

Figure 2 shows the XRD patterns of $MAPbI_3$ perovskite layer spin-coated at various temperatures. XRD pattern with COD entry no. 96-433-5369, which is based on the publication by Stoumpos *et al.*

[9], was used as a reference. XRD peaks from $MAPbI_3$ perovskite crystal are denoted with letter “P” along with its Miller indices. These XRD peaks are in good agreement with literatures. The most intense peak is at 14.1°, which indicates the diffraction from (110) and (002) planes. The presence of small peaks at 23.5° and 30.1° indicates the diffraction from (211) and (213) planes, which belongs to β-$MAPbI_3$ phase [3,10]. The β-phase is phase of $MAPbI_3$ that presents in room temperature.

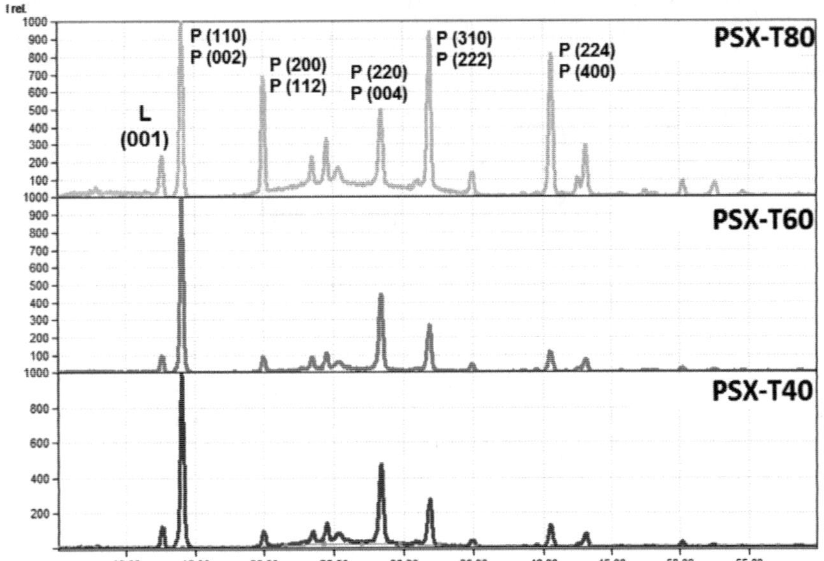

Figure 2. XRD patterns of $MAPbI_3$ spin-coated at 40 °C (PSX-40 sample), 60 °C (PSX-60 sample), and 80°C (PSX-80 sample).

In addition to peaks originated from the perovskite crsytal, there is also a peak which does not belong to perovskite crystal. A peak (denoted with letter “L”) observed at 2θ = 12.65° is the diffraction peak from (001) plane of PbI_2 [11]. The presence of this L-peak in all XRD patterns shows that there is a small amount of unconverted PbI_2 left in perovskite films. As the precursor solution temperature increasing from 40°C, 60°C to 80°C, there is a decrease in L-peak’s intensity from 4967 cps, 4738, and 4402 cps. Table 1 shows the percentages of PbI_2 in perovskite layer made at various precursor solution temperatures. As the temperature increasing from 40 °C to 60 °C, the PbI_2 content in perovskite decreases from 4.70% to 3.97%. However, when the temperature is increased from 60°C to 80°C, there was an increment in PbI_2 percentage.

Table 1. PbI_2 contents in $MAPbI_3$ perovskite layer (based on Match! Software calculation).

No.	Precursor temperature	PbI_2 content in perovskite
1	40 °C	4.70%
2	60 °C	3.97%
3	80 °C	9.74%

These results may be related with better PbI_2 solubility at higher temperature, which is related to the increment of (110) and (002) peak intensity from 19,589 cps to 24,555 cps for an increase of precursor temperature from 40°C to 60°C. However, the same peak intensity from sample of 80°C is just 10,366 cps, which may indicate much thinner layer due to higher viscosity of precursor solution at this temperature.

Figure 3 shows the SEM images of perovskite layer surface which were prepared at various temperatures. Perovskite layers spin-coated from precursor solution with temperature at 40°C and 60°C have a mesh-like crystal structure. The perovskite crystals tend to grow at one direction, as in agreement with their XRD patterns that exhibit one intense peak at 14.1°.

(a) (b) (c)

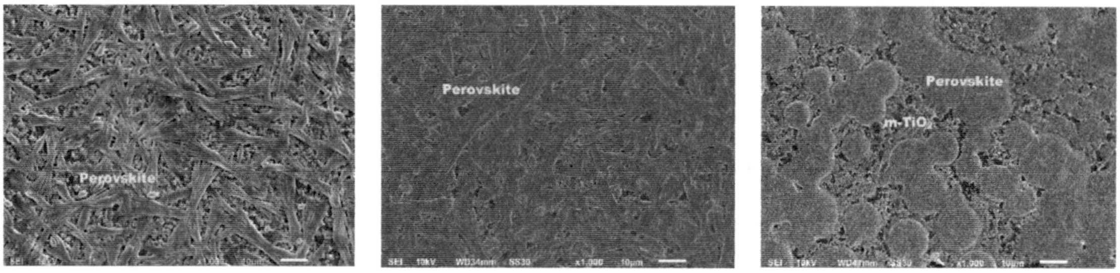

Figure 3. SEM images of perovskite layers spin-coated at (a) 40°C, (b) 60°C, and (c) 80°C.

On the contrary, perovskite layer spin-coated from precursor solution with temperature at 80°C have flake-like morphology, which may indicate higher dimensionality of crystal structure. This shape explains why the XRD pattern of sample PSX-T80 exhibits several intense peaks from lattice plane with larger Miller indices. However, PSX-T80 shows uncovered surface area that may cause a problem on the cell performance.

3.2. Influence of precursor concentration on perovskite crystal formation

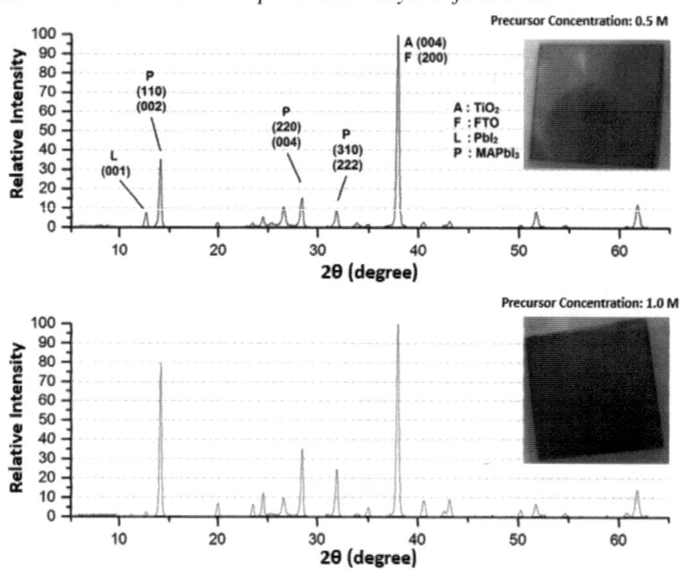

Figure 4. XRD patterns of perovskite layer spin-coated from precursor solution with concentration of (a) 0.5 M (PSX-C05 sample) and (b) 1.0 M (PSX-C10 sample).

Figure 4 shows the XRD patterns of perovskite layer samples. In the patterns, a peak appeared at 37.9° which belongs to the diffraction from (200) plane of FTO and (004) plane of TiO_2 anatase phase. The appearance of this peak might be happen if the perovskite layer was very thin compared to the X-ray penetration depth.

IOP Conf. Series: Journal of Physics: Conf. Series **877** (2017) 012046 doi:10.1088/1742-6596/877/1/012046

As the precursor solution concentration was decreased from 1.0 M to 0.5 M, the intensity of (110) peak at 14.1° also decreases but the PbI_2 peak increases. The PbI_2 content in PSX-C05 sample and PSX-C10 sample are estimated to be about 5.71% and ~0.1%. PSX-C05 sample has brown color, while PSX-C10 has black color. Nevertheless, both samples show brown color if viewed in transmittance mode. The observation of high PbI_2 content in low concentration of precursor solution may be explained as follows. Precursor solution with lower concentration contains more solvent than the higher concentration one. With the same volume of precursor solution dispensed onto the substrate and the same spinning rate of spin coating process, solvent in low concentration precursor will evaporate slower than the high concentration precursor. Delaying solvent evaporation would give enough time for PbI_2 to crystallize as the temperature drops. Therefore, precursor with 1.0 M concentration is better than precursor with 0.5 M concentration because it has higher conversion to perovskite crystals.

3.3. Solar cell performance

Figure 5 shows the J-V curves of PSC where the precursor solution was set at 60°C. PSC samples made from precursor with concentration of 0.5 M (PSC-05 sample) and 1.0 M (PSC-10 sample) exhibit power conversion efficiency of 4.65% and 4.34%. As the precursor concentration was decreased from 1.0 M to 0.5 M, there is an increment in efficiency, J_{sc}, and V_{oc} by 0.31%, 1.83 mA/cm^2 and 0.22 V, respectively. The Filling Factor (FF) value seems, however, not good enough. This may be due to the presence of uncovered mesoporous TiO_2 regions and unconverted PbI_2, where it becomes Ohmic and recombination sites during the J-V curve measurement. For PSC made by 1.0 M precursor concentration, the precursor might be viscous enough so that the solution rather difficult to penetrate TiO_2 pores. Therefore, the perovskite crystals are dominantly deposited on the surface of mesoporous TiO_2 layer. This condition would prolong electrons paths to travel to from photo-excited perovskite to FTO layer and increase the recombination rate, but that condition creates more perfect junction structure. A solar cell having perfect junction structure would have high FF value, which is also necessary in order to convert photon into electricity efficiently.

(a) (b)

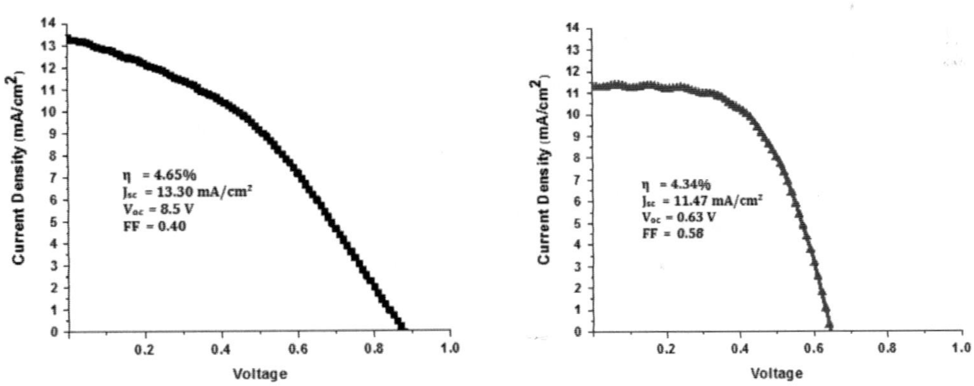

(a) (b)

Figure 5. J-V curves of PSC made in this work from (a) 0.5 M and (b) 1.0 M precursor solutions with temperature solutions at 60°C.

4. Conclusion

The present experimental result shows how the precursor solution parameters, i.e. its temperature and concentration, in one-step solution process affect the formed perovskite crystal characteristics and also the fabricated PSC. The temperature of precursor solution affects the formed perovskite layer crystallinity and morphology as well as the unconverted PbI_2 amount. Perovskite layers made by spin coating precursor at 40°C and 60°C produce a mesh-like structure, while the layer at higher precursor temperature (80°C) produce a flake-like structure. In this work, perovskite layer with the smallest PbI_2 content was obtained using precursor solution temperature at 60°C. At low precursor concentration (0.5 M), the amount of unconverted PbI_2 increases up to 5.71%. Those formed perovskite layer has influence on their PSC characteristics indicated by their J-V curves. Although PSC made from 0.5 M precursor concentration has slightly higher efficiency, J_{sc}, and V_{oc}, but it has bad FF value. This may be due to imperfect mesh-like structure with considerable amount of uncovered TiO_2 region and unconverted PbI_2. PSC from precursor with 1 M precursor concentration shows much better FF value that indicates the formation of better junction structure. Although the perovskite crystlas are more deposited on the top of the mesoporous TiO_2 layer, rather inside its pores, the formation of junction structure seems better in this case.

Acknowledgements
Authors are very grateful for the kindness of Prof. M Okazaki of Osaka University and his laboratory members for using their experimental facilities during our research visit. This research was supported by Program Hibah Kerjasama Luar Negeri 2015/2016 from DIKTI.

References
[1] Chen Q, Marco N D, Yang Y M, Song T, Chen C, Zhao H, Hong Z, Zhou H and Yang Y 2015 *J. Nanotod.* **10** 355-396
[2] Snaith H J 2013 *J. Phys. Chem. Lett.* **4** 3623- 30
[3] Baikie T, Fang Y, Kadro J M, Schreyer M, Wei F, Mhaisalkar S G, Graetzel M and White T J 2013 *J. Mater. Chem. A* **1** 5628- 41
[4] Kojima A, Teshima J, Shirai Y and Miyasaka T 2009 *J. Am. Chem. Soc.* **131** 6050- 51
[5] NREL 2015 Research Cell Efficiency Records. http://www.nrel.gov/ncpv/images/efficiency chart.jpg
[6] Gao P, Graetzel M and Nazeeruddin M K 2014 *J. Energy Environ. Sci.* **7** 2448- 63
[7] Cao D H, Stoumpos C C, Malliakas C D, Katz M J, Farha O K, Hupp J T, and Kanatzidis M G 2014 *APL Mater.* **2** 091101; doi: 10.1063/1.4895038
[8] Bi D, El-Zhory A M, Hagfeldt A and Boschloo G 2015 *ACS Photonics* **2** 589- 94
[9] Stoumpos C C, Malliakas C D and Kanatzidis M G 2013 *Inorg. Chem.* **52** 9019- 38
[10] Song Z, Watthage S C, Phillips A B, Tompkins B L, Ellingson R J and Heben M J 2015 **27** 4612- 19
[11] Khatiwada D, Vankatesan S, Adhikari N, Dubey A, Mitul A F, Mohammad L, Iefanova A, Darling S B and Qiao Q 2015 *J. Phys. Chem. C* **119** 25747–53

Study of Natural Convection Passive Cooling System for Nuclear Reactors

Habibi Abdillah[1], Geby Saputra[2], Novitrian[3], Sidik Permana[4]

Physics Department, Institut Teknologi Bandung, Jl. Ganesha No.10, Kota Bandung, Jawa Barat 40132

[1]habibi.abdillah@students.itb.ac.id, [2]geby.saputra@fi.itb.ac.id, [3]novit@fi.itb.ac.id, [4]psidik@fi.itb.ac.id

Abstract. Fukushima nuclear reactor accident occurred due to the reactor cooling pumps and followed by all emergencies cooling systems could not work. Therefore, the system which has a passive safety system that rely on natural laws such as natural convection passive cooling system. In natural convection, the cooling material can flow due to the different density of the material due to the temperature difference. To analyze such investigation, a simple apparatus was set up and explains the study of natural convection in a vertical closed-loop system. It was set up that, in the closed loop, there is a heater at the bottom which is representing heat source system from the reactor core and cooler at the top which is showing the cooling system performance in room temperature to make a temperature difference for convection process. The study aims to find some loop configurations and some natural convection performances that can produce an optimum flow of cooling process. The study was done and focused on experimental approach and simulation. The obtained results are showing and analyzing in temperature profile data and the speed of coolant flow at some point on the closed-loop system.

1. Nuclear Passive Cooling System

Nuclear reactor accident in Fukushima, Japan, becomes one of the important issues to increase the level of nuclear reactor safety especially due to the natural disaster which affected the nuclear reactor system. Lack of electricity to cool the reactor and spent nuclear fuel for longer time caused by station black out (SBO) condition due to the giant earthquake which was happened.

Fukushima nuclear reactor accident occurred due to the reactor cooling pumps and followed by all emergencies cooling systems could not work, that affected to the increase of reactor temperature and made the reactor core melted with the fuel and water. Therefore, the system which has a passive safety system that rely on natural laws such as natural convection passive cooling system which can work without any active system or electricity dependence, is strongly recommended for the next advance nuclear power plant.

2. Natural Circulation Theory

In natural convection, the cooling material can flow due to the different density of the material due to the temperature difference. Like all other substances, fluid composed of molecules. But in this paper, the macroscopic behavior seen from the fluid and the fluid is considered as a substance that is a continuous distribution.

In the case of fluid dynamics, control volume approach was used. There are three laws of physics that can explain the phenomenon of fluid dynamics, namely the conservation of mass, Newton's second law of motion, and the law of conservation of energy or the first law of thermodynamics.

In natural convection, fluid flow occurs due to the buoyancy of the fluid produced by changes in fluid density. Fluid density changes can be caused by differences in fluid temperature. The relationship between the density changes with temperature difference can be written as follows

$$\rho = \rho_0 (1 - \beta \Delta T) \tag{1}$$

with ρ_0 is the initial density of the fluid, ρ the fluid density in a layer that is heated, β is thermal expansion coefficient, and ΔT is the temperature difference between the fluid that is heated by the initial fluid temperature. Equation (4) is an equation derived from Boussinesq approach which considers density changes only affected by temperature differences. Due to the density change will occur buoyant force per unit volume (f_a) expressed as

$$\vec{f}_a = (\rho_0 - \rho)\vec{g} \tag{2}$$

by substituting equation (4) into the equation (5), it will obtain the equation

$$\vec{f}_a = \rho_0 \beta \vec{g} \Delta T \tag{3}$$

3. Methods

In this study, we used two methods, simulation and experimental methods.

3.1. Simulation

Simulations performed using software that uses finite element principles. In the simulation, the closed loop is formed as follows.

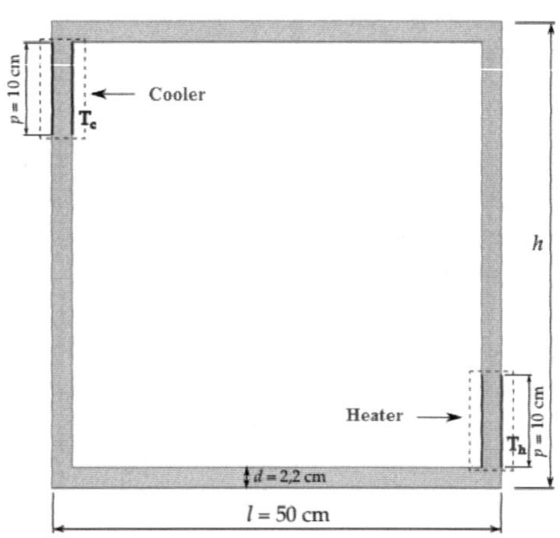

Figure 1. Natural Convection loop system

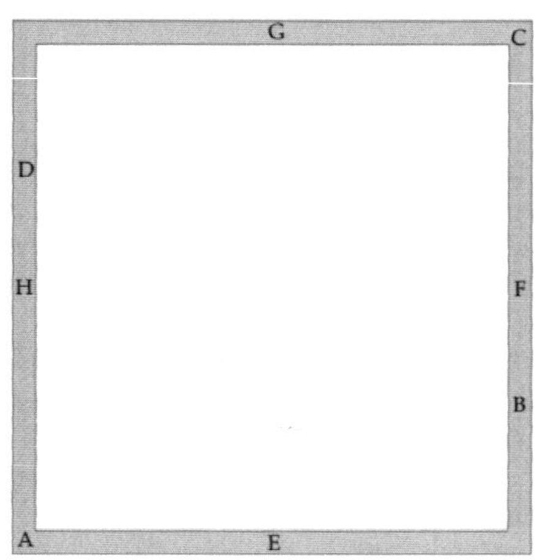

Figure 2. Position of sensor

The loop is filled with water, on the lower right side, there is a hot wall and on the top left, there is a cold wall. This simulation will look for the value of the speed and temperature at multiple points. Measurement points can be seen in the following figure.

Points A, B, C, and D is the temperature measurement point and point E, F, G, and H is the velocity of fluid flow measurement point. Simulations carried out using three different models. Here is a table that explains the configuration of the three models.

Table 1. Configuration Simulation Model

Model	l (cm)	h (cm)	T_h (°C)	T_c (°C)	ΔT (°C)
I	50	50	40	15	25
II	50	80	40	15	25
III	50	100	40	15	25

Simulation is run for a span of 1800 second, the simulation results can be seen in the following graph.

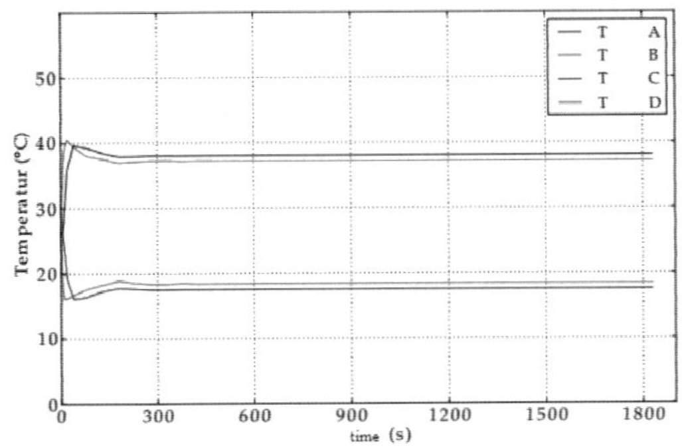

Figure 3. Graph of Temperature of water versus Time for height 50 cm

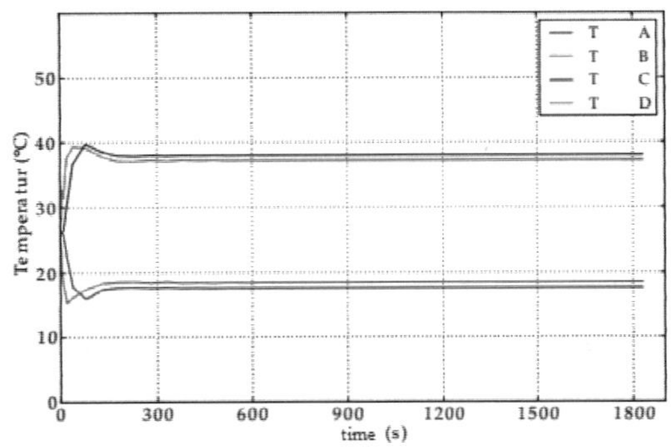

Figure 4. Graph of Temperature of water versus Time for height 80 cm

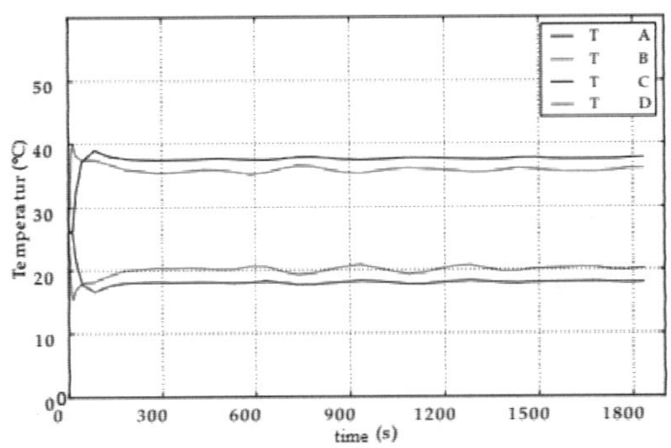

Figure 5. Graph of Temperature of water versus Time for height 100 cm

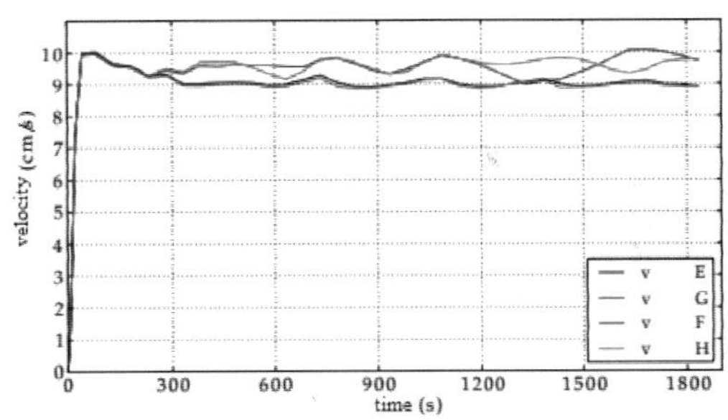

Figure 6. Graph of Velocity of water versus Time for height 100 cm

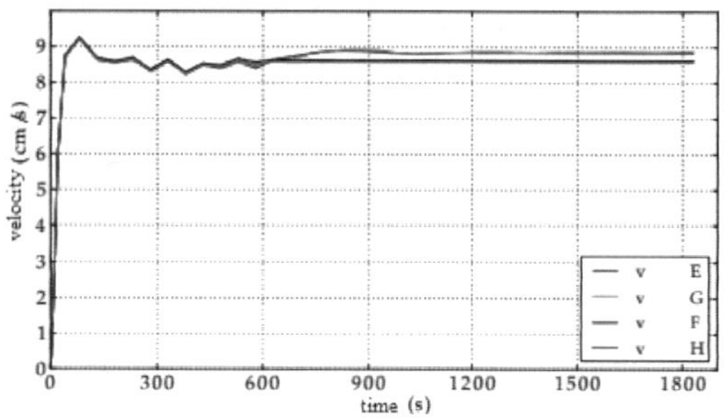

Figure 7. Graph of Velocity of water versus Time for height 80 cm

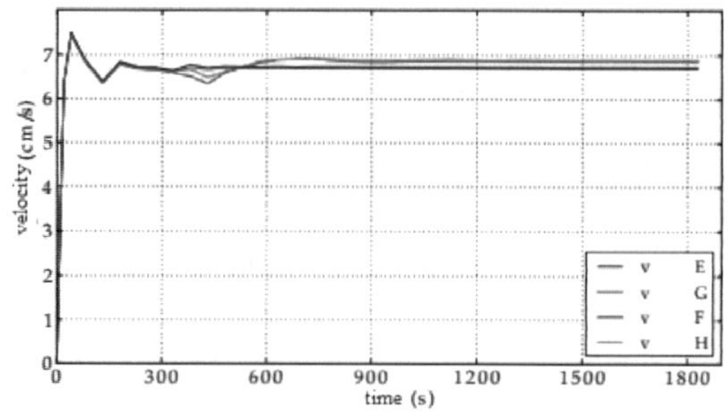

Figure 8. Graph of Velocity of water versus Time for height 50 cm

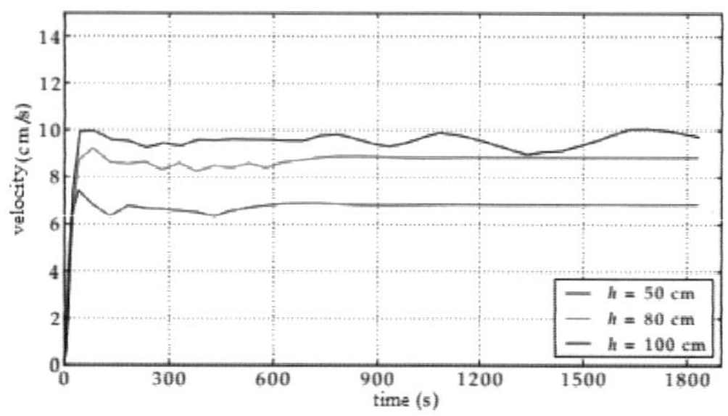

Figure 9. Comparison of Velocity of water versus Time for height 50 cm, 80 cm, and 100 cm

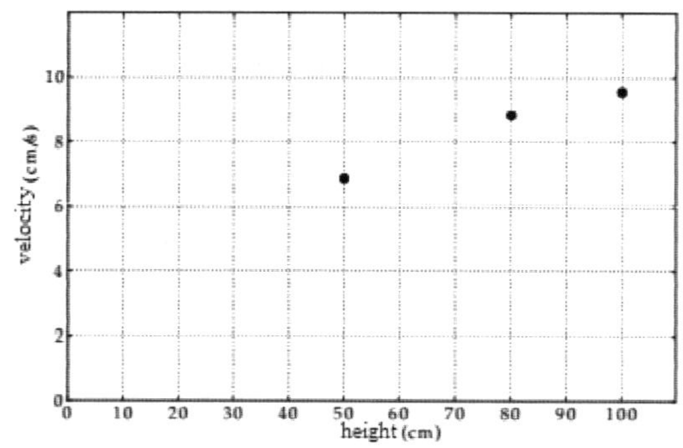

Figure 10. Graph of velocity versus height of loop

3.2. Experiment

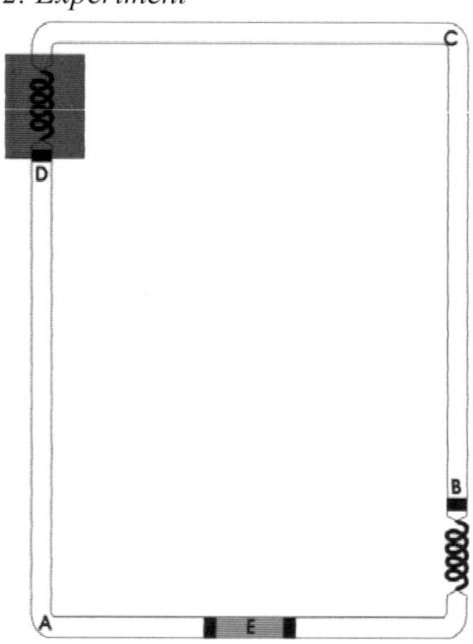

Figure 11. Design of Apparatus of Natural Convection Loop

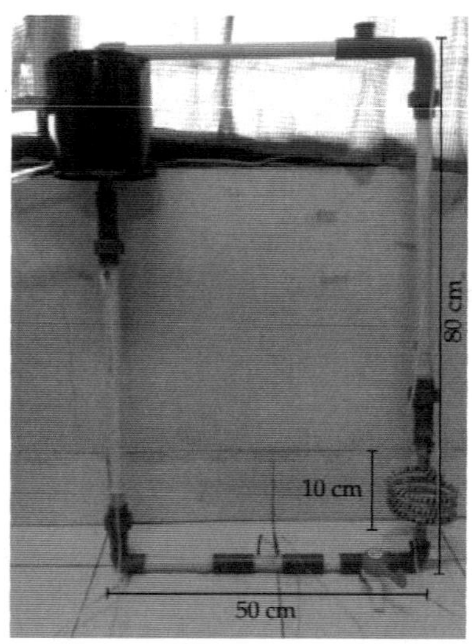

Figure 12. Apparatus of Natural Circulation Loop

The experiment begins with the installation of a temperature sensor and flow sensor on a loop system apparatus. A temperature sensor placed at points A, B, C, and D and a flow sensor placed at point E (see **Figure 11**).

Temperature and flow sensor then connected with a data acquisition system. After that, put in a water circulation system. The heater is then activated and put the ice as a cooling material into the cooler. After that enable computer and data acquisition system to function receives the data from the sensors.

The results of the experimental data collection is as follows.

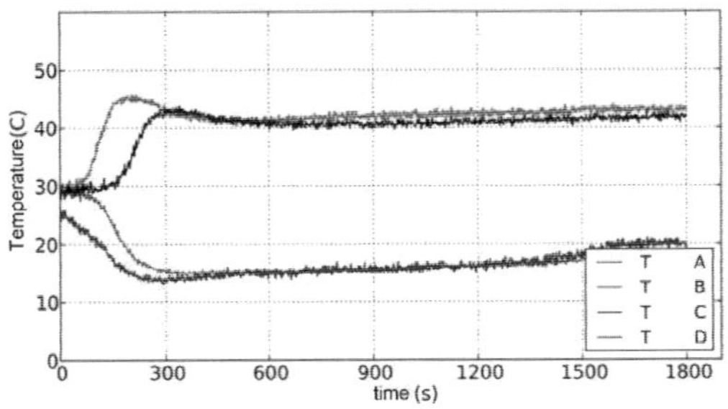

Figure 13. Graph of Temperature versus time from experiment data (height of loop 80 cm)

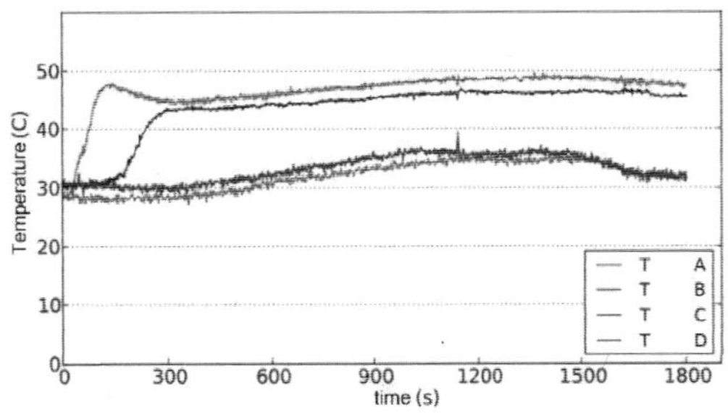

Figure 14. Graph of Temperature versus time from experiment data (height of loop 100 cm)

4. Results

From the simulation results, it can be seen that the height of loop apparatus affects the transient time to get the stationary temperature. The higher the loop apparatus, the longer the time of the transient (see Figure 3 - Figure 5). And when compared with experimental results (see Figure 13 and Figure 14), while transient temperature experimental result is greater than the time transient simulation results. This happens because at the beginning of the simulation cooling and heating temperature is constant from the beginning of the simulation, while the experiment heating and cooling in the loop apparatus takes time to reach the desired temperature.

Based simulations showed that the higher the loop apparatus, the higher the velocity of water flows in the apparatus. This is shown in Figure 9 and Figure 10. The simulation results have not been compared with experimental results. In the experiment we did, fluid flow sensor that we have not been able to detect the velocity of water flow in the loop apparatus. Our hypothesis is that our flow sensor less sensitive, so it can't detect small flow velocity.

5. Future plan experiment

From the experiments we have done, we plan to improve our flow sensor system. Furthermore, we have plan to increase the level of loop apparatus at a maximum of 5 meters to find out more about the effect of height of loop apparatus on the flow of water in the apparatus.

Acknowledgements

The authors thank the RISTEKDIKTI which provides funds to this research with contract number 583g/I1.C01/PL/2016 and the members of the Nuclear Physics Laboratory.

References

[1] Adam Bejan, Allan D. Kraus. 2003. Heat Transfer Handbook. John Wiley and Sons : New Jersey. Page 531-532.

[2] Alan T. McDonald, Philip J. Pritchard, Robert W. Fox. 2004. Introduction to Fluid Mechanics. John Wiley and Sons : New Jersey. Page 184-187.

[3] D. W. Spitzer. 2001. Flow Measurement. ISA : North Carolina. Page 475-478.

[4] Edwin N. Lightfoot, R. Byron Bird, Warren E. Stewart. 2002. Transport Phenomena Second Edition. John Wiley and Sons : New Jersey. Page 490-492.

[5] James Siregar. 2012. Two-Dimensional Simulation of Fluid Flow On Vertical Natural Circulation Equipment With Closed Loop. Physics Department of Institut Teknologi Bandung, Bandung. Page 1 and 16, Appendix A.

Comparison of Pretreatment Methods on Vetiver Leaves for Efficient Processes of Simultaneous Saccharification and Fermentation by *Neurospora* sp.

E Restiawaty[1,2*], **A Dewi**[2]

[1]Study Program of Bioenergy and Chemurgy Engineering, Research Group of Chemical Engineering Process Design and Development, Faculty of Industrial Technology, Institut Teknologi Bandung, Jln. Let. Jen. Purn. Dr. (HC) Mashudi No.1 Sumedang, West Java 45363, Indonesia
[2] Biosciences and Biotechnology Research Centre, Institut Teknologi Bandung, Jln. Ganesha 10 Bandung 40132, Indonesia

*E-mail: erestiawaty@che.itb.ac.id

Abstract. Lignocellulosic biomass is a potential raw material for bioethanol production. *Neurospora* sp. can be used to convert lignocellulosic biomass into bioethanol because of its ability to perform simultaneous saccharification and fermentation. However, lignin content, degree of polymerization, and crystallinity of cellulose contained in lignocellulosic biomass can inhibit cellulosic-biomass digestion by *Neurospora* sp, so that a suitable pretreatment method of lignocellulosic biomass is needed. The focus of this research was to investigate the suitable pretreatment method for vetiver leaves (*Vetiveria zizanioides* L. Nash) used as a raw material producing bioethanol in the process of simultaneous saccharification and fermentation (SSF) by *Neurospora* sp.. Vetiver plants obtained from Garut are deliberately cultivated to produce essential oils extracted from the roots of this plant. Since the vetiver leaves do not contain oil, some of harvested leaves are usually used for crafts and cattle feed, and the rest are burned. This study intended to look at other potential of vetiver leaves as a source of renewable energy. Pretreatments of the vetiver leaves were conducted using hot water, dilute acid, alkaline & dilute acid, and alkaline peroxide, in which each method was accompanied by thermal treatment. The results showed that the alkaline peroxide treatment is a suitable for vetiver leaves as indicated by the increase of cellulose content up to 65.1%, while the contents of hot water soluble, hemicellulose, lignin, and ash are 8.7%, 18.3%, 6.8%, and 1.1%, respectively. Using this pretreatment method, the vetiver leaves can be converted into bioethanol by SSF process using *Neurospora* sp. with a concentration of bioethanol of 6.7 g/L operated at room temperature.
Keywords: vetiver leaves, chemical pretreatment, SSF, Neurospora sp.

1. Introduction

Based on the feedstocks of bioethanol production, there are three generations of bioethanol [1, 2], i.e. (a) the first generation, which is produced from sugar plants and starch; (b) the second generation, which is produced from lignocellulosic biomass, such as household waste, wood, and grass; and (c) the third generation, which is produced from (micro)algae. This research focuses on second-generation bioethanol production process because Indonesia has abundant sources of lignocellulosic biomass derived from agroindustrial waste. Unlike the first generation bioethanol, the feedstock of second-generation bioethanol production process is not used for food, so that this process does not compete with the food industry [3].

One of agroindustrial wastes that is potentially used as a source of lignocellulosic biomass in Indonesia is a waste of vetiver essential oil industry. Indonesia is one of vetiver oil-producing country in the world [4]. Vetiver essential oil obtained from the roots of vetiver grass (*Vetiveria zizanioides* L. Nash). This essential oil is used in various industries for perfumes, cosmetics, aromatherapy, food and flavouring productions. Vetiver grass is usually planted at the beginning of the rainy season. Harvesting can be done after the plant was 12-16 months. Before the root of vetiver grass can be harvested, the leaves of vetiver grass is usually cut every few months to encourage the growth of seedlings and to reduce the fire hazard in the dry season. Amount of cut vetiver leaves are more than 10 tons/Ha. Some vetiver leaves are used as animal feed (especially young leaves), crafts, and as compost. A large amount of left residues, however, are not used and usually just burned at the edge of the farm field. Therefore, the utilization of lignocellulosic content of vetiver leaves residues into a high value-added chemical such as bioethanol is very important to be studied. The process of conversion of lignocellulosic biomass into ethanol generally consists of four processes, namely, biomass pretreatment, saccharification, fermentation and product separation. Kuhiran and Punnapayak (2000) reported that 13% of bioethanol yield from dry vetiver leaves was obtained after separation process using one-cycle column distillation. They run a series of conversion process from vetiver leaves into bioethanol started from alkali treatment for the leaves. The next processes were the saccharification and fermentation, which were carried out simultaneously. This process is known as a simultaneous saccharification and fermentation (SSF). For the purpose of saccharification to obtain fermentable sugars, they used cellulose enzyme derived from *Trichoderma reesei*. Simultaneously, they used yeast for the glucose fermentation into bioetanol in the same reactor. The SSF process was carried out for seven days at 40°C and pH 5.0 [5]. The cost and energy used in SSF process is much lower than the conventional process that separates saccharification and fermentation using two different reactors. In addition, the fermentable sugars formed in the SSF process will soon be fermented into ethanol so as to avoid inhibition of the hydrolysis process [6].

Biomass pretreatment plays an important role in the conversion of lignocellulosic biomass into bioethanol. This pretreatment aims to increase the pore size, reducing the content of lignin, and reduce the crystallinity of cellulose. The pretreatment can be performend by using chemical and heat treatment. Typically, acid pretreatment is used to hydrolyze the hemicellulose coating, whereas alkali pretreatment is mainly used to remove lignin. Therefore, pretreatment is highly important to allow cellulase access the cellulose fibers in saccharification process [1]. The pretreatment products can certainly affect the product yield of bioethanol.

This study aimed to investigate the effect of pretreatment method chosen to increase bioethanol yield from vetiver leaves through the SSF process by *Neurospora* sp. *Neurospora* sp. is a fungus that can be obtained from traditional Indonesian fermented food, known as oncom. These fungi have the ability to produce the enzymes needed for saccharification and fermentation processes simultaneously, so that the process is no longer required the addition of (hemi-)cellulose enzymes. However, lignin content, degree of depolymerization (DP), and the crystallinity of cellulose can inhibit cellulosic-biomass digestion by *Neurospora* sp., so that the suitable pretreatment method of lignocellulosic biomass is strongly needed. This study started by investigating the contents of lignocellulose in vetiver leaves, then applied some method of acid or alkaline pretreatment, or any combination thereof, as well as thermal treatment. The pretreated vetiver leaves from different pretreatment methods were applied into SSF process and the bioethanol yields were then determined.

2. Materials and Methods

2.1. Microorganism
Neurospora sp. used in this study was isolated from the fungus on oncom, a traditional Indonesian fermented food. Stock cultures were maintained on potato dextrose agar slants at 4°C. *Neurospora* sp is kept and cultivated in Bioconversion Laboratory, Biosciences and Biotechnology Research Centre, Institut Teknologi Bandung, Indonesia.

2.2. Lignocellulosic biomass
Vetiver grass (*Vetiveria zizanioides* L. Nash) used as lignocellulosic biomass feedstock was obtained from Leles Garut, West Java, Indonesia with the growing age of 6 months. The grasses were washed and dried in sunlight. Dried vetiver leaves were milled using a blender and then sheaved to obtained the particle size of +120/-230 mesh.

2.3. Pretreatment methods

2.3.1. Hot water pretreatment. 100 mL of water was added into 500 mL Erlenmeyer flask containing 10 g of milled vetiver leaves. The mixture was then autoclaved at 130°C and 24.5 psi (relative to atmosphere) for 1 h. Before the autoclaved leaves can be used as feedstock for SSF process, they were filtered, washed with distilled water, sterilized using autoclave (121°C for 20 min), and dried at 60°C for 16 h.

2.3.2. Dilute acid pretreatment. About 7.83 $g_{dry\ solid\ (DS)}$ of milled vetiver leaves were slurred with 1L of 1.5%-v/v dilute sulfuric acid. The mixture was autoclaved at 130°C and 24.5 psi (relative to atmosphere) for 30 min. After acid pretreatment, pH of vetiver slurry was adjusted to pH 5.0 with 10 M NaOH [7]. After pH adjustment, the vetiver leaves were filtered, washed, sterilized, and dried at 60°C for 16 h.

2.3.3. Alkaline & dilute acid pretreatment. This method has two steps of treatments. In the first step, 10 g of milled vetiver leaves were mixed with 1.0 M NaOH and then incubated at room temperature for 24 h. Before going to the second step, the leaves were washed, filtered, and dried overnight at 60°C for 16 h. In the second step, three grams of pretreated vetiver leaves were hydrolyzed with 2.0%-v/v of sulfuric acid at 111°C for 30 min using autoclave. Before applying to SSF process, the treated leaves were filtered, washed, sterilized, and dried again at 60°C for 16 h [8].

2.3.4. Alkaline peroxide pretreatment
This method was adopted and modified from the one described by Saha and Cotta (2007) [9]. 15 g milled vetiver leaves were suspended in 1 L of 7.5%-v/v H_2O_2. Alkaline level was adjusted to 11.5 using NaOH. The samples were shaken (160 rpm) and incubated at 35°C for 48 h. The samples were then filtered and solid residues were washed up to neutral pH. Similar with other methods, before the samples applied to SSF process, they were sterilized and dried at 60°C for 16 h.

2.4. Simultaneous saccarification and fermentation (SFF) using Neurospora sp.
Neurospora sp. was initially cultivated in a 50 mL vial containing 10 mL potato dextrose broth (PDB) medium. The culture was shaken (150 rpm) and incubated at room temperature for 48 h in aerobic condition. The aerobic condition was intended for the growth of *Neurospora* sp. After gaining the proper weight of biomass, 0.1 g of pretreated vetiver leaves was fed into fermenter. In order to increase the yield of bioethanol, the fermentation was continued in anaerobic condition for 24 h.

2.5. Analytical method
Lignocellulosic contents of vetiver leaf were analyzed according to the Chesson-Datta method [10]. SSF products were analyzed using high performance liquid chromatography (HPLC) equipped with an

ion moderated partition chromatography column, Aminex HPX-87H (Bio-Rad, CA). The HPLC was used in combination with a Waters (2414) refractive index detector. The flow rate of the mobile phase (5mM H_2SO_4) was adjusted at 0.6 ml/min and the temperature at 60°C.

3. Results and Discussion

3.1. The influence of pretreatment methods on the composition of lignocellulosic vetiver leaves

Vetiver leaves are an attractive feedstock for bioethanol production, because they content high hemicellulose and cellulose. The mass fraction of hemicellulose and cellulose on fresh vetiver leaves (no pretreatment) were 0.35 and 0.31, respectively as shown in Table 1. Besides polysaccharides, lignin, hot water soluble (HWS), and ash are also detected in the lignocellulose content analysis of vetiver leaves. Lignin, which surrounds the cellulose, can inhibit enzymatic hydrolysis of cellulose. Besides lignin contents, enzymatic hydrolysis of lignocellulosic biomass is inhibited by crystallinity of cellulose, DP, surface area, and moisture contents [11 - 14]. In this research, several methods of pretreatments have been conducted in order to enhance the digestible contents of vetiver leaves as biomass feedstock in simultaneous saccharification and fermentation. Lignocellulose composition measured after pretreatment are shown in Table 1.

Table 1. Lignocellulosic contents of vetiver leaves with different pretreatment methods.

Lignocellulose Contents	No Pretreatment	Hot Water	Dilute Acid	Alkaline & Dilute Acid	Alkaline Peroxide
HWS (%-w/w)	14.01 ± 0.65	7.33 ± 0.65	10.33 ± 0.58	38.33 ± 0.58	8.67 ± 0.58
Hemicellulose (%-w/w)	34.55 ± 0.58	35.67 ± 0.58	9.00 ± 1.00	16.33 ± 0.58	18.33 ± 0.58
Cellulose (%-w/w)	31.39 ± 0.26	34.33 ± 0.58	54.67 ± 0.58	40.67 ± 0.58	65.07 ± 0.12
Lignin (%-w/w)	17.58 ± 0.44	20.50 ± 0.58	23.33 ± 0.44	2.33 ± 0.58	5.46 ± 0.25
Ash (%-w/w)	2.47 ± 0.23	2.17 ± 0.29	2.67 ± 0.58	2.34 ± 0.58	2.47 ± 0.23

The examined pretreatment methods in this study are hot water, dilute acid, alkaline & dilute acid, and alkaline peroxide methods. All those pretreatment methods begin with mechanical treatment by milling the dried vetiver leaves for reducing the particle size. Mechanical treatment may cause an increase of surface area, reduction of DP, and shearing of the biomass. All of these things can lead an increase of hydrolysis product yield of lignocellulosic biomass [15]. The lignocellulosic composition of pretreated vetiver leaves with hot water does not show a significant difference to the composition of untreated vetiver leaves (see Table 1). Temperature used in hot water pretreatment is 130°C, which is not high enough temperature to solubilize hemicellulose and lignin. Some research reported that hemicellulose starts to dissolve at temperature more than 150°C [16], while lignin can be solubilized if the lignocellulose biomass is heated up around 180°C [17]. Other research reported that liquid hot water pretreatment conducted at the range of temperature 220-280°C could increase the enzymatic hydrolysis of 2-5-fold for their lignocellulosic biomass [18].

Dilute acid pretreatment was intended to solubilize the hemicellulose, so that the (hemi-)cellulase of *Neurospora* sp. could freely access the (hemi-)cellulose. Solubilized hemicelluloses can be hydrolyzed in dilute acid treatment. Presented in Table 1 that the composition of hemicellulose after dilute acid pretreatment is lower than the one without pretreatment. Products of hydrolysis of hemicellulose can be monomer, hydroxymethylfurfural (HMF), furfural, and the products of other volatiles [15]. HMF, furfural, soluble lignin compounds, and other inhibitors produced from pretreatment are not desired in the fermentation step [19]. In this study, the pretreated vetiver leaves were washed before being fed to the bioreactor so that the remained pretreatment chemicals and the inhibitors produced during pretreatment do not interfere the SSF process. Some of hydrolysis products were brought along with washing water. Only the insoluble compounds in cold washing water are remained on the pretreated sample. As a result, the mass fraction of HWS after dilute acid treatment in Table 1 is lower than the one without pretreatment. The loss of some compounds of this hydrolysis product causes the proportion of cellulose and lignin increased.

Alkaline & dilute acid pretreatment is a combination between alkaline and acid pretreatments. This method aims to improve the anaerobic digestibility. In this method, an alkaline pretreatment was previously applied to the vetiver leaves beffore a dilute acid pretreatment was applied. Compared with the previous method, in this method, there is an additional alkaline pretreatment. Alkaline pretreatment can cause swelling of the biomass, which can later be accessed by enzymes or other biocatalyst agent [15]. Alkaline pretreatment allows the dissolution of lignin and change the state of crystalline cellulose [20]. As explained earlier that the dilute acid would hydrolyze the soluble hemicellulose so that the combination of these two pretreatment methods can eliminate lignin and hemicellulose gradually. As a result, the cellulose content increased after alkaline & dilute acid pretreatment compared with the content without pretreatment as can be shown in Table 1.

In the last method, an oxidizing compound, hydrogen peroxide, is used. It is intended to remove the hemicellulose and lignin so that biomass can be digested by a biocatalyst. At a pH of 11.5 - 11.6, hydroxyl ions could be expected to do delignification. Gould (1984) reported that approximately 50% lignin can be dissolved by applying this method for about 20 hours [21]. In this study, the use of alkaline peroxide method can eliminate approximately 70% lignin in biomass (see Table 1).

3.2. Simultaneous saccharification and fermentation of pretreated vetiver leaves using Neurospora sp.
The pretreated vetiver leaves were fed into the bioreactor as a carbon source and substrate for *Neurospora* sp. to produce bioethanol. *Neurospora* sp. is a biological agent that is capable of producing enzymes for (hemi-)cellulose hydrolysis and converting fermentable sugars into bioethanol [22]. The ability of *Neurospora* sp. to convert the pretreated vetiver leaves into bioethanol was assessed for each pretreatment methods that previously done.

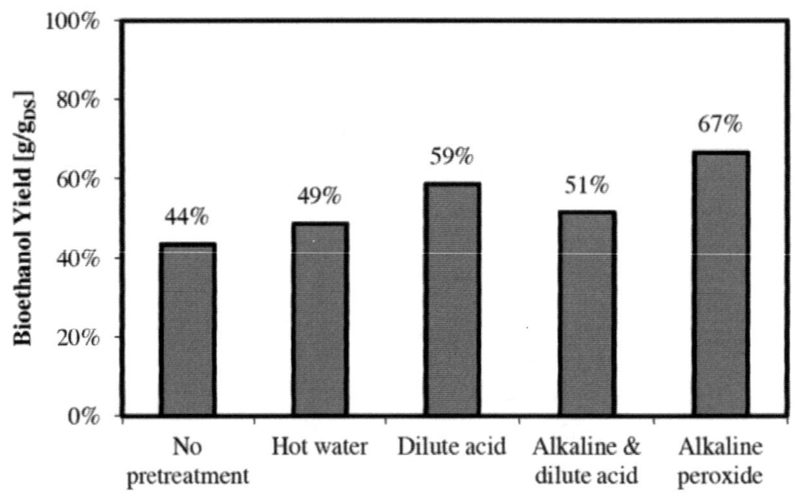

Figure 1. Yields of bioethanol from different pretreated vetiver leaves by *Neurospora* sp. Bioethanol yield is in $g_{bioethanol}/g_{dry\ solid\ (DS)\ of\ vetiver\ leaves}/L_{fermentation\ mixture}$.

The bioethanol yields (per liter) from fermentations of each lignocellulosic vertiver coming from the different treatments are given in Figure 1. Results indicate that the type of pretreatment method can affect the performance of *Neurospora* sp. in producing bioethanol. Alkaline peroxide method provides the highest yield of bioethanol, i.e. 67% (g/g$_{DS}$), followed by a dilute acid, alkaline & dilute acid, and hot water pretreatment methods, subsequently. Production of inhibitor compound and the loss of fermentable sugars frequently occur during pretreatment, hence the yield of bioethanol is still low. As reported by other researcher that the use of an acid during the pretreatment can lead to the formation of volatile compounds that can inhibit the fermentation process, while the use of an alkaline can produce other inhibitor compounds which can interfere the fermentation process [15]. The use of oxidative compounds can cause non-selective oxidation so that lot of sugar is lost [15].

4. Conclusions

Several methods for pretreatment of lignocellulosic biomass, such as pretreatments with hot water, dilute acid, alkaline and dilute acid, and alkaline peroxide have been done for vetiver leaves. The success of pretreatment stage in reducing the lignin content and also increasing the polysaccharides and oligosaccharides contents significantly affects the performance of *Neurospora* sp. in digesting the pretreated vetiver leaves and carrying out saccharification and fermentation. In this study, pretreatment methods using alkaline peroxide could reduce lignin content of vetiver leaves by 69% and increase the total cellulose and hemicellulose content of 26% compared to the lignocellulose contents of the leaves without any pretreatment. The vetiver leaves after alkaline peroxide pretreatment can be converted by *Neurospora* sp. into bioethanol with a yield of 67% (g/g_{DS}).

Acknowledgments

The authors would like to gratefully acknowledge to the Asahi Glass Foundation for the financial support.

References

[1] Naik S N, Goud V V, Rout P K and Dalai A K 2010 Production of first and second generation biofuels: A comprehensive review *Renew. Sustainable Energy Rev.* **14** 578-97

[2] Jambo S A, Abdulla R, Azhar S H M, Marbawi H, Gansau J A and Ravindra P 2016 A review on third generation bioethanol feedstock *Renew. Sustainable Energy Rev.* **65** 756-69

[3] Dogaris I, Mamma D, and Kekos D 2013 Biotechnological production of ethanol from renewable resources by *Neurospora crassa*: an alternative to conventional yeast fermentations *Microbiol. Biotechnol.* **97** 1457-73

[4] Hanief M M A, Mushawwir H A W and Mahfud 2013 Ekstraksi minyak atsiri dari akar wangi menggunakan metode steam-hydro distillation dan hydro distillation dengan pemanas microwave *J Teknik Pomits* **2** 219-23

[5] Kuhiran M and Punnapayak H 2000 Leaves of vetiver grass as a source of ethanol *An abstract of a poster paper presented at ICV-2* Phetchaburi, Thailand 18-22 January 20

[6] Öhgren K, Bura R, Lesnicki G, Saddler J, Zacchi G 2007 A comparison between simultaneous saccharification and fermentation and separate hydrolysis and fermentation using steam-pretreated corn stover *Process Biochem.* **42** 834-39

[7] Saha B C, Iten L B, Cotta M A and Wu Y V 2005 Dilute acid pretreatment, enzymatic saccharification and fermentation of wheat straw to ethanol *Process Biochem.* **40**: 3693-700

[8] Mai Chiang 2009 Comparison of hydrolysis conditions to recover reducing sugar from various lignocellulosic materials *J. Sci.* **36** (3): 384-94

[9] Saha B C and Cotta M A 2007 Enzymatic saccharification and fermentation of alkaline peroxide pretreated rice hulls to ethanol *Enzyme Microb. Tech.* **4** (41): 528–32

[10] Datta R 1981 Acidogenic fermentation of lignocellulose – acid yield and conversion of components *Biotechnol, Bioeng.* **23** (9): 2167-70

[11] Chang V S and Holtzapple M T 2000 Fundamental factors affecting enzymatic reactivity Appl. Biochem. Biotechnol. **84** 5-37

[12] Koullas D P Christakopoulos P Kekos D Macris B J Koukios E G 1992 Correlating the effect of pretreatment on the enzymatic hydrolysis of straw *Biotechnol. Bioeng.* **39** 113-16

[13] Laureano-Perez L Teymouri F Alizadeh H Dale B E 2005 Understanding factors that limit enzymatic hydrolysis of biomass *Appl. Biochem. Biotechnol.* **124** 1082-99

[14] Puri V P 1984 Effect of crysrallinity and degree of polymerization of cellulose on enzymatic saccharification *Biotechnol. Bioeng.* **26** 1219-22

[15] Hendriks A T W M and Zeeman G 2009 Pretreatments to enhance the digestibility of lignocellulosic biomass *Bioresour. Technol.* **100** 10-18

[16] Garrote G Domingguez H and Parajo J C 1999 Hydrothermal processing of lignocellulosic materials *Holz Roh Werkst.* **57**, 191-202

[17] Bobleter O 1994 Hydrothermal degradation of polymers derived from plants *Prog. Polym. Sci.* **19** 797-841.

[18] Weil J R Sarikaya A Rau S L Goebz J Lasisch C M Brwer M Hendrickson R and Tadisch M R 1998 Pretreatment of corn fiber by pressure cooking in water *Appl. Biochem. Biotechnol* **73** 1-17.

[19] Laser M Schulman D Allen S G Lichwa J Antal Jr M J and Lynd L R 2002 A comparison of liquid hot water and steam pretreatments of sugar cane bagasse for bioconversion to ethanol *Bioresour. Technol.* **81** 31-44.

[20] Gregg D and Saddler I N 1996 A techno-economic assessment of pretreatment and fractionation steps of a biomass-to-ethanol process *Appl. Biochem. Biotechnol.* **57** 711-27.

[21] Gould J M 1984 Alkaline peroxide delignification of agricultural residued to enhance enzymatic saccharification *Biotechnol. Bioeng.* **26** 46-52.

[22] Dogaris I Vakontios G Kalogeris E Mamma D and Kekos D 2009 Induction of cellulases and hemicellulases from Neurospora crassa under solid-state cultivation for bioconversion of sorghum bagasse into ethanol *Ind. Crops Prod.* **29** 404-11.

IOP Conf. Series: Journal of Physics: Conf. Series **877** (2017) 012049 doi:10.1088/1742-6596/877/1/012049

Comparative Studies on UO_2 Fueled HTTR Several Nuclear Data Libraries

Anni N. Hidayati[1], Puguh A. Prastyo[1],Abdul Waris[2]*, and Dwi Irwanto[2]

[1]Department of Physics, Faculty of Mathematics and Natural Sciences, Institut Teknologi Bandung, Jl. Ganesa 10 Bandung 40132, INDONESIA
[2]Nuclear Physics & Biophysics Research Division, Department of Physics,Faculty of Mathematics and Natural Sciences, Institut Teknologi Bandung, Jl. Ganesa 10 Bandung 40132, INDONESIA

E-mail: (*)awaris@fi.itb.ac.id

Abstract. HTTR (High Temperature Engineering Test Reactor) is one of Generation IV nuclear reactors that has been developed by JAERI (former name of JAEA, JAPAN). HTTR uses graphite moderator, helium gas coolant with UO_2 fuel and outlet coolant temperature of 900°C or higher than that. Several studies regarding HTTR have been performed by employing JENDL 3.2 nuclear data libraries. In this paper, comparative evaluation of HTTR with several nuclear data libraries (JENDL 3.3, JENDL 4.0, and JEF 3.1) have been conducted.. The 3-D calculation was performed by using CITATION module of SRAC 2006 code. The result shows some differences between those nuclear data libraries result. K-eff or core effective multiplication factor results are about 1.17, 1,18 and 1,19 (JENDL 3.3, JENDL 4.0, and JEF 3.1) at Begin of Life, also at the End of Life (after two years operation) are 1.16, 1.17 and 1.17 for each nuclear data libraries. There are some different result of K-eff but for neutron spectra results, those nuclear data libraries show the same result.

1. Introduction

Nuclear power plants, which produce low-carbon electricity at stable and competitive costs, constitute an element of the solution to global warming and a means of delivering power to emerging and developed countries. Further development of nuclear technology is needed to meet future energy demand[1]. Learning from the Three Mile Island, Chernobyl, and Fukushima Daiichi accidents, since 1990s the studies on the Generation IV reactors have been conducted which are expected to be operated since 2030. One of Generation IV reactors is HTTR of High Temperature Test Reactor in Japan. In this paper, the neutronic analysis of HTTR using some nuclear data libraries will be discussed.

The high-temperature engineering test reactor (HTTR) is a block-type high-temperature gas-cooled reactor (HTGR) of generation IV nuclear reactor, designed for 950°C gas temperature. The high outlet gas temperature may increase the fission product release from fuel because of increased fuel temperature[2]. Thermal output of this reactor is 30 MW. First critically point reached at November 10th, 1988. As a high temperature reactor, HTTR is expected to give some co-generation product such as hydrogen production and purification of coal[3].

2. Objective

The objective of this study is comparing several nuclear data such as JENDL3.3, JENDL4.0 and JEFF3.1 on UO_2 Fueled HTTR 30 MWT.

3. Design and Methodology

General specification of HTTR are listed below:

Table 1: Specification of the HTTR[2]

Parameter	Specification
Thermal power	30 MW
Outlet coolant temperature	950 ^oC
Inlet coolant temperature	395 ^oC
Primary coolant pressure	4 MPa
Core structure	Graphite
Coolant	Helium Gas
Equivalent core diameter	2.3 m
Effective core height	2.9 m
Average power density	2.5 W/cm^3
Burnup period	660 days
Number of fuel blocks	150
Number of fuel columns	30
Number of pairs of CRs :	
In core	7
In reflector	9

This reactor has low average power density because it is not a commercial nuclear power plant. And the maximum average burnup in a fuel element must not exceed 33.000 MWd/t[2]. Because of that. HTTR has a unique core design called double heterogeneity design which uses different levels of ^{235}U enrichment. Those distribution of ^{235}U enrichment are:

Table 2: ^{235}U enrichment (%) and burnable poison (BP) configuration plan of HTTR core

Layer	radial				BP
	1	2	3	4	
1	6.7	7.9	9.4	9.9	2.0
2	5.2	6.3	7.2	7.9	2.5
3	4.3	5.2	5.9	6.3	2.5
4	3.4	3.9	4.3	4.8	2.0
5	3.4	3.9	4.3	4.8	2.0

BP or Burnable Poison that used in this core calculation is B_4C/C. which has 2.0% and 2.5% of natural boron concentration. There are two step of calculation method that used in this case, first is fuel block calculation using SRAC 2006 PIJ[4], hexagonal assembly with asymmetric pin rods geometry and for the core calculation using CITATION, 3D triangular (XT,YT,H) geometry with 30 MWT power and two years operation period. Several nuclear data libraries are used in this calculation and this is the comparison between JENDL 3.3, JENDL 4.0, and JEF 3.1:

Table 3: comparison between nuclear data libraries

Nuclear library	JENDL 3.3	JENDL 4.0	JEFF 3.1
Number of nuclides	396	450	412
year of release	2002	2011	2006

4. Result and Discussion

All nuclear data libraries that used in this calculation give some different result of K-eff or core effective multiplication factor. The result is: The result shows that all nuclear data give

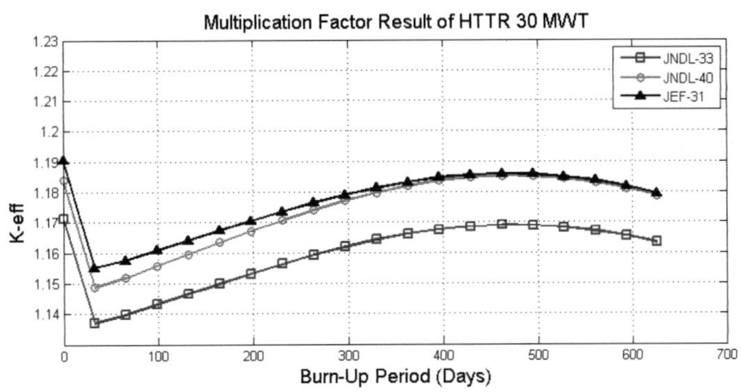

Figure 1: effective multiplication factor result

different value at begin of operation and the highest value is 1.19 (JEFF 3.1), but at the end of operation there is same value for K-eff beetween JENDL 4.0 and JEFF 3.1. At the second step of operation, drop K-eff are occurred. It is because of BP concentration still high at this step. So that neutron population will increase after second step and will give effect to the multiplication factor. And this is neutron spectrum neutron result:

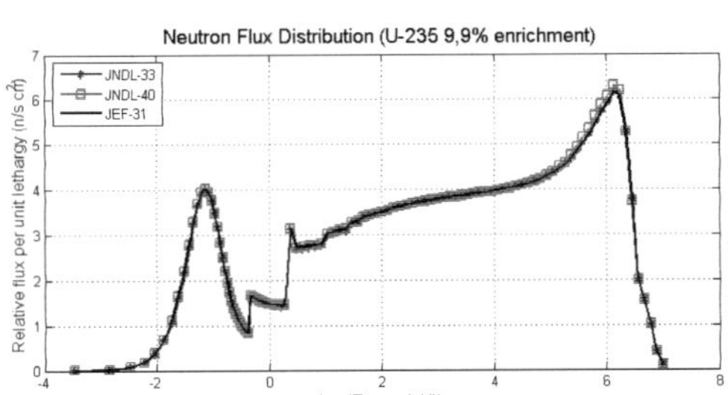

(a) ^{235}U 9.9% enrichment

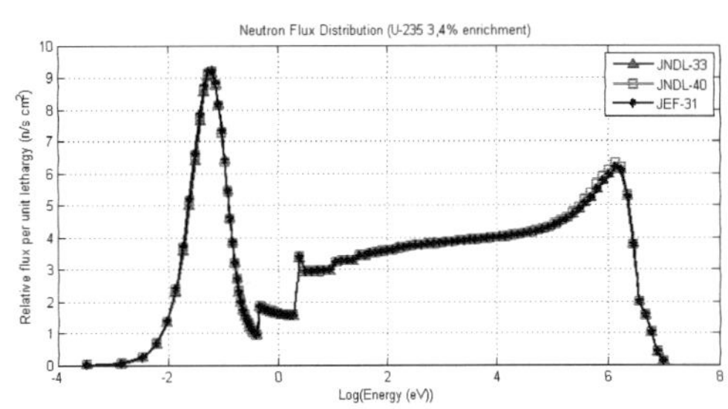

(b) ^{235}U 3.4% enrichment

Figure 2: Neutron Flux Distribution for highest and lowest ^{235}U enrichment in HTTR core

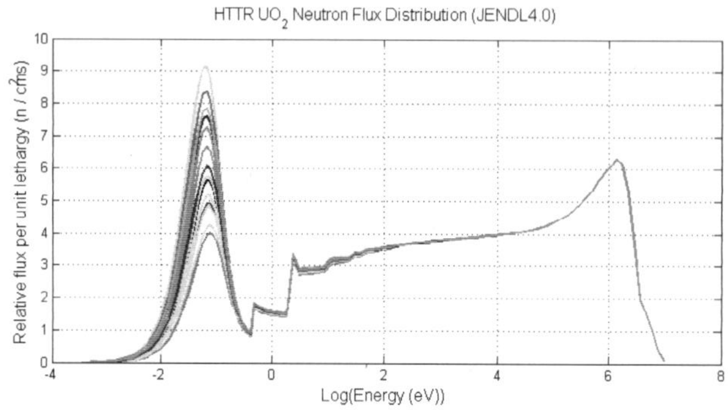

Figure 3: Neutron Flux Distribution for all ^{235}U enrichment in HTTR core (JENDL 4.0)

5. Conclusion

Several nuclear data such as JENDL3.3, JENDL4.0 and JEFF3.1 have been compared through SRAC 2006 calculation on UO_2 fueled HTTR 30MWT. There are some different multiplication factor result between JENDL3.3, JENDL4.0 and JEFF3.1. the lowest K-eff between those nuclear data is 0.0120 and the highest is 0.0193 or about 1.2% and 1.93%. And in the end of operation period, K-eff value between JENDL4.0 and JEFF3.1 are pretty close. There are no different result of neutron spectra between those nuclear data. And the spectra are dominant in the thermal range for all ^{235}U enrichment. It will be better if the neutronics calculation use more nuclear data libraries for comprehensive comparison.

References

[1] Technology Roadmap Update for Generation IV Nuclear Energy System, January 2014. Gen IV International Forum.

[2] IAEA-TECDOC-1382. 2003. Evaluation of High Temperature gas Cooled Reactor Performance: Benchmark Analysis Related to Initial Testing of the HTTR and HTR-10. Nuclear Power Technology Development Section, International Atomic Energy Agency

[3] Clief P, Sandro. 2015. Studi Desain GFR (Gas Coolef Fast Breeder Reactor) Berbahan Bakar Uranium-Plutonium Carbide Untuk Produksi Listrik dan Gas Hydrogen.Thesis Magister, Departemen Fisika ITB.

[4] K. Okumura. 2007. SRAC2006: A Comprehensive Neutronics Calculation Code System. Reactor Physiscs Group,apan Atomic Energy Research Institute, Japan.

Review of Membrane Oxygen Enrichment for Efficient Combustion

Danu Ariono* and Anita Kusuma Wardani

Department of Chemical Engineering, ITB, Jl. Ganesha 10, Bandung 40132, Indonesia

*Corresponding author, Email: danu@che.itb.ac.id

Abstract. Oxygen enrichment from air is a simple way of increasing the efficiency of combustion process, as in oxy-combustion. Oxy-combustion has become one of the most attracting combustion technologies because of its potential to address both pollutant reduction and CO_2 capture. In oxy-combustion, the fuel and recycled flue gas are combusted with oxygen enriched air (OEA). By using OEA, many benefits can be obtained, such as increasing available heat, improving ignition characteristics, flue gas reduction, increasing productivity, energy efficiency, turndown ratio, and flame stability. Membrane-based gas separation for OEA production becomes an attractive technology over the conventional technology due to the some advantages, including low capital cost, low energy consumption, compact size, and modularity. A single pass through membrane usually can enrich O_2 concentration in the air up to 35% and a 50% concentration can be achieved with a double pass of membrane. The use of OEA in the combustion process eliminates the presence of nitrogen in the flue gas. Hence, the flue gas is mainly composed of CO_2 and condensable water that can be easily separated. This paper gives an overview of oxy-combustion with membrane technology for oxygen enrichment process. Special attention is given to OEA production and the effect of OEA to the efficiency of combustion.

1. Introduction

Coal is a combustible sedimentary rock with a high amount of carbon and hydrocarbons. Coal contains the energy stored by plants that lived hundreds of millions of years ago. Therefore, coal becomes the primary fuel, especially for electricity generation. Currently, about 40% of the world's electricity is generated with coal [1]. Demand for coal resources is expected to continue to increase worldwide because of increasing energy demand in the future. However, coal as an energy source has a number of negative environmental impacts, including the release of particle matter (PM), oxides of sulphur and nitrogen (SO_x and NO_x), carbon monoxide (CO), and trace metals such as mercury [1-4]. Therefore, Coal combustion presents challenges for reducing emissions of air pollutants and carbon dioxide (CO_2). In response to these challenges, a number of technologies have been developed with further prospective developments toward 'near zero emission' power plants. One of the promising technologies for emissions reduction is by using oxy-combustion.

Oxy-combustion has become one of the most attracting combustion technologies in the last decade [5]. In Oxy-Combustion, the fuel and recycled flue gas are combusted with oxygen enriched air (OEA) with the result that the flue gas consists mainly of CO_2 and water vapor. The net flue gas contains about 80–95% CO_2 depending on the fuel used and the particular oxy-fuel combustion process [6]. The water can be removed easily by condensation and the remaining CO_2 can be purified relatively inexpensively. This process is having numerous environmental benefits as well as improved energy efficiency and productivity. Using oxygen enriched air (OEA) as the feed to the power plant ensures that the flue gas contains nearly pure CO_2, which eliminates the expensive CO_2/N_2 separation necessary for capture and sequestration [7, 8].

To establish oxy combustion, it needs oxygen enrichment process. There are three methods that usually used in oxygen enrichment: distillation, adsorption, and membrane technology. Some reviews relating to oxygen enrichment have been published. However, the reviews more focused on cryogenic distillation process. None of them focus on the membrane oxygen enrichment for oxy combustion. Therefore, this paper provides an overview of oxy-combustion with membrane technology for oxygen enrichment process. Special attention is given to OEA production and the effect of OEA to the efficiency of combustion.

2. Oxy combustion

Oxy combustion is recognized as an effective process for coal-fired power plants to continue operation in a carbon-constrained world. In 1982 oxy-coal combustion was originally proposed to produce CO_2 for Enhanced Oil Recovery (EOR) [2, 6, 9, 10]. But lately, interest has been paid to oxy-coal combustion as a means to reduce pollutant emission and create a CO_2 gas stream that can easily be removed. The core concept of oxy-combustion is the use of a high purity oxidant stream for the combustion process. The aim is to obtain a carbon dioxide-rich stream that is ready for sequestration, after removing water vapor and other impurities. Replacing combustion air with a mixture of oxygen and recycled flue gases has a significant impact on the combustion, such as increase efficiency, lower emissions, improve temperature stability and heat transfer, increase productivity, improve ignition characteristics, and low risk [5, 11, 12].

$$Coal + O_2 + 3.76N_2 \rightarrow \underbrace{CO_2 + H_2O}_{\text{unavoidable}} + \underbrace{3.76N_2}_{\text{unnecessary}} + \underbrace{(CO, unburned\ coal, NO_x)}_{\text{undesirable}} + \underbrace{Heat}_{\text{needed}} \quad (1)$$

Combusting the fuel using oxygen enriched air at near-stoichiometric conditions creates a flue gas consisting mainly of CO_2 (80-95 % on a dry basis), water vapour and minor amounts of noble gases and, depending on fuel composition [13]. Due to its high concentrations, CO_2 can be recovered relatively easily and economically with the flue gas compression train [2]. After the water in the flue gas is condensed and the small amounts of impurities such as SO_x, NO_x, O_2, noble gases, metals and particulates are removed to meet transport and storage requirements [3, 8, 14]. In addition, for safety and environmental requirements, the CO_2 can be sent to storage.

3. Oxygen enrichment

One of the most important processes in oxy combustion is oxygen enrichment from air. Some technologies have been developed to meet high concentration of oxygen to improve combustion efficiency. Three methods that currently exist to separate oxygen from the air include distillation, adsorption, and membranes [15, 16]. Distillation is the most mature of the three technologies and allows for both high purities (>99%) and large scale productions [13, 17]. Adsorption is able to reach purities of up to 95% oxygen but the requirement of solvents limits its size capacity due primarily to capital costs [18]. Membrane technology is the most recent method for oxygen enrichment, which includes polymeric and high temperature ion transport membranes. Polymeric membranes can produce oxygen enriched air of various concentrations, meanwhile ion transport membranes can produce purities of close to 100% [19].

A commercial-scale coal-fired oxy-combustion power plant would require thousands of tons of oxygen each day. Cryogenic distillation is one of technology that commercially available today to produce such large quantities of O_2 economically [20]. Cryogenic distillation is a technology that has been around since the early 1900s when the Linde process was developed [21]. Gases such as oxygen and nitrogen, the primary constituents of air, decrease in temperature and condense when throttled because they have a positive Joules–Thompson coefficient. By controlling how much the temperature decreases through pressurizing and throttling the air, oxygen and nitrogen can be separated by phase as they have different boiling temperatures [19].

Table 1. Oxygen production technology alternatives [18, 22]

Process	Status	Purity (vol%)
Cryogenic distillation	Mature	> 99
Pressure Swing Adsorption	Semi-mature	95
Polymer membrane	Semi-mature	> 99
Ion transport membrane	Developing	> 99

Beside cryogenic distillation, oxygen enrichment can be obtained by pressure wings adsorption (PSA). In PSA, the regeneration of the adsorbent is performed by reducing the total pressure of the system, the total pressure of the system "swings" between high pressure in feed and low pressure in regeneration. Zeolites A and X are the most common used adsorbate for air separation [18, 23-25]. Compressed air enters the adsorption bed at a high pressure. Since nitrogen is adsorbed appreciably more than oxygen at high pressure, oxygen-enriched air can be easily obtained at the outlet. The bed is then blown down to a lower pressure to desorb the nitrogen in zeolite [25-27].

Both cryogenic distillation and PSA are energy intensive, which limits further development of the oxy combustion technique. Therefore, membrane process becomes one of the attractive technologies to replace both processes. Membrane offers some benefits, such as simple separation, continuous nature of the process, and operation at near ambient conditions [16, 18]. With an induced pressure difference, an O_2/N_2 selective polymeric membrane can partly separate oxygen from air by having properties that allow oxygen molecules to pass through the membrane easier than nitrogen molecules. However, the membrane cannot fully separate oxygen from the air; it can only be used to increase the concentration of oxygen as some nitrogen molecules will also flow through the membrane [19].

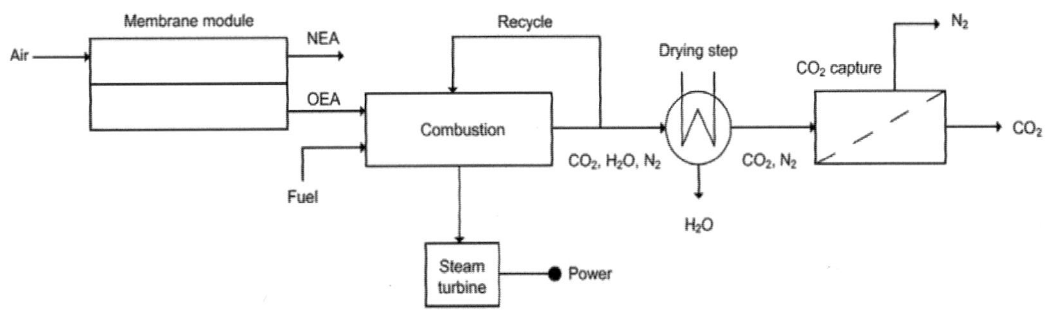

Figure 1. Oxy combustion with membrane as oxygen enrichment process, adapted from [28, 29]

Recently, some studies have reported the oxygen production by membrane processes. Smith and Klosek [18] used inorganic oxide ceramic materials to produce oxygen by the passage of oxygen ions through the ceramic crystal structure. These systems operated at high temperatures, over 11008°F. Oxygen molecules were converted to oxygen ions at the surface of the membrane and transported through the membrane by an applied electric voltage or oxygen partial pressure difference, then reform oxygen molecules after passing through the membrane material. The oxygen ions travel through the inorganic oxide ceramic materials at very high flow rates and produce nearly pure oxygen on the permeate side of the membrane.

Meanwhile, Clark and Rowan [30] studied an efficient method to enrich the oxygen concentration of the air stream by passing the incoming air through a series of polymeric membranes where there is a preferential diffusion rate for oxygen through the membrane over the other gases in the air. A single pass through membranes can enrich the oxygen concentration in the air to 35% and a 50% concentration can be achieved with a double pass of membrane. Later, Murali et al. [28] also used polymeric membrane based separation for O_2 production. Production of pure O_2 is difficult because a little quantity of N_2 always permeates through the membranes, resulting in oxygen enriched air (OEA) rather than pure O_2.

Besides by polymeric membrane, oxygen enrichment can be obtained by using an integrated oxygen ion transport membrane (ITM). Oxy-fuel combustion with ITM is a thermodynamically attractive concept that seeks to mitigate the penalties associated with CO_2 capture from power plants. ITM utilizes a high-temperature mixed-conducting (ionic and electronic) ceramic membrane to separate oxygen from air, and in some cases counter-currently oxidize a fuel within the same unit. ITM systems operate at relatively high temperatures, typically 800-1000°C [9, 31, 32].

Ion transport membranes (ITMs) offer promising oxygen production technology with high purity (up to 99%) without adversely affecting the efficiency of the oxy-fired plants. The separation rate of such ITMs can be increased by replacing the conventional inert sweep gas with a reactant/diluent mixture (e.g. CO_2, CH_4) as this reduces the permeate partial pressure on the permeate side of the membrane, which, along with the temperature, governs the permeation flux [33].

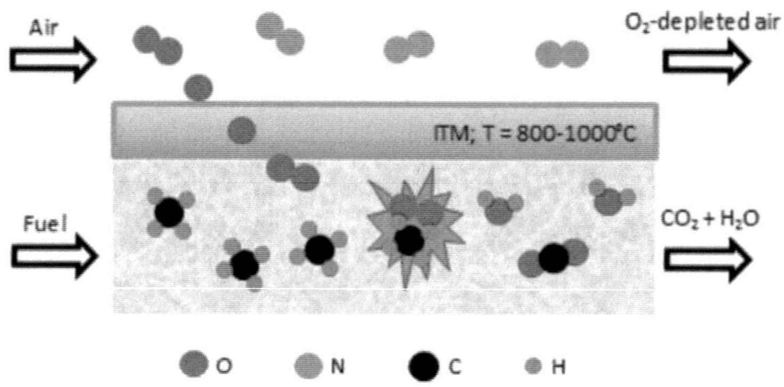

Figure 2. ITM counter-currently oxidize a fuel within the same unit, adapted from [6, 34, 35]

Ceramic membranes are selected as materials of ITM to maximize the oxygen permeation rate while keeping the membrane temperature within the operating limits [36]. Perovskites (ABO_3) are the most studied ITM materials due to their additional ability to conduct electrons, which allows oxygen transport to take place without an external circuit to transport electrons from one surface to the other. There are three types of perovskites for ITM, including BSCF (Ba0.5Sr0.5Co0.2Fe0.8O3), which has high oxygen fluxes but poor stability at low temperatures, LSCF (La0.6Sr0.4Co0.2Fe0.8O3−δ) which has lower fluxes but better stability, and BCFZ (BaCoxFeyZrzO3−δ) which has produced very stable membranes, operated for up to 2200 hours [9, 37].

4. Effect of OEA on oxy combustion

Using oxy-combustion for specific applications may improve efficiency, depending on the exhaust gas temperature and percentage of oxygen in the combustion air. In the past decades, intense research efforts have been directed to the development and improvement of ceramic and polymeric membranes for oxygen separation from air at high temperature operations. The result showed that by increasing oxygen flux, the combustion process is improved slightly and the maximum temperature inside the reaction zone is increased a little bit [10, 38].

Increasing the concentration of oxygen in combustion air also allows a reduction in fuel usage. This energy savings is due to the fact that nitrogen is not necessary for the combustion process and energy is lost as the inert nitrogen is heated during combustion and exhausted to the atmosphere [39]. As the oxygen concentration in the combustion air increases, more energy is available from the combustion thus less fuel is required. When oxygen concentration increases from 21% to 40%, energy saving of 60% can be achieved with the assumption that the exhaust temperature is constant, while in the real case it increases and thus there will be reduced savings.

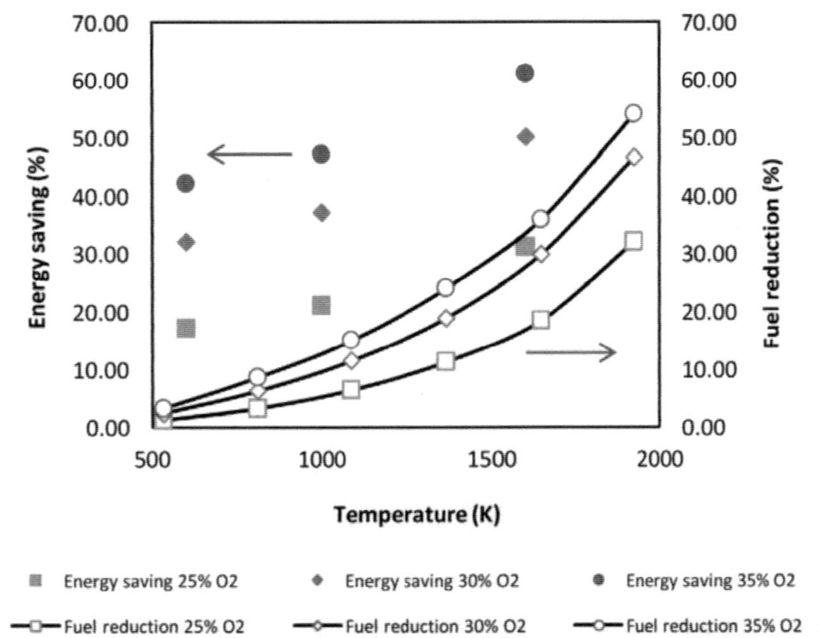

Figure 3. Effect of oxygen concentration and temperature on energy saving and fuel reduction, adapted from [11, 40]

The oxy-fuel efficiency can be increased by an optimisation of membrane operation parameters, such as oxygen separation degree, partial pressure ratio, and total pressure ratio. For example, the overall efficiency can be increased to 40.8% by reducing the mass flow in the recirculation which reduces the mean partial pressure ratio to 17 [35]. A reduction of the partial pressure ratio is only possible by increasing the area of the membrane. With this partial pressure ratio the upper temperature difference of the air preheater reaches a limit. Therefore, a further reduction of the partial pressure ratio requires changes to other parameters as well.

5. Commercial oxy combustion

There are a number of pilot-scale facilities around the world, typically ranging in size from about 0.3-3.0 MW that has been used to characterise the combustion performance of coals under oxy-fuel combustion conditions. A number of demonstration projects ranging in size from 30 MW to 300 MW have been proposed [41-43].

Furthermore, large oxy-combustion plant analyses and configurations also have been completed. The Babcock & Wilcox Company (B&W), through its Power Generation Group, and Air Liquide (AL) have been actively involved in the development of oxy-coal technologies for power generation throughout the past decade [4, 44]. During 2007 and 2008, B&W successfully demonstrated oxy-combustion by using eastern bituminous, sub-bituminous and lignite coals. The oxy-combustion emission levels were all lower than emissions under air-blown combustion for all fuels. From 2009, B&W has been actively pursuing projects and partners for demonstrating CCS with oxy-combustion in the range of 50MW to 150MW. This size provided for commercial scale testing and also generated up to 1 million tons/year of CO_2 [45].

Other company that develop oxy combustion is Alstom. They started with lab scale (<3 MW) in 1990s. In 2012, they successfully demonstrated large pilot plant by using pulverized coal with capacity of 150-400 MW [46]. Nowadays, by using scale up modelling, they try to design full scale oxy combustion plant that will be operated in 2018/2019 with capacity of 600-1100 MW. Besides, there are some other large scale demonstration projects of oxy combustion that have or will be developed, such as Vattenfall's Jaenschwalde 30 MW Project (Germany), CIUDEN/Endessa's Compostilla OxyCFB300 Project (Spain), KEPCO's Young Dong Project (South Korea), FutureGen 2.0 Project (USA), and White Rose Project (UK).

6. Conclusion

Oxy-combustion has become one of the most attracting combustion technologies where the fuel and recycled flue gas are combusted with oxygen enriched air with the result that the flue gas consists mainly of CO_2 and water vapor. Some benefits can be obtained by oxy-combustion, such as increase efficiency, lower emissions, improve temperature stability and heat transfer, increase productivity, improve ignition characteristics, and low risk. Three branches of technologies have been developed to meet high concentration of oxygen to improve combustion efficiency, include distillation, adsorption, and membranes. A single pass through membranes can enrich the oxygen concentration in the air to 35% and a 50% concentration can be achieved with a double pass of membrane.

By increasing oxygen flux, the combustion process is improved slightly and the maximum temperature inside the reaction zone is increased a little bit. Besides, increasing the concentration of oxygen in combustion air allows a reduction in fuel usage. When oxygen concentration increases from 21% to 40%, energy saving of 60% can be achieved.

Reference

[1] Farzan H, Vecci S J, Châtel-Pélage F, Pranda P and Bose A C 2005
[2] Shaddix C and Molina A 2011: Woodhead Publishing)
[3] Tan Y, Croiset E, Douglas M A and Thambimuthu K V 2006 *Fuel* **85** 507-12
[4] Moorman S and McDonald D 2011 *Zaragoza*
[5] Alsayed M F M 2008: An-Najah National University)
[6] Abraham B, Asbury J, Lynch E and Teotia A 1982 *Oil Gas J.;(United States)* **80**
[7] GHG I 2005 *Int. Energy Agency Greenhouse Gas R&D Programme*
[8] Buhre B J P, Elliott L K, Sheng C D, Gupta R P and Wall T F 2005 *Prog. Energy Combust. Sci.* **31** 283-307
[9] Lockwood T 2014 *IEA Clean Coal Centre, Report No. CCC/240, ISBN: 978-92-9029-561-7, August 2014.,*
[10] Scheffknecht G, Al-Makhadmeh L, Schnell U and Maier J 2011 *Int. J. Greenhouse Gas Control* **5, Supplement 1** S16-S35
[11] US D E 2005 (Wahington DC: Energy Efficiency and Renewable Energy)
[12] Ochs T L, Summers C A, Oryshchyn D B, Gross D, Patrick B, Gross A, Dogan C, Simmons W and Schoenfeld M 2004 *Oxy-fuel combustion systems for pollution free coal fired power generation*: Citeseer)
[13] Anheden M, Yan J and De Smedt G 2005 *Oil Gas Sci. Tech.* **60** 485-95
[14] Skeen S A, Yablonsky G and Axelbaum R L 2009 *Combust. Flame* **156** 2145-52
[15] Koros W J and Mahajan R 2000 *J. Membr. Sci.* **175** 181-96
[16] Coombe H S and Nieh S 2007 *Energy Convers. Manage.* **48** 1499-505
[17] Belaissaoui B, Le Moullec Y, Hagi H and Favre E 2014 *Sep. Purif. Tech.* **125** 142-50

[18] Smith A R and Klosek J 2001 *Fuel Process. Technol.* **70** 115-34
[19] Burdyny T and Struchtrup H 2010 *Energy* **35** 1884-97
[20] Gale J, Hendriks C, Turkenberg W, Tranier J-P, Dubettier R, Darde A and Perrin N 2011 *Energy Procedia* **4** 966-71
[21] Cengel Y A and Boles M A 1994 *Sea* **1000** 8862
[22] Adeosun A, Muthiah A and Abu-Zahra M R 2013 *Int. J. of Thermal Env. Eng.* **5** 113-21
[23] Zahra M, Jafar T and Masoud M 2008 *World Academy Sci. Eng.Tech.* **47** 99-103
[24] Yoshida S, Ogawa N, Kamioka K, Hirano S and Mori T 1999 *Adsorption* **5** 57-61
[25] Chiang A S T, Hwong M Y, Lee T Y and Cheng T W 1988 *Ind. Eng. Chem. Res.* **27** 81-5
[26] Rao V R, Farooq S and Krantz W B 2010 *AlChE J.* **56** 354-70
[27] Grande C A 2012 *ISRN Chem. Eng.* **2012**
[28] Murali R S, Sankarshana T and Sridhar S 2013 *Sep. Purif. Rev.* **42** 130-86
[29] Favre E, Bounaceur R and Roizard D 2009 *Sep. Purif. Tech.* **68** 30-6
[30] Clark G A and Rowan M 1996: DTIC Document)
[31] Dillon D, Panesar R, Wall R, Allam R, White V, Gibbins J and Haines M 2004
[32] Mancini N D and Mitsos A 2011 *Energy* **36** 4721-39
[33] Habib M A, Ahmed P, Ben-Mansour R, Badr H M, Kirchen P and Ghoniem A F 2013 *J. Membr. Sci.* **446** 230-43
[34] Habib M A, Nemitallah M and Ben-Mansour R 2013 *Energy Fuels* **27** 2-19
[35] Stadler H, Beggel F, Habermehl M, Persigehl B, Kneer R, Modigell M and Jeschke P 2011 *Int. J. Greenhouse Gas Control* **5** 7-15
[36] Nemitallah M A, Habib M A and Mezghani K 2015 *Energy* **84** 600-11
[37] Mancini N D and Mitsos A 2011 *Energy* **36** 4701-20
[38] Nemitallah M A, Habib M A and Mansour R B 2013 *J. Membr. Sci.* **432** 1-12
[39] Ming X, Borgnakke D S, Campos M A, Olszewski P, Atreya A and Borgnakke C 2003 *University of Michigan, ACEEE Summer Study on Energy Efficiency in Industry*
[40] Ming X, Borgnakke D S, Campos M A, Olszewski P, Atreya A and Borgnakke C 2013 *University of Michigan, ACEEE Summer Study on Energy Efficiency in Industry*
[41] Wall T, Liu Y, Spero C, Elliott L, Khare S, Rathnam R, Zeenathal F, Moghtaderi B, Buhre B, Sheng C, Gupta R, Yamada T, Makino K and Yu J 2009 *Chem. Eng. Res. Des.* **87** 1003-16
[42] McCauley K 2009
[43] Davidson R M and Santos S O 2010 *Oxyfuel combustion of pulverised coal*: IEA Clean Coal Centre London, UK)
[44] McCauley K, Farzan H, Alexander K, McDonald D, Varagani R, Prabhakar R, Tranier J and Perrin N 2009 *Energy Procedia* **1** 439-46
[45] Varagani R, Pranda P, Gautier F, Tranier J-P, Perrin N, Farzan H, Vecci S J, McDonald D K and McCauley K J 2007
[46] Levasseur A A, Chapman P J, Nsakala N and Kluger F 2009

The Calculation Study of Electronic Properties of Doped RE (Eu, Er and Tm)-GaN using Density Functional Theory

Aflah Zaharo and Acep Purqon

Physics of Earth and Complex System, Institute of Technology Bandung, Bandung, 40132, Indonesia

Email : aflahzaharo1@gmail.com

Abstract. The calculation of the structure and electronic properties of Rare Earth (RE) at the wurtzite Gallium Nitride (GaN) based on DFT has completed. GGA approximation used for exchange correlation and Ultra soft pseudo potential too. The stability structure of GaN is seen that difference lattice parameter 11% lower than another calculation and experiment result. It is shown the stability structure GaN have direct band gap energy on Gamma point hexagonal lattice Brillouin zone. The width Eg is 2.6 eV. When one atom Ga is substituted with one atom RE, the bond length is change 12 % longest. An in good agreement with theoretical doping RE concentration increases, the edge of energy level shifted towards to make the band gap narrow which is allow the optical transitions and help to improve the optical performance of GaN. The RE doped GaN is potentially applicable for various color of LED with lower energy consumption and potentially energy saving application

1. Introduction

Lighting technology is widely used in daily life. The huge growth of population and advanced life demand equipped developing with qualified lighting and saved. New materials using RE over significant progress in the development of optoelectronic technology and photonic like element Neodymium (Nd) which can emit a laser and phosphorus in the solid state, Er or Tb for displays and lighting, Er or Pr used in fiber optics for telecommunications [1], so that the use of RE also potentially be applied to the manufacture of LEDs. While blue LED GaN-bassed is made for the first time in 1971 by Jaques Pankove at RCA laboratories. In the mid-1990s, a researcher named Shuji Nakamura of Nichia Corp company, demonstrated a blue LED GaN-bassed for the first time with high brightness (high brightness GaN-based blue LED) [2].The discovery cause the revolution of lighting from incandescent bulbs in the era of Thomas Edison into LED based that certainly provides a high level of efficiency and tremendous saving both in energy and cost.

The main material of LED lights currently is based on semiconductors category III-nitride, one of them is a Gallium Nitride (GaN). GaN is a semiconductor that has wide and direct band gap (direct wide band gap) [3]. With the characteristic, GaN is a semiconductor that is appropriate for optoelectronic applications, one of them is LED. The first approach is the color mixing by configuring three single colors LED (Red-Green-Blue) and the second approach is to combine LEDs with conversion of emission color materials in whole or in part [4].

2. The Calculation Study Electronic Properties of GaN RE using DFT

2.1. Many Body Problem

The problem of quantum systems with many particles number is called Many Body Problem. Then the equation is done by completing the configuration of the ground state of the electrons first, then calculate the energy of the system configuration, and finally completed the calculation of core movement. Separation of the movement of electrons and nuclei calculation is known as the Born-Oppenheimer approach [5]. Hamiltonian becomes:

$$\hat{H} = -\frac{1}{2}\nabla^2 + \sum \frac{1}{|r_i - r_j|} + \hat{V}_{ext} \tag{1}$$

$$\hat{V}_{ext} = -\sum \frac{Z'}{|r - R'|} \tag{2}$$

2.2. Hohenberg-Kohn Theory

The big idea of the DFT is that the wave functions of many particles, can be defined as a function of the electron density. It comes from the idea that for a certain value, it will represent a wave function well (unique). In general theory of Hohenberg-Kohn stated:x

First, a wave function of the ground state uniquely defined by a function of the density in the ground state. In mathematical equation becomes:

$$\psi_0(\vec{r_1}, \vec{r_2}, \ldots, \vec{r_N}) \Rightarrow \psi[n_0(r)] \tag{3}$$

It is most essential in the DFT. As a consequence, the value expectations of the ground state (Ground State) for all is also a function of [5]

Secondly, the expected value of the total energy of the ground state which is expressed in:

$$E_{v,0} = E[n_0] = \langle \psi[n_0] | \hat{H} | \psi[n_0] \rangle \tag{4}$$

Energy value has a principle of variation which,

$$E[n_0] \leq E[n'] \tag{5}$$

Applies,

$$E[n_0] = \langle \psi[n_0] | \hat{H} | \psi[n_0] \rangle \leq \langle \psi[n'] | \hat{H} | \psi[n'] \rangle \tag{6}$$

This principle can be utilized, that minimizes the expected the total energy value eV an external potential (V_{ext}) can be done with a functional derivative of the expectation value of the electron densities

2.3. Total Energy

DFT is a method for calculating the electronic structure of a system of atoms, molecules, or material. In general, the total energy of an electron systems are defined:

$$E_v = \langle \psi[n] | \hat{H} | \psi[n] \rangle \tag{7}$$

And Hamiltonian $\hat{H}$ can be written:

$$\hat{H} = \hat{T} + \hat{U} + \hat{V} \tag{8}$$

In this case $\hat{T}$ the kinetic operator, $\hat{U}$ the operator of potential interactions between electrons, and is the operator of potential interactions between the electrons with the outside. Total electron energy becomes:

$$E[n] = T_s[n] + U[n] + V_{ext}[n] + E_{xc}[n] \tag{9}$$

2.4. Kohn-Sham Equation

The most stable state is obtained when the system is in a state of minimum energy. Hohenberg Kohn theory of the equation (12), the minimum total energy directly correlated with minimum electron density. So that minimum energy can be obtained using the equation lagrange multiplier by lowering the equation (9) to the electron density n (r). By entering these constraints, the minimization of the total energy equation becomes:

$$\frac{\partial}{\partial_n(r)}\left(E[n]-\sum_i \lambda_i \int \phi_i * \phi_i d^3r\right)=0 \tag{10}$$

2.5. Self Consistent Field (SCF)

Total electron density is the sum of the square of the orbital all off the function. KS then to solve this equation, we need an educated guess electron density value which is then used to solve the equation KS. SCF cycle is shown in the algorithm in Figure.1.

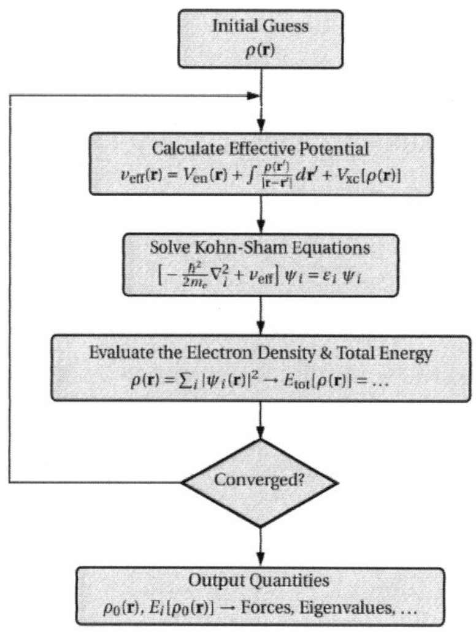

Figure.1. SCF Cycle

2.6. Pseudopotential

Pseudo potential method gives the idea that electrons can change and interact are just part of the outer electrons (valence electrons). With a core electron approaches that have been fixed, then this method provides a solution by replacing the external potential V_{ext}^{pseudo} at the core electrons with pseudo potential and wave function with pseudo wave function. So the equation that needs to be resolved:

$$\left(-\frac{1}{2}\nabla + v_{ee} + v_{ext}^{pseudo} + v_{exc}\right)\phi_i^{pseudo} = \lambda_i \phi_i^{pseudo} \tag{11}$$

2.7. *Exchange Correlation Energy*

2.7.1 Local Density Approximation (LDA). LDA is the basis and the simplest of all approaches to energy E_{xc}. The main idea of the LDA is the charge distribution in a region are uniform and localized. E_{xc} so that energy can be written:

$$E_{xc}^{LDA}[n] = \int n(r) v_{xc}[n(r)] d^3r \tag{12}$$

2.7.2 Generalized Gradient Approximation (GGA)The difference from the concept of LDA, is information that is used not only on data from the electron density n (r) at position r, but also retrieve information from the gradient changes electrons density in the position. Then the E_{xc} energy becomes:

$$E_{xc}^{GGA}\left[n_\alpha, n_\beta\right] = \int (n_\alpha, n_\beta, \nabla n_\alpha, \nabla n_\beta) d^3 r \tag{13}$$

3. Method

This study uses a program package to PHASE / 0 [6]. The program is able to perform the calculation for the extended case through super cell scheme. Program PHASE / 0 capable of calculating the energy value of a system in an atomic scale, also called SCF calculation, to determine the condition of the ground state of the system. In addition to the calculation of SCF, PHASE / 0 is also capable of performing calculations to see either the electronic structure of Band Structure and Density of States (DOS), the spin density distribution and electron density.

Firstly, determine the model structure of solids that will be calculated (the position of atoms, crystalline form, and the lattice constants). Calculate the charge density by using SCF cycle. SCF cycle can be seen in Figure.2

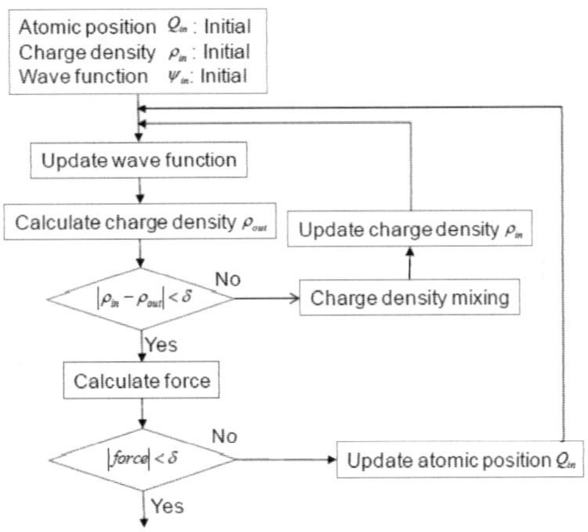

Figure.2 Working Cycle on Phase

This study uses a wurtzite crystal structure, because it is the most stable structure among other GaN structures. Figure.3 showed the smallest cell unit wurtzite GaN crystal structure consisting of four atoms, two Ga atoms and two N atoms has a cell unit vector unit is 1 = (a, 0,0), a2 = (a / 2, √ (3/2), 0) and a3 = (0,0, c) where a and c is the length of the lattice parameter wurtzite. GaN atomic coordinate position is (1 / 3.2 / 3.0) and (2 / 3.1 / 3.1 / 2), while the position of the N atom is (1 / 3,2 / 3, u) and (2 / 3,1 / 3,1 / 2 + u), where u is the internal parameters. In ideal wurtzite, is a constant value u (u = 3/8) and the value. To obtain the value of the optimal parameters for a, c, c/a and u, then the geometrical structure optimization process unit cell of GaN.

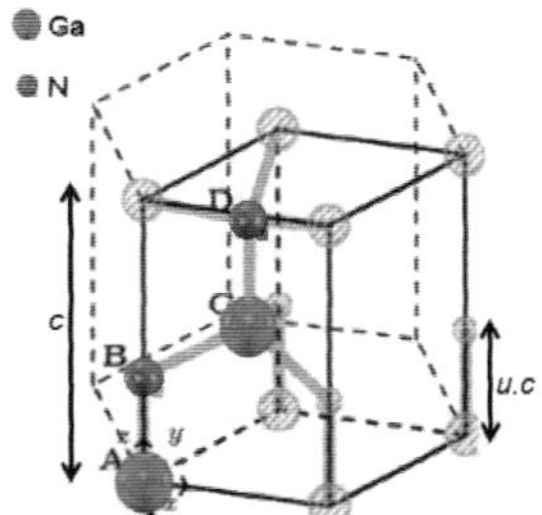

Figure.3 Wurtzite unit cell geometry and consists of two gallium atoms (red) and twonitrogen atoms (blue), with lattice constants a and c as well as the internal parameter u.

In RE: GaN single Ga atom is substituted by one atom RE, so the calculation of RE: GaN using $Ga_{15}N_{16}RE$ configuration. In Figure.4 super cell shows a model by substituting a single atom with one atom RE: GaN. This study uses super cell size of 32 atoms, with the configuration $Ga_{1-x}RE_xN$ in super cell 32 atoms (x = 0.0625).

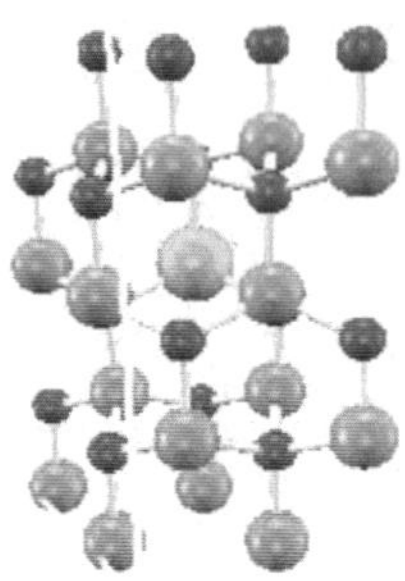

Figure. 4 Geometry wurtzite GaN super cell with substitutions 1 Ga atom by atom RE

E_{xc} energy type approach is used in this calculation use cut off wave function 25.00 Ry and cut off the charge density is 225.00 Ry. This calculation uses meshing methods for the integration of the brillouin zone with values for super cell k-point 2x2x2. SCF calculations done in a relaxed position with the convergence of energy limit 1.0e-03 Ha tree / Bohr and bounds in the style of the geometry optimization of 1.0e-03 Ha tree / Bohr.

4. Result

4.1. Geometry Structure Optimization

In this study used the crystal structure is wurtzite GaN. The procedures performed in this geometry optimization process following the procedures in reference C. Stamp et al [8]. Figure.5 show visualization of the modeling results unit cell GaN.

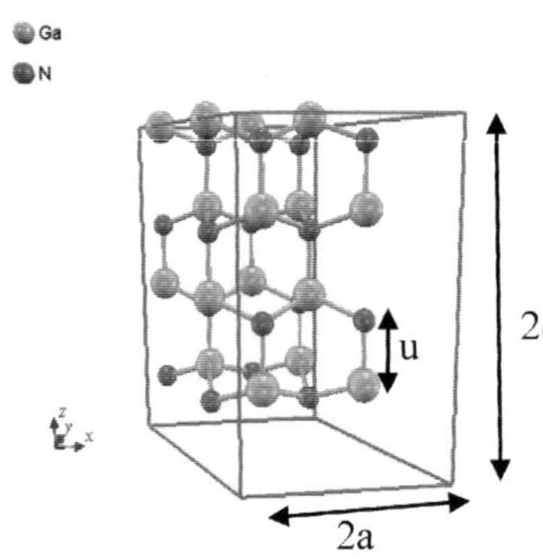

Figure .5 GaN modelling results using PHASE / 0 in wurzite structure optimization forum for lattice parameter values a and c, c / a ratio and internal parameter u for wurtzite GaN

4.2. Pure GaN

In the plot shown Figure.5 total energy as a function of the lattice parameters a and c, the value of c / a ratio as well as the internal parameter u. Tabel.1 shows the comparative results of the calculation of the geometrical structure optimization unit cell GaN with experimental data and other calculations. The

IOP Conf. Series: Journal of Physics: Conf. Series **877** (2017) 012051 doi:10.1088/1742-6596/877/1/012051

calculation in this study demonstrates the value corresponding compared with experimental data and other calculations, even for internal parameter u showed identical results with experimental data.

Table 1. Comparison of the calculation results wurtzite GaN structure optimization geometry a and c, c / a, internal parameter u with the results of other calculations and experimental data

Parameter	This work	Others Calculation	Experiment
a (A°)	3.142	3.124 - 3.245[a]	3.180 - 3.192[b]
c (A°)	5.123	5.0 - 5.228[a]	5.166 - 5.185[b]
c/a	1.630	1.622 - 1.632[a]	1.624 - 1.627[b]
U	0.377	0.375 - .376[a]	0.375 - 0.377[b]

[a]Ref [3,8] [b]Ref [8]

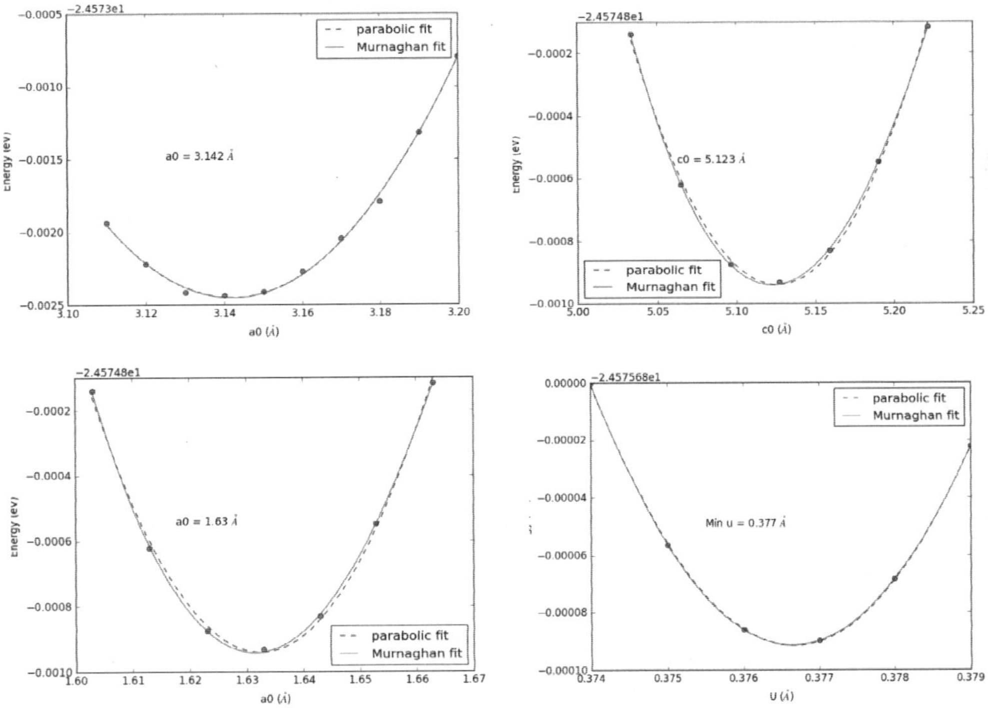

Figure.6 Total energy of the system as a function of cell units and value lattice parameters a and c, the ratio of c = a and u internal parameters for wurtzite GaN

4.3. GaN Doped RE

Calculation of total energy of the system, and the energy band structure and DOS of the system using the super cell 2x2x2 containing 32 atoms consisting of 16 atoms of gallium and 16 atoms of nitrogen. Figure.7 an energy band structure of pure GaN. From these images are shown GaN having a band gap energy directly (direct-band gap) in both the conduction band minimum (CB_{min}) and the valence band maximum (VB_{max}) are at a point with a difference of 2.6eV

The results of calculation of the energy band gap is smaller by about 26% compared with experimental data [8, 2]. This value also occurs in other theoretical calculations that use the same approach to energy exchange, ie GGA [7]. This occurs because of the underestimation factor calculation that occurs in DFT calculations using the GGA as reported in other work by Simanovskii et al [5].

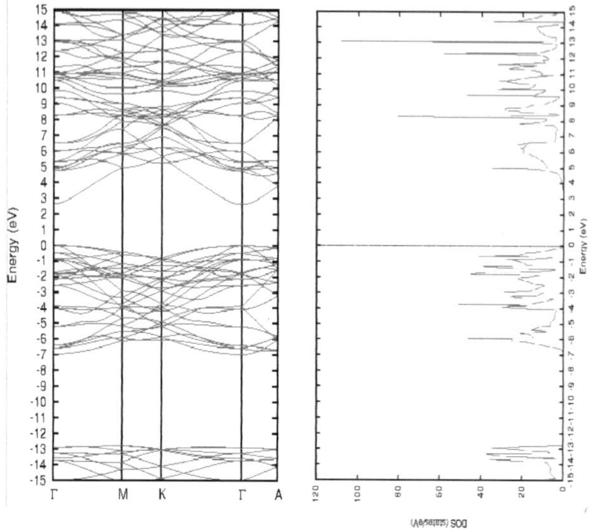

Figure.7 Band gap energy band structure (left) and total DOS for pure GaN (right) .Fermi level is set at the level of zero

The pattern of the electronic structure of GaN energy band gap of the calculation results, proving that the material properties of GaN has a direct wide band gap. These properties indicate GaN material particularly suitable for optoelectronic applications, such as LEDs

Table 2 Comparison of the energy band gap calculation results of theoretical calculations using the GGA with experimental data

Parameter	This Work	Others Calculation	Experiment
Eg (eV)	2.6	2.3[a]1.9[b] 1.7[c]	3.474-3.507[d]

GaN crystal structures form four bonds between gallium and nitrogen atoms form a tetrahedral coordination. The results of structural optimization in the previous step shows in Table 3

Table 3 Comparison of the long bond-length RE-N on structure optimization RE: GaN

No	Super cell	d_{RE-N} (°A) This Work	d_{RE-N} (°A) Others Calculation	d_{RE-N} (°A) experiment
1	$Ga_{15}N_{16}Eu$	2.1435-2.2009	2.260-2.304[a] 2.00-2.21[b]	2.3-2.5[d]
2	$Ga_{15}N_{16}Er$	2.1267-2.1887	2.152-2.178[a] 2.14-2.20[b]	2.23 ± 0.06[c]
3	$Ga_{15}N_{16}Tm$	2.1273-2.1281	2.15-2.30[b]	

In previous calculations show the structure of GaN without doping RE with the wide energy gap of 2.6 eV. Further calculations to investigate the state of the electronic structure in terms of the energy band gap structure as well as the state of DOS Table 4 and Figure. 8, 9 and 10.

Table 4 Comparison of wide band gap (band gap) and the impurity band due to the influence of RE dopant in GaN

No	Dopant	Atom Number	Wide of band gap (eV)	Impurity level VB_{maks} (eV)
1	*Eu*	63	2.6	1.22
2	*Er*	68	2.63	0.51
3	*Tm*	69	2.64	0.35

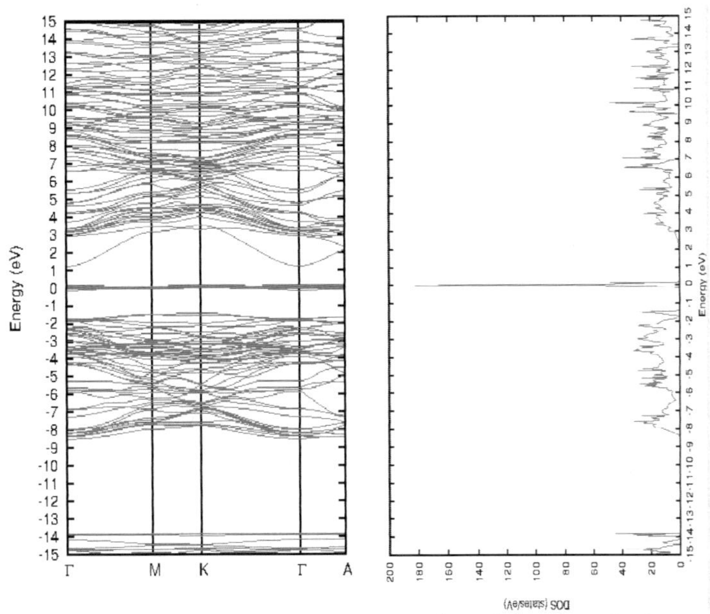

Figure.8 Energy band structure and DOS for $Ga_{15}N_{16}Eu$

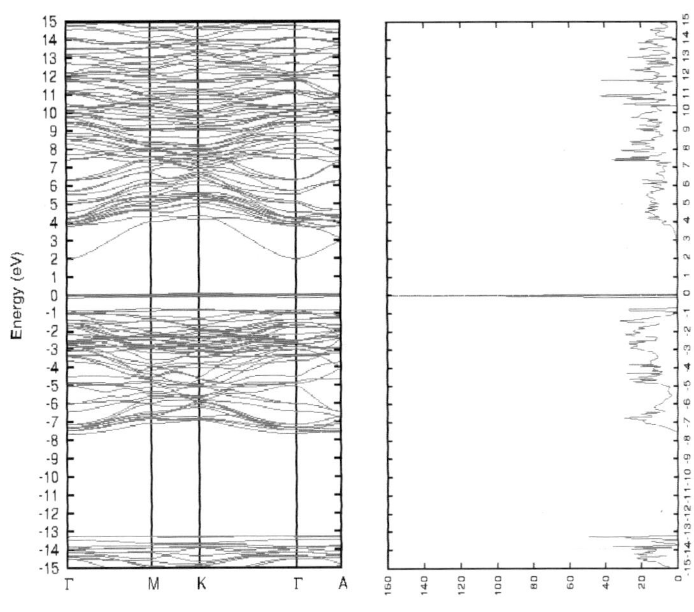

Figure.9 Energy band structure and DOS for $Ga_{15}N_{16}Er$

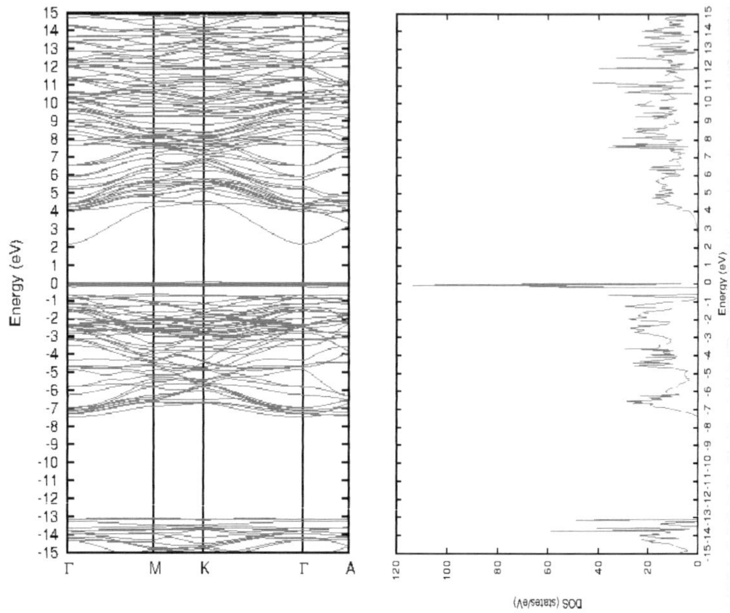

Figure.10 Energy band structure and DOS for $Ga_{15}N_{16}Tm$

Figure.8, Figure 9, Figure 10 showed the same trend, namely a change in energy band structure in the form of indirect band gap formation and an impurity band. RE doping in GaN also resulted in the formation of an impurity band is located between CB_{min} and VB_{max}. Pita impurity gives an advantage, since the narrowing of the energy band gap and this allows the optical transitions, which helps improve the optical performance of GaN.

5. Summary

The results of electronic structure calculations RE ions in GaN showed a structure stability RE: GaN in tetrahedral coordination with relaxation RE-N bond lengths varying between 2.1-2.2°A, this result is consistent with the data structure of the results of other calculations and experimental data EXSAF. RE: GaN also indicates a change in the electronic structure, which resulted in the band gap energy into indirect (indirect band-gap) and the formation of an impurity band with varying sizes for each dopant RE. Impurity band that forms resulted in a narrowing of the energy band gap which facilitates the optical transitions in materials that are expected to improve optical performance on GaN. Addition of RE dopant concentration in the GaN shows a shift in the boundary of the valence band is getting towards the low energy area, thus making the material more stable at room temperature.

References

[1] C. J. Humphreys (2008). *Solid State Lighting*. MRS Bulletin, 33, 459-470.
[2] D. M. Simanovskii, H. A. Schwettman, H. Lee, and A. J. Welch. (2003), Midinfrared Optical Breakdown in Transparent Dielectrics, Phys. Rev. Lett, 91, 107601
[3] C. Stamp and C. G. Van de Walle. Density Functional Calculations for III-V Nitrides Using the Local Density Approximation and the Generalized Gradient Approximation, Phys. Rev. B., 59, 5521-5535.(1999).
[4] E. F. Schubert. The Blue Laser Diode: The Complete Story. Cambridge, UK.(2003)
[5] Hohenberg, P., Kohn, W. *Inhomogeneous Electron Gas*, Phys. Rev. B, 136, B864-B871. (1964).
[6] Dirk Vogel, Peter Kruger, and Johannes Pollmann. *Structural and Electronic Properties of Group-III Nitrides*, Phys. Rev. B, 55, 12836{12839}. (1997).
[7] Center for Research on Innovative Simulation Software (CISS), the Institute of Industrial Science

(IIS), the University of Tokyo. First-*principles Electronic Structure Calculation Program PHASE/0 2014.* (2014).

[8] I. Vurgaftman, J. R. Meyer and L. R. Ram-Mohan. *Band Parameters for IIIV Compound Semiconductors and Their Alloys*, J. Appl. Phys, 89, 5815. (2001)

[9] Rubio, Angel and Corkill, Jennifer L. and Cohen, Marvin L. and Shirley, Eric L. and Louie, Steven G., *Quasi particle band structure of AlN and GaN*, Phys. Rev. B, 48, 11810{11816}.(1993).

Effect of Acetylene Black Content to Half Cells Li-ion Battery Performance Based on $Li_4Ti_5O_{12}$ using Li_2CO_3 as Lithium Ion Source with Hydrothermal Mechanochemical Process

B Priyono[1], Faizah[1], A Z Syahrial[1], A Subhan[2]

[1]Departement of Metallurgical and Materials Engineering, Universitas Indonesia, Depok, 16424, Indonesia
[2]Research Center for Physics – LIPI, PUSPIPTEK, Tangerang, Banten, 15310, Indonesia

Email: bambang.priyono@ui.ac.id

Abstract. Lithium titanate ($Li_4Ti_5O_{12}$)/LTO is a promising candidate to be used as anode electrode in Li-ion battery, to replace graphite in Li-ion battery application. Crystal structure of lithium titanate/LTO is more stable or undergoes less strain than graphite during intercalation and de-intercalation process Li^+ ions. However, although lithium titanate has good stability, the material has low electrical conductivity and lithium ion diffusion. The purpose of this research is to synthesis the spinel LTO using combinated hydrothermal and mechanochemical processes from xerogel TiO_2. Then, to increase the conductivity, in the half-cell battery assembly process it was added acetylene black conductive (AB) additive with various from 10%, to 15% in wt. The LTO obtained were characterized using scanning electron microscope (SEM), X-Ray Diffraction (XRD) and Brunauer–Emmett–Teller (BET). The XRD showed a rutile as minor phase, while SEM showed homogeneous distribution of particle with an average particle size of 0.35 μm. The BET showed that the surface area of LTO formed is 2.26 m^2/g. The assembled coin half cells used this $Li_4Ti_5O_{12}$ as a cathode and lithium metal foil as the anode were tested using electrochemical impedance spectroscopy (EIS), cyclic voltammetry (CV) and charge discharge (CD). The conductivity value obtained from EIS corresponds to the contents of AB. Meanwhile, the CV and CD testing showed that higher percentage of AB causing the decrease of battery specific capacity. The highest specific capacity at the rate of 10C is obtained at the mixture of 10wt% AB with the value of 40.91 mAh/g.

1. Introduction

To the world concern on the environmental pollution and global warming, usage of renewable energy is considered as the most promising solution. This kind of energy needs the storage ultimately. One of useful storages is lithium ion batteries (LIBs), which could be applied for the hybrid electric vehicles (HEVs), plug-in hybrid electric vehicles (PHEVs), and storage systems for others renewable energy resources [1]. Most LiBs were made with graphite anodes since it is relatively cheap, discharge capacity 375 mAh/g [2] and the intercalation and de-intercalation of lithium ion is very stable [3]. However, it has limitation of formation SEI layer during the first charge or discharge cycle and lower rate capability [4].

The use of $Li_4Ti_5O_{12}$ (LTO) to replace graphite has been developed for optimal results. During intercalation, LTO based anode shows an almost zero strain properties with good thermal stability and can endure of thousands of cycles without significant capacity loss [5]. However, LTO has a low electrical conductivity (10^{-9} S.cm^{-1}) which is not favorable to high rate capability application [6].

The focus of this study is to increase the electrical conductivity of LTO using the addition acetylene black (AB) during assembly process of Li-ion battery half cells. The LTO in this work was prepared by mixing anatase TiO_2 xerogel with Li_2CO_3 using hydrothermal mechanochemical method. Combination of these methods is expected to produce $Li_4Ti_5O_{12}$ with nanoparticle size with good electrochemical performance. Influence of the AB content (wt.%) in the LTO slurry mixture will be investigated.

2. Experimental

2.1 Materials synthesis and electrode fabrication

$Li_4Ti_5O_{12}$ powder was prepared through the following steps, i.e: sol-gel, calcination hydrothermal mechanochemical and sintering process. The sol-gel process started with the mixing the primary solution (titanium butoxide + ethanol pH 3) with secondary solution (ethanol pH 3 + distilled water) to form a clear gel using a magnetic stirrer. The obtained gel was dried in the ambient temperature to get xerogel TiO_2. Calcination process was done at temperature of 300 °C for 2 h to obtain anatase TiO_2. Then, hydrothermal process was done at temperature of 120 °C for 15 h. The high energy ball mill was used to mix the hydrothermal result with Li_2CO_3. Finally, the sintering process was then performed at 750 °C in 1 hour to form LTO spinel phase and ready to be used for the active electrode material.

The electrochemical cycling performance of LTO powder was carried out with a coin cell using a lithium metal foil as anode. The cell had a configuration of Li metal, electrolyte and $Li_4Ti_5O_{12}$ and polypropylene was used as the separator. A combination of acetylene black as conductive agent with various 10 wt.%, 12 wt.%, and 15 wt.%. Polyvinylidene fluoride (PVDF) binder homogeneously mixed with DMAC was prepared into viscous slurries for efficient deposition. The slurries were deposited on current collectors copper foil by Doctor Blade with 6 cm/minutes coating speed. The electrode was then dried under vacuum at 80 °C for 2 h before electrochemical testing. Cell assembly was performed in a glove box filled with pure argon.

2.2 Characterization and Testing

The characterization of LTO powder to know the morphology of sample was observed by scanning electron microscope. The particle size of sample was measured with Image J software. Phase identification was examined by X-ray diffraction analysis (XRD) using Cu Kα radiation. BET (Brunaeur-Emmet-Teller) method was used to analyze the surface area of LTO by using Quantachrome Nova instrument.

The Cyclic voltammetry (CV) of coin cell was carried out at scan rate of 100 mVs^{-1}. Electrochemical impedance spectroscopy (EIS) tests was performed to show impedances value. The charge and discharge (CD) testing was conducted to evaluate the cell electrochemical performance from low current (C/3) to high current (10C) discharge rate.

3. Results and discussion

3.1 The structure of $Li_4Ti_5O_{12}$ powder

The morphology of the LTO powder produced from hydrothermal mechanochemical reaction synthesis was examined using scanning electron microscope (SEM). Figure 1. shows representative SEM image of LTO powder with 20000X (A) and 5000X (B) magnification. The grain size is almost uniform distribution, but there is still almost evenly agglomerates.

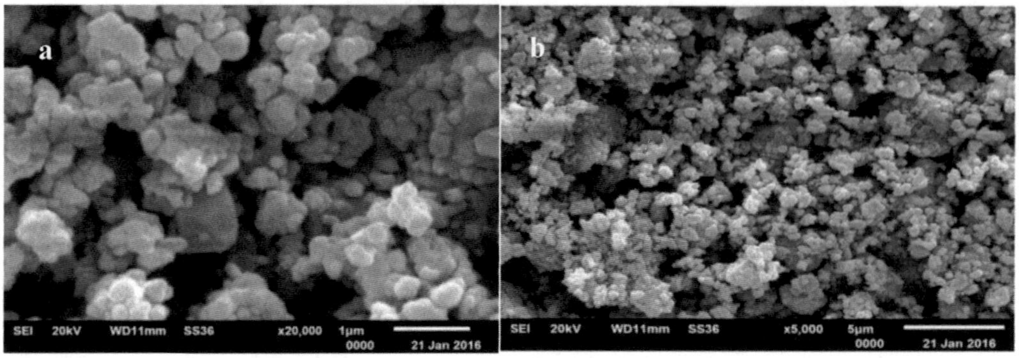

Figure 1. SEM image of LTO powder with magnification (a) 20.000X and (b) 5.000X

The LTO powder has diameter approximately 0.34 μm as seen in Figure 1. (A). The average diameter size of agglomerates was approximately 2.27 μm as seen in Figure 1. (B) That small particle size can improve the surface area of this LTO solid.

The structure of the synthesized powder was characterized using a X-ray diffraction (XRD). The data is processed using XRD results X'Pert High Score Plus software installed in the instrument and matches to JCPDS 26-1198 reference numbers for the LTO spinel phase and (JCPDS 01-076-0318) TiO_2 rutile [7].

X-ray diffraction patterns of LTO powder after sintering for 2 h is shown in Figure 2. The peaks at the XRD pattern take place at 2θ= 18.42°, 35.44°, 43.30°, 62.89° and 66.19° correspond to the characteristic of crystal peaks of LTO. The TiO_2 rutile phase obtained at 2θ position are 27.49° and 54.37°. The presence of little amount of TiO_2 rutile compound is considered as an impurities

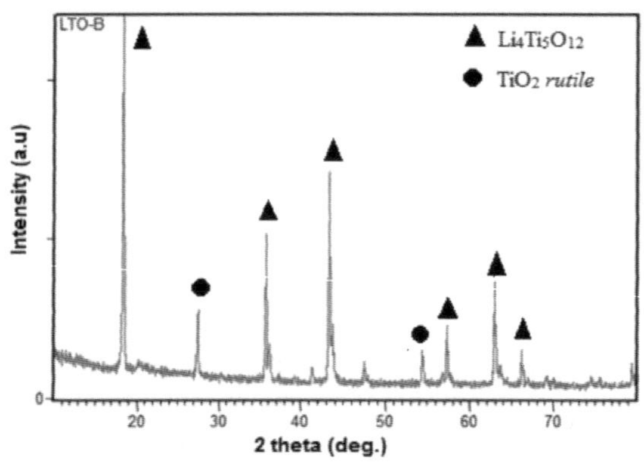

Figure 2. X-ray diffraction patterns of LTO powder after sintering for 2 h

The existence of these impurities caused by the presence of lithium kation loss [7] and upon mixing with insufficient homogeneity [8]. Peak intensity indicates that rutile content in the amount approximately of 15%. The surface area of LTO powder was 2.26 m^2g^{-1} analysed by BET measurement.

3.2 EIS characterization of LTO electrodes

Half-cell measurement of LTO and a lithium counter electrode using electrochemical impedance spectroscopy (EIS) can show the charge transfer resistance (Rct) electrode as seen in Figure 3. as follows.

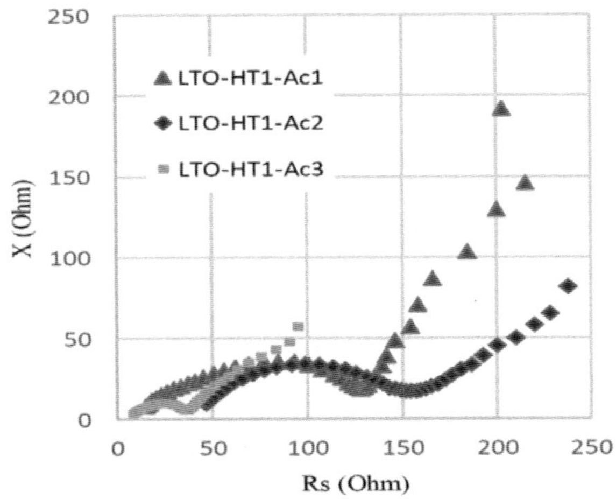

Figure 3. Complex impedance plots of the half cell with $Li_4Ti_5O_{12}$ anode prepared by hydrothermal-mechanochemical synthesis calcined at 750 °C for 3 h

The electrode-polarization resistance increased with added acetylene black for the cell with the $Li_4Ti_5O_{12}$ derived from the standard hydrothermal reaction. The results can be seen in **Table 1** shows electrolyte resistance (Re) and resistance charge transfer (Rct) to determine the resistance or impedance at the half cell batteries.

Table 1. Values of Re and Rct of ion exchange the experimental with variation acetylene Black Content

Samples	Re (Ω)	Rct (Ω)
LTO-HT1-Ac1	15.57	113.09
LTO-HT1-Ac2	46.90	107.85
LTO-HT1-Ac3	9.00	27.56

The electrolyte resistance were inversely proportional with the addition of acetylene black contents. Since resistance is the opposite of the conductivity, therefore the conductivity is proportional with the addition of acetylene black contents.

3.3 Cyclic Voltammetry (CV) characterization of LTO electrodes

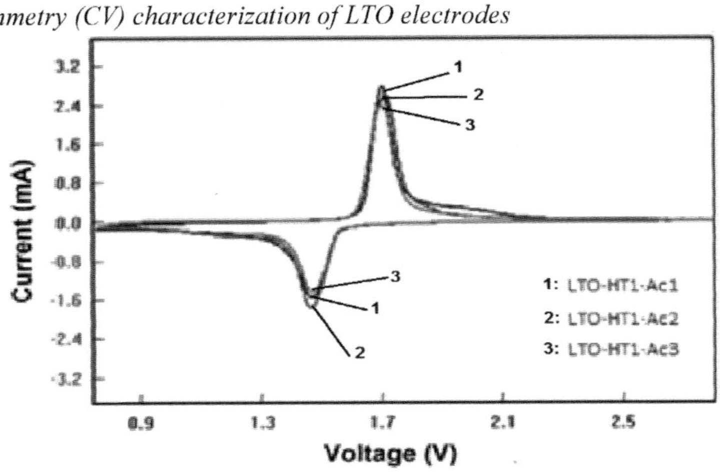

Figure 4. CV curve of half-cell, sweep rate 100 mVs^{-1}

The CV curve shows the initial three cycles of half-cell with potential sweep rate of 100 mV/s is shown in Figure 4. The specific capacity of LTO-HT1-Ac1 is 115.78 $mAhg^{-1}$ that is higher than others samples. The capacity LTO-HT1-Ac2 and LTO-HT1-Ac3 are 100.50 $mAhg^{-1}$ and 99.80 $mAhg^{-1}$ respectively. They showed decreased capacity because in this AB content range, the increase of conductivity has no effect to the specific capacity and the less proportion of LTO as the active material cause decreasing capacity on the battery.

Cycles behavior corresponded to kinetic reaction of highest peak LTO with average potential of 1.55 V which correlated to two phases that is $Li_4Ti_5O_{12}$ and $Li_7Ti_5O_{12}$ phases in the intercalation and de-intercalation process with reaction as follows [7]:

$$Li_4Ti_5O_{12} + 3Li+ + 3e^- \leftrightarrow Li_7Ti_5O_{12} \quad (3.1)$$

All sample had same anodic and cathodic working potential. From the peak potential that indicated acetylene black content did not affect performance of cell battery as seen in the result of cyclic voltammetry test. The working potential is average potential from oxidation and reduction curve [9] can be calculated with equation as below.

$$\boldsymbol{E^0} = \frac{\boldsymbol{E_{p,c}+E_{p,a}}}{\boldsymbol{2}} \quad (3.2)$$

The synthesis obtained from hydrothermal combined with mechano-chemical process generated higher surface area and less aggregation of the $Li_4Ti_5O_{12}$. The smaller grain size suggest a shorter diffusion distance, while a higher surface area means more interfacial sites and both are beneficial for higher reaction rate performance.

3.4 Charge-Discharge (CD) characterization of LTO electrodes

The charge and discharge (CD) test were conducted to evaluate the cell electrochemical performance at different current rate is shown in Figure 5. below. In Figure 5., it shows the discharge capacity of the half-cell at various acetylene black content at 0.3C to 10C with LTO powder synthesis after calcination at 750 °C for 3 h. The capacity from this synthesis techniques has not reached the theoretical value of 175 $mAhg^{-1}$ yet.

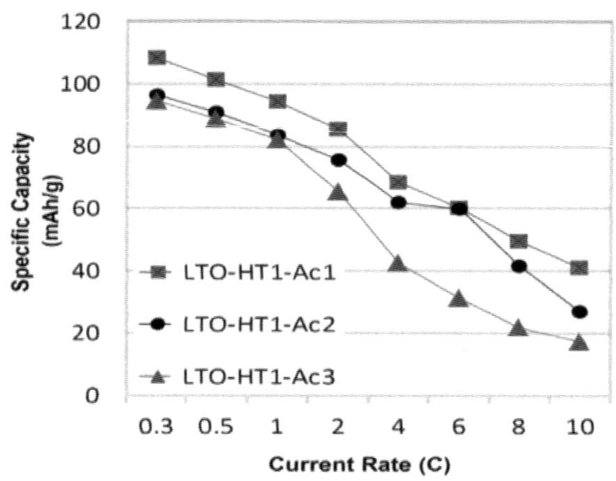

Figure 5. The discharge capacity of half-cell with current rate 0.3V to 10C

In the first cycle, the discharge capacity was 108.26 $mAhg^{-1}$ with 10 wt% AB content. At 10C discharge rate, the LTO-HT1-AC1 has the highest capacity compare to the other with the value of 40.91 $mAhg^{-1}$, LTO-HT1-AC2 was 26.88 $mAhg^{-1}$ g and LTO-HT1-Ac3 was 17.23 $mAhg^{-1}$ respectively. From the data obtained, indicating that the discharge rate with the addition of AB will degrade the battery capacity. From columbic efficiency generated by the current ratio variation rate and of AB obtained in Figure 6. The current efficiency obtained was approximately in average almost 100%.

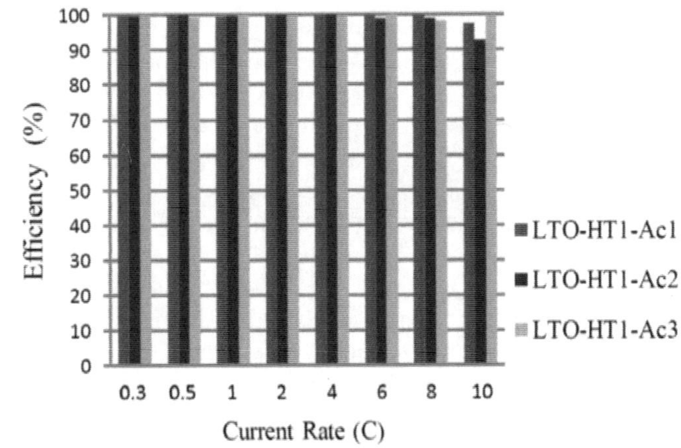

Figure 6. Columbic efficiency of half-cell during charge and discharge cycling

This is because the used of the active material (LTO) has a zero-strain properties. Columbic efficiency added of the AB content does not have a significant influence to change the columbic efficiency of lithium ion batteries.

4. Conclusion

The present $Li_4T_5O_{12}$/LTO has been successfully synthesized using hydrothermal-mechanochemical process and obtained a quite high surface area and minimum aggregation of the solid state. Conductivity of the assembled battery with various of acetylene black AB content from 10%, 12% to 15wt% shows that conductivity is proportional with the amount of AB contents in the slurry mixture.

The capacity of the assembled battery with various AB content from 10%, 12% to 15wt% shows the highest capacity at the 10wt%, but decrease at addition of more AB 12% and 15%. Thus, 10% is the best performance for the range of AB content from 10wt% to 15wt%. In this range, the conductivity is inversely proportional with capacity of the half-cell battery.

Finally, quite high capacity at the high current-rate 10C is reached at 10wt% AB, i.e: 40.91 $mAhg^{-1}$ compared to other AB contents. Accordingly, this LTO compound which could withstand under the high current rate of 10C is certainly well above the anode requirement for Indonesia electric car program which is set to be 4C.

Acknowledgements

The authors would like to thank the Direktorat Riset dan Pengabdian Masyarakat Universitas Indonesia (DRPM-UI) for the financial support under the grant of Hibah PITTA under contract number: 2106/UN2.R12/HKP.05.00/2016.

References

[1] X. Su, Q. Wu, X. Zhan, and J. Wu, "Advanced titania nanostructures and composites for lithium ion battery," pp. 2519–2534, 2012.
[2] Z. Jiang, C. Li, S. Hao, K. Zhu, and P. Zhang, "An easy way for preparing high performance porous silicon powder by acid etching Al–Si alloy powder for lithium ion battery," *Electrochimica Acta*, vol. 115, pp. 393–398, Jan. 2014.
[3] S. Hong, H. Hong, B. Cho, and B. Na, "Effect of heat treatment on electrochemical characteristics of spinel lithium titanium oxide," vol. 27, no. 1, pp. 91–95, 2010.
[4] Y. Zhang, C. Zhang, Y. Lin, D. Xiong, D. Wang, X. Wu, and D. He, "Influence of Sc^{3+} doping in B-site on electrochemical performance of $Li_4Ti_5O_{12}$ anode materials for lithium-ion battery," *Journal of Power Sources*, vol. 250, pp. 50–57, 2014.

IOP Conf. Series: Journal of Physics: Conf. Series **877** (2017) 012052 doi:10.1088/1742-6596/877/1/012052

[5] R. P. Maloney, H. J. Kim, and J. S. Sakamoto, "Lithium Titanate Aerogel for Advanced Lithium-Ion Batteries," *Applied Material & Interfaces*, vol. 4, pp. 2318–2321, 2012.
[6] Y. Li, Z. Wang, D. Zhao, and L. Zhang, "Gd doped single-crystalline $Li_4Ti_5O_{12}/TiO_2$ nanosheets composites as superior anode material in lithium ion batteries," *Electrochimica Acta*, vol. 182, pp. 368–375, 2015.
[7] Y.-J. Gu, Z. Guo, and H.-Q. Liu, "Structure and electrochemical properties of $Li_4Ti_5O_{12}$ with Li excess as an anode electrode material for Li-ion batteries," *Electrochimica Acta*, vol. 123, pp. 576–581, Mar. 2014.
[8] D. Yoshikawa, Y. Kadoma, J. Kim, K. Ui, N. Kumagai, N. Kitamura, and Y. Idemoto, "Electrochimica Acta Spray-drying synthesized lithium-excess Li_4+xT_5−xO_{12}−ı and its electrochemical property as negative electrode material for Li-ion batteries," *Electrochimica Acta*, vol. 55, no. 6, pp. 1872–1879, 2010.
[9] K. A. Narayana, "Electrode and Electrolyte Additive for Lifetime Extension in Lithium-Ion Batteries," 2014.

IOP Conf. Series: Journal of Physics: Conf. Series **877** (2017) 012053 doi:10.1088/1742-6596/877/1/012053

Understanding Kendal aquifer system: a baseline analysis for sustainable water management proposal

A Lukman [1,2], M D Aryanto [2], A Pramudito [2], A Andhika [2], D E Irawan [1]

[1] Faculty of Earth Sciences and Technology, Institut Teknologi Bandung, Indonesia
[2] PT. Supra Indodrill, Indonesia

dasaptaerwin@outlook.co.id

Abstract. North coast of Java has been grown as the center of economic activities and major connectivity hub for Sumatra and Bali. Sustainable water management must support such role. One of the basis is to understand the baseline of groundwater occurrences and potential. However the complex alluvium aquiver system has not been well-understood. A geoelectric measurements were performed to determine which rock layer has a good potential as groundwater aquifers in the northern coast of Kaliwungu Regency, Kendal District, Central Java province. Total of 10 vertical electrical sounding (VES) points has been performed, using a Schlumberger configuration with the current electrode spacing (AB/2) varies between 200 - 300 m and the potential difference electrode spacing (MN/2) varies between 0.5 to 20 m with depths target ranging between 150 - 200 m. Geoelectrical data processing is done using Ip2win software which generates resistivity value, thickness and depth of subsurface rock layers. Based on the correlation between resistivity value with regional geology, hydrogeology and local well data, we identify three aquifer layers. The first layer is silty clay with resistivity values vary between 0 - 10 ohm.m, then the second layer is tuffaceous claystone with resistivity value between 10 - 60 ohm.m. Both layers serve as impermeable layer. The third layer is sandy tuff with resistivity value between 60 - 100 ohm.m which serves as a confined aquifer layer located at 70 - 100 m below surface. Its thickness is vary between 70 to 110 m. The aquifer layer is a mixing of volcanic and alluvium sediment, which is a member of Damar Formation. The stratification of the aquifer system may change in short distance and depth. This natural setting prevent us to make a long continuous correlation between layers. Aquifer discharge is estimated between 5 - 71 L/s with the potential deep well locations lies in the west and southeast part of the study area. These hydrogeological settings should be used as the main starting point in managing water supply in this area.

1. Introduction

Java island has a nearly 2000 km coast line, stretching from east to west, which also the location of many industrial areas. With very established road network and flat lowland terrain, Kendal Regency (Mid Java Province) is one of the favorable area. With such condition, Kendal Regency needs a detail groundwater resource mapping to fulfill the daily water needs of the area. Research location located in the northern coast of Kendal Regency about 14 Km from Semarang to the northwest (figure 1). Our specific is to evaluate the hydrogeological setting of this coastal alluvium system in the research area with the specific interest in identifying groundwater resources beneath the location.

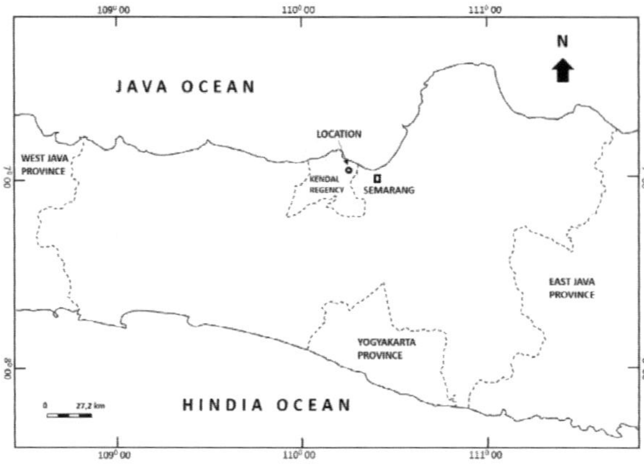

Figure 1. Research location map, located in the northern part of Kendal Regency about 14 Km to the northwest from Semarang.

2. Research Area

2.1. Research Location Description

Administratively, the research location is located in Wonorejo Distric, Kendal Regency about 5 to 6 Km from the coast line. It consist of flat coastal and swamp plain with very low slope gradient. Such condition are controlled by soft overlaid rock lies below the ground, eroded and deposited by river-tides-waves current energy. The locals grow rice field and fish pond in the northern part of the area, the measurements done in two major part in the wide area described with yellow dots and the narrow area described with orange dots (see map in figure 2). This research talk about the result of field measurements in the orange dots area, there are ten total geoelectrical measurement points.

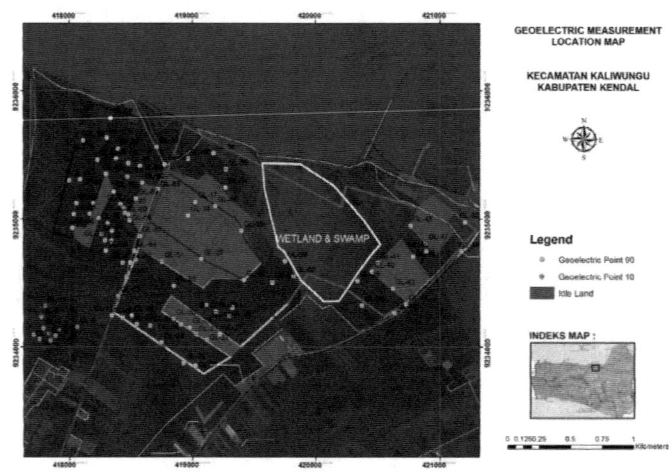

Figure 2. Research location description, near the northern coast, dominated by wetland and swamp.

2.2. Regional Condition

Two formations were identified based on previous studies: Damar formation (Qtd) and alluvium formation (Qa). Both were deposited in the Quarternary period, however, the Damar fm is older than alluvium fm. [5] from Geological Survey of Indonesia (figure 3). Damar fm consists of tuffaceous claystone, volcanic breccia, tuffaceous sandstone, tuff, conglomerate and a few laharik deposit in various area. This formation is distributed from the volcanic highland toward the coast, forming a volcanic-river-coastal depositional system. Aside to the hard and compact breccias and lavas, this formation has high groundwater potential stored in the porous volcanic sand layers.

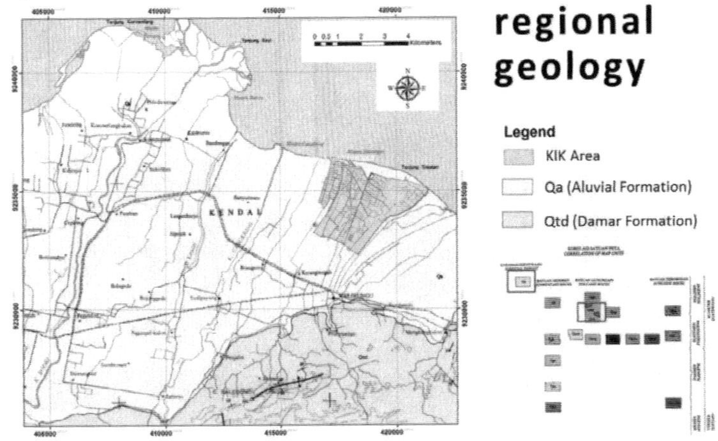

Figure 3. Regional geological map of the research location [5].

The younger alluvium fm consists of gravel, sand, silt and clay material which was deposited in the river and riverbanks as well as swampy area in the coast of Kendal. This formation is composed of unconsolidated carboniceous silty clay material that usually produce unpermeable layer. The maximum thickness of this formation can reach more than 50 m.

The hydrogeology of the area consists of two aquifer systems: alluvium system (swampy and lake deposits) and volcanic system (tuff, tuffaceous sandstone and volcanic breccia). The first system is a shallow aquifer with low permeability, whereas the second system has higher permeability. Based from previous studies, we could come up with three classes of aquifer productivity [4] (figure 4):

- moderate to high (green): fracture zones, deep groundwater table, various well yields and spring discharge;
- high (light blue): moderate transmissivity, shallow water level, wells yield 5 to 10 l/s;
- very high (dark blue): moderate to high transmissivity, very shallow water level, wells yield more than 10 l/s.

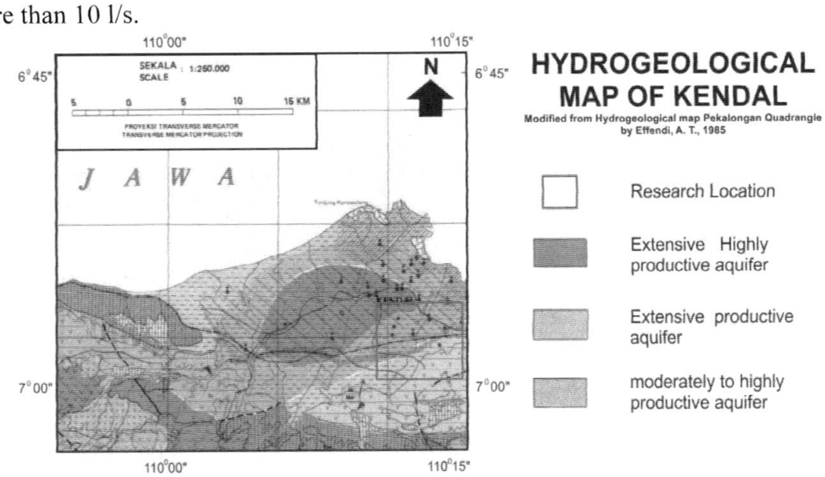

Figure 4. Hydrogeological map of research location [4].

There were 27 geotechnical boreholes (drilled in 2013) with depth varied between 45 to 70 m, contains: standard penetration test (SPT), physical properties of soil, and groundwater level. According

IOP Conf. Series: Journal of Physics: Conf. Series **877** (2017) 012053 doi:10.1088/1742-6596/877/1/012053

to the those data, generally, silty clay soil with a bit of sandy materials are found at 0-45 m of depth with SPT value range between 13 to 20, which are common for coastal-swampy alluvium deposit. The groundwater level is located at 10 m below the surface, flowing in shallow aquifers (figure 5).

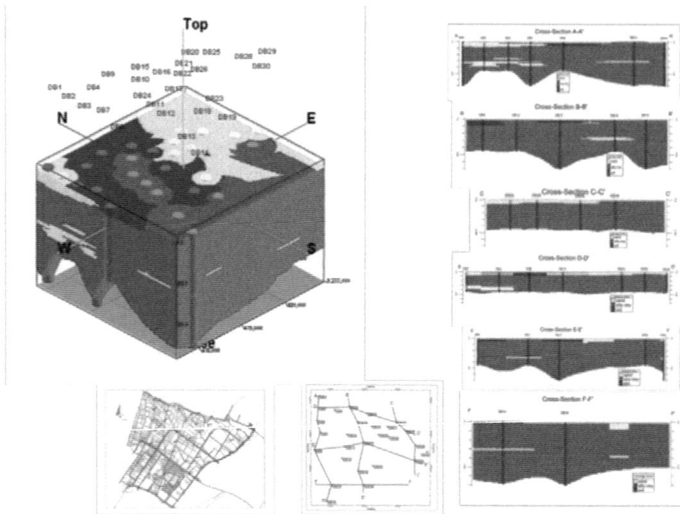

Figure 5. Drilling log correlation from previous geotechnical survey (drilled in 2013), it is known that at the shallow depth in the subsurface of research area dominated by clay with a little sand intercalation which probably act as shallow local aquifer.

To support deeper resistivity interpretation, we used one deep drill hole with depth of 125 m, located south of the area (figure 6). From the drilling log below, it shows that tuffaceous claystone with a bit sandy intercalations at shallow depth, which are equivalent with our resistivity values. Sandy tuff layers intercalated with volcanic breccia known as Damar fm, are found deeper. From the log, we identify the potential of deep aquifer.

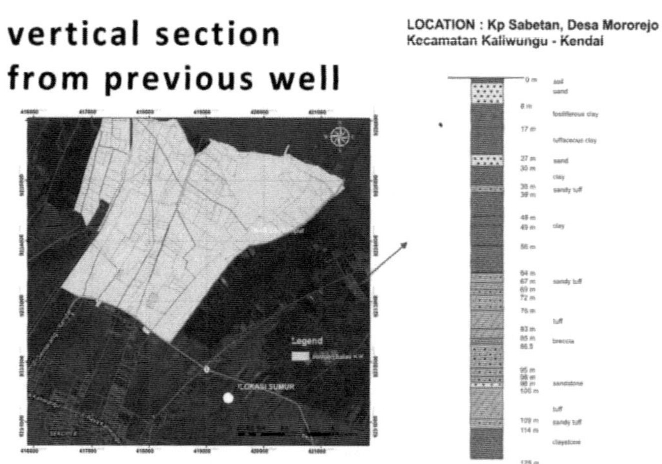

Figure 6. Drilling log from deep production well at the southern part of the research location, showing the same pattern with local geotechnical drilling result and some sandy tuff (probably as a member of Damar Formation) at the deeper part as the main aquifer layer in the location.

3. Material and Methods

3.1. Methods

3.1.1 Field Measurement Process. Geoelectrical measurement was conducted using four electrodes: two current electrodes (A and B) and two potential electrodes (M and N). Schlumberger configuration used in the study with distance of potential electrodes (MN/2) 0.5 to 10 m and distance of current electrodes (AB/2) 150 to 200 m (see figure 7). The principles of the method is to inject electrical current in below the surface through current electrodes (AB). Electrical currents propegate in the layers with different resistivity values, in which recorded by the receiving electrodes. Such values, the apparent resistivity, are guided by Ohm's Law and they are controlled by the injected current values, potential differences and geometrical factor. It can be calculated using equation 1 [3].

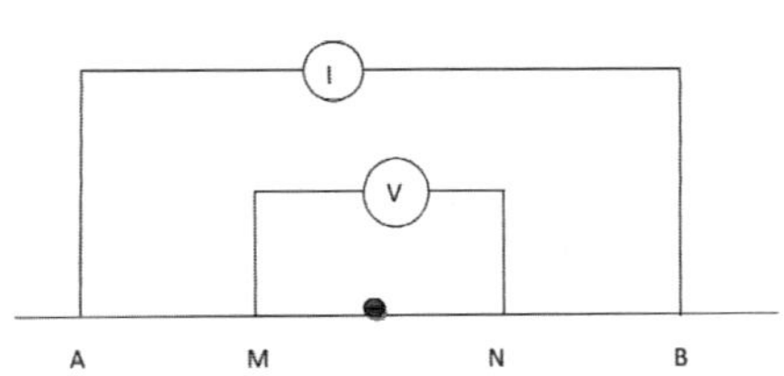

Figure 7. Schlumberger Configuration of Electrodes [3].

equation (1) :

$$\rho = K\frac{\Delta V}{I} \tag{1}$$

Where ρ is the value of apparent resistivity (Ohm.m), I is electrical currrent (mA), ΔV is differences of potentials (mV), and K is the geometrical factor. The value of geometrical factor of Schlumberger configuration can be calculated using equation 2.

$$K = \pi \frac{\left(\frac{AB}{2}\right)^2 - \left(\frac{MN}{2}\right)^2}{2\left(\frac{MN}{2}\right)} \tag{2}$$

Where :
K = geometrical factor
AB = Distance between current electrodes (m)
MN = Distance between potential electrodes (m)

Geoelectrical measurements in the area using the following equipments:

- Two sets of Resistivity meter (Naniura NRD 300 HF), each with 2 pieces of potential electrodes, 2 pieces of current electrodes, and four cable rolls @ 500 m;
- 1 unit GPS (Garmin 76 Csx)
- 1 set AVO meter, TDS meter anda water level detector

3.1.2 Data Processing. Resistivity values are plotted onto apparent resistivity curve log vs AB/2. The curve is used as a temporary interpretation in the field. Depth of penetration is proportional with a

IOP Conf. Series: Journal of Physics: Conf. Series **877** (2017) 012053 doi:10.1088/1742-6596/877/1/012053

third of AB/2 distance. The algorithm used in the interpretation is forward modeling and inversion modeling. This will determine the actual resistivity value, thickness and depth from the surface.

The resistivity values are derived from the fluid content and type of layer (eg: sand or clay). They also resembles the fluid salinity values as well. Moreover, resistivity of rock is related to porosity, mineral composition and water saturation. Sedimentary rock with high porosity has lower resistivity than igneous or metamorph rocks. Resistivity value of sedimentary rock is between 10 Ohm.m until 1000 Ohm.m. Clay has lower resistivity value than sandstone. Water has resistivity value between 10 to 100 Ohm.m, on the other hand, sea water has resistivity value of 0.2 Ohm.m. Low resistivity at sea water is because of high salt content [2].

Microsoft Excel spreadsheet software was used to process apparent resistivity values from equation 1 and equation 2 (Table 1). Values of AB/2, MN/2 and apparent resistivity were then processed with IP2win software to obtain actual resitivity value, depth and thickness of earth surfaces with correlate the inversion model (line) with the apparent resistivity path (dots) in the software. Furthermore, aquifer layer can be interpreted based on its resistivity value. Inversion from software IP2win will provide correlation between sounding points to produce 2D and 3D model using Rockwork 16 software.

Table 1. Schlumberger data sheet example to get apparent resistivity value from field measurement.

SCHLUMBERGER DATA SHEET

Client	**KIK**	**DATE**	**14-1-2016**
Province	**Central Java**	**INSTRUMENT**	**Naniura**
Location	**Kawasan Industri Kendal**	**X**	**417750**
		Y	**9234062**
Line No.	**GI-2**		

AB/2	**MN/2**	**K**	**I**	**V**	**Rho.App**
1	0.3	4.76	653	1479	10.79
1.5	0.3	11.31	593	246.6	4.7
2	0.3	20.47	582	48.2	1.7
2.5	0.3	32.25	520	25.5	1.58
3	0.3	46.65	645	12.7	0.92
4	0.3	83.3	605	3.8	0.52
5	0.3	130.43	554	1.8	0.42
5	1	37.7	555	7.3	0.5
6	1	54.98	840	6.6	0.43
8	1	98.96	1058	3.3	0.31
10	1	155.51	1268	4.3	0.53

IOP Conf. Series: Journal of Physics: Conf. Series **877** (2017) 012053 doi:10.1088/1742-6596/877/1/012053

4. Result and Analysis

Based on our interpretation, first layer from 0 to 70 m mostly consists of alluvium material with resistivity value at 0 – 10 Ohm.m and it is interpreted as alluvium sediment with clay to sand grain size according to the soil investigation borehole data as key interpretation. Very low resistivity values are expected to be sea water intrusion.

Second layer from depth 70 to 200 m mostly consists of more rigid rock material with medium to high resistivity value range at 10 – 200 Ohm.m. It is interpreted as sandy-clay and sandy-tuff lithology of Damar fm as also shown by existing deep water well. The interpreted resistivity and its lithological classification is shown in the following Table 2.

Table 2. Processed geoelectrical data in the research location.

No	line Number	Coordinates (UTM)		Depth		Thickness	Resistivity	Lithology
		X	y	Top	Base			
1	GL-1	417700	9234096	0	99.84	99.84	0.65	Silt Clay
				99.84	197.51	97.67	20.2	Tuffaceous claystone
				197.51	200	2.49	273.52	Breccia
2	GL-2	417750	9234062	0	37.51	37.51	0.45	Silt Clay
				37.51	58.52	21.01	44.66	Sandy tuff
				58.52	159.44	100.92	22.19	Tuffaceous claystone
				159.44	200	40.56	28.92	Tuffaceous claystone
3	GL-3	417839	9234050	0	50.47	50.47	0.52	Silt Clay
				50.47	93.63	43.16	16.55	Silt Clay
				93.63	150.81	57.18	19.22	Silt Clay
				150.81	200	49.19	67.21	Sandy tuff
4	GL-4	417883	9234088	0	52	52	0.65	Silt Clay
				52	81.32	29.32	45.32	Sandy tuff
				81.32	124.32	43	21.29	Tuffaceous claystone
				124.32	186.96	62.64	48.29	Sandy tuff
				186.96	200	13.04	293.12	Breccia
5	GL-5	417910	9234130	0	160	160	0.58	Silt Clay
				160	186	26	35.23	Sandy tuff
				186	200	14	17.31	Silt Clay
6	GL-6	417829	9234106	0	67.61	67.61	0.63	Silt Clay
				67.61	88.41	20.8	46.99	Sandy tuff
				88.41	118.98	30.57	71.04	Sandy tuff
				118.98	195.25	76.27	94.59	Sandy tuff
				195.25	200	4.75	194.85	Sandy tuff
7	GL-7	417783	9234146	0	35	35	0.41	Silt Clay
				35	64	29	44.43	Sandy tuff
				64	102	38	27	Tuffaceous

								claystone
				102	162	60	122	Sandy tuff
				162	200	38	156	Sandy tuff
8	GL-8	417764	9234234	0	99	99	0.5	Silt Clay
				99	124	25	18.49	Silt Clay
				124	145	21	69.86	Sandy tuff
				145	165	20	190	Sandy tuff
				165	200	35	122	Sandy tuff
9	GL-9	418063	9234158	0	93	93	0.53	Silt Clay
				93	123	30	35.85	Sandy tuff
				123	177	54	33.64	Sandy tuff
				177	200	23	59.13	Sandy tuff
10	GL-10	417797	9234320	0	20	20	0.5	Silt Clay
				20	165	145	2.53	Silt Clay
				165	184	19	70.24	Sandy tuff
				184	200	16	89.73	Sandy tuff

North to south section shows layers of clay (light green – dark green) at shallow depth and sandy-clay and sandy-tuff (yellow) from Damar fm with good porosity and permeability. The second layer is predicted as an aquifer (Figure 8).

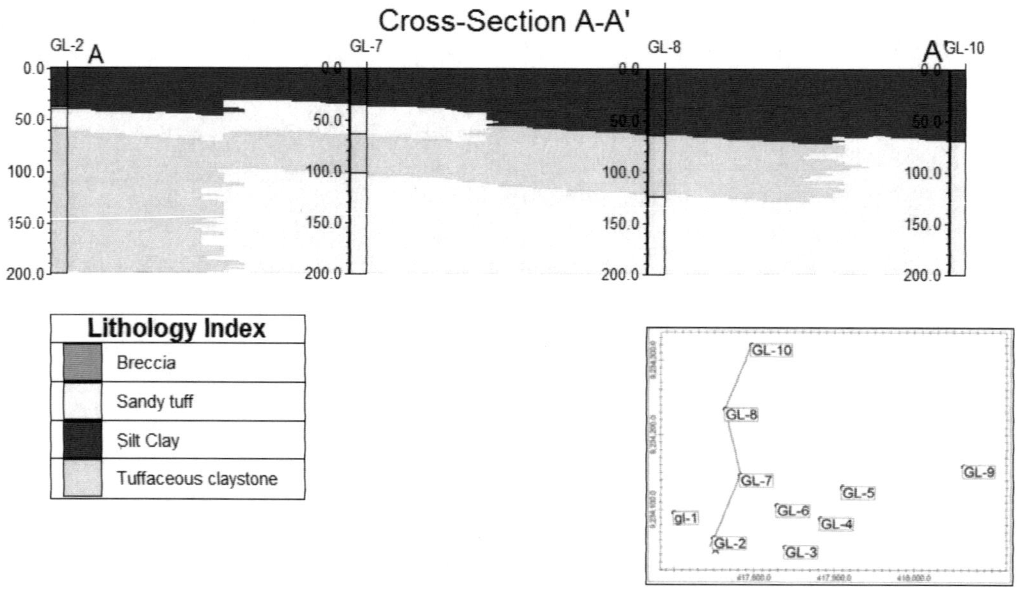

Figure 8. North to south cross section.

The east-west section presents similar condition as previous section. The first layer is dominated by silty clay and tuffaceous claystone as an aquitard layer and the second one at the deeper part is dominated by sandy tuff which is predicted as potential aquifer.

From the correlation we identified some sand lenses with north-south trend. Those lenses are interpreted as channel sand from the older river system in the area. The width is from 4 to 7 m with 2 - 4 m of thickness. We believe that the sand lenses are good aquifer for communal well water.

5. Discharge Estimation

Estimation of the discharge of proposed well is conducted by assuming the hydraulic conductivity range from 3.10^{-10} to 6.10^{-6} m/sec [1] which shown in table below. The following discharge estimation is a very rough due to the data availability, therefore pumping test later on should be used as the main guidance.

Table 3. Hydraulic Conductivity of Sedimentary Rocks.

Sedimentary Rocks	
Rock Type	**Hydraulic Conductivity (m/sec)**
Karst and reef limestone	1×10^{-6} to 2×10^{-2}
Limestone, dolomite	1×10^{-9} to 6×10^{-6}
Sandstone	3×10^{-10} to 6×10^{-6}
Siltstone	1×10^{-11} to 1.4×10^{-8}
Salt	1×10^{-12} to 2×10^{-10}
Anhydrite	4×10^{-13} to 2×10^{-8}
Shale	1×10^{-13} to 2×10^{-9}

The discharge estimation is following Sichardt Formula described below.

$$Q\ maks = 2\pi r b \frac{\sqrt{k}}{15} \quad (3)$$

Where:

Qmaks = maximum discharge of water extracted (m^3/s)
r_e = effective radius (m)
b = aquifer thickness (m)
k = hydraulic conductivity

Disharge estimation calculation give result of discharge range depend on the hydraulic condutivity value used. Other than that the calculation also use several assumption as follows:

- Aquifer Thickness is 70 meter
- Effective radius is 10 meter
- Hydraulic conductivity value 3.10^{-10}m/sec and 6.10^{-6} m/sec

Table 4. Discharge estimation

Hydraulic Conductivity	Q (m^3/s)	Q (l/s)
3.10^{-10} m/sec	0.005076	5.076
6.10^{-6} m/sec	0.071786	71.786

The different value of discharge depends on the following conditions:

- Porosity and permeability rate of aquifer.
- Fragment size of sandstone
- Matrix precentage of the sandstone.

6. Conclusion

Based on our preliminary results, the geoelectrical method has successfully identify the lateral and vertical distribution of the aquifer. We find volcanic sand from Damar fm forms good aquifer in more than 70 m of depth. We also find several ancient north-south river channel which could be used as a strong lead for further water supply exploration in Kendal coastal area. Disharge estimation calculation give result of discharge range depend on the hydraulic conductivity value assumption used, discharged estimation ranged between 5.076 L/s to 71.786 L/s, for further analysis it's needed to do the pumping test to get the real discharge value.

References

[1] Domenico P A & Schwartz F W 1990 *Physical and Chemical Hydrogeology* Vol 23 (New York, Chichester, Brisbane, Toronto, Singapore: John Wiley & Sons)
[2] Loke M H 2001 *2-D and 3-D inversion and modeling of surface- and borehole resistivity data* (Connecticut, USA : United States Geological Survey)
[3] Telford W M, L P Geldart, R E Sheriff 1990 *Applied Geophysics* Vol 2 (Cambridge : Cambridge University Press)
[4] Effendi A T 1985 *Hydrogeology Map of Pekalongan, Central Java* (Bandung : Geological Survey Departement)
[5] Thaden R E, Sumadirdja H, Richards P W 1975 *Geology Map of Magelang & Semarang, Central Java*: (Bandung : Geological Survey Departement)

IOP Conf. Series: Journal of Physics: Conf. Series **877** (2017) 012054 doi:10.1088/1742-6596/877/1/012054

Ab-Initio Calculation of Electronic Structure of Lead Halide Perovskites with Formamidinium Cation as an Active Material for Perovskite Solar Cells

E D Indari[1], T D K Wungu[2], and R Hidayat[1]

[1] Physics of Photonics and Magnetism Research Division, Physics Program Study, Faculty of Mathematics and Natral Sciences, Bandung Institute of Technology
Jl. Ganesha 10, 40132, Bandung – INDONESIA

[2] Nuclear Physics and Biophysics Research Division, Physics Program Study, Faculty of Mathematics and Natural Sciences, Bandung Institute of Technology
Jl. Ganesha 10, 40132, Bandung – INDONESIA

e-mail: rahmat@fi.itb.ac.id

Abstract. Organic lead halide perovskite material based solar cells show impressive power conversion efficiencies, which can reach above 19 percent for perovskite solar cell with methyl-ammonium cations. These efficiencies are originated from efficient photoexcitation and charge carrier transport and not observed in conventional perovskite crystals. In this preliminary research work, we therefore performed Density Functional Theory (DFT) calculation of formamidinium lead iodide (FAPI), an alternative to methyl-ammonium lead iodide (MAPI), to predict their electronic structure and density of state (DOS). The calculation result at the most stable lattice parameters show a good agreement with the experiment results. The obtained band gap energy is 1.307 eV. The valence band is dominantly formed by the 5p orbitals of I^- anions, while the conduction band is dominantly formed by the 6p orbitals of Pb^{2+} cations. The DOS of valence band of this perovskite seems smaller compared to the case of methyl-ammonium lead iodide perovskite, which then may explain the observation of smaller power conversion efficiencies in perovskite solar cells with this formamidinium cations.

1. Introduction

After its first synthesis in 1978, perovskite CH3NH3PbI3 (methyl-ammonium lead iodide or shortly MAPI) has been widely utilized for solar cells application [1]. In addition of ease and low cost fabrication, the achieved efficiencies have already reached more than 19 % [2]. A number of reports explain that these high performances are originated by long exciton diffusion lengths, very high light absorption coefficients which increase with temperature, and high open circuit voltages [3] [4] [5].

Owing to the success of $CH_3NH_3PbI_3$, scientists also explore the possibility to utilize other elements and/or molecules. One of the recently developed ones is taking an advantage of $CH(NH_2)_2$ (formamidinium lead iodide or shortly FAPI) which has been investigated by several groups to have different performance to MAPI [3] [6] [7]. In this particular study, as the band gap value plays an important role to the solar cells performance, we aim to investigate the difference between the FAPI and MAPI systems with a focus in the electronic structure computationally by state of the art Density Functional Theory (DFT) method.

2. Computational Details

All computations were performed with Quantum Espresso software package based on DFT, plane waves, and pseudopotentials. Generalized Gradients Approximations (GGA) and Ultrasoft Pseudopotentials (US PPs) were utilized. Scalar relativistic exchange correlation functionals constructed by Perdew-Burke-Ernzerhof (PBE) were applied to all of the constituent atoms. Stand on the choices, the valence orbitals comprised of 2s and 2p of C, 1s of H, 2s and 2p of N, 5d, 6s, and 6p of Pb, as well as 5s and 5p of I. In order to limit computations reach infinity, the kinetic energy cut-off for wavefunctions were limited to 40 Ry.

The Brillouin zones were divided by Monckhorst Pack k-mesh with 4x4x3 grid 3x3x3 in the case of MAPI (both tetragonal and pseudocubic structure) and FAPI respectively for both structure optimizations and electronic structure calculations. Due to the presence of Pb atom, a smearing type of Methfessel-Paxton were employed [8]. The declared lattice constants were selected from experimental results reported by other groups [3] [9] [10]. The stable positions of organic molecules were declared by performing a pre-optimization and applying DFT technique with B3LYP (Becke, three parameter, Lee-Yang-Parr) exchange correlation functional [11]. The structures of each system are depicted in Figure 1.

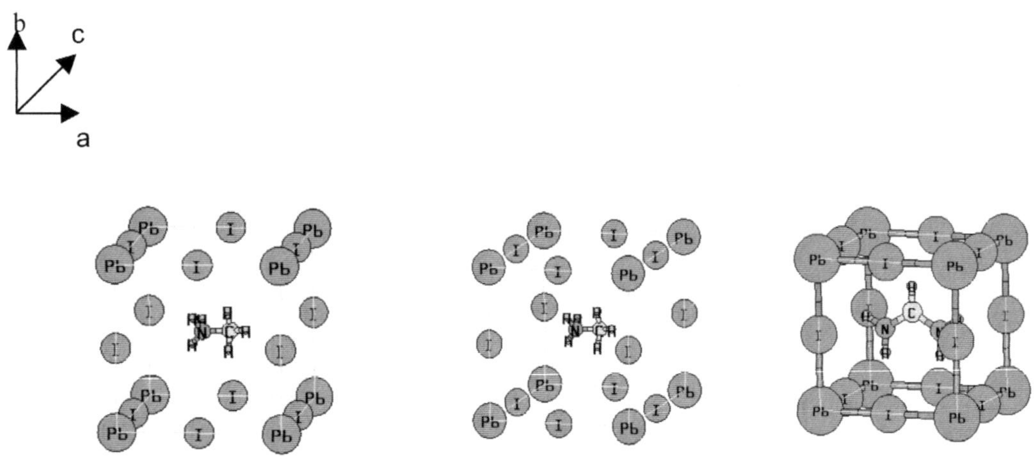

(a)	(b)	(c)
a=8.856 Å c=12.655 Å	a=6.2621 Å c=6.3275 Å	a=6.3620 Å

Figure 1. Crystal structure of tetragonal **(a)** and pseudocubic **(b)** MAPI and **(c)** FAPI obtained from experimental results [3] [9] [10].

3. Results and Discussion

The electronic structure of each system was calculated after performing optimization calculation by which most stable lattice constants were obtained. To validate as well as to compare the calculation, we first computed the MAPI system. Jishi et al reported that MAPI exhibits tetragonal structure at room temperature [9]. However, the predicted electronic structure does not provide enough information to be interpreted as shown in band structure (see Figure 2). We think that the profile is caused by not close enough distance between neighbouring atoms and accordingly there is no enough interaction observed. Hereby, as also proposed by Baikie et al, we computed the MAPI system with pseudocubic structure with ***a**** and ***c**** as the lattice constant ($a^*=a^{1/2}$ while $c^*=c/2$) [10]. The result of both tetragonal and pseudocubic structure of both systems are shown in Figure 2 and 3 respectively.

From Figure 3, it is then known that the band gap energy of pseudocubic MAPI is 1.232 eV at A high simmetry point which is lower than experiment result (1.51 eV) with an error of 18.42 %. Besides, the band gap energy of FAPI is 1.307 eV at R high symmetry point which is also lower than the reported experimental reports [3]. These lower values are due to the use of GGA exchange correlation functional and smaller lattice constants as depicted in Figure 1b for the case of pseudocubic MAPI [10].

In addition, PDOS (Projected Density of State) computation which exhibits both the DOS (Density of State) values and the projection along each atomic orbital which are also performed to both pseudocubic MAPI and FAPI.

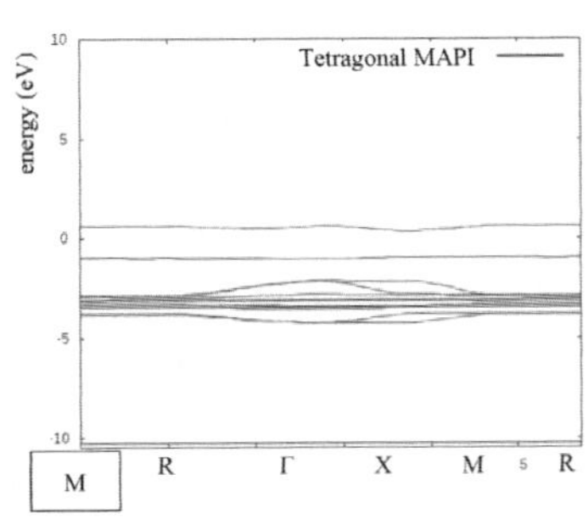

Figure 2. Band gap of tetragonal MAPI.

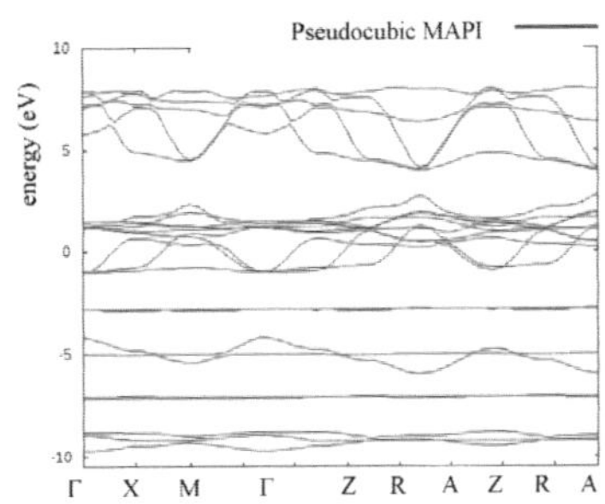

Figure 3. Band gap of pseudocubic MAPI.

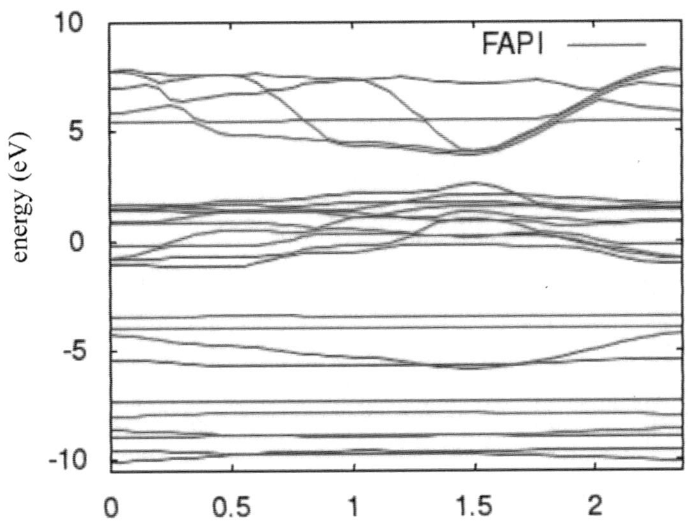

Figure 4. Band gap of FAPI.

Figure 5 shows that the valence bands of pseudocubic MAPI are dominated by 5p orbitals of I^- while conduction bands are dominated by 6p orbitals of Pb^{2+}. In addition, it is also known that the organic molecule cation gives small contributions into both conduction and valence orbitals. Based on the value of DOS of pseudocubic MAPI and FAPI (Figure 5 and 6), it is clear that FAPI exhibits lower

DOS, which can be associated to lower amount of charge carriers at the valence band to be extracted into electricity. Hereby, this finding may explain smaller power conversion of FAPI based solar cells compared to the MAPI ones.

The charge distribution of valence and conduction bands over different atoms are further analyzed by looking at Lowdin charge profiles as shown in table 1 and 2. From both tables, it is known that there are partial charger transfers from the organic molecules to the inorganic domains. Carbon atoms in MAPI system accept charge carriers from their surrounding which is the contrast mechanism to the FAPI system. Owing to the symmetry of the organic molecules, the charge transfers experienced by symmetrical N and H atoms are the same (see Figure 1c).

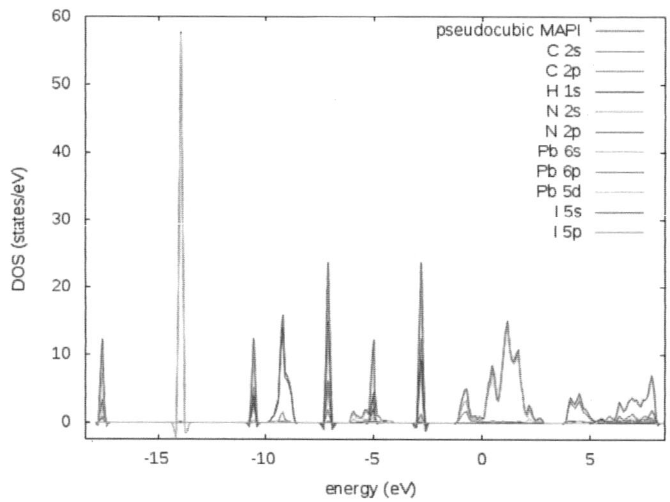

Figure 5. PDOS of pseudocubic MAPI.

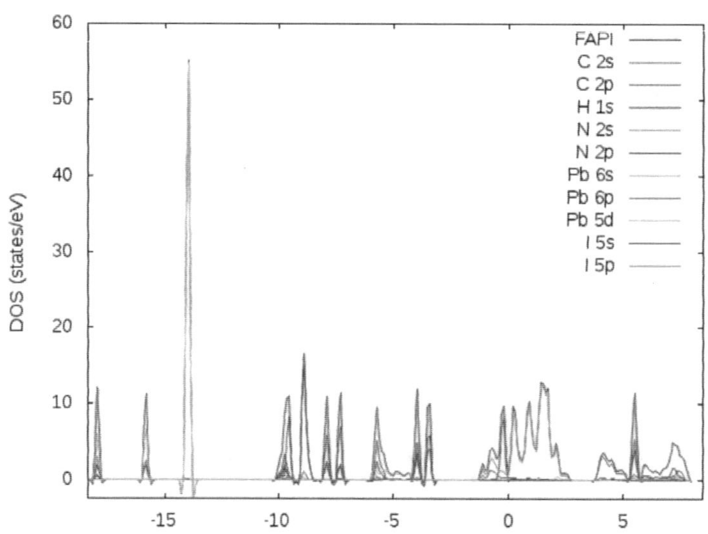

Figure 6. PDOS of pseudocubic FAPI.

Table 1. Lowdin charge orbitals of MAPI with spilling parameter of 0.0074.

MAPI atom	Valence electron number	Total charges	s	p	d
C	4	4.3365	1.0484	3.2881	0.0000
N	5	5.3442	1.2816	4.0627	0.0000
H	1	0.7757	0.7757	0.0000	0.0000
H	1	0.7751	0.7751	0.0000	0.0000
H	1	0.7756	0.7756	0.0000	0.0000
H	1	0.6437	0.6437	0.0000	0.0000
H	1	0.6351	0.6351	0.0000	0.0000
H	1	0.6343	0.6343	0.0000	0.0000
CH_3NH_3	15	13.9202	6.5695	7.3508	0.0000
Pb	14	13.4807	1.9681	1.5140	9.9985
I	7	7.4500	1.8755	5.5745	0.0000
I	7	7.1705	1.9414	5.2290	0.0000
I	7	7.6078	1.8510	5.7568	0.0000

Table 2. Lowdin charge orbitals of FAPI with spilling parameter of 0.0075.

FAPI atom	Valence electron number	Total charges	s	p	d
C	4	3.7564	0.9171	2.8393	0.0000
N	5	5.3681	1.2274	4.1406	0.0000
N	5	5.3681	1.2274	4.1406	0.0000
H	1	0.6599	0.6599	0.0000	0.0000
H	1	0.7551	0.7551	0.0000	0.0000
H	1	0.6599	0.6599	0.0000	0.0000
H	1	0.6656	0.6656	0.0000	0.0000
H	1	0.6656	0.6656	0.0000	0.0000

$CH(NH_2)_2$	15	17.8987	6.7780	11.1205	0.0000
Pb	14	13.4787	1.9656	1.5147	9.9984
I	7	7.2321	1.9372	5.2949	0.0000
I	7	7.5691	1.8726	5.6965	0.0000
I	7	7.4141	1.8869	5.5271	0.0000

4. Conclusion

In summary, we found out that the smaller power conversion performed by FAPI as the active material of hybrid perovskite solar cells may be explained by smaller value of the DOS. Both valence and conduction bands of MAPI and FAPI systems are dominated with 5p orbitals of I^- ions and 6p orbitals of Pb^{2+}. Partial charge transfers occurs in both system, from the organic molecules and inorganic domains. In case of organic molecules, symmetrical partial charge transfer occurs in FAPI which is not the case of MAPI. In contrast to MAPI system, atom C losses charge carriers instead of accepting them.

Acknowledgments
This research was supported by Program Hibah Kerjasama Luar Negeri 2015/2016 from DIKTI. We also thank Prof. Tjia May On and Dr. Fadjar Faturrahman for theoretical as well as technical discussions.

References

[1] Wang Y, Gould T, Dobson J F, Zhang H, Yang H, Yao X and Zhao H 2014 *Phys. Chem. Chem. Phys* **16** 1424–29
[2] Carmona C R, Gratia P, Zimmerman I, Grancini G, Gao P, Graetzel M and Nazeeruddin M K 2015 *Energy Environ. Sci.* **8** 3550-56
[3] Weller M T, Weber O J, Frost J M and Walsh A 2015 *J. Phys. Chem. Lett.* **6** 3209-12
[4] Green M A, Baillie A H and Snaith H J 2014 *Nature Photonics* **8** 506-14
[5] Ball J M, Lee M M, Hey A and Snaith H J 2013 *Energy Environ. Sci.* **6** 1739-43
[6] Aharon S, Dymshits A, Rotem A and Etgar L 2015 *J. Mater. Chem. A* **3** 9171-78
[7] Aguiar J A, Wozny S, Holesinger T G, Aoki T, Patel M K, Yang M, Berry J J, Al-Jassim M, Zhou W and Zhu K 2016 *Energy Environ. Sci.* **9** 2372-82
[8] Methefessel M and Paxton A T 1989 *Physical Review B* **40** 3616-21
[9] Jishi R A, Ta O B and Sharif A A 2014 *J. Phys. Chem. C* **118 (49)** 28344-49
[10] Baikie T, Fang Y, Kadro J M, Schreyer M, Wei F, Mhaisalkar S G, Graetzel M and White T J 2013 *J. Mater. Chem. A* 5628-41
[11] H. B. Schlegel and M. J. Frisch 1991 *Theoretical and Computational Models for Organic Chemistry* NATO-ASI Series C vol. 339, ed J. S. Formosinho, I. G. Csizmadia, and L. G. Arnaut (Kluwer Academic, The Netherlands,) p 5-33

Resistivtiy Structure in Tangkuban Parahu Area Drived from CSAMT Data

Enjang Jaenal Mustopa[1], Nurhasan[1], Wahyu Srigutomo[1], and Doddy Sutarno[1]

[1] Earth Physics and Complex System Research Divison, Faculty of Mathematics and Natural Sciences, Bandung Institute of Technology, Indonesia

E-mail: enjang@fi.itb.ac.id

Abstract. Tangkuban Parahu is located in Bandung, West Java, one of a volcano area. Controlled-source audio frequency magnetotelluric (CSAMT) survey has been conducted in the area using transmitter located in 3-5 km from the survey area. The objective of the survey was to determine a subsurface resistivity structures that may correlate to a geothermal reservoir in unknown geothermal potential area. In the paper, we applied a two-dimensional (2D) inversion scheme to interpret of the CSAMT data to obtain several resistivity sections describing the subsurface resistivity structures. The results of 2D inversion indicated several subsurface low resistivity anomaly (1-10 ohm.m) in the depth of 1000-1500 meter that may correlate to the existing a geothermal reservoirir confirmed by several surface manifestations in the area.

1. Introduction

Tangkuban Parahu is located in a volcanic region in Bandung, West Java, Indonesia. The area is geographically situated in the South latitude of 06°7'17"– 6°7'17" and in the East longitude of 107°36'12"- 107°40'16". The topographic elevation of the field is ranging from 800 to 2200 m above sea level. A Controlled-source audio frequency magnetotelluric (CSAMT) measurement, in the operating frequency range of 0.25 Hz to 8192 Hz, has been carried out in the area to acquire 42 soundings with the irregular space stations, varied from 200 to 500 m between sounding, covering the Tangkuban Parahu and its environs. The location of transmitter was about 3 – 5 km in the south from the survey area (Figure 1).

The objective of the survey was to demonstrate the capability of the CSAMT method to determine a subsurface resistivity structures in the deeper part and to delineate it that may correlate to a geothermal reservoir in unknown geothermal potential area. Further more, if it is possible in the future, the area can be exploited to produce a geothermal energy for generating the electricity after doing several geophysical, geological and geocemical exploration. For the reason, the geothermal energy is one of the alternatives in Indonesia.

The configuration of CSAMT measurement is shown in Figure 2. The CSAMT method utilizes a grounded electric dipole generating electromagnetic (EM) wave source has been developed by Goldstein and Strangway [1]. The EM wave polarization of the fields can be selected by the orientation of the transmitting antenna and the signal strength. The CSAMT measurement must be carried out at a distance greater than 3-5 skin depths from the transmitter site where the plane wave approximation is valid. Since the measurements were made using the electric field along the transverse, we assumed that the data correspond to the transverse magnetic (TM) mode [2].

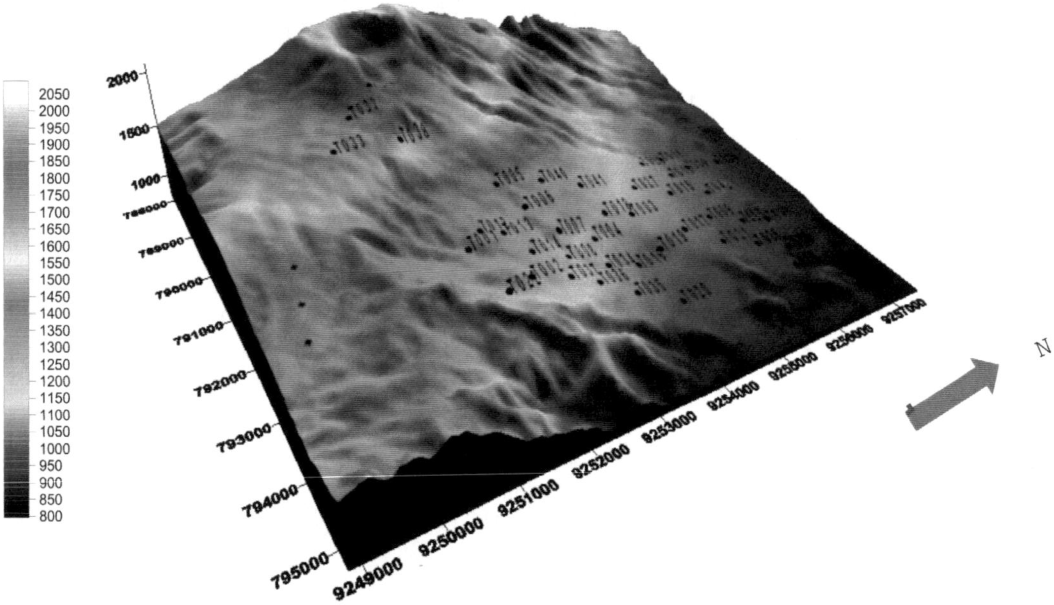

Figure 1. Map of Tangkuban Parahu area

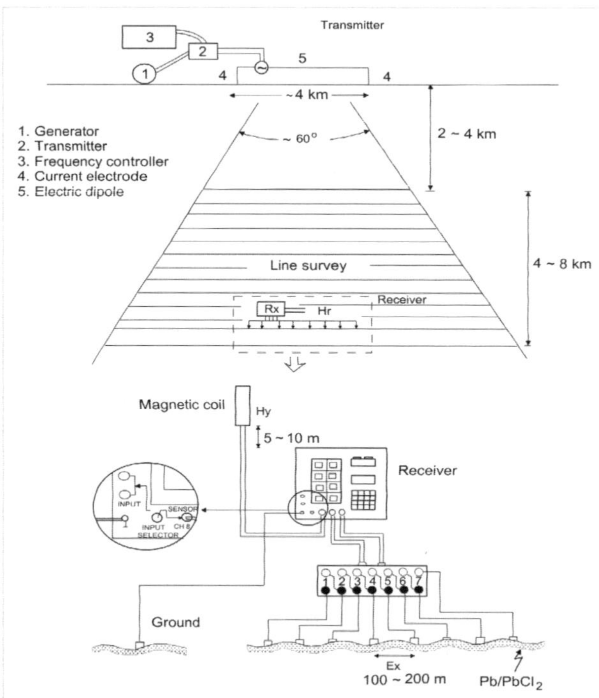

Figure 2. Configuration of CSAMT measurement

The location of the sounding sites of the CSAMT survey in Tangkuban Parahu is shown in Figure 3. The transmitter (Tx) was located about 3 to 7.5 km in the South from the sounding site area that was in Cikole, Lembang. The orientation of the electric field was parallel to the electric current dipole in every sounding site. On the other hand, the magnetic coil is located about 7 meters from the receiver and perpendicular to the electric field (TM mode).

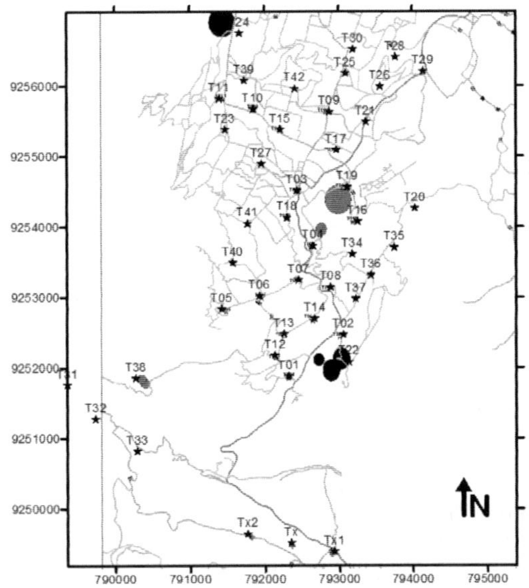

Figure 3. Location of the sounding sites of the CSAMT survey

The typical structure of geothermal system is schematically presented in Figure 4. The cooler upper zones are characterized by alteration of electrical conductive layer that is formed at temperatures above 70 °C. At higher temperatures, illite of less conductive layer becomes interlayered with smectite. The proportion of illite increases with the temperature, forming about 70% of the mixed-layer clay at 180 °C. Above this temperature, the smectite content continued to decrease, and pure illite commonly appears at greater than 220 °C[3]

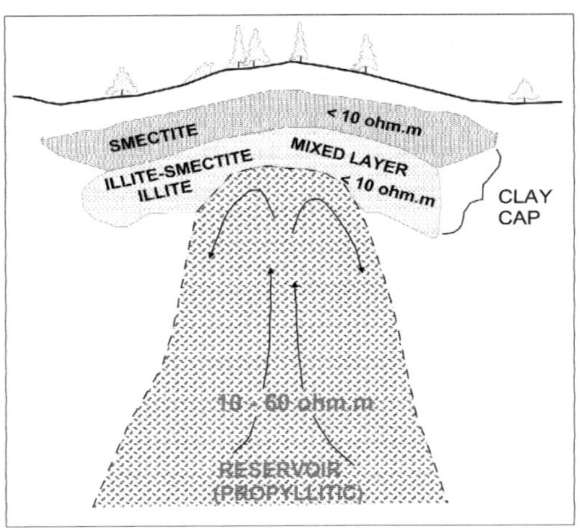

Figure 4. A generalized geothermal system [3]

2. Basic theory

The basic concept used in the CSAMT method can be derived from the solution of Maxwell's equations, these are

$$\nabla \times \mathbf{E} = -\mu \frac{\partial \mathbf{H}}{\partial t} \tag{1}$$

$$\nabla \times \mathbf{H} = \mathbf{J} + \frac{\partial \mathbf{D}}{\partial \mathbf{t}} \tag{2}$$

$$\nabla \cdot \mathbf{D} = 0 \tag{3}$$

$$\nabla \cdot \mathbf{B} = 0 \tag{4}$$

where **E** is the electric field intensity (V/m), **B** is the magnetic induction (tesla), **H** is the magnetic field intensity (A/m), ω is the angular frequency, μ_o is the magnetic permeability (H/m).

The general solutions of a tangential component of electric field (E_ϕ) and a radial component of magnetic field (H_r) are expressed in cylindrical coordinates as [4]

$$E_\phi \approx \frac{Idl \sin \phi}{\pi \sigma r^3} \tag{5}$$

$$H_r \approx \frac{Idl \sin \phi}{\pi \sqrt{\mu \sigma \omega r^3}} e^{\frac{-i\pi}{4}} \tag{6}$$

The apparent resistivity on the earth surface derived from Maxwell's equation in a far-field zone is

$$\rho_{axy} = \frac{1}{\mu \omega} \left| \frac{E_x}{H_y} \right|^2 \tag{7}$$

Although the CSAMT measurements were carried in a far-field zone, the data still contain some near field and transition zone data when the frequency from transmitter signal is low (deep survey). Therefore, it needs to make a correction technique for a near-field and transition effect to use the CSAMT data before applying the MT interpretation scheme [5].

3. Interpretation of CSAMT data

A 2D inversion technique based on a non-linier conjugate gradient method [7] was applied to invert the selected CSAMT data. Since the measurements were made using the electric field along the transverse and the magnetic field normal to the traverse, it is assumed that the data correspond to the TM-mode.

Employing a 2D inversion scheme to the CSAMT data in Tangkuban Parahu, we constructed the 2D resistivity section for several profiles. The resistivity section obtained from 2D inversion of Line-2 indicates that the distribution of the low resistivity anomaly (<10 ohm.m) in the depth of 700-1000 m is distributed in the northeast from t-19 to t-28 (Figures 5). This anomaly is may considered as hydrothermal alteration zone in a geothermal reservoir structure. On the contrary, the low resistivity anomaly is not found in the southeast. From the figure also shows that the electrical discontinuity is found below the sounding site t-07, it may correspond to the existing fault in the area.

It is identified from the 2D resistivity section of Line-8 of NW-SE direction (the position of line shown in the left of the Figure 6) that the distribution of the low resistivity anomaly is also found below sounding sites t-17 to t-35. It is also considered as the hydrothermal alteration zone. Furthermore, the electrical discontinuity also discovered below the sounding site t-10, it is considered as a fault in this zone.

Finally, from the resistivity section of Line-9 in NW-SE direction (Figure 7) indicates that the subsurface resistivity of this line is mainly composed of 3 layers. The first layer (overburden) is high resistivity (50 to 500 ohm-m) that is between 800 to 1000 meters thick overlying the lower resistivity zone (1 – 20 ohm.m) in the second layer that is approximately 1000-1500 meters thick. These anomalous features may be due to faulting, fracturing, and hydrothermal alteration along the significant fault.

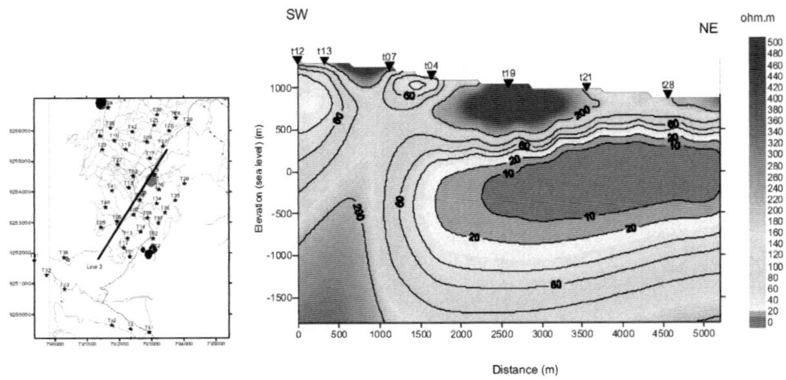

Figure 5. Resistivity section derived from 2D inversion result at Line-3

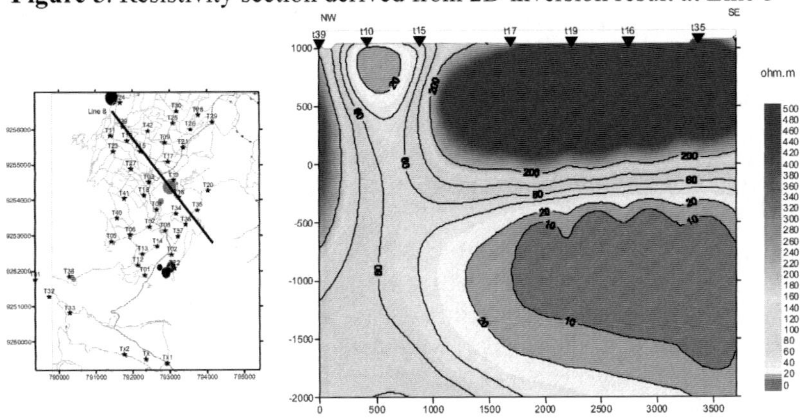

Figure 6. Resistivity section derived from 2D inversion result at Line-8

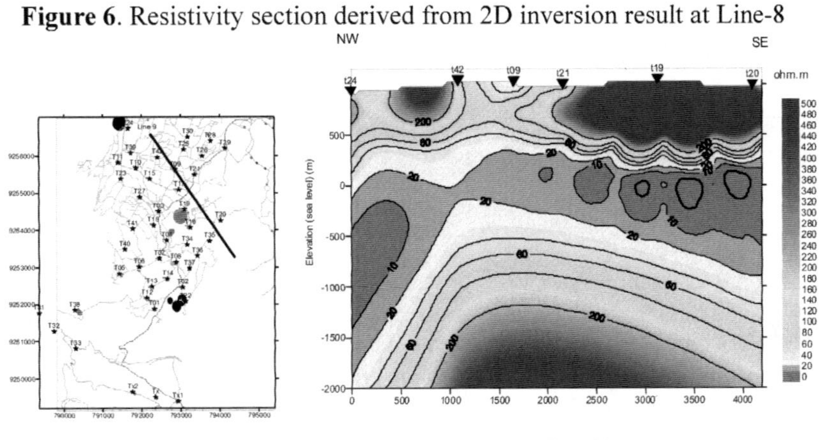

Figure 7. Resistivity section derived from 2D inversion result at Line-9

To understand the resistivity distribution in term of 3D geological structure, we constructed the 3D model of the low resistivity anomaly (<10 ohm.m) as shown in the Figure 8. The figure shows that the conductive (second) layer, which correlates to the layer of a hydrothermal alteration zone, is distributed in the northeastern part of the Tangkuban Parahu area. The region is probably a hydrothermal alteration zone where the geothermal reservoir may exist in that area.

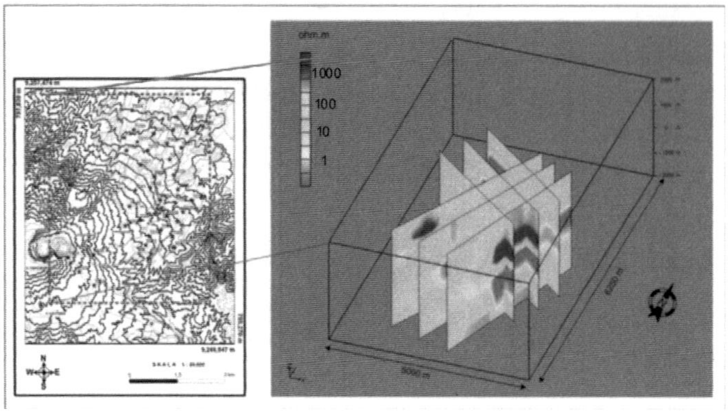

Figure 8. 3D view of resistivity structure in Tangkuban Parahu area

4. Conclusions

The CSAMT method applied at Tangkuban Parahu area was successful to locate an anomalously low resistivity zone that may indicate the existing of a potential geothermal reservoir. Employing the 2D inversion of CSAMT data generates several geoelectrical cross sections are found in the some locations that the resistivity structure is similar to a geothermal reservoir structure. The subsurface resistivity structures composed mainly by three types of resistivity feature obtained from 2D inversion results of CSAMT data. The overburden (first layer) has a resistivity of 30 – 150 ohm-m and thickness of 800 to 1000 m. The intermediate (second) layer has an extremely low resistivity of 3 – 15 ohm.m with 1000 – 1500 m thick. The low resistivity layer as an anomalous feature can be considered as a hydrothermal alteration zone (impermeable layer/cap rock). Finally, the basement (third) layer that is relatively more resistive than the second layer with resistivity of 30 – 100 ohm-m.

Further more, if it is possible in the future, the area can be exploited to produce a geothermal energy for generating the electricity after doing several geophysical, geological and geocemical exploration. For the reason, the geothermal energy is one of the alternatives in Indonesia.

Acknowledgment

The authors are grateful to *Penelitian Unggulan Perguruan Tinggi Ristek-Dikti* for funding of the research in the year of 2016.

References

[1] Goldstein M A and Strangway D W 1975 Audio frequency magnetotelluric with grounded electric dipole source *Geophysics* vol. 40 669-683.
[2] Sasaki Y, Yoshihiro Y and Matsuo K 1992 Resistivity imaging of controlled-source audiofrequency magnetotelluric data *Geophysics* 57 952-955.
[3] E. Anderson, D. Crosby and G. Ussher, Bulls-eye! -Simple Resistivity Imaging to Reliably Locate the Geothermal Reservoir, *Poc. WGC 2000*, 909-914, 2000.
[4] Kaufman A A and Keller G V 1983 *Frequency and transient sounding method* Elsevier Science Publ. Co. Inc.
[5] Yamashita M, Hallof P G and Pelton W H 1985 *CSAMT* case histories with a multichannel CSAMT system and near-field data correction *55th SEG Annual Convention* 276-278
[6] Rodi W and Mackie R L *2001* Nonlinear conjugate gradients algorithm for 2-D magnetotelluric inversion *Geophysics* **66** 174-187.
[7] Sasaki Y, Yoshihiro Y and Matsuo K 1992 Resistivity imaging of controlled-source audio frequency magnetotelluric data *Geophysics* **57** 952–955.

IOP Conf. Series: Journal of Physics: Conf. Series **877** (2017) 012056 doi:10.1088/1742-6596/877/1/012056

Simulation of Fluid Flow and Heat Transfer in Porous Medium Using Lattice Boltzmann Method

Imam Wijaya[1,a)], Acep Purqon[2]

[1,2]Earth Physics and complex system Research Division,
Institut Teknologi Bandung, Jalan Ganesa 10, Bandung 40132, Indonesia

a) robiah.wijaya239@gmail.com

Abstract. Fluid flow and heat transfer in porous medium are an interesting phenomena to study. One kind example of porous medium is geothermal reservoir. By understanding the fluid flow and heat transfer in porous medium, it help us to understand the phenomena in geothermal reservoir, such as thermal change because of injection process. Thermal change in the reservoir is the most important physical property to known since it has correlation with performance of the reservoir, such as the electrical energy produced by reservoir. In this simulation, we investigate the fluid flow and heat transfer in geothermal reservoir as a simple flow in porous medium canal using Lattice Boltzmann Method. In this simulation, we worked on 2 dimension with nine vectors velocity (D2Q9). To understand the fluid flow and heat transfer in reservoir, we varied the fluid temperature that inject into the reservoir and set the heat source constant at 410^0C. The first variation we set the fluid temperature 45^0C, second 102.5^0C, and the last 307.5^0C. Furthermore, we also set the parameter of reservoir such as porosity, density, and injected fluid velocity are constant. Our results show that for the first temperature variation distribution between experiment and simulation is 92.86% match. From second variation shows that there is one pick of thermal distribution and one of turbulence zone, and from the last variation show that there are two pick of thermal distribution and two of turbulence zone.

1. Introduction

Fluid flow and convection heat transfer of porous medium have been studied by many researchers in different field of science and engineering including geothermal engineering, civil engineering and mechanics, chemistry and petroleum engineering, hydrology, and nuclear especially in cooling management. In geothermal engineering, heat transfer and fluid flow are very important thing to understand, because it has correlation with electrical energy produced by the reservoir. The property of fluid flow and heat transfer in porous medium had been studied and modeled it with some numerical methods. The numerical method used to simulate is divided into three groups i.e. macroscopic modeling (FDM, NS, FEM), microscopic modelling (MD, DSMC), and meso-scopic modelling (LBM, LGA, SPH). The meso-scopic modeling is a relatively new method that could bridge between two approaches macro-scale and micro-scale. In the meso-scopic modelling (LBM) stated that the behavior of group particles into the behavior of a single particle. Therefore, we do not need to declare the behavior of each particle. The behavior of a group particle is stated by distribution function. The distribution function fulfilled for this case is Boltzmann Distribution.

LMB has several advantages, two of these are it does not need computer with high of RAM and it does not need to solve Poisson equation for each iteration, therefore the running process is more faster. In this study, we simplify the process of fluid flow and heat transfer in geothermal reservoir as a simple flow in porous medium canal with varying of fluid temperature injected, then we compare the result of simulation with experiment results by Olimpia Banete (2014).

2. Lattice Boltzmann Method

2.1. Basic Concept of Lattice Boltzmann Method

LBM is a kind of numerical method that can be used to simulate fluid flow. This method can bridge between two approaches macro-scale and micro-scale. In this method the behavior of group particles is stated into the behavior of a single particle. Therefore, we do not need to declare the behavior of each particle. The behavior of a group particle stated by distribution function. The distribution function fulfilled for this case is Boltzmann Distribution. The function is expressed by equation:

$$f_a(x+e_a\Delta t,t+\Delta t)=f_a(x,t)-\frac{\Delta t}{\tau}(f_a(x,t)-f_a^{eq}(x,t)) \tag{1}$$

where $f_a(x,t)$ is Boltzmann distribution function to calculate density and velocity field, τ is time relaxation that has correlated with kinematic viscosity v, $f_a(x+e_a\Delta t,t+\Delta t)$ is velocity distribution function at intervals time $t+\Delta t$, and $f_a^{eq}(x,t)$ is distribution function in equilibrium state.

$$f_a^{eq}(x,t)=w_a\rho\left[1+\frac{e_a.u^{eq}}{c_s^2}+\frac{(e_a.u^{eq})^2}{2c_s^4}-\frac{(u^{eq})}{2c_s^2}\right] \tag{2}$$

In this simulation, we used lattice model D2Q9, the figure of model as follow.

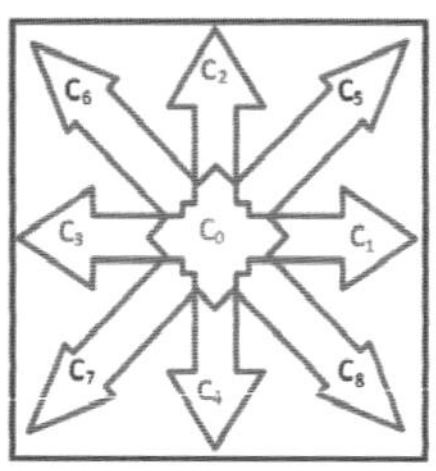

Figure 1. Lattice D2Q9 structure (Sukop,2007)

For this model, we define e_a as :

$$e_a=\begin{cases}(0,0)\rightarrow,a=0\\ \varphi_a c\left[\cos\frac{(a-1)\pi}{4},\sin\frac{(a-1)\pi}{4}\right]\rightarrow,a\neq 0\end{cases}\quad \text{With } \varphi_a\begin{cases}1\rightarrow,a=1,2,3,4\\ \sqrt{2}\rightarrow,a=5,6,7,8\end{cases} \tag{3}$$

Density and velocity define as

$$\rho=\sum\nolimits_a f_a u_i=\frac{1}{\rho}\sum\nolimits_a f_a c_{ai}u_i \tag{4}$$

2.2. LBM for heat transfer in pore medium

Niathiarasu et all. (1997) assumed if the Boussinesq in limit value are fulfilled and there is thermal equilibrium near fluid and solid, then the equations used to describe the behavior of incompressible fluid flow and convection proses in porous medium as follow:

$$\nabla . u = 0' \tag{5}$$

$$\frac{\partial u}{\partial t} + (u.\nabla)\left(\frac{u}{\varepsilon}\right) = -\frac{1}{\rho f}\nabla(\varepsilon p) + v_e \nabla^2 u + F \tag{6}$$

$$\sigma \frac{\partial T}{\partial t} + u.\nabla T = \nabla.(\alpha_m \nabla T) \tag{7}$$

where coefficient σ is ratio of thermal capacity for solid and liquid, u is discharge, p is pressure, v_c is effective viscosity, F is total external force experienced by porous medium. Guo and Zhao (2005) defined the total of external force experienced by the medium as follow:

$$F = -\frac{\varepsilon v}{K}u - \frac{\varepsilon F_\varepsilon}{\sqrt{K}}|u|u + \varepsilon G \tag{8}$$

where v is fluid viscosity, G is gravitational force. The gravitational force G is defined as follow :

$$G = -g\beta(T - T_0) + a \tag{9}$$

where g is gravitational acceleration, β is thermal expansion coefficient, T_0 initial temperature, a acceleration because of external force. Ergum (1952) developed a geometry function F and permeability K as a function of porosity based on their experiment. The function is defined as follow:

$$F_\varepsilon = \frac{1.75}{\sqrt{150\varepsilon^3}} \quad \text{and} \quad K = \frac{\varepsilon^3 d_p^2}{150(1-\varepsilon)^2} \tag{10}$$

2.3. LBM for velocity field

Guo and Zhao (2002) developed the equation for fluid flow in porous medium defined as follow:

$$f_a(x + e_a\delta_t, t + \delta_t) - f_a(x,t) = -\frac{1}{\tau_v}\left[f_a(x,t) - f_a^{eq}(x,t)\right] + \delta_t F_a \tag{11}$$

Where $\delta_t F_a$ is total force experienced by porous medium or total force experienced in the system. Guo and Zhao (2002) defined the distribution equation in equilibrium state for DnQb model as follow:

$$F_a = w_a \rho\left(1 - \frac{1}{2\tau_v}\right)\left[\frac{e_a.F}{c_s^2} + \frac{uF:(e_a e_a - c_s^2 I)}{\varepsilon c_s^4}\right] \tag{12}$$

Based on the equations above (Guo and Zhao), F has non-linear correlation with u. Guo and Zhao (2002) also defined the density and fluid velocity as follow:

$$\rho = \sum_a f_a \quad \text{and} \quad u = \frac{v}{c_0 + \sqrt{c_0^2 + c_1|v|}} \tag{13}$$

where v is

$$\rho v = \sum_a e_a f_a + \frac{\delta_t}{2}\varepsilon \rho G \tag{14}$$

c_0 and c_1 are defined as

$$c_0 = \frac{1}{2}\left(1 + \varepsilon\frac{\delta_t}{2}\frac{F_\varepsilon}{\sqrt{K}}\right) \quad \text{and} \quad c_1 = \varepsilon\frac{\delta_t}{2}\frac{F_\varepsilon}{\sqrt{K}} \tag{15}$$

2.4. Temperature Field Equation

He et all 1998 found the distribution function for equilibrium state. The equation is defined

$$g^{eq}=\frac{\rho e}{(2\pi RT)^{D/2}}\exp\left(-\frac{e^2}{2RT}\right)\left[\frac{e^2}{DRT}+\left(\frac{e^2}{DRT}-\frac{2}{D}\right)\frac{(e.u)}{RT}+\frac{(e.u)^2}{2(RT)^2}-\frac{u^2}{2RT}\right]+$$

$$\frac{\rho e}{(2\pi RT)^{D/2}}\exp\left(-\frac{e^2}{2RT}\right)\left[\left(\frac{e^2}{DRT}-\frac{D+4}{D}\right)\frac{(e.u)^2}{2(RT)^2}-\left(\frac{e^2}{DRT}-\frac{D+2}{D}\right)\frac{u^2}{2RT}\right] \quad (16)$$

For D2Q9 model, Seta et al 2006 used equation 17 until 19 to describe the distribution of density in equilibrium state which is the value of the function is discrete.

$$g_0^{eq}=-\frac{2\rho eu^2}{3c^2} \quad (a=0) \quad (17)$$

$$g_{1-4}^{eq}=\frac{\rho e}{9}\left[\frac{3}{2}+\frac{3}{2}\frac{e_a.u}{c^2}+\frac{9}{2}\frac{(e_a.u)^2}{c^4}-\frac{3}{2}\frac{u^2}{c^2}\right] \quad (a=1,2,3,4) \quad (18)$$

$$g_{5-8}^{eq}=\frac{\rho e}{36}\left[3+\frac{6e_a.u}{c^2}+\frac{9}{2}\frac{(e_a.u)^2}{c^4}-\frac{3}{2}\frac{u^2}{c^2}\right] \quad (a=5,6,7,8,9) \quad (19)$$

Internal energy has correlation with temperature, $e=3RT/2$ where R is ideal gas constant (R=8.314 J/moll-K).

2.5. Boundary Condition

In this simulation, we used many kind of boundary condition, such as bounce back boundary condition and inlet velocity boundary condition. The particles bounce when those hit the wall in back boundary condition. This is a simple boundary condition to make a solid formation in the system.

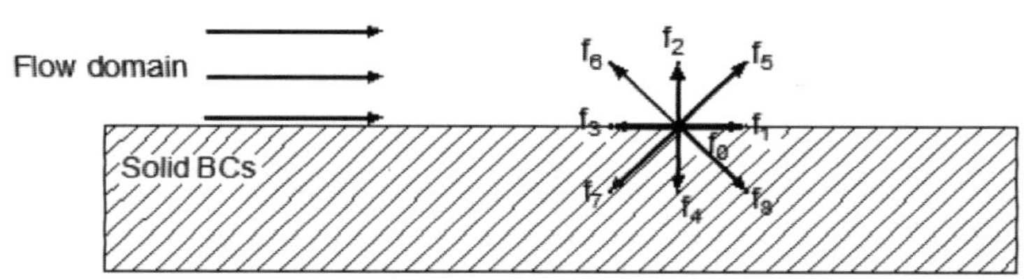

Figure 2. Bounce Back boundary Condition (Muhammed,2011)

By used inlet velocity boundary condition this boundary condition, we could take some assumption that there is no velocity in y-direction. The velocity in x-direction equal to 0 when the particle hit the wall. After the streaming process, there are unknown distribution function f_1, f_5, f_8 and also the density. To get the distribution function we used non-equilibrium equation (4). Temperature distribution function for lattice model D2Q9 defined

$$T=\sum_{g=0}^{8} g_a = g_0+g_1+g_2+g_3+g_4+g_5+g_6+g_7+g_8 \quad (20)$$

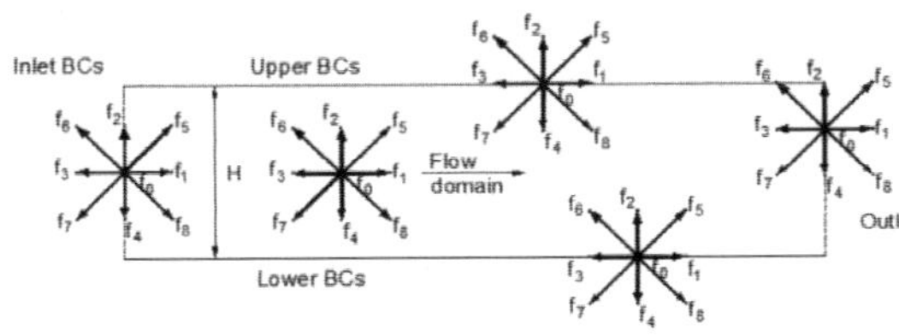

Figure 3. velocity Boundary Condition

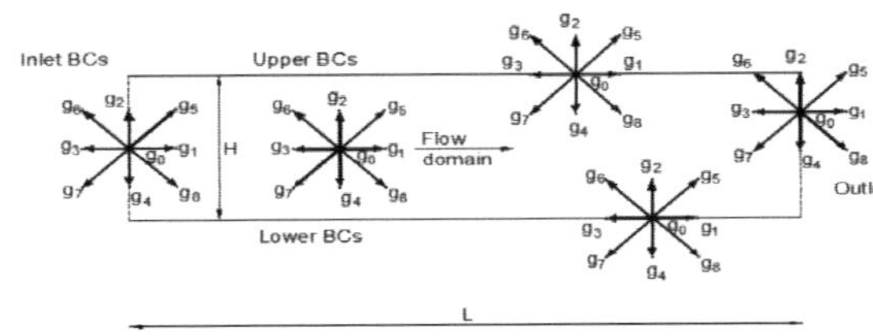

Figure 4. Temperature Boundary Condition

3. Model and Simulation Algorithm

The simulations were performed in this study divided in to two categories, the first categories is single phase fluid flow simulation for poissellie flow and heat convection, and second is simulation of fluid flow and heat transfer in porous medium.

3.1. Single Phase Fluid Flow

In the first category, we designed the simulation which the fluid flows in a pipe with both ends open to each other, otherwise it is assumed that there is no friction between the fluid and the pipe and there is no interaction between fluid particles. The fluid in the system is non-viscous and incompressible fluid. In this simulation we used Lattice Boltzmann Method D2Q9 to described the fluid flow in the cylinder. We used some assumption i.e. the interaction between particles only collision, there is no external force, and flow occurs just because of the differences pressure between inlet and outlet. We used bounce back boundary condition on the cylinder wall and assumed no slip. While in inlet and outlet of the cylinder we used periodic boundary conditions, therefore the fluid can flow continuously. The dimension of the models we used 400x100 lattice unit. The Reynolds number is 100, density is 1, and the inlet pressure is 0.005. The time relaxation is defined by this formula $\tau = 3v/2$ with $v = 2v_{maks}/\mathrm{Re}$. From this simulation we plot the velocity profile, then we compare the result to the theory.

3.2. Natural Convection in the Water

The second category is simulation of natural convection in the water. In this simulation, we used two distribution function, there are distribution function for liquid particle and distribution function for temperature, and we also used two kind of lattice model; there are D2Q9 for particle and D2Q5 for temperature. Both of distribution function we stacked together in the system. We assumed that the interaction between particles on the system is only collision, no external forces except gravity, and flow occurs because of the differences density between cold and hoot water. We used three kind of boundary conditions; there are no-slip boundary condition, periodic boundary condition, and streaming boundary condition. In this simulation we composed the model into dimension 200x100 lattice unit, the Rayleight number 20000. In this condition, heat source comes from the bottom wall of the system. We compared the result of this simulation with another simulation used FDM that has been done by Septian Setyoko to see the pattern of distribution temperature.

3.3. Fluid Flow and Herat Transfer in Porous Medium

In this simulation, firstly we simulate the fluid flow and heat transfer in porous medium with heat source comes from left wall, then we simulate fluid flow and heat transfer with various fluid temperature injected into the system and we set the heat source (comes from bottom wall) at constant temperature. From this simulation we plot the profile of temperature distribution.

3.4. Simulation Algorithm

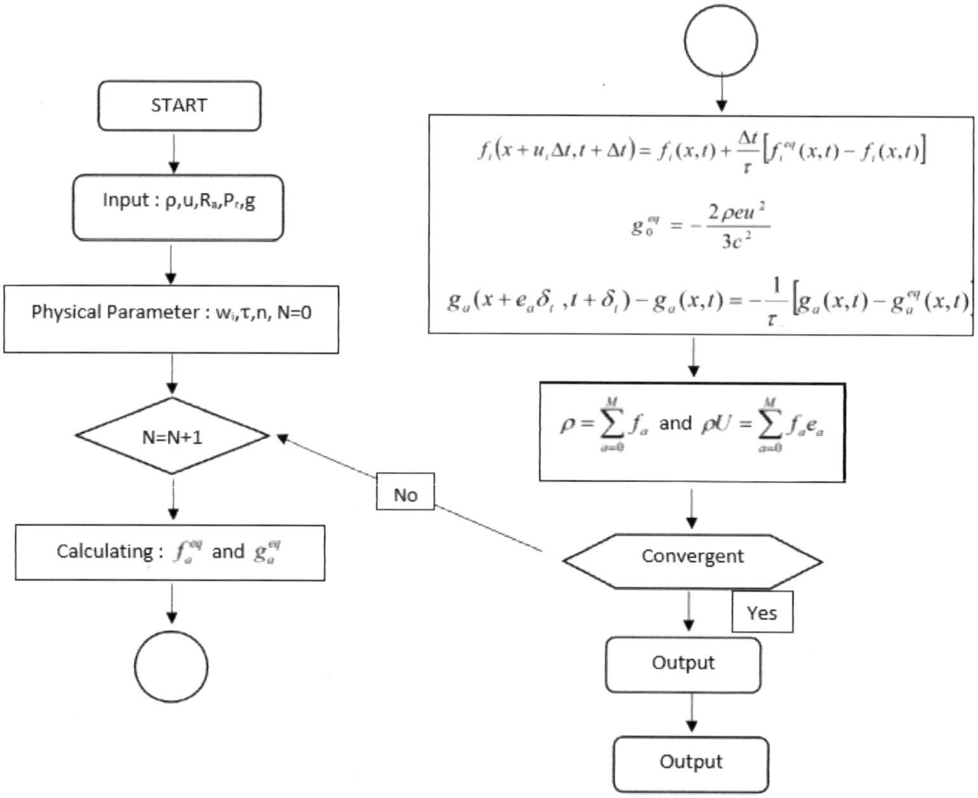

Figure 5. Simulation Algorithm

4. Results and Discussion

4.1. Single Phase Fluid Flow

From the simulations for the Poiseuille flow, we got the result as follows.

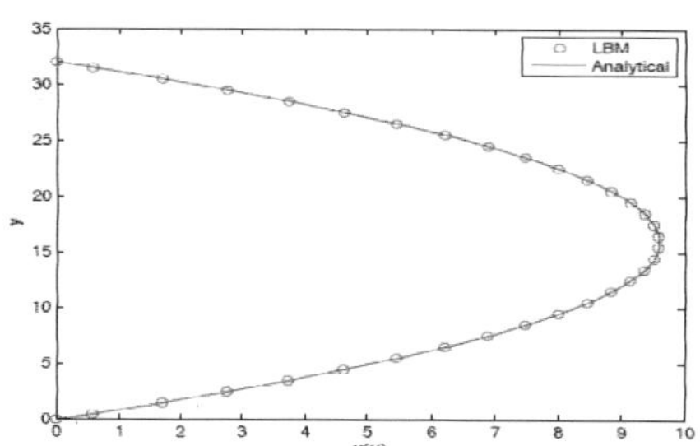

Figure 6. Velocity profile for poiseuille flow in the pipe (analytic and simulation) .

In Figure 5 we known that from simulation and the analytic result are the same. The fluid velocity profile has the shape of hyperbolic functions where the highest rate velocity located in the center of the pipe. Further validation process by comparing the visualization result from the simulation using LBM and Finite Difference method that has been done by Septian Setyoko (2012). The pattern of the temperature distribution results can be seen in the picture below.

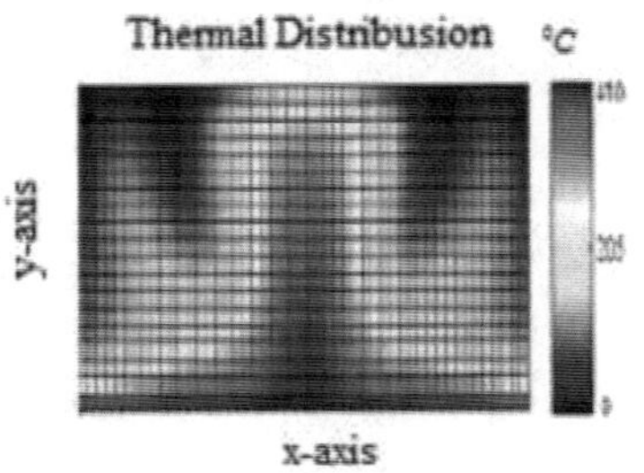

Figure 7a. Temperature distribution using FDM

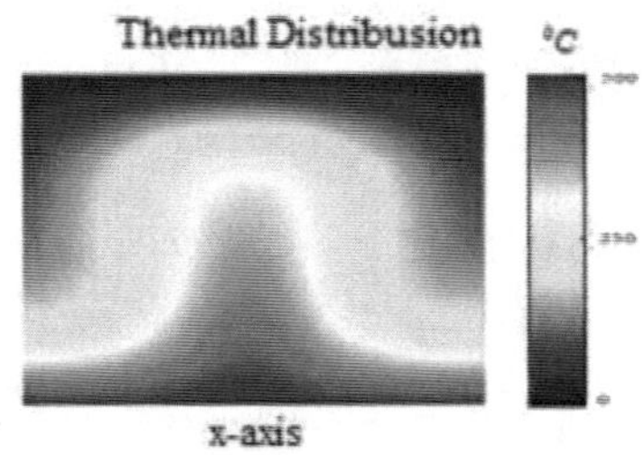

Figure 7b. Temperature distribution using LBM

From the simulation, using FDM and LBM we got similar profile of temperature distribution. The big difference is likely due to differences in physical parameters used in the simulation. Even though we can use the LBM to approach heat convection process in a porous medium.

4.2. Fluid Flow and Herat Transfer in Porous Medium

After the simulations with analytical validation was done, then we simulate heat transfer in a porous medium when the heat source comes from left wall and also simulate heat transfer in a porous medium with variation temperature fluid injected where the source comes from the bottom. Firstly we simulate fluid flow and heat transfer where the heat source comes from the left of the wall. The temperature on the left wall we set at 45^{0}C, while the upper and lower walls we set and we maintained into 0^{0}C.

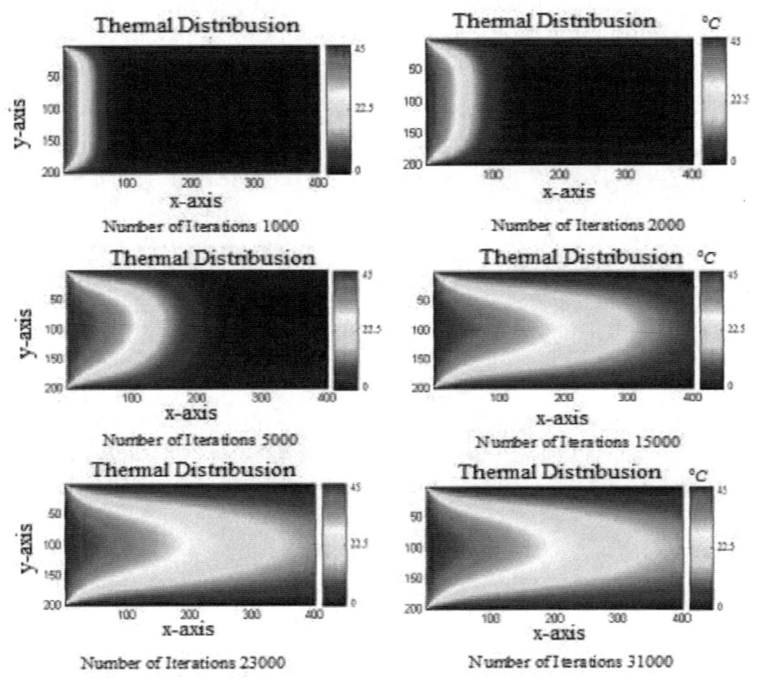

Figure 8. Temperature distribution profile in porous medium as convection procces used LBM with temperature fluid flow 45 ^{0}C and upper bottom walls temperature 0 ^{0}C

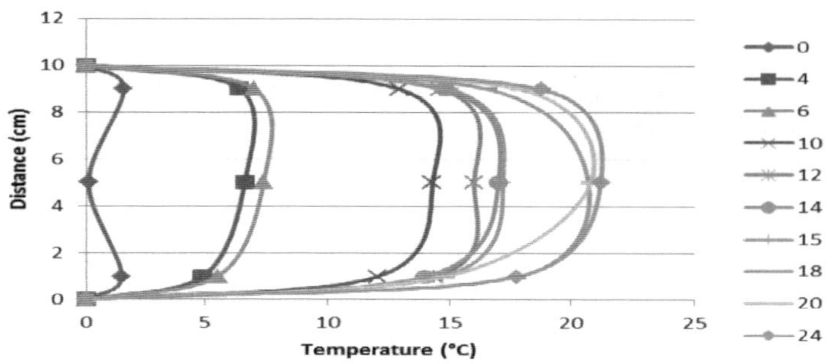

Figure 9. Experiment result of thermal temperature distribution profile in porous medium as convection process by Olimpia Banete.

From the simulations and experiments conducted by Olimpia Banete, we have the similar profile. The profile has a parabolic shape with a peak on the right. If we compare it, the peak has a value almost the same, 22.5 ^{0}C for the experiments and 21^{0}C for the simulation, but there is a striking difference around the walls. In the simulation, temperature distribution around the wall rise rapidly, while the experimental relatively slow to rise.

Secondly, we simulate fluid flow and heat transfer where the heat source comes from the bottom wall, in this case, there is a hot fluid injection from the left wall. We set the temperature at bottom wall constant at 410 ^{0}C and injection fluid temperature at 102.5 ^{0}C with the rate injection 0.02m /s. Their rate of injection in the left wall intended to approximate the behavior of the temperature distribution at the time of the geothermal reservoir injection process. This process is very important since it has relationship to the energy that would be generated by the reservoir. From the simulation, temperature distribution profile is obtained as follows.

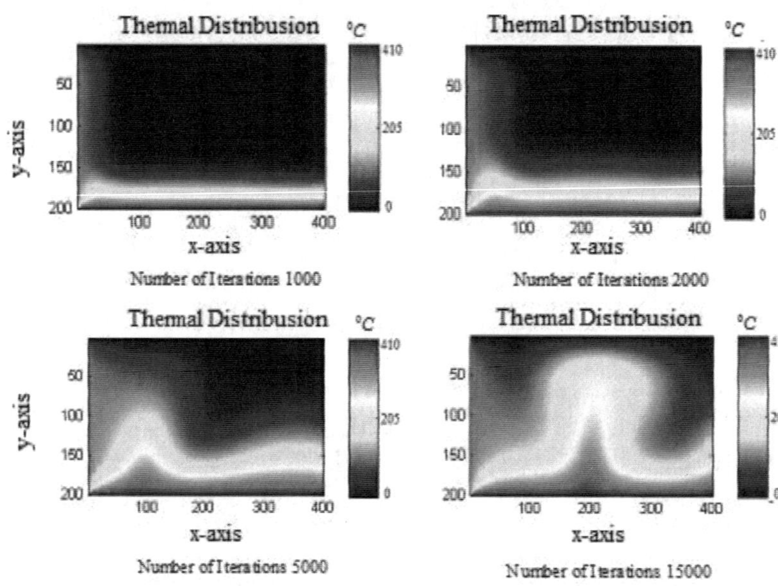

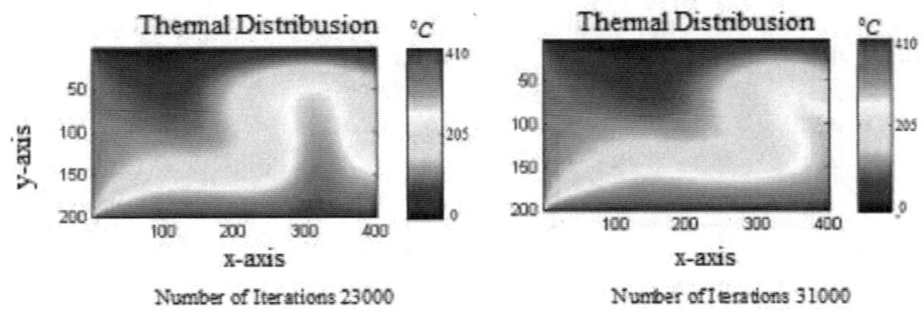

Figure 10. temperature distribution profile with heat source 410 0C and injection fluid temperature 102.5 ^{0}C. y-axis and x-axis is dimension of system .

From figure above, in 1000 number of iteration, the temperature does not rise significantly from the bottom and sides. In 5000 number of iteration, there is a peak of temperature distribution due to the injection process and the heating continuously from the bottom walls. The results show that the peak temperature distribution getting higher and shifting toward to the right. The high peak temperature distribution affects the pattern of fluid flow in the system. At low temperature, fluid flows in a laminar pattern, while at the high temperatures the fluids flow would be turning into turbulent pattern. it could be seen from the velocity vector of fluid flow in the following figure.

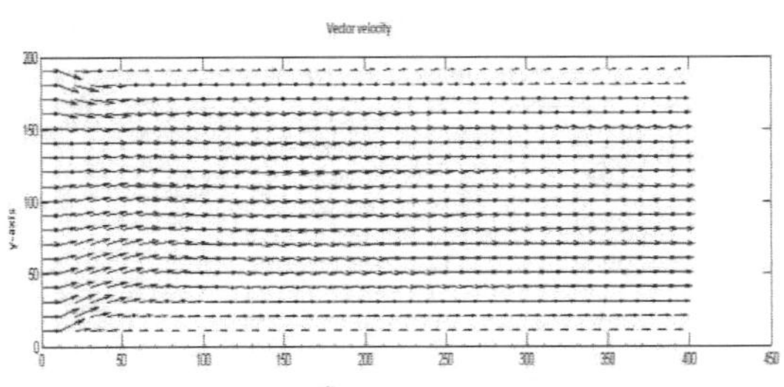

Figure 11a. Vector velocity in low temperature

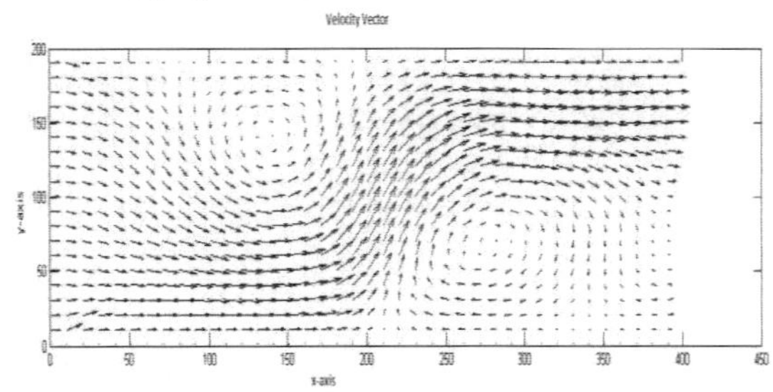

Figure 11b. Vector velocity in low temperature

For the third simulation we set the bottom temperature into 410 ^{0}C and injection fluid temperature 307.5 ^{0}C with the rate of injection 0.02m/s. The simulation process obtained the temperature distribution profile as show in figure 11. From the simulation, the temperature increases rapidly and the profile almost different from the previews doe to the differences fluid temperature injection with the previews simulation is significant, around 200 degrees. At 15000 number of iteration, there are two peak temperature. The first peak occurs due to a combination temperature from heat source and fluid temperature injection, while the second peak temperature comes from warming the fluid that has a higher temperature than the initial temperature.

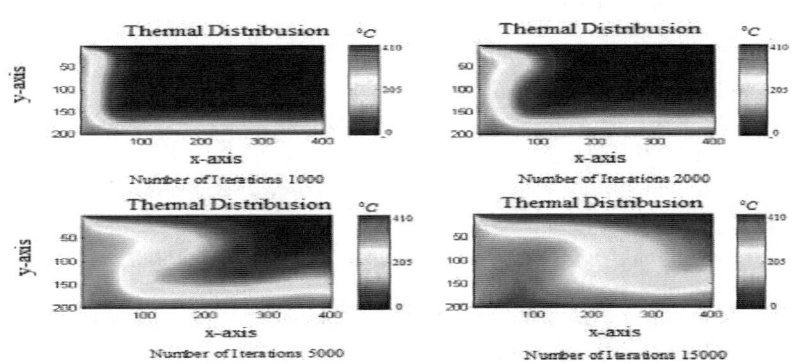

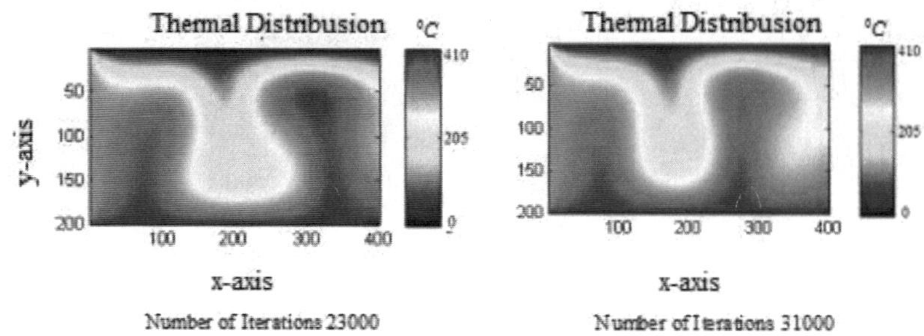

Figure 12. Temperature distribution profile with heat source 410 ^{0}C and injection fluid temperature 102.5 ^{0}C.

5. Conclusion

From this simulation, we could be conclude that the model composed by LBM can be used to solve fluid flow and heat transfer in a porous medium problems. The first simulation simulate fluid flow in a pipe, from the simulation that there are similarities with the theory of fluid flow profile for Poiseuille flow. The second simulation simulate natural water convection which is the heat source comes from the bottom of the wall. The results using LBM have the same profile with the Finite Difference Method, it can be concluded that LBM could also be used to simulate the heat flow in the fluid. In the simulation of fluid flow and heat transfer in a porous medium provides the results of the temperature distribution profile similar to the experiment, with the rate reaching 92.8%, it show in the temperature distribution profile.

References

[1] Nobovati,Aydin, and W.L.Edward., 2009, *General Model for the Permeability of Fibrous Porous Media Based on Fluid Flow Simulation using the Lattice Boltzmann Method.* Comp: part A **40** 860-869.

[2] Ramstad, Thomas.,2010, *Simulation of Two Phase Fluid Flow in Reservoir Rock Using a Lattice Boltzmann Method. J.*SPE 124617

[3] Pan.C, Hilpert.M, Miller.C.T.,2004, *Lattice Boltzmann Simulation of Two Phase Flow in Porous Media.* Water Resource Research, Vol.40, W01501,doi:10.1029/2003WR002120.

[4] Latra-Koko M, Rothman DH.,2005, *Static Contact Angle in Lattice Boltzmann Models of Immicible Fluid.*Physical Review E.2005; **72**(4):046701.

[5] Huber, Christian.,2007, *Lattice Boltzmann Modeling for Melting with Natural Convection.*

[6] Sukop,M.C, Thorne,D.T, *Lattice Boltzmann Modelling An Introduction for Geoscientists and Engineers*, Miami, Florida USA.

[7] Wolf,G.D.A.,2005, *Lattice Gas Cellular Automata and Lattice Boltzmann Models an Introduction*, Alfred Wegener Institute for polar and marine research, Germany.

[8] Huang, Haibo, Li, Zhicao, Lio,Shuaishuai.,2008, *Shan-and-Chen Type Multiphase Lattice Boltzmann Study of Viscous Couppling Effect for Two-Phase Flow in Porous Media*. Int.J. Numer.Meth.Fluids. doi:10.1002/fld.1972.

[9] Q.Zou, and X.He, *Pressure and Velocity Boundary Condition for the Lattice Boltzmann*, J.Phys.Fluids 9, 2591-2536(1996).

[10] S.Chen,D.Martirez, and R.Met, *on Boundary Condition in Lattice Boltzmann Methods*, J.Phys.Fluid 8,2527-2536(1996).

[11] Aaltosalmi,Urpo., 2005, *Fluid Flow in Porous Media With the Lattice Boltzmann Method*, University of Jyvashyla, Finland(Thesis).

[12] Banete, Olimpia.,2014, *Towards Modeling Heat Transfer Using Latic Boltzmann Method For Porous Medium*, Laurentian University,Canada.

Synthesis and Characterization of $BaFe_{12}O_{19}$/Poly(aniline, pyrrole, ethylene terephthalate) Composites Coatings as Radar Absorbing Material (RAM)

Nia Sasria[1,2], H Ardhyananta[1], R Fajarin[1] and Widyastuti[1]

[1] Department of Materials and Metallurgical Engineering, Faculty of Industrial Technology, Sepuluh Nopember Institute of Technology (ITS), Surabaya, Indonesia

Email niasasria15@gmail.com

Abstract. This research shows the processing and design of radar absorbing material (RAM) based on barium hexaferrite (BaM) and poly(aniline, pyrrole, ethylene terephthalate) (PAni,PPy,PET). BaM was prepared by sol gel method with Ni-Zn doping at mole fraction of 0.4 to obtain soft magnetic material. BaM/(PAni,PPy) composites were synthesized by in-situ polymerization method at ~0 °C. (BaM/PET) composite was prepared by melt compounding at 220°C. The composites were coated on A-grade AH36 steel using Dallenbach Layer, Salisbury Screen and Jaumann Layer methods with thickness of 2, 4, and 6 mm. The composites were evaluated using XRD, SEM, FTIR, VSM, LCM-meter and VNA. Results showed that doped BaM showed $BaNi_xZn_xFe_{12-2x}O_{19}$ structure. BaM/(PAni,PPy,PET) composites possessed globular morphology with M–O and C–H bonds. $BaNi_xZn_xFe_{12-2x}O_{19}$ exhibited the value of Ms and Hc, 56.6 emu/g and 60 Oe respectively. High electrical conductivity of 1.77744 x 10^{-5} S/cm was achieved of BaM/PAni composite. The maximum reflection loss (R_L) was reached at -48.720 dB and 8.1 GHz for BaM/PAni composite coating with 6 mm thickness at Jaumann Layer. These results indicated that BaM/PAni composite was a soft magnetic material with a high R_L value that is suitable for RAM, which used in stealth technology on naval vessels.

1. Introduction

Radar absorbing material (RAM) is a material used in stealth technology to camouflage structure of vehicle from radar detection. The principle of RAM same with microwave absorber that has been widely used to prevent and minimize the reflection of electromagnetic waves (GEM) of a large structure such as fighter aircraft, naval vessels and tanks. RAM can absorb the GEM by two ways, absorbs the magnetic field by magnetic materials and reduce the incident/reflected wave energy by dielectric materials [1]. Therefore, in RAM application it will need two main materials, magnetic and dielectric materials. Barium hexaferrite (BaM) is utilized as magnetic material. Hexaferrite has large anisotropic crystal and resonance location which can be modified in wide frequency range using ion substitution in hexaferrite. Besides, it is a soft magnetic material with relatively large permeability. Therefore, it is potential for RAM. Barium Heksaferit ($BaFe_{12}O_{19}$) possesses high saturation magnetization (Ms = 72 emu/g), high coercive force (Hc = 6700 Oe) and high Curie temperature (720 K) [2]. While, RAM has high magnetization with low coercivity. Therefore, metal ion substitution in BaM is required to RAM application. Ion substitution of Fe with Ni, Zn, Cu, Mn, Cu and others. The substitution causes intrinsic coercivity to decrease and magnetic saturation also decreases considerably at the same time [3]. Agustianto reported that substitution of Ni^{2+} and Zn^{2+} ions in BaM achieved minimum coersivity and magnetic saturation was 1200 Oe and 84.89 emu/g for $X_{Ni\text{-}Zn} = 0.6$ sample in temperature sintered 950 °C [4]. Conducting polymers as dielectric materials, polypyrrole (PPy) and polyaniline (PAni) have been regarded as a new class of materials in the last few years due to their high electrical conductivity, easy preparation, and excellent environment stability [5]. Wang reported that BaM/PPy composite has a reflection loss of -21,5 dB at 10,7 GHz with the thickness of 3 mm [5].

RAM effectiveness can be increased by the variations of thickness layer and coating method. The thicker the layer of RAM will increase the absorbtion. The coating methods are Dallenbach Layer, Salisbury Screen and Jaumann Layer. The three of those methods have layer constructions which differents so it will affect the absorbtion. Castro reported that PET(PU-PAni/DBSA) with Salisbury Screen method obtained optimum reflection loss of -23 dB at 9 GHz when the thickness of PET was 0.25 cm [6].

This paper deals with processing and design of the RAM having Dallenbach Layer, Salisbury Screen and Jaumann Layer, based on BaM and BaM/(PAni,PET,PPy) composites were produced by sol gel, in-situ polymerization and melt compounding method, respectively. These methods have several advantages such as shorter processing time and more general chemical reagents. The variations that were used in this research were the polymer type, coating method and coating thickness.

2. Experimental analysis

2.1. Materials

All chemicals used to prepare the BaM/(PAni,PPy,PET) composites were reagent grade and used without further purifications. Barium nitrate ($Ba(NO_3)_2$), ferric nitrate ($Fe(NO_3)_3.9H_2O$), zinc nitrate ($Zn(NO_3)_2.4H_2O$), nickel nitrate ($Ni(NO_3)_2.6H_2O$), citric acid ($C_6H_8O_7.H_2O$), ammonium hydroxide (NH_4OH), and aniline ($C_6H_5NH_2$) poly(ethylene terephthalate) (PET), pyrrole (C_4H_4NH), dodecyl benzene sulfonic acid (DBSA), ammonium persulfate ($(NH_4)_2S_2O_8$), polyol, isocyanate, acetone were supplied by Merck. The paint used in this research was Epoxy type from Jotun. Demineralized water was used during all synthesis and treatment process.

2.2. Preparation of Ni-Zn dopped Barium hexaferrite (BaM)

Barium hexaferrite (BaM) powder was produced by mixing $Ba(NO_3)_2$ and ($Fe(NO_3)_3.9H_2O$ solutions with constant stirring for 15 minutes. Afterwards, $Ni(NO_3)_2.6H_2O$ and $Zn(NO_3)_2.4H_2O$ solutions were added with mol fraction of 0.4 as doping materials followed by citric acid as reducing agent and then continued stirring for 15 minutes. After that, ammonia was added to obtain pH~7 and stirred for 1 hour. The mixture was then stirred at 80 °C for 3 hours to produce wet BaM gel and heated to 250 °C for 3 hours to dry it. To optimize its polymerization process, dry BaM gel was crushed into fine powder and sintered at 950 °C for 3 hours.

2.3. Preparation of BaM/(PAni,PPy,PET) composites

BaM/(PAni,PPy) composites powder (30:70 wt%) were prepared by in-situ polymerization method. The monomers were added with BaM and DBSA as oxidator. After that, ammonium persulfate (APS) aqueous solution was added and polymerized for 8 hours. It was then filtrated and washed using ethanol and aquades. The composites were dried at room temperature for 12 hours. While, BaM/PET composite powder (30:70 wt%) was produced by melt compounding above 200 °C.

The composites powder were mixed with epoxy paint (10:90 wt%) and painted (in accordance to ASTM 5068) on A-grade AH36 steel vessel as demonstrated in Figure 1, using Dallenbach Layer, Salisbury Screen and Jaumann Layer methods with thickness of 2, 4, and 6 mm. Polyurethane (PU) foams was used as band gap replacement in the layer constructions of Salisbury Screen and Jaumann Layer methods. PU was produced by reacting polyol and isocyanate (30:70 wt%) at room temperature. The coated steels were then dried at room temperature for two days to obtain the final product.

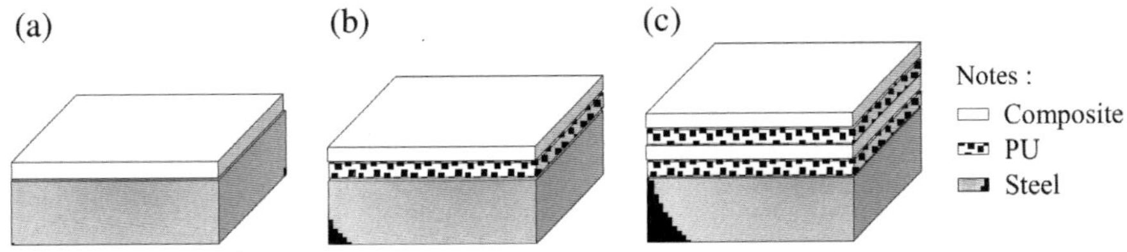

Figure 1. Illustration design of (a) Dallenbach Layer, (b) Salisbury Screen, (c) Jaumann Layer

2.4. Characterization

Phase characteristic and crystal structure of the powder samples were determined using X-ray diffraction (XRD) X'Pert PRO PANalytical utilizing Cu-Kα radiation (k= 1.54059 Å) at diffraction angle (2θ) ranging from 20 to 80°. The crystal size of non-doped and doped BaM were calculated by using Scherrer formula. Scanning electron microscope (SEM) FEI S50 with maximum magnification up to 20000 times and energy dispersion X-ray (EDX) of the samples were used to analyze the morphology and chemical composition of the samples. Functional groups of the composites powder were observed using fourier transform infrared (FTIR) Thermo Scientific in the range of 4000 – 400 cm^{-1}. Magnetic properties (magnetic saturation and coercivity) of non-doped and doped BaM powders were measured using vibrating sample magnetometer (VSM) OXFORD 1.2H. The magnetization measurements were observed in an external field up to 40 kOe at room temperature. Electrical conductivity properties of the samples were obtained using inductance capasitance resistance (LCM-meter) HIOKI over the frequency range of 10^{-1} – 10^{5} Hz at room temperature. The samples for the electrical conductivity measurement were made by pressing the polymer and composites into a slice with 10 mm diameter and about 0.5 mm thickness. While, microwave absorbing properties of the coated steels were analyzed using vector network analyzer (VNA) ADVANTEST R3770 (300 kHz – 20 GHz) in the X-Band frequency of 8–12 GHz.

3. Result and discussion

3.1. Structural characterization

Figure 2 shows the XRD pattern of non-doped ($BaFe_{12}O_{19}$) and doped BaM ($BaNi_xZn_xFe_{12-2x}O_{19}$) that both samples have the barium hexaferrite single-phase with no intermediate phase.

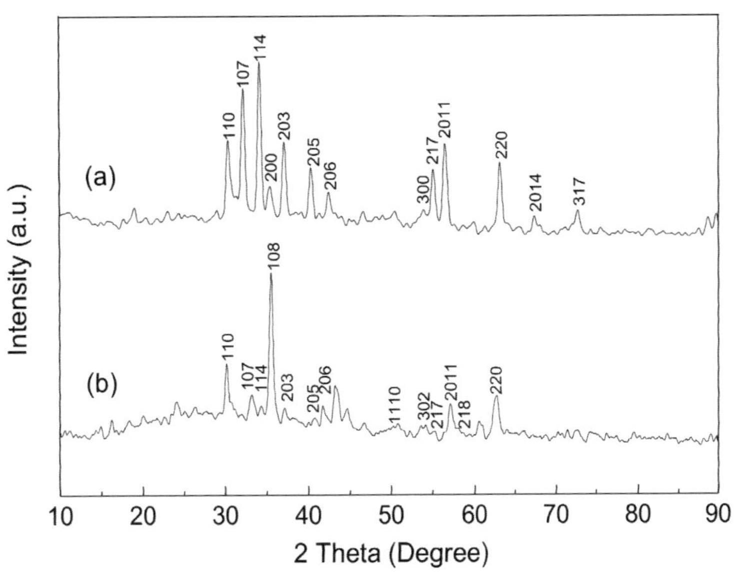

Figure 2. XRD at (a) Non-doped BaM, (b) Doped BaM

IOP Conf. Series: Journal of Physics: Conf. Series **877** (2017) 012057 doi:10.1088/1742-6596/877/1/012057

It was perfectly indexed to the crystal plane of hexagonal $BaFe_{12}O_{19}$ (JCPDS Card No. 01-074-1121 and 00-043-0002). The line profile, as described by Jiang et al. [7], Kanagesan et al. [8], Xie et al [9] and Sharma et al. [10], was shown in Figure 3. It was fitted for the observed 14 peaks with the following miller indices: (1 1 0), (1 0 7), (1 1 4), (1 0 8), (2 0 0), (2 0 3), (2 0 5), (2 0 6), (3 0 0), (2 1 7), (2 0 1 1), (2 2 0), (2 0 1 4) and (3 1 7).

Table 1. Crystal size of Barium hexaferrite (BaM)

Materials	Height [cts]	B	θ	Crystal Size [nm]	Unit Cell Volume [A^3]
Non-doped BaM	318.30	0.0669	17.056	61.228	696.99
Doped BaM	232.58	0.0669	17.089	61.238	697.56

The average crystallite size and unit cell volume of both samples have value as seen in Table 1. Furthermore, it indicated that Ni^{2+} and Zn^{2+} ions have entered $BaFe_{12}O_{19}$ lattice and become $BaNi_xZn_xFe_{12-2x}O_{19}$ particles. Subtitution of Ni^{2+} and Zn^{2+} ions caused the intensity of $BaFe_{12}O_{19}$ diffraction peaks declined and BaM peak phase shift to right as seen in Figure 2b. It showed that Ni^{2+} and Zn^{2+} ions have influenced to the phase growth of $BaFe_{12}O_{19,}$ in which the radius of Ni^{2+} and Zn^{2+} ions were larger than Fe^{3+}. So that, Ni^{2+} and Zn^{2+} ions substituted Fe^{3+} ion caused the crystal size was greater because the radius size of Ni-Zn ions were larger than Fe^{2+} ion that was substituted.

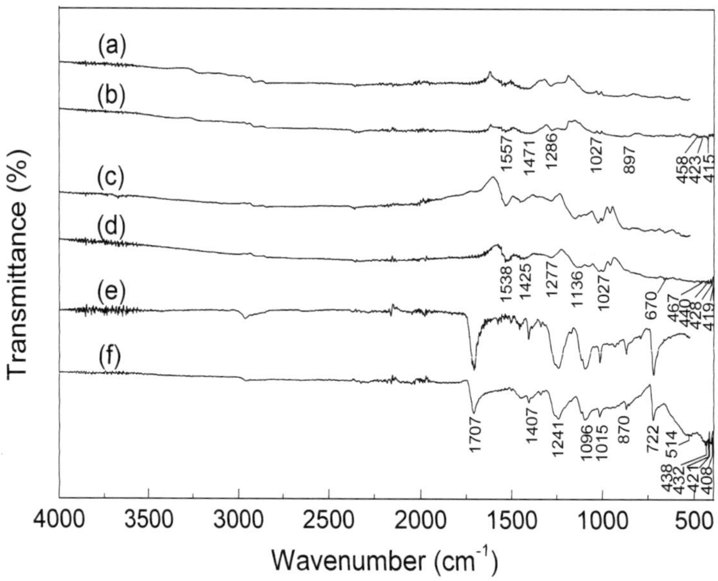

Figure 3. FTIR at (a) PAni, (b) BaM/PAni, (c) PPy, (d) BaM/PPy, (e) PET, (f) BaM/PET

Figure 3 shows that all composites have M–O and C–H bonds, wavenumbers of 514-408 cm^{-1} corresponding to M-O bond. It indicated interaction between metals (Fe, Ba, Ni, Zn) and oxygen. The peaks at 897-670 cm^{-1} were the out-of-plane deformation of C–H bonds in benzenoid ring which can be attributed to the presence of polymer. The vibrational mode of N=Q=N (Q is quinoid ring) bonds and C=C (pyrrole ring) in the doped PAni and PPy chain was also found at 1557 and 1538 cm^{-1} as described by Quillard et al. to be "electronic-like band" because their electronic structure is based on the overlap of alternating nitrogen and C_6 rings [11], as seen in Figure 3a-d. This electronic structure produced conductive properties in polymers. Therefore, it can be utilized as dielectric material to absorb energy of electromagnetic wave, known as radar absorbing material (RAM). The peaks at 1241 and 1407 cm^{-1} (Figure 3e and f), showed the stretched CH_3COOR and C-H ester bond in BaM/PET composite.

3.2. Morphology

Figure 4a shows SEM images of non-doped BaM. It exhibits pores and shape like sponge. Moreover, doped BaM showed polygonal morphology with increased pore size as seen in Figure 4b. Substitution of Ni-Zn ions caused the pore size was greater because the radius size of Ni-Zn ions were larger than Fe^{2+} ion that was substituted, this result also accord with XRD test results in Table 1. While, SEM images of BAM/(PAni,PPy,PET) composites on Figure 4c, d, and e, indicated same globular morphology. It was observed that BaM particles in BaM/(PAni,PPy) composites exhibits hexagonal and polygonal shape. Some aglomerates were found at surface of BaM/PAni composite (Figure 4c) which was caused by residual product of polymerized solution as concluded by Jiang et al. [12]. Figure 5a shows the presence of Ni-Zn elements inside BaM, while C, N, O, S, Ba, Fe, Ni and Zn were found on BaM/(PAni,PPy,PET) composites as seen in Figure 5b. It showed that BaM/(PAni,PPy,PET) composites have been succesfully synthesized, supported by FTIR spectra of polymers and composites on Figure 3.

Figure 4. SEM at (a) Non-doped BaM 2000x (b) Doped BaM 2000x (c) BaM/PAni 5000x (d) BaM/PPy 5000x (e) BaM/PET 20000x

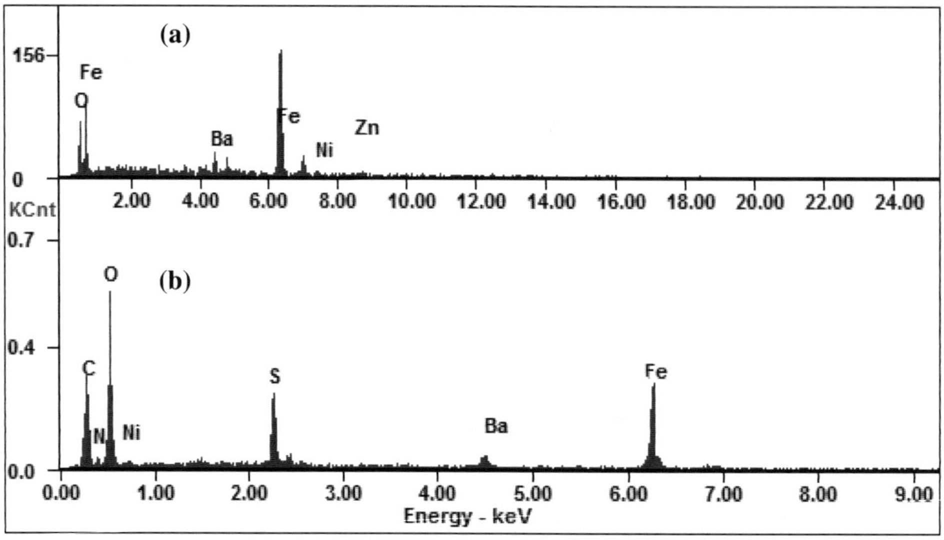

Figure 5. EDX at (a) Doped BaM (b) BaM/PAni

3.3. Magnetic property

VSM graphs on Figure 6 exhibits broad and narrow curve which indicated hysteresis loop of non-doped ($BaFe_{12}O_{19}$) and doped BaM ($BaNi_xZn_xFe_{12-2x}O_{19}$). Based on this curve and Table 2, it showed that non-doped BaM had magnetic saturation (Ms) and coercivity (Hc) value of 53.5 emu/gr and 13160 Oe. Whereas, doped BaM had Ms and Hc value of 56.6 emu/gr and 60 Oe. It can be concluded that doped BaM possessed soft magnetic properties based on its higher Ms and lower Hc value than non-doped one. This characteristic was caused by doping addition which reduced high energy loss in demagnetization process [13]. Therefore, it could be more efficient in absorbing the energy of electromagnetic wave or radar.

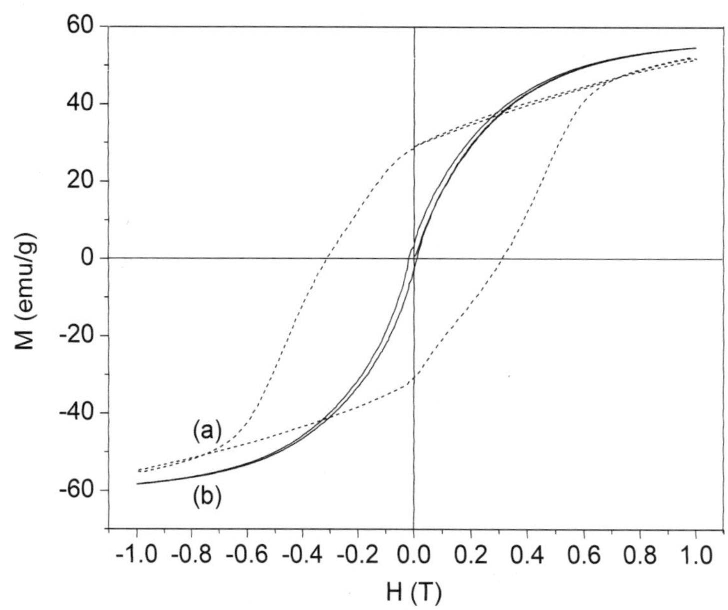

Figure 6. VSM at (a) Non-doped BaM, (b) Doped BaM

Table 2. Magnetic properties of Barium hexaferrite (BaM)

Materials	Ms (emu/g)	Hc (T)	Hc (Oe)
Non-doped BaM	53.5	0.329	13160
Doped BaM	56.6	0.0015	60

3.4. Electrical conductivity

The results of LCR-meter analysis are shown in Table 3, that a high electrical conductivity of 1.77744 x 10^{-5} S/cm was achieved by BaM/PAni composite. Whereas, BaM/(PPy,PET) had a smaller electrical conductivity than BaM/PAni, which is adequate to reduce the reflected wave energy. It indicates that the DBSA dopant has an important influence on conductivity. This may be attributed to the fact that the acid concentration has an important effect on the oxidative polymerization [14]. It is known that the presence of the acid is needed in the polymerization step and also the acid protonates the polymer [5]. Conducting polymers are inherently conducting due to the presence of a conjugated π electron system in their structure [9]. The appearance of these conjugated π electron caused by the effect of the acid that protonates the polymer.

Table 3. The electrical conductivity of polymers and composites

Materials	Conductivity (S/cm)
PAni	8.79809 x 10^{-8}
BaM/PAni	1.77744 x 10^{-5}
PPy	1.5007 x 10^{-7}
BaM/PPy	8.68413 x 10^{-6}
PET	1.47397 x 10^{-9}
BaM/PET	1.0354 x 10^{-5}

3.5. Microwave absorbtion properties

Figure 7 shows the R_L value of BaM/(PAni,PPy,PET) composites in the X-Band frequency of 8–12 GHz. All of samples have excellent absorbing performances for electromagnetic wave (EMW), which are related to the coating thickness of the absorber.

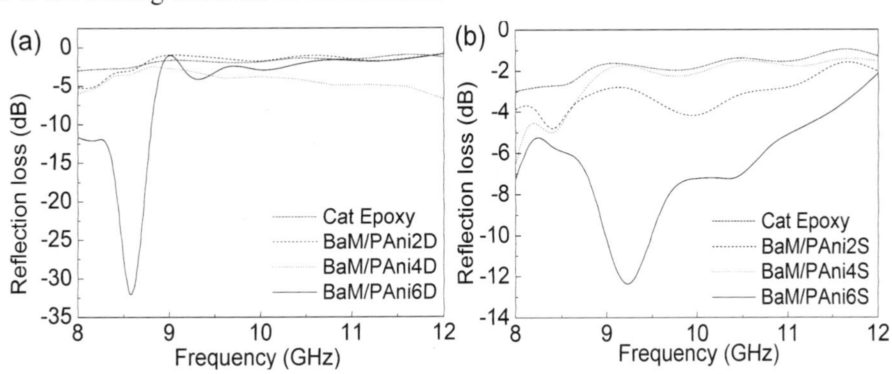

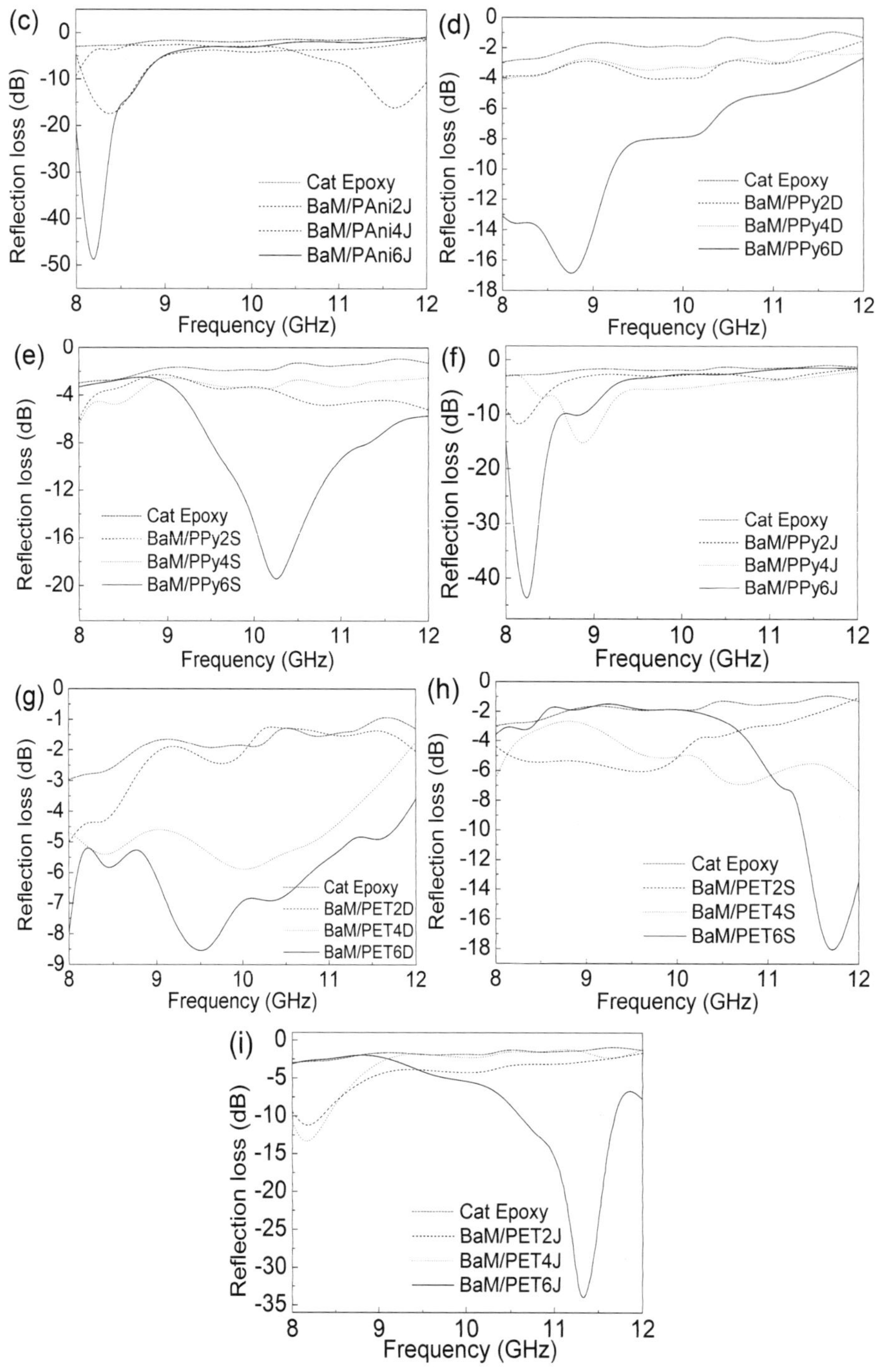

Figure 7. VNA at (a) BaM/PAni Dallenbach Layer, (b) BaM/PAni Salisbury Screen, (c) BaM/PAni Jaumann Layer, (d) BaM/PPy Dallenbach Layer, (e) BaM/PPy Salisbury Screen, (f) BaM/PPy Jaumann Layer, (g) BaM/PET Dallenbach Layer, (h) BaM/PET Salisbury Screen, (i) BaM/PET Jaumann Layer

Table 4. Reflection loss BaM/Polymer composites with 6 mm thickness

Coating Materials	Coating Method	Thickness (mm)	Frequency (GHz)	Reflection Loss (dB)
BaM/PAni	Dallenbach Layer	6	8,57	-32,029
	Salisbury Screen	6	9,2	-12,333
	Jaumann Layer	6	8,1	-48,720
BaM/PPy	Dallenbach Layer	6	8,8	-16,768
	Salisbury Screen	6	10,25	-19,435
	Jaumann Layer	6	8,312	-40,808
BaM/PET	Dallenbach Layer	6	9,6	-8,438
	Salisbury Screen	6	11,6	-16,959
	Jaumann Layer	6	11,28	-32,717

When the thickness is equal to 2 mm, the R_L value was lower than 4 mm but the maximum R_l value was reached at 6 mm thickness of BaM/PAni composite at Jaumann Layer of -48.720 dB at 8.1 GHz, as seen in Table 4. Whereas, BaM/PPy and BaM/PET composite at Jaumann Layer with 6 mm thickness have the R_l value of -40.808 dB 8.312 GHz and -32.717 dB 11.28 GHz, respectively. It can be concluded that absorption increase with increase in thickness of material [15]. It is caused when the layers of material are getting thicker, the absorber particles in the coating will also increase. As a result, the absorbtion process will increase. Park reported that -10 dB absorbing bandwidth means the frequency bandwidth having over 90% reflection loss characteristics [16].

There are many kinds of electromagnetic absorbers. Salisbury screens and the Dallenbach layer are resonant absorbers in the narrow band frequency. There are also Jaumann absorbers of multilayer structures with reflection film as broadband absorbers in the broadband frequency [16]. It was caused the maximum R_L value was obtained BaM/PAni composite at Jaumann Layer. This result also in accordance with the conductivity test results in Table 3, that BaM/PAni has highest conductivity value of 1,77744 x 10^{-5} S/m. The reflection loss of sample is related to dielectric loss and magnetic loss besides the coating thickness of the absorber [17]. It is well known that conducting polymers can effectively shield or absorb EMW generated from an electric source, whereas EMW from a magnetic source can be effectively shielded or absorbed only by magnetic material that was BaM material which has soft magnetic properties based on test results in Table 2. Thus, incorporation of magnetic constituents and conducting polymeric materials opens new possibilities for the achievement of good shielding or absorbing effectiveness for various electromagnetic sources [7]. Furthermore, the coming EMW are reflected to the receiver radar and become weaker, which means there has been an absorption of EMW by RAM composites.

Conclusions

Barium hexaferrite (BaM) was doped with Ni-Zn ion. Both of non-doped and doped BaM were single-phase $BaFe_{12}O_{19}$. BaM/Poly(aniline, pyrrole, ethylene terephthalate) (BaM/(PAni,PPy,PET)) composites were successfully synthesized. BaM and BaM/(PAni,PPy,PET) composites possessed globular morphology with M–O and C–H bonds. BaM ($BaNi_xZn_xFe_{12-2x}O_{19}$) showed the highest value of Ms and Hc, 56.6 emu/g and 60 Oe respectively. While, non-doped BaM had Ms and Hc value of 53.5 emu/gr and 13160 Oe. It can be concluded that doped BaM possessed soft magnetic properties. A high electrical conductivity of 1.77744 x 10^{-5} S/cm was achieved of BaM/PAni composite, which is adequate for rèduce the reflected wave energy. The maximum reflection loss (R_L) reached to -48.720 dB at 8.1 GHz of BaM/PAni coating with 6 mm thickness at Jaumann Layer. These results indicated that BaM/PAni composite was a soft magnetic material with a high R_L value that was suitable for radar absorbing material (RAM).

Acknowledgments
The author thanks to ITS (Sepuluh Nopember Institute of Technology), PSTBM (Pusat Sains dan Teknologi Bahan Maju) – BATAN Tangerang and PPET (Pusat Penelitian Elektronika dan Telekomunikasi) – LIPI Bandung (Lembaga Ilmu Pengetahuan Indonesia) for support of this research.

References
[1] Jun L W, Eui L S and Gon K C 2005 *Fukugo Zairyo Shinpojiumu Koen Yoshishu* **30** 25-6
[2] Rashad M M and Ibrahim I A 2011 *J. Magn. Magn. Mater.* **323** 2158–64
[3] Li W, Qiao X, Li M, Liu T and Peng H X 2013 *Mater. Res. Bull.* **48** 4449-53
[4] Agustianto R, Dwi F R M, Kristiputra R D, Marsha A A and Widyastuti 2015 *Proc. Basic Sci. Int. Conf.* vol 5, ed L Hakim et al. (Malang: Brawijaya University) pp 8-11
[5] Wang Y, Huang Y, Wang Q, He Q and Chen L 2012 *Appl. Surf. Sci.* **259** 486–93
[6] Castro F L d and Rezende M C 2008 *Mater. Res.* **11** 245-9
[7] Jiang J, Ai L-H, Qin D-B, Liu H and Chao L L 2009 *Synth. Met.* **159** 695–9
[8] Kanagesan S, Jesurani S, Sivakumar M, Thirupathi C and Kalaivani T 2011 *J. Sci. Res.* **3** 451-6
[9] Xie Y, Hong X, Gaob Y, Li M, Liu J, Wang J and Lu J 2012 *Synth. Met.* **162** 677-81
[10] Sharma R, Agarwala R C and Agarwala V 2008 *Mater. Lett.* **62** 2233–6
[11] Quillard S, Louarn G, Lefrant S and Macdiarmid A G 1994 *Phys. Rev. B.* **50** 12496-508
[12] Jiang J, Ai L-H and Yuan L-L 2010 *Mater. Lett.* **64** 888–90
[13] Kurniawati C A, Rosyidah and Zainuri M 2013 *J. Tek. Pomits* **1** 1–3
[14] Gosh M, Barman A, Meikap A K, De S K and Chatterjee S 1999 *Phys. Lett. A.* **260** 138-48
[15] Kaur H and Aul G D 2014 *Int. J. Sci. Res.* **3** 1141-5
[16] Park K Y, Lee S-E, Kim C-G and Han J-H 2006 *Compos. Sci. Technol.* **66** 576–84
[17] Li L, Chen K, Liu H, Tong G, Qian H and Hao B 2013 *J. Alloy Compd.* **557** 11-7

Preliminary Study of 20 MWth Experiment Power Reactor based on Pebble Bed Reactor

Dwi Irwanto, Sidik Permana and Syeilendra Pramuditya

[1]Nuclear Physics and Biophysics Research Division, Department of Physics, Institut Teknologi Bandung, Jl. Ganesha No 10, Bandung, 40132, Indonesia

*Email: dirwanto@fi.itb.ac.id

Abstract. In this study, preliminary design calculations for experimental small power reactor (20 MWt) based on Pebble Bed Reactor (PBR) are performed. PBR technology chosen due to its advantages in neutronic and safety aspects. Several important parameters, such as fissile enrichment, number of fuel passes, burnup and effective multiplication factor are taken into account in the calculation to find neutronic characteristics of the present reactor design.

1. Introduction

Pebble bed reactors were developed in Germany at the Juelich Research Center. The reactor produced heat by passing helium gas through the reactor core consisting of uranium fuel pebbles. A steam generator was used to generate electricity through a conventional steam electric plant.

The HTGR concept has outlet temperature of around 1,000 °C. This high output temperature enable applications such as desalination of sea water, hydrogen production, oil recovery and many other important industrial processes. There are two types of HTGR, prismatic block and pebble bed type which consist of pebble ball as the fuel. The Pebble Bed Reactor (PBR) has three fuel loading scheme: Multipass, Once Through Then Out and Peu a Peu or accumulative fuel loading scheme which have good neutronic aspect for both using uranium or thorium fuel [1-4].

Indonesia is now preparing to build Experiment Power Reactor based on PBR technology. The purposes of the present study are to analyze characteristics of selected parameters for a 20MWt Pebble Bed Reactor design that significantly contributed to the neutronic aspect, such as number of uranium enrichment, number of fuel passes, fuel loading scheme and burnup.

2. Calculation Methods

A method of neutron transport calculations to treat randomly distributed spherical fuel elements, coated particles, pebble bed fuels, plutonium spot, etc. with the continuous energy Monte Carlo method has been developed based on the statistical geometry model [5]. In this method, the position of the fuel particles is determined probabilistically from the nearest neighbor distribution (NDD) in the process of the random walk. The NDD determined from the packing fraction of fuel particles in the fuel region. This method can deal with reactor systems which have double heterogeneous structure consisting of coated fuel particles and fuel pins or fuel pebble.

In pebble bed reactor calculation, double heterogeneity need to be treated. It consists of two kinds of spherical fuel elements i.e. coated fuel particle (CFP) distribution in the fuel ball and the other one

IOP Conf. Series: Journal of Physics: Conf. Series **877** (2017) 012058 doi:10.1088/1742-6596/877/1/012058

is the fuel balls distribution in the reactor core. These calculation methods are implemented in MVP/MVP-BURN [6-7]. For nuclear data library, JENDL 3.3 are use. [8]

Table 1. Reactor Design Specification

Design Specification		
Reactor power	20	MWth
Fuel	TRISO	
Active core radius	90	cm
Active core height	180	cm
Reflector width	100	cm
^{235}U enrichment	7-20	%

At the present study, simplification to the actual reactor was done by created a cylindrical core loaded by pebble fuel ball using a reference design of HTR-10 [9]. Specification of the reactor core design and coated fuel particle are shown in Table 1 and Table 2, respectively.

Table 2. Coated Fuel Particle Specification

Fuel Kernel		
Material	UO_2	
Radius of the kernel	0.250	mm
Density	10.4	g/cm^3
Coatings		
First Buffer Layer (PyC)		
Thickness	0.09	mm
Density	1.1	g/cm^3
Second Layer (PyC)		
Thickness	0.04	mm
Density	1.9	g/cm^3
Third Layer (SiC)		
Thickness	0.035	mm
Density	3.18	g/cm^3
Forth Layer (PyC)		
Thickness	0.04	mm
Density	1.9	g/cm^3

3. Calculation Results and Discussion

Calculation are performed to study characteristic and connections of several important neutronic parameters for 20 MWt power of the present reactor design. Effect of uranium enrichment to effective multiplication factor are investigated at different pebble fuel speed through the reactor core. At the present study, 0.5, 0.75 and 1.0 cm/day pebble fuel speed are taken into account. Parametric survey for 7-20% uranium enrichment are performed and calculation results are shown in Fig. 1.

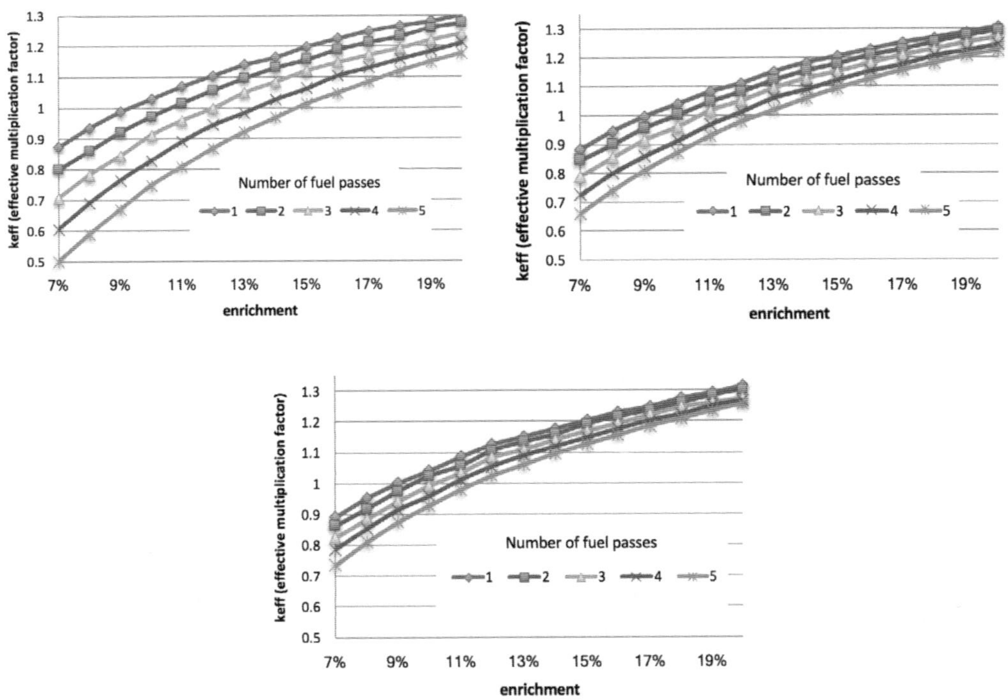

Figure 1. Characteristic of pebble fuel speed with five passes fuel loading scheme and minimum uranium enrichment needed for different fuel speed (0.5, 0.75 and 1.0 cm/day)

From the obtained data, for 0.5 cm/day fuel speed and 5 passes of pebble fuel ball through the reactor core, it will need at least 15% of uranium enrichment. Smaller number of fuel passes will need smaller uranium enrichment as shown in Fig.1. If one increased pebble fuel speed to 0.75 cm/day, minimum uranium enrichment needed will be at least 13%. For pebble fuel speed of 1.0 cm/day, minimum number of uranium enrichment is 12%. From these calculation results, correlation and optimum configuration for uranium enrichment, number of fuel passes and pebble fuel speed through the reactor core can be obtained.

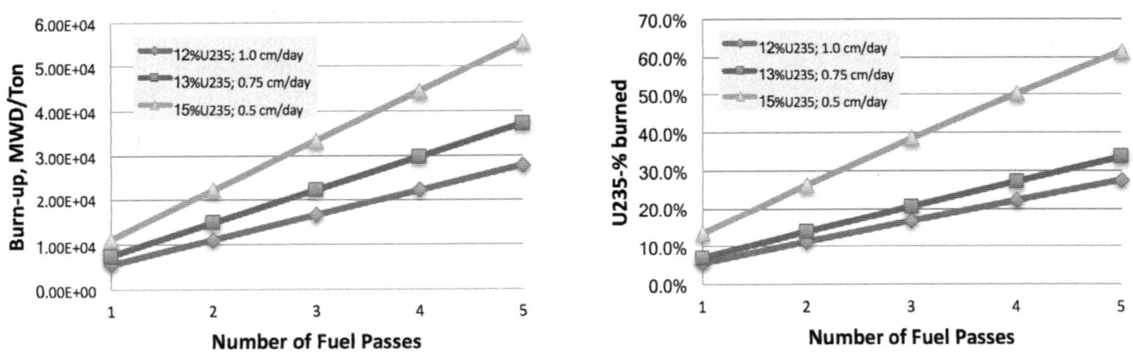

Figure 2. Characteristic of number of uranium loaded to the pebble fuel, pebble fuel speed with five passes fuel loading scheme and maximum burnup with minimum uranium enrichment needed for different fuel speed (0.5, 0.75 and 1.0 cm/day)

Burnup level is also one of the important parameters in reactor design. For the present reactor design, burnup and number of burned U235 for each configuration are shown in Fig.2. For fuel speed of 0.5 cm/day with 15% of uranium enrichment and 5 fuel passes, burnup is 56 GWD/T. For 0.75 cm/day fuel speed with 13% uranium enrichment and 5 fuel passes, burnup is 37 GWD/T. For 1.0 cm/day with 12% of uranium enrichment and 5 fuel passes, burnup is 28 GWD/Ton. If one need higher burnup, for example 80 GWD/T, one could increase number of pebble ball fuel passes through the reactor core or slowing down the fuel speed.

4. Conclusion

Parametric calculations for small power 20 MWt Pebble Bed Reactor have been performed in the present study. Uranium enrichment, number of pebble ball fuel passes and pebble fuel speed through the reactor core are important neutronic parameters taken into account. From the calculation performed in the present study, highest burnup could be obtained for combination of 0.5 cm/day pebble fuel speed, 15% uranium enrichment and 5 fuel passes. It is can be also understood from the calculation general characteristic and correlation between fuel speed, uranium enrichment and number of fuel passes.

Acknowledgment

The authors express their gratitude for the support from The Ministry of Research Technology and Higher Education of The Republic of Indonesia through the scheme of Penelitian Unggulan Perguruan Tinggi (PUPT) RISTEKDIKTI 2016.

References

[1] D. Irwanto, T. Obara, K. Yukitaka and I. Yamanaka, 2010 *Trans Am Nucl Soc* 103 p.791
[2] D. Irwanto and T.Obara, 2011 *J. Nucl. Sci. Technol*, vol. 48, No.11, p. 1385-1395
[3] D. Irwanto and T.Obara, 2012 *J. Nucl. Sci. Technol*, vol. 49, No.2, p.222-229
[4] D. Irwanto and T. Obara, 2013 *Ann. Nucl. Energy*, vol. 60, p. 383-395
[5] T. Mori, K. Okumura, Y. Nagaya, and H. Ando. 2000 *Trans. Am. Nucl. Soc.* 83, pp.283–284
[6] Y. Nagaya, et al, 2004 Japan Atomic Energy Agency
[7] K. Okumura, et al, 2005 Japan Atomic Energy Agency
[8] K.Shibata, et al, 2002 *J. Nucl. Sci. Technol*, Vol39, No.11, p. 1125-1136
[9] International Atomic Energy Agency, 2003 *Evaluation of High Temperature Gas Cooled Reactor Performance: Benchmark Analysis Related to Initial Testing of the HTTR and HTR-10*

IOP Conf. Series: Journal of Physics: Conf. Series **877** (2017) 012059 doi:10.1088/1742-6596/877/1/012059

Thermoeconomical Productivity Analysis in Manufacturing Sector in Indonesia

Widya Liana Aji[1,a] and Acep Purqon[1,b]

[1] Department of Physics, Institut Teknologi Bandung, Jl. Ganesha No. 10 Bandung, Indonesia, 40132

[1,a] widyalianaaji@gmail.com

Abstract. Negative temperature is a phenomenon interesting to study. In negative temperature regime, Boltzmann distribution is inverted where many particles occupy the higher energy states than the lower one. Iyetomi proposed a negative temperature case in Japan and applied it to the labor productivity distribution where the particle and energy state are replaced by worker and labor productivity, respectively. In this paper, we investigate the negative temperature concept to the labor productivity distribution in manufacturing sector in Indonesia which is divided by three industry groups according to BPS (Center of Statistical Agency of Indonesia), i.e. large and medium industries, small industry, and micro industry. For all industry groups, food industry possesses maximum productivity. The results represent that the negative temperature of large and medium industries is around ten times lower than negative temperature of micro industry indicating large and medium industries is lack demand of worker, while the negative temperature of small industry is among the temperature negative of large and medium industries and micro industry.

1. Introduction

Manufacturing industry in Indonesia is one of the economic sectors that undergoes fast economic growth. This sector has the largest contribution to Indonesia's Gross Domestic Product [3] i.e. important indicator of economic condition in Indonesia. Moreover, manufacturing industry also absorbs many workers. Hence, productivity of worker in the industry possesses important contribution to the national economy.

Iyetomi initiated that labor productivity distribution in Japan which consists of manufacturing sector and nonmanufacturing sector could be approximated with negative temperature distribution [1]. At negative temperature, particles possess unusual distribution over energy states. He adopted negative temperature concept in order to give new insight of the macroeconomy. In this circumstance, zero temperature in negative temperature regime indicated the highest productivity of worker.

In this study, negative temperature in the distribution of workers over their productivity was investigated in manufacturing sector in Indonesia. Empirical results of the distribution led us to know the possibility of negative temperature in its sector as proposed by Iyetomi. The negative temperature provided an overview to the amount of demand of workers to increase the productivity in certain industry.

2. Negative Temperature

The probability of finding particles at certain energy state is related to the Boltzmann distribution which describes that the probability varies exponentially as a function of $E_i / k_B T$. The relationship between probability of particles to occupy energy states and its energy states is given by

$$P_i \propto \exp\left(-\frac{E_i}{kT}\right), \tag{1}$$

where P_i denotes the probability at the i-th energy state E_i at a temperature T and k_B denotes the Boltzmann's constant.

In common physical system, P_i decreases exponentially as E_i increases at positive temperature. On the other hand, the number of particles decrease as the energy in system raises. Under certain condition, probability over all energy states are identical which is indicated by similar number of particles for all energy states. This condition occurs at infinity temperature.

In contrast, If we substitute $T < 0$ K on (1), P_i increases exponentially as E_i raises. At negative temperature, most of particles occupy the higher energy states than the lower energy states led to decreasing entropy. Therefore, the number of particles increase exponentially as the energy raises.

At negative temperature, P_i is divergent as E_i approaches infinity [2]. Therefore, negative temperature requires upper limit energy indicated by -0 K. At that circumtance, all particles are at the highest energy state as the energy in system is maximum. The entropy becomes zero indicating all particles occupy the highest energy state. The probability would be maximum at that energy state.

3. Labor Productivity Distribution in Japan

Negative temperature concept could be applied to the labor productivity distribution as Iyetomi proposed. Based on his work, empirical results of the labor productivity distribution in Japan consist of manufacturing sector and nonmanufacturing sector (Figure 1) where those sectors consist of around 480,000 firms (Figure 1).

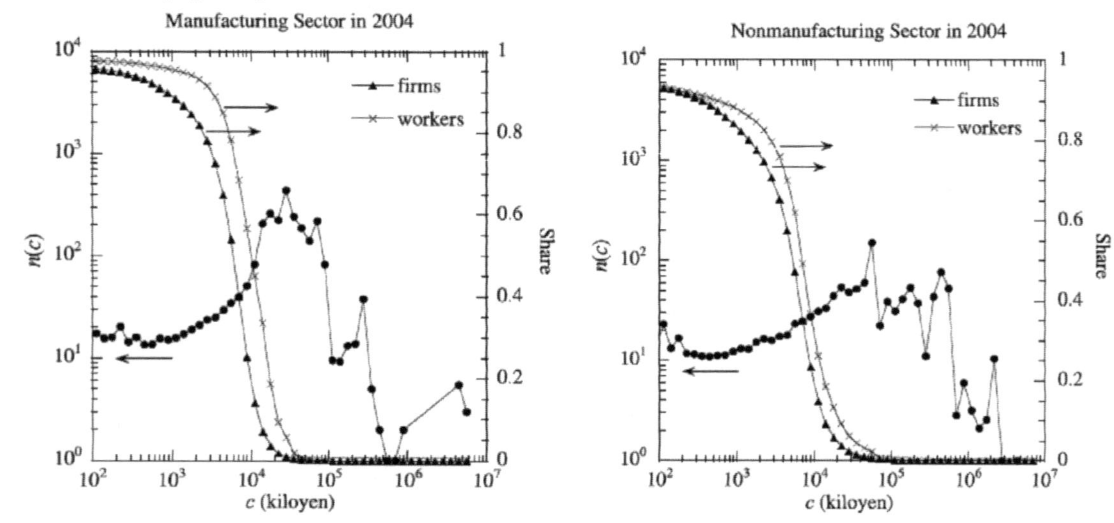

Figure 1. Empirical results of the labor productiviy distribution in Japan in 2004. Both axes are in logarithmic scale [1].

According to Iyetomi's research, the distribution of both sectors has sharp cups around $c = c^*$. In this circumtance, c^* indicates maximum productivity of workers. The number of workers $n(c)$ of both distributions tend to monotonic increase at $c < c^*$. In contrast, at the region $c > c^*$, $n(c)$ decrease of both sectors.

Iyetomi initiated Boltzmann distribution of workers over productivity:

$$n(c) = N\lambda \exp\left(-\frac{c}{T}\right), \tag{2}$$

where T is economic temperature indicating how large the demand of workers diminishes away from its maximum. The maximum condition is accomplished where all workers have maximum productivity. $N\lambda$ on (2) is related to equilibrium condition of two subsystems.

The equilibrium condition based on iyetomi's work consists of two condition

$$T_A = T_B, \tag{3}$$

and

$$N_A\lambda_A = N_B\lambda_B. \tag{4}$$

The first condition (3) describes the equlibrium of demand of workers, while the second condition (4) reveals the equilibrium of the number of workers between two subsystems. If $T_A > T_B$ then the demand of workers flow from subsystem A to subsystem B. $N_A\lambda_A > N_B\lambda_B$ indicated the workers flow from A to B and vice versa.

Recall from Iyetomi, the average number of workers increase as the productivity is raised. In this circumtancce, workers are regarded as particles and their productivity are considered as energy states of particles. Therefore, he allows us to adopt the concept of negative temperature formed with maximum productivity c^*.

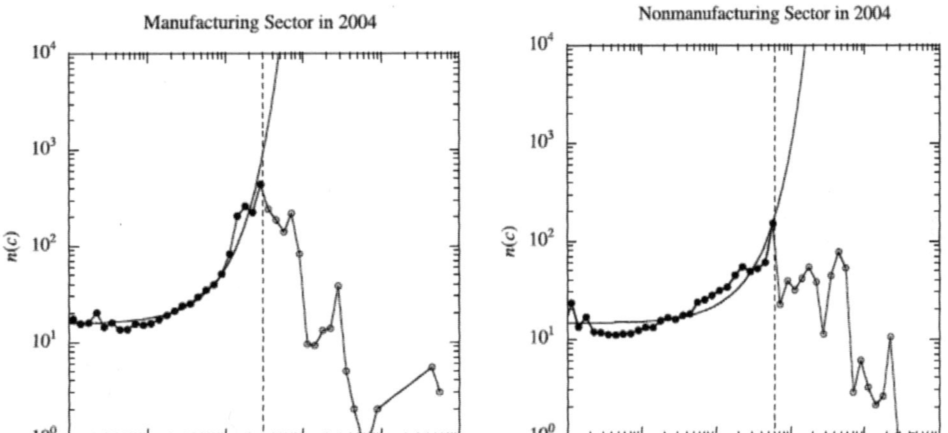

Figure 2. Fitting of labor productivity distribution in term Boltzmann distribution. Both axes are in logarithmic scales [1].

Least square fitting exponential to the Boltzmann form (Figure 2) of both distributions represents the exponential growth of the distribution of workers over productivity. Hence, economic temperature is negative. T and $N\lambda$ parameter could be determined by least square method based on Iyetomi's work.

4. Labor Productivity Distribution in Manufacturing Sector in Indonesia

4.1. Productivity data

We collected data of the productivity *c* and data of the number of workers *n(c)* compiled by BPS i.e. Center of Statistical Agency in Indonesia (www.bps.go.id). According to BPS, manufacturing sectors in Indonesia are divided by three groups:

- Large and medium industries (>19 people)
- Small industry (1-19 people)
- Micro industry (1-4 people)

In this study, we classified *c* and *n(c)* of large and medium industries in 2008-2013, while *c* and *n(c)* of micro and small industries are in 2010-2015.

Large and medium industries consist of 24 subsectors based on KBLI i.e. Indonesia's standard industrial classification in 2009, there are:

- Food
- Drink
- Tobacco
- Textil
- Clothes
- Leather
- Wood
- Paper
- Printing
- Coal and crude oil
- Chemical
- Pharmacy
- Rubber
- Nonmetallic mineral product
- Base metal
- Metal product
- Computer
- Electric product
- Mechanic
- Motor Vehicle
- Other transportation
- Furniture
- Other manufacturing
- Reparation services

Micro and small industries consist of 23 subsectors in which coal and crude oil industrial subsector were not included.

Labor productivity is defined as total output divided by the number of paid labor [4]. We could write the productivity of subsector *i* as c_i:

$$c_i = \frac{O_i}{n_i}, \tag{5}$$

where O_i and n_i represent the total output and the number of workers of subsector *i*. We should calculate c_i of micro and small industries by using (5). In contrast, we did not need to calculate c_i of large and medium industries since their productivity data were already provided in BPS.

4.2. Empirical results

We plotted *n(c)* respect to *c* for each industry groups to obtain the distribution of workers over their productivity. Each distribution has the maximum peak indicating maximum productivity of workers c^* (Figure 3). c^* of large and medium industries, small industry, and micro industry are around 270 million rupiah, 96 million rupiah, and 31 million rupiah, respectively. Food industry has the maximum productivity over all distributions in 2013.

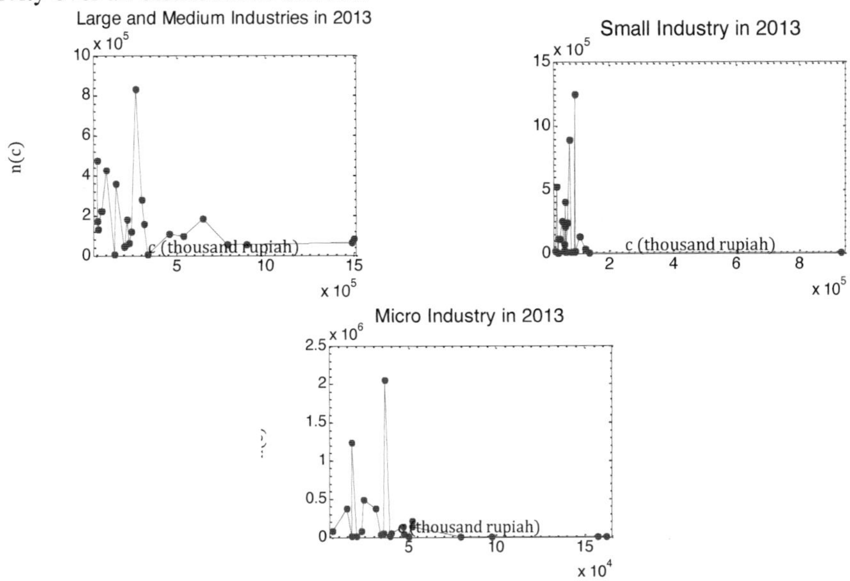

Figure 3. Empirical results of the distribution of workers over productivity for each industry groups in 2013.

We did not use logarithmic scale to plot *n(c)* since we lack of data from BPS. Moreover, we avoid to apply logarithmic scale on the distributions in order to indicate the difference of distribution as they are in $c < c^*$ and also in $c > c^*$. The distributions at $c < c^*$ have shorter interval than those at $c > c^*$. The average number of workers at $c < c^*$ is greater than those at $c > c^*$. For all distribution, the number of workers decrease drastically at $c > c^*$.

4.3 Exponential fitting of the labor productivity distribution

According to the distributions as shown in Figure 3, we could not fit the distributions with least square method. The distributions above are fluctuate. However, at each distribution, there is one highest peak indicating maximum productivity. In case of negative temperature, the maximum productivity indicated the upper limit of energy. In this circumstance, perhaps it allowed us to apply inverted Boltzmann distribution. On the other hand, exponential growth fitting could be applied to the distributions.

In this study, we formed exponential growth curve to the distributions as:

$$n(c) = a\exp(bc), \tag{6}$$

where

$$a = N\lambda, \tag{7}$$

and

$$b = -\frac{1}{T}. \tag{8}$$

a and *b* are guessed such that they could appropriate with the distributions. Therefore, we obtained $N\lambda$ and T parameters of each distributions.

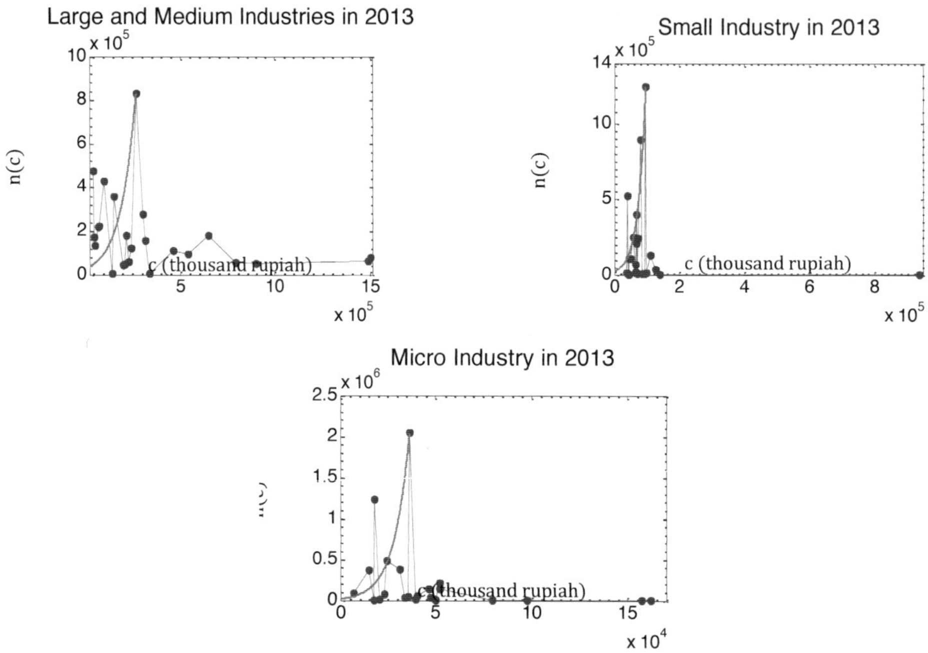

Figure 4. Exponential fitting curves appropriated to the distributions of each industry groups in 2013.

Exponential growth curve of each distribution indicated how many number of workers required to increase productivity in certain industry group (Figure 4). Exponential curve increasing rapidly indicates the great amount of the demand of workers. Exponential growth curve of large and medium industries are slower to increase than it of micro and small industries. Micro industry has the highest increment of exponential curves, while exponential curve of small industry is among micro industry and the large and medium industries.

4.4 T and Nλ parameters

The results of T and $N\lambda$ for each industry group are shown in Figure 5, 6, and 7 as initiated by Iyetomi where the blue color denotes T and the green color denotes $N\lambda$.

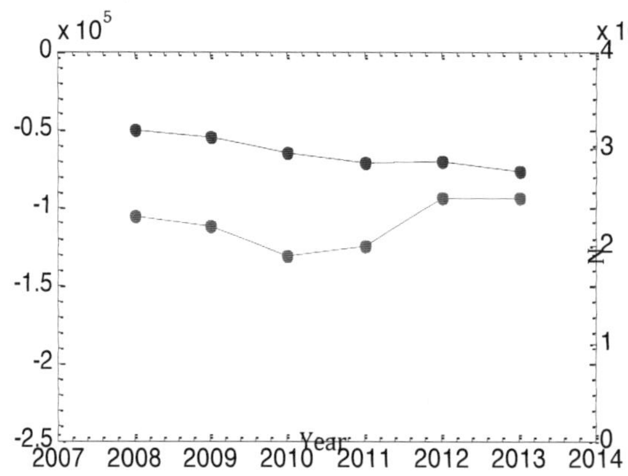

Figure 5. T and $N\lambda$ parameters of large and medium industries in the period 2008-2013.

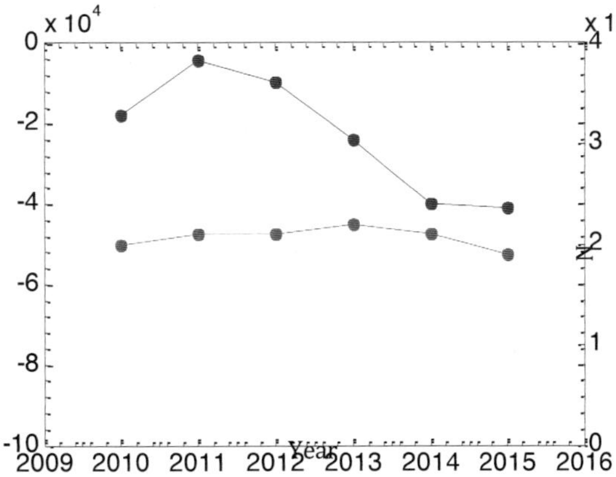

Figure 6. *T* and *Nλ* parameters of small industry in the period 2010-2015.

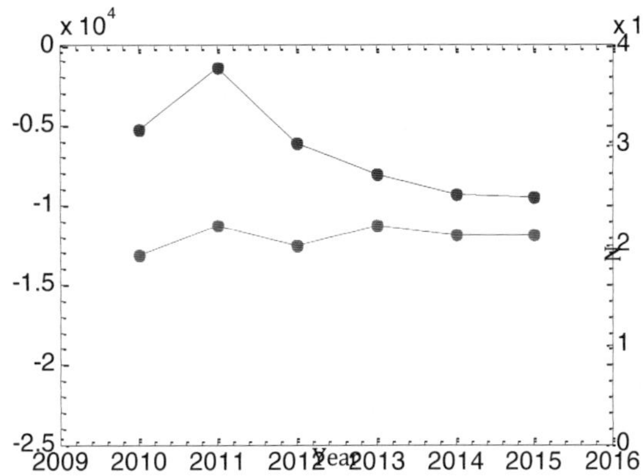

Figure 7. *T* and *Nλ* parameters of micro industry in the period 2010-2015.

T of large and medium industries are lower around ten times than it of micro industry indicating the demand of workers in micro indutry is greater than it in large and medium industries. These results are quite perceptible. Micro industry requires more amount of workers than large and medium industries in order to increase their productivity since the resounces are more limited in micro industry than those in large and medium industries. The large and medium industries will optimize their resources before they increase the demand of workers to increase their productivity.

T of small industry is among *T* of micro industry and *T* of large and medium industries. Economic temperature of micro and small industries tends to increase in the period 2010-2011. In contrast, their economic temperature tend to decrease in period 2011-2015. In 2011, both economic temperatures are the highest indicating the highest demand of workers of both industries.

In this study, we could not determined the results of *T* exactly, moreover in respect of *Nλ*. In Figure 5, 6, and 7 we obtained *Nλ* tends to equilibrium. In the right of fitting procedure by using the large amount of productivity data, however, the result of *Nλ* might be different.

5. Conclusion

We knew that the data of labor productivity in Japan is around 480.000 data, while data in this study are only 24 data for large and medium industries and 23 data for small and medium industries. We worked with 24(23) data to represent labor productivity distribution in Indonesia, particularly in manufacturing sector. The distributions of each industry groups had one highest peak indicating maximum productivity of workers. Therefore, there might be a negative temperature where the Boltzmann distribution is inverted. For each industry group, the results of T are quite understandable and they could describe the demand of workers adequately. In contrast to the results of $N\lambda$ which they could not be determined by our fitting procedure.

In the future research, we require to use more productivity data with the productivity data of firm for each industry groups in manufacturing sector as initiated by Iyetomi. Hence, the distribution of $n(c)$ over c will not fluctuate in order to use well the fitting procedure.

References

[1] H. Iyetomi, *Labor Productivity Distribution with Negative Temperature*, Progress of Theoretical Physics Supplement, DOI: 10.1143/PTPS.194.135.

[2] S. Braun et.al, *Negative Absolute Temperature for Motional Degrees of Freedom*, Journal of Quantum Gases 52 (2013) 339, arXiv: 1211.0545 [cond-mat.quant-gas].

[3] Yusri Alfian, *Analisis Peranan Sektor Industri Pengolahan terhadap wilayah di Propinsi Sumatra Utara*, Tesis (2002), Universitas Sumatra Utara.

[4] www.bps.go.id

IOP Conf. Series: Journal of Physics: Conf. Series **877** (2017) 012060 doi:10.1088/1742-6596/877/1/012060

Thorium Fuel Utilization Analysis on Small Long Life Reactor for Different Coolant Types

Sidik Permana [1]
[1]Nuclear Physics and Biophysics Research Division, Institut Teknologi Bandung
Jl. Ganesha 10, Bandung 40132, Gedung Fisika FMIPA ITB Indonesia

E-mail: psidik@fi.itb.ac.id

Abstract. A small power reactor and long operation which can be deployed for less population and remote area has been proposed by the IAEA as a small and medium reactor (SMR) program. Beside uranium utilization, it can be used also thorium fuel resources for SMR as a part of optimalization of nuclear fuel as a "partner" fuel with uranium fuel. A small long-life reactor based on thorium fuel cycle for several reactor coolant types and several power output has been evaluated in the present study for 10 years period of reactor operation. Several key parameters are used to evaluate its effect to the reactor performances such as reactor criticality, excess reactivity, reactor burnup achievement and power density profile. Water-cooled types give higher criticality than liquid metal coolants. Liquid metal coolant for fast reactor system gives less criticality especially at beginning of cycle (BOC), which shows liquid metal coolant system obtains almost stable criticality condition. Liquid metal coolants are relatively less excess reactivity to maintain longer reactor operation than water coolants. In addition, liquid metal coolant gives higher achievable burnup than water coolant types as well as higher power density for liquid metal coolants.

1. Introduction

Nuclear reactor utilization have been imrpoved to increases their reactor fuel capability for decades and one of the innovative reactor design to fullfil the requirement is reactor type for small power reactor and for long operation which can be deployed for small population densities and remote area as well as less transmission or grid. Long life reactor operation have been proposed by the IAEA as SMR program for a small and medium reactor to be used for some specific purposes which is basically based on thermal reactor and fast reactor types [1]. Beside uranium utilization, thorium fuel resources for SMR can be used as a "partner" fuel with uranium fuel. Thorium fuel technology has being developed well not only for conventional reactor but also provided the thorium breeder reactor programs to meet the global nuclear energy contribution as well as Uranium fuel technology. Better capability of thorium fuel is mainly coming from the superiority of fissile U-233 which shows higher eta-value in comparing with other fissile materials such as U-235 and Pu-239, except for higher energy regions, Pu-239 becomes more superior than others although U-233 still shows high eta-value. In addition, some advantages of thorium are estimated such as better breeding capability in thermal and epi-thermal neutron energy region, better fuel stability and better proliferation resistance level in comparison to uranium cycle [2-6]. Plutonium production in the spent fuel based on thorium fuel is very small compared with uranium fuel cycle, because of its nuclide chain, that affect to very low plutonium production as well as very low MA production as long-lived heavy nuclides. A comparative analysis on small long life reactor based on thorium fuel cycle for several reactor coolant types and several power output has been evaluated. In addition, several key parameters are analysed to evaluate its effect to the reactor performance such as reactor criticality, excess reactivity, reactor burnup achievement and power density profile.

Table 1 General Parameters

Parameters	Values/Type
Thermal Power Output [MWt]	100 - 500
Coolant	H2O, D2O, Na and Pb-Bi
Fuel	Oxide Type
Blanket	Nat Thorium
Core	U233-Th
Fuel Pellet Radius [cm]	0.6
Volume Fraction (Fuel/Clad/Cool) [%]	45 / 15 / 40
Core Operation Time [Years]	10

2. Methods and Parameters

To maintain the reactor criticality condition, some approaches have been done such as reactor size as geometrical reactor core approach, required fissile material and fuel cycle types as well as coolant types. Active core arrangement and fuel shuffling and refuelling schemes are also would be optimized for reactor operation, especially for long-life reactor core operation. One of the efficient and effective reactor operations is by adopting longer reactor operation with one batch fuel system, which can be called as a long-life reactor operation without any refuelling and shuffling processes [7-9]. The idea for designing a long-life reactor core operation for different reactor coolant types which based on water-cooled liquid metal cooled types as well as for several power reactor output. General parameters are shown in **Table 1** as a basic reactor design parameters. Small power outputs of 100 to 500 MWt are adopted as small reactor power types based on the IAEA definition.

The reactor system are based on thorium oxide fuel for different coolant types such as light water and heavy water coolants as representative of thermal reactor coolant types or water-cooled type as well as for liquid metal-cooled reactor types, it has been used some coolants such as sodium-cooled and lead-bismuth-cooled as coolant. Natural thorium is loaded into blanket regions and some required fissile content of U-233 and thorium are loaded into core regions as driver fuel regions. To analyse the reactor core optimization, an established code of SRAC-CITATION and JENDL-33 have been used for core optimization analysis and nuclear data library for two-dimensional core arrangement of radial and axial directions (R-Z) [10]. Reactor operation period is arranged to be 10 years as long-life reactor core operation. Adopting sufficient inner blanket region as well as core fuel regions in order to increase inner fuel conversion capability has done maintaining the reactor criticality condition for long-life core operation. Excess reactivity is used to analyse the surplus or additional neutron production during reactor operation, which related to reactor criticality. This neutron excess or excess reactivity can be used as one of the safety parameter for reactor operation which is considering the criticality safety point of view. Burnup level has been evaluated to estimate the reactor capability to burn the fissile material which can contribute to the reactor operation as well as reactor power production during operation period.

3. Results and Discussions

In the present study, small power output of nuclear reactor with several coolant materials are evaluated based on thorium fuel with some additional fissile material of U-233 in the core regions as driver fuel for long-life reactor operation for about 10 year reactor operation. In this section, some important parameters are shown and discussed such as reactor criticality, excess reactivity, burnup level and power density for different coolant types and several power output levels.

3.1. Reactor criticality

As a basic nuclear reactor analysis for the operation, the evaluation of reactor criticality becomes a key factor and it should be maintain more than unity to keep the reactor in operation. The study intend to maintain the reactor operation up to 10 years period which means the reactor should be maintain to be critical condition during intended period of 10 years operation without any refuelling and shuffling processes. Some basic approaches have been done for obtaining the reactor to be in a critical condition such geometrical arrangement and fuel type optimizations as well as fuel content optimization. Several coolants have been adopted for the evaluation based on different typical reactor that are exist and in operation for research and commercial reactor facilities which is generally divided into two general reactor types that is thermal reactor and fast reactor types. Typical coolant types of thermal reactors are light water (H2O coolant) and heavy water (D2O) coolants. Liquid sodium or sodium and liquid Pb-Bi coolants are adopted for typical fast reactor types.

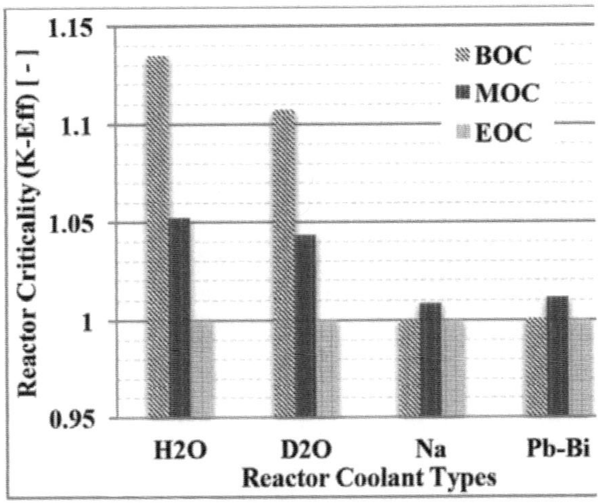

Figure 1 Reactor criticality for different coolant types

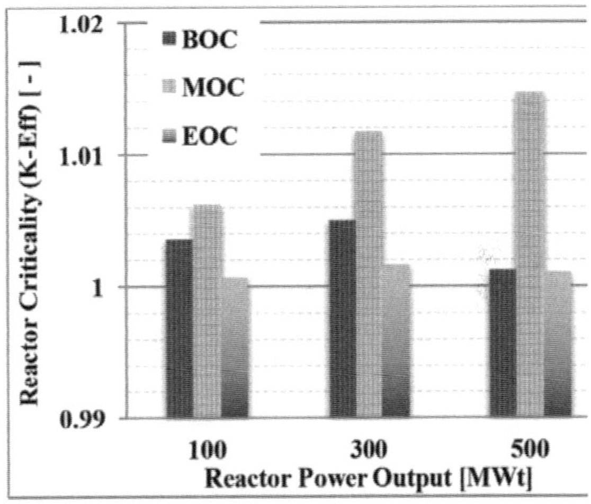

Figure 2 Reactor criticality for different power outputs

BOC : Beginning of cycle (0 year)
MOC: Middle of cycle (5 years)
EOC : End of cycle (10 years)

Obtained results of reactor criticality are shown in Figures 1 and 2 for different coolant types and power reactor outputs. Several coolant types have been employed as well as some different power outputs for the evaluation to analyse its effect to the reactor criticality condition during reactor operation. Figure 1 shows all reactor systems based on different coolant types obtain reactor operation for 10 year operation times which shows reactor criticality are always higher than unity. Light water coolant obtains higher criticality condition from the beginning of operation in comparison to other coolants. Water coolant types gives higher criticality than liquid metal coolants (Liquid sodium and liquid Pb-Bi coolants) at BOC and MOC, while at the end of operation or EOC almost all coolants obtain same level of criticality. Liquid metal coolant of coolant for fast reactor system gives less criticality at the BOC and increases at the MOC and it reduces for EOC. This condition shows that fast reactor system or liquid metal coolant system obtain almost stable criticality condition since the beginning of cycle and small increases at MOC. While water cooled types (light water and heavy water coolants) are obtained high at BOC and it reduces with increasing the time. Different power outputs give some different criticality conditions are shown in Figure 2 for sodium coolant type. All power output can maintain its reactor operation up to 10 years period. Based on the obtained results shows that the highest criticality condition is expected coming from the highest power output, which is shown at MOC period.

IOP Conf. Series: Journal of Physics: Conf. Series **877** (2017) 012060 doi:10.1088/1742-6596/877/1/012060

3.2. Excess reactivity

Criticality condition is not only used as reactor operation condition, but also to evaluate the safety condition based on excess reactivity from criticality of the reactors. Excess reactivity is defined as some surplus neutron population or criticality that exceed more than unity. This excess criticality will be used to estimate the safety from criticality point of view. Less excess reactivity or in ideal condition when the criticality reaches to a unity is the best condition. When abnormality condition occurs, some control rod should be used to compensate the excess criticality to push the reactor sub critic and if the excess reactivity is less, then it is easier to make the reactor shutdown and less control rod action. Less excess reactivity is also estimated the reactor stability from neutron population during reactor operation. Obtained results of excess reactivity are shown in Figures 3 and 4 for different coolant types and power reactor outputs.

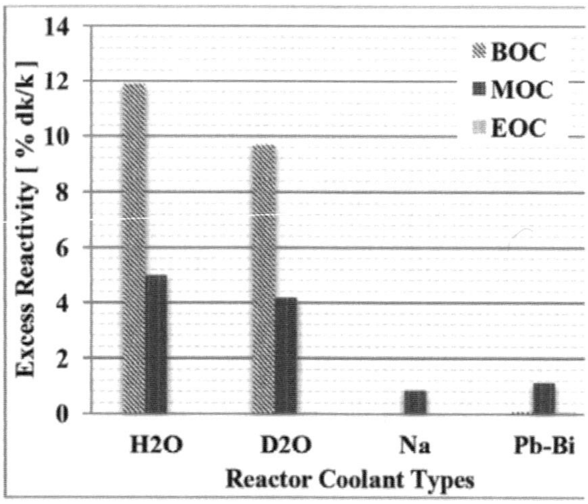

Figure 3 Excess reactivity for different coolant types

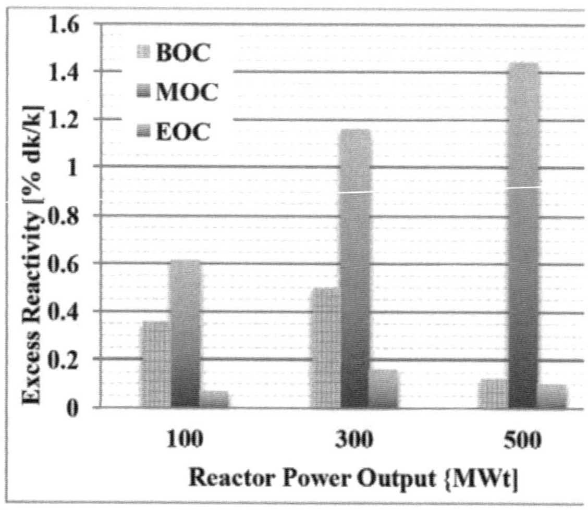

Figure 4 Excess reactivity for different power outputs

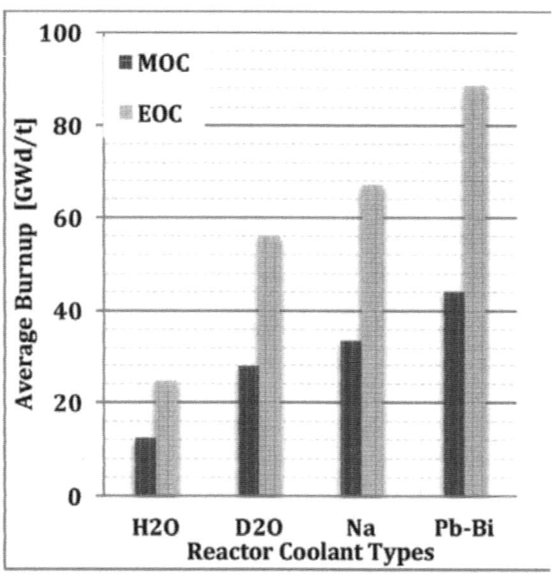

Figure 5 Average burnup for different coolant types

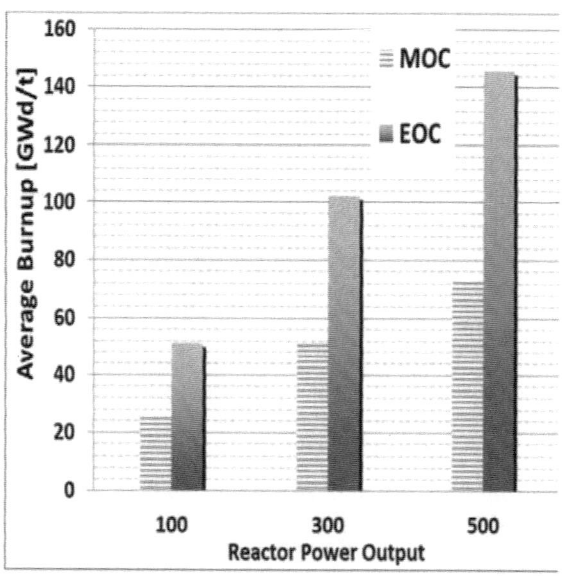

Figure 6 Average burnup for different power outputs

Figure show that larger excess reactivity are obtained by water cooled type (H2O and D2O coolants) and liquid metals coolant types (Na and Pb-Bi coolants) are obtained less. Higher excess reactivity of water-cooled type is estimated from the BOC period that more neutron populations are needed to maintain the reactor operation along 10 years periods. This reactivity condition is a typical type of water-cooled reactor for long-life operation. In case of liquid metal coolant, which is working as coolant for fast reactor system are relatively less excess reactivity to maintain longer reactor operation period for both sodium and Pb-Bi coolants. This condition can be estimated from the neutron spectrum characteristic that liquid metal has harder spectrum with relatively higher fissile content in core region. While blanket region can maintain internal fuel conversion by capturing produced neutron from core regions, which affects to a reduction factor of neutron population. At the same time some converted fertile materials in the blanket regions becomes fissile materials, which can be used for reactor operation longer. In case of water-cooled reactor, blanket regions does not significantly absorb the neutron from core regions and slowly reduce the excess reactivity at the BOC and gradually criticality reduces because of fissile material are reduces from caused by fission reaction. In case of different power output, it shows that higher power output obtains higher excess reactivity, which is shown at MOC period of operation. Based on the obtained results, it gives the excess reactivity less than 12% for water-cooled coolant types (H2O and D2O coolants) and less than 1.5% for liquid metal coolant types (Na and Pb-Bi coolants)

3.3. Reactor burnup

Obtained results based on the analyses of the different coolant and power output effects to the reactor burnup are shown in **Figures 5 and 6.** Liquid metal coolant gives higher achievable burnup than water coolant types which obtains higher than 60 GWd/t and more than 80 GWd/t along 10 years operation period for liquid sodium and liquid Pb-Bi coolants, respectively. While water coolant type obtains less than 60 GWd/t of reactor burnup for 10 years operation period.This phenomenon can be estimated from the neutron spectrum point of view that harder neutron spectrum work better to make more fission reaction which means more fuel material in the reactors are burnt during reactor operation for the same period of time. Liquid Pb-Bi gives harder spectrum than liquid sodium and those liquid metal coolants are relatively harder in term of neutron spectrum. Higher power output gives better burnup value, and it shows a gradual trend of increasing burnup, which is depending on the increase of reactor power. More power reactor output means the reactor produces more power from fission reaction in the reactor cores and more fission reactor affects to the increasing burnup.

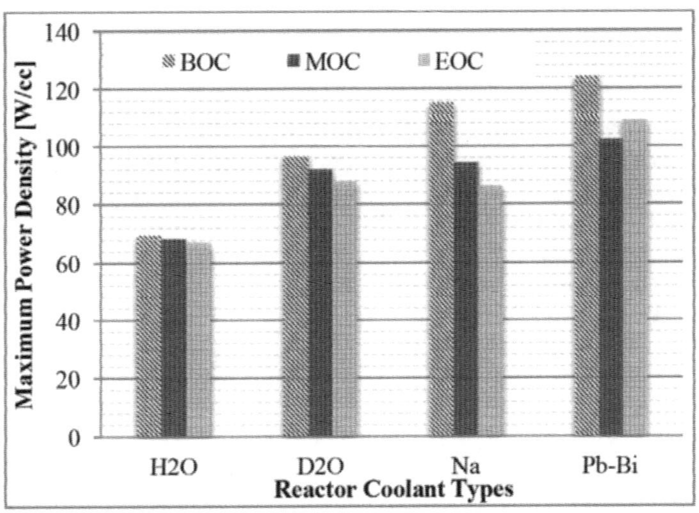

Figure 7 Maximum Power Density for different coolant types

3.4. Reactor power density

To evaluate the performance of reactor power density, a maximum power density value is adopted to show the optimum power density is produced during reactor operation. Obtained results of the study are shown in **Figure 7** for different coolant cases. It shows that the highest maximum power density is obtained by liquid Pb-Bi coolant followed by liquid sodium coolant and water coolant types. In General, power density profile is decreasing for longer reactor operation which is shown in the results, except for liquid Pb-Bi which shows some increasing power at the EOC than MOC, although it still less than BOC. Liquid metal obtains higher maximum power density than water coolants. This condition is estimated from the capability to make more fission reaction in the reactor, which can produce more power output as well as power density during reactor operation.

4. Conclusions

A small long-life reactor based on thorium fuel cycle for several reactor coolant types and several power output has been evaluated in the present study. Several key parameters are used to evaluate its effect to the reactor performances such as reactor criticality, excess reactivity, reactor burnup achievement and power density profile. An established code of SRAC-CITATION and JENDL-33 as nuclear data library have been used for core optimization analysis based on two dimensional reactor core arrangement of radial and axial directions (R-Z) for 10 years period of reactor operation period is arranged to be 10 years. Water-cooled types give higher criticality than liquid metal coolants (Liquid sodium and liquid Pb-Bi coolants). Liquid metal coolant of coolant for fast reactor system gives less criticality especially at BOC and increases in small changing, which shows that fast reactor system or liquid metal coolant system obtain almost stable criticality condition. Larger excess reactivity is obtained by water-cooled type (H2O and D2O coolants) than liquid metals coolant types (Na dan Pb-Bi coolants). This reactivity condition is a typical type of water-cooled reactor for long-life operation. In the case of liquid metal coolant for fast reactor system are relatively less excess reactivity to maintain longer reactor operation period for both sodium and Pb-Bi coolants. In addition, liquid metal coolant gives higher achievable burnup than water coolant types as well as higher power density for liquid metal coolants.

Acknowledgments

We would like to acknowledge and extend our gratitude to desentralisasi research program of ministry of research, technology and higher education and the Asahi foundation for the grant and international conference publication

References

[1] V. Kuznetsov 2002, *Progress in Nuclear Energy*, Vol. 50, No. 242
[2] P. Sidik, N. Takaki, H. Sekimoto 2007, *J. Nucl. Sci. Technol*, **44**[7], 946-957
[3] T. K. Kim and T. J. Downar 2002 *Nuclear Technology*, **138**[1], 17-29
[4] L. Michael, G. Otto 1998 *Nucl. Eng. Des*. 180[2], 133
[5] U. Turan, 2000, *Prog. Nucl. Energy*, 37, 137
[5] L. B. Freeman et al. 1989 *Nucl. Sci. Eng*. 102, 341 (1989)
[6] P. Sidik, N. Takaki, H. Sekimoto 2008, *J. Nucl. Sci. Technol*, **45**[7], 589-600
[7] Sekimoto, H, Zaki. S. 1991 *Proceeding of the international conference on fast reactors and related fuel cycle*, Kyoto, October 28- November 1.
[8] Zaki, S., Sekimoto H 1995 *Annals of Nuclear Energy* **22** 711
[9] Zaki, S., Sekimoto H 1995 *J. Nucl. Sci. Technol*. **32 9** 834-845
[10] K. Okumura et al. 1996 JAERI-Data/Code96-015, Japan Atomic Energy Research Institute (JAERI)

IOP Conf. Series: Journal of Physics: Conf. Series **877** (2017) 012061 doi:10.1088/1742-6596/877/1/012061

Heat and Mass Transfer Model in Freeze-Dried Medium

Sayahdin Alfat[1] and Acep Purqon[2]

[1] Graduate Computational Science, Faculty of Mathematics and Science, Institute of Technology Bandung, Jl. Ganesha 10 Bandung 40132, Indonesia
[2] Physics of Earth and Complex Systems, Institute of Technology Bandung, Jl. Ganesha 10 Bandung 40132, Indonesia

E-mail: sayahdin.alfat@yahoo.com

Abstract. There are big problems in agriculture sector every year. One of the major problems is abundance of agricultural product during the peak of harvest season that is not matched by an increase in demand of agricultural product by consumers, this causes a wasted agricultural products. Alternative way was food preservation by freeze dried method. This method was already using heat transfer through conduction and convection to reduce water quality in the food. The main objective of this research was to design a model heat and mass transfer in freeze-dried medium. We had two steps in this research, the first step was design of medium as the heat injection site and the second was simulate heat and mass transfer of the product. During simulation process, we use physical property of some agriculture product. The result will show how temperature and moisture distribution every second. The method of research use finite element method (FEM) and will be illustrated in three dimensional.

1. Introduction

Today, Indonesia is one of the countries producing and exporting coffee beans in the world. In Indonesia itself, the coffee beams is a major commodity and the coffee itself is the largest agricultural commodity exchange earners after oil palm, rubber and cocoa [1]. However, this product have some problems, one problem encounters is that the low value of exports because still low quality [2]. A modern solution, home freeze drying may be the best preservation approach. This approach is an artificial drying to biology product and an approach that is often used for the drying process [3].

Basically, the technology of freeze drying was developed during World War II. This technology as a way to keep vaccines for the wounded from spoiling due to uncertain refrigeration while transported to war zones. The technology was refined and used extensively during the NASA space program, providing varieties of freeze dried food for astronauts. This technology has been applied with success to diverse biological material, such as meet, seafood products, coffee, tea, vegetable, fruits, milk products, juices, dairy products, cells and bacteria and is now standard practice in the production of penicillin, protein hydrolyzes, hormones, blood plasma and vitamin preparations. Especially for food, this technology have many advantages such as minimize the decline in food nutrition, there is not shrink on food, and also freeze drying does not reduce aroma and flavor of food. The model or approach use heat and mass transfer to reduce water content in the food.

Heat and mass transfer are complex and complicated process. This phenomenon are found in other process such as; roasting [7, 5], cooking [9] and drying [14, 6, 10]. Understanding the heat and mass transfer in the products will help to improve drying process parameters and quality. There are advantages of heat and mass transfer In food product. First, heat and mass transfer treatment can reduce microbial growth [8]. Second, understanding the heat and mass transfer can be minimize consumption energy during treatment process [7].

Mathematical modeling is a useful method to describe heat and mass transport during drying food. Several theoretical model have been proposed in the study of convection food drying [10]. Luikov's couple systems equation is heat and mass transfer equation, but thermo-physical properties is constant [11, 13].

The discrete model is a part of numerical method and it have been widely used in solving real engineering problem. One of the discrete model is finite element method (FEM), it is most powerful numerical method to show mechanism of heat ad mass transfer. This is one numerical method and can show mechanism of heat and mass transfer, can show distribution heat and mass inside product and also can be used to complex domain [4].

The work was divided on two objectives namely main and specific objective. The main objective of this paper is to design a numerical method for heat and mass transfer as a part of coffee drying. For specific objective, we want to investigate heat and mass flow between surface and center of product. However, distribution profile of temperature and moisture content of the product will be explained.

This paper have six section. Section 2, we will explain some of the assumption used during the simulation process. The computational model and some boundary condition are also shown in this section. Section 3, we will give parameters of coffee bean and air. For section that explain the numerical method that is section 4. As for section 5, the numerical results of heat and mass transfer will be shown and it is discussed with physical analysis. Finally, section 6 is closed by the main conclusion and the possible future works as the extension in this research.

2. The Macroscopic governing equation

2.1. General distribution model

On freeze-drying medium, drying is usually described as a couple process of heat and mass transfer. In this study, the Luikov's couple system that simplified by Husain et, all [11] was used to simulate heat and mass transfer. Initial temperature and moisture content (T_0, M_0) of the agricultural product is uniform on the product (see fig 1).

In the model which will be presented here, the following consideration such as:

- There is no heat generation inside the product;
- The gas phase is water vapour, which behaves as an ideal gas;
- Thermo-physical properties of product are dependent on moisture;
- Heat transfer into the product through the entire surface by convection;
- Moisture content of product flowing out through the entire surface by diffusion;
- No chemical reaction takes place during drying, i.e. thermal and chemical properties of material, air and moisture are constant within the range of temperatures considered;
- Uniform distribution of air through the dryer;
- Gravity and shrinkage are negligible.

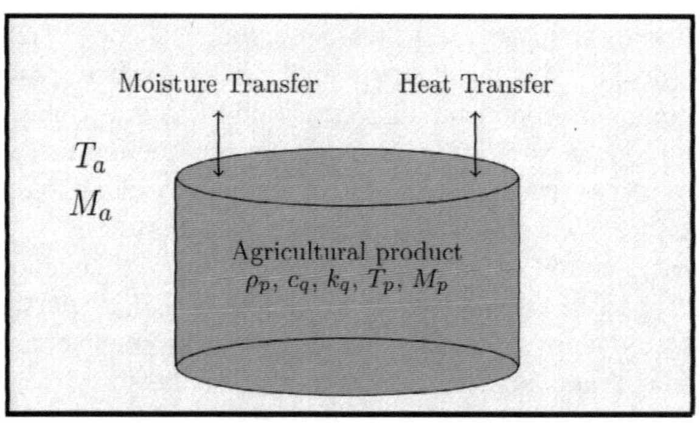

Figure 1. Schematic of physical model

2.2. Mathematical model

To simulate heat and mass transfer in freeze-dried medium, we use Luikov's equation that has been simplified by Husain, et, all [11]. As for the heat and mass transfer equation that are used as follows:

$$\rho_p c_t \partial_t T = \nabla \cdot (\kappa_t \nabla T) + L\rho_p \frac{\partial M}{\partial t} \qquad x \in \Omega \tag{1}$$

$$\partial_t M = \nabla \cdot (D \nabla M) \quad x \in \Omega \tag{2}$$

where T is temperature of coffee (K), ρ_p is coffee bean density (kg/m^3), c_t is the specific heat of coffee bean (J/kgK), κ_t is the thermal conductivity of coffee bean (W/ms), L is the latent heat of vaporization of water (J/kg), M is moisture content of coffee bean (%) and D is diffusion coefficient (m/s^2).

2.3. Boundary condition and initial condition

According assumption in subsection of general distribution model. So, the boundary conditions associated with this system of equations were written is a generalized form as:

$$\kappa_t \left(\frac{\partial T}{\partial n}\right) + h_t(T - T_\infty) = 0 \qquad \text{on } \Gamma_N \tag{3}$$

$$D\left(\frac{\partial M}{\partial n}\right) + h_m(M - M_\infty) = 0 \qquad \text{on } \Gamma_N \tag{4}$$

$$D\left(\frac{\partial M}{\partial n}\right) = 0 \qquad \text{on } \Gamma_D \tag{5}$$

$$\kappa_t \left(\frac{\partial T}{\partial n}\right) = 0 \qquad \text{on } \Gamma_D \tag{6}$$

where n is the normal to the surface; T_∞ and M_∞ are the temperature and moisture of the air, respectively; h_t and h_m are the convection heat transfer coefficients (W/m^{2}K) and the surface mass transfer coefficient (kg/m^{2}s), respectively; and Γ_N and Γ_D are Neumann boundary and Dirichlet boundary, respectively.

For initial condition (t = 0), temperature (T_0) and moisture content (M_0) of agriculture product are uniform:

$$T(x, y, z, 0) = T_0(x, y, z) \qquad (x, y, z) \in \Omega \tag{7}$$

$$T(x, y, z, 0) = T_0(x, y, z) \qquad (x, y, z) \in \Omega \tag{8}$$

On the model, one of the assumptions used were air is considered as an ideal gas; if the air pressure decreases, air density will decrease according to the ideal gas law:

$$\rho_a = \frac{p_a}{RT_a} \tag{9}$$

where ρ_a is the air density (kg/m^3), p_a is the air pressure (Pa) and R is the air constant (8.314 J/Kmol).

In heat and mass simulation, heat and mass transfer coefficient for freeze dried medium were calculate using Nusselt number. The Chilton-Colburn analogy holds [5]:

$$Nu = 0.664 Re^{0.5} Pr^{0.33} \tag{10}$$

with

$$Re = \frac{\rho_a v_a L}{\mu}, Pr = \frac{c_a \mu_a}{\kappa_a}$$

$$Sc = \frac{\mu_a}{\rho_a D}, Le = \frac{v_a}{D}$$

3. Thermo-physical properties of coffee bean

In this section, we will show thermo-physical properties of the agricultural product that is used on simulation. Here, the kind of the agricultural products is coffee bean, then the parameters that are used entirely a physical parameter coffee.

Table 1. Thermo-physical properties of coffee bean.

Thermo-physical properties	Value	*Ref*
Density, ρ_p (m^2/s)	$78.845M + 1723$	[7]
The thermal conductivity, κ_t (W/mK)	$0.0116M + 0.062$	[7]
The thermal expansion coefficients, α $(1/K)$	$5x10^{-9}M + 6x10^{-8}$	[7]
The convection heat transfer coefficients, h_t $(W/m^2\,K)$	$\frac{3.26\rho_a c_a v_a}{Pr^{2/3}} Re^{-0.65}$	[10]
The surface mass transfer coefficient, h_m $(kg/m^2\,s)$	$\frac{h_t Pr^{2/3}}{\rho_a c_a Sc^{2/3}}$	[10]
The specific heat, c_t (J/kgK)	$\frac{\kappa_t}{\rho_p \alpha}$	[10]
The latent heat of vaporization of water, L (J/kg)	$2.3x10^6$	[7]

Table 2. Thermo-physical properties of air .

Thermo-physical properties	Value	*Ref*
The kinematic viscosity, v (m^2/s)	$16.97\,x\,10^{-6}$	[16]
The air velocity, v_a (m/s)	1	-
The specific heat, c_a (J/kgK)	$1.005\,x\,10^3$	[16]
The thermal conductivity, k_a (W/mK)	0.0271	[16]
The dynamic viscosity, μ (Ns/m^2)	$vx\mu$	[16]
The diffusion coefficient, D (m^2/s)	$2.17x10^{-5}\left(\frac{101.325}{P_a}\right)\left(\frac{T_\infty}{273.16}\right)^{1.88}$	[5]

4. Numerical methods

In simulation heat and mass transfer of coffee bean, we use finite element method with continuous piecewise linear (P1-elements) function to solve the governing equation (1 - 2). As for the software is already in use is FreeFEM++, which was created especially for the finite element method.

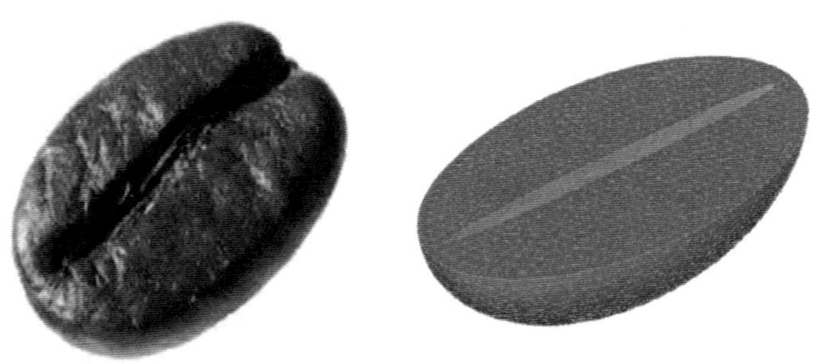

Figure 2. Grid profile of computational domain use GMSH software

To solve heat and mass transfer equation [1-2], we have to do a semi discretization in time. Let $T^k, M^k, \rho_p^k, c_t^k, \kappa_t^k, D^k$ be approximation of $T, M, \rho_p, c_t, \kappa_t, D$. we set $\Delta t > 0$ as a constant time step at

$t_k = k\Delta t\ (k = 0,1,2,\dots)$. The numerical solution of heat and mass transfer equation (T^k, M^k) are recomputed from (T^{k-1}, M^{k-1}) with the implicit scheme:

$$\rho^k c_t^k \left(\frac{T^k - T^{k-1}}{\tau_1}\right) = \nabla \cdot \left(\kappa_t^k \nabla T^k\right) + L\rho^k \left(\frac{M^k - M^{k-1}}{\tau_1}\right) \quad (11)$$

$$\rho^k c_t^k \left(\frac{T^k - T^{k-1}}{\tau_1}\right) = \nabla \cdot \left(\kappa_t^k \nabla T^k\right) + L\rho^k \left(\frac{M^k - M^{k-1}}{\tau_1}\right) \quad (12)$$

$$\kappa_t^k \frac{\partial T^k}{\partial n} = h_t\left(T_\infty - T^k\right) \quad \text{on } \Gamma_N \quad (13)$$

$$D^k \frac{\partial M^k}{\partial n} = h_m\left(M_\infty - M^k\right) \quad \text{on } \Gamma_N \quad (14)$$

$$\kappa_t^k \frac{\partial T^k}{\partial n} = 0, D^k \frac{\partial M^k}{\partial n} = 0 \quad \text{on } \Gamma_D \quad (15)$$

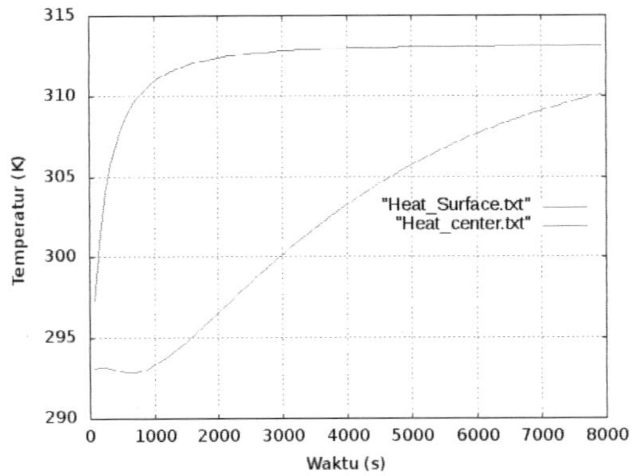

Figure 3. Predicted heat transfer between center and surface of coffee bean; pressure was kept at 20 kPa.

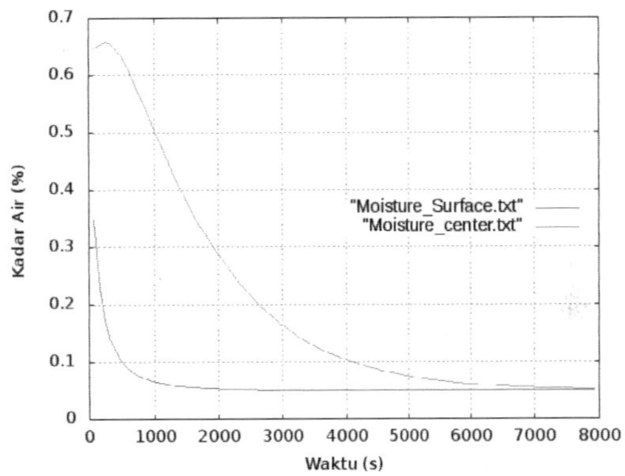

Figure 4. Predicted mass transfer between center and surface of coffee bean; pressure was kept at 20 kPa.

5. Numerical results and discussion

The effects of drying temperature and pressure on the freeze drying medium are shown in Figs. 3 and 4, in which drying curves (moisture ratio vs. time) and temperature evolution in the sample centre under different drying conditions are plotted.

According figure 3 - 4, show that there are significant different temperature and moisture content between surface and centre of coffee. In figure 3, the rate of temperature very fast in the surface of coffee. The fastest rate of temperature increase occurs between time interval 0 – 1000 s. On this state, the temperature is 311 K. In centre of coffee bean, no significant changes of temperature. The evolution temperature significantly increase in the centre of product after 1000 s. However, temperature increase slowly in the surface of product. Temperature increase should be offset by a decrease in the moisture content of the coffee bean. If we see picture 4, this show similarities. the drying rate of coffee surface is very fast. At t = 2000 s, moisture content is 0.05 % on surface. But in the same time, moisture content is 0.29 % on centre.

Physically, Drying phenomena occurs heat flow or transfer on coffee bean during time *t*. Beside that, thermo-physical parameters and pressure factor take effect on drying process. In this simulation of heat and mass transfer, time interval is 8000 s and air pressure is 20 kPa. The use of low air pressure can cause drying process more quickly.

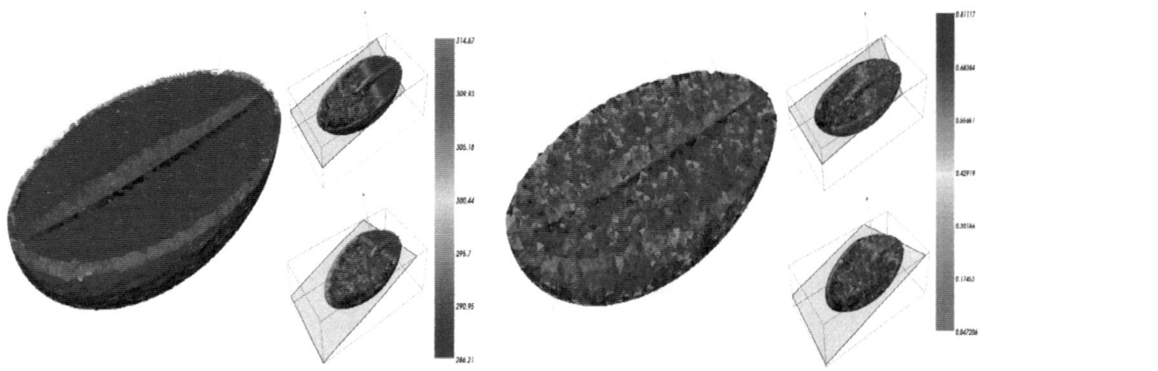

(a) (b)

Figure 5. Snapshot of distribution (a) temperature and (b) moisture content at time t = 40 s.

Qualitatively, the graph output is satisfies to illustrate drying process of coffee bean. However, display distribution profile of heat and mass transfer on drying process is better. On figure 5 – 6 shown distribution profile of temperature and moisture content inside coffee bean at time *t*. Distribution profile of temperature and moisture content at t = 40 s (see fig. 5a), temperature is 313 K in surface of coffee and 293 K in centre of coffee. In the same time, there is moisture content decrease, but the rate of decline was not significant. Moisture content in surface and centre of coffee is 0.04 % and 0.6 %, respectively. Moisture content distribution is almost uniform inside coffee bean at t = 8000 s, it is like that shown in the figure 6. As well as to the temperature distribution is almost same but it is not maximum in the centre of product. If we see figure 5 - 6, show that geometry of coffee bean is not smooth, this is because the number of triangle or grid slightly and also type of element is continuous piecewise linear

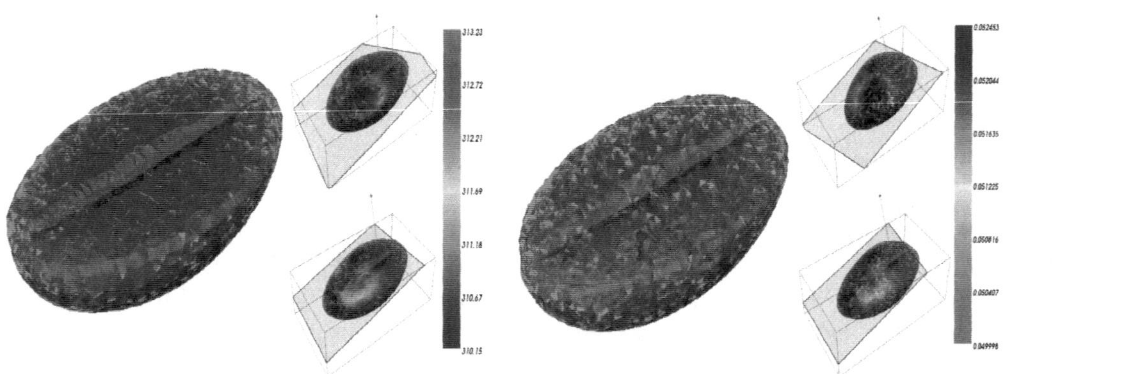

(a) (b)

Figure 6. Snapshot of distribution (a) temperature and (b) moisture content at time t = 8000 s.

6. Conclusion and future work

The finite element method (FEM) has been implemented to solve heat and mass transfer model in freeze dried medium. The method already successful to simulate drying phenomena. The method is also very easy to simulate phenomena on complex domain case. From this result also, we get that increase of temperature in the surface is very fast than in the centre of coffee. For drying process, in the surface is very vast that in the centre of coffee.

Acknowledgements
I would like to acknowledge the financial support in the form of *Bantuan Kerja Mahasiswa* (BKM) from Institute of Technology Bandung (ITB) so that the research have done. I also express appreciations to the entire FreeFEM++ authors; Frèdèric Hecht, Sylvian Auliac, Oliver Pironneau, Jasques Morice, Antoine Le Hyaric and Kohji Ohtsuka, for their help when first use this program.

References

[1] Anonim 2015 *Rencana Strategis Kementrian Pertanian Tahun 2015-2019 24-25.* (Jakarta: Kementerian Pertanian)
[2] B. Dradjat, A. Agustian and A. Supriatna 2007 *Ekspor dan daya saing kopi biji Indonesia di pasar internasional: implikasi strategis bagi pengembangan biji kopi organis* (Pelita Perkebunan) p 159-179.
[3] K. Haghighi, and L.J. Segerlind 1988 *Modeling simultaneous heat and mass transfer in an isotropic sphere-a finite element approach* American Society of Agricultural Engineers Vol. 31 p 629-637.
[4] Sh. Rafiee, M. Kashaninejad, A.R. Keyhani and A. Jafari 2009 *Pistachio nut (ohadi variety) mass transfer simulation during process of drying using finite element method* J. Agriculture Science Technology Vol. 11 p 137-146.
[5] F. Nadi, G.H. Rahimi, R. Younsi, T. Tavakoli, and Z. Hamidi Esfahani 2012 *Numerical Simulation of Vacuum Drying by Luikov's Equation* (Drying Technology Taylor & Francis) p 197-206
[6] Allan L. Phillips 1965 *Drying coffee with solar-heated air* Solar Energy Vol. 9 Issue 4 p 213-216
[7] Angelo Fabbri, et all 2011 *Numerical Modeling of Heat and Mass Transfer During Coffee Roasting Process}* Journal of Food Engineering Vol. 205 p 264-269
[8] Jafar F and Farid M *Analysis of heat and mass transfer in freeze-drying* Drying Technology Vol. 21 p 249-263
[9] Haiqing Chen, et all 1999 *Modeling Coupled Heat and Mass Transfer for Convection Cooking of Chicken Patties* J. of Food Engineering Vol. 24 p 139-146
[10] Hector J. Ciro-Velasquez, Luis C. Abud-Cano and Luis. R. Perez-Alegria 2010 *Numerical Simulation of Thin Layer Coffee Drying by Control Volumes* Dyna Vol. 163 p 270-278
[11] Husain A, Sun C.C and Clayton J.T 1973 *Simultaneous Heat and Mass Diffusion in Biological Materials* Journal Agricultural Engineering Research Vol. 18 p 343-354
[12] Luikov, A.V 1966 *{Heat and mass transfer in capillary-porous bodies* (New York: Pregamon Press
[13] A. Kovács, E. Lakatos, G. Milics dan M. Neményi 2010 *Finite Element Modeling of Coupled Heat and Mass Transfer of a Single Maize Kernel Based on Water Potential Using COMSOL Multiphysics Simulation* The Proceedings of The COMSOL Conference
[14] Miklos Nemenyi, et all 2000. *Investigation of Simultaneous Heat and Mass Transfer within The Maize Kernels During Drying* J. Computers and Electronics in Agriculture p 123-135.
[15] Michael J. Moran and Howard N. Shapiro 2004 *Termodinamika Teknik, Edisi 4* (Jakarta: Erlangga) p 229
[16] http://www.engineeringtoolbox.com/air-properties-d_156.html

Understanding Passive Layer Formation for Further Corrosion Management in Gas Production Pipes

R. K. Santoso[1,2], **S. D. Rahmawati**[1], **A. Gadesa**[1], **D. Wahyuningrum**[2]

[1]Petroleum Engineering Study Program, Institut Teknologi Bandung, Indonesia
[2]Chemistry Study Program, Institut Teknologi Bandung, Indonesia

[2]Corresponding author: ryan.kurniawan@students.itb.ac.id

Abstract. Corrosion is a critical issue during the development of a gas field, especially wet gas or retrograde gas field. Corrosion affects the management system of a field and further impacts the amount of investment. Therefore, accurate prediction of corrosion rate is needed to plan an effective preventive action before going further to the development phase. One of the important parameters that should be noticed to create an accurate prediction is the formation of the passive layer. In CO_2-H_2S environment, there will be three possibilities of passive layer: FeS, $FeCO_3$ or no passive layer. In this study, we create mathematical models to determine the formed passive layer in each segment of the gas production tubing and pipeline. The model is built using Faraday's Law and Thermodynamic approach to account the passive layer formation at different temperature, pH, corrosion rate and partial pressure of CO_2 and H_2S. From the simulation, it was found that there were three boundary conditions: no scale-FeS boundary, no scale-$FeCO_3$ boundary and FeS-$FeCO_3$ boundary. The first two boundaries evolved over a time as the concentration of Fe^{2+} ions was increasing. However, FeS-$FeCO_3$ boundary remained steady as it was not affected by the addition of Fe^{2+} ions. Using sample case study, few variations were noticed at production pipeline and tubing. It was caused by the gas composition, which contained high CO_2 and very low H_2S. Boundary conditions only changed slightly over two days period.

1. Introduction

Corrosion is defined as the destructive result of chemical reaction between a metal or metal alloy and its environment [18]. It occurs when a metal or metal alloy meets corrosive agents in aqueous solution at certain pressure and temperature. In gas field, corrosion is usually caused by CO_2 gas (sweet corrosion) and H_2S gas (sour corrosion). The existence of corrosion in production system could affect the amount of production and lead to safety issues. Therefore, early prevention of corrosion is needed before bringing a gas field into production.

In order to plan an effective corrosion management, one should first understand the fluid characteristic and environment condition (ambient temperature, location, etc) of a gas field. Moreover, an accurate prediction of future corrosion phenomenon is also critical. It can be assessed by creating a model that describes the corrosion phenomenon of a field accurately. Many corrosion models have been published so far by researchers [2] [12] [27] [35] [62] [65] [68] [72] and well-known companies to estimate the future corrosion rate in a production system. From all those models, a critical parameter that influences the assumptions inside the models and determines the success of the prediction is the type of passive layer.

IOP Conf. Series: Journal of Physics: Conf. Series **877** (2017) 012062 doi:10.1088/1742-6596/877/1/012062

Passive layer is known as the corrosion reaction product, formed on the surface of a metal or metal alloy where corrosion reaction occurs. The existence of passive layer could be helpful to protect the metal from further corrosion or harmful because it can block the gas flow. The type of passive layer is determined by the gas composition and thermodynamic condition (pressure and temperature) of production system. As the type of passive layer in a system determines the development of any corrosion models and accuracy in prediction of the future corrosion phenomenon, it is critical to have a mathematical model that can predict the type of passive layer in a gas production system.

As a part of our big research in creating a new corrosion-erosion model that is applicable in any environmental conditions of gas fields, we first developed a mathematical model to predict the type of passive layer using Faraday's law and thermodynamics approach. This model could simulate the formed passive layer based on the partial pressure of CO_2 and H_2S in gas production system. The model was applicable in vertical, incline and horizontal flow condition. We also used sample case study from to perform our model calculations.

2. Mathematical Model

The prediction model is built using Faraday's law and thermodynamic approach. Faraday's law is used to estimate the amount of Fe^{2+} ion in half-cell reaction. Thermodynamic approach is utilized then as the main foundation to create the model. The model is divided into three parts: no scale-mackinawite (FeS) boundary equation, no scale-siderite ($FeCO_3$) boundary equation and mackinawite-siderite boundary equation. We also assume that

- corrosion occurs uniformly in a specific segment of production pipe
- water is uniformly distributed in a specific segment of production pipe
- reaction has reached its equilibrium condition at certain pressure and temperature

2.1. No Scale-Mackinawite Boundary

In H_2S environment, there is a possible formation of mackinawite scale on the surface of the metal or metal alloy. This scale can be formed if the solubility limit of mackinawite is surpassed. The solubility of mackinawite can be expressed as

$$Fe^{2+}{}_{(aq)} + S^{2-}{}_{(aq)} \Leftrightarrow FeS_{(s)} \quad (1)$$

Therefore, the solubility limit can be represented using solubility product constant of mackinawite

$$Ksp_{FeS} = \left[Fe^{2+}\right]\left[S^{2-}\right] \quad (2)$$

The amount of Fe^{2+} ion in solution is related to the amount of metal loss (corrosion rate). Thus, $[Fe^{2+}]$ can be estimated using

$$\left[Fe^{2+}\right] = \frac{CR \times A \times \rho_{metal} \times \Delta t}{MW_{metal} \times HU \times V_{pipe}} \quad (3)$$

where

CR : corrosion rate, L T^{-1}
A : inner pipe area, L^2
ρ_{metal} : metal density, M L^{-3}
MW_{metal} : molecular weight of metal
HU : liquid hold up, fraction
V_{pipe} : pipe inner volume, L^3
Δt : exposure time, T

Based on ASTM G102 [1], in half-cell reaction, molecular weight of a metal is determined using (equivalent molecular weight)

$$MW_{metal} = \frac{1}{\sum_{1}^{n} \frac{f_i}{\left(a_i / n_i \right)}} \qquad \textbf{(4)}$$

where

f_i : mass fraction of an atom, fraction
a_i : molecular weight of an atom
n_i : valence

The boundary condition to differentiate the area of no-scale and mackinawite can be calculated using

$$Ksp_{FeS} = \frac{\left(\frac{CR \times A \times \rho_{metal} \times \Delta t}{MW_{metal} \times HU \times V_{pipe}} \right) \times k_{1H_2S} \times k_{2H_2S} \times k_{3H_2S} \times P_{H_2S}}{\left[H^+\right]_{(aq)}^2} \qquad \textbf{(5)}$$

The value of $K_{1\,H_2S}$ can be estimated using several empirical correlations: Weiss, 1970; Willhelm et al., 1977; Roberts, 1985; Carroll & Mather, 1989; Suleimenov & Krupp, 1994; De Bruyn et al., 1995; Rinker & Sandall, 2000; Nordsveen et al., 2003; Fernández-Prini et al., 2003; and Sander et al., 2011. The value of $K_{2\,H_2S}$ can be estimated using several empirical correlations: Wright & Maass, 1932; Ringborn, 1953; Pohl, 1962; Ellis & Golding, 1967; Tsonopoulos et al., 1976; Rao & Hepler, 1977; Broderius et al., 1977; Millero, 1986; Kharaka et al., 1989; and Suleimenov & Seward, 1997. The value of $K_{3\,H_2S}$ can be estimated using several empirical correlations: Ringborn, 1953; Maronny, 1959; Muhammad & Sundarahm, 1961; Pohl, 1962; Kolthoff, 1969; Giggenbach, 1971; Ellis & Giggenbach, 1971; Stephens and Cobble, 1971; Kharaka et al., 1989; and Flaschka et al., 1991. The value of Ksp_{FeS} can also be estimated using several empirical equations: Berner, 1967; Doyle, 1968; Tewari et al., 1978; Morse et al., 1987; Davison, 1991; Bagander & Carman, 1994; Davison et al., 1999; Bening et al., 2000; Rickard, 2006; and Sun et al., 2008.

2.2. No Scale-Siderite Boundary

In CO_2 environment, siderite will be formed as the passive layer. The solubility of siderite can be expressed as

$$Fe^{2+}{}_{(aq)} + CO_3{}^{2-}{}_{(aq)} \Leftrightarrow FeCO_{3(s)} \qquad \textbf{(6)}$$

Therefore, the solubility limit can be mathematically expressed as

$$Ksp_{FeCO_3} = \left[Fe^{2+}\right]\left[CO_3{}^{2-}\right] \qquad \textbf{(7)}$$

The correlation to identify the area of no-scale and siderite can be expressed mathematically as

$$Ksp_{FeCO_3} = \frac{\left(\frac{CR \times A \times \rho_{metal} \times \Delta t}{MW_{metal} \times HU \times V_{pipe}} \right) \times k_{1CO_2} \times k_{2CO_2} \times k_{3CO_2} \times P_{CO_2}}{\left[H^+\right]_{(aq)}^2} \qquad \textbf{(8)}$$

The value of $K_{1\,CO_2}$ can be estimated using several empirical correlations: Bohr, 1899; Weiss, 1974; Wilhelm et al., 1977; Edwards et al., 1978; Oddo & Tomson, 1982; Yoo et al., 1986; Crovetto, 1991; Zheng et al., 1997; Fernández-Prini et al., 2003; and Sander et al., 2011. The value of $K_{2\,CO_2}$ can be estimated using several empirical correlations: Hansson, 1973; Mehrbach, 1973; Millero, 1979; Plath et al., 1980; Oddo & Tomson, 1982; Johansson & Wedborg, 1982; Dickson & Millero, 1987; Goyet & Poisson, 1989; Roy et al., 1993; and Prieto & Millero, 2002. The value of $K_{3\,CO_2}$ can be estimated using several empirical correlations: Barnes et al., 1966; Hansson et al., 1973; Mehrbach, 1973; Millero, 1979; Plath et al., 1980; Oddo & Tomson, 1982; Johansson & Wedborg, 1982; Dickson & Millero, 1987; Goyet & Poisson, 1989; and Roy et al., 1993. The value of Ksp_{FeCO_3} can also be estimated using several empirical equations: Smith, 1918; Latimer, 1952; Singer & Stumm, 1970; Reiterer et al., 1981; Ptacek & Blowes, 1981; Draun, 1991; Greenberg & Tomson, 1992; Bruno et al., 1992; Ptacek & Reardon, 1993; and Benezeth et al., 2009.

2.3. *Mackinawite-Siderite Boundary*

The solubility limit for mackinawite and siderite is simply expressed as the ratio of Ksp_{FeS} and Ksp_{FeCO_3} [52].

$$Ksp_{FeS/FeCO_3} = \frac{k_{1H_2S} \times k_{2H_2S} \times k_{3H_2S} \times P_{H_2S}}{k_{1CO_2} \times k_{2CO_2} \times k_{3CO_2} \times P_{CO_2}} \tag{9}$$

3. Model Implementation

The model is used to predict the type of passive layer in gas pipes through hydrogen sulfide partial pressure vs carbon dioxide partial pressure in log-log plot [52]. The execution step for the model follows the algorithm in **Figure 1**.

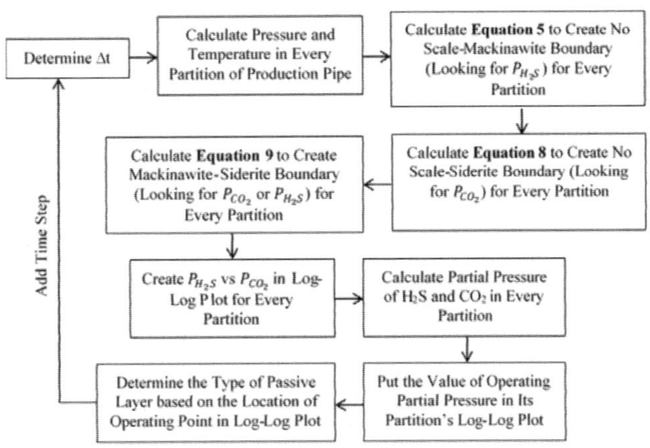

Figure 1. Algorithm for Execution

The critical parameters during the model's execution are pressure, temperature, composition and pH changing in each partition of the pipe. One should measure those parameters accurately to obtain good predictions.

4. Case Study

To perform the model's simulation, we used an example of wet gas field composition. Its fluid is characterized by the significant amount of CO_2. The detail fluid's composition of the case study can be seen at Table 1. The production data is shown at Table 2.

Table 1. Gas Composition Data

	Composition	Mole (%)
1	Hydrogen Sulfide (H_2S)	0.05
2	Carbon Dioxide (CO_2)	31.67
3	Nitrogen (N_2)	0.33
4	Methane (C_1)	56.92
5	Ethane (C_2)	0.11
6	Propane (C_3)	0.03
7	Water (H_2O)	10.90

Table 2. Production Data

	Parameter	**Value**	**Unit**
		Production Tubing	
1	Reservoir Pressure	1100	psig
2	Reservoir Temperature	325	F
3	Tubing Length	8152.9	ft
4	Tubing Internal Diameter	6.625	inch
5	Inclination (from Vertical)	0	degree
6	Wellhead Pressure	500	psig
7	Wellhead Temperature	255.3	F
8	Ambient Temperature	85	F
9	Gas Rate	57.13	MMSCFD
10	Tubing Material (Assumed)	API 5L-X 65	-
		Production Flowline	
1	Wellhead Pressure	500	psig
2	Wellhead Temperature	255.3	F
3	Ambient Temperature	85	F
4	Flowline Length	1875.4	ft
5	Flowline Outer Diameter	8	inch
6	Inclination (from Horizontal)	0	degree
7	Outlet Pressure	402.28	psig
8	Outlet Temperature	249.0	F
9	Gas Rate	57.13	MMSCFD
10	Tubing Material (Assumed)	API 5L-X 65	-

In this study, we assumed steady rate for two days period. We generated the pressure and temperature along tubing and flowline using commercial software with Beggs-Brill correlation for vertical and horizontal flow and black-oil model. The corrosion rate value was also assumed to be steady for two days, even there was passive layer on the surface of the metal. The corrosion rate was calculated using deWaard (1995) model. In the proposed model, we calculated all constant using the newest empirical correlation, for instance to calculate the value of $K_{1\,H_2S}$ we used Sander et al. (2011).

5. Results and Discussions

From the simulation using commercial software, the pressure and temperature profile in production tubing and flowline are shown at Figure 2.

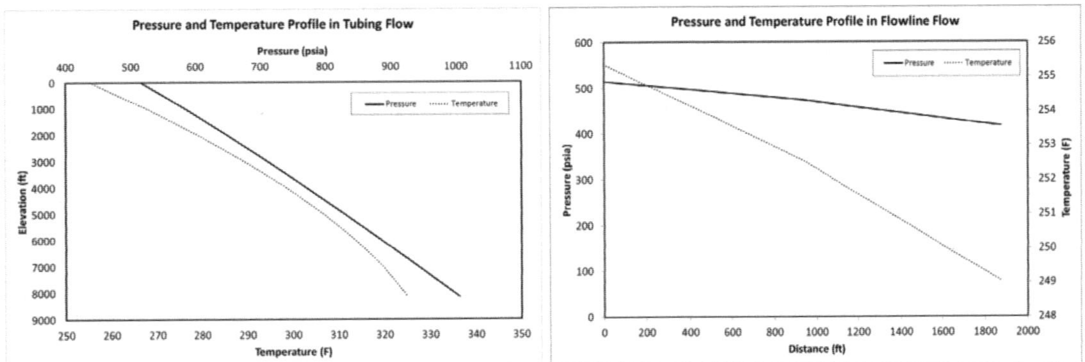

(a) (b)

Figure 2. Pressure and Temperature Profile in: Tubing Flow (**a**) and Flowline Flow (**b**)

Based on the environment condition in tubing and flowline, the corrosion rate can be seen at Figure 3.

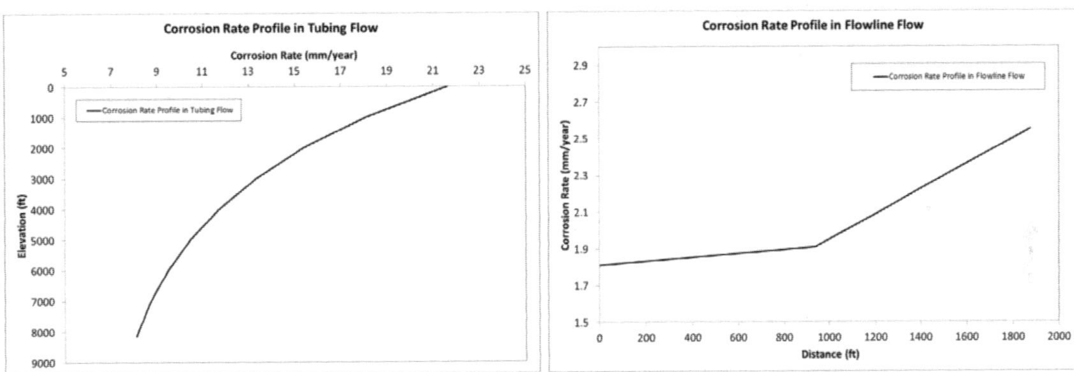

(a) (b)

Figure 3. Tubing's Corrosion Rate Profile (**a**) and Flowline's Corrosion Rate Profile (**b**)

Corrosion rate occurs worst at tubing because it has bigger pressure and temperature than in flowline. Based on the deWaard prediction, corrosion rate has its biggest value at the outlet location of the pipes.

Using the input of corrosion rate and environmental condition, the proposed model can be executed. It can be recognized that there are three boundaries, so that there are three regions [52]: no-scale, mackinawite and siderite region. The arrangement of the boundaries does not change at various conditions, only the value of the boundaries will be different at different conditions. For boundary profile in tubing and flowline, it can be seen at Figure 4.

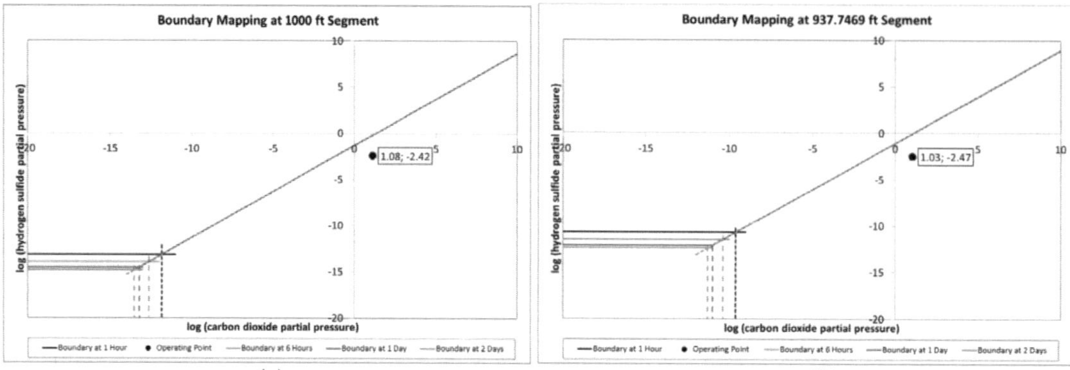

(a) (b)

Figure 4. Boundary Profile in Tubing at 5,000 ft Depth **(a)** and Boundary Profile in Flowline at 937.7469 ft Length **(b)**

At each segment, no scale-mackinawite and no scale-siderite boundaries change over time. As time goes longer, the value of those boundaries become smaller. It is caused by the addition of Fe^{2+} ion in the solution (In Equation 5 and 8, P_{H_2S} and P_{CO_2} are reciprocal with the amount of Fe^{2+} ion) . However, mackinawite-siderite boundary remains constant since it does not depend on the amount of Fe^{2+} ion.

It can also be seen that at Figure 4, the no scale-mackinawite and no scale-siderite boundary lie on negative side of log-log plot. It means that no scale condition could only occur at low operating pressure or low composition of H_2S and CO_2. Since the operating pressure of the pipe is relatively high, there is no doubt that scale is formed. Moreover, the existence of big amount of CO_2 makes it sure that siderite layer is formed.

When the whole segments at Figure 4 are compared, it can be seen that there is only small change in boundary and operating point values. It is caused by the significant amount of CO_2 in the stream, for about 30%. Therefore, it can be guaranteed that only siderite layer will exist over time in all segments of the production pipes.

Finally, for the case study, we can say that to calculate the corrosion rate, one should involve the existence of siderite passive layer. It is also needed to be noticed that siderite build-up is possible to occur in the future.

6. Conclusion

From the discussion above, it can be concluded that a new boundary model to determine the type of passive layer that is formed inside the gas production pipes has been successfully developed. The model is built using Faraday's law and thermodynamic approach.

The novelty of this study is the easy prediction of type of passive layer using the model. By only plugging in the working pressure and temperature at certain position to the boundary map, one can easily know the type of the passive layer formed at certain time.

In the model, no scale-mackinawite and no scale-siderite boundary are influenced by pressure, temperature, pH, partial pressure of carbon dioxide or hydrogen sulfide and concentration of Fe^{2+} ion (corrosion rate value). Therefore, these boundaries will evolve over time. However, mackinawite-siderite boundary only depends on pressure, temperature and partial pressure of carbon dioxide and hydrogen sulphide. Thus, this boundary will not change over time.

From the case study, it can be concluded that there will be only siderite passive layer formed on the inner surface of the pipes. It is caused by the huge amount of carbon dioxide in the flow.

Acknowledgement
This research was funded by DIKTI 2016.

References

[1] 1989 ASTM G 102 (Philadelphia: American Society for Testing and Materials) p 416-422

[2] Andrzej Anderko and Robert D. Young 1999 *Corrosion 99* No. 31 (Houston: NACE International) p 1-19

[3] Bågander, L.E.; and Carman, R. In-situ determination of the apparent solubility product of amorphous iron sulphide. *Applied Geochemistry*, 1994, **9**, 379–386.

[4] Barnes, H. L.; Helgeson, H. C.; and Ellis, A. J. Handbook of Physical Constants-Revised Edition. The Geological Society of America Memoir 97, 1966.

[5] Benning, L. G.; Wilkin, R. T.; nad Barnes, H. L. Reaction pathways in the Fe-S below 100 °C. *Chem. Geol.* 2000, **167**, 25-51.

[6] Berner, R. A. Thermodynamic stability of sedimentary iron sulfides. *Am. J. Sci.* 1967, **265**, 773-785.

[7] Bénézeth, P.; Dandurand, J. L.; and Harrichoury, J. C. Solubility product of siderite ($FeCO_3$) as a function of temperature (25–250 °C). *Chemical Geology Volume* **265**, Issues 1–2, 15 July 2009, Pages 3–12.

[8] Bohr, C. Definition und Methode zur Bestimmung der Invasionsund Evasionscoefficienten bei der Auflösung von Gasen in Flüssigkeiten. Werthe der genannten Constanten sowie der Absorptionscoefficienten der Kohlensäure bei Auflösung in Wasser und in Chlornatriumlösungen, *Wied. Ann.*, 1899. **68**, 500–525.

[9] Braun, R.D. Solubility of iron(II) carbonate at temperatures between 30 and 80°. *Talanta* **38**, 1991, 205–211.

[10] Bruno, J.; Wersin, P.; and Stumm, W. On the influence of carbonate in mineral dissolution: II. The solubility of $FeCO_3$(s) at 25 °C and 1 atm total pressure. *Geochim. Cosmochim. Acta.* 1992, **56**, 1149–1155.

[11] Broderius, S. J.; and Smith Jr., L. L. Direct Determination and Calculation of Aqueous Hydrogen Sulfide. Department of Entomology, Fisheries, and Wildlife: St. Paul, Minnesota, 1977.

[12] Carlos A. Palacios, Corpoven S. A., Valoy Chaudary 1996 *Fourth Latin American and Caribbean Petroleum Engineering Conference* No. SPE 36127 (Port of Spain: Society of Petroleum Engineers, Inc.) p 501-511

[13] Carroll, J.; and Mather, A. E. The solubility of hydrogen sulfide in water from 0 to 90 °C and pressures to 1 MPa. Geochim. Cosmochim. Acta 1989, 53, 1163-1170.

[14] Crovetto, R. Evaluation of solubility data for the system CO_2-H_2O from 273 K to the critical point of water, *J. Phys. Chem. Ref. Data*, 1991, **20**, 575–589.

[15] Davison, W. The solubility of iron sulfides in synthetic and natural waters at ambient temperature. *Aquatic Sciences*, 1991, **53**, 309–329.

[16] Davison, W.; Phillips, N.; and Tabner, B.J. Soluble iron sulfide species in natural waters: reappraisal of their stoichiometry and stability constants. *Aquatic Sci.*, 1999, **61**, 23–43.

[17] De Bruyn, W. J.; Swartz, E.; Hu, J. H.; Shorter, J. A.; Davidovits, P.; Worsnop, D. R.; Zahniser, M. S.; and Kolb, C. E. Henry's law solubilities and ´Setchenow coefficients for biogenic reduced sulfur species obtained from gas-liquid uptake measurements, J. Geophys. Res., 1995b, 100D, 7245–7251.

[18] Denny A. Jones 1996 *Principles and Prevention of Corrosion* 2nd edition (New Jersey: Prentice-Hall, Inc.) p 5

[19] Dickson, A.G.; and Millero, F.J. A comparison of the equilibrium constants for the dissociation of carbonic acid in seawater media. *Deep-Sea Res.*, 1987, **34**, 1733-1743.

[20] Doyle, R. W. Identification and solubility of iron sulfide in anaerobic lake sediment. *Am. J. Sci.* 1968, **266**, 980-984.

[21] Edwards, T. J.; Maurer, G.; Newman, J.; and Prausnitz, J. M. Vapor liquid equilibria in multicomponent aqueous solutions of volatile weak electrolytes, *AIChE J.*, 1978, **24**, 966–976.

[22] Ellis, A. J.; and Golding, R. M. Spectrophotometric determination of acid dissociation constant of hydrogen sulfide. J. Chem. Soc. 1959, 127-130.

[23] Ellis, A. J.; and Giggenbach, W. Hydrogen sulfide ionization and sulfur hydrolysis in high temperature solution. Geochim Cosmochim. Acta 1971, 35, 247-260.

[24] Fernández-Prini, R.; Alvarez, J. L.; and Harvey, A. H. Henry's constants and vapor-liquid distribution constants for gaseous solutes in H2O and D2O at high temperatures, J. Phys. Chem. Ref. Data, 2003, 32, 903–916.

[25] Flaschka, H. A.; Barnard, A. J.; and Jr. Starrock, P. E. Quantitative Analytical Chemistry, 2nd Edition; Willard Grant Press: Boston, MA, 1980.

[26] Greenberg, J. and Tomson, M. Precipitation and dissolution kinetics and equilibria of aqueous ferrous carbonate vs temperature. Appl. Geochem. 1992, **7**, 185–190.

[27] G. H. Al-Aithan, F. M. Al-Mutahar, J. R. Shadley, S. A. Shirazi, E. F. Rybicki and K. P. Roberts 2014

Corrosion 2014 No. 3854 (San Antonio: NACE International) p 1-15

[28] Giggenbach, W. Optical spectra of highly alkaline sulfide solutions and the second dissociation constant of hydrogen sulfide. Inorg. Chem. 1971, 10, 1333-1338.

[29] Goyet, C.; and Poisson, A. New determination of carbonic acid dissociation constants in seawater as a function of temperature and salinity. *Deep-Sea Res.*, 1989, **36**, 1635-1654.

[30] Hansson, I. A new set of acidity constants for carbonic acid and boric acid in seawater. *Deep-Sea Res.*, 1973, **20**, 461 478.

[31] Hogfeldt, E.; IUPAC. Stability Constants of Metal-Ion Complexes, Part A: Inorganic Ligands; Pergamon Press: Oxford, U.K., 1982.

[32] Jensen, D. L.; Boddum, J. K.; Tjell, J. C.; and Christensen, T. H. The solubility of rhodochrosite ($MnCO_3$) and siderite ($FeCO_3$) in anaerobic aquatic environments. *Applied Geochemistry* **17** (2002) 503–511.

[33] Johansson, O.; and Wedborg, M. On the evaluation of potentiometric titrations of seawater with hydrochloric acid. *Oceanologica Acta*, 1982, **5**, 209-218.

[34] Kharaka, Y. K.; Perkins, E. H.; Gunter, W. D.; Debral, J. D.; and Bamford, C. H. Solmineq 88: A Computer Program for Geochemical Modeling of Water Rock Interactions; Alberta Research Council: Menlo Park, CA, 1989.

[35] Kolthoff, I. M.; Sandell, E. B.; Meehan, E. J.; and Bruckenstein, S. Quantitative Chemical Analysis, 4th Edition; Macmillan: London, 1969.

[36] Latimer, W.M. The oxidation states of the elements and their potentials in aqueous solutions. *Prentice Hall Chemistry Series*. 1952. Prentice Hall, Englewood Cliffs, NJ.

[37] Maronny, G. Constants de dissociation de l'hydrogen sulfure. Electrochim. Acta 1959, 1, 58-69.

[38] Mehrbach, C. Measurement of the apparent dissociation constants of carbonic acid in seawater at atmospheric pressure. *Limnol. Oceanogr.*, 1973, **18**, 897-907.

[39] Millero, F.J. The thermodynamics of the carbonic acid system in seawater. *Geochim. Cosmochim. Acta*, 1979, **43**, 1651-1661.

[40] Millero, F. J. The thermodynamics and kinetics of the hydrogen sulfide system in natural waters. Mar. Chem. 1986, **18**, 121-147.

[41] Muhammad, S. S.; and Sundarahm, E. V. The spectrophotometric determination of the dissociation constants of hydrogen sulfide. J. Aci. Ind. Res. 1961, sect. B 20, 16-18.

[42] Morse, J. W.; Millero, F. J.; Cornwell, J. C.; and Richard, D. The chemistry of the hydrogen sulfide and iron sulfide systems in natural waters. *Earth Sci. Rev*. 1987, **24**, 1-42.

[43] Nordsveen, M.; Nesic, S.; Nyborg, R.; and Stangeland, A. A mechanistic model for carbon dioxide corrosion of mild steel in the presence of protective iron carbonate films Part 1: theory and verification. Corrosion 2003, 59, 443-457.

[44] Oddo, J. E.; and Tomson, M.B. Simplified Calculation of $CaCO_3$ Saturation at High Temperatures and Pressure in Brine Solutions, *J. of Petroleum Technology*, 1982, pp.1583-1590.

[45] Plath, D. C.; Johnson, K. S.; and Pytkowicz, R. M. The solubility of calcite--probably containing magnesium--in seawater. *Marine Chemistry*, 1980, **10**, 9-29.

[46] Pohl, H. A. Solubility of iron sulfides. J. Chem. Eng. Data 1962, 7, 295-306.

[47] Prieto, F. J. M.; and Millero, F. J. The values of $pK_1 + pK_2$ for the dissociation of carbonic acid in seawater. *Geochimica et Cosmochimica Acta Vol. 66*, 2002, **No. 14**, pp. 2529–2540.

[48] Ptacek, C.J.; and Blowes, D.W. Influence of Siderite on the pore-water chemistry of inactive mine-tailings impoundments. In: Alpers, C.N., Blowes, D.W. (Eds.), *Environmental Geochemistry of Sulfide Oxidation*. 1994, American Chemical Society, Washington, DC, pp. 172–189. (Chapter 13).

[49] Ptacek, C.J.; and Reardon, E.J. Solubility of siderite ($FeCO_3$) in concentrated NaCl and Na_2SO_4 solutions at 25 °C. *Water-Rock Interaction*, 1993, 181–183.

[50] Rao, S. R.; and Hepler, L. G. Equilibrium constants and thermodynamics of ionization of aqueous hydrogen sulfide. Elsevier Scientific Publishing Company: Amsterdam, 1977, 293-299.

[51] Reiterer, F.; Johannes, W.; and Gamsjager, H. V. Semimicro determination of solubility constants: copper(II) carbonate and iron(II) carbonate. *Mikrochim. Acta.*, 1981, **1**, 63–72.

[52] Richard Woollam, Kavitha Tummala, Jose Vera and Sandra Hernandez 2011 *Corrosion 2011* No. 11076 (Houston: NACE International) p 1-14

[53] Rickard, D. The solubility of FeS. *Geochimica et Cosmochimica Acta,* 2006, **70**, 5779–5789.

[54] Ringborn, A. Solubility of Sulfides, A Report to the Analytical Section of IUPAC, 1953.

[55] Rinker, E. B.; and Sandall, O. C. Physical solubility of hydrogen sulfide in several aqueous solvents, Can. J. Chem. Eng., 2000, 78, 232–236.

[56] Roberts, B. E. Vapor liquid equilibrium calculations for dilute aqueous solutions of CO2, H2S, NH3 and NaOH to 300 °C. Can. J. Chem. Eng. 1985, 63, 294-300.

[57] R. K. Santoso, I. Fauzi and S. D. Rahmawati 2015 *SPE Asia Pacific Oil and Gas Conference and Exhibition* No. SPE 176192 (Bali: Society of Petroleum Engineers, Inc.) p 1-17

[58] Roy, R. N.; Roy, L. N.; Vogel, K. M.; Porter-Moore, C.; Pearson, T.; Good, C. E.; Millero, F. J.; and Campbell, D. M. The dissociation constants of carbonic acid in seawater at salinities 5 to 45 and temperatures 0 to 45°C. *Mar. Chem.* 1993, **44**, 249–267.

[59] Sander, S. P.; Abbatt, J.; Barker, J. R.; Burkholder, J. B.; Friedl, R. R.; Golden, D. M.; Huie, R. E.; Kolb, C. E.; Kurylo, M. J.; Moortgat, G. K.; Orkin, V. L.; and Wine, P. H. Chemical Kinetics and Photochemical Data for Use in Atmospheric Studies, Evaluation No. 17, JPL Publication 10-6, 2011, Jet Propulsion Laboratory, Pasadena, available at: http://jpldataeval.jpl.nasa.gov (last access: 22 June 2016).

[60] Sander, R. Compilation of Henry's law constants (version 4.0) for water as solvent. Atmos. Chem. Phys., 2015, 15, 4399–4981.

[61] Singer, P.C.; and Stumm, W. The solubility of ferrous iron in carbonate-bearing waters. *J. Am. Water Works Assoc.* 1970, **62**, 198–202.

[62] Smith, H.J. Equilibrium in the system: ferrous carbonate, carbon dioxide and water. *J. Am. Chem. Soc.* 1918, **40**, 879–883.

[63] S. Nesic, S. Wang, H. Fang, W. Sun and K. K-L. Lee 2008 *Corrosion 2008 Conference and Expo* No. 08535 (Houston: NACE International) p 1-16

[64] Suleimenov, O. M.; and Krupp, R. E. Solubility of hydrogen sulfide in pure water and in NaCl solutions, from 20 to 320 °C and at saturation pressures. Geochim. Cosmochim. Acta 1994, 58, 2433-2444.

[65] Suleimenov, O. M.; and Seward, T. M. A spectrophotometric study of hydrogen sulfide ionization in aqueous solutions to 350 °C. Geochim. Cosmochim. Acta 1997, 61, 5187-5198.

[66] Sun, W.; Nešić, S.; Young, D.; and Woollam, R. C. Equilibrium Expressions Related to the Solubility of the Sour Corrosion Product Mackinawite. *Ind. Eng. Chem. Res.* 2008, **47**, 1738-1742.

[67] Stephens, H.P.; and Cobble, J.W. Thermodynamic properties of the aqueous sulfide and bisulfide ions and the second ionization constant of hydrogen sulfide over extended temperatures. Inorg. Chem., 1971, 10: 619--625.

[68] Tewari, P.H.; Wallace, G.; and Campbell, A.B. The solubility of iron sulphides and their role in mass transport in Girdler- Sulphide heavy water plants. *Report of the Atomic Energy Canada AECL-5960*, 1978, 1–34.

[69] T. Shibata 1996 *Corrosion* November 1996 813

[70] Tsonopoulos, C.; Coulson, D. M.; and Inman, L. B. Ionization Constants of Water Pollutants. J. Chem. Eng. Data, 1976, 21 (2), pp 190–193

[71] Weiss, R. F. Carbon dioxide in water and seawater: The solubility of a non-ideal gas, *Mar. Chem.*, 1974, **2**, 203–215.

[72] Weiss, R. F. The solubility of nitrogen, oxygen and argon in water and seawater. Deep Sea Res. 1970, 17, 721-735.

[73] Wei Sun and Srdjan Nesic 2007 *Corrosion 2007 Conference and Expo* No. 07655 (Houston: NACE International) p 1-26

[74] Wilhelm, E.; Battino, R.; and Wilcock, R. J. Low-pressure solubility of gases in liquid water, Chem. Rev. 1977, 77, 219–262.

[75] Wright, R. H.; and Maass, O. The solubility of hydrogen sulfide in water from the vapour pressures of the solutions. Can. J. Res. 1932, 6, 94-101.

[76] Yoo, K. P.; Lee, S. Y.; and Lee, W. H. Ionization and Henry's law constants for volatile, weak electrolyte water pollutants, *Korean J. Chem. Eng.*, 1986, **3**, 67–72.

[77] Zheng, D. Q.; Guo, T. M.; and Knapp, H. Experimental and modeling studies on the solubility of CO_2, $CHClF_2$, CHF_3, $C_2H_2F_4$ and $C_2H_4F_2$ in water and aqueous NaCl solutions under low pressures, *Fluid Phase Equilib.*, 1997, **129**, 197–209.

Carbon material@Chitosan composite as catalyst on the synthesis of FAME from used-cooking oil with electro-catalytic process

Rudy Syah Putra[1,2*], Yudi Antono[1], Kharis Pratama[1]

[1]Department of Chemistry and [2]New and Renewable Energy Research Group, Faculty of Mathematics and Natural Sciences, Universitas Islam Indonesia, Jl. Kaliurang km. 14 Yogyakarta 55584, Indonesia

[*]rudy.syahputra@uii.ac.id

Abstract. The conversion of fatty acid methyl ester (FAME) from soybean oil with a carbon@chitosan composite as alkaline catalyst using electrolysis process had been investigated. The carbon was added onto chitosan through sol-gel method. Carbon material@chitosan, featured with high electrical conductivity and large surface area and Scanning electron microscopy equipped with an energy dispersive spectroscope (EDS) detector was performed to characterize the microstructures as-prepared alcolgels composite. The evaluation of the synthesis process was followed by GC-MS, determining the fatty acid methyl ester (FAME) ratio at different operation variables (e.g oil:MeOH molar ratio at 1:6, THF:MeOH ratio at 1:1 v/v, 10 V and 60 mins). The results showed that the incorporation of carbon resulted in an observable change in the porous structure and an obvious increase in the conductivity strength. When compared with graphite@chitosan composite as catalyst, the carbon@chitosan composite exhibits remarkably FAME yields of 100% in 20 wt.% catalyst loading. The application of those processes was also evaluated when using used-cooking oil as a feedstock of biodiesel production.

1. Introduction

Fossil fuel is now widely recognized as unsustainable resource. The limitations of fossil fuel had demanded to look for other energy sources that can be renewable. With a global production of almost 60 million tonnes and a global vegetable oil market share of more than 35% by weight in 2012, palm oil is the most produced vegetable oil in the world [1]. Therefore, the amount of cooking oil consumption impacted directly the increase of use-cooking oil as raw material for biodiesel production. In addition, utilization of used-cooking oil as biodiesel diminishes the problem of contamination, because reusing of these greases can reduce the burden of the government in disposing of the waste, maintaining public sewers, and treating the oil wastewater [2].

The chitosan-catalyzed transesterification of soybean oil had used chitosan gel (i.e cryogel) as alkaline catalyst at high temperature (100-150 °C) was investigated by Kayser *et al.* (2014) [3] led to high FAME yields (up to 90%). Another researcher, Guan and Kusakabe (2009) [4] had used electrolysis process to afford FAME in high yields (> 97%) at room temperature reaction. In addition, the combined use of electrolysis and chitosan had been investigated by Putra *et al.* (2015) [5] with low FAME yields (up to 59.05%). Low yield on these synthesis processes due to lack of chitosan basicity which was influenced by low catalyst surface area and due to lack of electrical conductivity in the reaction. However, direct use of chitosan as heterogeneous alkaline catalyst combined with electrolysis process had enhanced the biodiesel synthesis. For example, the esterification and transesterification reaction occurring in one step reaction through the formation of H^+ and OH^- ions either in the electrolysis process or involved the catalytic reaction in the chitosan [5]. Therefore, in this study the increasing of FAME conversion from used-cooking oil is proposed using a carbon material@chitosan composite as hydrogel catalyst. The purposes of these studies are to provide high added-value lightweight material with outstanding textural properties, i.e. high surface area, open porosity and electrical conductivity, respectively.

2. Materials and Method

2.1. Chemical and equipment

Used-cooking oil was provided from home industry activities and soybean oil was obtained from supermarket. High grade of methanol, NaCl and co-solvent THF were purchased from Merck, Germany and chitosan (pharmaceutical grade) was purchased from UPT BPPTK LIPI Gunung Kidul, Yogyakarta. Fig. 1 shows an electro-catalytic cell (70 ml) contains two graphite electrodes (⌀ 0.8 mm x 130 mm) that were separated by a distance of 15 mm. The qualities of pre-treatment of used-cooking oil were determined by physicochemical analysis and the results were shown in the Tab. 1.

The process involves direct filtration of oil using filter paper following absorption process [2]. A 50 g of used-cooking oil was heated until 100°C after filtration to remove solid particle from oil. After that, a 5 g of activated carbon was poured into the oil and then stirred for 80 minutes. The mixture was then filtered and the filtrate was measured for physicochemical properties including density, refractive index, water content, and acid number (AN), respectively.

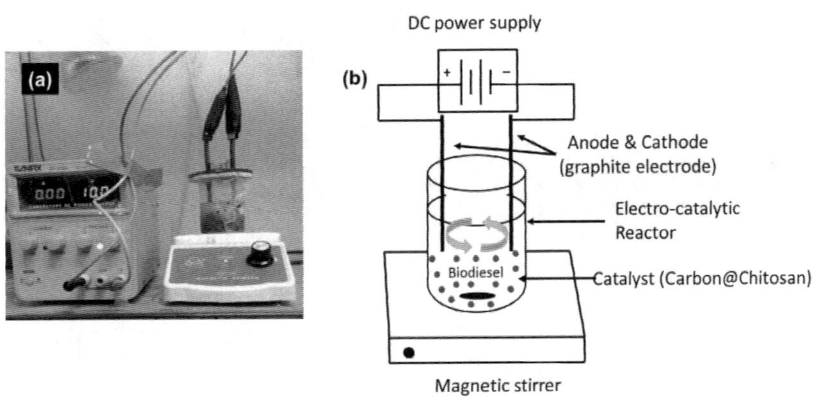

Figure 1. Electro-catalytic reactor set up (a) and illustrated functional apparatus in the reactor (b).

Table 1. Physicochemical properties of oil that was used as a feedstock in the experiment

Oil	Acid number	FFA, %	Refractive index	ρ, Kg/L	Moisture, %
Used-cooking	1.308	0.860	1.461	0.914	0.185
Soybean	0.378	0.263	1.469	0.920	0.217

2.2. Preparation of spherical carbon material@chitosan composite with sol-gel method

In order to evaluate the scope of carbon material@citosan composite as catalyst, uniform-size spherical hydrogel beads were prepared by adaptation and modified of reported procedures based on the alkaline coagulation of an acidic viscous chitosan solution added using a peristaltic pump equipped with a dropping silicon pipe [6] as illustrated in Fig. 2. Graphite was used as a carbon material that was composited with chitosan as a preliminary study before testing with carbon. The first step was the formation of chitosan hydrogel by reaction of a polymer solution with a gelling agent, 1 g chitosan (from shrimp shells, > 61.4% deacetylated) was dissolved in 50 mL acetic acid glacial solution (1% v/v) and stirred for 15 mins. After that 0.5 g graphite was added into the mixture and continued to be stirred for 24 h. The solution mixture was then slowly precipitated into an aqueous NaOH solution (1 M) until appropriate of graphite@chitosan-hydrogel composite was formed. The carbon@chitosan-hydrogel composite was washed with deionized water until constant pH was reached (pH 6 ~ 7). The resulting graphite@chitosan-hydrogel composite was stored in deionized water before used it as alkaline catalyst in biodiesel synthesis. Similar procedure was also applied when carbon@chitosan composite prepared.

The micro-structures of the produced graphite@chitosan composite samples were characterized using a desktop scanning electron microscope (SEM, Phenom ProX, Phenom-World B.V., The Netherlands) equipped with an energy dispersive spectroscope (EDS) detector. In this regards, the corresponding dried gel beads were prepared by alcogel, a water-to-ethanol solvent exchange (5 steps EtOH/water 100:0) was carried out prior to analysis [7].

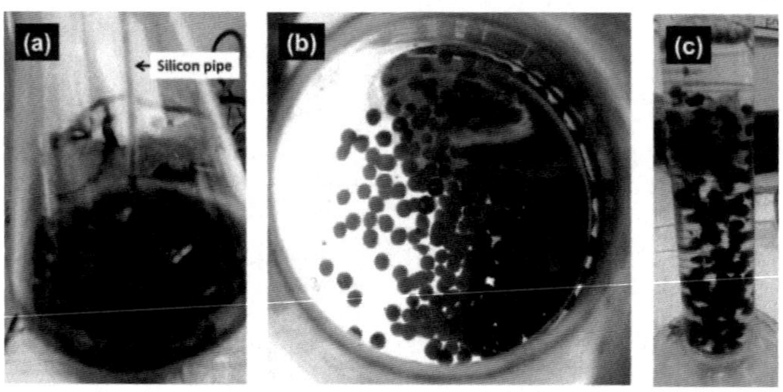

Figure 2. Experimental set-up used for the preparation of spherical graphite@chitosan with an average diameter of 3.0 ± 0.1 mm (a), and macroscopic view of milk-black colour graphite@chitosan beads after maturing (b, c).

2.3. FAME synthesis

Preliminary, soybean oil was used as a feedstock to obtain optimal condition before implemented to used-cooking oil for FAME synthesis. Effect of catalyst amount (10 wt.% and 20 wt.%) were evaluated for the conversion of FAME from soybean oil. The reaction mixture was agitated using magnetic stirrer. The electrolytic cell was filled with 70 ml of reaction mixture containing soybean oil, methanol, THF as a co-solvent, and NaCl as a supporting electrolyte. The methanol-to-oil molar ratio was 6:1 and co-solvent-to-methanol molar ratio was 1:1, and NaCl concentration was added on the basis of oil weight of the entire reaction mixture, respectively. The electro-catalytic was carried out at room temperature (25 °C) using a constant voltage of 10 V for 60 mins. Fig. 3 shows a step of optimization process for biodiesel production in this study.

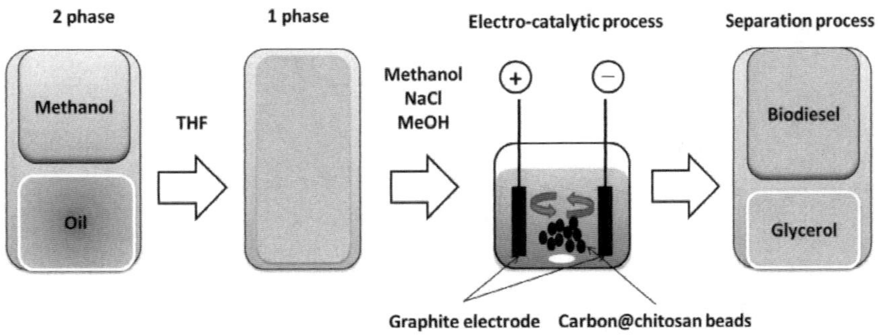

Figure 3. Steps on biodiesel production using electrolysis process and carbon@chitosan beads as catalyst.

2.4. FAME analysis

The biodiesel phase was separated from glycerol and washed with deionized water to remove the residual inorganic components. Concentration of produced FAME and unreacted oils remaining in the product were analyzed using GC-MS (GC-MS QP2010 Plus, Shimadzu Corporation, Japan) equipped with an electron ionization detector (EID) and a capillary column 30 m × 0.25 mm × 0.25 μm (Restex 5MS). The sum of FAME and unreacted glycerides were represented by peaks separated in the chromatograph. The biodiesel yield in the product was calculated by total sum of percentage area of chromatographic peak of FAME in each product.

3. Results and Discussion

3.1. Composite characterization

The morphology of graphite@chitosan composite was investigated by SEM as shown in Fig. 4. According to the SEM images, Fig. 4a represents the SEM morphology of the prepared graphite@chitosan composite. The SEM image of the composite showed that graphite particles were covered by chitosan gel. Figure 4b shows that the surface of the carbon@chitosan composite was relatively homogeneous and smooth being similar to those from previous reports [8]. In addition, the energy dispersive spectroscope (EDS) spectra of the graphite@chitosan composite (Figs. 5a and 5b) confirmed the presence of carbon material in the matrix composite.

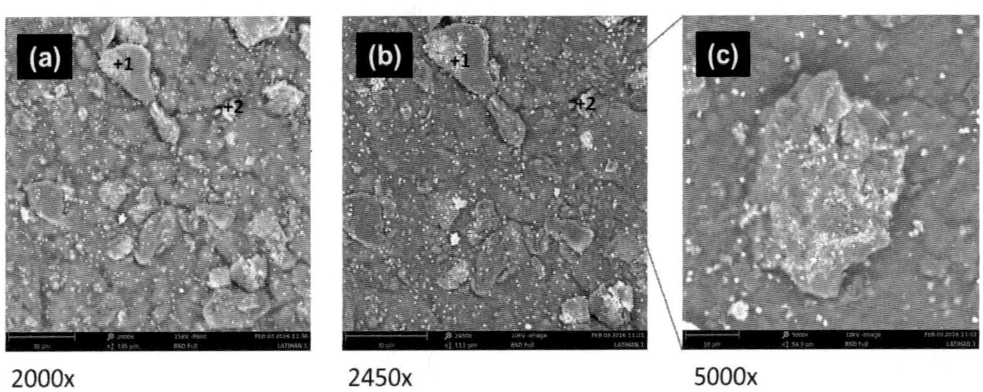

Figure 4. SEM images of graphite@chitosan composite with different magnification.

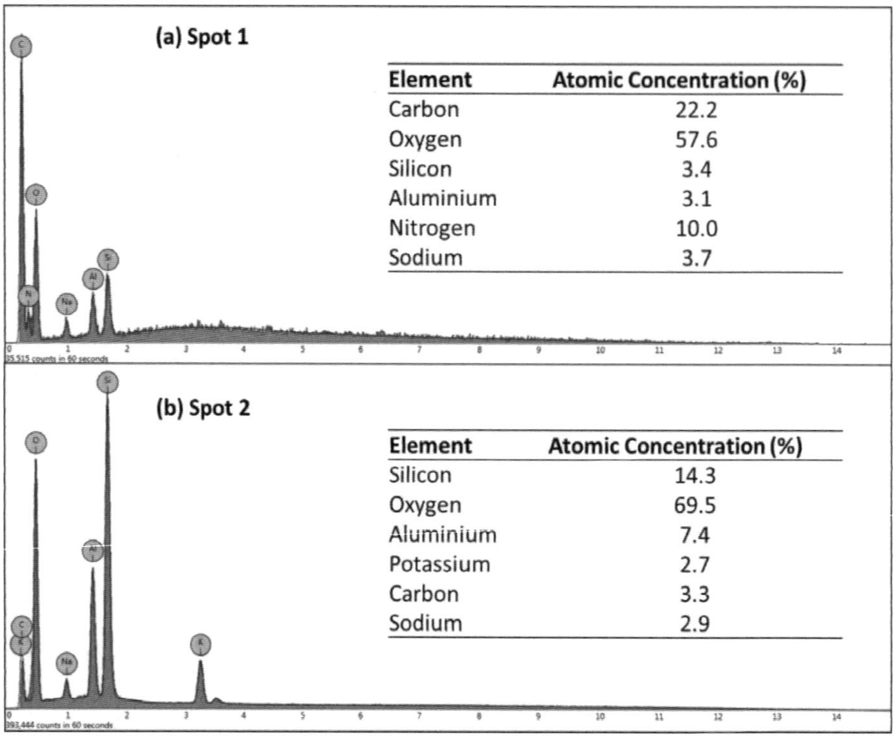

Figure 5. EDS spectra obtained from spot 1 and 2 in the SEM images of graphite@chitosan composite.

3.2 Washing effect of graphite@chitosan composite catalyst

One of the most critical aspects during the evaluation of the carbon material@chitosan composite catalyst is the meticulous washing protocol of the matured beads, thereby ensuring the removal of trapped hydroxyl ions (OH^-) from alkaline solution in the sol-gel process, which otherwise might influence the expected base catalysis by the free amino groups $-NH_2$ on the chitosan backbone, upon slow diffusion-controlled leaching of OH^- during the reaction [6]. In order to demonstrate this hypothesis, different batches of beads at different pH were prepared by tuning the wash procedure. The exact pH of the filtrate in each case was measured using a pre-calibrated pH meter. The general correlation of the pH of the filtrate with the internal pH of the beads was checked by extensive trituration-dissolution of the beads and measuring the pH of the resulting solution. Tab. 1 shows the correlation between pH of graphite@chitosan composite before and after washed to the FAME product by electro-catalytic process.

Neutral composite afforded 73.0% (Table 1, entry 4) of FAME conversion as well as alkaline composite before washed (Table 1, entry 1), which showed low conversion in the absence of the catalyst in the electro-catalytic process (Table 1, entry 5). In general, no significant improvement was observed by increasing two-fold washing cycle (Table 1, entry 3). However, it was high when neutral pH achieved. Those results indicating that simply the accessible free primary amino groups presented in these hydrogel beads of the native chitosan are active enough to promote satisfactorily the formation of the required methoxide ion (CH_3O^-) intermediate [5,6] under the present hydrogel conditions.

Table 2. Correlation between pH of graphite@chitosan composite before and after washed to the FAME product by electro-catalytic process

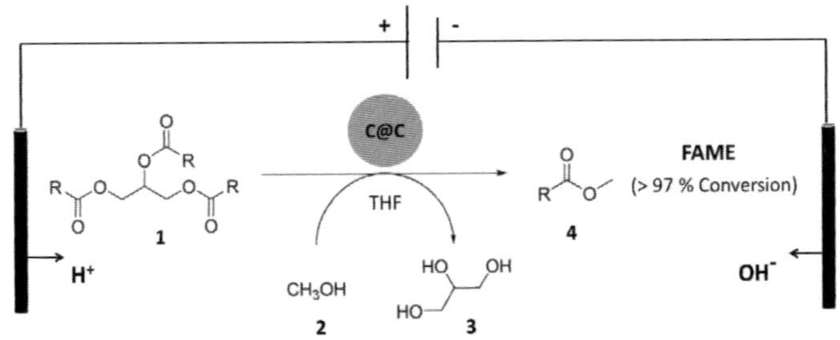

Entry	pH	Time (minute)	FAME, %
1	11.98	6o	76.1[a]
2	10.13	60	13.3
3	8.15	60	47.8
4	6.91	60	73.0[b]
5	-	60	42.2[c]

Reaction condition: soybean oil used as a feedstock; molar ratio methanol to oil, 6:1; volume ratio THF to methanol, 1:1; mass ratio of catalyst to oil, 10 wt.%; DC voltage in electro-catalytic, 10 V. Result of the composite [a]before washed and [b]after 3rd washing cycle. [c]Control experiment in which no catalyst was employed in electro-catalytic process at 18.2 V.

3.3 Electrical conductivity of carbon material@chitosan composite catalyst

Another critical aspect of the carbon material@chitosan composite is the electrical conductivity properties of the matured beads, thereby ensuring enhanced the electrical conductivity of the solution during the FAME conversion. Figure 6 shows the effect of carbon material@chitosan composite on the electrical conductivity of catalyst property. In the electro-catalytic process, a direct proportional to the hydrogen and hydroxyl ions formed at the anode and cathode as well as NaCl concentration enhanced electric current and conductivity of the reaction mixture during electro-catalytic process. Roughly, the electric current increased with the increase in the electrolysis voltage, which resulted in more hydrogen and hydroxyl ions formed on the anode and cathode [4,5]. The hydroxide ions are continuously produced at the cathode and might react immediately with nearby methanol and the oil transesterification reaction happened in methanol phase [4]. As a result, transesterification progressed mainly with the concentrated hydroxide ion that existed near the cathode. While, the electron transfer emerged by the formation of hydrogen ion that existed near the anode was derived indirectly as a catalyst of esterification process of oil (reaction scheme was showed in Tab.1).

The electrical conductivity during electro-catalytic process was also enhanced by the presenting of carbon material in the chitosan gel beads as a composite material. The electrical conductivity was measured in the aqueous media using conductivity meter at room temperature. It was showed that the electrical conductivity of carbon@chitosan higher than graphite@chitosan beads (Fig. 6).

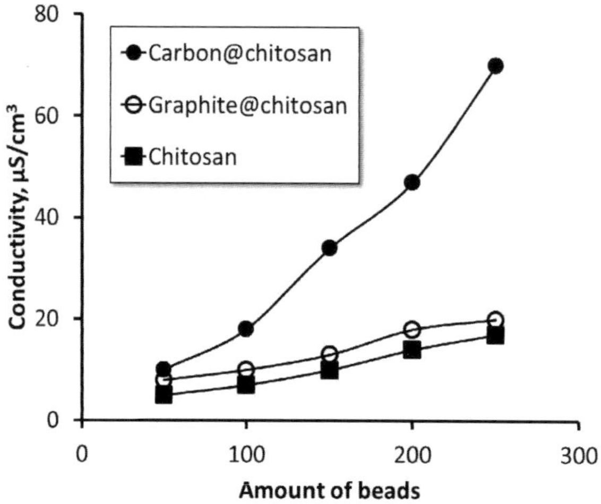

Figure 6. Effect of carbon material on the electrical conductivity of composite catalyst.

3.4 FAME conversion from soybean oil and used-cooking oil

For low quality used-cooking oil containing high proportion of free fatty acid (FFA) and water, homogeneous alkaline catalyst are markedly sensitive to FFA, leading to saponification problem, catalyst losses, lowering the efficiency and economics of the overall process [8-10]. Therefore, a pre-treatment process to reduce water content (<0.1 wt.%) and FFAs (<2.0 wt.%) are necessary in order to avoid an undesirable side reactions, such as saponification, which could lead to serious problem of product separation and low fatty acid methyl ester (FAME) yield. Tab. 1 shows that the pre-treatment process of used-cooking oil which used activated carbon as an adsorbent was able to reduce of water content (< 0.1%) and FFA (< 0.2%) in the oil significantly, which the results were similar to those from previous report [2]. In this study, effects of catalyst loading on the biodiesel yield was preliminary investigated for the transesterification of soybean oil before apply in used-cooking oil. The use of soybean oil which was more unsaturation per triglyceride expected led to such a high FAME yields [3].

Tab. 1 shows FAME conversion from oil (i.e soybean oil and used-cooking oil) did not reflect with the increasing of catalyst loading, but it was more affected by the carbon material@chitosan type as a composite catalyst. When compared with grahite@chitosan composite, the carbon@chitosan composite exhibits remarkably FAME yields of 100% in 20 wt.% catalyst loading when using soybean oil as feedstock. However, there were no good results when the process was used in the conversion of biodiesel from used-cooking oil.

Table 3. FAME conversion from oil using carbon material@chitosan type

Catalyst type	wt.% loading	FAME yield, %	
		Soybean oil	Used-cooking oil
Carbon@chitosan	10	0.00[a]	-[b]
	20	100	33.6
Graphite@chitosan	10	73.0	37.6
	20	0.00[a]	0.00[a]

Reaction condition: molar ratio methanol to oil, 6:1; volume ratio THF to methanol, 1:1; DC voltage in electro-catalytic, 10 V; pH composite washed 6.91. [a]Experiment was repeated (n = 2) in which no FAME yield in the product confirmed. [b]There is no experiment employed in this regards.

4. Conclusion

The incorporation of carbon material resulted in an observable change in the porous structure and an obvious increase in the electrical conductivity strength of composite as catalyst. Carbon@chitosan catalyst composite exhibits remarkably FAME yields of 100% in 20 wt.% of catalyst loading when using soybean oil as feedstock compared with grahite@chitosan composite as catalyst. In contrast, those results were not applied well on the FAME conversion from used-cooking oil.

Acknowledgements

Authors would like thank to Ministry of RISTEK-DIKTI, the Republic of Indonesia for financial support through PKM-P program 2015. Furthermore, additional support from AIPT program 2016 of Universitas Islam Indonesia is gratefully acknowledged.

References

[1] MPOB 2013 Malaysian Palm Oil Statistics 2012 Malaysian Palm Oil Board, Ministry of Plantation Industries & Commodities, Malaysia
[2] Putra RS, Julianto, TS, Hartono, P, Puspitasari, RD and Kurniawan A 2014 *Int. Journal of Renewable Energy Development (IJRED)* 3 (1):33-35
[3] Kayser H, Pienko F, Maria PD 2014 *Fuel* 116:267-272
[4] Guan G, Kusakabe K 2009 *Chem Eng* 153:159-163
[5] Putra RS, Hartono P, Julianto T 2015 *Energy Procedia* 65:309-316
[6] Kuhbeck D, Saidulu G, Reddy KR, and Dıaz DD 2012 *Green Chem* 14:378-392
[7] Kaysera H, Müllera CR, García-González CA, Smirnova I, Walter Leitnera W, Domínguez de María P 2012 *Applied Catalysis A: General* 445-446:180-186
[8] Di Serio M, Tesser R, Pengmei L, Santacesaria E 2008 *Energy Fuels* 22:207-17
[9] Melero JA, Iglesias J, Morales G 2009 *Green Chem* 11:285-308
[10] Patil PD, Deng S 2009 *Fuel* 88:1302-1306

IOP Conf. Series: Journal of Physics: Conf. Series **877** (2017) 012064 doi:10.1088/1742-6596/877/1/012064

Comparative Study on Various Geometrical Core Design of 300 MWth Gas Cooled Fast Reactor with UN-PuN Fuel Long-life without Refuelling

Ratna Dewi Syarifah[1], **Zaki Su'ud**[2], **Khairul Basar**[2] **and Dwi Irwanto**[2]
[1]Department of Physics, Faculty of Mathematics and Natural Sciences, Institut Teknologi Bandung, Jl. Ganesa 10 Bandung 40132, INDONESIA
[2]Nuclear Physics and Biophysics Research Division, Physics Department, Faculty of Mathematics and Natural Science, Institut Teknologi Bandung, Jalan Ganesha 10 Bandung 40132, INDONESIA

E-mail: syarifah.physics@gmail.com

Abstract. Nuclear power has progressive improvement in the operating performance of exiting reactors and ensuring economic competitiveness of nuclear electricity around the world. The GFR use gas coolant and fast neutron spectrum. This research use helium coolant which has low neutron moderation, chemical inert and single phase. Comparative study on various geometrical core design for modular GFR with UN-PuN fuel long life without refuelling has been done. The calculation use SRAC2006 code both PIJ calculation and CITATION calculation. The data libraries use JENDL 4.0. The variation of fuel fraction is 40% until 65%. In this research, we varied the geometry of core reactor to find the optimum geometry design. The variation of the geometry design is balance cylinder; it means that the diameter active core (D) same with height active core (H). Second, pancake cylinder (D>H) and third, tall cylinder (D<H). The reflector radial-axial width is 50 cm. The power is 300 MWth. First calculation, we calculate survey parameter for UN-PuN fuel with fissile contain from Plutonium waste LWR for each geometry. The minimum power density is around 72 Watt/cc, and maximum power density 114 Watt/cc. After we calculate with various geometry core, when we use the balance geometry, the k-eff value flattest and more stable than the others.

1. Introduction

The ten countries have joined together to form the Generation IV International Forum (GIF) to develop future generation nuclear energy system. It need to provide manageable nuclear waste, effective fuel utilization, competitive economics, recognized safety performance and secure nuclear energy systems and nuclear materials [1].

One type of Generation IV reactor is Gas Cooled Fast Reactor (GFR). The GFR system features a fast-spectrum helium-cooled reactor and closed fuel cycle. The fast spectrum makes it possible to utilize available fissile and fertile materials two orders of magnitude more efficiently than thermal spectrum gas reactor with once-through fuel cycles [2]. Comparative study on various geometrical core designs of 300 MWth Gas Cooled Fast Reactor with UN-PuN fuel has been done. The neutronic analysis of GFR with three geometry type, i.e. pancake cylinder, balance cylinder and tall cylinder has been done with SRAC2006 code calculation. Beforehand, analysis neutronic of GFR with SRAC2006 code system with modified CANDLE burn up scheme has been analysed [3-8]. The neutronic analysis for long-life 200MWth GFR also has been analysed [9]. In this research, the neutronic calculation calculate in order to the reactor can be operate long-life without refuelling for the fuel.

2. Design Concept and Calculation Methods

Table 1 shows parameter design of fuel pin and specification of reactor core. It use nitride fuel (UN-PuN) fuel which is the plutonium obtained from waste fuel from Light Water Reactor (LWR), so it can reduce the amount of plutonium in the world.

Table 1. Parameter design of fuel pin and specification of reactor core

Parameter	Specification
Power	300 MWth
Fuel material	UN-PuN
Cladding material	Silicon Carbide (SiC)
Coolant material	Helium
Fuel volume fraction	40%-60%
Cladding volume fraction	10%
Coolant volume fraction	30%-50%
Pin pitch	1.45 cm
Reactor core type	• Pancake cylinder (D>H) • Balance cylinder (D=H) • Tall cylinder (D<H)
Volume core	• Pancake cylinder = 4190476.19 cm^2 • Balance cylinder = 4158289.14 cm^2 • Tall cylinder = 4106666.67 cm^2
Variation of fissile contain per region	8% : 10% : 12%

The fuel pin design geometry use hexagonal cell geometry which is presented in **Figure 1**. The fuel pin is arranged into heterogeneous core configuration with three variation fuel (Fuel 1, Fuel 2 and Fuel 3). The variation of fissile contain per region show in Table 1, F1=8%, F2=10%, and F3=12%.

There are three reactor core geometry designs has been calculated (see **Figure 2**). The dimension (width and height) of the cylinder is show in **Table 2**.

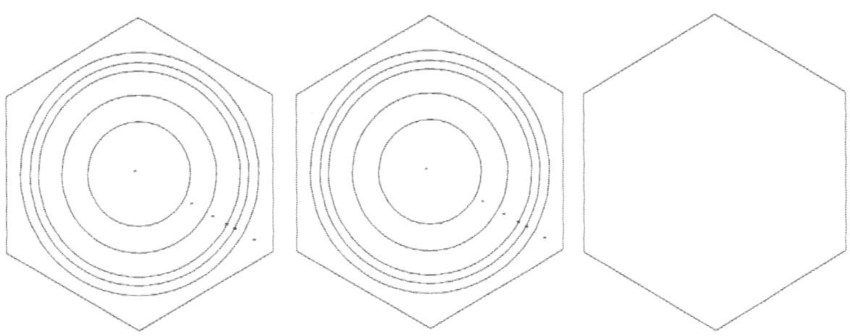

Figure 1. Fuel pin hexagonal geometry design (output result of PIJ Calculation)

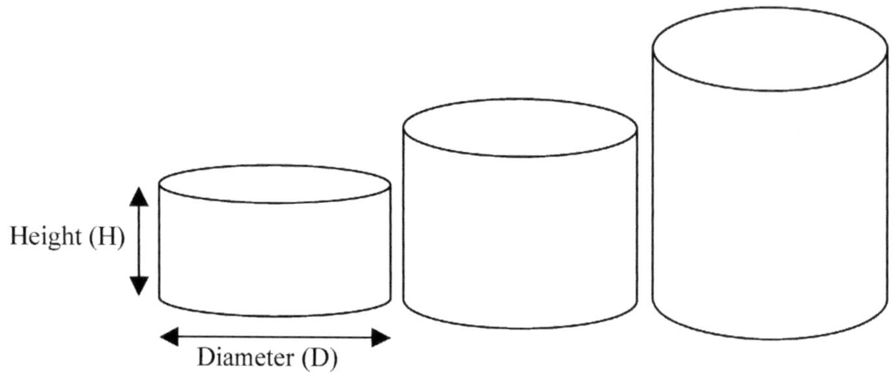

Figure 2. Scheme of various core geometry design (pancake cylinder, balance cylinder and tall cylinder)

Table 2. Various core geometry design of the reactor

Type of Geometry Core	Width active core	Height active core
Pancake cylinder	200 cm	100 cm
Balance cylinder	159 cm	159 cm
Tall cylinder	140 cm	140 cm

The neutronic analysis use SRAC2006 code system which is developed by JAERI (now JAEA) to calculate the neutronic analysis of the reactors. And the data libraries use JENDL 4.0[10].

3. Results and Discussion

Figure 3 shows effective multiplication factor (k-eff) value of gas cooled fast reactor with various core geometry design. For all core geometry design, they can be operated plus than 20 years. For all geometry have same trend of the k-eff graph. The balance cylinder has k-eff value flattest than the other. It means when we use balance cylinder, the k-eff value is more stable than the other. The maximum k-eff value of balance cylinder is 1.0088.

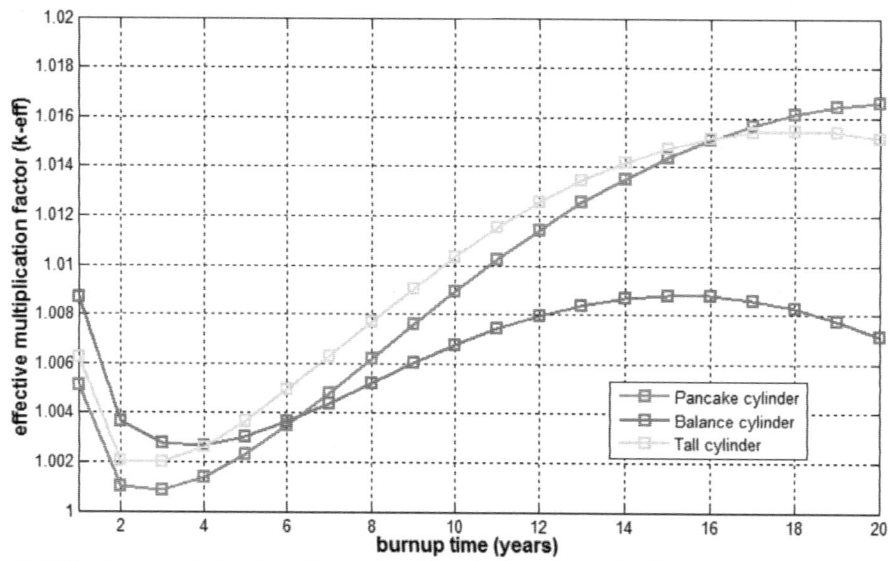

Figure 3. Effective multiplication factor (k-eff) value of gas cooled fast reactor with various core geometry design

Table 3 show the parameter and power density results of the reactor. From the table pancake cylinder has power density value greatest than the others. In this case, due to it has fuel volume fraction the greatest one. The balance and tall cylinder has average power density almost the same.

Table 3. The parameter and power density results of the reactor

Type of geometry	Volume fraction (%)			Power density (Watt/cc)	
	Fuel	Cladding	Coolant	Average	Peak
Pancake cylinder	54.5	10	35.5	66.35	114
Balance cylinder	46	10	44	51.34	75
Tall cylinder	49.5	10	40.5	51.73	72

Figure 4 show excess reactivity value of gas cooled fast reactor with various core geometry design. From **Figure 4** all geometry has excess reactivity value less than 1.8%, means that the reactor is stable. For balance cylinder has excess reactivity value less than 1%. The maximum excess reactivity value is 0.87%.

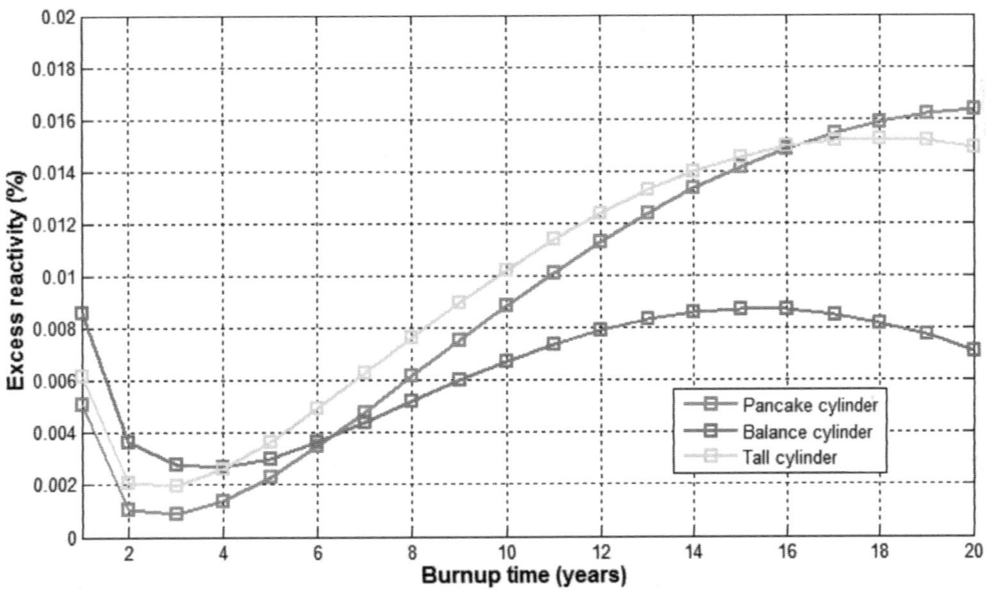

Figure 4. Excess reactivity value of gas cooled fast reactor with various core geometry design

4. Conclusion

Comparative study on various geometrical core designs of 300 MWth Gas Cooled Fast Reactor with UN-PuN fuel has been done. For all core geometry design (balance, pancake, tall cylinder) can be operated more than 20 years without refueling. It has same trend of graph in k-eff value and excess reactivity value. The maximum k-eff value is 1.0088 and the maximum excess reactivity value 0.87%. This design is more stable than the others.

Acknowledgment

This work was supported by Decentralization Research Grand 2016, DIKTI, Ministry of Research Technology and Higher Education of the Republic of Indonesia which are gratefully acknowledged.

References

[1] GIF (The Generation IV International Forum) 2002 *A technology Roadmap for Generation IV Nuclear Energy System* (U.S DOE Nuclear Energy Research Advisory Committee) p 1-22
[2] GIF (The Generation IV International Forum) 2009 *GIF R&D Outlook for Generation IV Nuclear Energy System* (the OECD Nuclear Energy Agency) p 6
[3] Fiber M Zaki S Abdul W Khairul B Menik A and Hiroshi S 2013 *Advanced Material Research* **772** 501
[4] Zaki S and Hiroshi S 2013 *Annals of Nuclear Energy* **54** 58
[5] Menik A Zaki S Fiber M Abdul W Khairurrijal Idam A Aziz F and Hiroshi S 2013 *Applied Mechanics and Materials* **261-262** 307
[6] Fiber M Menik A Zaki S Abdul W Khairul B Ferhat A Sidik P and Hiroshi S 2014 *AIP Conference Proceedings* **1584** 105
[7] Fiber M Zaki S Abdul W Khairul B Menik A and Hiroshi S 2014 *AIP Conference Proceedings* **1615** 47
[8] R.D. Syarifah, Z. Su'ud, K. Basar and D. Irwanto, 2016 *J. Phys.: Conf. Ser.* **776** 012103
[9] R.D. Syarifah, Y. Yulianto, Z. Su'ud, K. Basar and D. Irwanto, 2016 *Matec Web Conf.* **82** 03008
[10] Okumura K 2002 *SRAC version 2002* (Japan Atomic Energy Research Institute) p 1-28

Preliminary Studies of Thermoluminescence Dosimeter (TLD) $CaSO_4$:Dy Synthesis

N Nuraeni [1,2], F Iskandar[1,3], A Waris[1], and F Haryanto[1]

[1]Department of Physics, Faculty of Mathematics and Natural Sciences, Institut Teknologi Bandung, Jl. Ganesha 10, Bandung 40132, Indonesia
[2]Work Safety and Dosimetry Division, National Nuclear Energy Agency (BATAN) Jl. Lebak Bulus Raya Kav. 49, Jakarta Selatan 12440, Indonesia
[3]Research Center for Nanosciences and Nanotechnology (RCNN), Institut Teknologi Bandung, Jl. Ganesha 10, Bandung 40132, Indonesia

E-mail: nunung.213009@gmail.com

Abstract. Thermoluminescence dosimeter (TLD) $CaSO_4$:Dy was synthesised by co-precipitation. The TLD was observed after radiation exposure to Strontium-90. The thermoluminescence intensity was read using a TLD Reader Harshaw 3500. The thermoluminescent response obtained was 59.29 nC. Then re-annealing was conducted with the temperature varied at 700, 800 and 900 °C. The thermoluminescent intensity obtained at temperatures of 700 °C, 800 °C and 900 °C was 66.12 nC, 169.45 nC, and 552.37 nC respectively. The sensitivity of the TLD increased in response to the re-annealing temperature rise. In addition to observing the thermoluminescence properties, a comparison was made between the TLD obtained from this experiment with an existing TLD in the market. Finally, also the glow-curve characteristics of the TLD were observed.

1. Introduction

Radiation exists in everyday life. Radiation comes from nature but there is also artificial radiation. Artificial radiation is widely used in the health sector (diagnosis and therapy) and industry [1]. Because of the increasing number of users of radiation it is necessary to monitor radiation on people involved in the use of radiation [2]. The type of radiation dosimeter most commonly used for medical physics, personnel monitoring and environmental monitoring is the thermoluminescence dosimeter (TLD). When a thermoluminescent (TL) material is exposed to ionising radiation some of the liberated charge carriers (electrons or holes) are trapped in lattice imperfections in the crystalline solid. TL is light emission from materials when heated, following previous absorption of energy from radiation [3]. The structure of the defects can be controlled by the preparation method.

The technology of TLD fabrication was developed along with the development of material synthesis. One of the fabrication methods is co-precipitation. This method has the advantage of involving a simple and rapid preparation process, it is easy to control the particle size and composition, and it offers several possibilities to modify the particle surface state and overall homogeneity [4,5].

This study aimed to investigate the effect of re-annealing at various temperatures on the crystal structure and thermoluminescence properties of calcium sulphate doped with dysprosium ($CaSO_4$:Dy). Furthermore, the glow-curve characteristics of each type of TLD were observed. Besides that, we compared the thermoluminescence response of $CaSO_4$:Dy obtained from this experiment to that of an existing TLD in the market.

2. Materials and method

2.1. Material Preparation (co-precipitation)

TLD $CaSO_4$:Dy was synthesized using a co-precipitation method [6] with the reaction as shown in Equation 1:

$$\mathbf{(CH_3COOH)_2Ca+(NH_4)_2SO_4+Dy_2(SO_4)_3 \rightarrow CaSO_4{:}Dy\downarrow+2CH_3COONH_4} \quad (1)$$

Calcium acetate ($(CH_3COO)_2Ca$ (AR grade, Merck) was dissolved in double distilled water, then mixed with a mixture of dysprosium sulphate and ammonium sulphate ($(NH_4)_2SO_4$ (AR grade, Merck) in the presence of ethanol. Dysprosium sulfate was used as doping with a concentration of 0.1 mol %, made from Dy_2O_3 (Sigma Aldrich) and an appropriate H_2SO_4 (AR grade Merck). The precipitate was then washed several times with distilled water and dried in an oven for 4 hours. After drying, the TLD was annealed at a temperature of 650 °C for 1 hour in a furnace.

After that, the obtained TLD $CaSO_4$:Dy was re-annealed at various temperatures (700 °C, 800 °C and 900 °C). An illustration of the TLD $CaSO_4$:Dy synthesis process is shown in Figure 1.

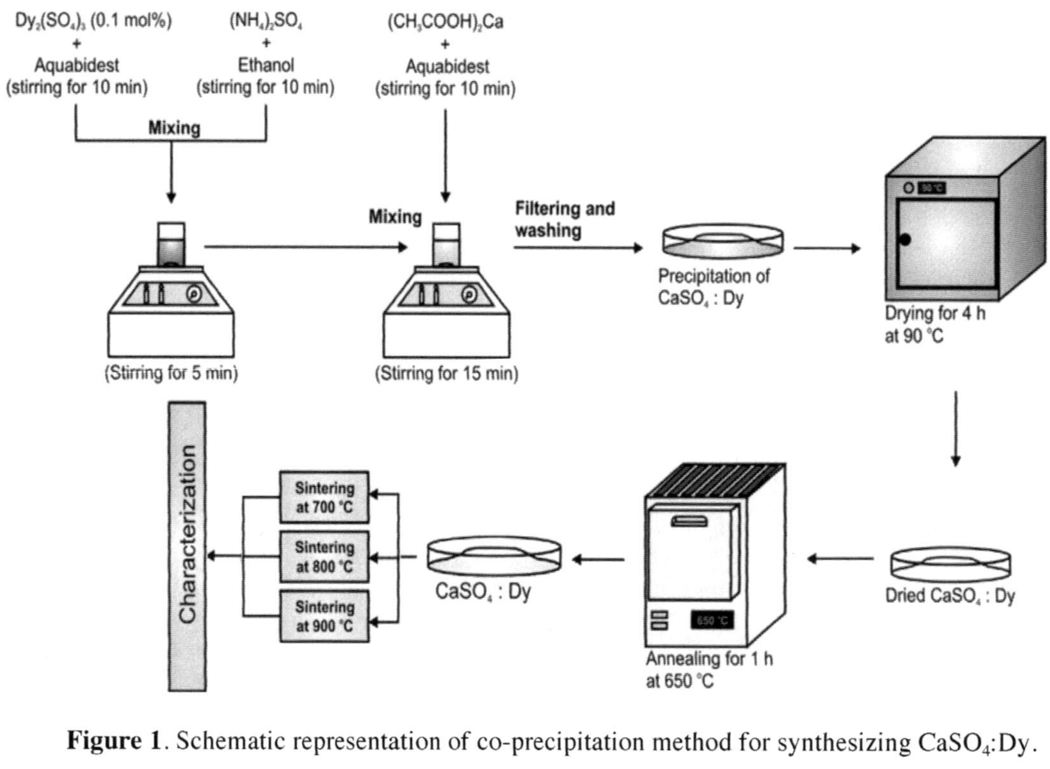

Figure 1. Schematic representation of co-precipitation method for synthesizing $CaSO_4$:Dy.

2.2. Characterization

The structure of the obtained result was characterized using an XR diffractometer (XRD, Philips Analytical PW 1710 BASED) before re-annealing and after re-annealing (700 °C). Then the thermoluminescence response was observed using a TLD Reader Harshaw 3500 at a maximum temperature of 260 °C after radiation exposure to Strontium-90 before re-annealing and after all re-annealing temperature variations (700 °C, 800 °C, 900 °C).

3. Results

From the XRD we could identify that our result was exactly $CaSO_4$ according to JCPDS no. 37-1496.

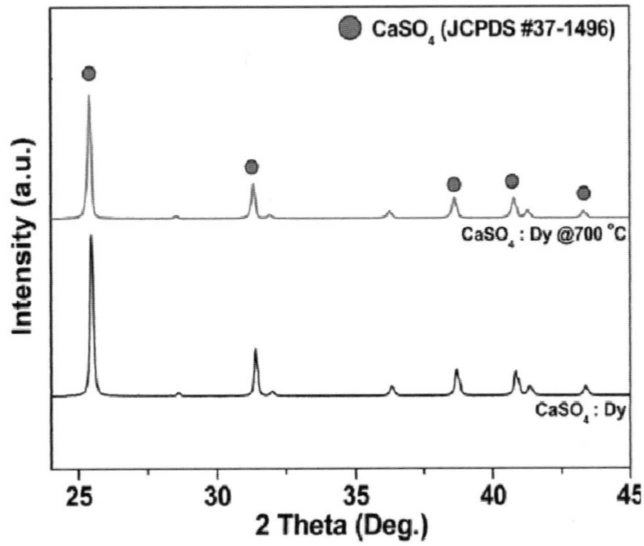

Figure. 2. XRD of pure $CaSO_4$:Dy and $CaSO_4$:Dy after re-annealing at 700 °C.

Using the Scherrer equation, the crystal size of the obtained $CaSO_4$:Dy from the experiment was 54.09 nm for the pure $CaSO_4$ while it was 50.91 nm:Dy for the $CaSO_4$:Dy re-annealed at 700 °C respectively. The crystal size decreased after the re-annealing treatment. This phenomenon shows that the pure $CaSO_4$:Dy was still more porous, so when the re-annealing treatment was applied, the crystal size of $CaSO_4$:Dy became smaller. The correlation between re-annealing temperature and thermoluminescence intensity of $CaSO_4$:Dy is shown in figure 3.

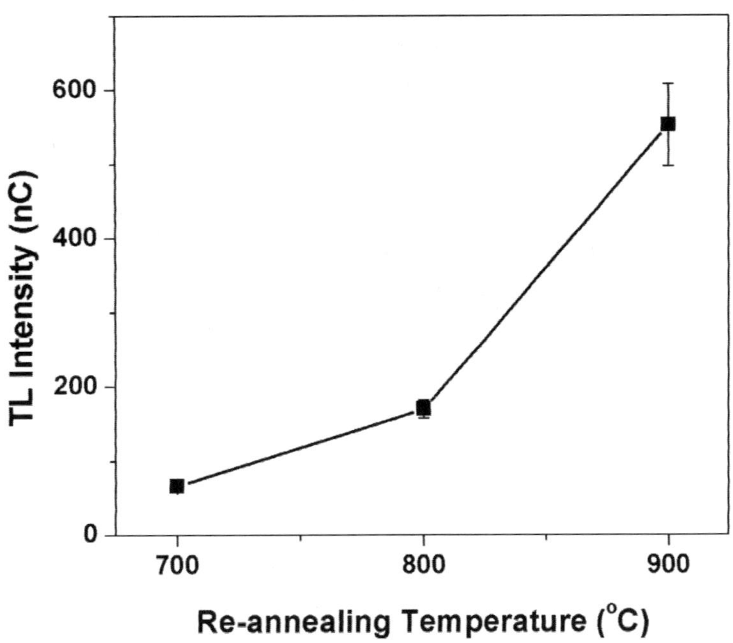

Figure 3. Correlation between re-annealing temperature vs. TL intensity of $CaSO_4$:Dy.

Besides applying the re-annealing temperature variations, the glow-curve characteristics of each type of TLD were observed. From the glow curve, the luminescence intensity distribution can be seen and the temperature range required for producing luminescence. A comparison of the glow curves from the experiment and from an existing TLD are shown in table 1.

Table 1. Comparison of Glow Curve Characteristics From TLD $CaSO_4$:Dy Obtained From Experiment With the Existing TLD

Type of TLD	Main peak (°C)	Remaining e^-	Glow curve
TLD $CaSO_4$:Dy experiment	160	No	Full curve
TLD 900	190	No	Full curve
TLD BARC	260	Yes	Truncated curve

The TLD $CaSO_4$:Dy from the experiment and the TLD BARC have a simple glow curve, but the TLD-900 has a complex glow curve. According to Azorin [4], a good TL material usually has a relatively simple glow curve and has its main peak at about 200 °C. The TLD $CaSO_4$:Dy from the experiment has a simple glow curve and all the electrons in the trap were read out. To determine the TL intensity performance of the TLD CaSO4:Dy from the experiment, a comparison with the existing TLD was made, as shown in Figure 4.

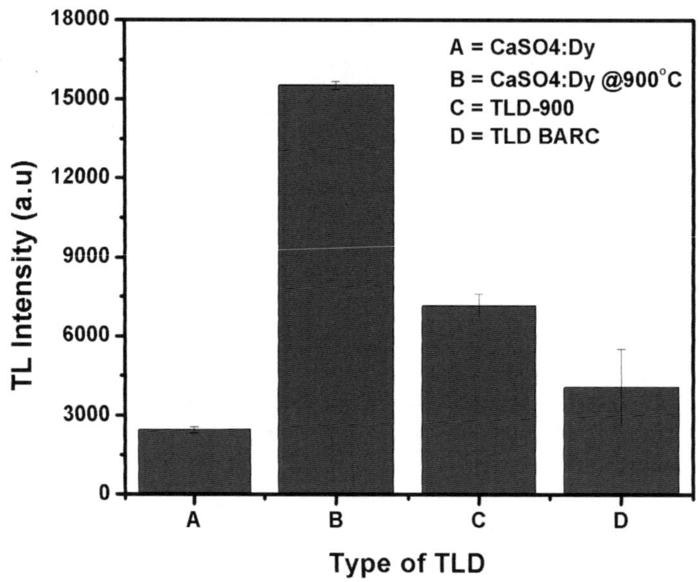

Figure 4. Comparison of TL intensity of TLD $CaSO_4$:Dy from the experiment vs. existing TLD.

4. Conclusion

Re-annealing caused a decrease in grain size or crystal size of $CaSO_4$:Dy. On the other hand, an increase of the re-annealing temperature increased the sensitivity of the TLD (high TL intensity). The TL response of the TLD obtained in the experiment was higher than that of an existing TLD in the market. The main peak of the obtained TLD in this experiment was between 160 and 200 °C. The TLD obtained in the experiment had good glow-curve properties; all radiation stored in the TLD was read out.

Acknowledgements
The authors thank to Work Safety and Dosimetry Division, Environment lab , Centre for Safety Technology and Radiation Metrology, National Nuclear Energy Agency (BATAN) and all colleagues in Energy and Environmental Material Laboratory, Department of Physics, ITB. The research was funded by the research grant of Riset Inovasi ITB Fiscal Year 2016.

References

[1] United Nations Scientific Committee on The Effect of Atomic Radiaiton, Source and Effect of Ionizing Radiation, Report to General Assembly with Scientific Annexes 2010 (UN Publications, New York)
[2] Kron T 1995 Thermoluminescence Dosimetry aud its application in medicine part 2: history and applications *Australian Physical & Engineering Science in Medicine* **18** 1-25
[3] Khan F.M 1987 The Physics of Radiation Therapy, (MD: Lippincott, Williams and Wilkins, Baltimore) pp.144-148
[4] Azorin J 2014 Preparation methods of thermoluminescent materials for dosimetric applications: An overview *Applied Radiation and Isotopes* 83 **Part C**187-191
[5] Nambi K S V, Bapat V N and Ganguly A K 1974 Thermoluminescence of $CaSO_4$ doped with rare earth *J. Phys. C: Solid State Phys.*, **7** 4403-15
[6] Salah N, P.D. Sahare, S.P. Lochab and Pratik Kumar 2006 TL and PL studies on $CaSO_4$: Dy nanoparticles *Radiation Measurements* **41**(1) 40-47

1-D DC Resistivity Inversion Using Singular Value Decomposition and Levenberg-Marquardt's Inversion Schemes

M Heriyanto and W Srigutomo

Physics of Earth and Complex System, Faculty of Mathematics and Natural Sciences, Institut Teknologi Bandung, Indonesia

mheriyanto37@gmail.com

Abstract. Exploration of natural or energy resources requires geophysical survey to determine the subsurface structure, such as DC resistivity method. In this research, field and synthetic data were used using Schlumberger configuration. One-dimensional (1-D) DC resistivity inversion was carried out using Singular Value Decomposition (SVD) and Levenberg-Marquardt (LM) techniques to obtain layered resistivity structure. We have developed software to perform both inversion methods accompanied by a user-friendly interface. Both of the methods were compared one another to determine the number of iteration, robust to noise, elapsed time of computation, and inversion results. SVD inversion generated faster process and better results than LM did. The inversion showed both of these methods were appropriate to interpret subsurface resistivity structure.

1. Introduction

DC Resistivity is one of geophysical exploration methods that can obtain subsurface resistivity structure. This method is frequently utilized for archeology surveys [15], mineral exploration [14], geothermal [13], and groundwater investigation [10,12].

Measurement principle of DC resistivity is injecting an amount of electric current into the earth to acquire potential difference on the surface thus apparent resistivity would be obtained based on Ohm's law. Regarding its measurement configuration, this method is divided into some types, one of examples is Schlumberger configuration. This configuration is usually used for obtaining resistivity variation to its depth in 1-D orientation. Generally, resistivity model is interpreted in 2-D and 3-D domain, nevertheless for shallow structure and not too complex geology condition, interpretation can be done one-dimensionally. The apparent resistivity data acquired in the measurement is not real, thus interpretation demands an inversion process to obtain subsurface resistivity model. Several inversion methods which can be applied to apparent resistivity data in 1-D case are stable iterative [11], ridge regression [5]. Levenberg-Marquardt [3,9] and singular value decomposition [1] are methods that often be used for solving geophysical problem but there is still a few of scientists quantitatively comparing the advantages and disadvantages of these two methods.

In this paper, we compared Levenberg-Marquardt (LM) to singular value decomposition (SVD) inversion method to determine each robustness of the methods.

2. Forward Modeling and Inversion

Within this section, theories concerning forward modeling and inversion of DC resistivity with Schlumberger's configuration are going to be explained.

2.1. Forward Modeling

As commonly used in forward modeling DC resistivity with Schlumberber's configuration [1,4]. The mathematical relation between Schlumberger's apparent resistivity data and subsurface layer's parameter can be described by Hankel's integral.

$$\rho_a = \left(\frac{AB}{2}\right)^2 \int_0^{\infty} T_1(\lambda) J_1\left(\frac{AB}{2}\lambda\right)\lambda d\lambda \tag{1}$$

where $\frac{AB}{2}$ is half the current electrode spacing in Schlumberger configuration, λ is electrode parameter, J_1 is the first-order Bessel's function of the first kind, $T_1(\lambda)$ is given by the recursive formulation [7] as follows:

$$T_{i-1}(\lambda) = \frac{T_i(\lambda) + \rho_{i-1}\tanh(\lambda t_{i-1})}{1 + T_i(\lambda)\tanh(\lambda t_{i-1})/\rho_{i-1}} \text{ ; i = NL, NL-1, ..., 1} \tag{2}$$

where ρ_i and t_i are resistivity and thickness of *i*th layer, NL is the consecutive number of layers. For the last layer, $T_{NL} = \rho_{NL}$. Using fast Hankel transforms, that apparent resistivity data for DC resistivity Schlumberger configuration can be calculated as [8]:

$$\rho_a = \sum_K T_1(\lambda_K) f_K \tag{3}$$

where K is the number of coefficient and f_K are the filter coefficients.

2.2. Inversion

Below is nonlinear relation between model parameter and data:

$$\mathbf{G}(\mathbf{m}) = \mathbf{d} \tag{4}$$

where **d** is vector of data with size of N x 1, **G** is nonlinear function relating data to model parameter and **m** vector of model parameters with size of M x 1. N is number of data and M is number of models. In this DC resistivity case, the data to be scrutinized is apparent resistivity (ρ_a) vs AB/2 and the model parameters are true resistivity (ρ) and layer's thickness (t).

Objective function to be minimized is [2]:

$$Y(\mathbf{m}) = \sum_{i=1}^{N}\left(\frac{G(\mathbf{m})_i - d_i}{\sigma_i}\right) \tag{5}$$

where σ_i is deviation of *i*th data.

2.2.1. Levenberg-Marquardt Scheme. Iteration equation of model update in this method is expressed as follow [2,3]:

$$\left(\mathbf{J}(\mathbf{m}^k)^T \mathbf{J}(\mathbf{m}^k) + \lambda\mathbf{I}\right)\Delta\mathbf{m} = -\mathbf{J}(\mathbf{m}^k)^T \mathbf{F}(\mathbf{m}^k) \tag{6}$$

where **I** is the identity matrix and λ is a damping factor. $\Delta\mathbf{m} = \mathbf{m}^{k+1} - \mathbf{m}^k$ with k is index of iteration. $\mathbf{F}(\mathbf{m}^k)$ is vector of differences between observed data and calculated data:

$$\mathbf{F}(\mathbf{m}) = \begin{bmatrix} f_1(\mathbf{m}) \\ \vdots \\ f_N(\mathbf{m}) \end{bmatrix} \tag{7}$$

where

$$f_i(\mathbf{m}) = \sum_{i=1}^{N} \left(\frac{G(\mathbf{m})_i - d_i}{\sigma_i} \right) \tag{8}$$

Meanwhile, **J(m)** is a sensitivity matrix or Jacobi written as below:

$$\mathbf{J}(\mathbf{m}) = \begin{bmatrix} \frac{\partial f_i(\mathbf{m})}{\partial m_1} & \cdots & \frac{\partial f_i(\mathbf{m})}{\partial m_M} \\ \vdots & \ddots & \vdots \\ \frac{\partial f_N(\mathbf{m})}{\partial m_1} & \cdots & \frac{\partial f_N(\mathbf{m})}{\partial m_M} \end{bmatrix} \tag{9}$$

Thus equation (6) can be written as follow:

$$\mathbf{m}^{k+1} = \mathbf{m}^k + \left(\mathbf{J}(\mathbf{m}^k)^T \mathbf{J}(\mathbf{m}^k) + \lambda \mathbf{I}\right)^{-1} \left[-\mathbf{J}(\mathbf{m}^k)^T \mathbf{F}(\mathbf{m}^k)\right] \tag{10}$$

Damping factor can be determined using golden section search (GSS) method [4] at each iteration.

2.2.2. Singular Value Decomposition Scheme. Iteration equation of model update in this method is based on a formulation by Meju (1994), where matrix **J** to be factorized into 3 matrices:

$$\mathbf{J} = \mathbf{U}\mathbf{S}\mathbf{V}^T \tag{11}$$

where **U** eigenvector of data with N x M size, **V** is eigenvector of model parameters with M x M size and **S** is diagonal matrix containing non-zero eigenvalue of **J** as many as R with R ≤ M which R indicates *R*th iteration. Then equation (11) is substituted into equation (6) thus yields:

$$\left(\mathbf{U}\mathbf{S}^2\mathbf{V}^T + \lambda \mathbf{I}\right)\Delta\mathbf{m} = -\mathbf{V}\mathbf{S}\mathbf{U}^T\mathbf{F}(\mathbf{m}^k) \tag{12}$$

where **S** is diagonal matrix containing singular values of matrix **J** $(\mu_1, \mu_2, \mu_3, \cdots, \mu_M)$.

$$\left(\mathbf{U}\mathbf{S}^2\mathbf{V}^T + \lambda \mathbf{I}\right) = \left(\mathbf{V}\mathrm{diag}(\mu_j^{\,2})\mathbf{V}^T + \lambda \mathbf{I}\right);\ \mathrm{j} = 1, 2, \ldots, \mathrm{M} \tag{13}$$

Ekinci and Demirci (2008) has derived equation (13) and obtained solution equation for this inversion problem:

$$\mathbf{m}^{k+1} = \mathbf{m}^k + \mathbf{V}\mathrm{diag}\left(\frac{\mu_j}{\mu_j^{\,2} + \lambda}\right)\mathbf{U}^T\left[-\mathbf{F}(\mathbf{m}^k)\right] \tag{14}$$

Damping factor is determined using formulation below [6]:

$$\lambda = \mu_L \Delta x^{\frac{1}{L}} \tag{15}$$

where L is test number performed and Δx is formulated as follow:

$$\Delta x_R = \frac{(x_{R-1} - x_R)}{x_{R-1}} \tag{16}$$

where x_{R-1} and x_R is misfit value of that the previous and the present iteration.

3. Applications and Results

In this section both inversion methods were applied to synthetic and field data. In order to facilitate analyzing quantitatively, we employed parameter of data's RMSE (root-mean-square error), models' RMSE, number of iteration and elapsed computing time.

RMS error is defined as follow:

$$\text{Data RMSE} = \sqrt{\frac{1}{N}\sum_{i=1}^{N}\left(d_i - G(\mathbf{m})_i\right)^2} \tag{17}$$

where d is observed data and $G(\mathbf{m})$ is calculated data obtained from calculation of equation (4). Then relative RMSE is defined as data RMSE which is divided by d at each data and multiplied with 100%.

$$\text{Relative RMSE} = \sqrt{\frac{1}{N}\sum_{i=1}^{N}\left(\frac{d_i - G(\mathbf{m})_i}{d_i}\right)^2} x100\% \tag{18}$$

$$\text{Model RMSE} = \sqrt{\frac{1}{M}\sum_{i=1}^{M}\left(m_i^{True} - m_i^{Est}\right)^2} \tag{19}$$

where m_i^{True} and m_i^{Est} is the real model and the estimated model at ith iteration.

3.1. Synthetic data

LM and SVD methods were applied to inversion with 3 test models of homogenous earth (model 1), 3-layered (model 2), and noisy 3-layered (model 3).

Forward modeling data calculated from those 3 models were then inverted using LM and SVD. The result, those 3 sets of synthetic data obtained small RMS errors, it can be seen on table 2. This is also indicated by curve of apparent resistivity vs. AB/2 of which observed data and calculated data have fit figures to each other (see figure 1a), 1b), and 1c)).

Table 1. True and estimated models: (a) homogenous model; (b) noise-free model; (c) noisy model (15% random noise)

		Calculated					Calculated					Calculated		
(a)	**True**	**SVD**	**LM**	**Unit**	**(b)**	**True**	**SVD**	**LM**	**Unit**	**(c)**	**True**	**SVD**	**LM**	**Unit**
ϱ_1	10	10	10	Ω m	ϱ_1	100	99.98	99.99	Ω m	ϱ_1	100	108.39	107.97	Ω m
ϱ_2	10	10	10	Ω m	ϱ_2	50	49.96	49.93	Ω m	ϱ_2	50	52.69	49.58	Ω m
ϱ_3	10	10	10	Ω m	ϱ_3	10	10	10	Ω m	ϱ_3	10	10.79	10.68	Ω m
t_1	10	5.34	5.78	m	t_1	5	5.01	5.01	m	t_1	5	5.45	5.87	m
t_2	10	4.63	4.03	m	t_2	20	20	20.01	m	t_2	20	18.91	19.65	m
										(+ noise)				

Table 2. Observed and calculated apparent resistivity data of model 1 (a), model 2 (b), model 3 (c).

AB/2 (m)	Apparent resistivity (Ω m)								
	Model 1			Model 2			Model 3		
		Calculated			Calculated			Calculated	
	Observed	SVD	LM	Observed	SVD	LM	Observed	SVD	LM
1	10	10	10	99.95	99.93	99.94	109.78	108.35	107.94
2	10	10	10	99.52	99.51	99.51	100.05	107.97	107.62
3	10	10	10	98.48	98.47	98.47	111.02	107.04	106.8
5	10	10	10	94.26	94.26	94.26	107.46	103.14	103.32
7	10	10	10	87.95	87.96	87.96	96.9	97	97.63
10	10	10	10	77.63	77.65	77.65	86.45	86.25	87.15
15	10	10	10	64.02	64.03	64.02	71.15	70.8	71.2
20	10	10	10	55.24	55.24	55.22	58.49	60.19	59.94
25	10	10	10	49.02	49.01	49	53.84	52.62	52.05
30	10	10	10	43.93	43.93	43.91	45.06	46.58	45.99
40	10	10	10	35.31	35.3	35.3	39.05	36.78	36.53
50	10	10	10	28.31	28.31	28.31	28.44	29.18	29.26
60	10	10	10	22.96	22.96	22.97	23.92	23.57	23.82
80	10	10	10	16.4	16.4	16.4	16.51	16.93	17.19
100	10	10	10	13.29	13.29	13.29	13.48	13.91	14.06
120	10	10	10	11.85	11.85	11.85	13.31	12.54	12.59
150	10	10	10	10.94	10.94	10.94	12.08	11.69	11.66
200	10	10	10	10.44	10.44	10.44	10.94	11.22	11.14
250	10	10	10	10.26	10.26	10.26	11.72	11.04	10.95
300	10	10	10	10.18	10.18	10.17	10.23	10.96	10.86
Data RMSE		**1.8e-10**	**5.4e-15**		**0.0089**	**0.0097**		**2.389**	**2.369**
Model RMSE		**1.42**	**1.46**		**0.0089**	**0.014**		**1.785**	**1.613**
Iter. No		**94**	**11**		**57**	**87**		**100**	**100**
Time (s)		**59.19**	**11.63**		**36.7**	**84.23**		**14.22**	**95.19**

The constructed homogenous model was a model with the same resistivity value over spaces and depth of 10 meters. Noisy 3-layered model, with random value of 15 % was constructed with 3 resistivity values and 2 layer's thicknesses and final apparent resistivity multiplied by 15% random value of apparent resistivity.

Inversion result of those two methods depended on initial model. If the initial model was too far from solution model thus its inversion process would obtain big calculation data and model RMSE values, also get trapped in local minimum. In model 1, the initial model for inversion were three resistivity values of 5 Ω m and depth values of 5 m. Meanwhile in model 2 and 3, the initial model were 3 resistivity values of 80 Ω m with layer's depth of 10 m.

Inversion result using model 1 obtained relative RMSE as much as 0 % for both SVD and LM, using model 2 resulted in 1.42e-6 % and 9.25e-6 %, for SVD and LM respectively, and using model 3 obtained

0.15 % and 16 %. LM method had a good inversion solution when being applied to model 1 and 3. This can be seen from the data RMSE and model RMSE as presented by table 2. Furthermore, considering the numbers of iteration, SVD inversion scheme spent less iteration than LM inversion scheme, especially when using model 2. Meanwhile, considering iteration time, SVD was relatively faster than the other one especially when using model 2 and 3 (see table 2).

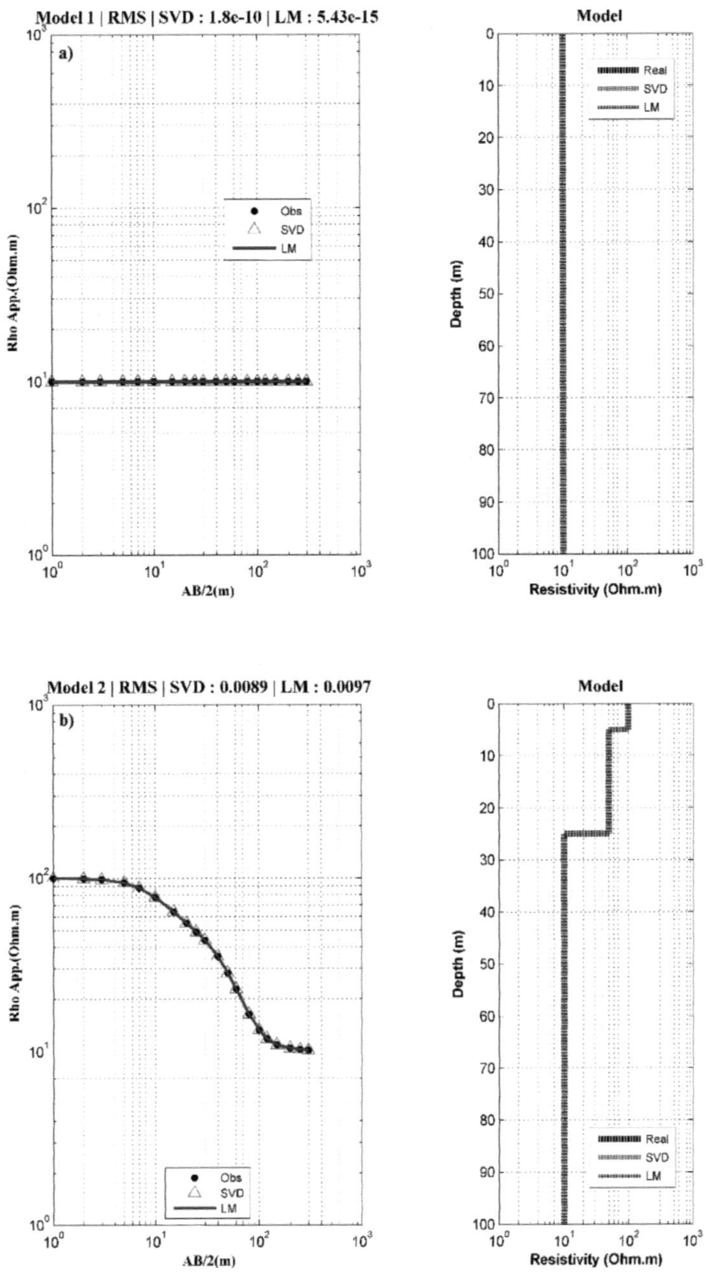

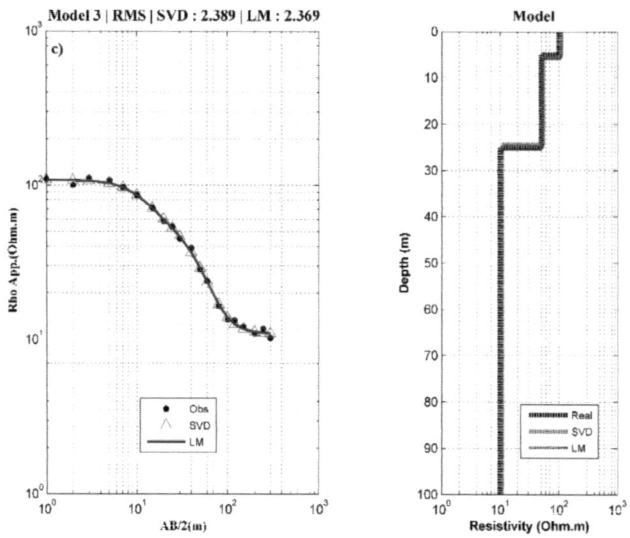

Figure 1. Left curve presents observed and calculated apparent resistivity data using a) model 1 b) model 2 and c) model 3. Black dot, red triangle, and blue line show observed, SVD-inverted, and LM-inverted data respectively. Right curve presents subsurface model. Black line, red line, and blue line show true, SVD-inverted, and LM-inverted model respectively.

3.2. Field data

After examining LM and SVD using synthetic data, we implemented those methods to field data. We used Kaleköy's data, located in northeastern part of Gökçeada [1]. Geological condition in the area has a thick sedimentary sequence, alluvium, and volcanic rock. The measurement was conducted using Iris Syscal-R1 + resistivity meter with maximum electrode space (AB) as long as 120 m.

Table 3. The estimated model of the field data

Model Parameter	Kaleköy		Gobi, China	
	SVD	**LM**	**SVD**	**LM**
ϱ_1	50.529	47.105	24.826	27.239
ϱ_2	18.862	10.665	69.653	77.219
ϱ_3	4.068	12.57	24.787	5.195
ϱ_4	219.527	388.252	474.424	343.836
t_1	2.659	3.964	6.565	11.209
t_2	10.577	38.927	38.886	64.411
t_3	16.029	15.941	582.457	103.027
Data RMSE	**0.686**	**1.6736**	**0.946**	**4.302**
Iter. No.	**100**	**100**	**100**	**100**
Time (s)	**66.147**	**110.53**	**47.407**	**104.151**

We selected the second field, Gobi area of West China [9]. Based on information provided by borehole measurement around the location, there were 4 layers of subsurface: Quaternary (0-4.63 m), Tertiary (4.63-48.34 m), Permian and Carboniferous (48.34-610 m), and Precambrian (basement).

The data inversion of both fields depended on initial model and model values would get trapped in local minimum misfit if initial models was less good. We gave the initial models consisting of 4 resistivity models and 3 thickness models. Kaleköy's inverted data had 0.686 and 1.6736 of data RMSE (see table 3), meanwhile relative RMSE as much as 0.08% and 0.64% for SVD and LM respectively while Gobi area's had 0.946 and 4.302 of data RMSE (see table 3), meanwhile relative RMSE as much as 0.07% and 1.47% for SVD and LM respectively. Hence, it can be seen that SVD inversion obtained smaller data RMSE and relative RMSE than LM inversion did, which indicated the calculated data had a good fit with observed (field) data.

The number of iteration in SVD and LM inversion using Kaleköy's data and Gobi's data were the same. Therefore, it could not become parameters to compare both inversion methods to each other. Considering computation time, LM inversion needed to take longer than SVD inversion (see table 3).

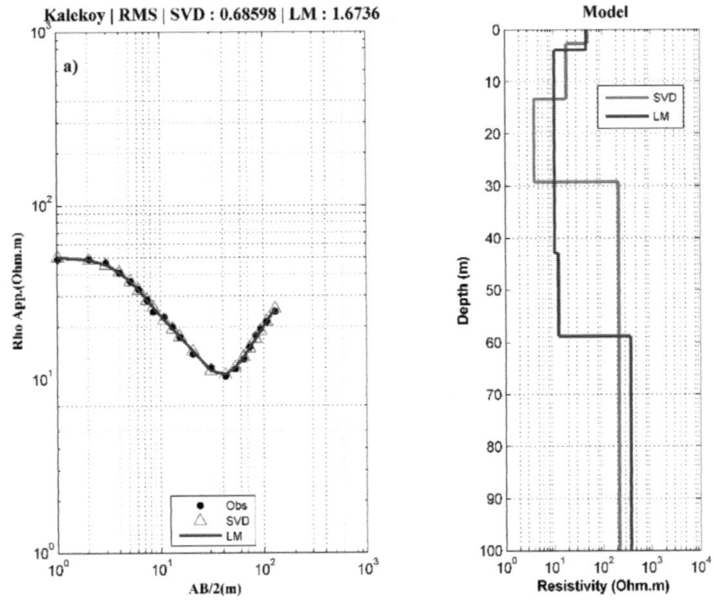

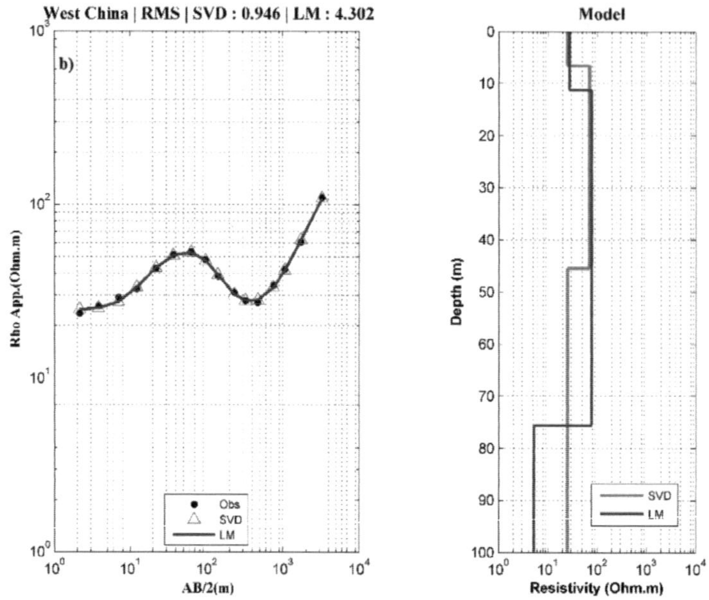

Figure 2. Inversion result of a) Kaleköy's data and b) Gobi area's data, West China. Left curve presents observed and calculated apparent resistivity data. Black dot, red triangle, and blue line show observed data, calculated SVD, and calculated LM data respectively. Right curve presents subsurface model recovered values. Red and blue line indicate result using SVD and LM model respectively.

4. Discussion

Considering to synthetic data from test model in table 1, it can be seen that with the same initial model, LM inversion method was good to be used to invert synthetic data from model 1 (homogenous) and model 3 (3-layered noisy), while SVD inversion method was good to be used to invert synthetic data model 2 (3-layered noise-free). This was probably because of good GSS's performance in determining damping factor when the constructed model was ideal (referred to synthetic data).

Computation time of LM inversion was longer than SVD inversion's in spite of the same number of iteration. This was caused by selecting optimum damping factor value using GSS method as each LM inversion's step must be evaluated within the range of damping factor value thus the inversion elapsed long computation time. Hence, it is recommended to use another method for determining the damping factor in order to make the inversion faster but still accurate in its results.

In both fields' data, with the same number of calculation iteration, the value of data RMSE obtained (see table 3) shows that SVD was faster than LM, has smaller number of iteration and data RMSE than LM does. This was caused by LM's dependence on damping factor selection with GSS method. In synthetic data case (ideal), GSS could determine optimum damping factor relatively easier than of that in field data case, which was relatively complex, because GSS's principle only save the ultimate output regardless detail of Jacobi matrix's sensitiveness. Meanwhile, in SVD, eigenvalue selection from Jacobi matrix was good enough because the method considered more the matrix's sensitiveness, nevertheless damping factor selection still used trial-and-error technique.

5. Conclusions

We have successfully examined Levenberg-Marquardt and singular value decomposition inversion method, and compared them to each other. The LM method was good to be used in ideal synthetic data. However, we should better to use SVD in the field data case because this method is faster than LM and producing smaller number of iteration, relative RMSE and data RMSE than LM does.

Acknowledgments

The authors gratefully thank to Prihandhanu M. Pratomo for constructive comments and suggestions for improving the manuscript, Firman I. Bismillah for reviewing and editing the manuscript's English structure, and all members of Modeling and Inversion Laboratory.

References

[1] Ekinci Y L and Demirci A 2008 A Damped Least-Squares Inversion Program for the Interpretation of Schlumberger Sounding Curves *J. of Appl. Sciences* 8 (**22**): 4070-4078

[2] Srigutomo W, Agustine E, and Zen M H 2006 Quantitative Analysis of Self-Potential Anomaly: Darivative Analysis, Least-Square Method and Non-Linier Inversion *Indonesian Journal of Physics* **17** No 2

[3] Rocky M and Srigutomo W 2015 Comparison of 1D Magnetotelluric Inversion using Levenberg-Marquardt and Occam's Inversion Schemes *AIP Conf. Proc.* **1656** 070014-1-070014-4

[4] Vedanti N, Srivastava R P, Sagode J, and Dimri V P 2005 An efficient 1D OCCAM'S Inversion Algorithm Using Analytically Computed First- and Second-order Derivatives for DC Resistivity Sounding *Computers & Geosciences* **31** 319-328

[5] Meju M A 1992 An Effective Ridge Regression Procedure for Resistivity Data Inversion *Computers & Geosciences* Vol **18** No 2/3 pp 99-118

[6] Arneson K and Hersir G P 1988 One Dimensional Inversion of Schlumberger Resistivity Soundings: Computer Program, Description and User's Guide: The United Nation University, Geothermal Training, Report 8, pp 59

[7] Koefoed O 1976 Error Propagation and Uncertainly in the Interpretation of Resistivity Data *Geophysical Prospecting* **24** 31-48

[8] Ghosh D P 1971 Inverse Filter Coefficient for the Computation of Apparent Resistivity Standard Curves for a Horizontally Layered Earth *Geophysical Prospecting* **29** 769-775

[9] Yin C 2000 Geoelectrical Inversion for One-dimensional Anisotropic Model and Inherent Non-uniqueness *Geophys. J. Int.* **140** 11-23

[10] Harja A, Srigutomo W, Mustofa E J, and Sutarno D 2007 CSAMT and DC-Resistivity Survey for Groundwater and Structural Investigation: Application to Eastern Bandung Basin, Indonesia *Proceesings of the 2007 Asian Physics Symposium* ISBN 978-979-17090-1-9

[11] Jupp D L B and Vozoff 1974 Stable Iterative Methods for the Inversion of Geophysical Data *Gephys. J. R. Astr. Soc.* **42**, 957-976

[12] Benson A K, Payne K L, and Stubben M A 1997 Mapping Groundwater Contamination using DC Resistivity and VLF Geophysical Methods-a Case Study *Geophysics* **62**, 80-86

[13] Verave R T, Mosusu N, and Irarue P 2015 1D Interpretation of Schlumberger DC Resistivity Data from the Talasea Geothermal Field West New Britain Province, Papua New Guinea *Proceedings World Geothermal Congress 2015*

[14] Srigutomo W, Trimadona, Pratomo P M 2016 2D Resistivity and Induced Polarization Measurement for Manganese Ore Exploration *Journal of Physics: Conference Series* **739**, 012138

[15] Noel M and Xu B 1991 Archaeological Inverstigation by Electrical Resistivity Tomography: a Premilinary Study *Geophys. J. Int.* **107**, 95-102

Neutronics Analysis of SMART Small Modular Reactor using SRAC 2006 Code

Rahmi N. Ramdhani[1], Puguh A. Prastyo[1], Abdul Waris[2,*], Widayani[2] and Rizal Kurniadi[2]

[1]Department of Physics, Faculty of Mathematics and Natural Sciences,
Institut Teknologi Bandung,
[2]Nuclear Physics & Biophysics Research Division, Department of Physics,
Faculty of Mathematics and Natural Sciences,
Institut Teknologi Bandung,
*E-mail: awaris@fi.itb.ac.id

Abstract. Small modular reactors (SMRs) are part of a new generation of nuclear reactor being developed worldwide. One of the advantages of SMR is the flexibility to adoptthe advanced design concepts and technology. SMART (System integrated Modular Advanced ReacTor) is a small sized integral type PWR with a thermal power of 330 MW that has been developed by KAERI (Korea Atomic Energy Research Institute). SMART core consists of 57 fuel assemblies which are based on the well proven 17x17 array that has been used in Korean commercial PWRs. SMART is soluble boron free, and the high initial reactivity is mainly controlled by burnable absorbers. The goal of this study is to perform neutronics evaluation of SMART core with UO_2 as main fuel. Neutronics calculation was performed by using SRAC 2006 code with JENDL 3.3 as nuclear data library.

1. Introduction

The world's energy sector is currently driven by a majority of fossil fuel power plants. In order to address and meet national and global requirements for clean, efficient, and consistent energy production designed to meet a growing demand for energy from developing nations, new and innovative nuclear reactor technologies will need to be developed. Part of this initiative includes the potential for smaller, modular nuclear reactors (SMRs) [1].

Korea Atomic Energy Research Institute (KAERI) has been developing a small-sized integral reactor called SMART for two purposes: seawater desalination and electricity generation [2,3]. The conceptual design and the basic design of SMART with a desalination system were completed in March of 1999 and in March of 2002, respectively [4]. SMART has been developed as a small-sized integral reactor, hence most of its major components is housed within the reactor vessel. Major components including 4 main coolant pumps, 12 steam generator cassettes, a self-pressurizeras well as active core are placed in a single reactor vessel as shown in Fig.1. This integral design eliminates large break LOCA (Loss of Coolant Accident). In its development, SMART also adopted the soluble boron free operation concept. Hence there is no need to control the boron concentration of the primary coolant and the system can be further simplified. However, an emergency boron storage tank is provided in case of scram failure [5].

Nuclear calculation for SMART reactor has been performed using CASMO-3/MASTER system and MCNAP code system. At first burnup period, effective multiplication factor for SMART core is 1.000162 calculated using CASMO-3/MASTER [6] system and 1.00641 with MCNAP code [7]. In this study, nuclear calculation for SMART reactor will be performed using CITATION module of SRAC 2006 code [8] that was developed by Japan Atomic Energy Agency (JAEA) with JENDL 3.3 as its nuclear data library [9].

2. Nuclear Core Design

As shown in Fig.2, the reactor core consists of 57 FAs with a 17 x 17 fuel rod array. The fuel assembly is based on the design of KOFA (Korean Standard Fuel Assembly) that was designed by KAERI/Siemens-KWU and used in the 900 MWe Westinghouse type Korean PWR's [3]. The fuel rods are arranged in rectangular lattices, as in KOFA but the FA's length has been reduced to 200 cm. The fuel rod contained uranium dioxide and the uranium enrichment for all the FAs is 4.95%.

Other than normal fuel rods, each FAs also consist of burnable absorber rods. There are two burnable absorber rods that are used in SMART's fuel assemblies, Gadolinia rods and Shim rods. Gadolinia rods are fuel rods with Gd_2O_3 as an integral part of the fuel matrix. Burnable absorber material in shim rods are B4C in Al_2O_3.

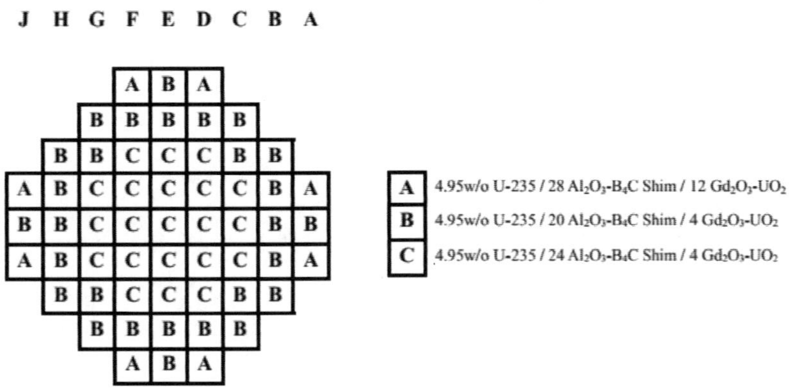

Figure 1. SMART core loading pattern [4].

In Fig. 1, SMART core loading pattern is presented, which consist of 3 fuel types. Since the neutron fluxes are higher in the core central region than in the peripheral region, more burnable absorber rods with a higher concentration of B-10 are used in fuel type C than in fuel type B. The number and concentration of the burnable absorber rods in each fuel type are selected so that reactivity of each assembly can be as flat as possible

Table 1. SMART major parameters

Parameter	Value
Thermal Output	330 MWt
Electric Output	100 MWe
Fuel Material	< 5 wt% UO2
Refueling Cycle	3 years
Active Length	200 cm
Cladding Material	Zircaloy-4
Cladding Thickness	0.57 mm
Pellet Diameter	8.05±0.01 mm
Pin Pitch	1.26 cm
Fuel rod to fuel rod	21.10 x 21.10
Material Absorber	Gd_2O_3-UO_2&Al_2O_3-B_4C

3. Methodology

3.1 Fuel assembly designs

There are three fuel assembly designs in this study. Each assembly has different configuration and specification. As mentioned above all the fuel assembly that was used in this study were developedbased on KOFA fuel assembly. In Fig. 2, we can see the difference in each fuel assemblies.

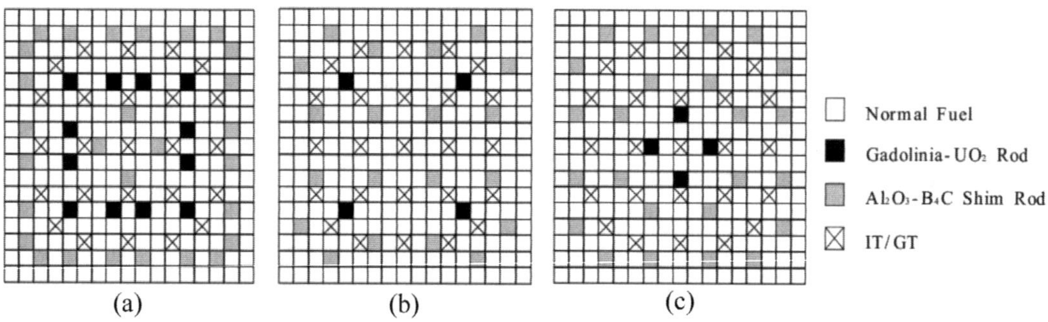

Figure 2. Fuel assembly designs used in this study (a) FA type A (b) FA type B (c) FA type C[7].

The number and concentration of burnable absorber rods in each FAs are selected so that reactivity of each FAs can be as flat as possible. The specification of each FA in this study can be seen in Table 2.

Table 2. Fuel Assembly Specification

Assembly Type	No. of Assembly	No. of Fuel Rods	No. AL_2O_3- B_4C rods	No. of Gd_2O_3-UO_2 rods
A	8	224	28	12
B	28	240	20	4
C	21	236	24	4

The fuel rod consisted of a uniform cylindrical pellet stacked together within a zircaloy-4 clad tube. Between the fuel stack and the cladding, a clearance was provided in order to accomodate the fuel swelling due to the accumulation of fission products. The gap was filled with helium gas to improve heat conduction from fuel to cladding.

3.2 Computational procedure

All the fuel assemblies and reactor core were modelled using SRAC (ver. 2006). The cross section for all materials (i.e. fuel, burnable absorber material, cladding, moderator and structural) were taken from JENDL 3.3 libraries. SRAC (*Standard Reactor Analysis Code system*) was developed by JAERI (*Japan Atomic Energy Research Institute*). This programme was used to analyze reactor design, especially for its neutronic analysis.

The study was carried out by referencing SMART fuel assembly and core design from previous studies. The fuel assembly burnup was calculated by using octant symmetric square with square array of pin rods geometry and the core calculation for 3D cylinder geometry.

4. Results and Discussion

The neutronic behaviour of the reactor was studied on three types of fuel assembly for 990 Effective full power days (EFPD). Figure 3 demonstrates the neutron infinite multiplication factor (k-infinity) for all assemblies which given in Fig. 2. The fuel assembly type C gives the highest infinite multiplication factor compared to others.

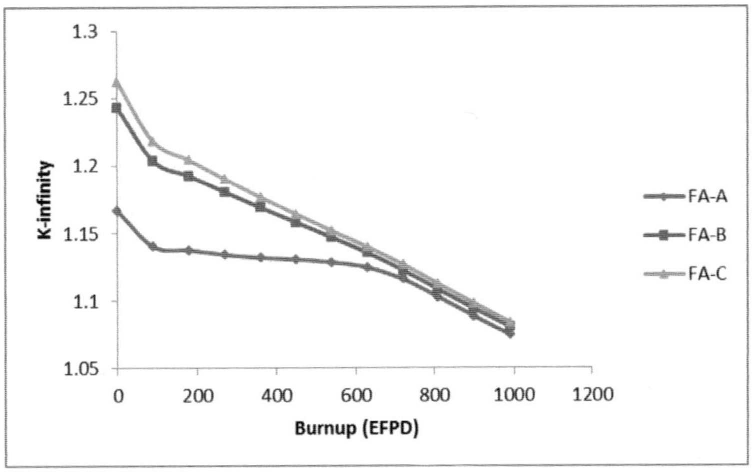

Figure 3. Infinite multiplication factors (k_{∞}) of fuel assemblies.

Figure 4 shows the conversion ratio for all FAs. As can be seen tfrom this figure, the conversion ratio for FA type C has the lowest value compared to other FAs with the same fuel enrichment. This profile shows that the trend of the infinite multiplication factor is inversely proportional with that of the conversion ratio.

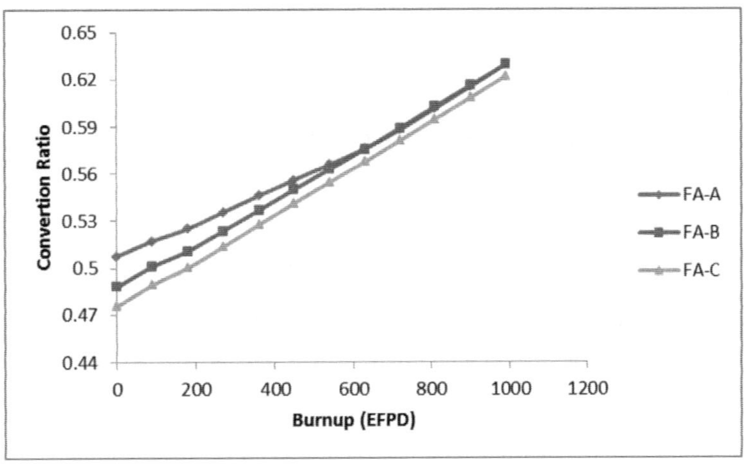

Figure 4. Convertion ratio of fuel assemblies.

Once each fuel assembly calculation finished, the next step is placing each fuel assemblies into its correct pattern according to Fig. 1. The core was surrounded by a stainless steel barrel. The 3D core calculation has been conducted by using with CITATION module in SRAC (ver. 2006). The calculation were carried out in order to find the effective multiplication factor (k-eff) and other parameters. The k- eff value for the first burn up periode which obtained from this simulation is 1.006119. Tabel 3 presents the comparation of k-eff values for the present research compared to different codes.

Table 3. Comparation of k_{eff} values

Burnup (days)	MASTER	MCNAP	CITATION
0	1.000162	1.00641	1.006119
90	1.000107	1.00186	1.00589

Another important nuclear parameters from this core calculation is average burnup. The number of fuel that was depleted in each burnup period is almost the same. The cycle average burn up from this calculation is 25.625 MWD/kgU. This means that the energy produced from this reactor design for 990 EFPD are equal to 2.5625 x 10^4 MWD for a ton of fuel.

Figure 5 shows the neutron spectrum for this reactor design. This graph shows that SMART reactor is a thermal reactor with slightly harder neutron spectrum. The harder neutron spectrum means that the spectrum shifts to high energy region [10-11].

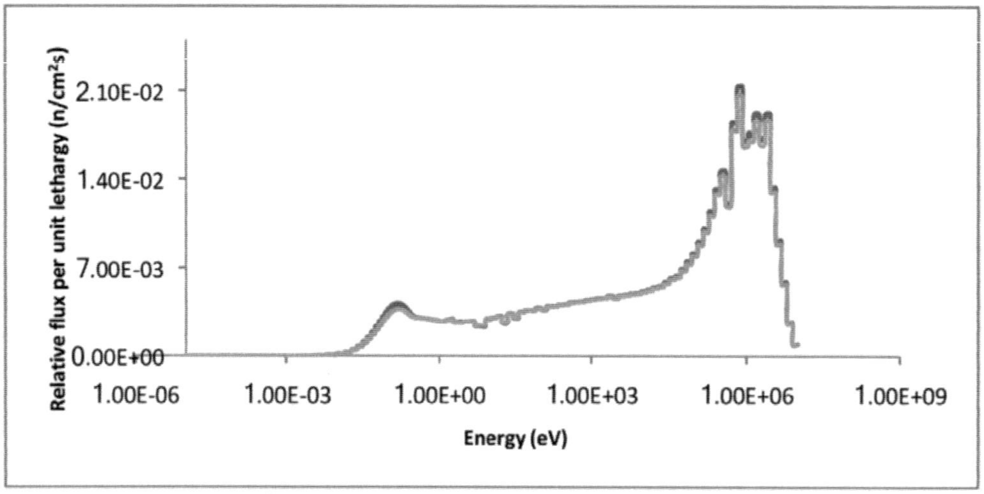

Figure 5. Neutron spectrum of SMART core

5. Conclusion

The study on SMART core design description and nuclear characteristics have been performed. Using 4.95wt% UO2 fuel, a 3-years cycle length is accomplished with the cycle average burn up from this design is 25.625 MWD/kgU. The obtained results show similar values compared to the references.

References

[1] Carelli M, Garrone G, Mancini M, Trucco M, and Ricotti M, 2010, Progress in Nuclear Energy, Economics Features of Integral, Modular, Small-to-Medium Size Reactors 52:403-14

[2] International Atomic energy Association (IAEA), 2005, Optimization of the Coupling of Nuclear Reactors and Desalination Systems, Final Report of a Coordinated Research Project (1999– 2003) *IAEA-TECDOC 1444*

[3] Seo J K, 1997, Advanced Integral Reactor (SMART) for Nuclear Desalination. IAEA-SM-347/40. Symposium of Desalination of Seawater with Nuclear energy. Taejon Korea May 26-30 1997

[4] Lee CC, Hwang D H, Zee S Q, and Chang M H, 2003, Nuclear and Thermal Hydraulic Design Characteristics of the SMART CoreInternational Conference on Global Environment and Advanced Nuclear Power Plants, Kyoto, Japan

[5] Kim S H, Kim K K, Yeo J W, Chang M H and Zee S Q, 2003, Design Verification Program of SMARTInternational Conference on Global Environment and Advanced Nuclear Power Plants, Kyoto, Japan
[6] Korea Atomic Energy Research Institute (KAERI), 1997, Design and Development of Small and Medium Integral Reactor Core
[7] Jeong J S, Shim H J, Kim C H, 2001, Monte Carlo Depletion Analysis of SMART core by MCNP Code, Spring Conference Proceedings, Korea Nuclear Society
[8] Okumura K, Kugo T, Kaneko K and Tsuchihashi K, 2007, SRAC2006: a Comprehensive Neutronics Calculation Code System, JAEA
[9] Shibata K, et. al., 2002, Journal of Nuclear Science and Technology. Japanese Evaluated Nuclear Data Library Version 3 Revision-3: JENDL-3.3, 39 pp 1125-1136
[10] Waris A, et.al., 2012, Energy Conversion and Management, Effect of void-fraction on characteristics of several thorium fuel cycles in BWR, 63 pp 11-16
[11] Waris A, et.al., 2013, Energy Conversion and Management. Preliminary study on plutonium and minor actinides utilization in thorims-nes minifuji reactor, 72 pp 27-32

IOP Conf. Series: Journal of Physics: Conf. Series **877** (2017) 012068 doi:10.1088/1742-6596/877/1/012068

Reducing The Light Reflected by Silicon Surface Using ZnO/TS Antireflection Coating

Andi Suhandi[1], Yuyu R. Tayubi[2], Firmanul C. Wibowo[3], Pepen Arifin[4] and Supriyatman[5]

[1,2]Department of Physics Education, Universitas Pendidikan Indonesia, Indonesia
[3]Department of Physics, Intitut Teknologi Bandung, Indonesia
[4]Department of Physics Education, Universitas Sultan Ageng Tirtayasa, Indonesia
[5]Department of Physics Education, Universitas Tadulako, Indonesia

E-mail: andi_sh@upi.edu

Abstract. Zinc Oxide (ZnO) thin films was coated on a texturized silicon (TS) surface using a spincoating technique. The TS layer was prepared by a wet etching method using 20 % KOH solution at temperature of 80°C for 5 minutes. To prepared precursor solution for ZnO layer, zinc acetate dehydrate, 2-methoxyethanol and monoethanolamine are used as a starting material, solvent and stabilizer, respectively. The XRD and SEM measurements confirmed that the thin films grown by spincoating technique have a single oriented crystal plane and homogenous surfaces. From photoluminescence measurement found that the optical band gap of grown films to be 3.44 eV. The optical reflectance of the grown films is characterized by UV-VIS spectrometry show that the presence of anti-reflection coating ZnO/TS is proven to reduce the reflection of solar radiation by silicon surface significantly.

1. Introduction

Until now solar cells are widely used and marketed for the purposes at the terrestrial are made of silicon material. There are five reason why silicon today is the material that has been chosen as the material for solar cells, namely: 1) Silicon has the potential for high efficiency, 2) Silicon is available limited, second most element of the earth's crust, 3) The involved materials and processes are non-toxic and do not harm the environment, 4) The silicon technology already exists and is reliable, and 5) Already exists a broad knowledge of the materials and the devices [1].

One of the problems encountered in the development of silicon solar cells is the high reflection of solar radiation by the surface of the silicon, which reached 35% of the incoming solar radiation. The high reflection of sunlight results in the conversion efficiency of silicon solar cells is limited [2]. To overcome these problems, on the surface of solar cells usually added anti-reflections coating (ARC), that is a thin film of semiconductor material that has a different refractive index to silicon materials. Zinc oxide (ZnO) is very potential for anti-reflections coating of silicon solar cells because it has a refractive index of about 2, has a large optical bandgap in the range of 3.3 to 3.7 eV, are transparent in the visible region and has a good adhesion properties and hardness. [3]

Texturing a silicon (TS) surface that acts as a light trapping thereby forming system ZnO/TS is believed to be much longer reduce the percent of reflection light by the surface of the silicon. Some advantages of texturized silicon surface, diantaranya: 1) At least second reflection, 2) The effective absorption length of the silicon layer will be reduced, the light way through the layer increases , 3)The area of the surface becomes bigger, 4) Total internal reflection on the inside of the front layer possible, and 5) Reflection can be reduced about 9/10 of the former reflection [1]. TS are an excellent antireflection coating (ARC) for solar cells. The ARC can enhance the efficiency of solar cells by increasing light trapping in the active region [4,5]. In the last couple of years, many researchers have increasingly focused on zinc oxide (ZnO) because of its multiple applications, such as laser diodes [6], transparent conductive contacts [7], and solar cells [8,9]. ZnO has a direct band gap (3.37 eV), a high exciton binding energy (60 meV) at room temperature (RT), and a wurtzite crystal structure [10]. Recently, ZnO has attracted attention as one of the materials for the ARC of solar cells because it has good transparency and an appropriate refractive index [11,12,13].

ZnO thin films are currently fabricated by chemical vapor deposition [14], pulsed laser deposition [15], Molecular beam epitaxy [16], chemical spray hydrolysis [17], or other high-temperature and in some cases capital and equipment intensive methods. Amongst the different available techniques, the sol-gel technique has the advantage of coating on large areas with easy control of the doping level, solution concentration and homogeneity, without using expensive and complicated equipment compared with other methods.

The results on the thin films of ZnO grown by solgel method have been presented. This technique of film preparation is a low-cost process and is attractive as the film properties can be tailored conveniently for a given application. This process, thus, becomes a preferred option for exploratory studies where a large number of candidate materials require screening for their compositions and properties prior to their applications in devices.

In the present work, we report the structural, morphology, and optical properties of ZnO/TS thin films prepared by sol-gel method using spin coating technique. Recently, few studies had reported on the growth of the ZnO thin film on texturized silicon substrate.

2. Experiment

Texturized silicon (TS) surface was prepared by a wet etching method using 20 % KOH solution at temperature of 80°C for 5 minutes. To prepared the precursor solution for ZnO layer, first, zinc acetate dehydrate is first dissolved in mixture of 2-methoxy ethanol and monoethanolamine at room temperature. The sol is prepared by dissolving 8.23 grams (0.75 moles) of zinc acetate and the molar ratio of monoethanolamine to zinc acetate is kept at 1:1. The resultant solution is stirred for 1hr at 60°C to yield a homogeneous, clear and transparent solution using a magnetic stirred. The seed layers are spin coated a day after the solution is prepared. Precursor solution is dropped onto silicon substrate, which are then spinned at 1500 rpm for 30 sec. After processing, the substrate is baked at 350°C for 10 min to evaporate the solvent and remove the organic component in the film. And then the film is heating at 700°C on the furnace to form the ZnO thin solid film on the silicon surface.

The crystal structur of ZnO/TS layer was characterized by X-Ray Diffraction (XRD), the surface morphology and the cross section of ZnO/TS layer were investigated by Scanning Electron Microscopy (SEM), and optical properties of ZnO/TS layer were determined from Photoluminescence (PL) Spectroscopy and UV-Vis Spectroscopy data.

3. Result and Discussion

3.1. Texture of silicon surface and surface morphology of ZnO layer on the silicon surface

Figure 1 show the cross section SEM image of Texturized Silicon.

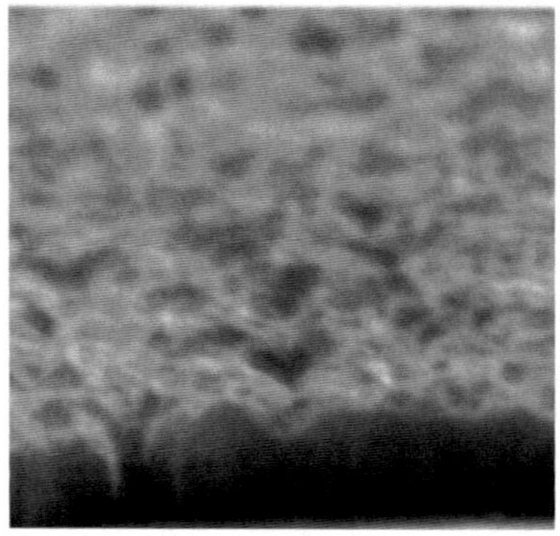

Figure 1. Cross section SEM image of Texturized Silicon

From the cross section SEM image appears that although not perfect, but it seemed to have been formed a texture on the silicon surface as a result of the wet etching process. The desired texture is pyramidal shape, but the obtained texture in this study have not been entirely pyramid shaped. This indicated that it is necessary to optimization in the etching process, both in percent of KOH solution, etching temperature, and duration of etching.

Figure 2 shows the SEM image of as grown ZnO film deposited on TS substrate. The film deposited in the optimized condition is smooth, dark gray in colour and uniformly covering the substrate with good adherence. The overall surface structure shows grains of spherical shape uniformly covering the substrate without any cracks.

Figure 2. Surfece morphology of ZnO/TS layer

3.2. Crystal structure of ZnO layer on the Texturized Silicon Surface

Figure 3 shows the X-Ray diffraction peak intensity for ZnO/TS thin film. ZnO thin films were deposited on a silicon surface both textured and non-textured have single orientation crystal planes. ZnO diffraction peaks occur at $2\theta = 34{,}40^{\circ}$ corresponding to the orientation of the crystal planes (002) that is oriented along the c-axis vertical to the surface of the silicon.

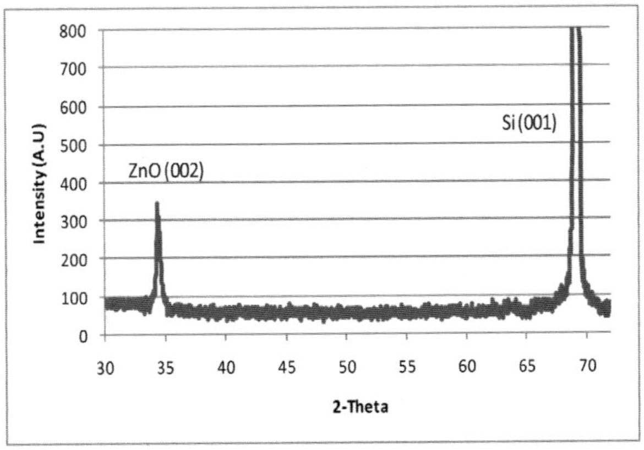

Figure 3. X-Ray diffraction peak intensity for ZnO/TS thin film

3.3. Optical Properties of ZnO layer on the Texturized Silicon Surface

Figure 4 shows the percent of light reflectanse versus wavelength obtained from UV-Vis spectroscopy measurement.

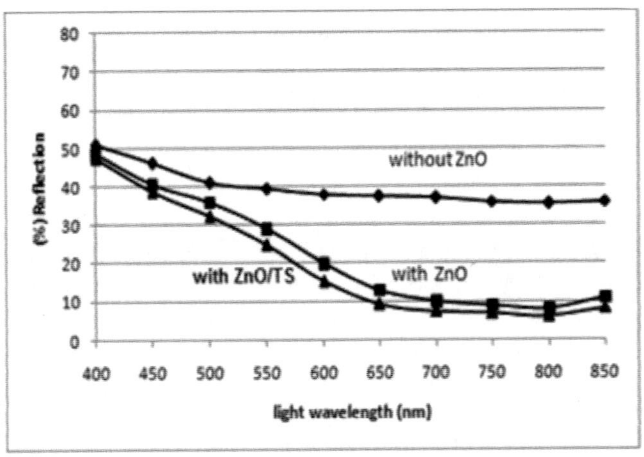

Figure 4. Percent of light reflectance versus wavelength

It showed that the addition of ZnO antirefrecting coating can degrade percent of light reflection by the silicon surface. Texturing a silicon surface making the system ZnO/TS can be further reduce percent of light reflection by the silicon surface. Presence of the texturized silicon surface can be caused the effective absorption length of the silicon layer will be reduced, the light way through the layer increases as shown in fig. 5.

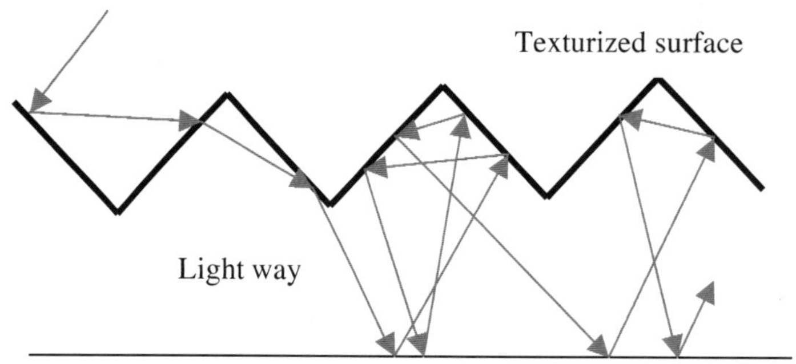

Figure 5. The increases of light way through the silicon substrate as effect of the presence of texturized silicon surface

Light trapped in the silicon material can be generate the hole-electron pair which can be produce photocurrent in the silicon solar cells. Hence the conversion efficiency of silicon solar cell can be increase.

4. Conclusion

The structural, surface morphology and optical properties of solgel spin coated ZnO/TS thin film have been investigated. The XRD spectra indicate that the film is of single crystalline structure. The SEM image showed that the film have a homogenous surface morphology. The values of the optical band gap of ZnO/TS film of about 3,44 eV. The optical reflectance of the grown films are characterized by UV-VIS spectrometry shows that the presence of anti-reflection coating ZnO/TS is proven to reduce the reflection of solar radiation by silicon surface significantly. It is thought that because of these properties, ZnO thin films can be used as an antireflection material in photovoltaic applications.

References

[1] Linder, S., 2016. *New Generation Silicon Solar Cells*, Engineering Physics, (Germany: TUM).

[2] Chen, D., 2001. Anti Reflection (AR) Coatings Made by Sol-gel Process : A Review, *Solar Energy Materials and Solar Cells*. **68**, pp. 313-336.

[3] Sevvanthi, P., Claude, A., Jayanthi, C., and Poiyamozhi, A., 2012. Instrumentation for fabricating an indigenous spin coating apparatus and growth of zinc oxide thin films and their characterizations, *Advances in Applied Science Research*, **3 (6)**, pp. 3573-3580.

[4] Yoo, J., 2010. Reactive ion etching (RIE) technique for application in crystalline silicon solar cells. *Sol. Energy,* **84**, pp. 730–734.

[5] San Vicente, G., Bayo´n, R., Germa´n, N., Morales, A., 2011. Surface modification of porous antireflective coatings for solar glass covers. *Sol. Energy,* **85**, pp. 676–680.

[6] Ye, Z.Z., Lu, J.G., Zhang, Y.Z., Zeng, Y.J., Chen, L.L., Zhuge, F., Yuan, G.D., He, H.P., Zhu, L.P., Huang, J.Y., Zhao, B.H., 2007. ZnO lightemitting diodes fabricated on Si substrates with homobuffer layers. *Appl. Phys. Lett.* **91**, pp. 113503–113505.

[7] Wong, E.M., Searson, P.C., 1999. ZnO quantum particle thin films fabricated by electrophoretic deposition. *Appl. Phys. Lett.* **74**, pp. 2939– 2941.

[8] Bahadur, L., Hamdani, M., Koenig, J.F., Chartier, P., 1986. Studies on semiconducting thin films prepared by the spray pyrolysis technique for photoelectrochemical solar cell applications: Preparation and properties of ZnO. *Sol. Energ. Mater.* 14, pp. 107–120.

[9] Sil, D., Chakrabarti, S., 2010. Photocatalytic degradation of PVC-ZnO composite film under tropical sunlight and artificial UV radiation: A comparative study. *Sol. Energy*, **84**, pp. 476–485.

[10] Pawar, R.C., Shaikh, J.S., Babar, A.A., Dhere, P.M., Patil, P.S., 2011. Aqueous chemical growth of ZnO disks, rods, spindles and flowers: pH dependency and photoelectrochemical properties. *Sol. Energy,* **85**, pp. 1119–1127.

[11] Yun-Ju, L., Ruby, D.S., Peters, D.W., Mckenzie, B.B., Hsu, J.W.P., 2008. ZnO nanostructures as efficient antireflection layers in solar cells. *Nano Lett.* **8**, pp. 1501–1505.

[12] Riaz, M., Fulati, A., Amin, G., Alvi, N.H., Nur, O., Willander, M., 2009. Buckling and elastic stability of vertical ZnO nanotubes and nanorods. *J. Appl. Phys*. **109**, pp. 034309–034314.

[13] Chen, J.Y., Sun, K.W., 2010. Growth of vertically aligned ZnO nanorod arrays as antireflection layer on silicon solar cells. *Sol. Energy Mater. Sol. Cells*, **94**, pp. 930–934.

[14] Gomez, H., Maldonado, A., Asomoza, R., Zironi, E.P., Canetasortega, J., Palaciosgomez, J., 1997. Characterisation of indium-dopped zinc oxide films deposited by pyrolytic spray with different indium compounds as doppants. *Thin Solid Films*, **Vol.293**, pp. 117-123.

[15] Ennaoui, Weber, A., Scheer, R., Lewerenz, H. J., 1998. Chemical-bath ZnO buffer layer for $CuInSe_2$ thin film soalr cells. *Solar Energy Materials and Solar Cells,* **Vol. 54**, pp. 277-286.

[16] Yang, Y., Yan, H., Fu, Z., Yang, B., Xia, L., Xu, Y., Zuo, J., Li, F., 2006. Photoluminescence and raman studies of electrochemically as-grown annealed ZnO films. *Solid State Communications*, **Vol. 138**, pp. 521-525.

[17] Banerjee, A. N., Kundoo, S., Saha, P., Chattopadhyay, K.K., 2003. Synthesis and characterization of nano-crystalline fluorin-doped thin oxide films by sol-gel method. *Journal of Sol-Gel Science and Technology,* **Vol.28**, pp. 105-110.

IOP Conf. Series: Journal of Physics: Conf. Series **877** (2017) 012069 doi:10.1088/1742-6596/877/1/012069

Optimation Pressure Loss to Oil Production System with Electrical Submersible Pump (ESP) at the Well A SW Field Bojonegoro, East Java

A M Ilmah[1], L Hendrajaya[1]

[1] Physics Departement, Institut Technology Bandung, Indonesia

Abstract. Study about oil and gas well A and SW fields have been conducted in Bojonegoro area, East Java Province. From the research, it is known that one of the constrain in oil and gas production process is a pressure drop, so that is necessary to set Electrical Submersible Pump (ESP) as an artificial lift pump at low pressure wells. It is needed to analyze the production well pressure as an effort to optimize the fluid production. It is in describe to relations between IPR (Inflow Performance Relationship) curve with TPR (Tubing Performance Relationship), are getting down point cut TPR curve with IPR curve, fluid rate its optimal, so need tubing's diameter variation and the number of stage pump ESP variation to can the most optimal fluid rate. Produced that to tubing variation are 2,441 ID is 876,729 bpd , 2,991 is 961,197 bpd , 3,476 id is 996 , 26 Bpd , while on the number of stage pump esp variation The most high is stage 400 with fluid rate 961,197 by comparison stage 338 it has value fluid rate 921,165 bpd , stage 200 it has fluid rate 827,889 bpd , and stage 169 has fluid rate 774,646 bpd .Of some variation tubing and stage pumps that has the most influence to fluid rate optimalize is the change number of stage pump..

1. Introduction

Bojonegoro is one of the cities located in the east java, which are production sites of oil and gas. One is in the SW field, which in the field there are several wells, one of which is well A. wells production in the field are generally scattered throughout the field. Each well has a production value of diverse circumstances which very good affect oil production. So the company perform a method that petroleum can be produced optimally By using a physical approach.

Formation fluid flow from the reservoir to the surface of the well passes through two stages, namely the flow of fluid from the reservoir to the bottom of the well through porous rock and then the flow of fluid from the surface to the bottom of a well past the vertical pipe media. Analyzes the relationship between fluid flow from the formation to the bottom of the well and the fluid flow from the bottom of the well to the surface of the well can be used to forecast oil production at a well.

Basically the fluid flow in oil production to make it to the surface pass through two phases: initial is the flow of fluid from the formation to bottom of well through porous media one of which is the reservoir rock , and continued with fluida flowing from the bottom of the well to the surface of the well through the media vertical pipes called tubing.

1.1. Fluid Flow of Formation To Basics wells

The ability of a well to produce can be described with IPR curve is represented as the relationship between the production rate of fluid flow to the bottom hole flow pressure (PWF) (Mahenda , 2014). Productivity Index Index is used to denote the ability of a well that is producing at certain conditions . It is expressed as the ratio between the production rate of the well in the bottomhole flowing pressure (PWF) certain differences bottomhole pressure at static state (Pr) and the well pressure during a flow (PWF) is expressed in equation (1)

$$q_0 = (\mathrm{Pr} - pwf)PI = \frac{KA}{\mu}(\mathrm{Pr} - Pwf) \tag{1}$$

IOP Conf. Series: Journal of Physics: Conf. Series **877** (2017) 012069 doi:10.1088/1742-6596/877/1/012069

When found the flowing fluid is a combination of fluid one phase and fluid two-phase , there will be conditions of reservoir pressure (Pr) is greater than the pressure bubble point (Pb) and the pressure flowing bottomhole (PWF) will decrease to less than Pb (Vogel , 1967) , so it can be determined by equation (2) as follows .

$$q_0 = q_{\max}\left(1 - 0{,}2\frac{pwf}{pr} - 0{,}8\left(\frac{pwf}{pr}\right)^2\right) \quad \textbf{(2)}$$

IPR curve will be found on a comparison between the flow rate at bottomhole flowing pressure as shown in Figure 1

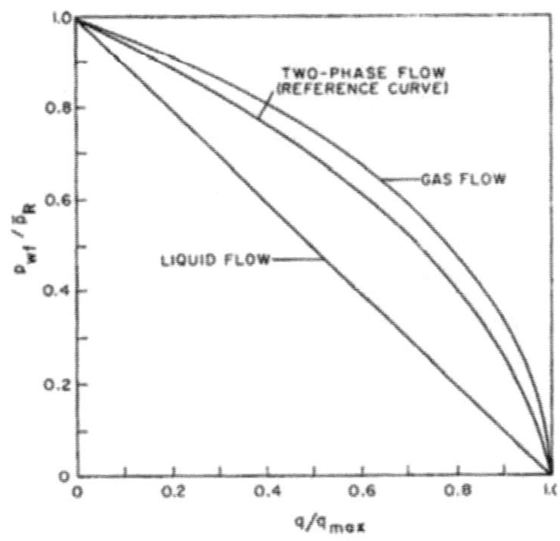

Figure 1. IPR Models (Vogel, 1968)

1.2. Flow in pipe

Fluid flows from the bottom of the well to the surface past the media pipeline , the vertical pipe is usually placed under the surface called tubing , in this case used the principle of a pressure gradient derived from the energy equation by using the principles of thermodynamics , analysis of pressure loss in vertical pipes can be approximated by the laws of thermodynamics which can be represented by Figure 2 .

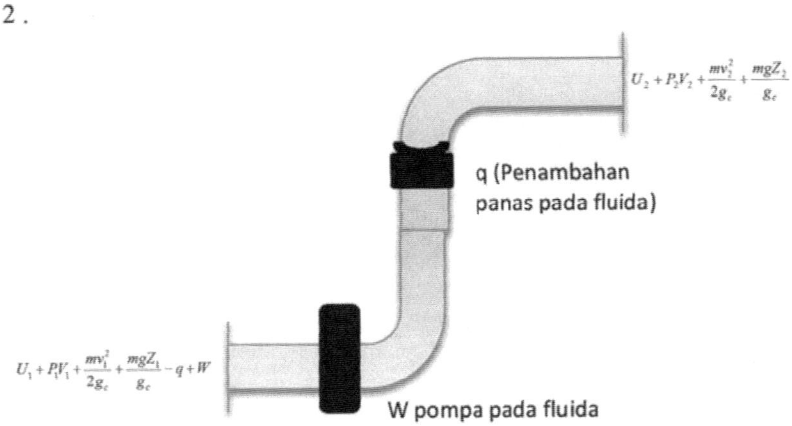

Figure 2. Equilibrium of energy

$$U_1 + P_1V_1 + \frac{mv_1^2}{2g_c} + \frac{mgZ_1}{g_c} - q - W_s = U_2 + P_2V_2 + \frac{mv_2^2}{2g_c} + \frac{mgZ_2}{g_c} \quad (3)$$

And equation (3) lowered into the equation (4)

$$\frac{dP}{dZ} = -\left(\frac{g}{g_c}\right)\rho - \frac{\rho v dv}{g_c dZ} - \frac{\rho dLw}{dZ} \quad (4)$$

and get the results of the analysis which can be seen in equation (4) , there are three components of the equation (4), namely :

a. Elevation component

$$\left(\frac{dP}{dZ}\right)_{eL} = \frac{g}{g_c}(\rho) \quad (4.a)$$

b. Velocity Component

$$\left(\frac{dP}{dZ}\right)_{acc} = \frac{\rho v dv}{g_c dZ} \quad (4.b)$$

velocity components are usually always ignored in the calculation of the pressure gradient.

c. Friction Component

$$\left(\frac{dP}{dZ}\right)_{f} = \frac{d.Lw}{dZ}(\rho) = \frac{f\rho v^2}{2g_c d} \quad (4.c)$$

Equation (4) can be expressed

$$dLw = \frac{P_1}{\rho} - \frac{P_2}{\rho};$$

with might be called due to friction . We can connect the equations (4.b) and (4.c) with dz we consider dL as assumptions vertical pipe, then becomes :

$$dp = \frac{fL\rho v^2}{2g_c d}$$

$$P_1 - P_2 = \frac{fL\rho v^2}{2g_c d}$$

$$\Delta p = \frac{fL\rho v^2}{2g_c d} \quad (5)$$

Pressure loss due to elavasi is 0 because the sect is a vertical flow , as well as the speed is very small because the cross-sectional area is constant , so it can be expressed in the analysis of energy loss due to the friction is :

$$h_f = \frac{fLv^2}{2g_c d} \quad (6)$$

With a conversion factor entered the equation to make it consistent dimensions.

Analysis of fluid flow in a vertical pipe can be represented by the TPR curve I Tubing Relathionship performance) .In the manufacture of TPR curve in need of equation (5)

- Where in the calculation of the pressure loss approximated by the equation (5)
- discharge pressure value can be approximated by the equation

$$P_1 = Pwh + \Delta P$$

In calculating the intake tubing pressure value can be approximated by

$$P_2 = P_1 - (\frac{\rho_{fsc} \times h}{808,314}) \times St$$

In calculating the intake bottomm of well can be approximated by

$$P_3 = P_2 + 0,433 \times Pwh$$

2. Methods

2.1 IPR and TPR Analysis

So from the relationship shown in Figure 2 is obtained that oil production in production wells can be predicted by a curve of intersection between the curves IPR with TPR curve. Basically the analysis of the relationship between TPR and IPR curve is often called nodal analysis, where the nodal analysis in this study is the connection between a porous medium with vertical media (tubing) with the connecting point is PWF.

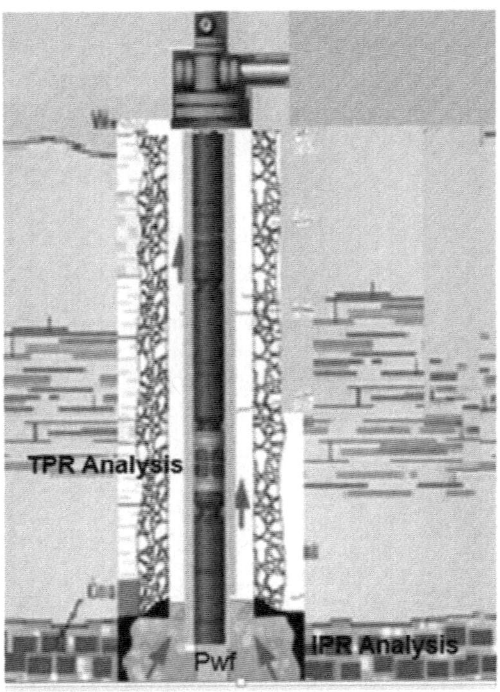

Figure 3. Analysis for Well Production

Nodal system analysis is required in order to:

1. Analyzing the behavior of the reservoir fluid flow at each well system components to determine the influence of each of these components of the well system as a whole.
2. Combine reservoir fluid flow behavior throughout the components so as to estimated production rate of the well.

In the analysis of the pressure loss in vertical pipes (tubing), the relationship between IPR and TPR important to do in the determination of oil production, caused by the intersection of the two curves can be seen the value of the rate of optimal fluid, the IPR curve called sebgai curve inflow derived from nature, ie the curve that is based with the ability reservor rocks, while the TPR curve (outflow) is the curve that is constituted with human behavior untukk obtain optimal flow rate of a fluid. In determining the outflow curve in this case been a nodal system for submerged pump wells (ESP) where the dot is located nodalnya bottom wells (Yohana, 2010).

• In determining the outflow curve can be influenced by several aspects, and aspects of the possible in influencing the productivity of wells that need to be further analyzed one of which will be discussed in this case is

a. Diameter Tubing influence on TPR
b. Stage Pumps influence on the curve TPR

3. Analysis and Discussion

Tabel 1. Well A Data

Parameter	
***Depth** (Ft)*	6218
SFL (Ft)	1000
WFL(Ft)	3900
Sg oil	0,8592
Sg water	1,125
WC (water Cut)(%)	0,24
API	33
Q liquid(Bpd)	836
Sg Oil	0,8592
T (Temperatur)(^0F)	180
Pb (Bubble Point) (psi)	683
tekanan Atmosfer (psi)	14,7
Densitas gas (lbm/cu ft) ρ	8,25
Viskositas liquid (cp) μ_{liquid}	0,34
Viskositas gas (cp) μ_{gas}	0,02
Surface strain	24
mid perforasi	5160
Pump Setting Depth	4900

Stage	400
Pwh	300
h(Feet/stage)	10
gc	32,2

Analysis of IPR

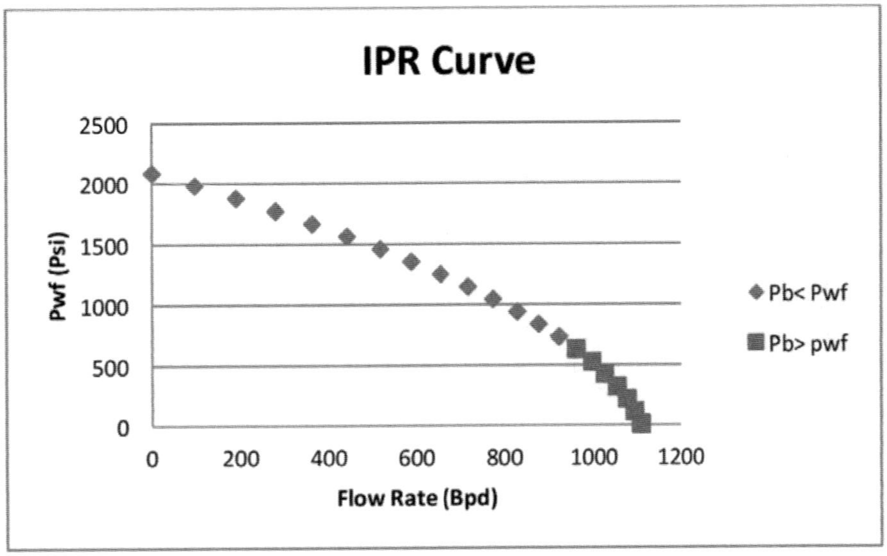

Figure 4. IPR Curve Well A

In this analysis it was found that there is a phase change from one phase into two phases , where the curve is red there is a phase change from one phase into two phases , namely phases of oil and gas this can result in fluid viscosity will change while the viscosity is very influential in the production fluid ,

TPR analysis

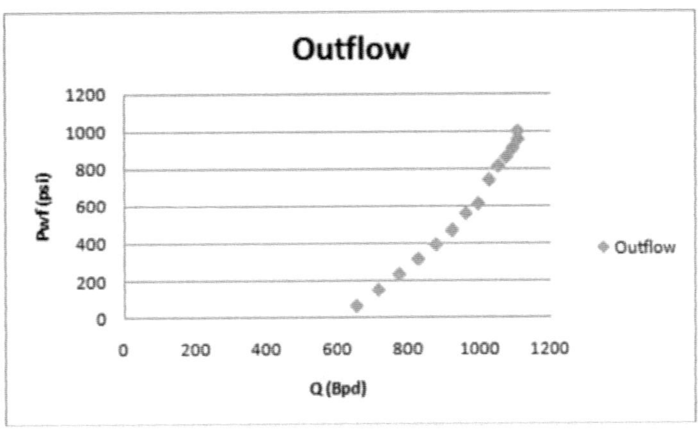

Figure 5. TPR Analysis

IOP Conf. Series: Journal of Physics: Conf. Series **877** (2017) 012069 doi:10.1088/1742-6596/877/1/012069

The image above is the calculation of outflows or vertical pipe analysis illustrated with TPR curve determination outflow curve can be influenced by several aspects , and aspects of the possible in influencing the productivity of wells that need to be further analyzed one of which will be discussed in this case is

a . Diameter Tubing influence on TPR

b . Stage Pumps influence on the curve TPR

a. Influence diameter tubing for the TPR

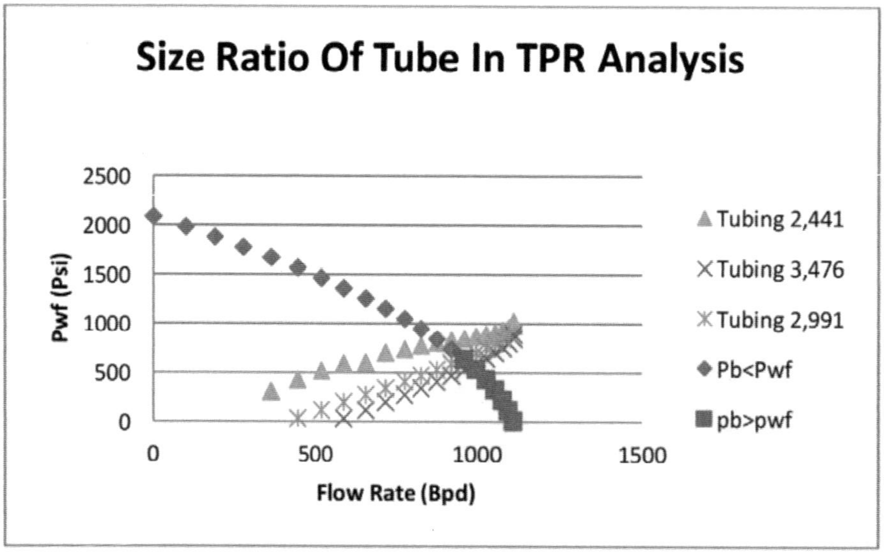

Figure 6. Size Ratio Of Tybe In TPR Analysis Curve

Table 2. Size Ratio Of Tube In TPR Analysis

tube size	Q (bpd)
2.441 ID	876,729
2,991 ID	961,197
3,476 ID	996,26

b . The influence of theTPR Analysis to stage pump

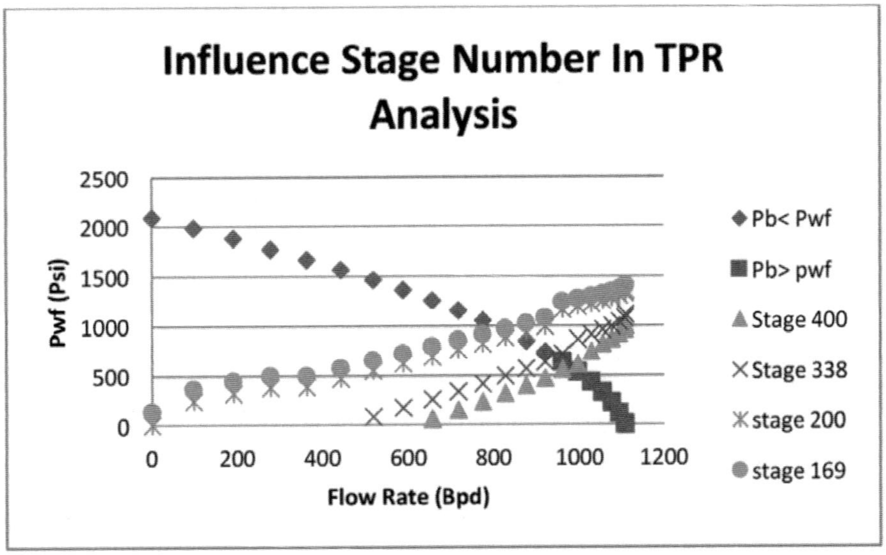

Figure 7. Influence Stage Number In TPR Analysis

Table 3. Influence Stage Number In TPR Analysis

Stage	Q (BPD)
400	961,1977
338	921,1655
200	827,889
169	774,646

4. Conclusion

So from this Analysis Based on the results of research conducted in wells A SW field and the description in previous chapters , it can be concluded that the ability of the rate of production at oil and gas wells A tubing can be optimized with its greatest diameter is 3.476 at 996.26 ID BPD , as well as the number of stages the ESP pumps were also crucial stage where an increasing number of more optimal production rates obtained namely by stage 400. in addition, the analysis also concluded that the pressure affects the friction pressure loss compared with the shape of the pipe cross-section (magnification) . At the appropriate type of pump ESP field characteristics obtained that initial pump installed in wells A far more optimal than the kind of pumps in others.

References

[1] Vogel,J.V,(1968): Inflow Performance Relationship for solution-gas drive wells",JPT Trans., AIME, 243
[2] *Mahenda, Althidia (,2014):* Pressure buildup test analysis with horner and standingmethods to get productivity condition of sgc-x well PT. Pertamina ep asset 1 field jambi, Unsri
[3] *Yohana, Kenes.,(2010) : Analisa Pengaruh watercut pada sistem produksi menggunakan analisa nodal dengan metode hagedorn and brown di lapangan JK* (in Indonesian), UIR Riau

Preliminary Study of 2-D Time Domain Electromagnetic (TDEM) Modeling to Analyze Subsurface Resistivity Distribution and its Application to the Geothermal Systems

Cahyo Aji Hapsoro*, Acep Purqon, and Wahyu Srigutomo
Earth Physics and Complex System, Faculty of Mathematics and Natural Sciences, Institut Teknologi Bandung, Bandung, Indonesia

* cahyoajihapsoro@gmail.com

Abstract. 2-D Time Domain Electromagnetic (TDEM) has been successfully conducted to illustrate the value of Electric field distribution under the Earth surface. Electric field compared by magnetic field is used to analyze resistivity and resistivity is one of physical properties which very important to determine the reservoir potential area of geothermal systems as one of renewable energy. In this modeling we used Time Domain Electromagnetic method because it can solve EM field interaction problem with complex geometry and to analyze transient problems. TDEM methods used to model the value of electric and magnetic fields as a function of the time combined with the function of distance and depth. The result of this modeling is Electric field intensity value which is capable to describe the structure of the Earth's subsurface. The result of this modeling can be applied to describe the Earths subsurface resistivity values to determine the reservoir potential of geothermal systems.

1. Introduction

Electromagnetics is one of the Geophysical exploration uses electromagnetic induction in subsurface that developed in the 1920 for the first time for detecting rocky conductivity [1]. Electromagnetic method is divided into two parts, which are frequency domain electromagnetics and time domain electromagnetics. Some of the frequency domain electromagnetics are Very Low Frequency (VLF), Very High Frequency (VHF), Ground Penetrating Radar (GPR), Magnetotelluric (MT), and Controlled Source-audio Magnetotelluric (CSMAT). Then, the kind of time domain is Time Domain Electromagnetics (TDEM) or Transient Electromagnetics (TEM).

Magnetotelluric method has been used to observe electric $\left(\vec{E}\right)$ and magnetic field $\left(\vec{B}\right)$ variation simultaneously [2,3]. Although has been used largely, frequency domain electromagnetics that the energy wave is transmitted by source, will decays by depth factor. The other problem for applying frequency domain electromagnetics methods is when frequency wave that propagates has high order to Megahertz [4]. In the early 1970s, there is EM methods developing significantly, such as time domain system called TEM or TDEM [1,5]. TDEM methods has some advantages compared by frequency domain, such as brief measurement procedure because when the current is doing once time loop from the transmitter is only need a minute. The other one is will be reached larger exploration than transmitter loop size and will be resulted better lateral resolution [6].

TEM or TDEM method has been used for exploration activity, such as for predicting the location of groundwater source [7], hydrogeology structure [8], aquifer potential based on the observed electric resistivity [9], and volcano resistivity structure [10]. 3-D TDEM solution is very important to solve and understand physical sense of observed earth responses [11]. For solving EM problem numerically is generally used by three methods, which are integral, differential, or both of them (hybrid). Finite Difference (FD) for time domain electromagnetic methods called Finite Difference Time Domain (FDTD) has been successfully used for solving EM problem. The advantage of FDTD is able to solve EM interaction with complex geometry, analyzing transient problem and modeling non-homogeneous geometry [12]. The other one is penetration ability of EM in dielectric and or magnetic material [13].

FDTD technique is efficiently able to solve curl operation in the time domain Maxwell's equation or equivalent integral equation [14]. In this paper will be modeled physical parameter which is electric intensity for analyzing subsurface resistivity distribution to the geothermal system.

2. Theoretical Background

The behavior of EM field at any frequency is able to describe briefly by Maxwell's equation [15] that can be expressed by the following equation:

$$\vec{\nabla} \cdot \vec{E} = \frac{\rho}{\varepsilon_0} \tag{1}$$

$$\vec{\nabla} \cdot \vec{B} = 0 \tag{2}$$

$$\vec{\nabla} \times \vec{E} = -\frac{\partial \vec{B}}{\partial t} \tag{3}$$

$$\vec{\nabla} \times \vec{B} = \mu_0 \vec{J} + \mu_0 \varepsilon_0 \frac{\partial \vec{E}}{\partial t} \tag{4}$$

By using calculus vector operation to equation (2) and (4), we will obtain general solution for EM wave that propagates in the-z direction and by assuming it is harmonically propagating, we will obtain diffusion equation expressed by:

$$\nabla^2 \vec{E} - \mu\sigma \frac{\partial \vec{E}}{\partial t} = 0 \tag{5}$$

And then, by comparing the intensity of electric and magnetic field will be obtained impedance that contains information of the electromagnetic fields in the subsurface. This impedance can be expressed by the following equation:

$$Z = -\frac{E_z}{H_y} \tag{6}$$

For the resistivity value is obtained by the following relation:

$$\rho_{a,i,j} = \frac{1}{\mu_0 \omega} \left| Z_{ij}(\omega) \right|^2 \tag{7}$$

Presence of source in transmitter causes EM field that propagates is should be solving by EM field equation with the source. If we consider that the EM wave propagates in the wave guide, so we will have electric and magnetic field components are able to express by the following equations:

For TM_z **(Transverse Magnetic) mode:**

$$E_x = \frac{1}{(\sigma + i\varepsilon\omega)} \frac{\partial^2 A_z}{\partial x \partial z}, \tag{8a}$$

$$E_y = \frac{1}{(\sigma + i\varepsilon\omega)} \frac{\partial^2 A_z}{\partial y \partial z}, \tag{8b}$$

$$E_z = \frac{1}{(\sigma + i\varepsilon\omega)} \left(\frac{\partial^2}{\partial z^2} + k^2 \right) A_z, \tag{8c}$$

$$H_x = \frac{\partial A_z}{\partial y}, \tag{9a}$$

$$H_y = -\frac{\partial A_z}{\partial x}, \tag{9b}$$

$$H_x = 0, \tag{9c}$$

For TE_z (Transverse Electric) mode:

$$E_z = -\frac{\partial F_z}{\partial y}, \tag{10a}$$

$$E_y = \frac{\partial F_z}{\partial x}, \tag{10b}$$

$$E_z = 0, \tag{10c}$$

$$H_x = \frac{1}{i\omega\mu}\frac{\partial^2 F_z}{\partial x \partial z}, \tag{11a}$$

$$H_y = \frac{1}{i\omega\mu}\frac{\partial^2 F_z}{\partial y \partial z}, \tag{11b}$$

$$H_z = \frac{1}{i\omega\mu}\left(\frac{\partial^2}{\partial z^2} + k^2\right)F_z, \tag{11c}$$

Equation (8) and (9) describes that only the magnetic field propagates in the wave guide perpendicularly. This condition is called transverse magnetic (TM_z). For equation (10) and (11) describes that only electric field propagates in the wave guides perpendicularly and called transverse electric (TE_z).

3. Methodology

For solving full EM wave, 3-D analysis is required to determine the relation propagation mode dispersion. The geometry of closed type wave guide that *z*-axis direction and parameters of propagation β is shown as figure 1. The *z*-axis direction represents electric and magnetic field propagation. TM polarization to *z*-axis direction will be obtained $H_z(x,y,z,t)=0$, and TE polarization to *z*-axis direction will be obtained $E_z(x,y,z,t)=0$.

Cubical cell unit in the coordinate system is shown by figure 2. For each magnetic field component is surrounded by four components of electric field. And then for each electric field components is surrounded by four magnetic field. Time-stepping Finite Difference for electric and magnetic field can be obtained by analyzing six coupled scalar partialy differential equations. By using rectangle coordinate system, $E_x(x,y,z,t)$, $E_y(x,y,z,t)$, and $E_z(x,y,z,t)$ that represents electric components. Then for $H_x(x,y,z,t)$, $H_y(x,y,z,t)$, and $H_z(x,y,z,t)$ that represents magnetic components.

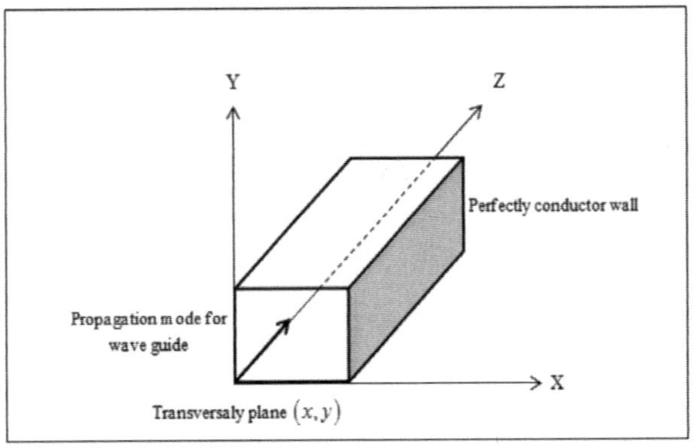

Figure 1. Geometry of closed-type wave guide

Δx, Δy, and Δz represents lattice special addition as long as x, y, and z coordinate. The Δt represents additional time and (i, j, k) represents the location of $(i\Delta x, j\Delta y, k\Delta z)$.

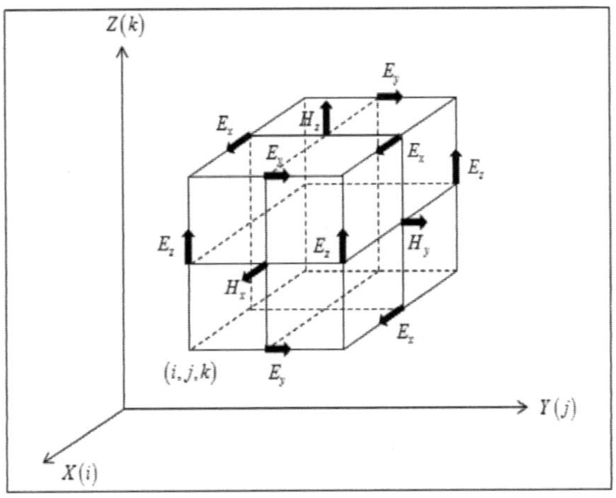

Figure 2. Geometry of closed-type wave guide

The selecting of spatial additional Δx, Δy, and Δz also time additional Δt are chosen because the stability accurate and algorithm. For make sure the stability of computational field, Δt is chosen to fulfill inequality that expressed by these following equation:

$$\Delta t \leq \frac{1}{c_{\max}} \left[\frac{1}{\Delta x^2} + \frac{1}{\Delta y^2} + \frac{1}{\Delta z^2} \right]^{-1/2} \tag{12}$$

where $c_{\max}$ is the maximum wave velocity.

4. Result and Discussion

FDTD modeling is made by three variations of source, which are sinusoidal source, single Gaussian source, and periodic Gaussian source. For each variation, the maximum and minimum amplitude is shown to analyze propagation of electric field in the subsurface. Each modeling is also given by time-step of 150 computational times to observe the attenuation of electric field's amplitude in the subsurface.

1. Sinusoidal Source

The electric current is generated from the transmitter as a sinusoidal function. The amplitude of electric current reaches the maximum value at the time-step is equal to zero. As the sinusoidal function, the electric current will decay and reaches the minimum value after reaches its maximum. This condition is shown by figure 3(a) and 3(b) that illustrate the changing of electric current intensity from maximum to minimum point. At time-step 150, the maximum and minimum amplitude is can be distinguished easily because the source of electric current is still generated by the transmitter as shown by figure 3(d).

For a comparing between maximum and minimum value, we separate the maximum and minimum amplitude of electric field into two graphs, as shown by figure 3(c). Maximum and minimum value is can be analyzed the resistivity structure in that position. This model also can be applied to the real observed data from the geothermal field, so we can analyze the resistivity structure in the real condition of geothermal system.

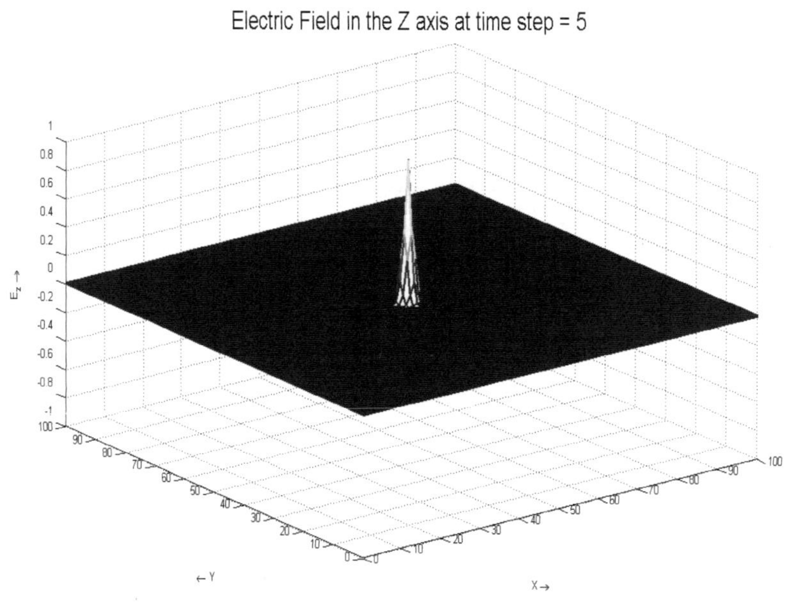

Figure 3(a). The pattern of electric field for the sinusoidal source at time-step = 0

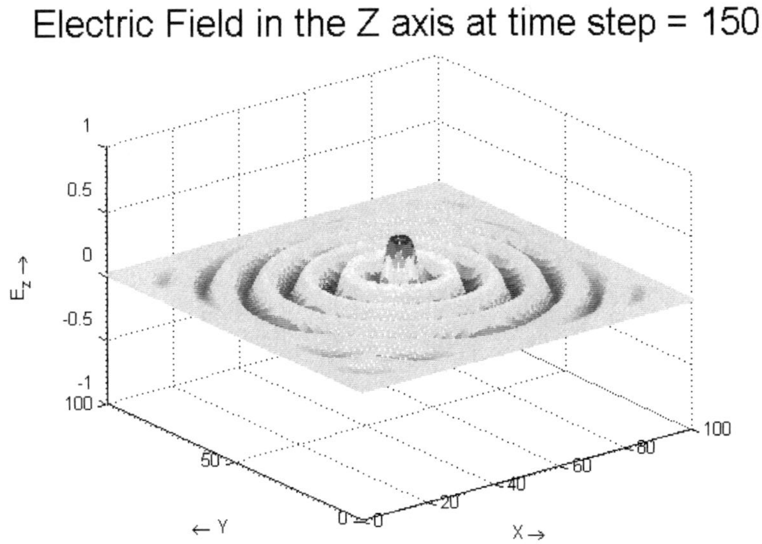

Picture 3(b). The pattern of electric field for the sinusoidal source at time-step = 150

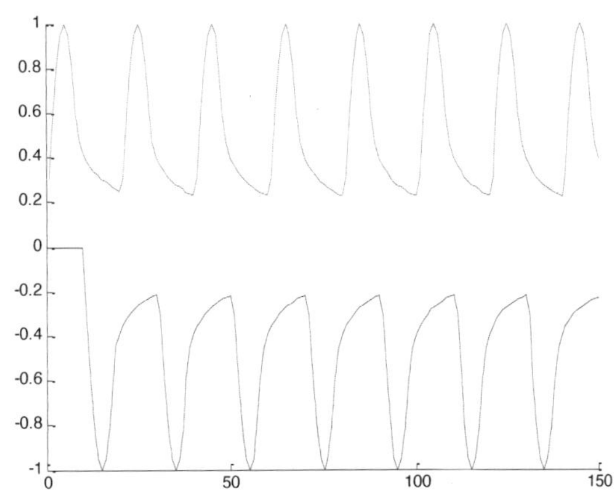

Figure 3(c). The maximum (red graph) and minimum (blue graph) amplitude of electric field at time-step = 150

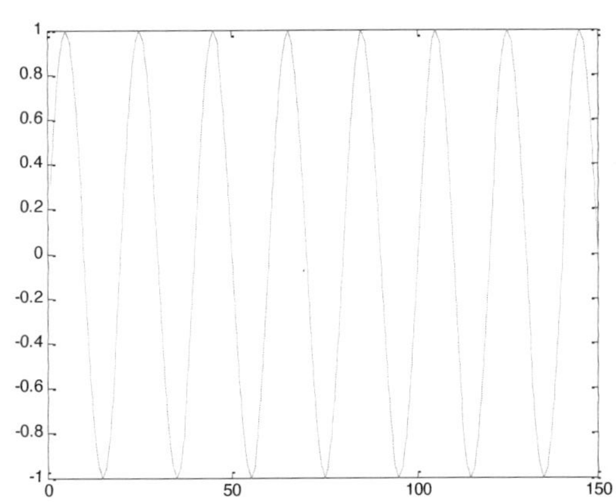

Figure 3(d). The maximum and minimum amplitude of source at time-step = 150

2. Single Gaussian Source

This variation source is decaying faster than the first one. It's happened because the source is generated only one time and almost be zero after reaches time-step 150, as shown figure 4(a). The maximum value will be attenuated to be almost zero or reaches the minimum point after reaches its maximum point as shown figure 4(b).

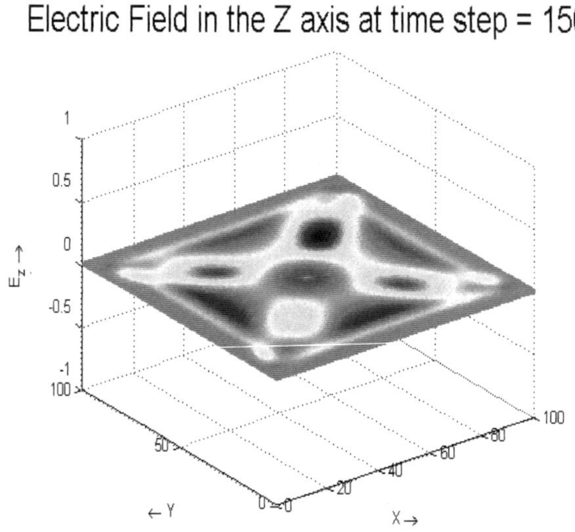

Figure 4(a). The pattern of electric field for the single gaussian source at time-step = 150

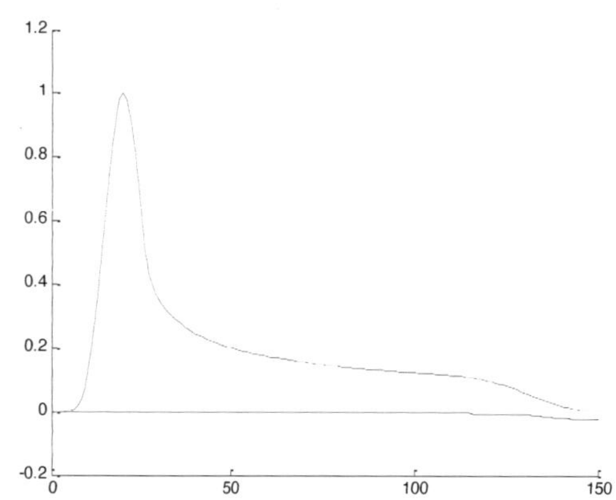

Figure 4(b). The maximum (red graph) and minimum (blue graph) amplitude of electric field at time-step = 150

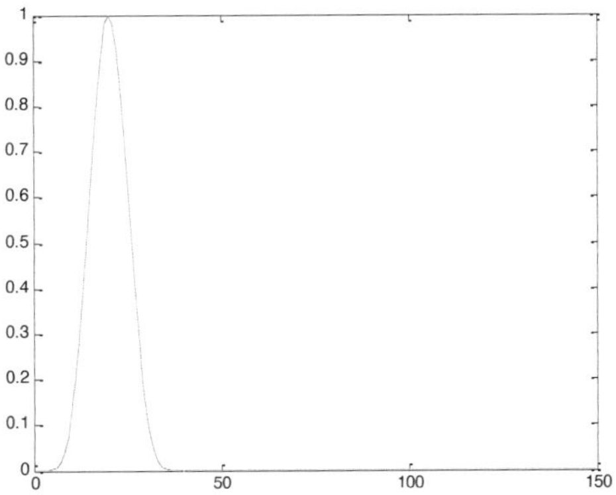

Figure 4(c). The maximum and minimum amplitude of source at time-step = 150

3. Periodic Gaussian Source

The amplitude of electric field is still observed at time step 150, compare to initial condition as shown by figure 5(a). This condition is occurred because electric current is generated by transmitter periodically, so we have only the maximum value that also generated periodically as shown by figure 5(b). If we compare figure 5(c) to picture 3(d), we can say that the maximum amplitude is generated periodically. But at the same time, which time-step is 150, sinusoidal source has propagated two times than periodic Gaussian source.

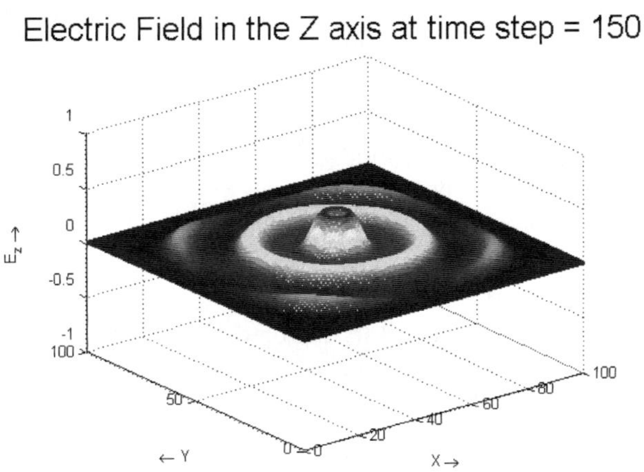

Figure 5(a). The pattern of electric field for the periodic gaussian source at time-step = 150

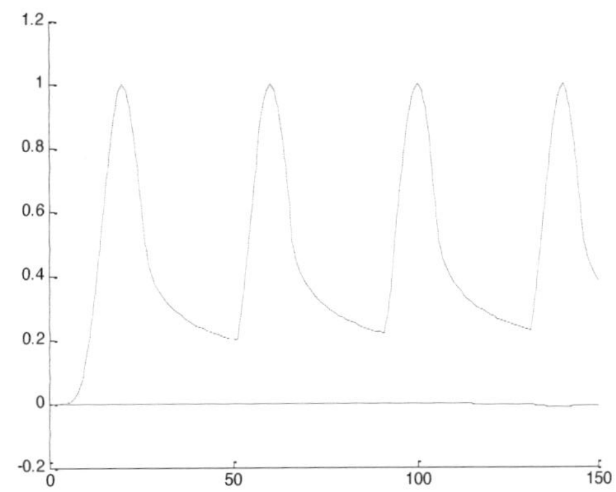

Figure 5(b). The maximum (red graph) and minimum (blue graph) amplitude of electric field at time-step = 150

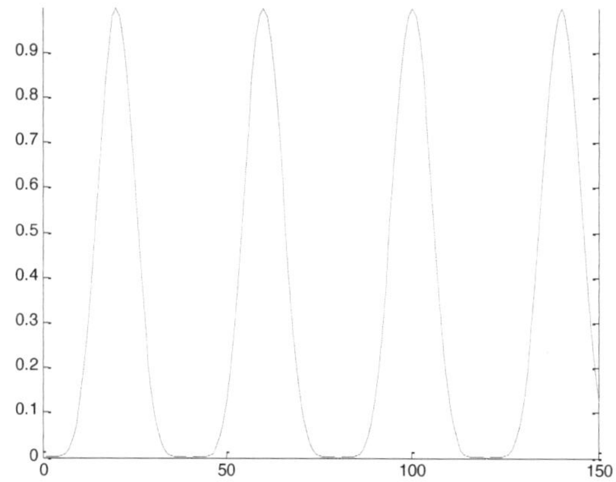

Figure 5(c). The maximum and minimum amplitude of source at time-step = 150

5. Conclusion

From this modeling that has been conducted, we can conclude that the amplitude of electric field will attenuate when it propagates in the earth. As shown by electric field propagating at time–step 150 we can see that the amplitude is decaying by the increasing of time. One of the reasonable for this attenuation is because there is an interaction between electric field and materials in the subsurface. This condition means that the resistivity structure in the subsurface is able to analyze by comparing the intensity or amplitude of electric and magnetic field. This information of electric field can be used for analyzing the resistivity structure by comparing to magnetic field that also propagates in the subsurface.

For the geothermal system, resistivity structure is very important to analyze the conductivity of rocks or geological structure in the layered earth. Conductivity gives us the information about the physical parameters such as the rock density, dimension of geothermal energy storage, and temperature of fluids or gases flow in the subsurface.

Propagation of EM field for TE mode is successfully conducted. We can analyze the source of geothermal prospect in the subsurface that can be explored for energy resources. This model also can be applied to the real observed data from the geothermal field, so we can analyze the resistivity structure in the real condition of geothermal system.

References

[1] Telford, W. M., Geldart, L. P., dan Sheriff, R. E. 1990. *Applied Geophysics 2nd Edition*. New York, Cambridge University Press
[2] Tikhonov, A. N. 1950. *On Determining Electrical Characteristics of The Deep Layers of The Earth's Chrust*. Geophysical Institute Academy of Science, USSR, Reprinted from Doklady, **73**, 2, 295-297
[3] Cagniard, L. 1953. *Basic Theory of The Magnetotelluric Method of Geophysical Prospecting*. Paris
[4] Zhdanov, M. S. 2010. *Electromagnetic geophysics: Notes from the past and the road ahead*. Geophysics, **75**, 5 (2010); 75A49-75A66
[5] French, R. B. 2002. *Time Domain Electromagnetic Exploration*. Oregon: Northwest Geophysical Associated, Inc
[6] Ward, S. H. 1990. *Geotechnical and Environmental Geophysics*. Vol 2-Environmental and Groundwater. Society of Exploration Geophysicists. Tulsa-Oklahoma
[7] Fitterman, D. V. dan Stewart, M. 1986. *Transient electromagnetic sounding for groundwater*. Geophysics, **51** (April, 1986), 995-1005
[8] Papadopoulos, dkk. 2004. *A TDEM survey to define local hydrogeological structure in Anthemountas Basin, N. Greece*. Jurnal of Balkani Geophysical Society, **7**, 1, Agustus 2004; 1-11
[9] Krivochieva, S. dan Chouteau, M. 2003. *Integrating TDEM and MT methods for characterization and delineation of the Santa Catarina aquifer (Chalco Sub-Basin, Mexico)*. Journal of Applied Geophysics, **52** (2003), 23-43

IOP Conf. Series: Journal of Physics: Conf. Series **877** (2017) 012070 doi:10.1088/1742-6596/877/1/012070

[10] Srigutomo, W. dkk. 2015. Time-domain electromagnetic (TDEM) baseline survey for CCS in Gundih area, Central Java, Indonesia. Proceedings of the 12th SEGJ International Symposium 2015

[11] Wang, T. dan Hohmann, G. W. 1993. *A Finite-Difference, Time-Domain Solution for Three-Dimensional Electromagnetic Modeling*. Journal of Geophysics, Vol. **58**, No.6; P. 797-809

[12] Pridmore, D.F. dkk. 1981. *An investigation of finite-element modeling for electrical and electromagnetic data in three dimensions*. Geophysics, **46**, 7, July 1981; 1009-1024

[13] Sha, W. dkk. 2007. *Application of The Simplectic Finite-Difference Time-Domain Scheme to Electromagnetic Simulation*. Journal of Computational Physics **225** (2007) 33-50

[14] Rao, S. M. 1999. *Time Domain Electromagnetics*. A Harcourt Science and Technology Company, Academic Press

[15] Nabighian, M. N. 1987. *Electromagnetic Methods in Applied Geophysics 2nd Edition*. Vol 1-Theory, Society of Exploration Geophysicists. Tulsa-Oklahoma

IOP Conf. Series: Journal of Physics: Conf. Series **877** (2017) 012071 doi:10.1088/1742-6596/877/1/012071

Calculation Study of Electric Properties on Molybdenum Disulfide By Using Density Functional Theory

Diah Angraina Fitri[1], Acep Purqon[2]

[1]Graduate Computational Science, Faculty of Mathematics and Science, Institute of Technology Bandung, Jl Ganesha 10 Bandung 40132, Indonesia

[2]Physics of Earth and Complex Systems, Institute of Technology Bandung, Jl Ganesha 10 Bandung 40132, Indonesia

diahwirna30@gmail.com

Abstract. Molybdenum disulfide (MoS_2) has some unique properties such as symmetry properties in honeycomb structure. In this work, we investigate the electronic properties of different number of layer by using first-principles calculation based on density functional theory (DFT) with calculation using PHASE/0. The Generalized Gradient Approximation (GGA) and ultrasoft pseudopotential method were used. In this study, we compare between monolayer and multilayer of MoS_2, and we investigate the effects of layer number on their electronic properties such as showing direct and indirect bandgap. The properties is important for applications in material devices and energy saving applications.

1. Introduction

The human need for energy growing increased, especially in the field of electrical energy. Electricity invented by Thomas Edison is very beneficial for people around the world. At the moment very growing use Light Emitting Diode (LEDs). LEDs has several advantages compared to ordinary bulbs such as more energy efficient and LEDs does not emit heat. However, LEDs lights also have shortcomings that led prices more expensive than regular bulbs. Molybdenum disulfide is one of material main to make a LEDs. Several studies investigating the use of semiconducting MoS_2 as a host material so as to emit light in multiple colors include red, green and blue. The use of LEDs produced from MoS_2 can provide high levels of efficiency and excellent savings in energy and costs. In previous theoretical research results using algorithms and computing power are predicting that indirect bandgap in bulk MoS2 into direct bandgap when molded into monolayer MoS_2 [1].
The purpose of the present study is to observe the electronic structure of MoS_2. We carry out first-principles calculation to investigate the electronic structure of monolayer and bilayer MoS_2 by using first-principles calculation based on density functional theory (DFT). The results of first-principles calculations are based on density functional theory (DFT) by comparing with the reference as a validation results of the calculation. One approach that is popular in the field of quantum mechanics is the DFT [2]. DFT is a calculation involving first-principles calculations on the effect of the interaction of materials or molecules that applying the local approach and density functional semilocal [3].

2. Model

We carry out first-principles calculation based on density functional theory. We have performed density functional theory (DFT) with exchange-correlation by generalized gradient approximation (GGA), The generalized gradient approximation (GGA) method more accurate than local density

IOP Conf. Series: Journal of Physics: Conf. Series **877** (2017) 012071 doi:10.1088/1742-6596/877/1/012071

approximation to calculation many electronic properties. We apply ultrasoft in pseudopotential method.

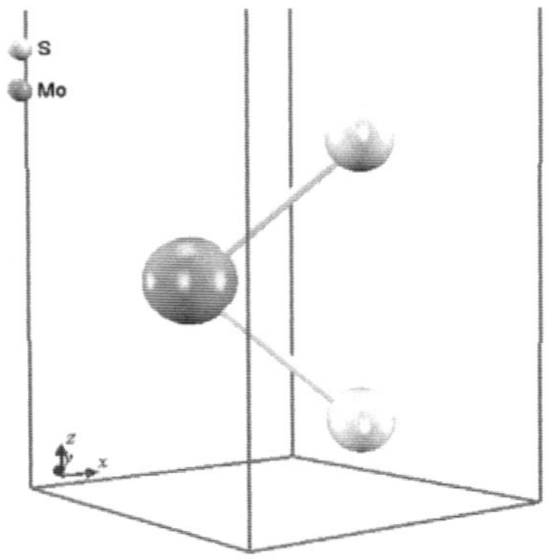

Figure 1. The crystal structure of monolayer MoS_2

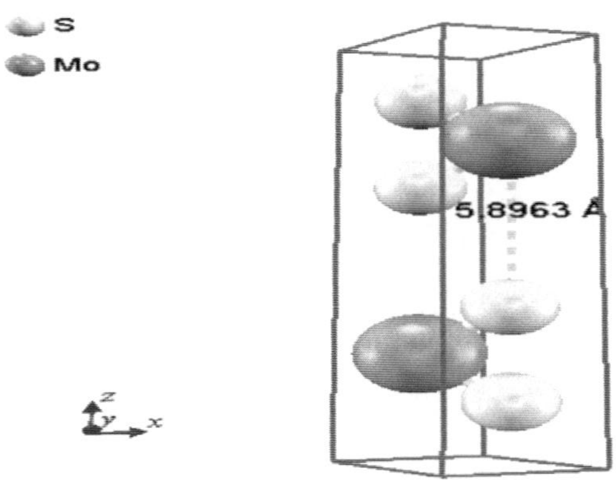

Figure 2. The crystal structure of bilayer MoS_2

The crystal structure of monolayer MoS_2 composed one atom of moybdenum's (Mo) and two atoms of sulphur (S). And also crystal structure bilayer MoS_2 composed two atoms of Mo and four atoms of S. In this calculation, monolayer MoS_2 obtained the bond length between Mo-S atoms is 2.4219 Å. In this result almost similar with experimental value is 2.383 Å [4]. Figure 2 show that bilayer MoS_2 has the bond length between Mo and S is 5.8963 Å. The result calculation almost close with experimental value is 6.0 Å [5].

We use the k-sampling 6x6x1 with mesh method and primitive unit cell with lattice constant, a = 3.16 Å [4] and c = 20 Å and 30 Å for monolayer and bilayer, respectively. We performed from four atoms in each layer is the rectangular lattice. We use cutoff energies of 25 rydberg and 250 rydberg for wave function and charge density grid, respectively. In optimization value of the kinetic energy cut off are wave function and charge density with a total value of state energy calculation result tend to convergence . In this calculation, self-consistent field (SCF) used limits the energy value convergence is 1.0 e-05 Hartree/Bohr.

3. Result and Discussion

3.1 Monolayer MoS_2

The cutoff wave function and charge density has been used in our calculation. After optimization, we find that the cutoff of kinetic energy to wave function and charge density are 25 rydberg and 250 rydberg, respectively. Due to the value of the kinetic energy of the charge density multiplied by 9 in accordance with tutorials on PHASE/0 manual. And also, the monolayer MoS_2 obtained k-point is 6x6x1. The lattice constant of MoS_2 has been used in our calculations. The calculation of the minimum value of cell unit in MoS_2 is done by varying lattice parameter of lattice constant (a). We find that the lattice constant of monolayer MoS_2 is 3.2183 Å. The result of our calculation is almost close to the experimental value is 3.16 Å [4].

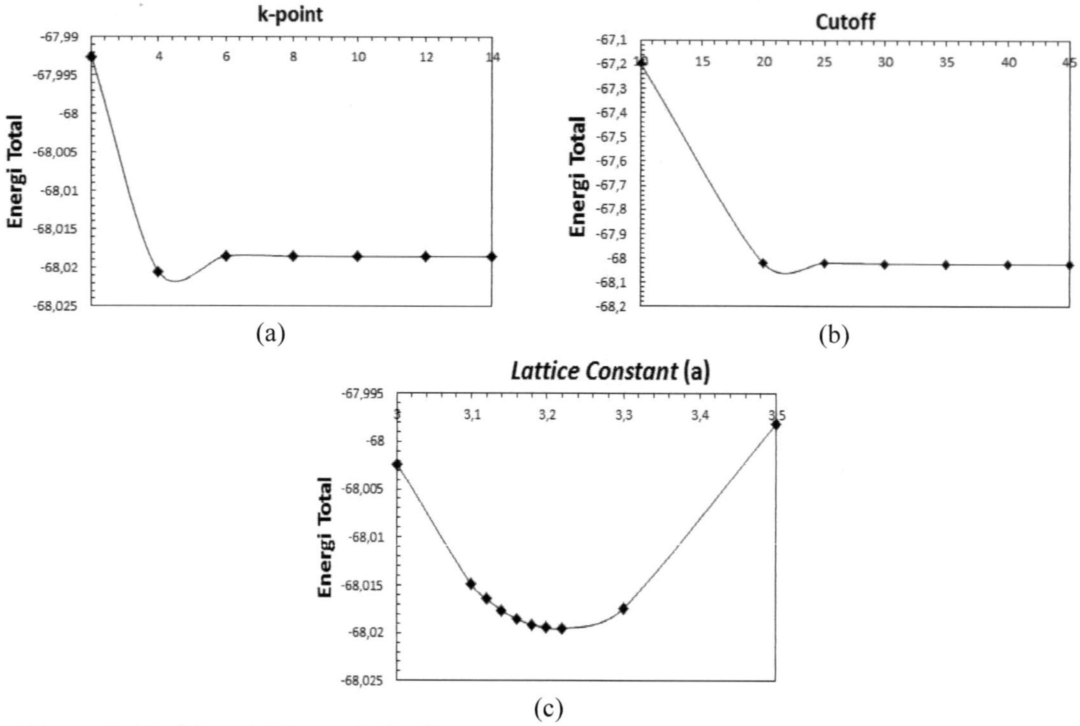

Figure 3. (a), (b), and (c) are relation between total energy and cutoff, k-point, and lattice constant, respectively

Figure 4 is band structure and density of state (DOS) of monolayer MoS_2. Monolayer MoS_2 has a direct bandgap energy located at K→K of k-point with the different in value gap is 1.63 eV. The result of our calculation is almost close to the experimental value is 1.79 eV [4].

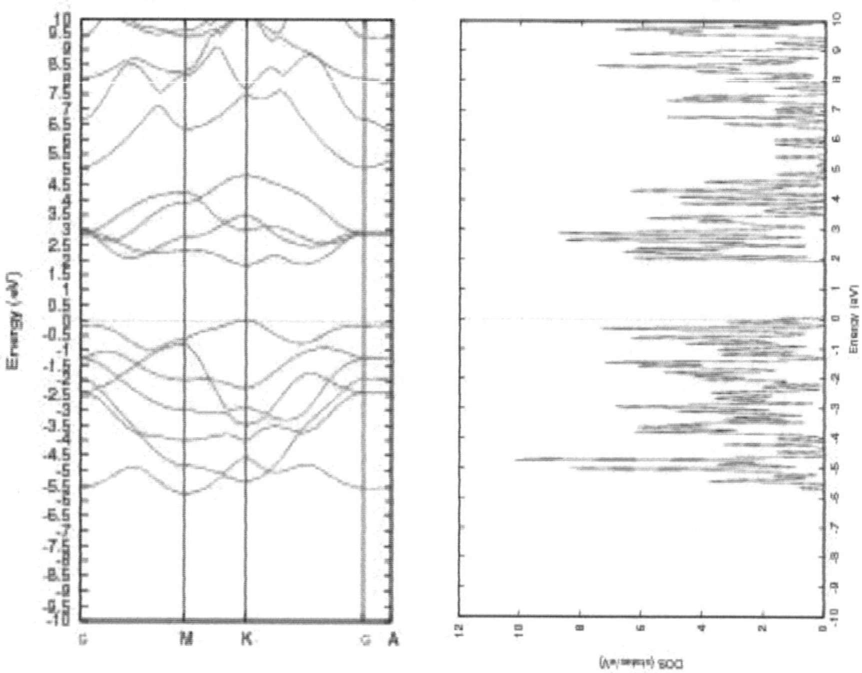

Figure 4. The band structure and density of state (DOS) of monolayer MoS_2

In this calculation. the pattern on the electronic structure of monolayer MoS_2 show that crystal structure of monolayer MoS_2 has a direct bandgap, there may be mentioned monolayer MoS_2 a direct bandgap of the semiconductor [4].

3.2 Bilayer MoS_2

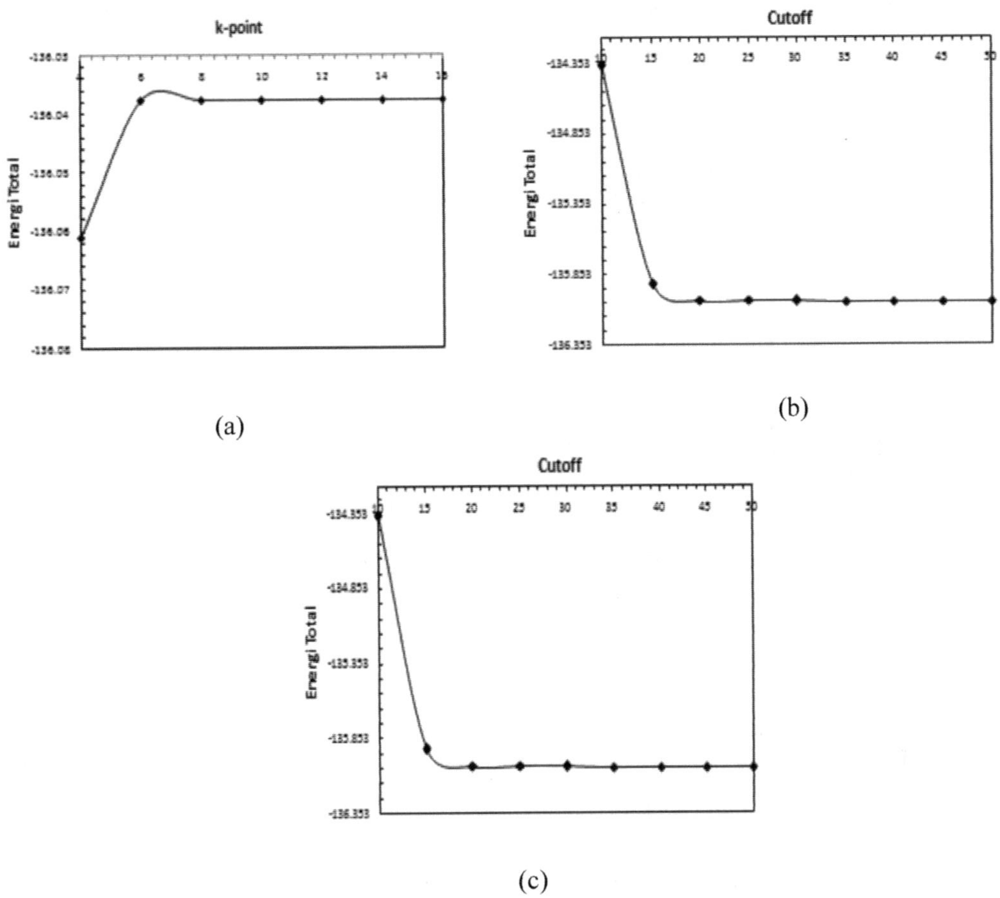

Figure 5. (a), (b), and (c) are relation between total energy and cutoff, k-point, and lattice constant of bilayer MoS_2, respectively

Figure 5 show that the value of k-point approach convergent is 8 x 8 x 1, the value cutoff wave function approaching convergence by 30 Rydberg then cutoff the charge density to 270 Rydberg had reached a value of convergence. Bilayer structure of MoS2 lattice constant achieves minimum energy at 3.26 Å. Optimization is considered complete when lattice constant has reached the minimum energy at a certain value. After optimization on all obtained the parameters for further calculations can be done.

Figure 6 show explain that structure of the left image is a band structure and the right of the image is the image density of states (DOS) in bilayer MoS_2. In the image, the band structure is the top and bottom of the conduction band is the valence band. The bilayer MoS_2 has an indirect bandgap that point A→ Γ with the difference in the value of 1.38 eV. The difference band structure within the conduction band and valence band become smaller. Due to the distance between atoms in the layer MoS_2 greater. In this calculation, the conduction band produced fewer than the valence band. The difference in input a few parameters in this calculation greatly affects the energy gap of MoS_2. The pattern of the electronic structure of the result of this calculation indicates that the material MoS_2 with a bilayer structure has the energy band gap properties that are not direct (indirect bandgap).

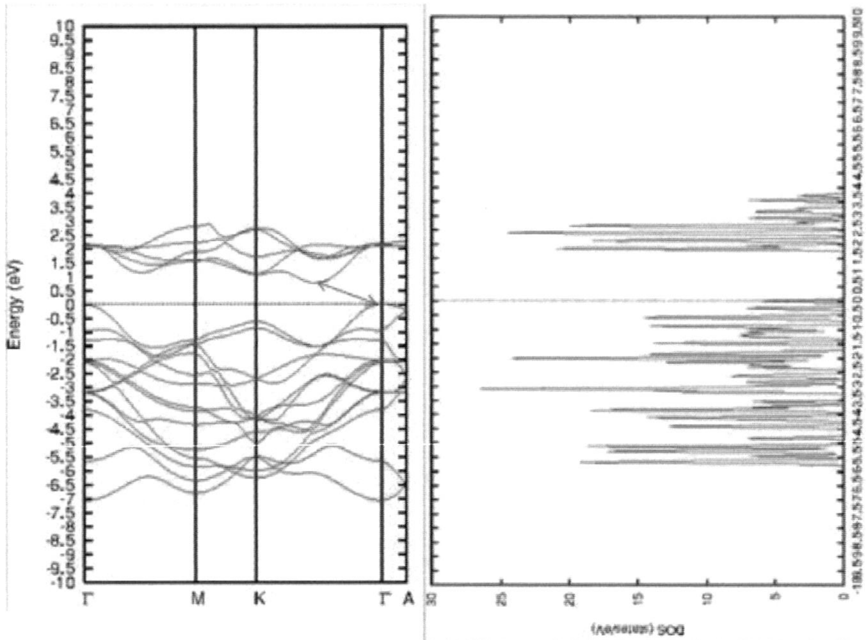

Figure 6. The band structure and density of state (DOS) of bilayer MoS_2

4. Conclusion

In this study, we compare between monolayer and bilayer of MoS_2, and we investigate the effects of layer number on their electronic properties such as showing direct and indirect bandgap. The properties is important for applications in material devices and energy saving applications.

References

[1] Li T and Galli G 2012 *Phys.Chem. C.* **111** 16192
[2] Capelle K 2006 *A Bird's-Eye View of Density -Functional Theory* arXiv:cond-mat
[3] Dion M, Rydberg H, Schrodinger E, Langreth D. C. and Lundqvist B. I. 2004 *Phys. Rev. Lett.* **92**
[4] Eugene S. Kadantsev and Pawel Hawrylak 2012 *Solid State Communication* **152** 909-913
[5] Liu K, Zhang L, Cao T, Jin C, Qiu D, Zhou Q, Zettl A, Yang P, Louie S. G, Wang F 2014 *Nature Communications* **5966** 1-6

Ferrous Ion and Medium Composition Effects on Acidogenic Phase in Biobutanol Production from Molasses

E Restiawaty[1,2*], **D Grinanda**[2]

[1]Study Program of Bioenergy and Chemurgy Engineering, Research Group of Chemical Engineering Process Design and Development, Faculty of Industrial Technology, Institut Teknologi Bandung, Jln. Let. Jen. Purn. Dr. (HC) Mashudi No.1 Sumedang, West Java 45363, Indonesia

[2]Biosciences and Biotechnology Research Centre, Institut Teknologi Bandung, Jl. Ganesha 10 Bandung 40132, West Java, Indonesia.

[*]E-mail : erestiawaty@che.itb.ac.id

Abstract. *Clostridium acetobutylicum* B530 has ability to convert sugar into biobutanol through two phases, i.e. acidogenic and solventogenic. This fermentation process is often hampered by high raw material cost and low product yield. In order to suppress the production cost, the molasses, a byproduct of sugar cane process production, was used as carbon source in this research. Molasses has nitrogen content in a small amount, thus could be negating the beef extract component, which is expected not to affect the growth of *C. acetobutylicum* B530 and also can reduce the production cost. In addition, a certain amount of Fe^{2+} (ferrous ion), a precursor in the formation of the enzyme ferredoxin, was added to the fermentation medium to contribute in the synthesis of acetyl-CoA, so that the formation of acidogenic products such as butyric acid and acetic acid is affected. This study aimed to investigate the effect of ferrous ion and the medium composition in acidogenic phase. The addition of 20 ppm $FeSO_4.7H_2O$ in the fermentation medium without beef extract can increase the concentration of butyric acid by 20% at a temperature of 35°C, while acetic acid concentration decreased by 6%. According to those results, it is expected that the product selectivity of butanol will increase in solventogenic phase. In addition, the removal of beef extract in the fermentation medium does not affect the kinetics of growth of *C. acetobutylicum* B530.
Keywords: *Clostridium acetobutylicum* B530, acidogenic phase, ferrous ion, molasses

1. Introduction

Development of alternative energy from biomass is needed to overcome the crisis of fossil-based natural resources. One of the promising biofuel developments is the production of biobutanol from agroindustry waste. Biobutanol has excellent properties as substitute fuel for gasoline, such as less hygroscopic, lower vapor pressure, and high energy content. Moreover, blending of butanol and gasoline is possible at any concentrations, and there is no need for adjustment of vehicle and engine technologies [1,2].

Biobutanol can be produced with biological process using solventogenic bacteria, such as clostridia. *Clostridium acetobutylicum* is biological agen that is often used in the production of acetone, butanol and ethanol (ABE). These bacteria have the ability to convert fermentable sugar into acids (acidogenic phase) followed by the reactions to produce ABE solvents (solventogenic phase).

The mole ratio of the solvent production (acetone:butanol:ethanol) is approximately 3:6:1 [3]. The acidogenic phase occurs when the *C. acetobutylicum* growth is in the exponential phase, while the solventogenic phase occurs when the cell growth slowed [4]. Metabolic pathways of ABE solvents production from glucose by *C. acetobutylicum* is illustrated in **Figure 1**.

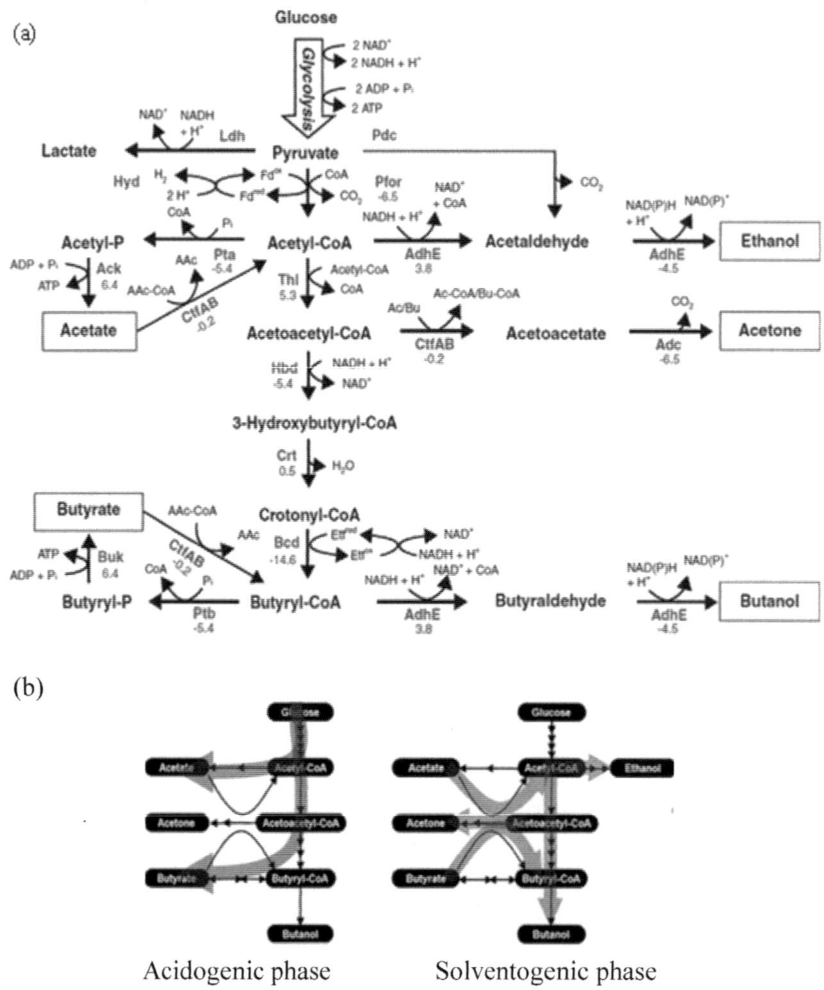

Figure 1. (a) Metabolism pathway of *C. acetobutylicum* to produce ABE solvents from glucose [5] and (b) acidogenic and solventogenic phases [6]

The acids produced in acidogenic phase will serve as precursor in solventogenic phase, so that the yield of each acid can affect the selectivity of the solventogenic products. The yield of ABE solvents can be affected by environmental conditions such as pH, temperature, and nutritional supplements. Previous studies reported that the addition of ferrous ion can increase the yeld of ABE solvents [7-9]. Ferrous ion is used as a cofactor for ferredoxin in the synthesis of acetyl-CoA. Acetyl-CoA plays an important role as precursor for the synthesis of all desired products, including acetogenic products. This study deals with the proper concentration of ferrous ion given in the fermentation process in order to obtain high yield of butyric acid which is expected to spur a high butanol selectivity. Fermentation stage observed in this study is limited to the exponential phase of *C. acetobutylicum* growth so it is possible to assess the effect of ferrous ion in acidogenic phase.

C. acetobutylicum can grow well in a reinforced clostridial medium (RMC). However, RCM has expensive constituent components, such as beef extract for nitrogen and vitamin sources, leading to not reliable process for commercial application. In addition. this study also examines the use of sugarcane molasses to replace the use of glucose and eliminate the use of beef extract in medium. Sugarcane molasses is a byproduct of the sugar industry and can be obtained at a cheap price compared to the price of glucose. Molasses has a high sugar content, i.e. 55-65%-w, so it is appropriate to be used as a carbon source. In addition, molasses also contains nitrogen and vitamins in small amounts. It is estimated that the use of molasses in the absence of beef extract in the medium does not affect the growth of *C. acetobutylicum*.

2. Material and Methods

2.1. Strain and culture maintenance

Clostridium acetobutylicum B530 used in this study is a gift from ARS (Agricultural Culture Collection), USA. The culture was grown in 50 ml vial bottle under anaerobic condition at 37°C. Before fermentation, inoculum was prepared by incubation at 37°C for 24 hours. Cultures were stored at 4°C and reseeded into fresh medium every 14 days.

2.2. Pretreatment of sugarcane molasses

Decalcification or addition of sulfuric acid was performed as cane molasses pretreatment in this study. Distillated water was added to 60 grams of cane molasses until the volume 1 L and then the mixture was acidified to pH 3.5 with H_2SO_4 96%. Following water bath at 60°C for 2 h, the mixture was centrifuged at 3000 rpm for 2 h using ultra speed centrifuged HERMLE Z 36 HK. Mixture was filtered using Whatman filter paper no. 1 (particle retention = 11 μm; diameter = 9 cm). Before being used for fermentation, the supernatant was neutralized to pH 6.5 with NaOH 4 N.

2.3. Fermentation mediums

The fermentation medium was composed of pretreated cane molasses 6%-w/v. The fermentation medium contained the following components per liter of distilled water: 10 g pepton (Bio Basic Inc., Canada), 5 g NaCl (Bio Basic Inc., Canada), 3 g yeast extract (Himedia, India), 3 g CH_3COOH (Bio Basic Inc., Canada), 1 g soluble starch (Bio Basic Inc., Canada), dan 0.5 g L-cystein HCl (Bio Basic Inc., Canada). 10 g beef extract (Oxoid, England) was added to certain experiment. The final pH was adjusted to 6.8. The variations of fermentation medium in this study are tabulated in **Table 1**.

Table 1. Variation of fermentation medium

	Carbon source	**Beef extract**	**$FeSO_4.7H_2O$ (ppm)**	**Incubation temperature (°C)**
Glu	Glucose	+	-	37
Mol#1	Molasses	+	-	37
Mol#2	Molasses	-	10	37
Mol#3	Molasses	-	20	37
Mol#4	Molasses	-	20	35
Mol#5	Molasses	-	30	35

\+ added to the medium
\- not given in the medium

2.4. Fermentation

Fermentation studies were conducted in 30 mL vials covered with rubber and stainless steel seal. 20 mL medium listed in **Table 1** were put in those vials and sterilized in autoclave at 121°C for 20 min. 10%-v/v of inoculum was applied into the sterilized fermentation medium. Inoculation was performed in the anaerobic chamber. Anaerobic chamber used in this research shown in **Figure 2**. Nitrogen gas was filtered and used for purging the anaerobic chamber up to no oxygen gas inside the chamber. The fermentations were incubated for 120 hours at certain temperatures (35 or 37°C). Samples were taken periodically to analyze the composition of fermentation products. Gasses formed during fermentation (CO_2 and H_2) were taken intermittently with sterile syringe. The observations of acidogenic stage were performed on the first 30 hours.

Figure 2. A simple anaerobic chamber connected with a nitrogen gas tube and equipped with a UV lamp.

2.5. Measurement of biomass density and cell dry weight
The optical cell density was determined at 600 nm using *spectrophotometer optima* SP-300. Cell dry weight was obtained from the following procedure. The samples of fermentation were centrifuged at 6000 rpm for 15 min. After removing the supernatant, pellets were subsequently put into oven at 60°C until the constant mass was obtained. The constant weight indicated the cell dry weight.

2.6. Morphology observation
Gram staining was performed to check the morphology of *C. acetobutylicum* B530. *C. acetobutylicum* B530 is a gram positive bacteria, so it will give purplish color. Morphology observation was performed using a Nikon Eclipse E200MV RS microscope with 100x focus.

2.7. Product quantification
The concentration of glucose, acetic acid, and butyric acid were measured using HPLC refractive index detector (RID) (Waters 2414) with Aminex HPX87H column (Biorad,CA) autosampler waters 2707. H_2SO_4 0.005 M was used as mobile phase with flow rate of 0.6 ml/min. The temperature operation was 60°C.

3. Results and Discussion

3.1. Effect of fermentation medium modification to the growth kinetics of C. acetobutylicum
The curve of *C. acetobutylicum* growth is depicted in **Figure 3**. *C. acetobutylium* growth in a medium containing glucose and beef extract (Glu) were observed faster than the growth in medium containing 6%-w/v molasses (with (Mol#1) or without (Mol#2) beef extract. The doubling time of *C. acetobutylicum* in a medium containing glucose and beef extract is higher than in ones containing molasses (see **Table 2**). It is quite acceptable because the molasses has a sugar content, which is more complex than glucose. *C. acetobutylicum* growths in molasses mediums with beef extract (Mol#1) or without beef extract (Mol#2) do not show a significant difference. The specific growth rate and the doubling time of *C. acetobutylicum* in both the fermentation medium also have similar values.

Sugarcane molasses contains various carbon sources and other nutrients, such as 30%-w/v sucrose, 13%-w/v fructose, 7.2%-w/v glucose, 0.4%-w/v lipid, and vitamins. Naturally, cane molasses contain ± 3%-w/v of nitrogen, so that the fermentation without beef extract component does not affects the growth of *C. acetobutylicum* B530 [11]. The results showed that molasses can be potentially as a carbon source and may negate beef extract in a medium. Further discussion after this section will look at the use of molasses as fermentation medium of *C. acetobutylicum*.

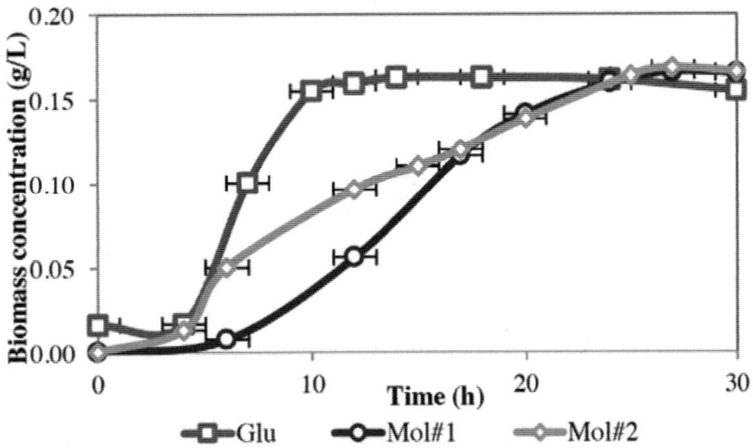

Figure 3. Growth profile of *C. acetobutylicum* B530 in different fermentation mediums (square for Glu, circle for Mol#1, dan diamond for Mol#2). The compositions of medium were mentioned in subsection *2.3*.

Table 2. The specific growth rate and doubling time of *C. acetobutylicum* B530 in different fermentation mediums

	μ (h^{-1})	τ (h)
Glu	0.369	1.879
Mol#1	0.245	2.824
Mol#2	0.247	2.807

3.2. pH changes during fermentation
The decreasing of pH was observed during molasses fermentation (with or without beef extract) as presented in **Figure 4**. This decrease in pH indicates the occurrence of acids production that took place on acidogenic phase. This pH decrease occurred in the exponential phase of *C. acetobutylicum* growth. The similar profiles of pH decrease indicate that the presence or absence of beef extract in molasses fermentation does not affect significantly in pH change of the medium during fermentation [5,6]. Low pH is one of the prerequisite conditions for *Clostridium sp.* to swith phase from acidogenic phase to solventogenic phase. Previous studies explained that optimum pH condition for culture to enter solventogenic phase is 4.5 [5,12,13].

3.3. Effect of iron ions in the production of acids in the acidogenic phase
Molasses fermentations by *C. acetobutylicum* were conducted in several concentrations of ferrous ion (10 ppm, 20 ppm, and 30 ppm) and different temperatures (35°C and 37°C) as mentioned in **Table 1**. The fermentation done in a medium without the addition of ferrous ion (Mol#1) was used as a base case. The acids detected as acidogenic products in this study were acetic acid and butyric acid. Acetic and butyric acid obtained from the fermentation of molasses on acidogenic stage for the medium variations of Mol#2 (10 ppm Fe^{2+}, 37°C), Mol#3 (20 ppm Fe^{2+}, 37°C), Mol#4 (20 ppm Fe^{2+}, 35°C),

and Mol#5 (30 ppm Fe^{2+}, 35°C) were compared with those obtained from the fermentation in Mol#1 (0 ppm Fe^{2+}, 37°C). The percentage of acid productivity on different media relative to the productivity on Mol#1 is given in **Figure 5**.

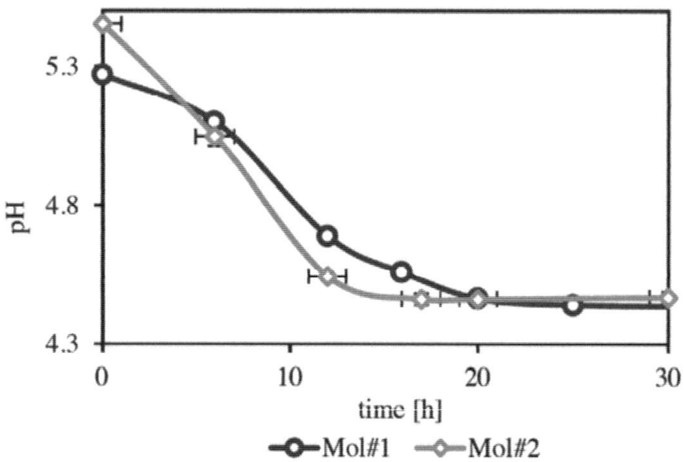

Figure 4. pH profile of molasses fermentation by *C. acetobutylicum* B530 in different mediums (circle for Mol#1, dan diamon for Mol#2)

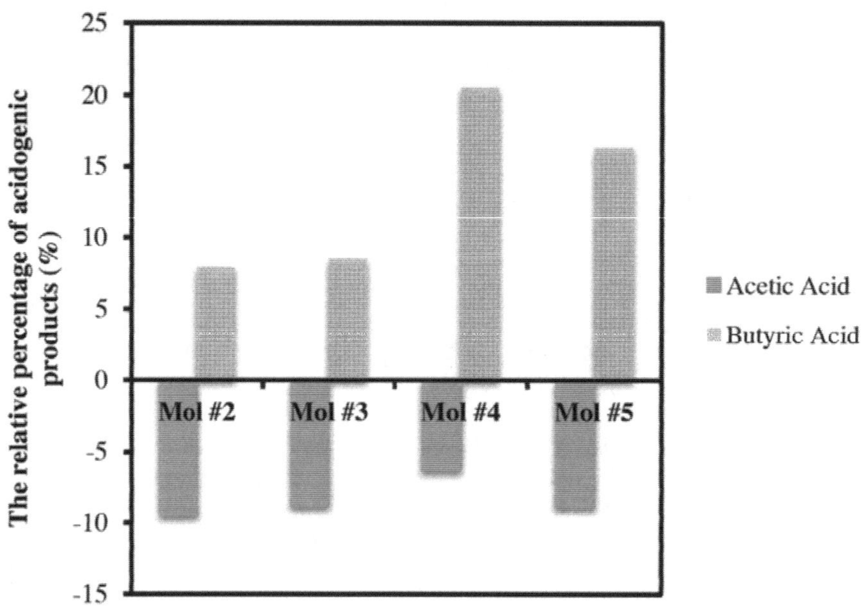

Figure 5. The percentages of acidogenic products from molasses fermentation using different medium (Mol#2, Mol#3, Mol#4, and Mol#5) relative to fermentation products using Mol#1 medium. The composition of medium is given in **Table 1**.

Result shows that the molasses fermentation conducted with the addition of 20 ppm of ferrous ion (at 35°C) is able to produce butyric acid by 20% higher than the base case (Mol#1), while acetic acid produced is 6% smaller. This result is expected to improve the selectivity toward biobutanol. The composition of those acidogenic products could influence the ratio of ABE solvent production. Ferrous ion can affect the metabolism of *C. acetobutylicum*. As shown in *C. acetobutylicum* metabolism pathway, acidogenic phase begins with the formation of acetyl-CoA, which is synthesized from the oxidation of pyruvic acid by the ferredoxin. Ferredoxin is an enzyme containing iron and sulfur as essential substances involved in the oxidation, therefore the addition of $FeSO_4.7H_2O$ will increase the production of this enzyme [14]. Furthermore, lowering the incubation temperature slightly could decrease the rate of cell metabolism, so that the performance of enzymes involved in acidogenic synthesis product will increase [12]. High productivity of butyric acid will enhance biobutanol production, because butyric acid plays an important role as precursor in the synthesis of biobutanol. The increase of acetic acid concentration as acidogenic product will provide more energy for biobutanol synthesis [9].

4. Conclusions

Sugarcane molasses can be used as a potential carbon and also vitamin sources in the fermentation by *C. acteobutylicum*. The addition of $FeSO_4.7H_2O$ into the fermentation medium can enrich the production of butyric acid in the acidogenic phase. 20 ppm $FeSO_4.7H_2O$ that was added into the molasses fermentation medium (at T = 35°C) can increase about 20% of butyric acid productivity compared to the molasses fermentation without ferrous ion (at T = 37°C). The increase in the production of butyric acid could be expected to increase the selectivity toward biobutanol in ABE fermentation.

Acknowledgments

The authors would like to gratefully acknowledge the ITB Research Program for the the financial supports. Gratitude to ARS (Agricultural Culture Collection), USA for providing of *C. acetobutylicum* B530 and also to PT. Indo Acidatama Tbk. for giving the sugarcane molasses.

References

[1] Schiel-Bengelsdorf B, Montoya J, Linder S and Dürre P 2013 Butanol fermentation *Environ. Technol.* **34** 1691-710

[2] Al-Shorgani NK, Ali E, Kalil MS and Yusoff WM 2012 Bioconversion of butyric acid to butanol by *Clostridium saccharoperbutylacetonicum* N1-4 (ATCC 13564) in a limited nutrient medium *BioEnergy Res.* **5** 287-93

[3] Green EM 2011 Fermentative production of butanol—the industrial perspective *Curr. Opin. Biotechnol.* **22** 337-43

[4] Amador-Noguez D, Brasg I A, Feng X J, Roquet N and Rabinowitz J D 2011 Metabolome remodeling dying the acidogenic-solventogenic transition in *Clostridium acetobutylicum Appl. Environ. Microbiol.* **77** 7984-97

[5] Lütke-Eversloh T and Bahl H 2011 Metabolic engineering of *Clostridium acetobutylicum*: recent advances to improve butanol production *Curr. Opin. Biotechnol.* **22** 634-47.

[6] Jang YS, Lee JY, Lee J, Park JH, Im JA, Eom MH, Lee J, Lee SH, Song H, Cho JH and Lee SY 2012 Enhanced butanol production obtained by reinforcing the direct butanol-forming route in *Clostridium acetobutylicum MBio.* **3** e00314-12.

[7] Monot F, Martin JR, Petitdemange H and Gay R 1982 Acetone and butanol production by Clostridium acetobutylicum in a synthetic medium *Appl. Environ. Microbiol.* **44** 1318-24.

[8] Durán-Padilla VR, Davila-Vazquez G, Chávez-Vela NA, Tinoco-Valencia JR and Jáuregui-Rincón J 2014 Iron effect on the fermentative metabolism of Clostridium acetobutylicum ATCC 824 using cheese whey as substrate *Biofuel Res. J.* **1** 129-33.

[9] Gheshlaghi RE, Scharer JM, Moo-Young M and Chou CP 2009 Metabolic pathways of clostridia for producing butanol *Biotechnol. Adv.* **27** 764-81.

[10] Fond O, Matta-Ammouri G, Petitdemange H and Engasser JM 1985 The role of acids on the production of acetone and butanol by Clostridium acetobutylicum *Appl. Microbiol. Biotechnol.* **22** 195-200.
[11] Syed Q 1994 Biochemical studies on anaerobic fermentation of molasses by Clostridium acetobutylicum (Doctoral dissertation, University of the Punjab, Lahore).
[12] Lee SY, Park JH, Jang SH, Nielsen LK, Kim J and Jung KS 2008 Fermentative butanol production by Clostridia *Biotechnol. Bioeng.* **101** 209-28.
[13] Ezeji TC, Qureshi N and Blaschek HP 2007 Bioproduction of butanol from biomass: from genes to bioreactors *Curr. Opin. Biotechnol.* **18** 220-7.
[14] Durán-Padilla VR, Davila-Vazquez G, Chávez-Vela NA, Tinoco-Valencia JR and Jáuregui-Rincón J 2014 Iron effect on the fermentative metabolism of Clostridium acetobutylicum ATCC 824 using cheese whey as substrate *Biofuel Res. J.* **1** 129-33.

Preliminary Study on Synthesis of Composite rGO/Ni by Microwave Assisted Method

Hafizh A. Fakhri[1], Miftahul Husnah[1], Akfiny Hasdi Aimon[1], and Ferry Iskandar[1,2*]

[1] Department of Physics, Institut Teknologi Bandung, Bandung 40132 Indonesia
[2]Research Center for Nanosciences and Nanotechnology, Institut Teknologi Bandung, Bandung 40132 Indonesia

*E-mail : ferry@fi.itb.ac.id

Abstract. Composite reduced graphene oxide-nickel (rGO/Ni) was successfully synthesized by wet chemical process via microwave assisted method. The nickel fraction was varied 0%, 5% and 20%. Crystallinity, morphology, chemical composition, chemical bonding, and conductivity were investigated by X-ray diffraction (XRD), scanning electron microscopy (SEM), elemental dispersive X-ray spectroscopy (EDS), Fourier transform infrared spectroscopy (FTIR), 4 point probes and electrochemical impedance spectroscopy (EIS). Based on the XRD patterns, the crystal sizes were 1.62 nm for 5% rGO/Ni and 1.64 nm for 20% rGO/Ni respectively. Furthermore, vibration of C=C aromatic stretching was detected in the FTIR spectra, which corresponds to rGO fingerprint. These results give a new perspective on synthesis of composite rGO/Ni.

1. Introduction

The lithium ion battery, which used as energy storage device, consists of three principle parts, i.e. anode, cathode and electrolyte [1]. Lithium iron phosphate ($LiFePO_4$) is a common material for the cathode. For anode, graphite is a general material for this purpose [2]. Several research groups have reported the improvement of lithium ion batteries [3-4]. Some of them replaced the graphite with graphene as an anode material [5]. Graphene is a carbon allotrope that has a two-dimensional structure [6]. Graphene has a high electrical and a good thermal conductivity, a wide surface and a high electrical mobility [7]. However, several researchers have reported synthesis of graphene by processes such as chemical solution, chemical vapor deposition, epitaxial growing, and micromechanical exfoliation [8-9]. From these methods, chemical solution method is an easier method with a high throughput. On the other side, transition metal is a type of inorganic material that has a good electrochemical behavior. Nickel is a transition metal that is abundant in Indonesia. To improve the anode performance, some researchers have composited graphene with a transition metal by microwave assisted method, hydrothermal method, or chemical vapor deposition method. Microwave assisted method have a rapid thermal process that easier to conduct than the other methods. In this study, graphene was composited with Ni (rGO/Ni) by microwave assisted method in order to improve the anode performance. Fourier Transform Infrared (FTIR), Scanning Electron Microscopy (SEM), Elemental Dispersive X -Ray Spectroscopy (EDS), and X-Ray Diffraction (XRD) were used to investigate the structural properties of the prepared sample. Further, 4 point probe measurement and electrochemical impedance spectroscopy (EIS) was used to study the electronic property of the sample. This result may give a new horizon in the improvement of anode material.

2. Experiment

All chemical materials were used without further purification. Graphite oxide and nickel nitrate hexahydrate were used as precursors of rGO/Ni. The graphite oxide was synthesized from graphite powder (Aldrich). Sulfuric acid (Merck) and phosphoric acid (Brataco) as concentrated acids and potassium permanganate (Merck) as oxidizer agent were used to oxidize the graphite. The volume of concentrated acid was 22.5 ml for sulfuric acid and 2.5 ml for phosphoric acid, respectively, for 1 gram graphite. During this process, the temperature was kept below 10 °C. After three gram of potassium permanganate was added, the compound was stirred at 50 °C for 40 minutes. Then distilled water was added to the solution. After that 30% of hydrogen peroxide was added to stop the oxidizing process. The solvent and solute were then separated and the solute was washed using 5% hydrochloric acid, alcohol, and distilled water. Finally, the washed solute was heated for 12 hours at 60 °C in an oven.

In the composite process, the graphite oxide was added with ethylene glycol and nickel nitrate as nickel source. In the sonication process, the compound of GO, nickel nitrate and ethylene glycol was sonicated using a sonicator for 2 hours. Then hydrazine hydrate as reduction agent was dropped into the compound. The compound was reduced using a microwave assisted method for 20 minutes. The solvent and solute were separated and then washed again using alcohol and distilled water. After that the solute was heated for 12 hours at 60 °C in an oven.

Samples were characterized by XRD, FTIR, SEM, EDS and 4 point probes. X ray diffraction was used to investigate the crystal structure of the samples by a Raigaku D/max-RA X ray diffractometer with Cu K radiation. The morphology of the samples was studied using scanning electron microscopy (Jeol JCM-6000 Bench top SEM). FTIR measurement from Alpha FTIR Spectrometer was used to study the chemical bonding of the sample. Furthermore, the electrical conductivity was characterized using the 4-point probes method using an Adventest R6240A DC Voltage Current Source/Monitor as the current source. Further, the EIS measurement was done using Gamry Instruments Reference 3000^{TM}.

3. Result and Discussion

Figure 1 shows the FTIR spectra of the GO sample and the 5% rGO/Ni sample. The GO spectra show some peaks which relates to O-C bond at 1037 and 1220 cm^{-1}(epoxide group), O=C bind at 1718 cm^{-1} (carbonyl group), C=C bond at 1613 cm^{-1} and O-H bond at 3123 cm^{-1} (hydroxyl group). These fingerprints indicate that GO was successfully synthesized by chemical process. Furthermore, Figure 1b shows the FTIR spectra of 5% rGO/Ni. Vibration of C=C aromatic bonding was found at ~1600 cm^{-1}, which corresponds to rGO. Vibration peaks of epoxide, carbonyl, and hydroxyl groups were not found in the spectra that indicate the sample (GO) was already reduced. Nickel bonding (< 500 cm^{-1}) was not found due to limitations of the instrument.

IOP Conf. Series: Journal of Physics: Conf. Series **877** (2017) 012073 doi:10.1088/1742-6596/877/1/012073

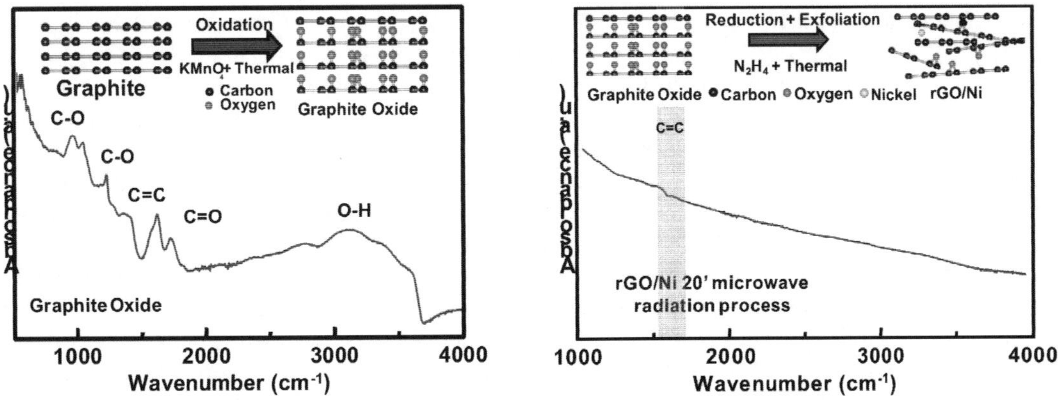

(a) (b)

Figure 1. Fourier transform infrared spectra of a) GO and b) Ni/rGO.

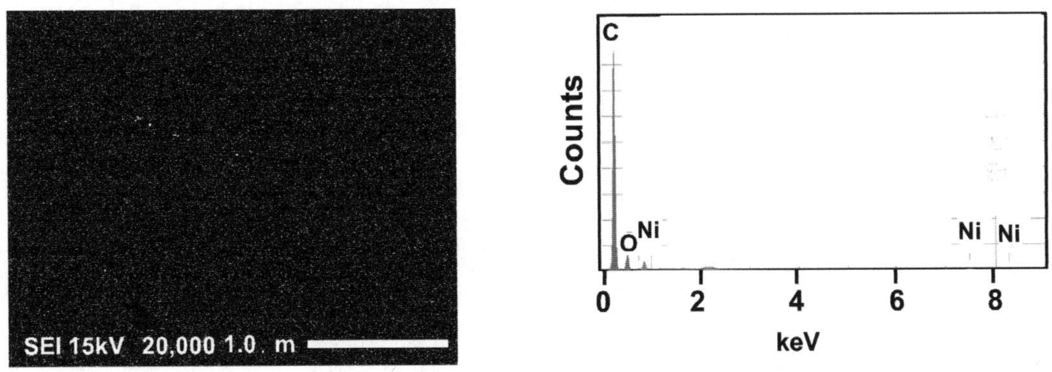

(a) (b)

Figure 2. a) SEM images of 5% rGO/Ni with 20.000 magnification and b) EDX characterization result of 5% rGO/Ni.

Figure 2a shows a SEM image of 5% rGO/Ni with 20,000 times magnification. The image shows the morphology of the sample as thin sheets. The atomic percentages from the EDX characterization were 86.61% for carbon, 11.91% for oxygen and 1.48% for nickel, respectively.

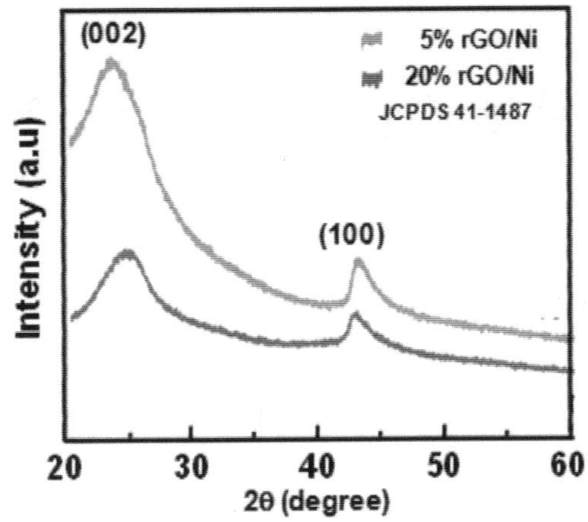

Figure 3. XRD pattern of 5% and 20% rGO/Ni.

The XRD patterns of the 5% and 20% rGO/Ni samples are shown in Figure 3, which correspond to graphite (JCPDS 41-1487) with diffraction peaks at $2\theta = 24.5°$ for (002) and at $2\theta = \sim 43°$ for (100) respectively. The graphite diffraction peak was widened at 20% sample due to stretching of the carbon layer distance. However, the nickel diffraction peaks could not be observed because the nickel was too small or its content in the sample was very slight compare to carbon content.

The electrical conductivity of the samples was measured using the 4-point probes method as shown in Figure 4a. Here, sample with 20% nickel had lower conductivity. It might be explained that bigger amount of Ni in the sample attracts the oxygen in the sample and rearrange into nickel oxide. However, this conductivity value represents the intrinsic property of the sample as single material. Hence, the EIS measurement was done to evaluate the interaction of the rGO/Ni composite as anode material with its environment. The EIS result of the samples was shown in Figure 4b with charge transfer resistance (Rct) value 3Ω, 0.86Ω, and 0.37Ω and surface resistance (Rsf) value 21.6 Ω, 9.36 Ω, and 3 Ω for 5%, 8% and 20% rGO/Ni composite, respectively. On contrary with conductivity result, the EIS result shows lower Rct and Rsf value for 20% nickel in the composite. This result shows that rGO/Ni composite can give a high charge transfer between the anode and the electrolyte and possibly to diminish the solid electrolyte interface (SEI) formation in the anode. Hence, rGO/Ni composite is potentially applied as an anode material.

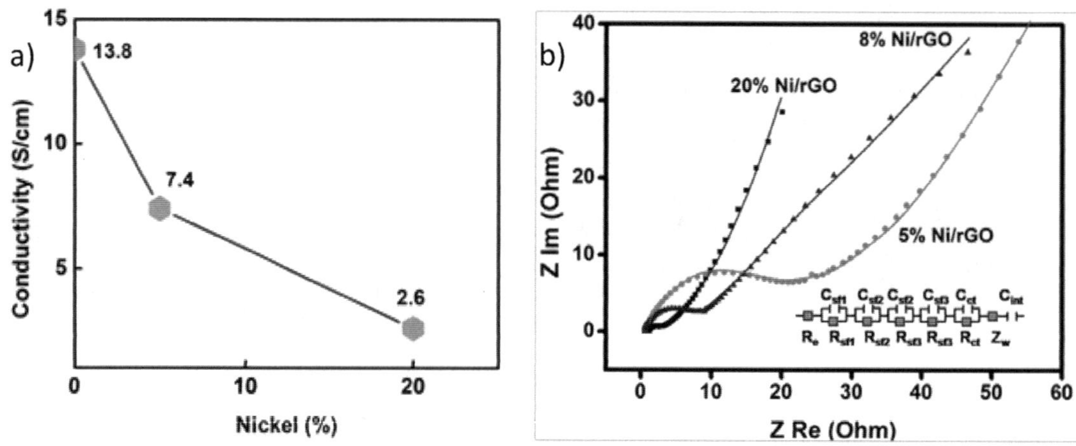

Figure 4. Four point probes result (a) and Nyquist plot of rGO/Ni composites.

4. Conclusion

Composite rGO/Ni was successfully synthesized by microwave assisted method with varying nickel content in the sample. FTIR result for rGO sample shows C=C aromatic stretching which confirms the formation of rGO sample. SEM image shows the sample in thin sheets formations which contain 86.61% carbon. Based on EIS results, rGO/Ni composite can give a high charge transfer between the anode and the electrolyte and has possibility to diminish the solid electrolyte interface (SEI) formation in the anode. Therefore, as a conclusion, rGO/Ni composite can give potential application for an anode material with microwave assisted synthesis method.

Acknowledgments

The authors would like to the Program Penelitian Unggulan Perguruan Tinggi (Desentralisasi) DIKTI 2016 research fund.

References

[1] M. Ren and Z. Zhou 2006 *Journal of Power Source.* **162** 1357.
[2] A. K. Padhi, K. S. Nanjundaswamy and J.B. Goodenough 1997 *Jurnal of Electrochemical*

Society. **144** 1188.
[3] H. Ni, J. Liu, and L. Z. Fan 2013 *Nanoscale.* **5** (5) 2164.
[4] W. J. Zhan, 2011 *Journal of Power Source.* **196** 2962.
[5] Abdolhosseinzadeh, S., H. Asgharzadeh, and H.S. Kim 2015 *Scientific Reports*. **5**.
[6] Zheng, Q. and J.-K. Kim 2015 *Graphene for Transparent Conductors; Synthesis, Properties and Applications,* Springer.
[7] Hummers Jr, W.S. and R.E. Offeman 1958 *Journal of the American Chemical Society*. **80** 1339-39.
[8] Berger, C., Z. Song, X. Li, X. Wu, N. Brown, C. Naud, D. Mayou, T. Li, J. Hass, and A.N. Marchenkov 2006 *Science*. **312** 1191-96.
[9] Geim, A.K. and K.S. Novoselov 2007 *Nat Mater.* **6** 183-191.

Recent Advances on Bioethanol Dehydration using Zeolite Membrane

I G B N Makertihartha[1,*], P T Dharmawijaya[1] and I G Wenten[1]

[1] Chemical Engineering Department, Institut Teknologi Bandung, Jl. Ganesha 10, Bandung, Jawa Barat, 40132, Indonesia

[*]Email: makertia@che.itb.ac.id

Abstract. Renewable energy has gained increasing attention throughout the world. Bioethanol has the potential to replace existing fossil fuel usage without much modification in existing facilities. Bioethanol which generally produced from fermentation route produces low ethanol concentration. However, fuel grade ethanol requires low water content to avoid engine stall. Dehydration process has been increasingly important in fuel grade ethanol production. Among all dehydration processes, pervaporation is considered as the most promising technology. Zeolite possesses high potential in pervaporation of bioethanol into fuel grade ethanol. Zeolite membrane can either remove organic (ethanol) from aqueous mixture or water from the mixture, depending on the framework used. Hydrophilic zeolite membrane, e.g. LTA, can easily remove water from the mixture leaving high ethanol concentration. On the other hand, hydrophobic zeolite membrane, e.g. silicate-1, can remove ethanol from aqueous solution. This review presents the concept of bioethanol dehydration using zeolite membrane. Special attention is given to the performance of selected pathway related to framework selection.

1. Introduction

Bioethanol is one of potential biofuel and has the potential to reduce pollution by replacing petroleum energy source [1]. Moreover, global environmental protection seen that industrial uses of fossil fuels released a large quantity of CO_2 is released leading to increasing atmospheric temperature [2]. Bioethanol is commonly produced from fermentation route from sugar or starch containing raw materials such as molasses and cornstarch [3]. Fermentation route produces aqueous solution with low ethanol concentration. To obtain absolute ethanol from said solution, dehydration step is required commonly using azeotropic distillation due to the formation ethanol-water azeotrope. However, pervaporation is reported to require less energy than azeotropic distillation. Cardona Alzate and Sanchez Toro [4] reported that further energy saving could be obtained in bioethanol production plant by using pervaporation compared with using azeotropic distillation. Pervaporation only uses 1/5 energy that consumed by using azeotropic distillation. Kaminski et al. [5] compared the cost of ethanol dehydration using vapor permeation, pervaporation, distillation and adsorption. The total operating cost reported are US$ 15.75/tonnes, US$ 12.6-16.6/tonnes, US$ 31.95-45.65/tonnes and US$ 36.3/tonnes, respectively.

Membrane technology has shown its excellent performance in separation process [6-12]. Compared with other approach, membranes tend to have lower energy requirement and other operational cost. One type of membrane based separation is pervaporation. Pervaporation uses membrane as a selective layer between liquid phase feed and vapor phase permeate. Only desired component can pass through membrane by vaporization. Vaporization of permeate is carried out using vacuum that kept on the permeate side while atmospheric or elevated pressure is used at feed side. Separation via pervaporation is not rely on vapor liquid equilibrium due to limiting transport resistance in the membrane [13, 14].

Zeolite has been used in various sectors such as gas/liquid separation [15-20], catalysis [21, 22], microelectronic, and corrosion resistant areas [23-26]. Zeolite owes uniform, well-defined pores in molecular size, high porosity, and excellent thermal and chemical stability. Zeolite is particularly promising in industrial scale with harsh condition.

There are some reviews in bioethanol purification using pervaporation [1, 5, 27, 28]. However, a brief explanation in zeolite membrane application and the progress in zeolite membrane performance are still needed. This paper will discuss general application of zeolite membrane for pervaporation in bioethanol production. Special attention will be given to progress on membrane flux and separation factor generated in order to compete with conventional distillation process.

2. Zeolite Membrane Concept in Bioethanol Production

Bioethanol is generally produced from fermentable raw materials containing sugar or starch, such as sugar cane molasses and cornstarch, cellulose, as well as industrial waste products [2, 29]. The production of bioethanol from fermentation route produces aqueous ethanol solutions. In order to obtain absolute ethanol from the aqueous ethanol solution, the dehydration process is carried out after fermentation process up to 99.5 wt% ethanol [30].

For the past few years, several researches are reported for the pervaporative dehydration of ethanol–water mixtures [31-35]. Kondo et al. [36] prepared NaA zeolite membrane on the surfaces of porous tubular supports composed ofmullite, alumina and/or cristobalite using the hydrothermal synthesis. They obtained pervaporation fluxes and separation factor of 2.35 kg/m^{2}h and 5100 respectively for pervaporation of 95 wt.% ethanol–water mixtures at 95 ◦C.

High permeation flux while maintaining high separation factor is the major challenge of zeolite membrane. There have been various efforts towards increasing permeation flux. Generally, permeation flux has the correlation with mass transfer resistant of porous support and membrane layer. Reduction of membrane layer or support is one effective way. Another approach involves using more porous support since the separation factor is only affected by the membrane layer. However, the preparation using such high porosity support is more complicated since high porosity support poorly hold seed layer on it. Algieri et al. [37] conducted pervaporation experiments with thin MFI zeolite membranes synthesized by in situ nucleation and secondary growth at 70 ◦C and 9.4 wt.% ethanol–water mixtures. They obtained high fluxes (2.1 kg/m2 h) and low separation factor (EtOH/H_2O = 1.3). The fluxes and separation factor for NaA type zeolite membrane varies from 0.23 to 5.60 kg/m2 h and 3600 to 10,000 relatively. The fluxes and separation factor for these zeolite membranes are generally higher than the silicalite and mordenite type zeolite membranes. This is because the pore diameters are larger for these zeolite membranes, which means that they are less size selective for water and they are also less hydrophilic compared to NaA type zeolite membrane [38]. Li et al. [39] prepared NaA zeolite membrane using cheap coarse macroporous support to not only reduce mass transfer resistance but also reduce the price of zeolite membrane.

Utilization of zeolite membrane for ethanol recovery from fermentation broth has been proposed widely. Earlier scheme involve distillation of broth up to azeotropic limit continued with hydrophilic membrane pervaporation up to 99.5 wt% [40]. A further utilization of pervaporation module proposes direct pervaporation using hydrophobic membrane to deplete ethanol from fermentation broth [27]. However, a column is still needed when hydrophobic membrane cannot deliver high enough purity. Figure 1 depict proposed diagram of pervaporation using both hydrophilic and hydrophobic membrane. The dephlegmator in between modules is used to cool the mixture down (using coolant) and condense water from the mixture resulting in higher ethanol percentage on overhead product [28].

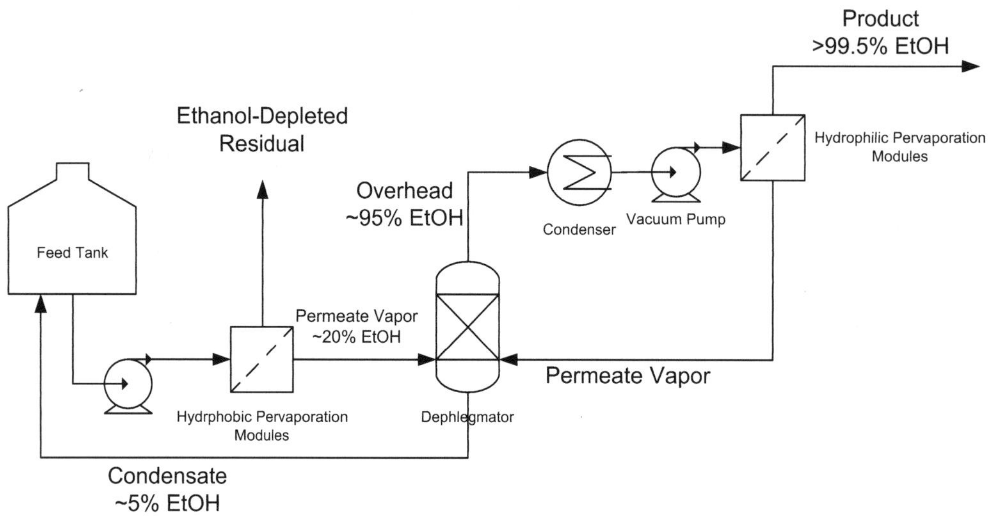

Figure 1. Proposed Scheme for Ethanol Recovery using Pervaporation adapted from [28]

3. Removal of Organic From Water

The first step after synthesizing ethanol via fermentation route is the recovering from fermentation broth. Fermentation broth may contain a lot of components depend on the condition of the reaction. Generally, the main component that hardly separated from ethanol is water since other components are larger in size than ethanol. Viable and dead whole cells, suspended solids, and other cell components have the potential to accumulate in membrane modules, blocking flow path hence reducing permeance [28]. Dissolved solids such as NaCl and Sugars such as glucose and xylose interact with water preventing it to pass through membrane. On the other hand, glycerol interacts with ethanol giving similar effect like sugar to water. However, dissolved solids may also precipitate when operating at higher temperature. pH may effect differently among zeolites used related to their stability but in general no direct effect caused by pH [28].

There have been a lot of researches that report pervaporative dehydration of ethanol-water mixtures [31-35]. Kondo et al. [36] fabricate LTA zeolite layer on porous tubular mullite, alumina and/or cristobalite via hydrothermal synthesis. The resulting membrane has flux and separation factor of 2.35 kg/m2h and 5100, respectively. Algieri et al. [37] was using MFI zeolite membrane prepared via in situ nucleation and secondary growth at 70 °C and 9.4 wt.% ethanol-water mixtures. High fluxes (2.1 kg/m^2h) but low separation factor ($EtOH/H_2O$=1.3) was obtained. Several report on pervaporation of ethanol –water mixture using hydrophobic zeolite membrane is presented in Table 1.

Table 1. Hydrophobic zeolite membranes reported for ethanol-water separation

Zeolite Type	Support	Ethanol in feed (wt.%)	T (K)	Membrane Performance		Ref
				Flux (kg/m^2h)	α ($EtOH/H_2O$)	
Silicate-1	SS	5	333	0.9	106	[41]
Silicate-1	Al_2O_3	5	333	1.8	89	[42]
B-ZSM-5	Al_2O_3	5	303	0.16	31	[43]
Ge-ZSM-5	SS	5	303	0.223	47	[44]
ZSM-5	α-Al_2O_3	5	333	7.6	51	[45]
ZSM-5	Al_2O_3	5	333	9.8	58	[46]
B-ZSM-11	Al_2O_3	5	333	1.51	35	[47]
ZSM-5	α-Al_2O_3	5	333	2.9	50	[48]
ZSM-5	α-Al_2O_3	10	383	51.6	5.2	[49]

Integration of membrane pervaporation with fermentation is also another approach to improve the performance of ethanol production [50]. In situ continuous ethanol extraction will boost fermentation efficiency since ethanol of about 10 wt% inhibits further fermentation [45]. Ikegami et al. [51] employed silicate membranes for pervaporation of fermentation broth to obtain ethanol. The result showed an increase in average glucose consumption rate of about ~20% compared with conventional batch reactor. The study also suggests addition of sugars will enhance ethanol selectivity through silicate membrane [52]. However, the selectivity of ethanol is still very low (~5). Nomura et al. [53] studied continuous pervaporation from fermentation broth using silicate membrane. The result shows good performance in terms of flux (0.1 kg/m^{2}h) and separation factor (up to 218) for pervaporation of fermentation broth and ethanol concentration of 98.2% in the permeate side. This result is higher compared with using only water/ethanol mixture.

In general, MFI membrane possesses the highest potential in extraction of ethanol from fermentation broth. MFI has 0.5 nm pore size which is higher than ethanol and water but smaller than most fermentation broth component [54]. Therefore, MFI selectivity towards ethanol is mainly due to surface properties not molecular sieving. MFI not only has the resistance to withstand fermentation broth but also shows higher performance when fermentation broth feed is used [53]. Hydrophobic zeolite membrane can be used as simple as detached pervaporation module filtering batch fermentation products or attached to the bioreactor and enable continuous production of ethanol. However, there still are more room to the development of hydrophobic zeolite membrane in terms of increasing performance and reducing cost [55].

4. Removal of Water from Organic

Removal of water from organic is used as the downstream process of ethanol purification to produce fuel grade ethanol. This method is preferably placed in downstream because of low water content in the feed stream (~10%) thus reducing the membrane load. Moreover, high hydrophilicity zeolite membrane tends to have high Al content making it more prone to dealumination from acidic environment. Meanwhile, fermentation may be carried out in acidic condition and/or produce organic acid as a byproduct [56].

Table 2. Hydrophilic zeolite membranes reported for ethanol-water separation

Zeolite Type	Support	Water in feed (wt.%)	T (K)	Membrane Performance		Ref
				Flux (kg/m^{2}h)	α (H_2O/EtOH)	
NaA	TiO_2 Tube	10	323	0.8-1.00	8500	[57]
NaA	α-Al_2O_3 tube	5	318	0.23	8300	[58]
NaA	α-Al_2O_3 tube	9.2	366	2.5	130	[59]
NaA	α-Al_2O_3 tube	10	348	5.6	>5000	[60]
NaA	α-Al_2O_3 tube	10	323	0.5	16000	[34]
NaA	α-Al_2O_3 tube	10	398	3.8	3600	[61]
NaA	α-Al_2O_3 tube	10	353	0.54	>10000	[62]
Silica	γ-Al_2O_3 disc	10	353	1	800	[63]
Mordenite	α-Al_2O_3 tube	10	423	0.16	139	[64]
Mordenite	α-Al_2O_3 tube	15	363	0.06	60	[65]
NaA	α-Al_2O_3 tube	10	348	8.38	>10000	[66]
NaA	α-Al_2O_3 tube	10	348	11.1	>10000	[67]

Several studies for the pervaporative dehydration of alcohol from aqueous mixtures using hydrophilic zeolite membrane are presented in Table 2 . The fluxes and separation factor for NaA type zeolite membrane varies from 0.23 to 5.60 kg/m^{2}h and 3600 to 10,000 relatively. The fluxes and separation factor for these zeolite membranes are generally higher than the silicalite and mordenite type zeolite membranes. This is because the pore diameters are larger for these zeolite membranes, which means that they are less size selective for water and they are also less hydrophilic compared to NaA type zeolite membrane [38].

The first commercialization of hydrophilic zeolite membrane for ethanol/water separation is using LTA membrane by Mitsui Engineering & Shipbuilding Co. in 1999 [68]. The first large scale pervaporation plant using zeolite NaA membrane is designed to dehydrate various solvents such as ethanol, IPA, and methanol. The plant is a major breakthrough in the development of zeolite membrane. It proves the reliability of LTA zeolite membrane on large scale operation. The highest reported water flux is 16 kg water/m^{2}h in the pervaporation of water/i-propanol mixtures at 120 ^{o}C (Caro dkk. 2000). The plant uses 16 modules; each consists of 125 pieces of tubular type NaA zeolite membrane. The plant adopts two vacuum systems for more effective dehydration. Firstly, the feed is dehydrated up to 1.5 wt. % water at 15 mmHg permeate pressure. Then, a final dehydrated product of 0.5 wt. % water is obtained at 8 mmHg permeates pressure [68].

5. Summary and Future Outlook

General aspects of zeolite membrane aided organic acid esterification have been reviewed. Zeolite membrane can be used to completely separate ethanol from fermentation broth. A lot of works have shown the effectiveness of zeolite based membrane in pervaporation to substitute conventional processes. While downstream hydrophilic zeolite membrane process has had its place in industrial scale, hydrophobic zeolite membrane still requires further work to perfect its performance. However, current performance of zeolite membrane is already sufficient for early industrial development of full membrane ethanol purification from fermentation broth.

In the future, this technology needs some improvement in the subject of membrane (materials, methods, and reproducibility), and cost optimization (mainly for industrial implementation). Economical study of zeolite membrane pervaporation for ethanol recovery should be done to review the effectiveness of current zeolite membrane and set a quantitative target for several aspects. The numbers of researches that study various aspects of this technology suggest that it is ready for further industrial implementation.

References

[1] Frolkova A and Raeva V 2010 *Theor. Found. Chem. Eng.* **44** 545
[2] Ikegami T, Yanagishita H, Kitamoto D, Negishi H, Haraya K and Sano T 2002 *Desalination* **149** 49
[3] Groot W and Luyben K C A 1987 *Biotechnol. Lett* **9** 867
[4] Alzate C C and Toro O S 2006 *Energy* **31** 2447
[5] Kaminski W, Marszalek J and Ciolkowska A 2008 *Chem. Eng. J.* **135** 95
[6] Wenten I G and Widiasa I N 2002 *Desalination* **149** 279
[7] Wenten I G, Julian H and Panjaitan N T 2012 *Desalination* **306** 29
[8] Purwasasmita M, Nabu E B P, Khoiruddin and Wenten I G 2015 *Journal of Engineering and Technological Sciences* **47** 426
[9] Himma N F, Anisah S, Prasetya N and Wenten I G 2016 *J. Polym. Eng.* **36** 329
[10] Wenten I G and Khoiruddin 2016 *Desalination* **391** 112
[11] Drioli E, Stankiewicz A I and Macedonio F 2011 *J. Membr. Sci.* **380** 1
[12] Strathmann H, Grabowski A and Eigenberger G 2013 *Ind. Eng. Chem. Res.* **52** 10364
[13] Mallada R and Menéndez M, *Inorganic membranes: synthesis, characterization and applications*. Vol. 13. Elsevier, 2008.
[14] Purwasasmita M, Kurnia D, Mandias F C, Khoiruddin and Wenten I G 2015 *Food and Bioproducts Processing* **94** 180
[15] Lai Z, Bonilla G, Diaz I, Nery J G, Sujaoti K, Amat M A, Kokkoli E, Terasaki O, Thompson R W and Tsapatsis M 2003 *Science* **300** 456
[16] Huang A, Liang F, Steinbach F, Gesing T M and Caro J r 2010 *J. Am. Chem. Soc.* **132** 2140
[17] Yuan W, Lin Y and Yang W 2004 *J. Am. Chem. Soc.* **126** 4776

[18] Pham T C T, Kim H S and Yoon K B 2011 *Science* **334** 1533
[19] Wang Z, Ge Q, Shao J and Yan Y 2009 *J. Am. Chem. Soc.* **131** 6910
[20] Ge Q, Wang Z and Yan Y 2009 *J. Am. Chem. Soc.* **131** 17056
[21] Zhong Y, Chen L, Luo M, Xie Y and Zhu W 2006 *Chem. Commun.* 2911
[22] Kim S-J, Yang S, Reddy G K, Smirniotis P and Dong J 2013 *Energy Fuels* **27** 4471
[23] Wang Z, Wang H, Mitra A, Huang L and Yan Y 2001 *Adv. Mater.* **13** 746
[24] Wang Z, Mitra A, Wang H, Huang L and Yan Y 2001 *Adv. Mater.* **13** 1463
[25] Lew C M, Li Z, Li S, Hwang S J, Liu Y, Medina D I, Sun M, Wang J, Davis M E and Yan Y 2008 *Adv. Funct. Mater.* **18** 3454
[26] Beving D E, McDonnell A M, Yang W and Yan Y 2006 *J. Electrochem. Soc.* **153** B325
[27] Caro J, Noack M and Kölsch P 2005 *Adsorption* **11** 215
[28] Vane L M 2005 *J. Chem. Technol. Biotechnol.* **80** 603
[29] Nigam P S and Singh A 2011 *Prog. Energy Combust. Sci.* **37** 52
[30] Kazemimoghadam M and Mohammadi T 2011 *Desalination* **281** 298
[31] Pera-Titus M, Llorens J, Tejero J and Cunill F 2006 *Catal. Today* **118** 73
[32] Zhou L, Wang T, Nguyen Q T, Li J, Long Y and Ping Z 2005 *Sep. Purif. Technol.* **44** 266
[33] Ahn H, Lee H, Lee S-B and Lee Y 2006 *Desalination* **193** 244
[34] Pera-Titus M, Mallada R, Llorens J, Cunill F and Santamaría J 2006 *J. Membr. Sci.* **278** 401
[35] Pera-Titus M, Llorens J, Cunill F, Mallada R and Santamaría J 2005 *Catal. Today* **104** 281
[36] Kondo M, Komori M, Kita H and Okamoto K-i 1997 *J. Membr. Sci.* **133** 133
[37] Algieri C, Bernardo P, Golemme G, Barbieri G and Drioli E 2003 *J. Membr. Sci.* **222** 181
[38] Wee S-L, Tye C-T and Bhatia S 2008 *Sep. Purif. Technol.* **63** 500
[39] Li H, Wang J, Xu J, Meng X, Xu B, Yang J, Li S, Lu J, Zhang Y, He X and Yin D 2013 *Journal of Membrane Science* **444** 513
[40] Richter H, Voigt I and Kühnert J-T 2006 *Desalination* **199** 92
[41] Lin X, Chen X, Kita H and Okamoto K 2003 *AlChE J.* **49** 237
[42] Lin X, Kita H and Okamoto K I 2001 *Ind. Eng. Chem. Res.* **40** 4069
[43] Bowen T C, Kalipcilar H, Falconer J L and Noble R D 2003 *J. Membr. Sci.* **215** 235
[44] Li S, Tuan V A, Falconer J L and Noble R D 2003 *Microporous Mesoporous Mater.* **58** 137
[45] Xia S, Peng Y, Lu H and Wang Z 2016 *Microporous Mesoporous Mater.* **222** 128
[46] Xia S, Peng Y and Wang Z 2016 *J. Membr. Sci.* **498** 324
[47] Chai L, Li H, Zheng X, Wang J, Yang J, Lu J, Yin D and Zhang Y 2015 *J. Membr. Sci.* **491** 168
[48] Chai L, Yang J, Wang J, Lu J, Yin D and Zhang Y 2016 *AlChE J.*
[49] Yu L, Korelskiy D, Grahn M and Hedlund J 2015 *Sep. Purif. Technol.* **153** 138
[50] Shabtai Y, Chaimovitz S, Freeman A, Katchalski-Katzir E, Linder C, Nemas M, Perry M and Kedem O 1991 *Biotechnol. Bioeng.* **38** 869
[51] Ikegami T, Yanagishita H, Kitamoto D, Haraya K, Nakane T, Matsuda H, Koura N and Sano T 1997 *Biotechnol. Tech.* **11** 921
[52] Ikegami T, Yanagishita H, Kitamoto D, Haraya K, Nakane T, Matsuda H, Koura N and Sano T 1999 *Biotechnol. Lett* **21** 1037
[53] Nomura M, Bin T and Nakao S-i 2002 *Sep. Purif. Technol.* **27** 59
[54] Baerlocher C, McCusker L B and Olson D H, *Atlas of zeolite framework types*. Elsevier, 2007.
[55] Meindersma G and De Haan A 2002 *Desalination* **149** 29
[56] Pera-Titus M, Llorens J and Cunill F 2009 *Chem. Eng. Process.* **48** 1072
[57] Pera-Titus M, Bausach M, Llorens J and Cunill F 2008 *Sep. Purif. Technol.* **59** 141
[58] Zah J, Krieg H M and Breytenbach J C 2006 *J. Membr. Sci.* **284** 276
[59] Tiscareno-Lechuga F, Tellez C, Menendez M and Santamarí J 2003 *J. Membr. Sci.* **212** 135
[60] Sato K and Nakane T 2007 *J. Membr. Sci.* **301** 151
[61] Pina M, Arruebo M, Felipe M, Fleta F, Bernal M, Coronas J, Menendez M and Santamaria J 2004 *J. Membr. Sci.* **244** 141
[62] Li Y, Chen H, Liu J, Li H and Yang W 2007 *Sep. Purif. Technol.* **57** 140

[63] Sekulić J, Ten Elshof J and Blank D 2005 *J. Membr. Sci.* **254** 267
[64] Navajas A, Mallada R, Téllez C, Coronas J, Menéndez M and Santamaría J 2007 *J. Membr. Sci.* **299** 166
[65] Casado L, Mallada R, Téllez C, Coronas J n, Menéndez M and Santamaría J 2003 *J. Membr. Sci.* **216** 135
[66] Yan H, Ma N, Zhan Z and Wang Z 2015 *Microporous and Mesoporous Materials* **215** 244
[67] Shao J, Zhan Z, Li J, Wang Z, Li K and Yan Y 2014 *Journal of Membrane Science* **451** 10
[68] Morigami Y, Kondo M, Abe J, Kita H and Okamoto K 2001 *Separation and Purification Technology* **25** 251

IOP Conf. Series: Journal of Physics: Conf. Series **877** (2017) 012075 doi:10.1088/1742-6596/877/1/012075

Preparation and characterization of polysulfone/PEG heterogeneous ion exchange membrane for reverse electrodialysis (RED)

D Ariono[1], Khoiruddin[1], D Prabandari[1], R Wulandari[1] and I G Wenten[1,*]

[1] Department of Chemical Engineering, Institut Teknologi Bandung, Jl. Ganesha 10, Bandung 40132, Indonesia

*igw@che.itb.ac.id

Abstract. Heterogeneous cation-exchange membrane is synthesized using solution casting method. The casting solution is prepared by dispersing finely ground cation-exchange resin particles in N,N-dimethylacetamide (DMAc) solutions of polysulfone (PSf) while polyethylene glycol (PEG400) is used as a modifier. The results show that the PEG400 can increase water uptake, conductivity, and ion-exchange capacity (IEC) of the heterogeneous cation-exchange membrane due to the hydrophilic nature of PEG400. The more hydrophilic membrane results in higher water uptake and wider access for functional sites. However, when the concentration of PEG400 is increased further, the IEC and conductivity tend to decrease. This tendency is more pronounced when the ion-exchange resin particle is increased from 50 to 60%-wt. It could be attributed to the washed out of some ion-exchange particle during membrane immersion due to lower bonding between membrane matrix and the particles.

1. Introduction

Ion-exchange membrane (IEM) has been widely used in a number of different applications such as desalination and seawater pre-concentration, production of chlorine-alkaline, production of organic acids and organic salts, production of ultrapure water, wastewater treatment, and energy generation [1-8]. One of the applications of IEM membrane in energy generation is in reverse electrodialysis (RED). RED unit is very similar to conventional electrodialysis wherein a number of anion and cation exchange membranes are stacked together in an alternating pattern between a pair of electrode (anode and cathode). Due to the gradient salinity of two different solutions (e.g. seawater or brine water with river water) fed into RED cells, the chemical potential is converted to electrical energy by transporting ions from high salinity into low salinity solution called concentrate and diluate, respectively [9]. This technology can be combined with desalination plant such as seawater reverse osmosis (SWRO) in an integrated SWRO-RED system [10, 11]. It is found from the study that the brine from SWRO plant provides a better source for energy recovery. Moreover, the discharged brine can be controlled for minimizing environmental impact due to dilution effect during the process of RED [12]. Development of IEM with improved properties is required to achieve a better performance since IEM is the key of the process. Conductivity and pemselectivity of IEM are crucial properties in order to get a high power density in RED.

In general, commercial IEMs can be classified into homogeneous and heterogeneous membranes with their typical properties and advantages [13]. The homogeneous membrane exhibits better electrochemical properties than the heterogeneous [14]. On the other hand, heterogeneous IEMs have better mechanical strength, easier preparation procedure, and lower production cost [15, 16]. Recently, a number of works have been reported in the development of the heterogeneous IEMs which has been driven by the need to obtain IEMs with better properties, lower production cost, and easier preparation procedure [13]. Several modifications methods are proposed to improve the characteristic of heterogeneous IEM or even to obtain a specific property. The modifications methods include additive blending, surface modification, and special treatment after fabrication of a base membrane.

Introduction of additive into membrane solution or the additive blending method is a simple method for improving membrane characteristics. It has been used in many works especially for incorporating inorganic particles [17-25]. In this method, additives are blended together with a polymer and solvent. This solution is then cast into a pre-formed membrane and the solvent is separated through solvent evaporation, solution immersion, or combined process. This blending method is also applied in membrane preparation using extrusion [26]. The properties of the additive such as electrical conductivity, hydrophilicity, and adsorption capacity can be utilized to produce a membrane with better electrochemical characteristics.

In this study, polysulfone-based cation-exchange membrane is prepared using phase inversion method and solution casting technique. The effects of polyethylene glycol (PEG400) and ion-exchange resin content on the membrane characteristics were studied and evaluated in terms of water content, ion-exchange capacity, and membrane conductivity.

2. Materials and method

2.1. Materials

Polysulfone (PSf) supplied by Solvay Advanced Polymers (Udel PSf P-3500 LCD) was used as a polymer binder. N,N-dimethylacetamide (DMAc) from Shanghai Jingsan Jingwei Chemical Co., Ltd was employed as the solvent. Strong acid cation-exchange resin (Amberlite^{TM} IR120-Na) was used in membrane preparation as functional groups. Polyethylene glycol (PEG400) was used as additive. Demineralized water is used throughout the experiment.

2.2. Membrane preparation

The heterogeneous cation-exchange membrane is prepared by phase inversion method and solution casting technique (figure 1). Prior to the membrane preparation, ion-exchange resins were dried in an oven at 60 °C for 24 hours, then pulverized into fine particles and sieved into the desired mesh size (-325+400). The polymer solution was prepared by dissolving PSf and PEG400 into DMAc and stirred until the solution was optically homogeneous. The concentration of PEG was varied from 0 to 45% by weight in a total polymer (PSf/PEG400). The concentration of polymer was maintained at 25% by weight (polymer/solvent). This was followed by dispersing a specific quantity of resin particles (50 and 60 %-wt in total solid or PSf/PEG/resin) and mixing until homogeneous. Afterwards, the mixture was casted onto a glass plate, dried at ambient temperature for 24 hours, and immersed in demineralized water. The thickness of the membranes was about 0.35 (±0.07) mm.

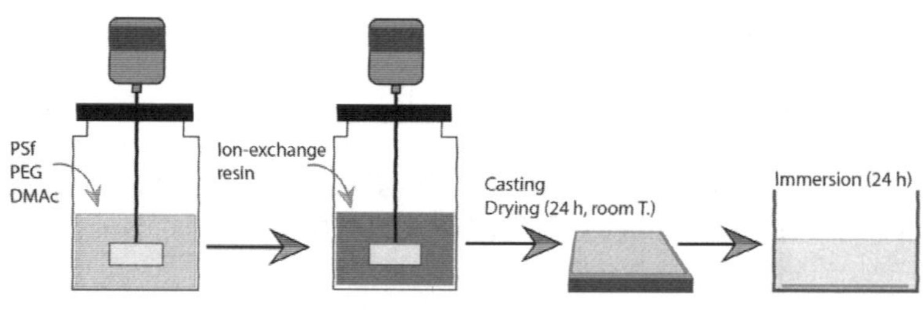

Figure 1. Membrane preparation steps

2.3. Characterization

Water uptake (Wu, g water/g dried membrane) was measured as weight difference between the dried (in oven 60 °C for 24 hours) and wetted membrane as follows:

$$Wu = \frac{W_{wet} - W_{dry}}{W_{dry}} \quad (1)$$

In equation (1), W_{wet} and W_{dry} are the weight of the wetted and dried membrane, respectively.

Ion exchange capacity (IEC) of the membrane was determined by using titration method. First, the membrane tip was immersed in a hydrochloric acid solution for 24 hours. Then, the membrane was immersed in a saturated NaCl solution for 48 hours at room temperature to change the H^+ ions with Na^+ ions. The H^+ ions in the solution are titrated with NaOH 0.01 M while phenolphtalein is used as an indicator. The IEC was calculated by using the following equation:

$$IEC = \frac{C \times V}{W_{dry}} \quad (2)$$

Where IEC is ion-exchange capacity ($meq.g^{-1}$), C is molar concentration of the titrant (M), V is titrant volume (ml), and W_{dry} is dry weight of the membrane sample (g).

The measurement of membrane conductivity, σ ($\Omega^{-1}.cm^{-1}$), was conducted by using LCR Meter instrumentation with a frequency range from 20 Hz to 2 MHz. To characterize the membrane conductivity, the membrane samples were immersed in the deionized water for 24 h until they were wet enough and saturated. Each sample was cut according to the electrode size used in the measurement cell. The measurement process was conducted at room temperature. The conductivity of the membrane was calculated by using the following equation:

$$\sigma = \frac{L}{R_m A} \quad (3)$$

Where L is membrane thickness and A is membrane effective area.

3. Results and discussion

The effects of PEG400 and resin concentration on Wu and IEC are shown in figure 2. Figure 2 shows that the water content is increased with increasing PEG concentration either for the membrane with 50%-wt and 60%-wt of resin. In addition, the membranes with 50%-wt resin have higher water content than that with 60%-wt of resin. Meanwhile, the IEC of the membrane is increased when the PEG concentration is 15%-wt and decreased again for higher PEG concentration. The effect of resin concentration on IEC has a similar trend with the water content.

The increasing water content is attributed to the hydrophilic nature of PEG. As the PEG concentration is increased, the hydrophilicity of the membrane is improved which results in an increase of water adsorption of the membrane. The increasing water content then improves the accessibility of the functional site inside the membrane matrix. As the result, the IEC of the membrane with 15%-wt of PEG is increased.

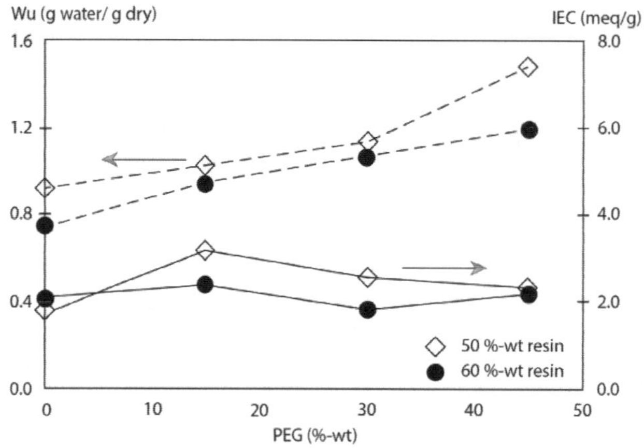

Figure 2. Wu and IEC of membrane with various PEG and ion-exchange resin concentrations.

However, the increase of water content is not always accompanied by the increase of IEC. When the PEG concentration is above 15%, it has negatives effect to IEC. This may due to washed-out of some resin particles during the membrane immersion. This washed out of the particles could be the results of lower bonding between resin particles and polymer matrix or binder. Since PEG replaces some portion of PSf, the structure of the membrane becomes weaker. The lower bonding between polymer and resin particles is also observed when the resin concentration is higher (60%-wt). Therefore, the prepared membrane with higher resin concentration exhibits less water content and lower IEC.

The membrane conductivity in various PEG400 and resin concentration is shown in figure 3. It can be observed that the highest conductivity is obtained when the PEG concentration is 15%-wt. This highest conductivity is associated with the high water content and high IEC of the prepared membrane. The adsorbed water inside the membrane contributes to electrical current transfer. The more adsorbed water also indicates the higher conductive part of the membrane instead of the inert or hydrophobic part. Therefore, the high water content results in high electrical conductivity. Meanwhile, IEC represents the amount of functional sites inside the membrane. These functional sites provide the active sites for bridging an electrical current transfer. As the functional sites are increased, the available active media involved in the electrical current transfer is also increased that yields in higher conductivity.

It is indicated in figure 2 and 3 that the electrical conductivity is affected by water content and IEC. However, the role of IEC in improving membrane conductivity is more dominant than the water content. The highest conductivity is obtained from the highest IEC (15%-wt of PEG and 50%-wt of resin). A relationship between IEC and membrane conductivity is elucidated in figure 4. It could be explained from the figure that the conductivity is increased with increasing IEC. It has been explained by the microheterogeneous model that there are two phases which have a contribution to membrane conductivity [27]. Those are a gel phase or the polymeric phase of the membrane and inter-gel phase or the solution phase filling a gap in the membrane matrix. Moreover, the gel phase could be further divided into (i) pure gel phase or the polymeric part with has the functional sites and (ii) inert phase. According to the model, the membrane conductivity (σ) is expressed by the following equation (if the conductive phase is randomly distributed) [27]:

$$\sigma = \sigma_1^{f_1} \sigma_2^{f_2} \tag{4}$$

Where σ_1 and σ_2 are conductivity of the gel and the inter-gel phase, respectively. Meanwhile, f_1 and f_2 are fractions of the first and the second phase, respectively.

It is indicated in equation (4) that both phases have the proportional effect on membrane conductivity. Since the measurement of electrical conductivity was conducted using membrane wetted with demineralized water, the functional sites inside the membrane had more contribution in electrical current transfer than the water content. The contribution of the inter-gel phase would be increased if the membrane used in a solution containing high concentration of ions.

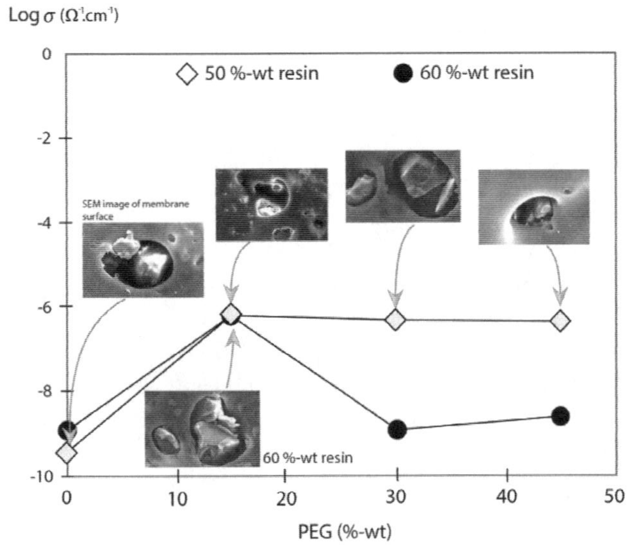

Figure 3. Conductivity of membrane with various PEG concentration.

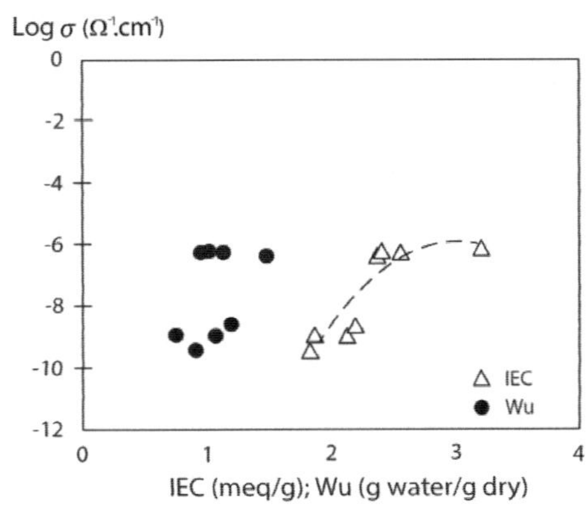

Figure 4. Membrane conductivity in various IEC and Wu.

4. Conclusion

The PSf-based heterogeneous cation-exchange membrane is synthesized using solution casting method and PEG400 is used as a modifier. The membrane is prepared in the various concentration of PEG and resin particles. The results show that the PEG400 can increase Wu, IEC, and conductivity of the heterogeneous cation-exchange membrane due to the hydrophilic nature of PEG400. The more hydrophilic membrane results in higher water content and wider access for functional sites. However, when the concentration of PEG400 is increased further, the IEC and conductivity tend to decrease. This tendency is more pronounced when the ion-exchange resin particle is increased from 50 to 60%-wt. It could be attributed to the washed out of some ion-exchange particle during membrane immersion due to lower bonding between polymer binder and the particles.

Acknowledgments
Financial support was obtained from Kementerian Riset Teknologi dan Pendidikan Tinggi, Republic of Indonesia, under a scheme of "Program Riset Unggulan ITB-DIKTI 2016 (Desentralisasi)" No. 077/SP2H/PL/LT/II/2016.

References

[1] Strathmann H, Grabowski A, and Eigenberger G 2013 *Ind. Eng. Chem. Res.* **52** 10364
[2] Huang C, Xu T, Zhang Y, Xue Y, and Chen G 2007 *J. Membr. Sci.* **288** 1
[3] Alvarado L and Chen A 2014 *Electrochim. Acta* **132** 583
[4] Khoiruddin, Hakim A N, and Wenten I G 2014 *Membr. Water Treat.* **5** 87
[5] Leong J X, Daud W R W, Ghasemi M, Liew K B, and Ismail M 2013 *Renew. Sust. Energ. Rev.* **28** 575
[6] Wenten I G, Khoiruddin, Arfianto F, and Zudiharto 2013 *Desalination* **314** 109
[7] Khoiruddin, Widiasa I N, and Wenten I G 2014 *J. Food Eng.* **133** 40
[8] Li X, Zhang H, Mai Z, Zhang H, and Vankelecom I 2011 *Energy Environ. Sci.* **4** 1147
[9] Długołęcki P, Gambier A, Nijmeijer K, and Wessling M 2009 *Environ. Sci. Technol.* **43** 6888
[10] Tufa R A, Curcio E, van Baak W, Veerman J, Grasman S, Fontananova E, and Di Profio G 2014 *RSC Adv.* **4** 42617
[11] Kim J, Park M, Snyder S A, and Kim J H 2013 *Desalination* **322** 121
[12] Wenten I G and Khoiruddin 2016 *Desalination* **391** 112
[13] Wenten I G and Khoiruddin 2015 *Journal of Engineering Science and Technology* **11** 916
[14] Shah B G, Shahi V K, Thampy S K, Rangarajan R, and Ghosh P K 2005 *Desalination* **172** 257
[15] Vyas P V, Shah B G, Trivedi G S, Ray P, Adhikary S K, and Rangarajan R 2001 *J. Membr. Sci.* **187** 39
[16] Kikhavani T, Ashrafizadeh S N, and Van Der Bruggen B 2014 *J. Appl. Polym. Sci.* **131** 39888
[17] Hosseini S M, Madaeni S S, and Khodabakhshi A R 2010 *J. Appl. Polym. Sci.* **118** 3371
[18] Hosseini S M, Madaeni S S, and Khodabakhshi A R 2010 *J. Membr. Sci.* **362** 550
[19] Hosseini S M, Madaeni S S, Heidari A R, and Moghadassi A R 2011 *Desalination* **279** 306
[20] Hosseini S M, Koranian P, Gholami A, Madaeni S S, Moghadassi A R, Sakinejad P, and Khodabakhshi A R 2013 *Desalination* **329** 62
[21] Parvizian F, Hosseini S M, Hamidi A R, Madaeni S S, and Moghadassi A R 2014 *J. Taiwan Inst. Chem. Eng.* **45** 2878
[22] Hosseini S M, Hamidi A R, Moghadassi A R, Koranian P, and Madaeni S S 2015 *Ionics* **21** 437
[23] Akli K, Khoiruddin, and Wenten I G 2016 *Journal of Membrane Science and Research* **2** 141
[24] Khoiruddin and Wenten I G 2016 *Journal of Engineering and Technological Sciences* **48** 1
[25] Heidary F, Kharat A N, and Khodabakhshi A R 2016 *J. Cluster Sci.* **27** 193
[26] Válek R and Zachovalová J 2015 *Desalin. Water Treat.* **56** 1
[27] Zabolotsky V I and Nikonenko V V 1993 *J. Membr. Sci.* **79** 181

IOP Conf. Series: Journal of Physics: Conf. Series **877** (2017) 012076 doi:10.1088/1742-6596/877/1/012076

Advances of zeolite based membrane for hydrogen production via water gas shift reaction

I G B N Makertihartha*, M Zunita, Z Rizki, and P T Dharmawijaya

Chemical Engineering Department, Institut Teknologi Bandung, Jl. Ganesha 10, Bandung, West Java, 40132, Indonesia

makertia@che.itb.ac.id

Abstract. Hydrogen is considered as a promising energy vector which can be obtained from various renewable sources. However, an efficient hydrogen production technology is still challenging. One technology to produce hydrogen with very high capacity with low cost is through water gas shift (WGS) reaction. Water gas shift reaction is an equilibrium reaction that produces hydrogen from syngas mixture by the introduction of steam. Conventional WGS reaction employs two or more reactors in series with inter-cooling to maximize conversion for a given volume of catalyst. Membrane reactor as new technology can cope several drawbacks of conventional reactor by removing reaction product and the reaction will favour towards product formation. Zeolite has properties namely high temperature, chemical resistant, and low price makes it suitable for membrane reactor applications. Moreover, it has been employed for years as hydrogen selective layer. This review paper is focusing on the development of membrane reactor for efficient water gas shift reaction to produce high purity hydrogen and carbon dioxide. Development of membrane reactor is discussed further related to its modification towards efficient reaction and separation from WGS reaction mixture. Moreover, zeolite framework suitable for WGS membrane reactor will be discussed more deeply.

1. Introduction

Hydrogen is the most abundant element on the planet. However, it can only be found naturally bonded with other elements and can be extracted from water, biomass, or hydrocarbons. Hydrogen combustion only produces water that it often referred as 'clean energy'. Hydrogen is already produced around the globe at the annual capacity of 5 billion cubic metres. There are many uses of hydrogen mainly for reactant of various syntheses such as ammonia production for fertiliser (~50%), oil refining (~37%), methanol production (8%), and in the chemical and metallurgical industries (4%) [1]. Over the past few years, many countries have invested in hydrogen mainly for energy use.

Water gas shift (WGS) reaction is an important process for production of H_2 from syngas [2, 3]. WGS is an endothermic and equilibrium limited reaction. To achieve good reaction performance, namely high reaction rate and CO conversion, most of today's WGS reaction employs two-step operation consists of high temperature shift (HT-WGS) and low-temperature shift (LT-WGS) [4]. The first step (HT-WGS) is operated using iron-based catalyst at high temperature range of about 350-400 °C to boost reaction rate but resulted in low equilibrium conversion. The second step (LT-WGS) generally operated using copper-zinc oxide catalyst at about 200-250 °C [5-7] to further boost the equilibrium conversion [8, 9]. This two-step conversion process is rather complex, high catalyst consumption and high steam recycling [10].

Due to the complexity of two-step conversion, membrane reactors that integrate hydrogen-selective membrane separation into reaction vessel have been investigated. The result shows that one-step operation shows great potential for CO deep conversion [11, 12]. Most WGS membrane reactors are using H_2 separation membranes [13-15]. The lack of H_2 due to removal results in high CO conversion at a given temperature, hence retentate stream consist mainly of CO_2 can be obtained. The process will be much simplified from two-step reaction process into a single stage shift reactor that produces high purity hydrogen without or with minimum further separation/purification step. This will also reduce the duty of the scrubbing system and high pressure CO_2 can be obtained which will minimize the cost of CO_2 capture [16]. Nevertheless, stability is one key issue of WGS membrane reactors. WGS reaction is conducted at relatively high temperature and pressure. Several harsh gasses such as steam and H_2S are also included. Therefore, WGS membrane reactors require high thermal, mechanical and chemical stability membrane. Pt and Pt alloys are the main selection of hydrogen selective membrane material due to their high performance. However, several disadvantages of using Pd-based membrane such as high cost and poor resistant to H_2S and CO make some researchers seek microporous inorganic membranes for hydrogen selective layer [10]. Pd-based membranes also have moderate durability and high brittleness that limit their practical applications[17]. SiO_2 membranes have been used for H_2 and other gas components separation and shows excellent performance [18].

Membrane reactor for water gas shift reaction is one interesting technique to boost the production of hydrogen efficiently. However, the selection of selective hydrogen layer is still the main challenge to commercialize it. The potential of zeolite based membrane for gas separation has been demonstrated well by various authors [19-21]. Moreover, the high price of Palladium based membrane for hydrogen selective layer make zeolite more interesting choice. This paper will discuss about zeolite membrane reactor for water gas shift reaction. The progress, challenges, and effort given to commercialize said technology will be discussed.

2. General hydrogen production

Since hydrogen is not available in gas state, hydrogen has to be extracted from various compounds. The common compounds used to extract hydrogen are water and hydrocarbons. Hydrogen extraction from water uses electrolysis technology while water gas shift reaction is employed to extract hydrogen from hydrocarbons. While electrolysis offer direct high purity hydrogen at various scale, water gas shift route is more interesting route nowadays due to its lower cost.

2.1. Electrolysis of water

Water electrolysis unit at least consists of an anode, a cathode, power supply and an electrolyte [22]. Current is then applied to maintain the electricity balance and electrons flow to the cathode where it consumed by hydrogen ions (protons) to form hydrogen. To maintain electrical charge and valence, hydroxide ions (anions) transfer via electrolyte to the anode where it gives away electrons. Electrolytes may vary from traditional ion solution, polymer electrolyte membrane, and molten salt electrolyte [23]. The half reactions occurring on the cathode and anode are:

Cathode: $$2H^+ + 2e^- \rightarrow H_2 \quad (1)$$

Anode: $$2OH^- \rightarrow \frac{1}{2}O_2 + H_2O + 2e^- \quad (2)$$

Overall: $$H_2O \rightarrow H_2 + \frac{1}{2}O_2 \quad (3)$$

As can be seen, pure hydrogen and oxygen gas are formed in different electrodes so it needs no further separation process. The practical limitation of electrolysis of water to produce hydrogen is energy price in terms of electricity. Electrolysis of water is now mainly built to store energy generated from solar, wind, wave, etc. in form of hydrogen for later use.

2.2. Water gas shift reaction

Hydrogen is widely produced via steam reforming and water gas shift reaction from hydrocarbons. Natural gas (mainly consist CH_4) is the major source of hydrocarbon for steam reforming and WGS reaction due to its cheaper price and high H/C ratio compared with other hydrocarbons. High temperature is needed to break hydrocarbons into syngas (CO, H_2) via reaction (4). Carbon monoxide is further reacted with steam shifted into carbon dioxide and more hydrogen via reaction (5).

$$CH_4 + H_2O \leftrightarrow CO + 3H_2 \quad (4)$$

$$CO + H_2O \leftrightarrow CO_2 + H_2 \quad (5)$$

To produce high purity hydrogen from said reactions, further separation process should be performed [1]. To the date, amine absorption is the most competitive technology to purify hydrogen via CO_2 removal due to its maturity. However, membrane systems have the potential to be more economical in terms of both relative capital investment and unit recovery cost [24]. The detailed aspects of water gas shift reaction will be discussed in the following section.

3. Water gas shift for hydrogen production

The Water Gas Shift (WGS) reaction is one alternative pathway to produce hydrogen that is commonly chosen. Traditionally, WGS reaction is performed to produce hydrogen for ammonia production from synthesis gas [25]. WGS is a reaction that uses carbon monoxide gained from combustion of hydrocarbons via steam addition. The equation of WGS is shown in equation (6). The WGS reaction is an exothermic catalytic reaction which conversion is limited by the equilibrium condition [26].

$$\mathrm{CO} + H_2O \rightarrow CO_2 + H_2\,; \quad \Delta H^0_{298} = -41{,}09\ kJ/mol \quad (6)$$

The conventional WGS reaction has many drawbacks that need to be improved. In order to achieve both fast catalytic reaction and considerable conversion, the traditional WGS reaction was performed in two steps, using high temperature (300-400°C) prior to low temperature reaction (200-300°C) [27]. Furthermore, the products need to process further to separate the hydrogen from CO_2 and excess reactants.

Answering those problems, a membrane reactor is applied in WGS reaction. The idea of membrane reactor is to remove one of the products (hydrogen) after the reaction occurred. By then, the equilibrium will be shifted which gives better performance in term of conversion and product recovery [28-30]. Furthermore, a membrane reactor in WGS reaction eliminates the second reactor and further separation related to the product purification. The concept of membrane reactors have been applied to other critical reaction processes besides the WGS reaction [31, 32].

Table 1. Performance of Various WGS Membrane Reactor

Research Type *)	Membrane Material	Catalyst	Configuration	Max CO Conversion (%)	Max H_2 Recovery (%)	Reference
E	Pd-Ag	Cu-Zn	Tubular	almost 100%	70%	[9]
E	Pd-based	Fe-Cr	Tubular**)	95	91,5	[27]
E	Pd-Ag	Pt-CeO2	Hollow Fiber	>60	ND	[33]
E+S	MFI Zeolite	Fe-Cr-Cu	Tubular	>95	>90	[16]
E+S	Pd-Ag	ND	Tubular	95,3	43	[34]
S	ND	ND	Shell and tube **)	> 95	ND	[35]
S	ND	ND	Tubular	>99	86,8	[36]

*) E = Experiment, S = Simulation
**) catalyst outside

The researches in membrane reactor have already been studied a lot via both experiments and simulations and several configurations have been proposed [29]. However, the tubular configuration is the most common with the process stream inside a tube [37, 16] while the reversal configuration has also been already performed [27]. The performance of several WGS membrane reactors is summarized in Table 1.

4. Zeolite membrane and modification

Among the materials that used in membrane reactor, zeolite becomes one favourable option due to its unique properties. Zeolite is known to be stable under harsh condition (e.g. chemical, high temperature and selective separation) [38, 39]. In addition, it also has unique properties that makes zeolite suitable in wide range of applications including membrane separation, integrated membrane reactor, electrode, chemical sensor and insulator[40].

The properties of zeolite membrane depend on the preparation method. Furthermore, the natural zeolites are not suitable for hydrogen separation [38]. Therefore, modifications are needed to customize and improve the properties of zeolite into specific required properties. The modifications can be a formation of composite materials or adjustment in synthesis technique. Feng [39] mentioned that the most attractive type of zeolite composites are the mixed membrane matrix (MMM) and metal organic frameworks (MOFs).

The most common method to prepare dense zeolite membrane is hydrothermal synthesis. Hydrothermal synthesis can be done with seeding (namely secondary growth method) and the heating procedure can be performed using microwave (microwave hearing). Zeolite membrane requires porous support to be attached to, which can be in the form of tubular, hollow fibre or any other membrane module configuration. The methods during synthesis process will determine how dense the zeolite membrane be thus affecting the separation performance [39].

Zeolite membrane and its derivatives have been widely studied in gas separation, especially for hydrogen. Several applications of zeolite membrane in hydrogen separation are summarized in Table 2.

Table 2. Zeolite Membrane for Hydrogen Separation

Type of Zeolite membrane	Compounds separated	Separation notes	Ref
Pd/Silicate-1 Zeolite	H_2/N_2	High Selectivity H_2/N_2	[41]
MFI Zeolite	H_2/CO_2+CO+H_2O	Very high selectivity for H_2O and moderate for CO and CO_2	[42]
MFI-ZSM5	H_2/He	Separation factor below 2	[20]
Natural Zeolite	$H_2/C1+C_2$ Hydrocarbon	High selectivity at ambient temperature and moderate to low at 500°C	[43]
Acrylate-zeolite MMM	H_2/CO_2	more than 2 times selectivity increase	[44]
Hydroxy Sodalite Zeolite	H_2/CH_4	17% selectivity increase	[45]

5. Application and performance of zeolite based membrane in hydrogen production via WGS reaction

WGS reaction is important reaction for CO conversion and H_2 production from fossil fuel and biomass via gasification and steam reforming [46]. Silica membrane has been integrated to WGS reactor to produce high purity H_2 [47-50]. Generally, WGS reaction membrane reactor is fabricated based on properties of capability to separate H_2 [14, 51, 52]. Therefore, WGS reaction reactor will be to achieve high CO conversion and high hydrogen purity using a single reactor. Thus, membrane used in WGS reaction used should have high H_2 permselectivity, so that it can improve H_2 yield and lower operation costs [42]. Moreover, the membrane should also has good thermal stability because the WGS reaction can be carried out in two condition, which are HT-WGS reaction (high temperature WGS reaction at 350-450°C) and LT-WGS reaction (low temperature WGS reaction at 190-250°C)[53, 54]. Zeolite is suitable for WGS reaction reactor is zeolite because it is hydrothermally stable and intrinsically resistant to sulphur compounds due to their crystalline structure [55, 46, 56, 57]. Scheme of zeolite-based membrane reactor for H_2 production is depicted in figure 1.

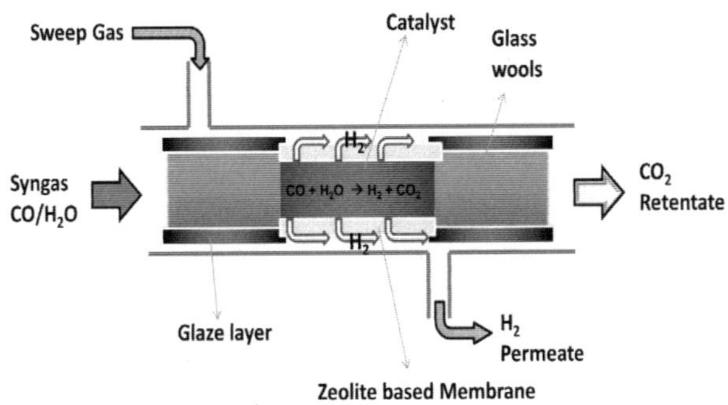

Figure 1. Scheme of zeolite-based membrane reactor for H_2 production

Table 3. Performance of Zeolite based Membrane in Production/Separation H_2 via WGS reaction and Another Reactions

Zeolite Membrane Type	Modification	Catalyst	H_2/CO_2 Separation Factor	CO Conversion (%)	H_2 Recovery (%)	Ref.
MFI	Synthesized on a porous α-alumina disc	Fe/Ce	31	>99.5	73.2	[58]
ZSM-5	Silicate bilayer	Fe/Cr/Cu	>16	>95	. 90	[16]
ZSM-5/silicate bilayer	Porous α-alumina support coated with an YSZ	N/A	25.3	N/A	>99	[59]
MFI	α-alumina hollow fiber support	Fe/Cr/Al	52	>95	>98.2	[60]
MFI	depositing molecular silica	Ce/Fe	68.3	81.7	40	[61]
ZSM-5 silicate layer	Porous α -alumina support coated with an intermediate YSZ barrier layer	Fe/Ce	23	>80	>20	[42]
MFI	modified by on-stream catalytic cracking deposition (CCD) using methyl diethoxysilane (MDES)	CuO/ZnO/Al_2O_3	42.6	>95.4	18	[10]
A-type	N/A	Pt, Pd, Au	N/A	>99	N/A	[62]
MFI	modified by the on-stream CCD process using MDES	Fe/Ce	45	>99	>60	[46]
MFI	Various high Si/Al	N/A	31	N/A	N/A	[63]

In recent decades, many researchers are developed and used zeolite membrane reactor to improve performance of reactor for H_2 production. Kim et al. [46] have modified MFI zeolite membrane with nanocrystalline Fe/Ce for HT-WGS reaction. The result showed hydrogen permeance of 0.9 × 10−7 mol s^{-1} m^{-2} Pa^{-1} with selectivity of ~31 and 25 for H_2/CO_2 and H_2/CO selectivity, respectively at above 500°C temperature. Several simulation or modeling experiments of WGS reaction has been studied to know the performance of WGS reactor. Ghasemzadeh et al. [64] studied the modeling of inorganic membrane reactor performance to WGS reaction. The results showed hydrogen recovery and CO conversion were improved in silica and Pd-Ag membrane by increasing the reaction pressure. Dong et al. [16] modelled the zeolite membrane reactor modified by silica layer to WGS reaction using ZSM-5/silicalite bilayer. The membrane can improve CO conversion and H_2 recovery from 89.8% and 28.5% to > 95% and > 90%, respectively. Tang and coworkers [61] have also modified MFI zeolite membrane by depositing silica and cerium-doped ferrite as catalyst, their research result showed H_2/CO_2 permselectivity at 68.3 with H_2 permeance about 2.94×10^{-7} mol/m^{2}sPa at operating temperature above 550°C. Moreover, the CO conversion is 81.7% which was significantly higher than the CO obtained in the traditional packed-bed reactor i.e. 62.5%. Performance of various zeolite based membrane reactors in production/separation H_2 from WGS reaction and another reaction is shown comprehensively in Table. 3.

6. Optimization of zeolite based membrane in hydrogen production via WGS reaction

The performance of zeolite based membrane should be optimized so that the H_2 production via WGS reaction will increase significantly. Moreover, the product should have high purity and the reaction carried away in milder condition than traditional WGS reaction reactor [58, 65, 34]. At high temperature, transport of small molecules is governed by gaseous diffusion leads to low H_2 selectivity, while the other gases will permeate simultaneously [66-68, 63, 69]. Therefore, many researchers developed studies about the modification of zeolite membrane pores. Accordingly, the pore size of MFI zeolite membranes are reduced by multiform modification such as catalytic cracking of a methyl diethoxysilane (MDES) precursor to deposit mono silica inside the zeolitic channels [70, 59, 10]. By reduction of zeolite membrane pores size, it is expected that the H_2 selectivity will increase [20, 47, 55, 71, 65, 62, 50].

To produce H_2 with high yield, catalyst plays an important role in WGS reaction [11, 72, 73, 70, 2]. Zeolite membrane provides support for catalytically active nano-sized particles so that the catalytic phase may contact reactants directly and reduce by-passing, short contact time, and as centre of reaction is helpful since the temperature of reaction increase can lead to promotion of reverse WGS reaction [74-76]. Hasegawa et al. [73] modified Y-type zeolite membranes by Pt catalyst. Pt catalyst was used to improve the CO conversion and selectivity by dispersing it in the zeolite membrane (ZSM-5) [77, 20, 55, 16, 78, 79]. To further optimize the performance of zeolite membranes, composite membrane was prepared by coating an Rh-loaded C-Al_2O_3 membrane with a defect-free SiO_2 layer hence SiO2/Rh/C-Al_2O_3 membrane can improve the CO oxidation and the H_2 production will increase rapidly[80, 81].

7. Conclusion

Based on several literatures reviewed, high hydrogen permselectivity material can improve hydrogen yield in WGS reaction thus lower operation cost. One promising material for said purpose is zeolite based membrane. Some profitable properties in zeolite membrane are stability under harsh condition (e.g. harsh chemical, high temperature). Pore size of zeolite membrane needs to be reduced by surface modification with coating materials such as silicate layer, YSZ (Yttria stabilized zirconia), methyl diethoxysilane (MDES), and Si/Al mixture.

References

[1] Lu G, Da Costa J D, Duke M, Giessler S, Socolow R, Williams R and Kreutz T 2007 *Journal of colloid and interface science* **314** 589-603
[2] Lee J Y, Lee D-W, Lee K-Y and Wang Y 2009 *Catalysis Today* **146** 260-4
[3] Mendes D, Garcia H, Silva V, Mendes A and Madeira L M 2008 *Industrial & Engineering Chemistry Research* **48** 430-9
[4] Ma D and Lund C R 2003 *Ind. Eng. Chem. Res.* **42** 711-7
[5] Nishida K, Atake I, Li D, Shishido T, Oumi Y, Sano T and Takehira K 2008 *Appl. Catal., A* **337** 48-57

[6] She Y, Zheng Q, Li L, Zhan Y, Chen C, Zheng Y and Lin X 2009 *international journal of hydrogen energy* **34** 8929-36
[7] Shishido T, Nishimura S, Yoshinaga Y, Ebitani K, Teramura K and Tanaka T 2009 *Catalysis Communications* **10** 1057-61
[8] Adams T A and Barton P I 2009 *International Journal of Hydrogen Energy* **34** 8877-91
[9] Gawade P, Mirkelamoglu B and Ozkan U S 2010 *J. Phys. Chem. C* **114** 18173-81
[10] Zhang Y, Wu Z, Hong Z, Gu X and Xu N 2012 *Chemical Engineering Journal* **197** 314-21
[11] Bi Y, Xu H, Li W and Goldbach A 2009 *Int. J. Hydrogen Energy* **34** 2965-71
[12] Tosti S, Basile A, Chiappetta G, Rizzello C and Violante V 2003 *Chem. Eng. J.* **93** 23-30
[13] Babita K, Sridhar S and Raghavan K 2011 *international journal of hydrogen energy* **36** 6671-88
[14] Gallucci F, Fernandez E, Corengia P and van Sint Annaland M 2013 *Chem. Eng. Sci.* **92** 40-66
[15] Smart S, Lin C, Ding L, Thambimuthu K and Da Costa J D 2010 *Energy Environ. Sci.* **3** 268-78
[16] Dong X, Wang H, Rui Z and Lin Y 2015 *Chemical Engineering Journal* **268** 219-29
[17] Jamshidi S, Babaluo A, Haghighi M and Noruzi A 2016 *Separation Science and Technology*
[18] Ockwig N W and Nenoff T M 2007 *Chem. Rev.* **107** 4078-110
[19] Adrover M E, Borio D and Pedernera M 2016 *International Journal of Hydrogen Energy*
[20] Antunes R, Borisevich O and Demange D 2016 *Chemical Engineering Research and Design* **109** 327-34
[21] Sun Y, Grigore M, Hla S, Morpeth L and Edwards J 2016 *International Journal of Hydrogen Energy* **41** 10335-45
[22] Zeng K and Zhang D 2010 *Progress in Energy and Combustion Science* **36** 307-26
[23] Oldham K and Myland J 2012 *Fundamentals of electrochemical science*: Elsevier)
[24] Spillman R W 1989 *Chemical engineering progress* **85** 41-62
[25] Smith R, Loganathan M and Shantha M S 2010 *International Journal of Chemical Reactor Engineering* **8**
[26] Bustamante F, Enick R, Cugini A, Killmeyer R, Howard B, Rothenberger K, Ciocco M, Morreale B, Chattopadhyay S and Shi S 2004 *AIChE Journal* **50** 1028-41
[27] Barbieri G, Brunetti A, Caravella A and Drioli E 2011 *RSC Advances* **1** 651-61
[28] Dragomirova R and Wohlrab S 2015 *Catalysts* **5** 2161-222
[29] Manrique Y J A 2012 In: *Department of Chemical Engineering*: Universidade do Porto)
[30] Mendes D, Mendes A, Madeira L, Iulianelli A, Sousa J and Basile A 2010 *Asia Pac. J. Chem. Eng.* **5** 111-37
[31] Casanave D, Ciavarella P, Fiaty K and Dalmon J-A 1999 Z *Chemical Engineering Science* **54** 2807-15
[32] Shu J, Grandjean B P and Kaliaguine S 1994 *Appl. Catal., A* **119** 305-25
[33] García-García F, Torrente-Murciano L, Chadwick D and Li K 2012 *Journal of membrane science* **405** 30-7
[34] Morpeth L D and Dolan M D 2015 *Chemical Engineering Journal* **276** 289-302
[35] Adrover M E, López E, Borio D O and Pedernera M N 2009 *Chemical Engineering Journal* **154** 196-202
[36] Boutikos P and Nikolakis V 2010 *Journal of Membrane Science* **350** 378-86
[37] Basile A, Curcio S, Bagnato G, Liguori S, Jokar S and Iulianelli A 2015 *International Journal of Hydrogen Energy* **40** 5897-906
[38] Eun Jee S, McGaughey A J and Sholl D S 2009 *Molecular Simulation* **35** 70-8
[39] Feng C, Khulbe K, Matsuura T, Farnood R and Ismail A 2015 *Journal of Membrane Science and Research* **1** 49-72
[40] Caro J and Noack M 2008 *Microporous and Mesoporous Materials* **115** 215-33

[41] Guo Y, Jin Y, Wu H, Zhou L, Chen Q, Zhang X and Li X 2014 *International Journal of Hydrogen Energy* **39** 21044-52
[42] Wang H, Dong X and Lin Y 2014 *Journal of Membrane Science* **450** 425-32
[43] An W, Swenson P, Wu L, Waller T, Ku A and Kuznicki S M 2011 *Journal of Membrane Science* **369** 414-9
[44] Khan A L, Cano-Odena A, Gutiérrez B, Minguillón C and Vankelecom I F 2010 *Journal of Membrane Science* **350** 340-6
[45] Nabavi M S, Mohammadi T and Kazemimoghadam M 2014 *Ceramics International* **40** 5889-96
[46] Kim S-J, Xu Z, Reddy G K, Smirniotis P and Dong J 2012 *Ind. Eng. Chem. Res.* **51** 1364-75
[47] Asaeda M, Sakou Y, Yang J and Shimasaki K 2002 *Journal of Membrane Science* **209** 163-75
[48] Boffa V, Blank D and Ten Elshof J 2008 *Journal of Membrane Science* **319** 256-63
[49] Kanezashi M and Asaeda M 2006 *Journal of Membrane Science* **271** 86-93
[50] Yoshida K, Hirano Y, Fujii H, Tsuru T and Asaeda M 2001 *Journal of chemical engineering of Japan* **34** 523-30
[51] Ramasubramanian K, Zhao Y and Winston Ho W S 2013 *AIChE Journal* **59** 1033-45
[52] Smart S, Lin C X C, Ding L, Thambimuthu K and Diniz da Costa J C 2010 *Energy Environ. Sci.* **3** 268-78
[53] Polychronopoulou K, M Kalamaras C and M Efstathiou A 2011 *Recent Patents on Materials Science* **4** 122-45
[54] Ratnasamy C and Wagner J P 2009 *Cat. Rev. - Sci. Eng.* **51** 325-440
[55] Dong J, Lin Y and Liu W 2000 *AIChE Journal* **46** 1957-66
[56] Ockwig N and Nenoff T 2007 *Chem. Rev.* **107** 4078-110
[57] Verweij H, Lin Y and Dong J 2006 *MRS bulletin* **31** 756-64
[58] Kim S-J, Yang S, Reddy G K, Smirniotis P and Dong J 2013 *Energy Fuels* **27** 4471-80
[59] Wang H and Lin Y S 2012 *Journal of Membrane Science* **396** 128-37
[60] Zhang Y, Sun Q and Gu X 2015 *AlChE J.* **61** 3459-69
[61] Tang Z, Kim S-J, Reddy G K, Dong J and Smirniotis P 2010 *Journal of Membrane Science* **354** 114-22
[62] Rosso I, Galletti C, Saracco G, Garrone E and Specchia V 2004 *Appl. Catal., B* **48** 195-203
[63] Sjöberg E, Barnes S, Korelskiy D and Hedlund J 2015 *Journal of Membrane Science* **486** 132-7
[64] Ghasemzadeh K, Zeynali R and Basile A 2016 *International Journal of Hydrogen Energy* **41** 8696-705
[65] Michalkiewicz B and Koren Z C 2015 *Journal of Porous Materials* **22** 635-46
[66] Bernardo P, Algieri C, Barbieri G and Drioli E 2006 *Catalysis today* **118** 90-7
[67] Bernardo P, Algieri C, Barbieri G and Drioli E 2008 *Separation and Purification Technology* **62** 629-35
[68] Moon J-H, Bae J-H, Han Y-J and Lee C-H 2010 *Journal of Membrane Science* **356** 58-69
[69] Sjöberg E, Sandström L, Öhrman O G W and Hedlund J 2013 *Journal of Membrane Science* **443** 131-7
[70] Hong M, Falconer J L and Noble R D 2005 *Ind. Eng. Chem. Res.* **44** 4035-41
[71] Liu B and Au C 2001 *Catalysis letters* **77** 67-74
[72] Bustamante F, Enick R, Cugini A, Killmeyer R, Howard B, Rothenberger K, Ciocco M, Morreale B, Chattopadhyay S and Shi S 2003 *AlChE J.* **58**
[73] Hasegawa Y, Sotowa K-I and Kusakabe K 2003 *Chemical engineering science* **58** 2797-803
[74] Igarashi H, Uchida H, Suzuki M, Sasaki Y and Watanabe M 1997 *Appl. Catal., A* **159** 159-69
[75] Kahlich M, Gasteiger H and Behm R 1997 *Journal of Catalysis* **171** 93-105
[76] Korotkikh O and Farrauto R 2000 *Catalysis Today* **62** 249-54
[77] Alshebani A, Pera-Titus M, Landrivon E, Schiestel T, Miachon S and Dalmon J-A 2008 *Microporous and Mesoporous Materials* **115** 197-205

[78] Kwan S M, Leung A Y L and Yeung K L 2010 *Separation and Purification Technology* **73** 44-50
[79] Masuda T, Fukumoto N, Kitamura M, Mukai S R, Hashimoto K, Tanaka T and Funabiki T 2001 *Microporous and mesoporous materials* **48** 239-45
[80] Hasegawa Y, Ueda A, Kusakabe K and Morooka S 2002 *Appl. Catal., A* **225** 109-15
[81] Sotowa K-I, Hasegawa Y, Kusakabe K and Morooka S 2002 *International journal of hydrogen energy* **27** 339-46

IOP Conf. Series: Journal of Physics: Conf. Series **877** (2017) 012077 doi:10.1088/1742-6596/877/1/012077

Synthesis and Characterization of La, Sc, Yb and Nd co-doped Gadolinium doped Cerium (GDC) Composite Electrolyte for IT-SOFC

Damisih, Jarot Raharjo, Masmui, Raffty Setya Aninda, Novita Ami Lestari

Center for Materials Technology, Agency for Assessment and Application of Technology, Bldg 224, PUSPIPTEK, Serpong, South Tangerang 15314, Indonesia

damisih@bppt.go.id

Abstract. Composite based on gadolinium doped cerium (GDC) co-doped Lanthanum (La), Scandium (Sc), Ytterbium (Yb) were investigated as electrolyte for solid oxide fuel cell (SOFC), namely GDC-La, GDC-Sc, GDC-Yb and GDC-Nd, respectively. The second co-doped La, Sc, Yb and Nd ensured the stability and high ionic conductivity of the GDC ceria-based electrolyte materials for SOFC. The GDC powder was synthesized via sol-gel technique. Then the La-GDC, Sc-GDC, Yb-GDC and GDC-Nd were subsequently prepared by mixing the GDC with La, Sc, Yb, and Nd respectively, through solid-state reaction in ballmill at 200rpm alumina balls. The composite electrolytes were then characterized in terms of its morphology, phase and thermal properties of the powders. Among the composite electrolytes investigated, GDC-Yb powder show higher purity and better dispersion than the others co-doped GDC. TGA analysis present that the addition of co-dopant led to improve thermal resistance and stability of solid electrolyte powders. The results confirm that GDC with co-dopant is promising alternative electrolyte for intermediate temperature solid oxide fuel cells (IT-SOFC).

1. Introduction

Fuel cells are electrochemical device that directly convert chemical energy from a fuel into electricity. This device produces electricity with no harmful byproduct for environment [1]. When hydrogen rich fuel such as natural gas passes through fuel cells, it electrochemically reacts with the oxygen to produce electricity, heat and water. These results in power generation that is essentially no present of nitrogen oxide (NOx), sulfur dioxide (SOx) or particulate matter that assists in reduction of air pollution and hazardous gas emission. Fuel cells are classified based on type of electrolyte used namely alkaline fuel cell (AFC), molten carbonate fuel cell (MCFC), phosphoric acid fuel cell (PAFC), proton exchange membrane fuel cell (PEMFC), and solid oxide fuel cell (SOFC). Among the various types of fuel cells, solid oxide fuel cell has been widely developed and its technology has been extensively adopted.

Solid oxide fuel cell technology is promising system for generating electricity due to its high conversion efficiency from chemical energy into electricity [2]. SOFC has many advantages over another type of fuel cells. First, due to its high operating temperature, no precious metals are required as catalyst. It potentially reduces the cost for its manufacturing. Second, SOFC has high resistance to sulfur gas. It is more flexible to various kinds of fuel where it can internally reform methane gas so that no reformer is needed. Third, SOFC is not experiencing carbon monoxide gas poisoning; even it can be used as fuel and this advantage makes SOFC can utilize gas from coal.

SOFC operates at high temperature and it causes instability for the system for long term period, high manufacturing cost and difficulties in selection of appropriate materials. Recently, the further research has been conducted on SOFC system which operates at medium temperature (600-800°C) [3]. Lowering of operating temperature in SOFC has been attracted much attention of some researchers due to it gives some advantages such as employing cheaper materials but still have good performance, high stability of electrolyte and rapid start up process [4].

Ceria-based electrolyte has received much more attention and considered as the best material for intermediate SOFC [5]. Aliovalent dopants such as Yb_2O_3, Gd_2O_3, La_2O_3 and Sc_2O_3 will react with CeO_2 and establish vacancies in crystal lattice as of facilitate oxygen ion to move. Accordingly, the ion of dopants will easily make a bonding with Ce^{3+}. Ion vacancies has important role towards ionic conductivity in ceria-based electrolyte [6]. Gadolinium doped cerium (GDC) is considered as one of the best choice among ceria-based electrolytes [7].

Strategy of co-dopant addition in solid electrolyte shows better conductivity than single-doped. Co-doped electrolyte materials might sort the oxygen vacancy become more organized so as ease of oxygen ion stepping and increase ionic conductivity. For example, N. Kim et al [8] reported the addition of five trivalent metal (Y, Sm, Nd, Pr and La) oxide as co-dopant materials for 20 mol% of GDC results indicated that the addition of co-dopant in GDC solid electrolyte can generally improve its electrical properties. It is also reported by Maricle et al [9] that electrolyte domain regime increased by two order of oxygen partial pressure through the addition of 3 mol% Pr into GDC. Although co-dopant addition into GDC has been studied by several researchers, it is still somewhat unclear how co-dopant addition affects the properties of SOFC electrolyte.

Co-doping approach with two or more aliovalent was known improving conductivity over single doping [10]. Based on this, in present work was focused on addition of co-dopant in GDC synthesis for solid electrolyte for SOFC. Oxides metal of La_2O_3, Sc_2O_3 and Yb_2O_3 were used as second dopant for GDC. The GDC powder was synthesized via sol-gel method and the dopant was added through solid state reaction in ballmill. The properties of co-doped GDC were then characterized in term of its morphology, compositional properties and thermal properties of the powders.

2. Experimental

First, gadolinium doped cerium powders with composition of $Gd_{0.10}Ce_{0.90}O_{1.95}$ (GDC10) were synthesized by using sol-gel method. Cerium nitrate hexahydrate and gadolinium nitrate hexahydrate were used as starting materials. While as solvent and metal chelating agent, it was used demineralized water and citric acid monohydrous respectively. At the beginning, starting materials was mixed in the required stoichiometric ratio after each homogeneously dissolved. Citric acid was added into the mixture of precursors to maintain the total molar ratio of metal to citric acid. The solution was stirred and then temperature was raised to 80°C. Stirring was still maintained around 1 hour until it turned into yellow color and foam was formed. Sample was kept at room temperature and then it was dried in an oven at temperature of 110°C for about 24 hours. After drying, sample was grinded by using mortar grinder into the fine powders and then it was calcined at temperature of 700°C.

Second, metal oxides consisting of La_2O_3, Sc_2O_3 and Yb_2O_3 as co-dopants were added into successfully synthesized GDC10 through solid state reaction. The co-dopant addition was amount of 1%wt of GDC using ball mill. The weighed powders were mixed in ball mill jar and ethanol was added as dispersant to maintain colloid system in the mixture. Aluminum balls were used as mixing media in the ball mill. Speed rotation was 240 rpm for approximately 4 hours with 30 minutes of resting time. After co-doping step, sample powders were dried at temperature of 110°C for 24 hours in an oven.

The co-doped GDC powders were then characterized using Scanning Electron Microscope (HITACHI SU3500, Japan) to and Energy Dispersive X-Ray (EDX) to analyze structural and compositional properties, respectively. Thermal Gravimetric Analysis (LabSys Evo TG-DTA, France)

IOP Conf. Series: Journal of Physics: Conf. Series **877** (2017) 012077 doi:10.1088/1742-6596/877/1/012077

was performed to investigate thermal properties of samples. In addition, ionic conductivity of selected sample was also measured.

3. Results and Discussion

Figure 1 shows SEM images of co-doped GDC powders of all samples after calcined at temperature of 700°C. It can be seen that the powders appearance looks like flakey shape at most and some of them at small amount shows to have porous nature if observed at higher magnification.

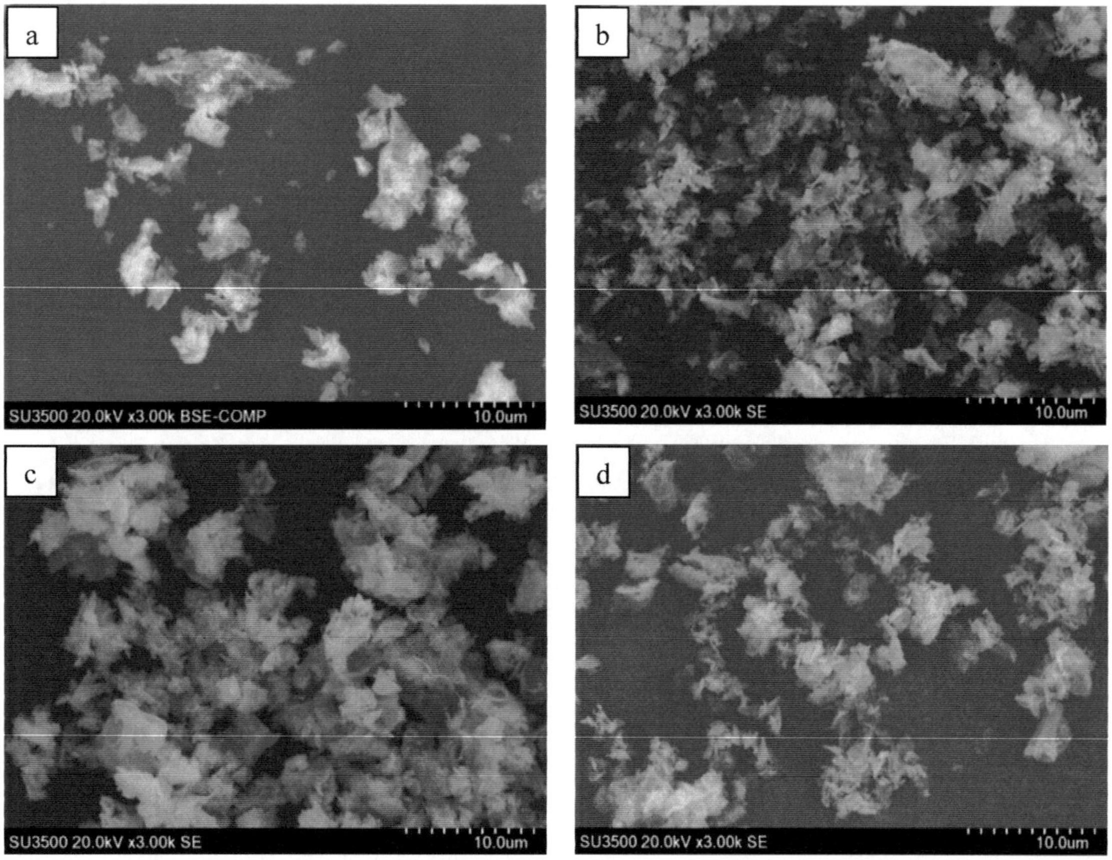

Figure 1. SEM image of co-doped GDC powders (a) GDC-La; (b) GDC-Sc; (c) GDC-Yb; and (d) GDC-Nd after calcined at temperature of 700°C

Figure 2 shows EDX spectrum results of GDC-La sample consisting of lanthanum, gadolinium and cerium element. The composition of lanthanum as second dopant or co-dopant is much smaller than gadolinium as main dopant, as expected. From EDX quantitative analysis of elements, lanthanum content on the sample powders in spectrum 3 was about 0.41wt% and 0.13at% as shown in Table 1. While gadolinium as first dopant was detected as much as 7.39%wt and 2.01at%. Another EDX spectrum shows almost similar results of quantitative element analysis. However, lanthanum was not found in all of EDX spectrums but only in spectrum 3, suggesting that co-dopant was still not homogeneously dispersed. Moreover, in this GDC-La sample powders, it was found peak indicating aluminum impurity that might come from alumina ball used as mixing media in ball milling process. These results suggested that the preparation of co-doped GDC powders need to be carefully performed

IOP Conf. Series: Journal of Physics: Conf. Series **877** (2017) 012077 doi:10.1088/1742-6596/877/1/012077

in order to obtain homogeneous dispersion of co-dopant with no impurities. N. Kim et al. [8] has prepared co-doped GDC powders with no impurities by using Pechini process including incorporation of selected five trivalent metal (Y, Sm, Nd, Pr and La) as co-dopant. Meanwhile, in this work, co-dopant was added by solid state method through ball milling process.

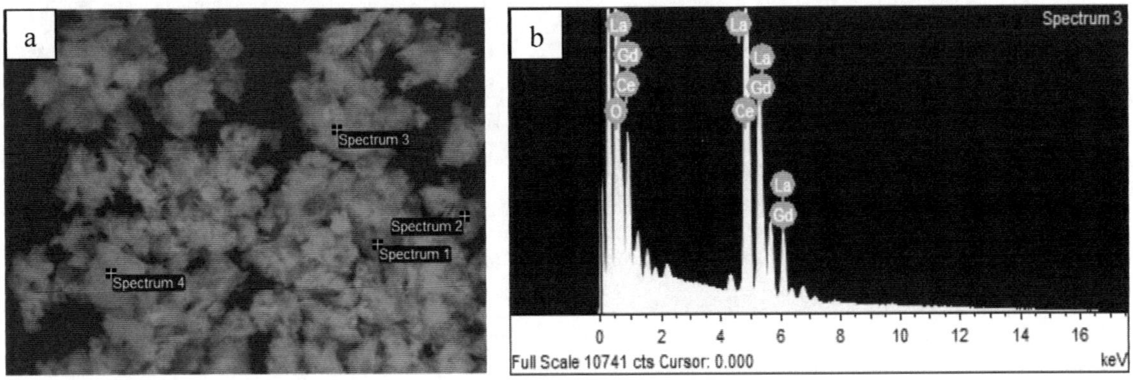

Figure 2. (a) EDX point analysis of GDC-La sample (b) EDX results of spectrum 3

Similar with GDC-La, in GDC-Sc and GDC-Nd sample powders were also found that the co-dopant dispersion was not sufficiently uniform. It can be seen from EDX elemental analysis that showing not all of point spectrum consisted of co-dopant element of Sc and Nd, respectively.

Table 1. EDX quantitative analysis of spectrum 3 from GDC-La sample

Element	Weight%	Atomic%
O K	29.41	78.68
La L	0.41	0.13
Ce L	62.79	19.18
Gd L	7.39	2.01
Totals	100.00	

In case of ytterbium element addition into GDC namely GDC-Yb, the dispersion of co-dopant seems relatively more uniform in the powders. It can be seen from EDX element analysis which was identifying peak spectrum of Yb in every investigated point.

The present of Yb co-dopant was confirmed by EDX analysis with the concentration of 0.71wt% and 0.16at% in point of spectrum 4 like shown in Table 2. The concentration of gadolinium element was in the amount of 7.44wt% and 1.83at%. Another points exhibited identical results of EDX elemental analysis. There is also no other impurities found in this GDC-Yb sample powders, but only gadolinium, cerium and ytterbium present. The purity of GDC-Yb powders seems to be higher than the others co-doped GDC.

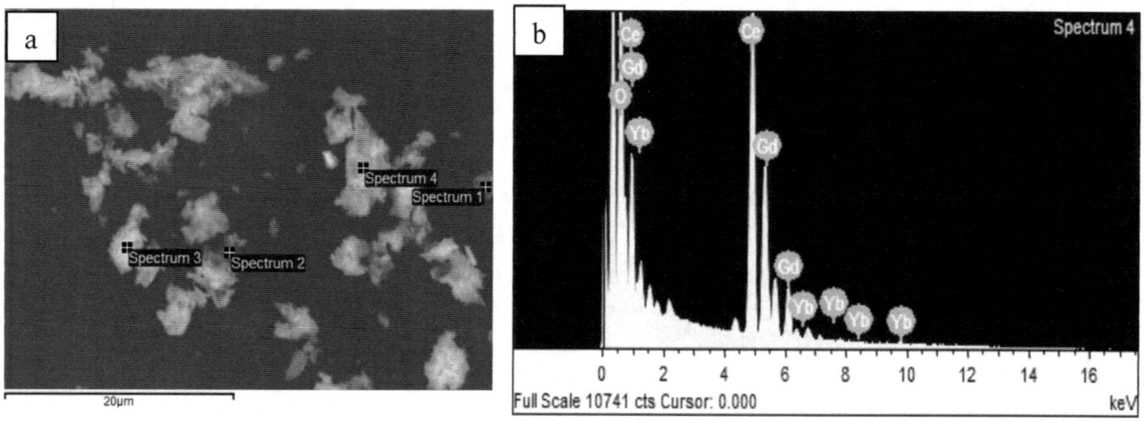

Figure 3. (a) EDX point analysis of GDC-Yb sample (b) EDX results of spectrum 1

Table 2. EDX quantitative analysis of spectrum 4 from GDC- Yb sample

Element	Weight%	Atomic%
O K	33.99	82.06
Ce L	57.87	15.95
Gd L	7.44	1.83
Yb L	0.71	0.16
Totals	100.00	

Figure 4 shows TGA plot for un-doped GDC and co-doped GDC at intermediate operating temperature of SOFC. From the results, it is found that there is no weight loss observed at those range of temperature. The graph is relatively flat indicating that no decomposition occurred. From the point of view of Intermediate Temperature SOFC (IT-SOFC) which operate at temperature of 600-800°C, this materials are regarding as promising solid electrolyte for SOFC [4]. It is comparably stable in which to address the challenge of degradation problem in SOFC technology. However, thermal stability is becoming the main issue when the device is subjected to high temperature operation.

It can be observed also that GDC with co-dopants addition presenting slightly higher level of TG% compared with un-doped GDC. It indicates that the addition of co-dopant led to improve thermal resistance and stability of solid electrolyte. From the graph in Figure 4, we can see also that GDC-Yb has slightly lower TG% compared with other co-doped GDC. According to N. Kim et al. [8], thermal properties that were represented by thermal expansion coefficient (TEC) value slightly changed around 5% with the addition of co-dopant. The change in TEC might be related with the change of lattice parameter. They also reported that co-dopants addition (Y, Sm, Nd, La) slightly changed the value of TEC, except the addition of Pr co-dopant which TEC rapidly increased.

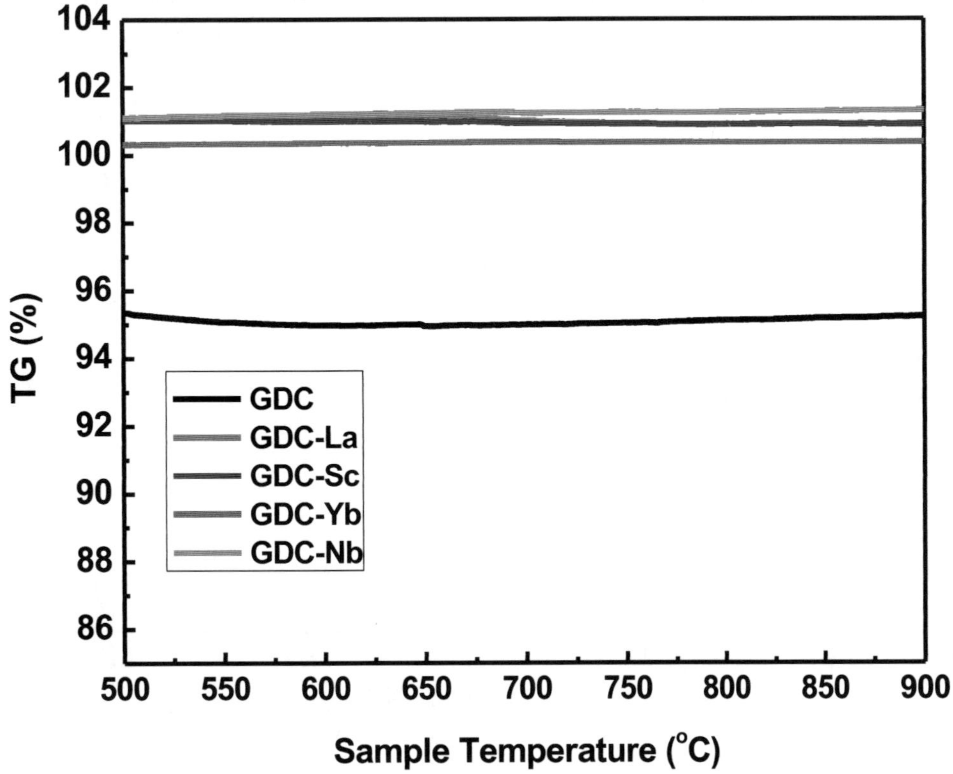

Figure 4. Thermal Gravimetric Analysis (TGA) plot of un-doped GDC and co-doped GDC at range of intermediate operating temperature of SOFC

The powders were pressed into a pellet and sintered at temperature of 1050°C and 1150°C. Then, it was performed ionic conductivity measurement where in this case, only GDC-Nd was selected. Figure 5 shows ionic conductivity of selected sample of GDC-Nd at different operating temperature and activation energy values of a certain temperature range. Total conductivity increase as temperature increase. The highest total ionic conductivity for GDC-Nd sample was about 0.040 Scm^{-1} at 800°C for GDC-Nd sintered at 1150°C, as shown in Table 3. While for GDC-Nd sample sintered at temperature of 1050°C shows highest total ionic conductivity of 0.026 Scm^{-1}. It implies that higher sintering temperature gives higher ionic conductivity where higher ionic conductivity could be related to the increase of number of equi-interaction energy site which reduces the activation energy and facilitates diffusion of ion oxygen in the crystal lattice [13]. It was reported by N. Kim et al [8] that electrical conductivity increased as much as 0.046 Scm^{-1} with the additon of Nd (1 mol%) into GDC. This is in accordance with the results we obtained. However, sintering temperature in our work was about 1150°C while they sintered the electrolyte pellet disc at higher temperature of 1400°C.

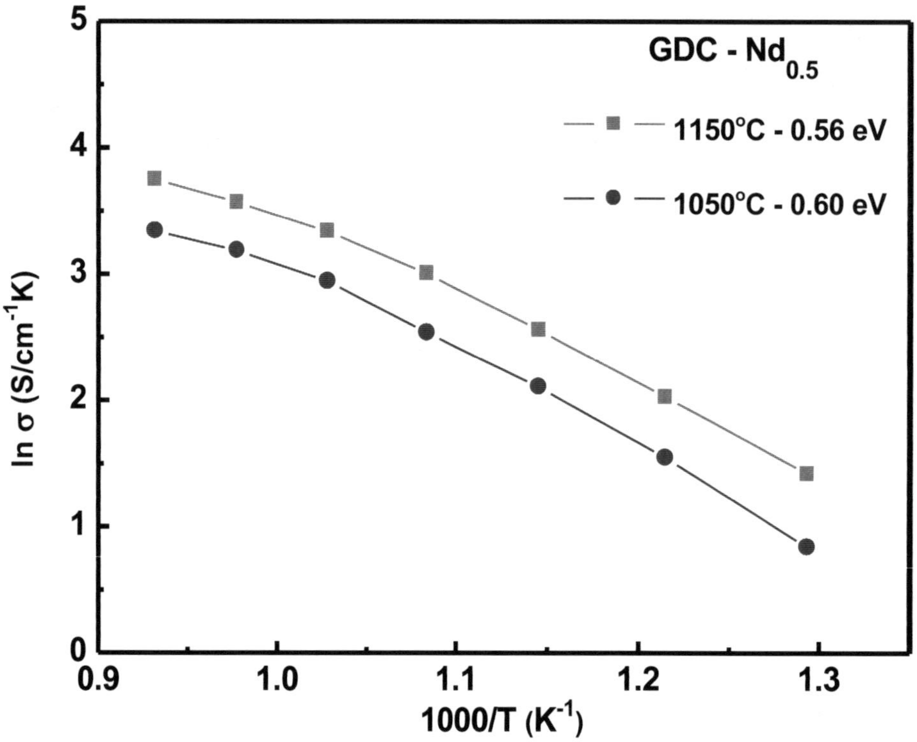

Figure 5. Ionic conductivity of GDC-Nd pellet at two temperature sintering of 1150°C and 1050°C

Table 3. Total ionic conductivity for GDC-Nd electrolyte

Sintering temperatur	Total ionic conductivity, σ (Scm^{-1})						
	800°C	750°C	700°C	650°C	600°C	550°C	500°C
1150°C	0.040	0.034	0.029	0.022	0.015	0.009	0.005
1050°C	0.026	0.024	0.019	0.017	0.009	0.006	0.003

4. Conclusion

Composite based on gadolinium doped cerium (GDC) co-doped lanthanum (La), scandium (Sc), ytterbium (Yb) and neodymium (Nd) were investigated as electrolyte for solid oxide fuel cell (SOFC), namely GDC-La, GDC-Sc, GDC-Yb and GDC-Nd respectively. GDC powders were successfully synthesized via sol-gel method and addition of metal oxide as co-dopant has been done with solid state reaction. SEM images show that the synthesized powders are sticking each other indicating particle agglomeration. The present of co-dopant on the powders were well confirmed by EDX analysis. GDC-Yb powders seem having higher purity than the others co-doped GDC and dispersed better. TGA analysis present that the addition of co-dopant led to improve thermal resistance and stability of solid electrolyte powders. The results confirm that GDC with co-dopant addition is a promising alternative electrolyte for intermediate temperature solid oxide fuel cells. Further investigations need to be performed later on.

References

[1] Sammes N 2006 Fuel Cell Technology Reaching Toward Commercialization (London: Springer Verlag)
[2] Ahn J, Omar S, Yoon H, Nino J, Wachsman E 2010 J Power Sources **195** 2131
[3] Zhu B N 2006 Int J Energ Res **30** 895
[4] Xia C, Li Y, Tian Y, Liu Q, Wang Z, Jia L, Li Y 2010 J Power Sources **195** 3149
[5] Fuentes R, Baker R, 2008 Int J Hydrogen Energy **33** 3480
[6] Sorrel C C et al 2005 Materials for Energy Conversion Devices (Cambridge: Woodhead Publishing)
[7] Wang S, Kobayashi T, Dokiya M, Hashimoto T 2000 J Electrochem Soc **147** 3606
[8] Kim N, Kim B, Lee D 2000 J Power Source **90** 139
[9] Maricle D L *et al* 1992 Solid State Ionic **52** 173
[10] Tian C, Ji B, Xie J, Bao W, Liu K, Cheng J, Yin Q 2014 J Rare Earth **32** 1162
[11] Fuentes R, Baker R, 2009 J Power Source 186 (2009) 268
[12] Prasada D, Son J, Kim B, Lee H, Lee J, 2010 J Ceram Process Res **11** 176
[13] Omar S, Wachsman E, Nino J 2008 Solid State Ionic **178** 1890